CAPTURE GAMMA-RAY SPECTROSCOPY AND RELATED TOPICS

To learn more about the AIP Conference Proceedings, including the Conference Proceedings Series, please visit the webpage
http://proceedings.aip.org/proceedings

CAPTURE GAMMA-RAY SPECTROSCOPY AND RELATED TOPICS

12th International Symposium

Notre Dame, Indiana   4 – 9 September 2005

EDITORS
Andreas Woehr
Ani Aprahamian
Institute for Structure & Nuclear Astrophysics
Notre Dame, Indiana

SPONSORING ORGANIZATIONS
Joint Institute for Astrophysics (JINA)
University of Notre Dame College of Science
University of Notre Dame Graduate School
University of Notre Dame Office of Research

Melville, New York, 2006
AIP CONFERENCE PROCEEDINGS ■ VOLUME 819

Editors:

Andreas Woehr

Ani Aprahamian

Institute for Structure & Nuclear Astrophysics
Department of Physics
University of Notre Dame
Notre Dame, IN 46556

E-mail: awoehr@nd.edu
 aprahamian.1@nd.edu

Authorization to photocopy items for internal or personal use, beyond the free copying permitted under the 1978 U.S. Copyright Law (see statement below), is granted by the American Institute of Physics for users registered with the Copyright Clearance Center (CCC) Transactional Reporting Service, provided that the base fee of $23.00 per copy is paid directly to CCC, 222 Rosewood Drive, Danvers, MA 01923, USA. For those organizations that have been granted a photocopy license by CCC, a separate system of payment has been arranged. The fee code for users of the Transactional Reporting Services is: 0-7354-0313-9/06/$23.00

© 2006 American Institute of Physics

Permission is granted to quote from the AIP Conference Proceedings with the customary acknowledgment of the source. Republication of an article or portions thereof (e.g., extensive excerpts, figures, tables, etc.) in original form or in translation, as well as other types of reuse (e.g., in course packs) require formal permission from AIP and may be subject to fees. As a courtesy, the author of the original proceedings article should be informed of any request for republication/reuse. Permission may be obtained online using Rightslink. Locate the article online at http://proceedings.aip.org, then simply click on the Rightslink icon/"Permission for Reuse" link found in the article abstract. You may also address requests to: AIP Office of Rights and Permissions, Suite 1NO1, 2 Huntington Quadrangle, Melville, NY 11747-4502, USA; Fax: 516-576-2450; Tel.: 516-576-2268; E-mail: rights@aip.org.

L.C. Catalog Card No. 2006920733
ISBN 0-7354-0313-9
ISSN 0094-243X

Printed in the United States of America

CONTENTS

Preface... xix

NUCLEAR STRUCTURE (ISOMERS AND GAMMA SPECTROSCOPY)

K-Isomers and Aspects of Nuclear Structure 3
    G. D. Dracoulis

In-Beam γ-Spectroscopy of Low-Spin Mixed-Symmetry States of
^{138}Ce with Gammasphere in Singles-Mode 11
    N. Pietralla, G. Rainovski, T. Ahn, M. P. Carpenter, R. V. F. Janssens,
    C. J. Lister, and S. Zhu

Nuclear Isomers: Stepping Stones to the Unknown........................ 16
    P. M. Walker

Violations of K-Conservation in ^{178}Hf 24
    A. B. Hayes, D. Cline, C. Y. Wu, J. Ai, H. Amro, C. Beausang,
    R. F. Casten, J. Gerl, A. A. Hecht, A. Heinz, R. Hughes, R. V. F. Janssens,
    C. J. Lister, A.O. Macchiavelli, D. A. Meyer, E. F. Moore, P. Napiorkowski,
    R. C. Pardo, Ch. Schlegel, D. Seweryniak, M. W. Simon, J. Srebrny,
    R. Teng, K. Vetter, and H. J. Wollersheim

Theoretical Description of K-Isomers 30
    Y. Sun

Isomer and In-Beam Spectroscopy of Medium-Spin States in ^{91,92}Zr 35
    P. H. Regan, N. J. Thompson, A. B. Garnsworthy, H. C. Ai, L. Amon,
    R. B. Cakirli, R. F. Casten, C. R. Fitzpatrick, S. J. Freeman, G. Gurdal,
    A. Heinz, G. A. Jones, E. A. McCutchan, J. Qian, V. Werner,
    S. J. Williams, and R. Winkler

NUCLEAR STRUCTURE: DEDICATED TO THE MEMORY OF D. D. WARNER

Dave Warner: Friend, Mentor, Colleague................................... 43
    A. Aprahamian

Emerging Collectivity in Nuclei, Shape Coexistence, and
Proton-Neutron Interactions 49
    R. F. Casten

Symmetry and Wigner Energy in Nuclear Matter......................... 57
    P. Van Isacker

Varying Shell Structure of *sd-pf* Nuclei 65
    T. Otsuka

Recent Advances in Studies of Mirror-symmetry in the A=40-56 Shell......... 71
    A. M. Bruce

FUNDAMENTAL PHYSICS

**Measurement of Parity-Violating Gamma-ray Asymmetry in
Compound Nuclei with Cold Neutrons** 81
    P.-N. Seo, J. D. Bowman, R. D. Carlini, T. E. Chupp, S. D. Covrig,
    M. Dabaghyan, S. J. Freedman, T. Gentile, M. T. Gericke, R. C. Gillis,
    G. L. Greene, F. W. Hersman, T. Ino, G. L. Jones, M. Kandes, B. Lauss,
    M. B. Leuschner, W. R. Lozowski, R. Mahurin, M. Mason, Y. Masuda,
    G. S. Mitchell, S. Muto, H. Nann, S. A. Page, S. I. Penttilä, W. D. Ramsay,
    S. Santra, E. I. Sharapov, T. B. Smith, W. M. Snow, W. S. Wilburn,
    V. Yuan, and H. Zhu

**Scattering of keV Neutrons and the Problem of
Quantum Entanglement** 86
    R. Moreh, R. C. Block, Y. Danon, and M. Neuman

Random Matrix Thermodynamics 91
    G. E. Mitchell, T. N. Nogueira, M. P. Pato, J.C. Sartorelli, and
    A. P. B. Tufaile

**Symmetry of Isoscalar Matrix Elements and Systematics in the sd
and Beginning of fp Shells** 96
    J. N. Orce, P. Petkov, V. Velázquez, C. J. McKay, S. R. Lesher, S. Choudry,
    M. Mynk, A. Linnemann, J. Jolie, P. von Brentano, V. Werner, S. W. Yates,
    and M.T. McEllistrem

Symmetry and Structure Tests in ^{18}O and ^{18}Ne 101
    S. Choudry, J. N. Orce, V. Varadarajan, S. Lesher, D. Bandyopadhyay,
    S. Mukhopadhyay, S. W. Yates, and M. T. McEllistrem

**Gamma-Ray Transitions in the Decay of the Superallowed Beta
Emitter ^{62}Ga** 105
    B. Hyland, C. E. Svensson, G. C. Ball, J. R. Leslie, D. Albers,
    C. Andreoiu, P. Bricault, R. Churchman, D. Cross, M. Dombsky,
    P. E. Garrett, C. Geppert, G. F. Grinyer, G. Hackman, V. Hanemaayer,
    J. Lassen, J. P. Lavoie, D. Melconian, A. C. Morton, C. J. Pearson,
    M. Pearson, A. A. Phillips, M.A. Schumaker, and J. J. Valiente-Dobón

DATA FOR NUCLEAR- AND ASTROPHYSICS APPLICATIONS

**Nuclear Structure and Decay Data: Current Status and
Future Perspectives** 113
    F. G. Kondev and J. K. Tuli

**An Updated Library of Reaction Rates for the Astrophysical
rp-Process** 118
    A. Sakharuk, T. Elliot, J. L. Fisker, S. Hemingray, A. Kruizenga,
    T. Rauscher, H. Schatz, K. Smith, F.-K. Thielemann, and M. Wiescher

**KADoNiS—The Karlsruhe Astrophysical Database of Nucleosynthesis
in Stars** 123
    I. Dillmann, M. Heil, F. Käppeler, R. Plag, T. Rauscher, and
    F.-K. Thielemann

Nuclear Data Resources for Capture γ-Ray Spectroscopy and Related Topics 128
    B. Pritychenko, M. W. Herman, P. Obložinský, A. A. Sonzogni, and V. Zerkin

Measurements of Neutron Capture Cross Sections for Gd Isotopes in the Energy Region from 10 keV to 90 keV 133
    G. N. Kim, W. C. Chung, T. I. Ro, T. Ohsaki, and M. Igashira

The Evaluated Gamma-ray Activation File (EGAF) 138
    R. Firestone, G. L. Molnár, Zs. Révay, T. Belgya, D. P. McNabb, and B. W. Sleaford

NUCLEAR MASSES AND SHELL MODEL

Nuclear Forecasting as Pattern Recognition: Can we Predict Nuclear Masses? 151
    A. Frank, J. C. López Vieyra, J. Barea, J. G. Hirsch, V. Velazquez, and P. Van Isacker

Mass Measurements with the CSS2 and CIME Cyclotrons at GANIL 159
    M. B. Gomez Hornillos, M. Chartier, W. Mittig, B. Blank, L. Caballero, F. Chautard, C. E. Demonchy, G. Georgiev, A. Gillibert, B. Jacquot, B. Jurado, N. Lecesne, A. Lépine-Szily, N. Orr, G. Politi, M. Rousseau, P. Roussel-Chomaz, H. Savajols, and A. C. C. Villari

Recent Achievements and Future of Mass Measurements with Fast Radioactive Beams 164
    M. Matoš

NUCLEAR ASTROPHYSICS

How Accurately Do We Know the Cross Section of the ^{7}Be (p, γ)^{8}B Reaction? 171
    M. Gai

Nuclear Astrophysics Experiments at CIAE 176
    W. Liu, Z. Li, X. Bai, G. Lian, B. Guo, S. Zeng, S. Yan, B. Wang, N. Shu, and Y. Chen

An Independent Measurement of the ^{12}C(α, γ)^{16}O Cross Section with the Karlsruhe 4π BaF$_2$ Detector 181
    R. Plag, M. Heil, F. Käppeler, and K. Wisshak

^{19}F(p, γ) ^{20}Ne: Putting a Lid on the CNO Cycle 186
    A. Couture, M. Beard, M. Couder, J. Görres, L. Lamm, P. LeBlanc, H.-Y. Lee, S. O'Brien, A. Palumbo, E. Stech, E. Strandberg, W. Tan, E. Uberseder, C. Ugalde, and M. Wiescher

Threshold States in ^{19}Ne and the CNO Breakout Reaction ^{15}O(α, γ)^{19}Ne 191
    W. P. Tan, J. Görres, J. Daly, M. Couder, A. Couture, H. Y. Lee, E. Stech, E. Strandberg, C. Ugalde, and M. Wiescher

Nuclear Isomers: Structures and Applications 196
    Y. Sun, M. Wiescher, A. Aprahamian, and J. Fisker

Radiative Capture Reactions and α-Elastic Scattering on ^{106}Cd for
the Astrophysical p-Process 201
    Gy. Gyurky, Z. Elekes, G. G. Kiss, Zs. Fülöp, E. Somorjai, Z. Máté,
    J. Görres, A. Palumbo, M. Wiescher, H.-Y. Lee, N. Özkan, R. T. Güray,
    G. Efe, D. Galaviz, A. Kretschmer, K. Sonnabend, A. Zilges, and
    T. Rauscher

NUCLEAR REACTIONS (REACTIONS WITH NEUTRON-RICH NUCLEI)

Neutron-Transfer Reactions with Exotic Neutron-Rich Beams:
Surrogates for Neutron-Capture Reactions 209
    J. A. Cizewski, K. L. Jones, S. D. Pain, J. S. Thomas, D. W. Bardayan,
    J. C. Blackmon, M. S. Smith, R. L. Kozub, and M. S. Johnson

Do Halos Exist on the Dripline of Deformed Nuclei? 216
    F. M. Nunes, B. Avez, and T. Dugnet

Doppler Shift as a Tool for Studies of Resonant (p,n) Reactions with
RIBs, Spectroscopy of ^{7}He 221
    P. Boutachkov, G. V. Rogachev, V. Z. Goldberg, A. Aprahamian,
    F. D. Becchetti, J. P. Bychowski, Y. Chen, G. Chubarian, P. A. DeYoung,
    J. J. Kolata, L. O. Lamm, G. F. Peaslee, M. Quinn, B. B. Skorodumov, and
    A. Wöhr

Model Calculations of Radiative Capture of Nucleons in MeV Region 226
    E. Běták

NUCLEAR STRUCTURE (0^+ STATES)

Study of 0^+ States in Deformed Nuclei 233
    S. R. Lesher, Z. Ammar, M. Merrick, C. D. Hannant, N. Warr,
    T. B. Brown, N. Boukharouba, C. Fransen, M. T. McEllistrem, and
    S. W. Yates

Pairing Versus Quadrupole Collectivity of Low-Lying 0^+ States in
Deformed Nuclei 239
    N. Lo Iudice, A. V. Sushkov, and N. Yu. Shirikova

Recent Progress on the 0^+ Dominance 244
    N. Yoshinaga, A. Arima, and Y. M. Zhao

Gamma-Ray Spectroscopy at TRIUMF-ISAC 249
    P. E. Garrett, C. E. Svensson, G. C. Ball, G. Hackman, E. F. Zganjar,
    C. Andreoiu, A. Andreyev, S. F. Ashley, R. A. E. Austin,
    D. Bandyopadhyay, J. A. Becker, S. Chan, H. Coombes, R. Churchman,
    R. S. Chakrawarthy, P. Finlay, G. F. Grinyer, B. Hyland, E. Illes,
    G. A. Jones, W. D. Kulp, J. R. Leslie, C. Mattoon, A. C. Morton,
    C. J. Pearson, A. A. Phillips, P. H. Regan, J. J. Ressler, F. Sarazin,
    M. A. Schumaker, J. Schwarzenberg, M. B. Smith, J. J. Valiente-Dobón,
    P. M. Walker, S. J. Williams, J. C. Waddington, L. M. Watters, J. Wong,
    and J. L. Wood

NUCLEAR ASTROPHYSICS (S-PROCESS NUCLEOSYNTHESIS)

Neutron Capture Reaction Rates for Stellar Nucleosynthesis 257
    A. Mengoni

The Weak s-Process and its Relation to Explosive Nucleosynthesis 265
    M. Heil and M. Pignatari

Measurement of Neutron Capture Cross Section of ^{62}Ni in the keV-Region... 273
    A. M. Alpizar-Vicente, T. A. Bredeweg, E.-I. Esch, U. Greife, R. C. Haight,
    R. Hatarik, J. M. O'Donnell, R. Reifarth, R. S. Rundberg, J. L. Ullmann,
    D. J. Vieira, J. B. Wilhelmy, and J. M. Wouters

^{102}Pd(n, γ) Cross Section Measurement Using DANCE.................... 278
    R. Hatarik, A. M. Alpizar-Vicente, T. A. Bredeweg, E.-I. Esch, U. Greife,
    R. C. Haight, J. M. O'Donnell, R. Reifarth, R. S. Rundberg, J. L. Ullmann,
    D. J. Vieira, and J. M. Wouters

Measurement of ^{139}La(n, γ) Cross Section............................. 283
    R. Terlizzi, U. Abbondanno, G. Aerts, H. Álvarez, F. Alvarez-Velarde,
    S. Andriamonje, J. Andrzejewski, P. Assimakopoulos, L. Audouin,
    G. Badurek, P. Baumann, F. Bečvář, E. Berthoumieux, F. Calviño,
    D. Cano-Ott, R. Capote, A. Carrillo de Albornoz, P. Cennini, V. Chepel,
    E. Chiaveri, N. Colonna, G. Cortes, A. Couture, J. Cox, M. Dahlfors,
    S. David, I. Dillmann, R. Dolfini, C. Domingo-Pardo, W. Dridi, I. Duran,
    C. Eleftheriadis, M. Embid-Segura, L. Ferrant, A. Ferrari,
    R. Ferreira-Marques, L. Fitzpatrick, H. Frais-Koelbl, K. Fujii, W. Furman,
    R. Gallino, I. Goncalves, E. Gonzalez-Romero, A. Goverdovski,
    F. Gramegna, E. Griesmayer, C. Guerrero, F. Gunsing, B. Haas, R. Haight,
    M. Heil, A. Herrera-Martinez, M. Igashira, S. Isaev, E. Jericha, Y. Kadi,
    F. Käppeler, D. Karamanis, D. Karadimos, M. Kerveno, V. Ketlerov,
    P. Koehler, V. Konovalov, E. Kossionides, M. Krtička, C. Lamboudis,
    H. Leeb, A. Lindote, I. Lopes, M. Lozano, S. Lukic, J. Marganiec,
    L. Marques, S. Marrone, P. Mastinu, A. Mengoni, P. M. Milazzo,
    C. Moreau, M. Mosconi, F. Neves, H. Oberhummer, S. O'Brien,
    M. Oshima, J. Pancin, C. Papachristodoulou, C. Papadopoulos, C. Paradela,
    N. Patronis, A. Pavlik, P. Pavlopoulos, L. Perrot, R. Plag, A. Plompen,
    A. Plukis, A. Poch, C. Pretel, J. Quesada, T. Rauscher, R. Reifarth,
    M. Rosetti, C. Rubbia, G. Rudolf, P. Rullhusen, J. Salgado, L. Sarchiapone,
    I. Savvidis, C. Stephan, G. Tagliente, J. L. Tain, L. Tassan-Got, L. Tavora,
    G. Vannini, P. Vaz, A. Ventura, D. Villamarin, M. C. Vincente,
    V. Vlachoudis, R. Vlastou, F. Voss, S. Walter, H. Wendler, M. Wiescher,
    and K. Wisshak

Measurement of the Resonance Capture Cross Section of ^{204,206}Pb and Termination of the s-Process.. 288
    C. Domingo-Pardo, S. O'Brien, U. Abbondanno, G. Aerts, H. Álvarez,
    F. Alvarez-Velarde, S. Andriamonje, J. Andrzejewski, P. Assimakopoulos,
    L. Audouin, G. Badurek, P. Baumann, F. Bečvář, E. Berthoumieux,
    F. Calviño, D. Cano-Ott, R. Capote, A. Carrillo de Albornoz, P. Cennini,
    V. Chepel, E. Chiaveri, N. Colonna, G. Cortes, A. Couture, J. Cox,
    M. Dahlfors, S. David, I. Dillmann, R. Dolfini, W. Dridi, I. Duran,
    C. Eleftheriadis, M. Embid-Segura, L. Ferrant, A. Ferrari,

R. Ferreira-Marques, L. Fitzpatrick, H. Frais-Koelbl, K. Fujii, W. Furman,
C. Guerrero, I. Goncalves, R. Gallino, E. Gonzalez-Romero,
A. Goverdovski, F. Gramegna, E. Griesmayer, F. Gunsing, B. Haas,
R. Haight, M. Heil, A. Herrera-Martinez, M. Igashira, S. Isaev, E. Jericha,
Y. Kadi, F. Käppeler, D. Karamanis, D. Karadimos, M. Kerveno,
V. Ketlerov, P. Koehler, V. Konovalov, E. Kossionides, M. Krtička,
C. Lamboudis, H. Leeb, A. Lindote, I. Lopes, M. Lozano, S. Lukic,
J. Marganiec, L. Marques, S. Marrone, P. Mastinu, A. Mengoni,
P. M. Milazzo, C. Moreau, M. Mosconi, F. Neves, H. Oberhummer,
M. Oshima, J. Pancin, C. Papachristodoulou, C. Papadopoulos, C. Paradela,
N. Patronis, A. Pavlik, P. Pavlopoulos, L. Perrot, R. Plag, A. Plompen,
A. Plukis, A. Poch, C. Pretel, J. Quesada, T. Rauscher, R. Reifarth,
M. Rosetti, C. Rubbia, G. Rudolf, P. Rullhusen, J. Salgado, L. Sarchiapone,
I. Savvidis, C. Stephan, G. Tagliente, J. L. Tain, L. Tassan-Got, L. Tavora,
R. Terlizzi, G. Vannini, P. Vaz, A. Ventura, D. Villamarin, M. C. Vincente,
V. Vlachoudis, R. Vlastou, F. Voss, S. Walter, H. Wendler, M. Wiescher,
K. Wisshak, and The n_TOF Collaboration

NEUTRON CAPTURE REACTIONS

The Radiative Strength Function Using the Neutron-Capture Reaction on ^{151,153}Eu 295
    U. Agvaanluvsan, A. Alpizar-Vicente, J. A. Becker, F. Bečvář,
    T. A. Bredeweg, R. Clement, E. Esch, C. M. Folden, III, R. Hatarik,
    R. C. Haight, D. C. Hoffman, M. Krtička, R. A. Macri, G. E. Mitchell,
    H. Nitsche, J. M. O'Donnell, W. Parker, R. Reifarth, R. S. Rundberg,
    J. M. Schwantes, S. A. Sheets, J. L. Ullmann, D. J. Vieira, J. B. Wilhelmy,
    P. Wilk, J. M. Wouters, and C. Y. Wu

Determination of Thermal Neutron Capture Cross Sections Using Cold Neutron Beams at the Budapest PGAA-NIPS Facilities 300
    T. Belgya, Zs. Révay, L. Szentmiklósi

Method for (n, γ) Cross Section Measurements on Unstable Isotopes 307
    S. Walter, M. Heil, F. Käppeler, R. Plag, and R. Reifarth

Neutron Capture Cross Sections of ^{236}U and ^{234}U 312
    R. S. Rundberg, T. A. Bredeweg, E. M. Bond, R. C. Haight, L. F. Hunt,
    A. Kronenberg, J. M. O'Donnell, J. M. Schwantes, J. L. Ullmann,
    D. J. Vieira, J. B. Wilhelmy, and J. M. Wouters

Neutron Capture Cross Section Measurements at n_TOF of ^{237}Np, ^{240}Pu and ^{243}Am for the Transmutation of Nuclear Waste 318
    D. Cano-Ott, U. Abbondanno, G. Aerts, H. Álvarez, F. Alvarez-Velarde,
    S. Andriamonje, J. Andrzejewski, P. Assimakopoulos, L. Audouin,
    G. Badurek, P. Baumann, F. Bečvář, E. Berthoumieux, F. Calviño,
    R. Capote, A. Carrillo de Albornoz, P. Cennini, V. Chepel, E. Chiaveri,
    N. Colonna, G. Cortes, A. Couture, J. Cox, M. Dahlfors, S. David,
    I. Dillmann, R. Dolfini, C. Domingo-Pardo, W. Dridi, I. Duran,
    C. Eleftheriadis, M. Embid-Segura, L. Ferrant, A. Ferrari,
    R. Ferreira-Marques, L. Fitzpatrick, H. Frais-Koelbl, K. Fujii, W. Furman,
    R. Gallino, I. Goncalves, E. Gonzalez-Romero, A. Goverdovski,

F. Gramegna, E. Griesmayer, C. Guerrero, F. Gunsing, B. Haas, R. Haight,
M. Heil, A. Herrera-Martinez, M. Igashira, S. Isaev, E. Jericha, Y. Kadi,
F. Käppeler, D. Karamanis, D. Karadimos, M. Kerveno, V. Ketlerov,
P. Koehler, V. Konovalov, E. Kossionides, M. Krtička, C. Lamboudis,
H. Leeb, A. Lindote, I. Lopes, M. Lozano, S. Lukic, J. Marganiec,
L. Marques, S. Marrone, T. Martinez, P. Mastinu, A. Mengoni,
P. M. Milazzo, C. Moreau, M. Mosconi, F. Neves, H. Oberhummer,
S. O'Brien, M. Oshima, J. Pancin, C. Papachristodoulou, C. Papadopoulos,
C. Paradela, N. Patronis, A. Pavlik, P. Pavlopoulos, L. Perrot, R. Plag,
A. Plompen, A. Plukis, A. Poch, C. Pretel, J. Quesada, T. Rauscher,
R. Reifarth, M. Rosetti, C. Rubbia, G. Rudolf, P. Rullhusen, J. Salgado,
L. Sarchiapone, I. Savvidis, C. Stephan, G. Tagliente, J. L. Tain,
L. Tassan-Got, L. Tavora, R. Terlizzi, G. Vannini, P. Vaz, A. Ventura,
D. Villamarin, M. C. Vincente, V. Vlachoudis, R. Vlastou, F. Voss,
S. Walter, H. Wendler, M. Wiescher, K. Wisshak, and The n_TOF
Collaboration

**Gamma-Ray Cross Section Standards in the MeV Energy Range and
^{56}Fe Inelastic Scattering** 323
R. O. Nelson, M. Devlin, N. Fotiades, J. A. Becker, P. E. Garrett,
W. Younes, D. Dashdorj, T. Ethvignot, and T. Granier

NUCLEAR STRUCTURE

Level Dynamics of Non-Yrast States 331
J. Jolie and S. Heinze

New Findings for Mixed-Symmetry States 340
V. Werner, N. Pietralla, P. von Brentano, U. Kneissl, H. H. Pitz,
A. Tonchev, M. W. Ahmed, C. Fransen, H. von Garrel, C. Kohstall, J. Li,
A. Linnemann, S. Müller, I. V. Pinayev, D. Savran, M. Scheck, F. Stedile,
W. Tornow, S. Walter, H. R. Weller, and Y. K. Wu

**Independent Evidence for M1 Scissors Resonances Built on the Levels
in the Quasicontinuum of ^{163}Dy** 345
M. Krtička, F. Bečvář, M. Heil, F. Käppeler, R. Reifarth, I. Tomandl,
F. Voss, and K. Wisshak

**Low-Energy Dipole Modes of Excitation Below the Neutron
Separation Energy** 350
A. P. Tonchev, C. Angell, M. Boswell, C. R. Howell, H. J. Karwowski,
J. H. Kelley, W. Tornow, and N. Tsoneva

**Search for Enhanced Alpha Preformation in the N=Z+1 Nuclei
^{113}Ba, ^{109}Xe, ^{105}Te** 355
A. A. Hecht, C. J. Lister, C. N. Davids, A. Heinz, N. Hoteling,
C. Mazzocchi, J. Palombo, D. Seweryniak, J. Shergur, M. Stoyer,
W. B. Walters, P. J. Woods, and S. Zhu

NEUTRON CAPTURE REACTIONS II AND NUCLEAR STRUCTURE

Photo-Induced Population of the $h_{11/2}$ Isomeric States in
(γ, n) Reactions..363
    C. T. Angell, H. J. Karwowski, J. H. Kelley, A. P. Tonchev, and W. Tornow

General Approach to Materials Classification Using Neutron
Analysis Techniques 368
    V. G. Solovyev and D. S. Koltick

Measurement of Gamma Rays from keV-Neutron Capture Reaction
on Zr-90, 94..373
    K. Ohgama, M. Igashira, and T. Ohsaki

Measurement of Capture Gamma Rays from the 46- and 84-keV
Neutron Resonances of ^{24}Mg...378
    T. Ohsaki, D. Ikenaga, and M. Igashira

Bound Excited States in ^{27}F 383
    Z. Elekes, Zs. Dombrádi, A. Saito, N. Aoi, H. Baba, K. Demichi,
    Zs. Fülöp, J. Gibelin, T. Gomi, H. Hasegawa, N. Imai, M. Ishihara,
    H. Iwasaki, S. Kanno, S. Kawai, T. Kishida, T. Kubo, K. Kurita,
    Y. Matsuyama, S. Michimasa, T. Minemura, T. Motobayashi, M. Notani,
    T. K. Ohnishi, H. J. Ong, S. Ota, A. Ozawa, H. K. Sakai, H. Sakurai,
    S. Shimoura, E. Takeshita, S. Takeuchi, M. Tamaki, Y. Togano, K. Yamada,
    Y. Yanagisawa, and K. Yoneda

APPLICATION AND DEVELOPMENT OF NUCLEAR TECHNIQUES

Thermal Neutron Capture Cross Sections of the Palladium Isotopes..........389
    R. B. Firestone, M. Krtiáka, D. P. McNabb, B. Sleaford, U. Agvaanluvsan,
    T. Belgya, and Zs. Révay

Atomic-Nuclear Coupling Experiments..396
    J. A. Becker

Distortion of Pulse-Height Spectra of Neutron Capture Gamma Rays 402
    A. Laptev, H. Harada, S. Nakamura, J. Hori, M. Igashira, T. Ohsaki, and
    K. Ohgama

NUCLEAR ASTROPHYSICS (R-PROCESS NUCLEOSYNTHESIS)

Nuclear-Physics Issues of r-Process Nucleosynthesis.......................409
    K.-L. Kratz

Neutron Captures and the r-Process 419
    K. Farouqi, K.-L. Kratz, B. Pfeiffer, T. Rauscher, and F.-K. Thielemann

Radioactive Beam Experiments Relevant for the Astrophysical
r-Process..423
    H. Schatz

Magnetic Dipole and Gamow-Teller Modes in Neutrino-Nucleus
Reactions: Impact on Supernova Dynamics and Nucleosynthesis 432
    P. von Neumann-Cosel, T. Adachi, A. Byelikov, H. Fujita, Y. Fujita,
    A. Heger, E. Kolbe, K. Langanke, G. Martínez-Pinedo, A. Richter,
    A. Shevchenko, and Y. Shimbara

NUCLEAR STRUCTURE (HIGH SPIN EXPERIMENTS)

Hyperdeformed Rotational Bands Observed in the Actinide Region 439
    A. Krasznahorkay, M. Csatlós, Y. Eisermann, T. Faestermann, G. Graw,
    J. Gulyás, D. Habs, M. N. Harakeh, R. Hertenberger, M. Hunyadi,
    H. J. Maier, Z. Máté, O. Schaile, P. G. Thirolf, and H. J. Wirth
High Spin Bands in the A~130 Nuclei: A "Non- Chiral" Explanation 447
    R. Kaczarowski
Nuclear Structure and Octupole Collectivity in Nucleus ^{122}Cs 454
    I. M. Govil
Rotational Damping, Ridges, and the Quasi-continuum of γ rays
in ^{152}Dy 459
    T. Lauritsen, R. V. F. Janssens, T. L. Khoo, I. Ahmad, M. P. Carpenter,
    F. G. Kondev, C. J. Lister, E. F. Moore, D. Seweryniak, S. Zhu, T. Døssing,
    B. Herskind, A. M. Heinz, D. G. Jenkins, R. M. Clark, P. Fallon,
    A. O. Macchiavelli, D. Ward, G. Lane, P. Chowdhury, A. Korichi,
    A. Lopez-Martens, and A. J. Larabee
Multinucleon Transfer Reactions to Study Single-Particle Evolution in
Se Isotopes 464
    P. H. Regan, G. A. Jones, Zs. Podolyák, N. Yoshinaga, K. Higashiyama,
    G. deAngelis, Y. H. Zhang, A. Gadea, C. A. Ur, M. Axiotis, D. Bazzacco,
    R. Broda, E. Bucurescu, E. Farnea, W. Gelletly, M. Ionescu-Bujor,
    A. Iordachescu, Th. Kröll, S. D. Langdown, S. Lenzi, S. Lunardi,
    N. Marginean, T. Martinez, N. Medina, R. Menegazzo, D. R. Napoli,
    B. Quintana, B. Rubio, C. Rusu, R. Schwenger, D. Tonev, J. J. Valiente
    Dobon, and W. von Oertzen

NUCLEAR STRUCTURE (THEORY)

Nuclear Mean Field from the Shell Model Point of View 475
    N. A. Smirnova, A. De Maesschalck, and K. Heyde
Softness of Doubly-Magic ^{78}Ni and Related Topics 483
    A. F. Lisetskiy, B. A. Brown, and H. Schatz
Regular and Chaotic Nuclear Vibrations 487
    P. Cejnar, M. Macek, P. Stránský, and M. Kurian
Continuum Shell Model, Reactions and Giant Resonances 493
    V. Zelevinsky and A. Volya

Correlation of Nuclear Level Densities and Masses.........................498
    T. von Egidy and D. Bucurescu
Effective Interactions and Operators in Nuclei within the No-Core
Shell Model 504
    B. R. Barrett, P. Navratil, I. Stetcu, and J. P. Vary

NUCLEAR REACTIONS

Precision Gamma-Ray Spectroscopy at the Institut Laue Langevin...........511
    H. G. Börner
Radiative Capture versus Coulomb Dissociation 518
    H. Esbensen
Spectroscopy Studies Using Two-Nucleon Knockout..........................523
    J. A. Tostevin
Influence of the Projectile Description on Breakup Calculations 528
    P. Capel and F. Nunes

NUCLEAR STRUCTURE

Distribution of the GT Strength Starting from the Ground State
of ^{14}N 535
    A. Negret, T. Adachi, C. Bäumer, A. M. van den Berg, G. P. A. Berg,
    P. von Brentano, D. Frekers, D. De Frenne, K. Fujita, Y. Fujita,
    E. W. Grewe, P. Haefner, K. Hatanaka, M. Hunyadi, M. A. de Huu,
    H. Johansson, E. Jacobs, Y. Kalmykov, K. Kawabata, A. Korff,
    K. Nakanishi, P. von Neumann-Cosel, T. Ogama, L. Popescu, S. Rakers,
    A. Richter, N. Ryezayeva, Y. Sakemi, A. Shevchenko, Y. Shimbara,
    Y. Shimizu, A. Tamii, M. Uchida, H. J. Wörtche, and M. Yosoi
Microscopic Calculations for Waiting-Point Nuclei........................540
    K. P. Drumev, C. Bahri, and J. P. Draayer
Level Densities of Iron Isotopes and Low-Energy Enhancement of
γ-Strength Function 545
    A. V. Voinov, S. M. Grimes, U. Agvaanluvsan, E. Algin, T. Belgya,
    C. R. Brune, M. Guttormsen, M. J. Hornish, T. N. Massey, G. E. Mitchell,
    J. Rekstad, A. Schiller, and S. Siem
Spin-Isospin Excitations from the Ground-State of ^{64}Ni....................550
    L. Popescu, T. Adachi, C. Bäumer, G. P. A. Berg, A. M. van den Berg,
    P. von Brentano, D. Frekers, D. de Frenne, K. Fujita, Y. Fujita,
    E. W. Grewe, P. Haefner, K. Hatanaka, M. Hunyadi, M. de Huu, E. Jacobs,
    H. Johansson, A. Korff, A. Negret, K. Nakanishi, P. von Neumann-Cosel,
    S. Rakers, N. Ryezayeva, Y. Sakemi, A. Shevchenko, Y. Shimbara,
    Y. Shimizu, H. Simon, Y. Tameshige, A. Tamii, M. Uchida, H. J. Wörtche,
    and M. Yosoi

POSTER SESSION

α-Stripping Reactions with Excotic Nuclei: $^{12}C(^{7}Be,^{3}He)^{16}O$ 557
    H. Amro, F. D. Becchetti, Y. Chen, H. Jiang, M. Ojaruega, H. C. Griffin,
    J. J. Kolata, B. B. Skorodumov, J. D. Hinnefeld, and G. Peaslee

Structural Ambiguities In ^{114}Cd 559
    D. Bandyopadhyay, P. E. Garrett, S. R. Lesher, C. Fransen,
    N. Boukharouba, M. T. McEllistrem, and S.W. Yates

$^{207}Pb(n,2n\ \gamma)^{206}Pb$ Cross-Section Measurements by In-Beam Gamma-Ray Spectroscopy 561
    P. Baumann, C. Borcea, E. Jericha, S. Jokić, M. Kerveno, S. Lukić,
    L. C. Mihailescu, A. Pavlik, A. J. M. Plompen, G. Rudolf, and the n_TOF Collaboration

Anomalous Neutron Radiative Capture in ^{197}Au Revisited 563
    M. M. Krtička, F. Bečvář, M. Heil, F. Käppeler, R. Reifarth, I. Tomandl,
    F. Voss, and K. Wisshak

New Method for the Determination of Accurate Gamma-Ray Intensities for the $^{14}N(n,\ \gamma)^{15}N$ High Energy Standard 565
    T. Belgya

Simultaneous Measurement of (n, γ) and (n,fission) Cross Sections with the DANCE 4π BaF$_2$ Array 568
    T. A. Bredeweg, M. M. Fowler, J. A. Becker, E. M. Bond,
    M. B. Chadwick, R. R. C. Clement, E.-I. Esch, T. Ethvignot, T. Granier,
    L. F. Hunt, R. A. Macri, J. M. O'Donnell, R. S. Rundberg,
    J. M. Schwantes, J. L. Ullmann, D. J. Viera, J. B. Wilhelmy, J. M. Wouters,
    C.-Y. Wu, and J. E. Yurkon

Classial Chaos in the Interacting Boson Model 570
    M. Macek and P. Cejnar

Measurement of Disintegration Rates and Absolute γ-ray Intensities 572
    D. J. DeVries and H. C. Griffin

New Band Mechanism of Doubly-odd Nuclei Around Mass 130 575
    K. Higashiyama and N. Yoshinaga

Measurement of Neutron Capture Cross Section for Natural Palladium 577
    J. Hori, H. Yashima, T. Oishi, W. Takahashi, M. Baba, and K. Nakajima

Systematic Measurement of keV-neutron Capture Cross Sections and Capture Gamma-ray Spectra of Sn Isotopes 579
    J. Nishiyama, M. Igashira, T. Ohsaki, G. N. Kim, W. C. Chung, and
    T. I. Ro

$^{18}F(\alpha,p)^{21}Ne$ Reaction: Neutron Source For r-Process In Supernovae 581
    H.-Y. Lee, M. Beard, H.-W. Becker, A. Couder, A. Couture, J. Görres,
    L. Lamm, P. LeBlanc, S. O'Brien, A. Palumbo, E. Stech, E. Strandberg,
    W. Tan, C. Ugalde, and M. Wiescher

Search for Multiphonon and Mixed-Symmetry States in ^{127}I 583
    S. Mukhopadhyay, J. N. Orce, S. N. Choudry, V. Varadarajan, A. Kumar,
    M. T. McEllistrem, and S. W. Yates

^{106}Cd and ^{112}Sn: Alpha-Induced Cross Section Measurements for the
Astrophysical P-Process 585
    A. Palumbo, J. Görres, H.-Y. Lee, M. Wiescher, W. Rapp, N. Özkan,
    R. T. Güray, G. Efe, Gy. Gyürky, Zs. Fülop, and E. Somorjai

$K=0^+$ Bands in Gadolinium Nuclei 587
    G. Popa, A. Aprahamian, A. M. Bruce, J. G. Hirsch, and J. P. Draayer

Nuclear Reaction and Structure Databases of the National Nuclear
Data Center 589
    B. Pritychenko, R. Arcilla, M. W. Herman, P. Obložinský, D. Rochman,
    A. A. Sonzogni, J. K. Tuli, and D. F. Winchell

Nuclear Structure Properties of Neutron Rich Ge-Br Isotopes in the
Astrophysical r-Process 591
    M. Quinn, A. Aprahamian, A. Woehr, P. Mantica, J. Pereira Conca,
    H. Schatz, S. Hennrich, and K.-L. Kratz

Thermal Neutron Capture and Thermal Neutron Burn-up of K
Isomeric State of ^{177m}Lu: A Way to the Neutron Super-Elastic
Scattering Cross Section 593
    O. Roig, G. Bélier, V. Méot, J. Aupiais, J.-M. Daugas, Ch. Jutier,
    G. Le Petit, A. Letourneau, F. Marie, P. Romain, and Ch. Veyssiere

γ-ray Detection Possibilities in the European Recoil Separator for
Nuclear Astrophysics 595
    D. Schürmann, A. Di Leva, L. Gialanella, D. Rogalla, F. Strieder,
    N. De Cesare, A. D' Onofrio, G. Imbriani, R. Kunz, A. Ordine, V. Roca,
    C. Rolfs, M. Romano, F. Schümann, F. Terrasi, and H.-P. Trautvetter

The Radiative Strength Function Using the Neutron-Capture
Reaction on ^{94,95}Mo 597
    S. A. Sheets, U. Agvaanluvsan, M. Krtička, G. E. Mitchell, J. A. Becker,
    F. Bečvář, and the DANCE Collaboration

Multichannel R-matrix Analysis of CNO Cycle Reactions 599
    E. C. Simpson, R. E. Azuma, M. Wiescher, A. Champagne, P. Bertone,
    H.-P. Trautvetter, J. Görres, and C. Ugalde

Spectroscopic Structure of Exotic ^{19}Na: Astrophysics Implication 601
    B. B. Skorodumov, G. V. Rogachev, P. Boutachkov, A. Aprahamian,
    J. J. Kolata, L. O. Lamm, M. Quinn, and A. Woehr

Development of a Prompt Gamma-ray Analysis Combined with
Multiple Gamma-ray Detection 603
    Y. Toh, M. Oshima, M. Koizumi, A. Osa, and A. Kimura

Systematic Studies of Odd Isotopes in the Vicinity of the Closed
Shell Z=50 605
    J. Honzátko, I. Tomandl, V. Bondarenko, T. von Egidy, H.-F. Wirth,
    R. Hertenberger, Y. Eisermann, G. Graw, D. Bucurescu, and
    V. Yu. Ponomarev

Mass of the lowest T=2 State of ^{32}S 607
    S. Triambak, A. García, E. G. Adelberger, G. J. P. Hodges, H. E. Swanson,
    S. A. Hoedl, S. K. L. Sjue, and A. L. Sallaska

The Complete (n, γ) Level Scheme of ^{124}Te 609
    T. von Egidy, H.-F. Wirth, I. Tomandl, and J. Honzatko

Nature of One- and Two-Phonon Mixed Symmetry States in ^{92}Zr and ^{94}Mo from High-Resolution Electron and Proton Scattering **611**
P. von Neumann-Cosel, N. T. Botha, O. Burda, J. Carter, R. W. Fearick, S. V. Förtsch, C. Fransen, H. Fujita, M. Kuhar, A. Lenhardt, R. Neveling, N. Pietralla, V. Yu. Ponomarev, A. Richter, E. Sideras-Haddad, F. D. Smit, and J. Wambach

Photographs 613
Programs 628
List of Participants 638
Author Index 653

PREFACE

The 12th International Conference on Capture Gamma-Ray Spectroscopy and Related topics (CGS-12) was hosted by the University of Notre Dame from Monday September 5 to Friday September 9, 2005. The University of Notre Dame du Lac founded in 1842 by a priest of the Congregation of Holy Cross, is an independent, national Catholic university located in a rather picturesque part of Indiana adjacent to the city of South Bend and approximately 90 miles southeast of Chicago. The conference was preceded by a one-day workshop on Nuclear Isomers organized by Professor Yang Sun on Sunday September 4, 2005. The conference program was lively and interactive with a large number of women and young scientists giving invited talks. The conference theme varied across current topics in Nuclear Structure, Nuclear Symmetries, Nuclear Reactions with stable as well as radioactive ion beams, Nuclear Astrophysics, Tools of Nuclear Science from instrumentation to facilities, as well as Applications of Nuclear Science. The last meeting of the series, CGS-11, was held in Pruhonice near Prague in the Czech Republic in 2002. Some of the other conferences were held in Santa Fe, USA (1999); Budapest, Hungary (1996); Fribourg, Switzerland (1993); Asilomar, USA (1990); Leuven, Belgium (1987); Knoxville, USA (1984); Grenoble, France (1981); Brookhaven, USA (1978); Petten, the Netherlands (1974); and Studsvik, Sweden (1969).

More than 150 scientists from 23 countries attended this conference. The program was rather densely packed for five and a half full days with many excellent contributions from the speakers as well as those who presented their posters making this conference a big success.

There are enormous changes in the field of interest for this conference since its inception. The first symposium of this series was held in Studsvik, Sweden in 1969. The closing down of the reactor at Studsvik unfortunately coincided with CGS-12. The main emphasis of the field has shifted towards Nuclear Astrophysics and towards other applications of nuclear science. New facilities have come on line specifically using neutrons with n_TOF at CERN, Geneva and DANCE at Los Alamos National Laboratory. Another pleasant surprise for us was the participation of significantly large numbers of women at this meeting.

The conference program at CGS-12 was scheduled to avoid any parallel sessions enabling all the participants to be a part of the entire program. In a special evening session – assisted by other spirits – we included a session on "Data for Nuclear- and Astrophysics Application" to address the needs of the collection, compilation and dissemination of nuclear data in our field. This was perhaps one of the liveliest sessions of the conference. An emerging trend is the shift towards web-based data bases and compilations. In addition to the existing nuclear data bases on nuclear structure and reaction cross-sections, new data bases addressing are emerging for the special needs of reaction-rates for nuclear astrophysics.

A special session of the conference was dedicated to the memory of our dear friend, colleague and one of the most influential founding members of neutron capture spectroscopy, **Dr. Dave Warner**. Dave Warner was also a member of the International Advisory Committee, one of the main invited speakers to CGS-12, and the first person to have registered on-line to attend CGS-12. Dave Warner was not only influential in developing the program in Nuclear Physics in the UK but also in Europe as well as the USA. He was on the advisory and executive committees of every facility in Europe, including the GSI-FAIR project, NuPecc, and the initial stages of planning for the Rare Isotope Accelerator in the US. Dave was a great friend, mentor, and colleague and he is missed more than we can say in words.

Professor Steven Yates of the University of Kentucky in Lexington funded a prize of $500 to be awarded to the best poster presented by a postdoctoral fellow or a graduate student. **The Founder's Award** was inaugurated in honor of the memories of Jean Kern, Raman Subramanian, and Gabor Molnar. All three of them have played crucial roles in establishing the CGS series of conferences as well as hosting one of the former symposia for CGS. The best posters were chosen by the selection committee consisting of Professors Art Champagne (University of North Carolina at Chapel Hill), Alejandro Frank (UNAM – Mexico City, Mexico) and Jan Jolie (University of Koeln, Germany). The committee had a tough time to decide on a winner and they suggested splitting the prize among two graduate students. The winners were Hye Young Lee and Smarjit Triambak both of them are presently graduate students at the University of Notre Dame Nuclear Structure Laboratory. Smarjit Triambak is presently located at the University of Washington in Seattle working with his advisor Alejandro Garcia, Hye Young Lee works in Nuclear Astrophysics at the University of Notre Dame with her advisor Michael Wiescher.

The CGS-12 welcoming reception was held in the courtyard of the University of Notre Dame Snite Museum of Art with live music provided by the **Nuclear Jazz Quarktet**. The collections of the Snite Museum of Art place it among the finest university art museums in the USA. The galleries were open for the conference participants and a special exhibit was held for CGS-12. The Fritz & Millie Kaeser Mestrovic Studio Gallery of the museum featured the **BRANCACCI PROJECT - PHASE ONE**. In this series of murals, Bill Sandusky, professor of art at Saint Mary's College (Notre Dame) had reinterpreted the fresco cycle painted in the Brancacci Chapel of Santa Maria del Carmine in Florence, Italy by Masaccio, Masolino and Filipino Lippi painted between about 1424 and 1480. Prof. Sandusky was available in the gallery on Sunday evening for a gallery talk and questions.

The conference excursion was to view the architectural styles of Chicago. We traveled by coach from Notre Dame to Chicago where we boarded "Chicago's First Lady". The boat took us on an architectural roundtrip on the Chicago river and we got a humbling view of the skyscrapers of the Windy City. The conference dinner was served on the boat and ended with a silhouette of the city of Chicago lit up at night from Lake Michigan.

The conference ended with an impressive performance "A Universe of Dreams" with the Ensemble Galilei with National Public Radio's Neal Conan. The performance consisted of music, poetry and stories with projected huge images from the Hubble Space Telescope.

We also would like to thank the members of the International Advisory Committee for their advice and guidance. They were: Juha Aysto (Univ. of Jyvaskyla, Finland), John Becker (LLNL, USA), Frantisek Becvar (Charles Univ., Czech Republic) Hans Boerner (ILL, Grenoble, France), Rick Casten (WNSL, Yale Univ., USA), Jolie Cizewski (Rutgers Univ., USA), I. M. Govil (Panjab University, India), William Gelletly (Univ. of Surrey, UK), Kris Heyde (Univ. of Gent, Belgium), Robert Janssens (ANL, USA), Jan Jolie (Univ. of Cologne, Germany), Rostislav Jolos (Dubna, Russia), Franz Kaeppeler (Karlsruhe, Germany), Shigeru Kubono (Univ. of Tokyo, Japan), Jan Kvasil (Charles Univ., Czech Republic), Alinka Lepine-Szily (Univ. of Sao Paulo, Brazil), Gary Mitchell (Duke Univ., USA), Yuri Oganessian (JINR, Dubna, Russia), W.P. Liu (CIAE, Beijing, China), Achim Richter (TU of Darmstadt, Germany), Klaus Schreckenbach (TU of Munich, Germany), John Sharpey-Schafer (iTHEMBA, South Africa), Peter von Brentano (Univ. of Cologne, Germany) Till von Egidy (Univ. of Munich, Germany), David Warner (Daresbury, UK), Hans Weidenmueller (Univ. of Heidelberg, Germany), Steve Wender (LANL, USA), Steve Yates (Univ. of Kentucky, USA).

The Program committee had the challenging job of rating the submitted abstracts and choosing the invited talks. We also thank them profusely. They were: Marialuisa Aliotta (Univ. of Edinburgh, UK), Yoram Alhassid (Yale Univ., USA), Carmen Angulo (Louvain La Neuve, Belgium), Bertram Blank (Univ. of Bordeaux, France), Mark Chadwick (LANL, USA), John D'Auria (Simon Fraser Univ., Canada), Jutta Escher (LLNL, USA), Paul Fallon (LBL, USA), Paul Garrett (Univ. of Guelph, Canada), Thomas Glasmacher (NSCL, USA), Bob Haight (LANL, USA), Paul Koehler (ORNL, USA) Reiner Kruecken (TU of Munich, Germany), Walt Kutchera (Vienna, Austria), Henryk Mach (Studsvik, Sweden), Paul Mantica (NSCL, USA), Alberto Mengoni (Univ of Bologna, Italy), Filomena Nunes (NSCL, USA), Grigory Rogachev (FSU, USA), Hendrick Schatz (NSCL, USA), Piet van Isacker (CAEN, France) and Dave Winchell (BNL, USA).

The smooth running of any conference is dependent on the session chairs and we would like to specially thank them. They were Rick Casten (WNSL, Yale Univ., USA), Alan Wuosmaa (Western Michigan, USA), Rick Firestone (LBL, USA), Grant Mathews (Notre Dame, USA), Art Champagne (North Carolina, USA), Jerry Wilhelmy (LANL, USA), Gabriela Popa (OSU in Mansfield, USA), Karl-Ludwig Kratz (Mainz, Germany), Hans Börner (ILL, France), Torben Laurtisen (ANL, USA), Denis De Frenne (Ghent, Belgium), James Kolata (Notre Dame, USA), Franz Kaeppeler (FZK, Germany), Norbert Pietralla (Suny, USA), Stefan Frauendorf (Notre Dame, USA), Filomena Nunes (NSCL, USA), and Alison Bruce (Brighton, UK).

We also would like to thank our fellow members of the local organizing committee who helped to contribute to the success of this meeting: Artur Teymurazyan, Michael Wiescher, Shawn O'Brien, P.J. LeBlanc, Umesh Garg, Stefan Frauendorf, and Philippe Collon.

The members of the advisory and program committees present at CGS-12 met for dinner in the Golden Dome of the University of Notre Dame. The outcome was a unanimous decision to have the next conference in Cologne, Germany for 2008. Professor Jan Jolie of the "Universität zu Köln" has agreed to organize the next conference (CGS-13). We are looking forward to the 13th International Symposium on Capture Gamma-Ray spectroscopy and Related Topics!

Dr. Andreas Woehr
and
Prof. Ani Aprahamian

NUCLEAR STRUCTURE
(ISOMERS AND GAMMA SPECTROSCOPY)

K-Isomers and Aspects of Nuclear Structure

G.D. Dracoulis

Department of Nuclear Physics, R.S.Phys.S.E., Australian National University, Canberra, A.C.T. 0200, Australia

Abstract. Gamma-ray spectroscopy measurements with Gammasphere and multi-nucleon transfer reactions with heavy beams provide a means of accessing deformed nuclei at stability, or on the neutron-rich side of the stability line. New results on the discovery of high-K isomers in the Er-Tm-Yb-Lu region are discussed with a focus on the energy systematics and the factors which may govern K-purity and therefore K-hindrances. These include random-mixing in regions of high level density, specific Coriolis mixing in configurations involving the $i_{13/2}$ neutron configuration, and chance degeneracies between an individual high-K state and a specific collective state which provide an opportunity to extract effective mixing matrix elements between states of very different-K.

Keywords: K-Isomers; reduced hindrances; chance degeneracies; interaction matrix elements.
PACS: 21.10.Tg;21.60.Ev;23.20.Lv;27.70.+q

INTRODUCTION

Recent studies have demonstrated the value of inelastic and deep-inelastic reactions with heavy-ion beams for the population and identification of nuclei near stability and into the neutron-rich region, nuclei which are generally inaccessible through fusion-evaporation reactions [1]. The presence of isomeric states adds an additional dimension of selectivity to spectroscopic studies, enhancing by orders of magnitude the sensitivity in reactions which inevitably populate a very broad range of nuclei. Nevertheless, while substantial progress has been made in identifying multi-quasiparticle intrinsic states in deformed nuclei and sometimes their associated collective structures, the progress has been essentially incremental since the cross-sections for transferring neutrons on to a given target diminishes rapidly with the number of transferred particles.

In the well-deformed nuclei near $A \sim 180$ the presence close to the Fermi surface of neutrons and protons with relatively high projections (Ω) of the particle angular momentum on the nuclear deformation axis, results in an abundance of multi-quasiparticle isomeric states. For lower proton numbers, the frequency of isomers may be reduced since the proton components are likely to be less favoured in energy terms, resulting in a competition between collective states built on lower-seniority configurations and multi-quasiparticle intrinsic states. The intrinsic states may be embedded in regions of relatively high level density rather than being yrast, raising questions as to the possible dilution of the K-quantum number due to random mixing, and other factors.

The main aspects I will cover here are some of the new energy systematics, anomalous hindrances and implications for K-purity revealed through the analysis of specific Coriolis mixing, and interaction strengths extracted from chance degeneracies between intrinsic and collective states. I will draw on a series of measurements which have produced a suite of new isomers and level schemes for the Lu isotopes $^{174-179}$Lu, the Yb isotopes $^{174-178}$Yb, a range of Tm isotopes and for several neutron-rich Er isotopes. Some results have been reported recently [2, 3, 4].

EXPERIMENTAL DETAILS

The main body of data comes from a series of measurements using 6.0 MeV per nucleon ^{136}Xe beams provided by the ATLAS Facility at Argonne National Laboratory, using a configuration primarily designed for observation of transitions in nuclei which have stopped in the target or backing. The experimental conditions included the use of nanosecond pulses, separated by about 825 ns, with beams incident on a variety of targets. The target thicknesses were chosen so as to integrate over the main yield of inelastic processes from $\sim 20\%$ above and down to the Coulomb barrier. Gamma-rays were detected with Gammasphere [5], with ~ 100 detectors in operation. Recent measurements on ^{176}Yb complement our earlier measurements on targets of ^{175}Lu, ^{176}Lu, and ^{174}Yb. As well as the nanosecond-pulsed beams a series of measurements using a macroscopically chopped beam with (beam on)/(beam off) conditions ranging from $10\mu s/30\mu s$ to $1ms/3ms$ and with out-of-beam dual coincidence events recorded in reference to a precision clock, have been carried out, allowing long-lifetimes to be isolated gated on specific γ-ray cascades.

$K^\pi = 8^-$ ISOMERS IN THE N = 106 ISOTONES

Although very weakly populated, it has been possible to characterise a long-lived isomer in the neutron-rich nucleus ^{174}Er, revealing its yrast sequence for the first time [6]. The population of the new isomer which is associated with clear characteristic Er X-rays, is approximately 1.5% relative to the well-known ^{176}Yb, 8^- isomer, consistent with two-proton transfer from the target. Intensity balances for the 163 keV transition which depopulates the ^{174}Er isomer give a total conversion coefficient which is only consistent with an E1 multipolarity, supporting the proposed $K^\pi = 8^-$ assignment and therefore, association with the two-quasineutron 8^- isomer known in the N = 106 isotones. Note that because of the saturation of deformation in mid-shell, differentiating between excitation of the target and nearby (unknown) isotopes can be challenging because of the similarity in transition energies expected.

Energy Systematics

The systematics incorporating the new scheme for ^{174}Er, are shown in Fig. 1. These isomers from the same 2-quasineutron configuration: ν^2 $7/2^-[514] \otimes 9/2^+[624]$; $K^\pi = 8^-$, have lifetimes ranging from seconds down to microseconds. It is remarkable that the isomer occurs from essentially the middle of the neutron shell in the neutron-rich nucleide ^{174}Er, in all well-deformed isotones up to Os, through the transitional region (^{184}Pt), into the region of oblate-prolate shape co-existence (^{186}Hg) and finally in the very neutron-deficient isotope ^{188}Pb, where its presence has been taken as evidence for a prolate sub-minimum in a nucleus with triple shape co-existence [7, 8]. The robust nature

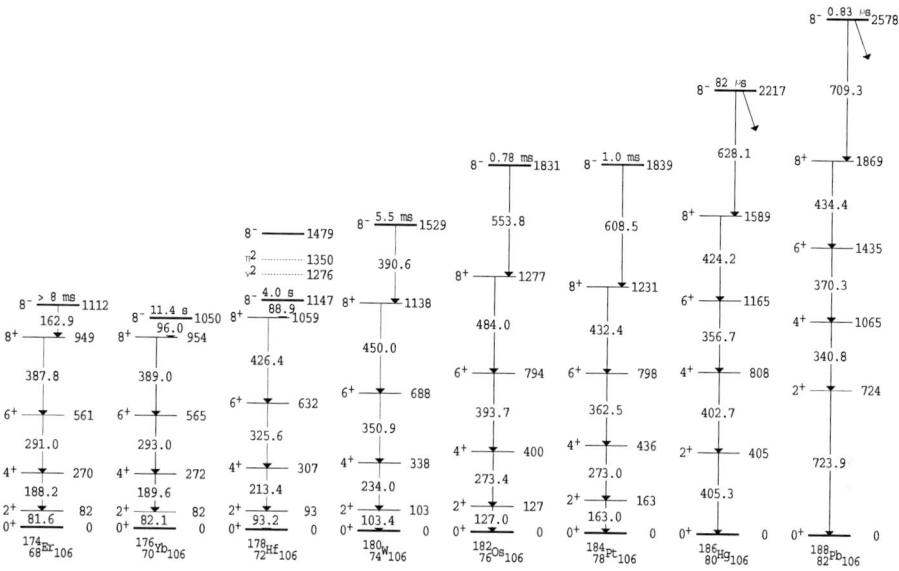

FIGURE 1. Energy systematics of the two-quasineutron $K^\pi = 8^-$ isomer for the N = 106 isotones including new results for ^{174}Er [6]. Note that two 8^- states occur in ^{178}Hf because of mixing with an alternative 2-quasiproton configuration. The unperturbed levels deduced previously are shown by dashed lines in that case. As indicated by the diagonal arrows, additional branches occur in ^{186}Hg and ^{188}Pb.

of the configuration is only disturbed in ^{178}Hf where strong mixing occurs with an 8^- state from the two-quasiproton configuration $\pi^2 7/2^+[404] \otimes 9/2^-[514]$, with the result that the lower experimental state is strongly mixed, being $64\%\nu^2 + 36\%\pi^2$ [9, 10, 11].

NORMAL AND ANOMALOUS K-HINDERED DECAYS

The isomerism arises in the 8^- cases from the forbidden nature of the E1 decay between the $K^\pi = 8^-$, 2-quasiparticle state and the $K = 0$ ground-state. Forbiddenness can be classified in terms of the shortfall in K-change, $\nu = \Delta K - \lambda = 7$, where λ is the multipolarity. Reduced values of the hindrance, $f_\nu = F^{1/\nu}$ with $F = \frac{\tau}{\tau_W}$, the ratio of the partial mean-lives to the Weisskopf estimates τ_W, are usually compared. Normal *reduced*-hindrances generally fall in the range $f_\nu = 30$-200.

"Forbidden" transitions presumably proceed through small admixtures of other K-components which will naturally be present, for example, when Coriolis mixing occurs. Other mechanisms which may effect the "goodness" of the nominal K-quantum number include

- statistical mixing which might be expected when high-K states are embedded in a region of high level density above the yrast line
- local (two-state) mixing due to chance degeneracies
- asymmetric shape oscillations - the breaking of the axial symmetry which defines the projection K

Transition matrix elements will also depend on orbital changes, and complex configuration changes, that is, re-arrangement of a large number of orbitals should further retard transition rates. These are all factors which need to be considered but in general they are not taken into account on an equal footing.

Unusually slow and unusually fast transitions in ^{174}Yb

Figure 2 shows part of the level scheme deduced for the stable nuclide ^{174}Yb from our recent measurements [3]. A feature of the scheme is the long-lived $K^\pi = 6^+$ isomer

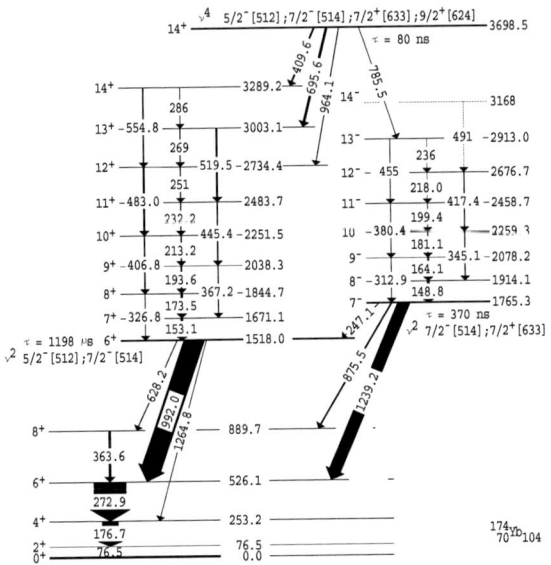

FIGURE 2. Partial level scheme for ^{174}Yb. (Adapted from ref. [3]

whose spin/parity had been the subject of considerable conjecture in the past, but which is largely confirmed by the new spectroscopic results and also by the properties of the rotational band based on the isomer. In addition the (previously unknown) 7^- configuration, which had been a competing assignment for the 1518 keV, 1.2ms isomer, is now identified, removing it as an alternative. The problematic factor remains the extremely hindered nature of the 1265 keV E2 transition from the $K^\pi = 6^+$ state to the 4^+ ground state band member. It has a reduced hindrance of $f_v = 352$, the largest in the region, corresponding to a strength of only 7×10^{-11} Weisskopf units. (This is notably the only case to date where such a transition is *more* hindered than is predicted on the basis of a tunneling prescription.) Its strength is in stark contrast to the strength of the 6-fold forbidden 964 keV E2 transition connecting the 14^+, 80 ns isomer and the 6^+ band. Although the forbiddenness is larger, the transition strength of 3×10^{-5} Weisskopf units is higher, and translates to a reduced hindrance of $f_v = 5.7$, an anomalously low value.

This could be partly because the v^4 configuration assigned to the 14^+ isomer involves two $i_{13/2}$ neutrons, the so-called t-band or Fermi-aligned configuration, the partner to the $K \sim 1$ aligned neutron configuration that is responsible for the backbending that occurs near N = 104-106 (see, for example, ^{179}W and ^{180}W [12, 13]). This configuration will have an average projection of $K = 8$ for the component from the aligned neutrons, with a specific form of K-mixing [14, 15, 16] although the present analysis of this K-mixing suggests that it is unlikely to be the dominant factor [3].

Instead, the reduction in hindrance is more likely attributable to the effect of semi-random mixing with background states, since the 14^+ isomer is significantly non-yrast and therefore is presumably located in a region of high level density. As shown in ref [3] the hindrance observed is consistent with the density-of-states formulation applied by Walker et al. [17] to isomers that are progressively further above the yrast line, although it should be noted that the fluctuation in the hindrance values is large.

The other notable point is that the 14^+ isomer arises from a 4-quasineutron configuration. This adds to a pattern that is emerging in nuclei near Z = 70. In the Hf cases (Z=72), multi-quasiparticle states (seniority >2) composed mainly of a single nucleonic species are less likely to compete against "balanced" configurations containing approximately equal numbers of quasiprotons and quasineutrons.

CORIOLIS K-MIXING AND CHANCE DEGENERACIES IN ^{174}LU

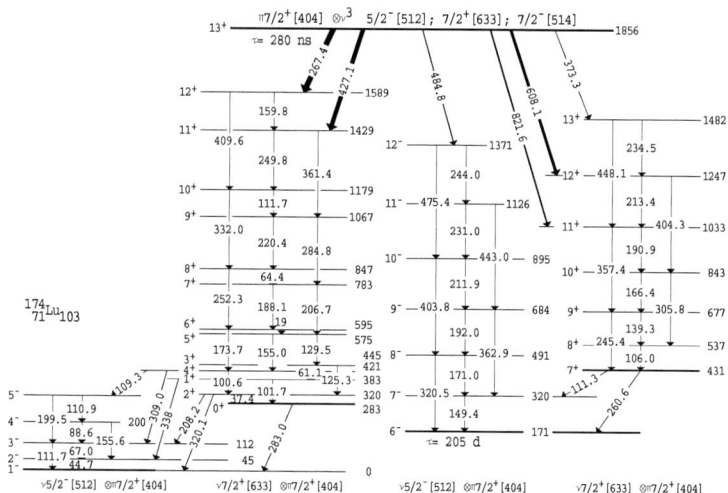

FIGURE 3. Partial level scheme and decay of a new isomer in ^{174}Lu [18].

Extensive results have also been obtained for the odd-odd nucleus ^{174}Lu [18] that may throw some light on the question of Coriolis-mixing. The partial level scheme given in Fig. 3 shows the multiple decays of a newly-identified $K^\pi = 13^+$, 280ns isomer in that nucleus.

Isomers and $K = \Omega_1 \pm \Omega_2$ Partners

The interesting facet is that the decay proceeds through rotational bands obtained from both parallel and anti-parallel couplings of the $7/2^+[404]$ proton and the $7/2^+[633]$ neutron which result in (nominal) $K^\pi = 0^+$ and 7^+ bands. The neutron Fermi level is close to the $7/2^+[633]$ orbital from the $i_{13/2}$ neutron, hence the bands will be Coriolis-mixed, in a precisely related fashion. Furthermore, the $K^\pi = 13^+$ isomer also contains the $7/2^+[633]$ neutron and will have some Coriolis mixing.

TABLE 1. Hindered transitions in the decay of the $K^\pi = 13^+$ isomer in ^{174}Lu [18].

E_γ (keV)	K_f	$M\lambda$	ν	f_ν nominal	corrected
267	0	M1	12	3.0	3.6
427	0	E2	11	1.9	2.1
485	6	E1	6	36	48
822	7	E2	4	17	31
608	7	M1	5	26	39
373	7	M1	5	28	43

As can be seen from table 1, the reduced hindrances for decays to the $K^\pi = 0^+$ band evaluated using the nominal K-values are very low, while those to the $K^\pi = 7^+$ band are borderline. Figure 4 shows the amplitudes of the K-distributions calculated using a band-

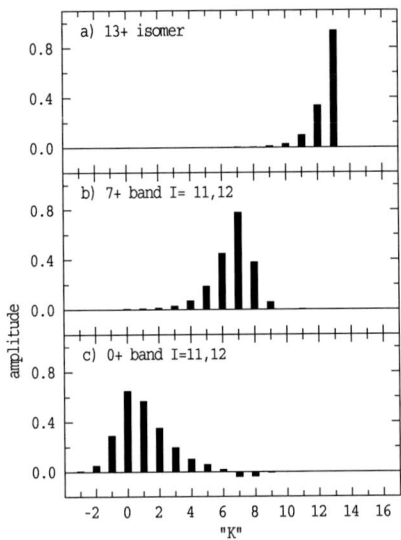

FIGURE 4. Distribution of mixed-amplitudes for the $K^\pi = 13^+$ isomer in ^{174}Lu and the $K^\pi = 0^+$ and $K^\pi = 7^+$ bands to which it decays, from a simplified particle-rotor model.

mixing model where the Coriolis mixing through the set of $i_{13/2}$ orbitals is included, and coupled to either the $7/2^+[404]$ proton orbital as a spectator, or to a $K^\pi = 19/2^+$

core from the v^2 5/2$^-$[512];7/2$^-$[514]⊗π7/2$^+$[404] coupling. This leads to a 14×14 matrix for the highest spins in each case, as formulated previously for a different case [19]. As can be seen from Fig. 4 there are significant components of $K < 13$ in the $K = 13^+$ band head, and a broad but non-symmetrical distribution around $K = 0$ and $K = 7$ for the states in the nominal 0^+ and 7^+ bands. When the calculated amplitudes are included in the derivation of a reduced hindrance by appropriate weighting of the inverse of the forbiddenness (1/v), the corrected values given in the last column of table 1 are obtained. The effective values for decays to the 7^+ band are significantly larger, and close to normal. Coriolis mixing therefore provides part of the explanation for the anomalously fast rates, but as is evident from the table, it does not solve the problem of the transitions to the 0^+ band.

Chance Degeneracies and Interaction Strengths

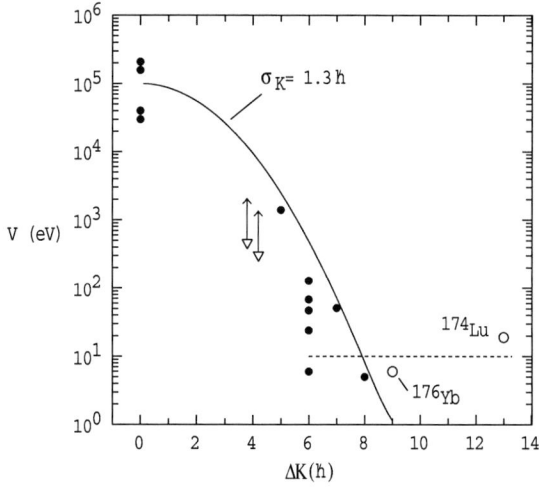

FIGURE 5. Interaction matrix elements as a function of the nominal K-difference deduced from cases of chance degeneracies [23]. The solid curve shows the expected dependence for two gaussian K-distributions with $\sigma = 1.3$ ℏ.

The cause can be isolated by extrapolating the energies of states in the 0^+ band. This extrapolation indicates that the 13^+ collective state would have fallen at about 1868 keV (with an uncertainty of about 0.5 keV), very close to the 13^+ intrinsic state. This is apparently another example (see [4, 20, 21]) of a chance degeneracy between a high-K state and a rotational state from a lower-K configuration, providing an opportunity to extract the interaction. The key factor is that although the mixing amplitudes may be very low, the admixture of a collective component into a wave function where other transitions are forbidden, has a very large effect on the apparent hindrances. The observed strength for the 427 keV E2 transition corresponds to a collective admixed amplitude of 1.6×10^{-3}, which can be induced through a mixing matrix element of only 19 eV. Figure 5 shows a collection of such matrix elements extracted from a variety of data satisfying the criterion [22] that the branching ratios should match the in-band ratios, as a function of the difference in K. The curve superimposed shows the expected interaction for a simple overlap of two Gaussian distributions separated by ΔK, each of the same full-width of 2.2σ, taking (arbitrarily) $V_0 = 100$ keV so that $V = V_0 \times \exp\{-(\frac{\Delta K}{2\sigma})^2\}$. The rapid fall-off in experimental values is a clear indication that the distributions in K-space are relatively narrow for both the initial and final configurations. Superimposed on the general dependence predicted will be the possibly large fluctuations in hindrance caused by configuration differences and Coriolis mixing.

There are five cases, for example, of $\Delta K = 6$ transitions, with matrix elements ranging from 6 eV to 128 eV and although not shown here, the distribution of these values is roughly correlated with seniority (and therefore K-value). Further examination may provide a clue to the factors causing this spread. The dashed line in the figure is arbitrarily drawn at 10 eV - presumably there will be some level below which interactions between nuclear states always occur, but it is the case that these values are at the atomic scale already, indicating, perhaps paradoxically since they are deduced from mixing, the essential purity of the K-quantum number.

ACKNOWLEDGMENTS

The results I have shown here are drawn from measurements and analyses in progress, to be published in due course. I am grateful to my colleagues for support, particularly to Greg Lane and Filip Kondev who have carried the substantial burden of the analysis.

REFERENCES

1. C. Wheldon, *et al*. Phys. Lett. **B 425**, 239 (1998)
2. G.D. Dracoulis *et al*. Phys. Lett. B **584**, 22 (2004)
3. G.D. Dracoulis *et al*. Phys. Rev. C **71**, 044326 (2005)
4. F.G. Kondev *et al*. Eur. Phys. Journ. A, **22**, 23 (2004)
5. R.V.F. Janssens and F.S. Stephens, Nucl. Phys. News. **6**, 9 (1996)
6. G.D. Dracoulis *et al*. to be published
7. G.D. Dracoulis *et al*., Phys. Rev. C **60**, 014303 (1999)
8. G.D. Dracoulis *et al*. Phys. Rev. C **69**, 054318 (2004)
9. F.W.N. de Boer *et al*., Nucl. Phys. A **263**, 397 (1976)
10. T.L. Khoo and G. Lovhoiden, Phys. Lett. B **67**,271 (1977)
11. S.M. Mullins, G.D. Dracoulis, A.P. Byrne, T.R. McGoram, S. Bayer, W.A. Seale and F.G. Kondev, Phys. Lett. B **393**, 279 (1997); erratum, Phys. Lett. B **400**, 401 (1997)
12. P.M. Walker *et al*., Nucl. Phys.A **568**, 397 (1994)
13. P.M. Walker *et al*. Phys. Lett. B **309**, 17 (1993)
14. S. Frauendorf, Nucl. Phys. A **557**, 259c (1993)
15. S. Frauendorf, Phys. Scr. **24**, 349 (1981)
16. S. Frauendorf, Nucl. Phys. A **677**, 115 (2000)
17. P.M. Walker *et al*., Phys. Lett. B **408**, 42 (1997)
18. F.G. Kondev *et al*. to be published
19. G.D. Dracoulis, C. Fahlander and M.P. Fewell, Nucl. Phys. **A383**, 119 (1982)
20. T.R. McGoram *et al*. Phys. Rev. C **62**, 031303(R) (2000)
21. F.G. Kondev *et al*. Phys. Rev. C **59**, R575 (1999)
22. T.R. Saitoh *et al*., Phys. Scr. **T88**, 67 (2000)
23. G.D. Dracoulis, T.R. McGoram and F.G. Kondev, to be published

In-Beam γ-Spectroscopy of Low-Spin Mixed-Symmetry States of ^{138}Ce with Gammasphere in Singles-Mode

N. Pietralla*, G. Rainovski*, T. Ahn*, M.P. Carpenter[†], R.V.F. Janssens[†], C.J. Lister[†] and S. Zhu[†]

NSL, SUNY at Stony Brook, NY 11794-3800
[†] *Physics Division, Argonne National Laboratory, Argonne, IL 60439*

Abstract. Gamma-rays from the nuclide ^{138}Ce have been measured with the Gammasphere-array following Coulomb excitation. Beams of ^{138}Ce ions were focussed on a carbon target at beam energies of 480 and 400 MeV. Gamma-ray yields and relative Coulomb excitation cross sections were measured in singles-mode. $M1$, $E2$, and $E3$ transition matrix elements from low-spin states were measured relative to the known $B(E2; 0_1^+ \to 2_1^+)$. The $2^+ \to 2_1^+$ $M1$ strength distribution from the lowest six 2^+ states up to 2.7 MeV enables us to identify the 2_4^+ state of ^{138}Ce as the dominant fragment of the one-phonon $2_{1,\text{ms}}^+$ mixed-symmetry state with F-spin quantum number $F_{\text{max}} - 1$. From its mixing with the nearby 2_3^+ state an F-spin mixing matrix element of 44 keV can be estimated.

Keywords: *radioactivity:* Coulomb excitation of ^{138}Ce on carbon target, $E_{\text{beam}} = 400, 480$ MeV; *measured:* γ-singles, $\gamma\gamma$-coinc., $I_\gamma(\theta)$, Gammasphere-array; *deduced:* E_x, J^π, multipole mixing ratio δ, COULEX cross section, EM matrix elements, mixed-symmetry state, F-spin mixing
PACS: 21.10.Re, 23.20.Js, 25.70.De, 27.60.+j

INTRODUCTION

Atomic nuclei are naturally occuring examples of strongly-correlated many-body quantum systems formed by two kinds of equivalent particles, protons and neutrons. Therefore, besides the study of the complicated nuclear forces, nuclear structure physics addresses three aspects that are of general interest for such systems. (*i*) The quantum nature of the system induces a shell structure. (*ii*) The many-body character induces collective phenomena due to the strong correlations between the particles. (*iii*) The equivalence of the two components (with respect to their interactions) induces isospin symmetry. In many ways these three aspects form the motivations for much of contemporary nuclear structure research (see, *e.g.*, [1]).

Particularly appealing objects of study are those nuclear structures that combine these three key-aspects, shell structure-dependence, collectivity, and the isospin degree of freedom, such as the isovector quadrupole excitations of the valence shell of heavy nuclei. These nuclear structures have been modeled [2] in terms of proton-neutron Mixed-Symmetry States (MSSs) in the framework [3] of the interacting boson model (IBM-2). The IBM-2 represents an effective phenomenological model for collective excitations of the nuclear valence shell and describes the proton-neutron degree of freedom through the inclusion of N_π proton bosons and N_ν neutron bosons where N_ρ

is taken as half the number of valence particles (or holes) of isospin ρ.

The strong forces between valence protons and neutrons leads to a coupling of collective proton and neutron excitations. This coupling can be quantified in the IBM-2 by the concept of F-spin [4]. Proton bosons and neutron bosons are considered as an F-spin doublet with projections $F_z = +1/2$ (proton boson) and $-1/2$ (neutron boson). F-spin for "elementary" bosons is analogous to isospin for "elementary" nucleons. In the F-spin limit [5], *i.e.*, if F-spin is a good quantum number, the boson wave functions with maximum F-spin $F_{max} = (N_\pi + N_\nu)/2$ are totally symmetric with respect to the mutual exchange of any two boson isospin labels and hence they are called Full-Symmetry States (FSSs). They correspond to wave functions of the IBM-1 where no distinction is made between proton bosons and neutron bosons. The strong coupling between proton and neutron bosons energetically favors the FSSs. This fact is considered one of the reasons why isoscalar collective models such as the IBM-1 are successful in describing many features of nuclear collective structures at low excitation energy [6]. MSSs are those boson states that do not have maximum F-spin, $F \leq F_{max} - 1$. They contain at least one pair of bosons consisting of one proton boson and one neutron boson that are coupled anti-symmetrically. Due to their isovector character their outstanding signature is the occurence of strong $M1$ transitions to FSSs with matrix elements of the order of 1 μ_N [7].

Four key questions arise. At what energy do the MSSs occur in heavy nuclei and what are their properties? How do their properties vary as a function of valence particle numbers and underlying shell model orbitals? To what extent is F-spin a good quantum number? How could knowledge on MSSs be extended to exotic nuclei?

Many authors have approached these questions previously in many ways, see, *e.g.*, [8, 9, 10, 11, 12, 13, 14, 15, 16, 17, 18] and references therein. Considerable progress was recently made on the investigation of MSSs of vibrational nuclei, *e.g.* [19, 20, 21, 22, 23, 24, 25, 26, 27, 28, 29]. We report here on a measurement of an F-spin mixing matrix element for the $N = 80$ nucleus ^{138}Ce which has been determined directly from the lowest-lying state with predominantly mixed-symmetry character, the one-phonon $2^+_{1,ms}$ state. No MSS of ^{138}Ce was previously known.

EXPERIMENT

In order to identfy the one-phonon $2^+_{1,ms}$ state of the vibrational nucleus ^{138}Ce we have performed a Coulomb excitation experiment at Argonne National Laboratory. A beam of ^{138}Ce ions was delivered by the ATLAS accelerator with an intensity of about 1 pnA. It has been extracted from natural cerium in which ^{138}Ce has a 0.25% abundance. The ion beams with energies of 480 MeV and 400 MeV bombarded a 1 mg/cm^2 thick carbon target for 15 h and 5 h, respectively. The γ-rays issued by the predominantly one-step Coulomb-excited projectiles were detected with the Gammasphere array which consisted of 98 HPGe detectors arranged in 17 rings around the beam axis. Gammasphere was used in singles mode at an average counting rate of about 4000 events per second. A total of 2.4×10^8 events of γ-ray fold 1 or higher were collected at a beam energy of 480 MeV in 15 h beam time.

Figure 1 shows a part of the γ-ray energy spectrum in coincidence with the $2^+_1 \to 0^+_1$

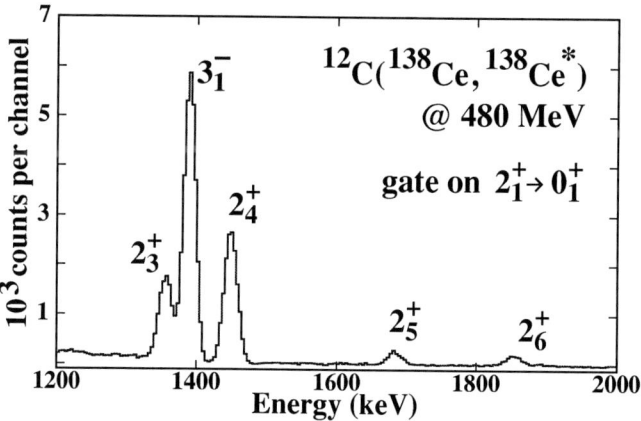

FIGURE 1. Doppler-corrected background-subtracted γ-ray spectrum observed with GAMMASPHERE in the Coulomb excitation reaction of a ^{138}Ce ion beam at 480 MeV on a 1 mg/cm^2 thick natural carbon target in coincidence with the 789-keV $2_1^+ \to 0_1^+$ transition of ^{138}Ce. The five transitions shown feed directly the 2_1^+ state and originate from the states indicated by the labels.

transition in ^{138}Ce. The velocity of the γ-ray emitting ^{138}Ce ejectiles amounted to $v/c \approx 6.9\%$. This induced a Doppler-broadening of the γ-ray lines leading to an effective energy resolution of about 1.4%. Two new γ-rays at 1354 keV and at 2143 keV were observed, the first one of them being in coincidence with the 789-keV transition from the 2_1^+ state to the ground state of ^{138}Ce. This γ-ray coincidence, the 2143-keV ground state transition, and the predominantly one-step population mechanism prove the existence of the previously unknown 2_3^+ state of ^{138}Ce at 2143 keV. Beside the 0_2^+, 3_1^-, and 4_1^+ states, the first six $2_{1,2,3,4,5,6}^+$ states up to an excitation energy of 2.7 MeV were observed.

Assignments of γ-ray multipolarities and spin quantum numbers are based on angular γ-ray intensity distributions. The 722-keV $2_2^+ \to 2_1^+$ transition is assigned 80(5)% E2 while the $2_{3,4}^+ \to 2_1^+$ transitions at 1354 and 1448 keV contain 59(4)% and 97(2)% M1 contribution, respectively. Measurement of the COULEX cross sections relative to the 2_1^+ state with a ground state transition strength of $B(E2; 2_1^+ \to 0_1^+) = 21.2(14)$ W.u. [30] yields information on the $B(E2; 2_i^+ \to 0_1^+)$ transition strength distribution. Observed decay branching ratios $I_\gamma(2_i^+ \to 2_1^+)/I_\gamma(2_i^+ \to 0_1^+)$ and deduced E2/M1 multipole mixing ratios for the $2_i^+ \to 2_1^+$ transitions also enable us to determine the $B(E2; 2_i^+ \to 2_1^+)$ and $B(M1; 2_i^+ \to 2_1^+)$ transition strength distributions.

DISCUSSION

The $B(M1; 2_i^+ \to 2_1^+)$ strength distribution up to 2.7 MeV is found to be dominated by the 2_4^+ state at 2.237 MeV with an absolute M1 matrix element of $|\langle 2_1^+ \| M1 \| 2_4^+ \rangle| = 0.78 \mu_N$. This state can be considered as the dominant fragment of the one-phonon $2_{1,\text{ms}}^+$ state of ^{138}Ce with $F = F_{\text{max}} - 1$. Its excitation energy corresponds within

5% to the excitation energy of the $2^+_{1,\text{ms}}$ state of the neighboring even-even $N = 80$ isotone ^{136}Ba which has been previously identified at 2.129 MeV from the transition strength $B(M1; 2^+_4 \to 2^+_1) = 0.26(3)\,\mu_N^2$ deduced from photon scattering data [19]. This corresponds to a larger $M1$ matrix element of $|\langle 2^+_1 \| M1 \| 2^+_4 \rangle|(^{136}\text{Ba}) = 1.14\,\mu_N$.

In contrast to the situation in ^{136}Ba, the nearby 2^+_3 state of ^{138}Ce at 2.143 MeV also acquires a considerable $M1$ strength with an $M1$ matrix element of $|\langle 2^+_1 \| M1 \| 2^+_4 \rangle| = 0.54\,\mu_N$. We interpret this situation as a fragmentation of the $2^+_{1,\text{ms}}$ one-phonon mode [13, 15, 31]. The $2^+_{3,4}$ states share the total $M1$ strength $\sum B(M1; 2^+_{3,4} \to 2^+_1) = 0.18\,\mu_N^2$ which is about 30% less than in ^{136}Ba. These two states are separated from the next 2^+ states by more than 230 keV. We, thus, consider a two-state mixing scenario

$$|2^+_3\rangle = \alpha|2^+_{\text{FSS}}\rangle + \beta|2^+_{1,\text{ms}}\rangle$$
$$|2^+_4\rangle = -\beta|2^+_{\text{FSS}}\rangle + \alpha|2^+_{1,\text{ms}}\rangle$$

between the $2^+_{1,\text{ms}}$ one-phonon MSS and a close-lying FSS[1]. Since the 2^+_1 state can be considered as a FSS and since $M1$ transitions between any two FSSs are forbidden, the ratio of the wave function probabilities can be obtained from the ratio of the $M1$ transition strengths to the 2^+_1 state

$$\frac{\beta^2}{\alpha^2} = \frac{B(M1; 2^+_3 \to 2^+_1)}{B(M1; 2^+_4 \to 2^+_1)} \tag{1}$$

which results in $\alpha^2 = 68(3)\%$ and $\beta^2 = 32(3)\%$. From the energy separation of 94 keV between the $2^+_{3,4}$ states a mixing matrix element of $V_{F-\text{mix}} = 44(4)$ keV can be concluded. A similar analysis for the data on the isotone ^{136}Ba results in a much smaller mixing matrix element of $V_{F-\text{mix}}(^{136}\text{Ba}) < 10$ keV. Since the neutron configuration is not expected to differ much for the isotones ^{136}Ba$_{80}$ and ^{138}Ce$_{80}$, it is suggested that this difference in size of the F-spin mixing matrix elements is related to the proton configurations. Ground state spins for proton-odd $N = 80$ isotones and the shell model indicate the $\pi(1g_{7/2})$ sub-shell closure for cerium isotopes at proton number $Z = 58$. While already the leading one-phonon 2^+ proton configuration requires promotion of protons to the $\pi(2d_{5/2})$ sub-shell in $^{138}_{58}$Ce the corresponding configuration for $^{136}_{56}$Ba can still be formed within the $\pi(1g_{7/2})$ sub-shell [32]. Thus, the one-phonon $2^+_{1,\text{ms}}$ state of ^{136}Ba is expected to consist of considerably simpler configurations than the more highly excited predominantly symmetric states that surround it at about 2 MeV excitation energy. This prevents strong mixing between the $2^+_{1,\text{ms}}$ state and nearby 2^+ states in ^{136}Ba in contrast to the situation in ^{138}Ce. This mechanism might be considered as a *shell-stabilization of mixed-symmetry structures*. This scenario is consistent with the observed reduction of $M1$ strength in ^{138}Ce with respect to ^{136}Ba.

---

[1] The one-phonon $2^+_{1,\text{ms}}$ state is the lowest MSS in a vibrational IBM-2 spectrum and might thus be surrounded only by FSSs or non-collective states outside of the IBM

We thank the ion-source group at ATLAS for the preparation of the excellent beams of ^{138}Ce ions. We gratefully acknowledge the support by the *NSF* under grant No. PHY 0245018 and by the U. S. Department of Energy, Office of Nuclear Physics, under contract No. W-31-109-ENG-38 and grant No. DE-FG02-04ER41334.

REFERENCES

1. NSAC Long Range Plan 2002,
   http://www.sc.doe.gov/production/henp/np/nsac/docs/LRP_5547_FINAL.pdf
2. F. Iachello, *Lecture Notes on Theoretical Physics*. Groningen, 1976;
   T. Otsuka, *Boson Model of Medium-Heavy Nuclei*. Ph.D. thesis, University of Tokyo, 1978.
3. F. Iachello and A. Arima, *The interacting boson model*, (Cambridge Univ. Press, Cambridge, 1987).
4. T. Otsuka, A. Arima, F. Iachello, Nucl. Phys. **A 309**, 1 (1978).
5. P. Van Isacker, K. Heyde, J. Jolie, and A. Sevrin, Ann. Phys. (NY) **171**, 253 (1986).
6. R.F. Casten and D.D. Warner, Rev. Mod. Phys. **60**, 389 (1988).
7. F. Iachello, Nucl. Phys. **A 358** (1981) 89c; Phys. Rev. Lett. **53** (1984) 1427.
8. D. Bohle, A. Richter, W. Steffen, A.E.L. Dieperink, N. LoIudice, F. Palumbo, and O. Scholten, Phys. Lett. **B137**, 27 (1984).
9. A. Richter, Prog. Part. Nucl. Phys. **34**, 261 (1995).
10. W.D. Hamilton, A. Irbäck, and J.P. Elliott, Phys. Rev. Lett. **53** (1984) 2469.
11. H. Harter, P. von Brentano, A. Gelberg, and R.F. Casten, Phys. Rev. C **32**, 631 (1985).
12. K.P. Lieb, H.G. Börner, M.S. Dewey, J. Jolie, S.J. Robinson, S. Ulbig, and Ch. Winter, Phys. Lett. **B215**, 50 (1988).
13. G. Molnár, R.A. Gatenby, S.W. Yates, Phys. Rev. C **37**, 898 (1988).
14. R. De Leo *et al.*, Phys. Lett. **B226**, 5 (1989).
15. J.R. Vanhoy, J.M Anthony, B.M. Haas, B.H. Benedict, B.T. Meehan, S.F. Hicks, C.M. Davoren, C.L. Lundstedt, Phys. Rev. C **52**, 2387 (1995).
16. P.E. Garrett, H. Lehmann, C.A. McGrath, Minfang Yeh, S.W. Yates, Phys.Rev.C **54**, 2259 (1996).
17. A. Giannatiempo, A. Nannini, A. Perego, P. Sona, D. Cutoiu, Phys. Rev. C **53**, 2770 (1996).
18. A. Leviatan, J.N. Ginocchio, Phys. Rev. C **61**, 024305 (2000).
19. N. Pietralla *et al.*, Phys. Rev. C **58**, 796 (1998).
20. N. Pietralla, C. Fransen, D.Belic, P. von Brentano, C. Frießner, U. Kneissl, A. Linnemann, A. Nord, H.H. Pitz, T. Otsuka, I. Schneider, V. Werner, I. Wiedenhöver, Phys. Rev. Lett. **83**, 1303 (1999).
21. N. Pietralla, C. Fransen, P. von Brentano, A. Dewald, A. Fitzler, C. Frießner, J. Gableske, Phys. Rev. Lett. **84**, 3775 (2000).
22. C. Fransen, N. Pietralla, P. von Brentano, A. Dewald, J. Gableske, A. Gade, A. Lisetskiy, and V. Werner, Phys. Lett. B **508**, 219 (2001).
23. N. Pietralla, C.J. Barton III., R. Krücken, C.W. Beausang, M.A. Caprio, R.F. Casten, J.R. Cooper, A.A. Hecht, H. Newman, J.R. Novak, and N.V. Zamfir, Phys. Rev. C **64**, 031301 (2001).
24. V. Werner *et al.*, Phys. Lett. **B 550**, 140 (2002).
25. C. Fransen *et al.*, Phys. Rev. C **67**, 024307 (2003).
26. C. Fransen, N. Pietralla, A.P. Tonchev, M.W. Ahmed, J. Chen, G. Feldman, U. Kneissl, J. Li, V.N. Litvinenko, B. Perdue, I.V. Pinayev, H.-H. Pitz, R. Prior, K. Sabourov, M. Spraker, W. Tornow, H.R. Weller, V. Werner, Y.K. Wu, and S.W. Yates, Phys. Rev. C **70**, 044317 (2004).
27. C. Fransen, V. Werner, D. Bandyopadhyay, N. Boukharouba, S.R. Lesher, M.T. McEllistrem, J. Jolie, N. Pietralla, P. von Brentano, and S.W. Yates, Phys. Rev. C **71**, 054304 (2005).
28. S.W. Yates, N. Orce *et al.*, contribution to this conference.
29. P. von Neumann-Cosel *et al.*, contribution to this conference.
30. S. Raman, C.W. Nestor, Jr, and P. Tikkanen, At. Data Nucl. Data Tab. **78**, 1 (2001).
31. D. Bandyopadhyay, C.C. Reynolds, C. Fransen, N. Boukharouba, M.T. McEllistrem, S.W. Yates, Phys. Rev. C **67**, 034319 (2003).
32. N. Lo Iudice and Ch. Stoyanov, Phys. Rev. C **65**, 064304 (2002).

Nuclear isomers: stepping stones to the unknown

P.M. Walker

Department of Physics, University of Surrey, Guildford GU2 7XH, UK

Abstract. The utility of isomers for exploring the nuclear landscape is discussed, including their role in superheavy-element research, and the possibility of observing neutron radioactivity. Emphasis is given to K isomers in deformed nuclei. Transition rates are examined in the $N_p N_n$ scheme for 2- and 3-quasiparticle K-isomer decays, and in connection with level densities for higher quasiparticle numbers.

Keywords: nuclear structure, isomers, limits of stability, high spin
PACS: 21.10.-k, 27.70.+q

INTRODUCTION

Nuclear isomers continue to make key contributions to the development and understanding of nuclear structure physics [1, 2]. Isomers are widely distributed in neutron number, proton number, excitation energy and angular momentum. With half-lives ranging from nanoseconds to years, they offer a variety of opportunities to explore unusual and extreme states of nuclei. The understanding of their occurrence and degree of stability promises additional nuclear structure insights, with the potential for novel applications.

Figure 1 shows a restricted set of isomers, with long half-lives and high excitation energies. It is clear that many such isomers are correlated with shell closures. These can be categorised as spin traps: they arise from special combinations of single-particle orbits in a spherical potential, with decay transitions that must carry high angular momentum (hence the long half-lives). Away from the closed shells are regions of deformed nuclei, where axial symmetry gives rise to the K quantum number. The associated K traps involve transitions with large changes in the orientation of the angular momentum (though not necessarily in the magnitude).

In Figure 1, only ^{242}Am represents shape isomers, where it is a nuclear shape change, rather than an angular-momentum change, that hinders the isomer deexcitation. A convenient listing of isomers, with their angular momenta and half-lives, can be found in ref. [3]. However, the tabulation is incomplete for isomers with half-lives < 1 ms.

The delay time for isomer decays provides an important experimental tool for separating them from the bulk of the radiation from nuclear reactions. This is all the more important when exploring the limits of nuclear stability, with non-selective reactions that lead to many final products. In the study of neutron-rich nuclei with deep-inelastic reactions, fragmentation reactions or fission, the additional selectivity provided by isomer decays can be especially valuable.

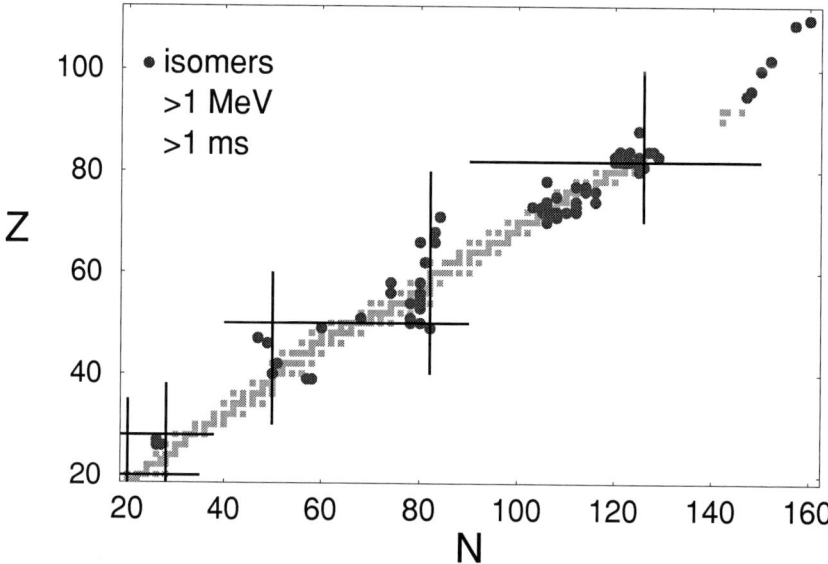

FIGURE 1. Nuclear chart: squares — naturally occurring nuclides; circles — isomers with half-lives > 1 ms and excitation energies > 1 MeV; and lines — closed shells (adapted from ref. [1]).

HIGH-K ISOMERS

The K value is the angular-momentum projection on the nuclear symmetry axis. This is not a directly measured quantity, but it is usually assumed to be equal to the total angular momentum (I) of the isomer (which is itself typically observed as the lowest-energy, lowest-spin member of a rotational band). To a first approximation, the K value is unaltered by collective rotation, which has it angular momentum perpendicular to the symmetry axis. However, inertial forces, principally the Coriolis force, cause K mixing and rotational alignment of individual nucleon orbits. It is these and other K-mixing effects that determine K-isomer decay rates.

The wide range of K-isomer half-lives is evident in Figure 2. It is also clear that the majority of K isomers are found in the A≈180 region. Indeed, it is only in that region that isomers involving > 3 quasiparticles are so far observed. Since the K values of isomers increase with quasiparticle number, it is through the properties of isomers in the A≈180 region that angular-momentum (and pairing) effects can best be studied.

Arguably the most famous K isomer is the 31-year, $K^\pi = 16^+$, 2.45-MeV excitation of ^{178}Hf, originally discovered over 35 years ago following reactor-neutron bombardment of natural hafnium [4]. The vast majority of the 31-year isomer decay strength is by (unobserved) conversion electrons associated with a 13-keV E3 transition to the I = 13

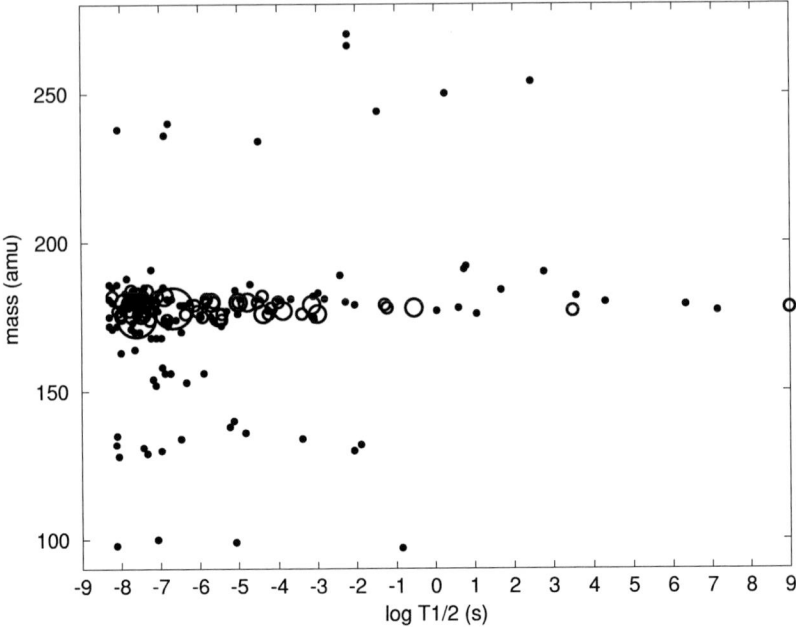

FIGURE 2. K-isomer mass number versus half-life (> 5 ns). The dots represent 2-quasiparticle isomers (excluding odd-odd nuclei) and 3-quasiparticle isomers, and the circles increase in size with quasiparticle number, up to 9 quasiparticles. Isomers in nuclides within four mass units of closed shells are omitted.

member of a K = 8 band. With decay by a $\lambda = 3$ transition, the isomer can be called a spin trap, and with a K change of 8 units the transition has a forbiddenness of $v = \Delta K - \lambda = 5$, classifying the isomer also as a K trap. This is an exceptional spin-trap/K-trap situation, which can explain the remarkable combination of high excitation energy and long half-life. Only recently have γ-ray emissions directly from the isomer been observed [5], in the form of low-intensity M4 and E5 branches, accounting for 0.015% and 0.006% of its decay, respectively. In comparison with Weisskopf single-particle estimates, these decays are hindered by factors, $F_W \sim 10^6 - 10^9$, while the reduced hindrances are much more uniform: $f_v = F_W^{1/v} \sim 100$.

This much is reasonably well understood, and the structure of the isomer can be explained in terms of the single-particle orbitals close to the Fermi surface: a broken neutron pair and a broken proton pair each contribute half of the 4-quasiparticle isomer's 16 $\hbar$. A puzzle remains in the experimental data, however. Unplaced transitions, illustrated in Figure 3, are observed with similar intensity to the E5 decay of the isomer. Considering that the source was prepared from an irradiation performed in 1980, and the illustrated measurements date from 2003 [5], it would appear that the unplaced tran-

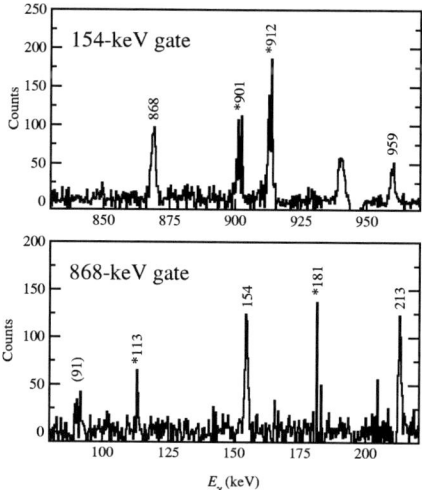

FIGURE 3. Gamma-ray coincidence spectra from a ^{178}Hf isomer source measurement [5] (see text). Known contaminant transitons are indicated by asterisks. Other labeled transitions are of unknown origin.

sitions, though of low intensity, are associated with a half-life of many years. Whether they might come from a higher-lying isomer in ^{178}Hf is speculation, but no trivial explanation has yet been found.

A more complete understanding of the γ-ray spectrum associated with the ^{178}Hf 31-year isomer decay is not only of academic interest. There has been much comment, see refs [1, 2], concerning the possibility of storing energy in nuclear isomers, potentially with controlled release of that energy. The ^{178}Hf isomer has been the most sought-after candidate, but the ability to trigger the release of the isomer's energy with < 100 keV photons has not yet been adequately demonstrated. Further knowledge of the nuclear structure of ^{178}Hf could prove valuable in this regard.

It may be noted that, while isomer triggering in ^{178}Hf remains unproven, the triggering of the only naturally occurring isomer, ^{180m}Ta, is well established [6], albeit with > 1 MeV photons; and the triggering pathway can be correlated in detail with the known level structure [7, 8]. This is an interesting example from an astrophysics perspective, as photo-activation could destroy the isomer in stellar s-process environments [6], if that is where the isomer is synthesised in the first place.

LIMITS OF NUCLEAR STABILITY

The drip lines, with associated proton and (potentially) neutron radioactivity, define the limits of nuclear stability in the N-Z plane, together with the fission and α decay of superheavy nuclei. Isomers can enhance our understanding of these limits, albeit complicated by their angular-momentum and excitation-energy dimensions.

In the superheavy domain, detailed measurements have now established high-K μs and ms isomers in $^{254}_{102}$No [9], highlighting the importance of conversion-electron detection. Going even heavier, a 6-ms isomer was recently found at ∼ 1 MeV in $^{270}_{110}$Ds [10], with a half-life that is longer than that of its ground state. Xu et al. [11] have calculated that comparable isomers should exist in a range of even-even superheavy nuclei, and in a real sense isomers can provide extra stability, which may be important for future experimental discoveries.

Approaching the proton drip line for lighter masses, the isomer ^{53m}Co provided the first example of proton radioactivity [12]. Peker et al. [13] subsequently pointed out that the angular-momentum barrier could lead to neutron radioactivity from high-spin isomers close to the neutron drip line. In this context, it is interesting that some high-spin isomers, such as the $K^\pi = 57/2^-$, 22-ns isomer in ^{175}Hf [14, 15], are unbound to neutron decay. However, the high angular momentum makes the probability very small for neutron emission, and it would be necessary to find much longer-lived isomers on the neutron-rich side of stability, to have a realistic chance of detecting their neutron decay. Such a possibility could perhaps apply to ^{187}Hf, where favored multi-quasiparticle isomers are predicted, as shown in Figure 4. Nevertheless, 7-quasiparticle excitations are needed in order to exceed the neutron separation energy. Experimentally, the ground state of ^{187}Hf has already been observed [17] from projectile-fragmentation reactions, but whether the required high angular momenta are within reach is not yet proven. However, the population of a spin 43/2 isomer in ^{215}Ra following ^{238}U fragmentation [18], and the inference that there must be a large and unexpected collective contribution to the angular momentum generated in the final isomeric fragment, gives encouragement. While, therefore, the objective of finding neutron decay from isomers remains challenging, the issue is not one of too-rapid decay in the absence of the Coulomb barrier, it is one of too-hindered decay in the presence of the angular-momentum barrier.

In the course of performing isomer calculations in the region of ^{187}Hf, it has been found that collective oblate rotation becomes an increasingly important mode, as the neutron richness increases [19]. In ^{190}Hf, the ground state itself is predicted to be oblate, and that shape remains favored up to high spin [20]. At the same time, coexisting multi-quasiparticle prolate states are predicted. Consequently, with sufficient primary beam intensity, it may be possible to identify the oblate collective states through their population following prolate isomer decays.

K-ISOMER DECAY RATES

Even though the configurations and excitation energies of K isomers can be calculated with some confidence, it remains problematical to predict their half-lives. The wavefunction overlaps are small, and even the relative importance of the different mechanisms for generating the overlaps is not well defined. The principal K-mixing mechanisms, necessary for non-zero "K-forbidden" transition rates, can be categorised as: (i) Coriolis (orientation) mixing; (ii) γ-tunneling (shape) mixing; (iii) statistical (thermal) mixing; and (iv) mixing due to chance near-degeneracies.

These mechanisms have been discussed in many different contexts, see for example

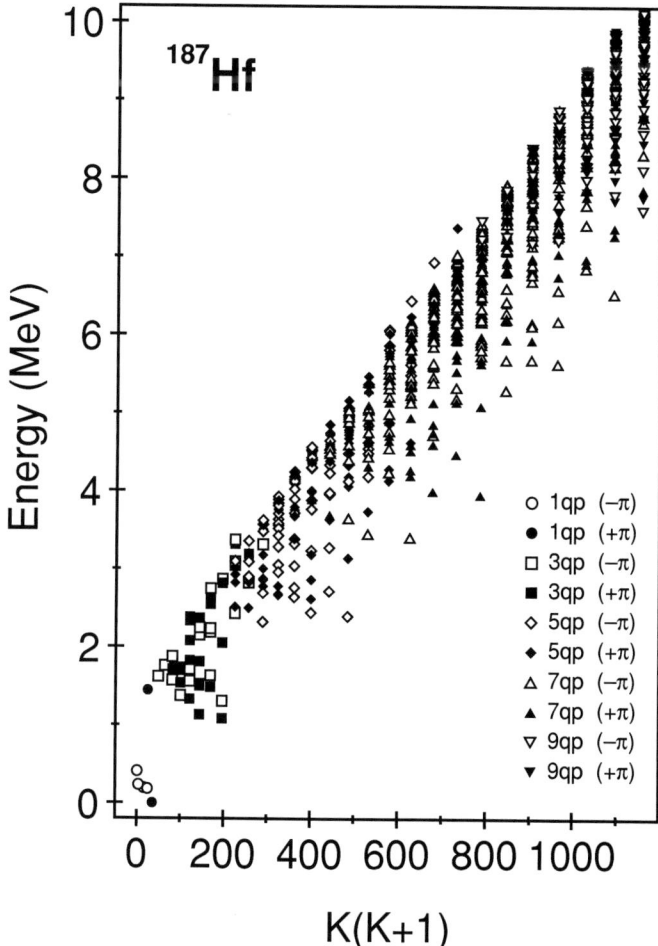

FIGURE 4. Multi-quasiparticle states in ^{187}Hf calculated with the Nilsson+BCS method of Jain et al. [16]. The neutron separation energy is approximately 4 MeV.

ref. [21], and here the presentation is restricted to extensions of the $N_p N_n$ scheme for 2- and 3-quasiparticle isomers [22], and statistical mixing considerations for higher quasiparticle numbers [23], in both cases concentrating on K-forbidden E2 transitions in the A≈180 region.

As described above, a simple measure of the goodness of the K quantum number is the hindrance per degree of K forbiddenness, often referred to as the reduced hindrance. For 2- and 3-quasiparticle E2 decays, there is found [22] to be a strong correlation of reduced hindrance with $N_p N_n$, the product of the valence nucleon numbers, which is itself a simple measure of collectivity. One may consider that for larger $N_p N_n$ values

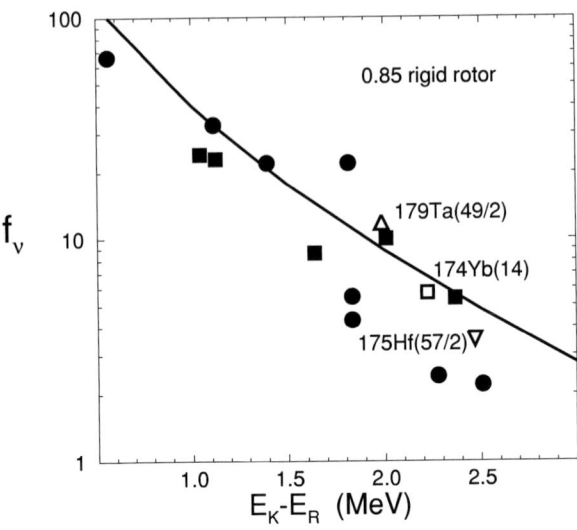

FIGURE 5. Variation of reduced hindrance with energy relative to a rotor with 85% of the rigid-body moment of inertia. Quasiparticle numbers are equal to 4 (circles — even-even nuclei), 5 (squares), 7 (triangle up), and 9 (triangle down). For the odd-mass nuclei, a pairing energy of 0.9 MeV has been added. The data are for E2 and E3 decays with $\Delta K > 5$. The full line represents the predicted level-density dependence [23]. The figure is adapted from ref. [28]. The labeled open symbols are discussed in the text.

the nuclei are more stably deformed and less susceptible to mixing effects, hence the correlation with reduced hindrance.

Recent data for a 6-ns ($31/2^+$) isomer in ^{191}Os [24] indicate an extension of the systematic reduced-hindrance behavior to a low value of $N_pN_n = 66$, with $f_v = 1.9$. This is in marked contrast to the value for the 6^+ isomer in ^{174}Yb, having $N_pN_n = 264$, and $f_v = 322$, which represents the other extreme of our present knowledge in the A≈180 region (and disagrees with the γ-tunneling model [25]). The remarkably high ^{174}Yb reduced hindrance has been discussed and confirmed recently by Dracoulis et al. [25], though not in the context of N_pN_n systematics. Data at even larger N_pN_n values can be anticipated in the near future, as 6^+, 2-quasi*neutron* isomers are predicted in the isotones ^{172}Er and ^{170}Dy [26], with $N_pN_n = 308$ and 352, respectively.

For higher quasiparticle numbers (> 3) the systematic behavior is distinctly different, perhaps as a consequence of blocked pairing correlations. Statistical mixing seems to be an important factor. This can be related to the level density, which may itself be determined from the excitation energy relative to a rigid-rotor reference [23]. If only a small range of angular momentum is considered (previously 4- and 5-quasiparticle isomers) then there is little sensitivity to the reference moment of inertia. However, including recent data from 7- and 9-quasiparticle isomers in ^{179}Ta [27] and ^{175}Hf [15], respectively, reduction of the reference moment of inertia by a factor of 0.85 is seen to be appropriate [28]. The resulting dependence of the reduced hindrance on the relative excitation energy is illustrated in Figure 5, showing that the general behavior is indeed systematic, but there are considerable fluctuations that presumably result from other significant degrees of freedom (see also ref. [21]). The new data point for a 4-quasiparticle isomer in ^{174}Yb [25] is in good accord with expectations.

SUMMARY

Isomers provide many insights into the nature of nuclear excitations and the limits of stability. The present paper has reviewed some of these aspects and pointed out that neutron decay from high-spin isomers may be observable. Extensions of the systematic behavior of K-isomer reduced-hindrance values have been presented, covering the full range of observed multi-quasiparticle decays.

ACKNOWLEDGMENTS

This work involves many collaborations: special thanks to Rodi Herzberg, Filip Kondev, Zsolt Podolyák, and Furong Xu. The UK EPSRC and Royal Society are thanked for financial support.

REFERENCES

1. P. M. Walker, and G. D. Dracoulis, *Nature* **399**, 35 (1999).
2. P. M. Walker, and J. J. Carroll, *Physics Today* **58-6**, 39 (2005).
3. G. Audi, O. Bersillon, J. Blachot, and A. H. Wapstra, *Nucl. Phys. A* **729**, 3 (2003).
4. R. G. Helmer, and C. W. Reich, *Nucl. Phys. A* **114**, 649 (1968).
5. M. B. Smith et al., *Phys. Rev. C* **68**, 031302(R) (2003).
6. D. Belic et al., *Phys. Rev. C* **65**, 035801 (2002).
7. P. M. Walker, G. D. Dracoulis, and J. J. Carroll, *Phys. Rev. C* **64**, 061302(R) (2001).
8. P. M. Walker, *Hyp. Int.*, **143**, 143 (2002).
9. R.-D. Herzberg et al., *Physica Scripta*, in press.
10. S. Hofmann et al., *Eur. Phys. J. A* **10**, 5 (2001).
11. F. R. Xu, E. G. Zhao, R. Wyss, P. M. Walker, *Phys. Rev. Lett.* **92**, 252501 (2004).
12. K. P. Jackson et al., *Phys. Lett. B* **33**, 281 (1970).
13. L. K. Peker, E. I. Volmyansky, V. E. Bunakov, and S. G. Ogloblin, *Phys. Lett. B* **36**, 547 (1971).
14. N. L. Gjørup et al., *Zeit. Phys. A* **337**, 353 (1990).
15. F. G. Kondev et al., to be published.
16. K. Jain et al., *Nucl. Phys. A* **591**, 61 (1995).
17. J. Benlliure et al., *Nucl. Phys. A* **660**, 87 (1999).
18. Zs. Podolyák et al., to be published.
19. F. R. Xu, P. M. Walker, and R. Wyss, *Phys. Rev. C* **62**, 014301 (2000).
20. F. R. Xu et al., to be published.
21. P. M. Walker, and G. D. Dracoulis, *Hyp. Int.* **135**, 83 (2001).
22. P. M. Walker, *J. Phys. G* **16**, L233 (1990).
23. P. M. Walker et al., *Phys. Lett. B* **408**, 42 (1997).
24. G. A. Jones et al., *J. Phys. G*, in press.
25. G. D. Dracoulis et al., *Phys. Rev. C* **71**, 044326 (2005).
26. P. H. Regan et al., *Phys. Rev. C* **65**, 037302 (2002).
27. F. G. Kondev et al., *Eur. Phys. J. A* **22**, 23 (2004).
28. P. M. Walker, *Acta Phys. Pol. B*, **36**, 1055 (2005).

Violations of K-Conservation in ^{178}Hf

A. B. Hayes[1], D. Cline[1], C. Y. Wu[1], J. Ai[2], H. Amro[2], C. Beausang[3], R. F. Casten[2], J. Gerl[4], A. A. Hecht[2], A. Heinz[2], R. Hughes[2], R. V. F. Janssens[5], C. J. Lister[5], A.O. Macchiavelli[6], D. A. Meyer[2], E. F. Moore[5], P. Napiorkowski[7], R. C. Pardo[5], Ch. Schlegel[4], D. Seweryniak[5], M. W. Simon[1], J. Srebrny[8], R. Teng[1], K. Vetter[6] and H. J. Wollersheim[4]

[1] *Nuclear Structure Research Laboratory**, Department of Physics, University of Rochester, Rochester, NY 14627
[2] *Wright Nuclear Structure Laboratory$^◊$, Yale University, New Haven, CT 06520*
[3] *Physics Department$^○$, University of Richmond, Richmond, VA 23173*
[4] *GSI, Gesellschaft für Schwerionenforschung, Planckstrasse 1, D-64291 Darmstadt, Germany*
[5] *Physics Division, Argonne National Laboratory†, Argonne, Illinois 60439*
[6] *Lawrence Berkeley National Laboratory‡, Berkeley, CA 94720*
[7] *Heavy Ion Laboratory,* [8] *Institute of Experimental Physics, Warsaw University‡, Warszawa, Poland*

Abstract. Coulomb excitation of $K^\pi=6^+$ ($t_{1/2}=77$ ns), 8^- ($t_{1/2}=4.0$ s) and 16^+ ($t_{1/2}=31$ y) ^{178}Hf isomers has led to the measurement of a set of $E\lambda$ matrix elements, coupling the isomer bands to the γ- and ground state bands. The resulting matrix elements, derived using a coupled-channel semiclassical Coulomb excitation search code, have been used to probe the K-components in the wave functions and revealed the onset and saturation of K-mixing in low-K bands, whereas K-mixing is negligible in the high-K bands. The implications can be applied to other quadrupole-deformed nuclei. An upper limit on the Coulomb depopulation yield of the 16^+ isomer was calculated based on the present set of matrix elements.

Keywords: isomer, K-isomer, K-mixing, Coulomb excitation
PACS: 27.70.+q, 23.20.-g, 25.70.De

INTRODUCTION

Studies of electromagnetic (EM) excitation and de-excitation of high-K isomeric states[1–5] have demonstrated significant violations of the K-selection rule in axially symmetric, quadrupole-deformed nuclei. The K-selection rule[6] does not allow EM transitions between two states $|I_iM_iK_i\rangle$ and $|I_fM_fK_f\rangle$ of an axially symmetric nucleus for which the forbiddenness $\nu \equiv |\Delta K| - \lambda$ is greater than zero, where λ is the multipole order and $\Delta K \equiv K_f - K_i$. The degree of hindrance of a K-forbidden transition can be expressed in terms of the "reduced hindrance" $f_\nu \equiv (B(\mathcal{M}\lambda)_{\text{W.u.}}/B(\mathcal{M}\lambda))^{1/\nu}$, where $B(\mathcal{M}\lambda)_{\text{W.u.}}$ is the Weisskopf single-particle estimate. The EM population of high-K states from the ground state band (GSB) is unlikely, either through highly hindered K-forbidden transitions or through multiple-step transitions of low or zero forbiddenness. For K-forbidden transitions, $f_\nu \gg 1$ is expected. The present work has used the hindrance of the K-forbidden transitions to probe K-admixtures in the ^{178}Hf rotational bands and has revealed the breakdown of K-selection as a function of spin.

EXPERIMENT

The $K^\pi=6^+$, 8^- and 16^+ isomer bands in ^{178}Hf were populated using a 650 MeV ^{136}Xe beam from ATLAS to bombard a thin 0.5 mb/cm^2 89% enriched ^{178}Hf target. Prompt γ decays were detected in coincidence with scattered ions by Gammasphere plus the CHICO PPAC array. Prompt γ-decay yields as a function of scattering angle θ_{scat} in the $K^\pi=6^+$ and 8^- isomer bands were measured as high as 10^{-3} of the $8^+\to6^+$ GSB strength. A remarkably high $19^+\to18^+$ yield, about 10^{-4} of the GSB strength, was unexpectedly observed in the known $K^\pi=16^+$ isomer band (Figure 1)[3,7]. In most cases, the coupled-channel semi-classical Coulomb excitation code GOSIA was used to adjust matrix elements for the relevant transitions and reproduce the experimental data. Typically, a single model-dependent parameter was obtained as the measure of the coupling strength between bands, including errors correlated with all other adjustable parameters. While the Alaga rule accurately reproduced yields due to K-allowed transitions, a spin-dependent mixing (SDM) reproduced yields due to K-forbidden paths[6] for low K.

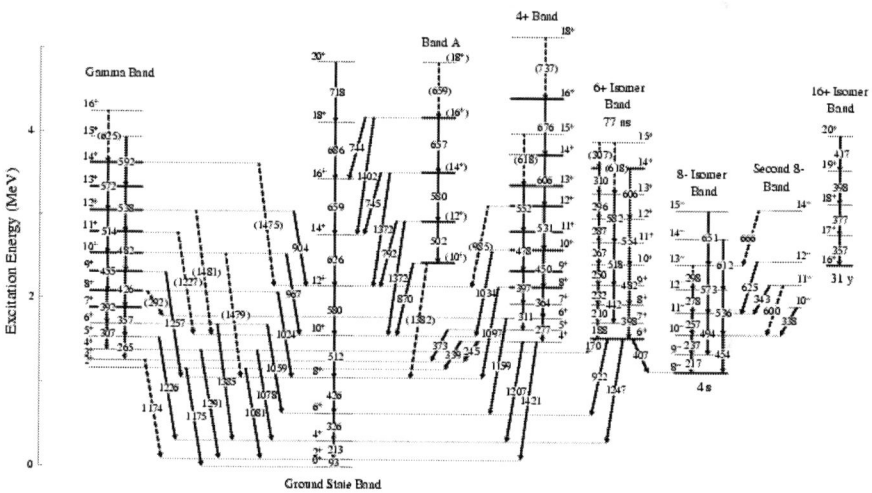

FIGURE 1. A partial level diagram for ^{178}Hf. Band A has been tentatively identified as previously unobserved states in the β-vibrational band.

A second experiment measured the Coulomb excitation function of the 16^+ isomer in order to extract model-independent $<I_{K=16}||E2||I_{GSB}>$ reduce matrix elements. A stack of five 1 mg/cm^2 natural Ta targets was irradiated by a ≈10 pnA, 858 MeV ^{178}Hf beam from ATLAS for E_{beam}=72% (target 5) to 88% (target 1) of the Coulomb barrier E_{Coul}. The targets were arranged normal to the beam, separated by 42 mg/cm^2 tantalum cylinders to collect scattered Hf ions over $40° <\theta^{lab,scat} <90°$, so that <<1% of the nuclei Coulomb-excited to the 16^+_{isom} state were lost or embedded in downstream targets. A 1.0(1) mg/cm^2 Ta scattering foil was used to scatter beam and target particles into a silicon detector mounted at $\theta_{scat}=45°$ to measure the absolute dose, 1.7(2) pmC, and the absolute activation cross sections (Figure 2). The GSB cascade following an isomer decay of the activated projectiles was used to count the activities five months later at Yale University's Wright Nuclear Structure Laboratory using two Ge "clover" detectors. The absolute efficiencies (≈3% for the 326 and 426 keV transitions in the GSB cascade) and the detection probabilities of relevant combinations of γ rays were calculated, including angular correlation and summing effects. Count rates were obtained from the >1-fold matrix by gating on the 326 keV $6^+ \to 4^+$ GSB transition and counting the coincident 426 keV $8^+ \to 6^+$ γ rays.

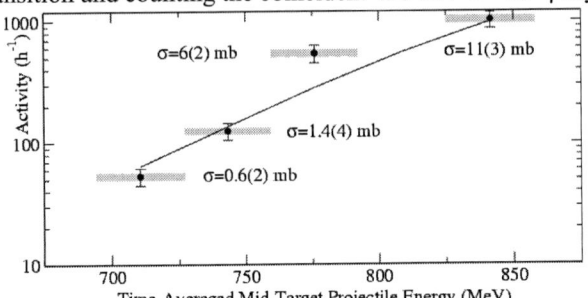

FIGURE 2. Measured and calculated activity after a direct fit of matrix elements (χ^2=3.5). The cross sections include errors in the total beam dose and measured target ablation. Target 2 (84% E_{Coul}) was not measured.

In the Xe beam experiment, it can be argued that the isomer bands could be populated through transfer reactions involving the ^{177,179}Hf contaminants in the target. An upper limit on ^{178}Hf(^{136}Xe,^{135}Xe)^{179}Hf transfer reactions was set using the only observed transition which could possibly be assigned to a ^{135}Xe transition (288 keV) in coincidence with a double gate on the ^{178}Hf GSB transitions. In the safe Coulomb excitation region, $25° < \theta_{scat} < 52°$, an upper limit on ^{177}Hf(^{136}Xe,^{135}Xe)^{178}Hf transfer was set at 10^{-5} of the ^{178}Hf GSB excitation. Assuming that the cross sections for ^{177}Hf(^{136}Xe,^{135}Xe)^{178}Hf (Q=-0.4 MeV) and ^{178}Hf(^{136}Xe,^{135}Xe)^{179}Hf (Q=-1.9 MeV) are similar, the upper limit on ^{177}Hf(^{136}Xe,^{135}Xe)^{178}Hf reactions in the 4.36% ^{177}Hf impurity is ~10^{-4} compared to Coulomb excitation of the ^{178}Hf GSB in the unsafe region, $52° < \theta_{scat} < 78°$, divided among several bands. Moreover, transfer to a 4 quasiparticle state (e.g. the 16^+_{isom} band) is very unlikely, since breaking a second pair of nucleons is a higher-order effect. Since no transfer is seen in the safe region, there should not be significant transfer near 52°, even in the unsafe region, where strong $K \leq 8$ isomer populations are already seen far above background in the double-gated data. In the ^{178}Hf beam activation experiment 16^+ isomer activation was observable at 72% E_{Coul}, consistent with the Coulomb excitation function (Figure 2). Nuclear interference is small at 88% E_{Coul} and $\leq 10^{-3}$ of the E2 contribution[8–10] for $E_{beam} \leq 80\% E_{Coul}$.

ANALYSIS

In both experiments, a possible excitation path was determined to be insignificant if it could not reproduce the measured data using reasonable reduced transition probabilities for the particular multipolarity and change in collective or single-particle structure, e.g., ≈1 W.u. for transitions between different quasiparticle configurations. Intrinsic matrix elements were fit to the ^{178}Hf(^{136}Xe,^{136}Xe)^{178}Hf data, connecting the

$K=0^+$ GS, γ-, 4^+, 6^+ and 8^- bands using Alaga and SDM systematics (Figure 3). The yield data for the 8^- bands were reproduced most accurately, and with the lowest B(E3) values (<4 W.u., reasonable in comparison to the 4 W.u. strength of the $3^-_{K=2} \to 0^+_{GSB}$ transition[11]) by two-step excitations to both 8^- bands through the γ-band in conjunction with single-step excitations from the GSB, both using Alaga rule coupling systematics for $K=5$ admixtures in the low-K bands. It was necessary to attenuate the $<8^-||E3||GSB>$ and $<8^-||E3||K^\pi=2^+>$ Alaga matrix elements smoothly by approximately an order of magnitude per unit spin as I_{GSB} decreases from 6 to 10 in order to keep the isomer cross section from growing unreasonably large compared to the other quasiparticle isomers in ^{178}Hf and to preserve the 4.0(2) s half life.

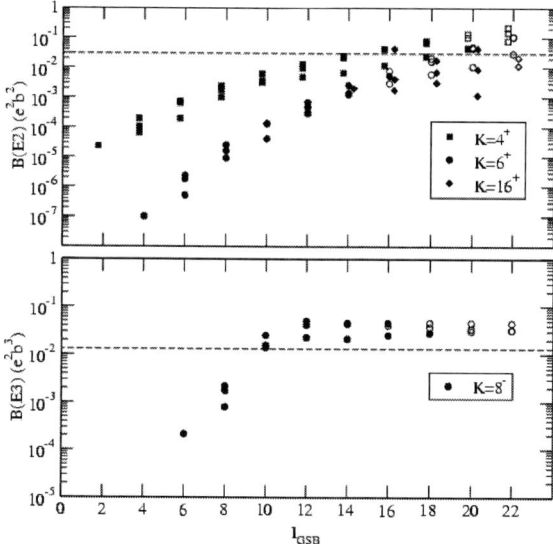

FIGURE 3. The three strongest reduced transition probabilities from each GSB level for GSB→K^π transitions. GSB→$4^+,6^+$ matrix elements follow SDM systematics. GSB→8^- matrix elements follow the Alaga rule, attenuated at low spin. Transitions to unobserved high-spin levels (hollow) are extrapolated to clarify the spin-dependence of the intrinsic matrix elements in the models used. Weisskopf estimates (dashed lines): B(E2↑)$_{w.u.}$=0.0297 e^2b^2. B(E3↑)$_{w.u.}$=0.0132 e^2b^3.

A simultaneous fit to the 16^+ isomer activity data and the prompt $19^+_{K=16}$ yields of the first experiment led to a coherent set of $<I_{K=16}||E2||I_{GSB}>$ matrix elements with B(E2;K=0→K=16)≤1.4 W.u. with upper limits, several lower limits and several diagonal (uncorrelated) errors. The fit was constrained by measured upper limits on both the GSB→16^+ feeding intensity (≈10^{-4} normalized to the $8^+_{GSB} \to 6^+_{GSB}$ yield in the Hf(Xe,Xe)Hf experiment), reasonable B(E2;GSB→K^π=16^+) values, etc. GSB→K^π=16^+ feeding is not insignificant, but the matrix elements which reproduce the yields are consistent with non-observation of feeding. In particular, the strength of the 1% $20^+_{GSB} \to 20^+_{K=16}$ γ-decay branch is 3 times smaller than the observable lower limit, while the observational lower limits for the energetically favored transitions are much higher, due to the unavailability of double-γ gates and clean single-γ gates. It was found that ≈75%—80% of the isomer activation comes directly from connections between the GSB and the $17^+_{K=16}$ and $16^+_{K=16}$ states for any set of matrix elements, as long as K-mixing between the GSB and the isomer band does not decrease with increasing spin and a liberal 10 W.u. upper limit is imposed on the B(E2;GSB→K=16) strengths.

INTERPRETATION

The systematic decrease with increasing spin of the hindrance of K-forbidden transitions is apparent from Figure 3. For each of the high-K isomer bands observed, reproduction of the measured yields requires that the interband B(Eλ) values increase with increasing spin and saturate at $\approx$1 W.u. for I$\geq$12 in the GSB and the γ-band. This saturation point represents the maximum mixing of K. For I$\geq$12, reduced hindrance values of K-forbidden transitions from low-K to high-K bands are as low as $f_v \sim 1$, showing that the K-selection rule has little predictive power at high spin—highly K-forbidden transitions have similar strength to allowed interband transitions.
Band interactions are reflected in the measured moments of inertia by an increase in slope of the moment of inertia I(ω), seen at I$\approx$6 and I$\approx$10 in the γ- and GS bands, respectively, while the B(Eλ) values (Figure 3) saturate at ~1 W.u. as low as I$\approx$8 and I$\approx$10 for transitions from the γ-band and the GSB respectively, in order to reproduce the measured γ-ray yields in the K^π=4$^+$, 6$^+$, 8$^-$ and 16$^+$ bands. Moreover, Coriolis alignment is expected to happen at much lower spin in low-K bands than in high-K bands[12] which are strongly deformation-coupled. The moments of inertia of the high-K bands are relatively constant in slope, with the exception of the 6$^+$ band at I$\approx$12, suggesting that the high-K bands are not K-mixed to the same degree as the low-K bands. The 16$^+$ band has a remarkably constant moment of inertia[7] up to I=22. In contrast with the K=0,2 transitions to the high-K isomer bands, the 16$^+_{isom}$→K^π=8$^-$ and 14$^-_{isom}$→K^π=8$^-$ γ decays are strongly hindered with 33$\leq$f$_v$$\leq$165(5) in all of the five known branches[13,14], showing that the onset of significant high-K admixtures in the 8$^-$ band must occur at I>18, if at all, whereas less hindered $f_v\sim$1 transitions from the γ- and GS bands are required to reproduce the present measured yields. That is, the strongly hindered decays of the K^π=16$^+$ and K^π=14$^-$ isomers to the 11$^-$<$I_{K=8}$<13$^-$ states are consistent with K being a good quantum number for the high-K bands, suggesting that mixing in the low-K bands is primarily responsible for the K-selection violations and that the EM matrix elements coupling to the high-K bands are sensitive probes of the K-distributions in the low-K bands. Coulomb excitation of a band with projection K, assuming that it is reasonably pure, would require admixtures K' in the low-K (nominally K_i) bands of $K-\lambda \leq K' \leq K+\lambda$. Hence, the mixing fractions of the 2$\leq$K'$\leq$6 components are depicted in Figure 3 as a function of spin by the B(E2;$K_i \to K$=4) values, the 4$\leq$K'$\leq$8 components by the B(E2;$K_i \to K$=6) values, etc.

The present results have revealed paths by which Coulomb depopulation of the ^{178}Hf 16$^+$ isomer could be achieved. The 16$^+$→GSB K-forbidden E2 paths would allow Coulomb depopulation of the isomer using heavy ions below the Coulomb barrier with a probability of $\leq$1% compared to the in-band excitations[15]. This path is not expected to be effective for photo depopulation, since photon absorption is dominated by E1 transitions, but this does not rule out photo depopulation via a 463 keV E1 transition to the known 15$^-_{K=14}$ state, for example, calculated to be a $\leq$0.1% effect using heavy ion bombardment and assuming B(E1;K^π=14$^-$→K^π=16$^+$)$\approx$1 W.u. for all transitions. The K^π=16$^+$ to K^π=8$^-$ transitions are highly forbidden, making this path ineffectual. While potential low-yield Coulomb depopulation paths have been discovered, intermediate states which might mediate x-ray photo depopulation were not found.

CONCLUSION

The present work has revealed the Coulomb excitation paths of the $K^\pi=4^+$, 6^+_{isom}, 8^-_{isom} and 16^+_{isom} rotational bands in ^{178}Hf by rapidly increasing K-mixing with increasing spin (I) in low-K bands, while the high-K bands remain very pure, even at the same spin levels where the low-K bands are completely mixed. The rapid increase in the K-forbidden interband B(Eλ) values coincides with the rotational alignment of low-K bands which has a noticeable effect on the moment of inertia above the $I\approx10$ levels of the γ-band and the GSB. Previous measurements of isomer decay branching ratios are inconsistent with significant mixing occurring in the high-K rotational bands, while high-K band heads are pure in K. It appears that higher-K components are admixed in the nominally low-K bands with increasing spin, until the reduced transition probabilities saturate near ~1 W.u. for I$\geq$12, signifying the total breakdown of the K quantum number.

ACKNOWLEDGMENTS

*Work supported by the National Science Foundation and the Air Force Office of Scientific Research. ⁰Work supported by the U.S. Department of Energy under grant number DE-FG02-91ER-40609. °Work supported by the U.S. Department of Energy under contracts DE-FG02-05ER41379 and DE-FG52-NA25929. †Work supported by the U.S. Department of Energy, Office of Nuclear Physics, under contracts W-31-109-ENG-38 (ANL) and DE-AC03-76SF00098 (LBNL). ‡Work supported by the Polish State Committee for Scientific Research under contract 5P03B04720.

REFERENCES

1. P. Chowdhury, et al., Nucl. Phys. A, 485:136, 1988.
2. P. M. Walker, et al., Phys. Rev. Lett., 65:416, 1990.
3. A. B. Hayes, et al., Phys. Rev. Lett., 89:242501, 2002.
4. P. M. Walker, G. D. Dracoulis and J. J. Carroll, Phys. Rev. C, 64:061302, 2001.
5. M. Loewe, et al., Phys. Lett. B, 551:71, 2003.
6. A. Bohr and B. R. Mottelson, Nuclear Structure, Vol. 2. Benjamin, Reading, 1975.
7. S. M. Mullins, et al., Phys. Lett. B, 393:279, 1997; Ibid. 400:401, 1997.
8. D. Cline, et al., Nucl. Phys. A, 133:445, 1969.
9. D. Cline, Annu. Rev. Nucl. Part. Sci., 36:683, 1986.
10. A. E. Kavka, Ph. D. Dissertation, Uppsala University, Uppsala, Sweden, 1989.
11. R. M. Ronningen, et al., Phys. Rev. C, 15:1671, 1977.
12. P. Ring and P. Schuck, The Nuclear Many-Body Problem, Springer-Verlag, New York, 1980.
13. M. B. Smith, et al., Phys. Rev. C, 68:031302R, 2003.
14. R. B. Firestone, Table of Isotopes, 8th Ed., Vol. 2. Wiley & Sons, New York, 1996.
15. D. Cline, A. B. Hayes and C. Y. Wu, In Proceedings of the 35th Winter Colloquium on The Physics of Quantum Electronics, Taylor and Francis, 2005. To be published.

Theoretical Description of K-Isomers

Yang Sun

Department of Physics and Joint Institute for Nuclear Astrophysics, University of Notre Dame, Notre Dame, Indiana 46545, USA

Abstract. A proper treatment of K-mixing is the key to understanding K-isomers. Here, we present a method based on the projected shell model. This method differs from the usual description of multi-quasiparticle states by introducing a transformation to the laboratory frame and a subsequent configuration mixing in that frame. It allows a quantitative study on the degree of K-violation through direct calculations of electromagnetic transitions.

Keywords: K-isomers, K-mixing, projected shell model
PACS: 21.60.Cs, 21.20.-k

INTRODUCTION

Many long-lived, highly-excited isomers in deformed nuclei owe their existence to the approximate conservation of the K quantum number [1, 2]. The selection rule for an electromagnetic transition would require that the multipolarity of the decay radiation, λ, be at least as large as the change in the K-value ($\lambda \geq \Delta K$). However, symmetry-breaking processes make possible transitions that violate the K-selection rule; such 'K-forbidden' transitions are hindered, rather than strictly forbidden. Much depending on the degree of the K-violation, decays from a high K state to low-lying low K states can have half-lives that range from nanoseconds to years.

It is thus clear that a proper description of K-violation in terms of K-mixing is at the heart of understanding K-isomers. A theoretical model that can treat K-mixing has preferably the basis states that are eigenstates of angular momentum I. Diagonalization of two-body interactions mixes these states and the resulting wavefunctions contain the information on the degree of K-mixing. In this kind of approach, the mixing and its consequences are discussed in the laboratory frame rather than in a body-fixed frame in which K is originally defined.

THE PROJECTED SHELL MODEL

The projected shell model (PSM) [3, 4] is a shell model that starts from a deformed basis. In the standard version of the PSM, the shell-model basis is constructed by considering a few quasiparticle (qp) orbitals near the Fermi surfaces and performing angular momentum projection (if necessary, also particle-number projection) on the chosen configurations. With projected multi-qp states as the basis states of the model, the PSM is designed to describe the rotational bands built upon qp excitations [3, 5]. The PSM has been rather successful in calculating the high-spin states of normally-deformed and superdeformed nuclei (see, for example, Refs. [6, 7]).

In many of the existing calculations, the PSM begins with deformed Nilsson single-particle states, with pairing correlations incorporated into these states by a BCS calculation. This defines a set of deformed qp states (with $a_\nu^\dagger$ and $a_\pi^\dagger$ being the creation operator for neutrons and protons, respectively) with respect to the qp vacuum $|0\rangle$. The PSM basis construction is then implemented in the multi-qp states with the following forms

- e-e nucleus: $\{|0\rangle, a_\nu^\dagger a_\nu^\dagger |0\rangle, a_\pi^\dagger a_\pi^\dagger |0\rangle, a_\nu^\dagger a_\nu^\dagger a_\pi^\dagger a_\pi^\dagger |0\rangle, a_\nu^\dagger a_\nu^\dagger a_\nu^\dagger a_\nu^\dagger |0\rangle, a_\pi^\dagger a_\pi^\dagger a_\pi^\dagger a_\pi^\dagger |0\rangle, \ldots\}$
- o-o nucleus: $\{a_\nu^\dagger a_\pi^\dagger |0\rangle, a_\nu^\dagger a_\nu^\dagger a_\nu^\dagger a_\pi^\dagger |0\rangle, a_\nu^\dagger a_\pi^\dagger a_\pi^\dagger a_\pi^\dagger |0\rangle, a_\nu^\dagger a_\nu^\dagger a_\nu^\dagger a_\pi^\dagger a_\pi^\dagger a_\pi^\dagger |0\rangle, \ldots\}$
- odd-ν nucleus: $\{a_\nu^\dagger |0\rangle, a_\nu^\dagger a_\nu^\dagger a_\nu^\dagger |0\rangle, a_\nu^\dagger a_\pi^\dagger a_\pi^\dagger |0\rangle, a_\nu^\dagger a_\nu^\dagger a_\nu^\dagger a_\pi^\dagger a_\pi^\dagger |0\rangle, \ldots\}$
- odd-π nucleus: $\{a_\pi^\dagger |0\rangle, a_\nu^\dagger a_\nu^\dagger a_\pi^\dagger |0\rangle, a_\pi^\dagger a_\pi^\dagger a_\pi^\dagger |0\rangle, a_\nu^\dagger a_\nu^\dagger a_\pi^\dagger a_\pi^\dagger a_\pi^\dagger |0\rangle, \ldots\}$

The omitted index for each creation operator contains information on the Nilsson orbitals. In fact, this is the standard way of building multi-qp states [8, 9, 10, 11]. However, the present model goes a step beyond by transforming these states from the body-fixed frame to the laboratory frame and mixing them in the laboratory frame through two-body residual interactions.

The angular-momentum-projected multi-qp states are thus the building blocks in the PSM wavefunction, which can be generally written as

$$|\psi_M^{I,\sigma}\rangle = \sum_{\kappa, K \leq I} f_\kappa^{I,\sigma} \hat{P}_{MK}^I |\phi_\kappa\rangle = \sum_\kappa f_\kappa^{I,\sigma} \hat{P}_{MK_\kappa}^I |\phi_\kappa\rangle. \quad (1)$$

The index σ labels states with same angular momentum and κ the basis states. $\hat{P}_{MK}^I$ is the angular-momentum-projection operator [3] and the coefficients $f_\kappa^{I,\sigma}$ are weights of the basis states.

The weights $f_\kappa^{I,\sigma}$ are determined by diagonalization of the Hamiltonian in the spaces spanned for various nuclear systems as listed above, which leads to the eigenvalue equation (for a given I)

$$\sum_{\kappa'} (H_{\kappa\kappa'} - E_\sigma N_{\kappa\kappa'}) f_{\kappa'}^\sigma = 0. \quad (2)$$

The Hamiltonian and the norm matrix elements in Eq. (2) are given as

$$H_{\kappa\kappa'} = \langle \phi_\kappa | \hat{H} \hat{P}_{K_\kappa K_{\kappa'}}^I | \phi_{\kappa'} \rangle, \qquad N_{\kappa\kappa'} = \langle \phi_\kappa | \hat{P}_{K_\kappa K_{\kappa'}}^I | \phi_{\kappa'} \rangle. \quad (3)$$

Angular-momentum-projection on a multi-qp state $|\phi_\kappa\rangle$ with a sequence of I generates a band. One may define the rotational energy of a band (band energy) using the expectation values of the Hamiltonian with respect to the projected $|\phi_\kappa\rangle$

$$E_\kappa^I = \frac{H_{\kappa\kappa}}{N_{\kappa\kappa}} = \frac{\langle \phi_\kappa | \hat{H} \hat{P}_{K_\kappa K_\kappa}^I | \phi_\kappa \rangle}{\langle \phi_\kappa | \hat{P}_{K_\kappa K_\kappa}^I | \phi_\kappa \rangle}. \quad (4)$$

In a usual approximation with independent quasiparticle motion, the energy for a multi-qp state is simply taken as the sum of those of single quasiparticles. This is the dominant term. The present theory modifies this quantity in the following two steps. First, the band energy defined in Eq. (4) introduces the correction brought by angular momentum projection and the two-body interactions, which accounts for the couplings between the rotating body and the quasiparticles in a quantum-mechanical way. Second, the corresponding rotational states are mixed in the subsequent procedure of solving the eigenvalue equation (2). The energies are thus further modified by the configuration mixing.

If the deformed states are axially symmetric, each of the basis states in (1), the projected $|\phi_K\rangle$, is a K-state. For example, an n-qp configuration gives rise to a multiplet of 2^{n-1} states, with the total K expressed by $K = |K_1 \pm K_2 \pm \cdots \pm K_n|$, where K_i is for an individual neutron or proton. In this case, shell model diagonalization, i.e. solving the eigenvalue equation (2), is completely equivalent to K-mixing. The amount of the mixing can be obtained from the resulting wavefunctions.

The above discussion is independent of the choice of the two-body interactions in the Hamiltonian. In practical calculations, the PSM uses the pairing plus quadrupole-quadrupole Hamiltonian (that has been known to be essential in nuclear structure calculations [12]) with inclusion of the quadrupole-pairing term

$$\hat{H} = \hat{H}_0 - \frac{1}{2}\chi \sum_\mu \hat{Q}_\mu^\dagger \hat{Q}_\mu - G_M \hat{P}^\dagger \hat{P} - G_Q \sum_\mu \hat{P}_\mu^\dagger \hat{P}_\mu. \qquad (5)$$

The strength of the quadrupole-quadrupole force χ is determined in such a way that it has a self-consistent relation with the quadrupole deformation ε_2. The monopole-pairing force constants G_M are

$$G_M = \left[G_1 \mp G_2 \frac{N-Z}{A}\right] A^{-1}, \qquad (6)$$

with "$-$" for neutrons and "$+$" for protons, which reproduces the observed odd–even mass differences in a given mass region if G_1 and G_2 are properly chosen. Finally, the strength G_Q for quadrupole pairing was simply assumed to be proportional to G_M, with a proportionality constant 0.16, as commonly used in the PSM calculations [3].

EXAMPLES

The nucleus ^{178}Hf has become a center of recent discission on K-isomers. The possibility to trigger the 2.45MeV, 31-year isomer decay by the application of external electromagnetic radiation has attracted much interest and potentially could lead to the controlled release of nuclear energy [13, 14]. Information on the detailed structure and transition of the isomeric and the surrounding states thus becomes a crucial issue. In the PSM calculation for ^{178}Hf [15], the model basis was built with the deformation parameters $\varepsilon_2 = 0.251$ and $\varepsilon_4 = 0.056$ (values taken from Ref. [16]). Fig. 1 shows the calculated energy levels in ^{178}Hf, compared with the known data [17]. Satisfactory agreement is achieved for most of the states, except that for the bandhead of the first 8^- band and the 14^- band, the theoretical values are too low.

It was found that the obtained states are generally K-mixed. One may still talk about the dominant structure of each band by studying the wavefunctions. We found that the 6^+ band has mainly a 2-qp structure $\{v[512]5/2^- \oplus v[514]7/2^-\}$, the 16^+ band has a 4-qp structure $\{v[514]7/2^- \oplus v[624]9/2^+ \oplus \pi[404]7/2^+ \oplus \pi[514]9/2^-\}$, the first (lower) 8^- band has a 2-qp structure $\{v[514]7/2^- \oplus v[624]9/2^+\}$, the second

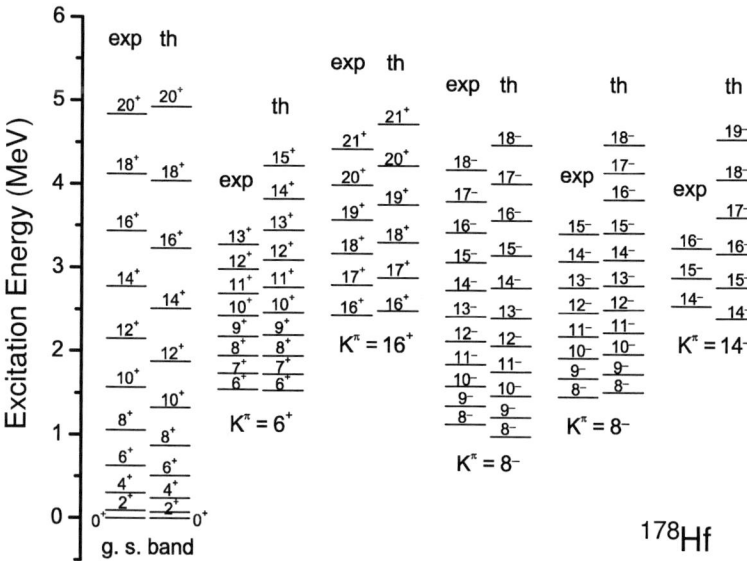

FIGURE 1. Comparison of the PSM calculation for the rotational bands in ^{178}Hf with data. This figure is adopted from Ref. [15].

(higher) 8^- band has a 2-qp structure $\{\pi[404]7/2^+ \oplus \pi[514]9/2^-\}$, and the 14^- band has a 4-qp structure $\{\nu[512]5/2^- \oplus \nu[514]7/2^- \oplus \pi[404]7/2^+ \oplus \pi[514]9/2^-\}$. These states, together with many other states (not shown in Fig. 1) obtained from the same diagonalisation process, form a complete spectrum including the high-K isomeric states. Electromagnetic transitions between any two of these states can be directly calculated by using the wavefunctions, as done in the previous publications (see, for example, Refs. [18, 19]). Results will be reported elsewhere.

FURTHER DEVELOPMENTS

The preceding discussions suggest that the projected shell model can be a useful tool for understanding K-isomers. So far, not much has been done along these lines except for a few selected applications [15, 20, 21]. Therefore, there is a large room for exploring the usefulness of the model. In addition, three straightforward extensions may be necessary in order to explain some of the current K-isomer data.

- *Enlargement of the configuration space* – There have been impressive high-K isomer data suggesting configurations with up to 10-quasiparticles. The current model space of the PSM does not have these configurations and must be enlarged.
- *High-order multipole transitions* – In some examples, violation of the K-quantum number can only be understood through electromagnetic transitions with high-order multipolarity. The current PSM code includes only the B(M1) and B(E2) parts. The

high-order transition rates must be considered.
- *The effective forces* – The pairing plus quadrupole Hamiltonian has been proven successful for many of the structure calculations. However, other effects such as the octupole correlation have also been considered important in some cases. Additional terms, such as the monopole-monopole and octupole-octupole interactions, can be included into the model if these are important for certain discussions.

SUMMARY

We have shown that the physics of K-mixing in multi-qp states is well-incorporated in framework of the projected shell model. The model follows the shell-model philosophy closely and may be viewed as a shell model constructed in a projected multi-qp basis. More precisely, the basis is first built in the qp basis with respect to the deformed-BCS vacuum; then rotational symmetry, violated in the deformed basis, is restored by angular-momentum-projection to form a basis in the laboratory frame; finally a two-body Hamiltonian is diagonalized in the projected multi-qp basis. The model can thus produce fully-correlated shell-model states and can generate well-defined wavefunctions, allowing one to compute, without any approximations, electromagnetic transition probabilities.

ACKNOWLEDGMENTS

This work is partly supported by NSF under contract PHY-0140324.

REFERENCES

1. A. Bohr and B.R. Mottelson, *Nuclear Structure* (W.A. Benjamin, Inc., New York, 1975).
2. P.M. Walker and G.D. Dracoulis, *Nature* **399**, 35 (1999).
3. K. Hara and Y. Sun, *Int. J. Mod. Phys. E* **4**, 637 (1995).
4. Y. Sun and K. Hara, *Comp. Phys. Commun.* **104**, 245 (1997).
5. Y. Sun and C.-L. Wu, *Phys. Rev. C* **68**, 024315 (2003).
6. Y. Sun, J.-y. Zhang and M. Guidry, *Phys. Rev. Lett.* **78**, 2321 (1997).
7. J.A. Sheikh, Y. Sun, and P.M. Walker, *Phys. Rev. C* **57**, R26 (1998).
8. K. Jain et al., *Nucl. Phys. A* **591**, 61 (1995).
9. V.G. Soloviev, *Nucl. Phys. A* **633**, 247 (1998).
10. J.-Y. Zeng, S.-X. Liu, L.-X. Gong, and H.-B. Zhu, *Phys. Rev. C* **65**, 044307 (2002).
11. F.-R. Xu, E.-G. Zhao, R. Wyss, and P.M. Walker, *Phys. Rev. Lett.* **92**, 252501 (2004).
12. M. Dufour and A.P. Zuker, *Phys. Rev. C* **54**, 1641 (1996).
13. B. Schwarzschild, *Physics Today* May issue, 21 (2004).
14. P.M. Walker and J.J. Carroll, *Physics Today* June issue, 39 (2005).
15. Y. Sun, X.-R. Zhou, G.-L. Long, E.-G. Zhao, and P.M. Walker, *Phys. Lett. B* **589**, 83 (2004).
16. R. Bengtsson, S. Frauendorf, and F.-R. May, *Atom. Data and Nucl. Data Tables* **35**, 15 (1986).
17. S.M. Mullins et al., *Phys. Lett. B* **393**, 279 (1997); *Phys. Lett. B* **400**, 401 (1997).
18. Y. Sun and J.L. Egido, *Nucl. Phys. A* **580**, 1 (1994).
19. P. Boutachkov, A. Aprahamian, Y. Sun, J.A. Sheikh, and S. Frauendorf, *Eur. Phys. J. A* **15**, 455 (2002).
20. C.M. Petrache et al., *Nucl. Phys. A* **617**, 249 (1997).
21. X.-R. Zhou, Y. Sun, G.-L. Long, and E.-G. Zhao, *Chin. Phys. Lett.* **19**, 1274 (2002).

Isomer and In-Beam Spectroscopy of Medium-Spin States in ^{91,92}Zr

P.H. Regan[1], N.J. Thompson[1,2], A.B. Garnsworthy[1,2], H.C. Ai[2],
L. Amon[2,3], R.B. Cakirli[2,3], R.F. Casten[2], C. R. Fitzpatrick[1,2],
S.J. Freeman[4], G. Gurdal[5], A. Heinz[2], G.A. Jones[1], E.A. McCutchan[2],
J. Qian[2], V. Werner[2], S.J. Williams[1], R. Winkler[2]

[1] Dept. of Physics, University of Surrey, Guildford, GU2 7XH, UK
[2] WNSL Yale University, 272 Whitney Avenue, New Haven CT 06520 , USA
[3] Dept. of Physics, Istanbul University, Istanbul, Turkey
[4] School of Physics and Astronomy, The University of Manchester, Manchester, M13 9PL, UK
[5] Clark University Worcester, Mass. 01610, USA

Abstract. Near-yrast states have been identified in the stable ^{91,92}Zr isotopes using the fusion evaporation reaction ^{82}Se(^{13}C,xn)$^{95-x}$Zr at an incident beam energy of 50 MeV. Gamma-ray spectroscopy of states above the reported $\tau = 6$ μs, $I^{\pi} = 21/2^{+}$ isomer in ^{91}Zr are reported for the first time with consequences for stimulating new shell model calculations which incorporate either the breaking of the N=50 shell closure and/or the inclusion of the negative parity $h_{11/2}$ neutron intruder orbital.

Keywords: ^{91,92}Zr, energy levels, γ spectroscopy, high-spin, fusion-evaporation reactions.
PACS: 21.60.Cs; 30.20.Lv; 25.70.Jj; 27.60.+j

INTRODUCTION

The stable Zr isotopes (Z=40) are a long-standing testing ground for the nuclear shell model, representing some of the least collective nuclei reported to date [1]. Their low-lying structure can be explained in terms of a shell model space truncated to allow proton excitations between the $g_{9/2}$ and $p_{1/2}$ orbitals, with neutron excitations allowed in the $g_{7/2}$, $d_{5/2}$, $d_{3/2}$ and $s_{1/2}$ positive parity orbitals. Negative parity states require the inclusion of the $h_{11/2}$ neutron intruder orbital within the valence space or proton one-particle-one hole excitations between the $g_{9/2}$ and $p_{1/2}$ orbitals. Shell model calculations for nuclei in this region have usually been performed assuming an ^{88}Sr (Z=38, N=50) closed core [2]. For seniority 3 excitations (i.e. 2 unpaired protons plus one unpaired neutron) the maximal spin positive-parity configuration is $I^{\pi} = 23/2^{+}$ arising from $[(\pi g_{9/2})^{2}{}_{8+}(\nu g_{7/2})_{7/2+}] = 23/2^{+}$. However, the single-particle energies for neutron-states in the N=51 ^{91}Zr suggest that the $g_{7/2}$ single particle level lies almost 2 MeV above the predominantly $d_{5/2}$ ground-state configuration [3]. Indeed, it is the simple coupling of this $d_{5/2}$ neutron ground state to the maximal $I^{\pi} = 8^{+}$ coupling of the

$(\pi g_{9/2})^2$ configuration which provides the basis for the favored $I^{\pi}=21/2^+$ isomer at $E_x=3.167$ MeV [4,5]. States of higher spin must therefore arise either from (i) increasing the seniority of the configuration to 5 unpaired particles and thus breaking the simple ^{88}Sr valence configuration (for example by promoting more 2 protons into the $g_{9/2}$ orbitals to form a $[(\pi g_{9/2})^4{}_{12+}(\nu d_{5/2})_{5/2+}] = 29/2^+$ configuration); or (ii) including excitations with the negative parity $h_{11/2}$ intruder orbital. In order to explain the structure of even higher spins states, the neutron N=50 core must be broken (e.g., as discussed in ref. [2]) and/or proton excitations across the Z=50 shell must be included in the shell model valence space. Such shell model calculations require significant experimental input of the structure of the near-yrast sequences in such nuclei. This is the aim of the current work, in which high-spin, near-yrast states have been identified in ^{91}Zr above the well established $I^{\pi}= 21/2^+$, $\tau=6$ μs isomer.

Experimental Details

Excited states in ^{91,92}Zr were populated following fusion evaporation reactions between a ^{13}C beam, with a laboratory energy of 50 MeV and a self-supporting ^{82}Se target of nominal thickness 200 μg/cm^2. The beam was provided by the Yale ESTU tandem [6] with a typical on-target beam current of 1pnA. The target had a flash of carbon on either side to reduce its sublimation under the beam bombardment. The reaction γ rays depopulating the nuclei of interest were observed using a modification of the YRASTBALL gamma-ray spectrometer [7]. In the current experiment, the detection system consisted of 10, four-element clover detectors, 6 at a central angle of 90° to the beam direction (5 of which were hardware Compton suppressed) and 4 more suppressed clovers at 140° to the beam. Two further germanium planar detectors were placed in the 90° ring for enhanced sensitivity for low-energy events. This resulted in a total of 4.5x10^7 γ–γ coincidences, measured over a 20 hour period, which were sorted in standard γ-ray energy coincidence matrices for subsequent off-line analysis using the RADWARE [8] suite of software.

FIGURE 1. Target chamber used in the current work. Note the Faraday cup situated 5 cm behind the target position.

Simulations of the fusion cross-section using the PACE code [9] predict a production cross-section of approximately 600 mb for the ^{91}Zr+4n evaporation channel, with the other significantly populated residual channels being ^{90}Zr+5n, ^{92}Zr+3n and ^{88}Sr+α3n. The estimated maximum input angular momentum at the beam energy used was ~26 ℏ using the standard semi-classical formulism. A tantalum Faraday cup was situated 5 cm behind the target position (see figure 1), which allowed the stopping of some of the residual recoils in view of the γ-ray array. The hardware master trigger condition was set such that at least 2 'prompt' γ rays had to be measured in separate clover detectors within approximately 30 ns of each other. The acquisition hardware master gate then allowed subsequent 'delayed' coincidences of up to approximately 0.5 μs later to be written in the same event, thereby allowing correlations between prompt excitations and delayed events from isomeric states in nuclei which were implanted in the Faraday cup.

Data Analysis and Preliminary Results

The upper panel on the left hand side of figure 2 shows the total projection of the γ–γ coincidence matrix obtained in the current work (note that no additional software time selection has been applied to this in the offline analysis and thus coincidences with a time differential of up to Δt = 0.5 μs were incremented into this matrix).

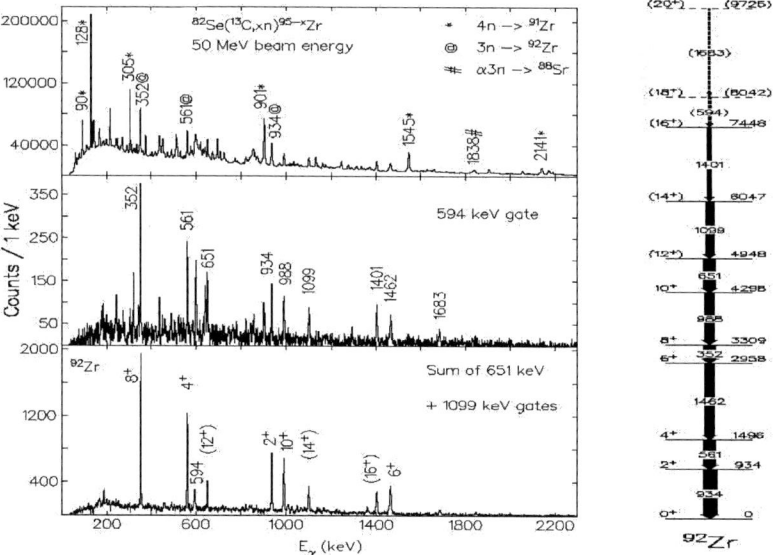

FIGURE 2. Total projection and γ-ray coincidence gates showing levels in ^{92}Zr from the current work.

Transitions previously identified in ^{92}Zr are clearly evident in the total projection [10-12]. The central panel of figure 2 shows the sum gates on the transitions depopulating the yrast (12$^+$) and (14$^+$) states in the nucleus. The near-yrast states reported in two recent fusion-fission induced studies of this nucleus [11,12] are evident, with an additional unreported transition at 594 keV. This transition is in coincidence with all the previously reported members of the yrast sequence in ^{92}Zr (as demonstrated by the spectrum shown in the central panel of figure 2) and our initial analysis suggest a stronger population than for the 1683 keV transition reported by Fotiades et al., [11]. However we stress that the ordering of the 1683 and 594 keV transitions is not definitive from the current analysis.

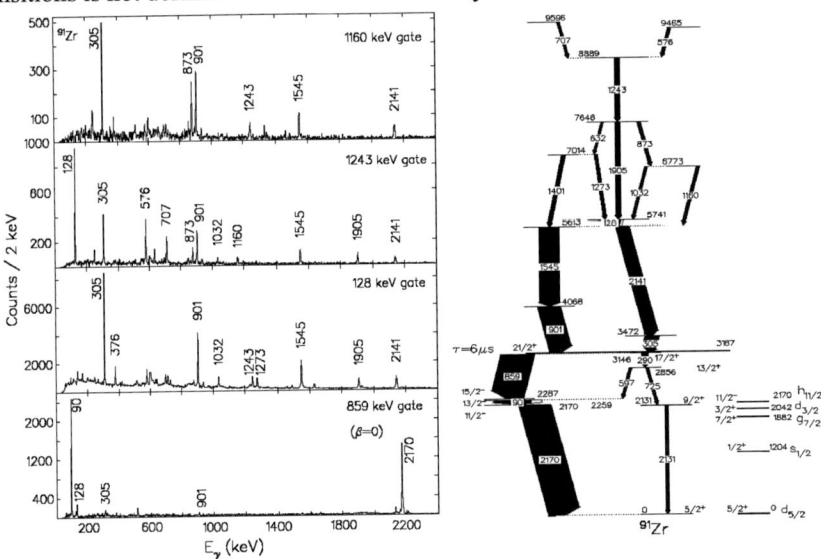

FIGURE 3. Spectra and preliminary partial level scheme for transitions above the $^{21}/_2^+$ isomer in ^{91}Zr as deduced from the current work. The states shown on the right of the level scheme can be associated with the main single particle strengths associated with the single-neutron valence states in ^{91}Zr. Their energies have been taken from reference [3].

Figure 3 shows the γ-ray coincidence spectra associated with ^{91}Zr obtained in the current work. The lower panel of the left hand side of figure 3 shows a γ-ray coincidence gate on the 859 keV transition reported in references [4,5] as being fed following the decay of the $^{21}/_2^+$ isomer in this nucleus. The Doppler correction used for this gate was set to β=0, appropriate for a recoil stopped in the Faraday cup. Note that all the other spectra shown in figures 2 and 3 assume a Doppler correction associated with the expected recoil velocity of v=1.2%c. The highly resolved nature of the 90 and 2170 keV transitions observed in this spectrum are consistent with the observation of *delayed feeding* from the $^{21}/_2^+$ isomer. The same spectrum also shows evidence for the previously unreported transitions at 128, 305 and 901 keV. These same γ rays are clearly observed in the total projection spectrum shown in the upper panel of figure 2. We suggest that these lines arise from prompt decays above the $^{21}/_2^+$ isomer in ^{91}Zr.

By placing coincidence gates on these transitions, the preliminary decay scheme for the states above the 6 μs isomer shown in figure 3 has been deduced. (Note that the ordering of the (305,2141) and parallel (901,1545) keV branches should be considered as tentative in the current data as the ordering of each branch is based on intensity measurements which are equal at the 2σ level within experimental uncertainties). A similar 'early-delayed' analysis on this nucleus following the population of this nucleus by the fusion-fission reaction between a ^{180}Hf beam and a ^{27}Al target also identifies the same transitions above the isomer in ^{91}Zr [13].

At present, no firm spin or parity assignments have been assigned to the levels identified above the $21/2^+$ isomer in the proposed level scheme. However, data on angular correlation and linear polarizations for these transitions are currently under analysis [14]. By making the standard assumption that fusion-evaporation reactions predominantly populate yrast and near-yrast states, it is very likely that states with spins above the maximum allowed using the simple ^{88}Sr valence space have been identified in the current work, including those associated with the breaking of the N=50 shell closure.

ACKNOWLEDGMENTS

This work is supported by the EPSRC (UK) and the US DOE by grants DE-FG02-91ER-40609 and DE-FG02-88ER-40417. PHR acknowledges support from the Yale University Flint Fund. NJT and ABG acknowledge financial support from Nexia Solutions (a subsidiary of BNFL).

REFERENCES

1. H. Mach et al., *Nucl. Phys.* **A523**, 197 (1991)
2. E.A. Stefanova et al., *Nucl. Phys.* **A69** 14 (2000); *Phys. Rev.* **C62**, 054314 (2000)
3. C. Baglin, *Nucl. Data. Sheets* **86**, 1 (1999)
4. B.A. Brown, P.M.S. Lesser and D.B. Fossan, *Phys. Rev. Lett.* **34**, 161 (1975)
5. B.A. Brown, P.M.S. Lesser and D.B. Fossan, *Phys. Rev.* **C13**, 1900 (1976)
6. H.R. McK. Hyder et al., *Nucl. Instr. Meth. Phys. Res.* **A268**, 285 (1988)
7. C.W. Beausang et al., *Nucl. Instr. Meth. Phys. Res.* **A452**, 431 (2000)
8. D. C Radford Nuc. Inst. Meth, **A361**, 297 (1995)
9. A. Gavron *Phys. Rev.* **C21**, 230 (1980)
10. B.A. Brown et al., *Phys. Rev.* **C14** 602 (1976)
11. N. Fotiades et al., *Phys. Rev.* **C65**, 044303 (2002)
12. D. Pantelica et al., *Phys. Rev.* **C72**, 024304 (2005)
13. G.A. Jones, *private communication, unpublished.*
14. N.J. Thompson and A.B. Garnsworthy, *private communication, unpublished.*

NUCLEAR STRUCTURE:
DEDICATED TO THE MEMORY OF D. D. WARNER

Dave Warner: Friend, Mentor, Colleague

A. Aprahamian

Institute for Structure & Nuclear Astrophysics, University of Notre Dame, Notre Dame, IN 46556

The morning of June 10, 2005 I had many urgent messages from colleagues on both sides of the Atlantic asking that I call back as soon as possible. The person that reached me on the phone first was Rick Casten. He was frantic and informed me of the unbelievable, that our friend Dave Warner had died on June 9, 2005 on his way home from work in the middle of the afternoon. My shock and disbelief was shared by many of our colleagues planning to attend CGS-12. Dave Warner had been there from the very beginning and played a very important role in exploiting neutrons in various aspects of Nuclear Structure Physics. His work at the ILL (Institut Laue Langevin in Grenoble, France) with GAMS 2 and 3 changed nuclear physics in a profound way by providing the world with spectroscopic studies of nuclei measured with high precision crystal spectrometers for the detection of gamma-rays as well as electrons. The first of these studies, the nucleus ^{168}Er became the benchmark nucleus with a complete level scheme up to an excitation energy of approximately 1.8 MeV [1]. The consequences of completeness have made a tremendous impact on nuclear structure Physics overall. Studies of the ^{168}Er nucleus was perhaps most instrumental in launching the new family of nuclear models based on algebraic or group theoretical approaches. The best known of this class of models is the **Interacting Boson Approximation** proposed by Iachello and Arima in the seventies [2-5]. One of the earliest and most crucial contributions that Dave Warner made to the testing and development of the IBA was the introduction of the consistent Q formalism for the IBA [6-8].

The latest complete spectroscopic study of a nucleus (^{162}Dy) was just published [9].

Dave Warner had many physics interests and led efforts in the United Kingdom as well as Europe and the USA for future facilities producing radioactive ion beams. He was a member of the NuPECC nuclear structure working group and the development of the Long Range Plan for Nuclear Science in Europe. He was part of the Isospin Laboratory international advisory committee amongst many others including GSI-Fair, Eurisol, the various GANIL scientific councils including SPIRAL and VAMOS and the CERN nuclear physics advisory council amongst others. Dave Warner's most recent science interest was in light and medium mass mirror nuclei.

Dave Warner was also a member of the International Advisory Committee for CGS-12. The email below is the response to my call to the Advisory Committee asking for suggestions of the science to be addressed at CGS-12.

Date: Thu, 24 Feb 2005 14:46:10 -0000
From: Dave Warner <d.warner@dl.ac.uk>
To: 'Ani Aprahamian' <aapraham@nd.edu>
Subject: RE: CGS-12 Int. Advisory Committee

Hi Ani,

The most exciting work in the world?? Well apart from my own. .....
(That must be the reply you'll get from just about everyone)

Anyway, some thoughts, most of which you've probably had:

Must be a big emphasis on rnb work:

MSU guys and nucleon knock-out work on N=28 shell closure round 42Si.
Coulex work: Seems to me a big current question is the differences between results coming from relativistic vs isol rnb coulex. eg in Mg32 region. Need contributions from Isolde, GANIL, MSU and GSI, maybe focusing on results from the same regions?

New facilities session. FAIR, SPIRAL2 and MAFF about to be funded.
Riken being built. plus status of RIA

Stable beams: Jyvaskyla work on heavy p-drip line and actinides

Theory: Large scale shell model. Phase changes and critical points.
Changing shell structure in n-rich.

No instrumentation session.

Those are first thoughts. Have to go to a meeting now. More later,
Cheers,
Dave

The conference took place at the University of Notre Dame Center for Continuing Education with over 150 participants from 23 countries. I wish Dave Warner could have been there with us. I owe him a large debt of gratitude for his friendship, the physics that he taught, and the colleague that he was. We will all miss him very much.

REFERENCES:

1. Casten, Warner, and Davidson, Phys. Rev. Lett. 45, 1077 (1980).
2. Arima and Iachello, Phys. Rev. Lett. 35, 1069 (1975).
3. Arima and Iachello, Ann. Phys. 99, 253 (1976).
4. Arima and Iachello, Ann. Phys. 111, 201 (1978).
5. Arima and Iachello, Phys. Rev. Lett. 40, 358 (1978).
6. Warner, D.D. and Casten, R.F., Phys. Rev. Lett. 48, 1385 (1982).
7. Warner, D.D. and Casten, R.F., Phys. Rev. C 28, 1798 (1983).
8. Casten, R.F. and Warner, D.D., Rev. Mod. Phys. 60, 389 (1988).
9. Aprahamian et al., Nucl. Phys. A 764, 42 (2006).

FIGURE 1: Dave Warner, Ani Aprahamian, and Alison Bruce at a picnic at Brookhaven National Laboratory. Dave Warner was a mentor to both Ani and Alison at Brookhaven while Rick Casten went to the University of Koeln as a Humboldt Fellow.

FIGURE2 : Dave Warner lecturing at the University of Surrey school, June, 2005.

FIGURE 3: I got you!

FIGURE 4. Stuart Pittel and Dave Warner at the "Mapping the Triangle" Conference in Jackson Lake LodgeGrand Teton National Park, Wyoming USA

FIGURE 5. Rick Casten, Jackie Mooney, and Dave Warner at Stufest on "Nuclear Physics, Large and Small" in Cocoyoc, Mexico (April, 2004).

FIGURE 6. Witek Nazarewicz and Dave Warner at Stufest in Cocoyoc, Mexico (April, 2004).

FIGURE 7. Cheers Dave…Oct. 31, 1983.

Emerging Collectivity in Nuclei, Shape Coexistence, and Proton-Neutron Interactions

R.F. Casten

Wright Nuclear Structure Laboratory, Yale University, New Haven, CT 06520 USA

Abstract. The growth of collectivity and phase transitional behavior in nuclei is one of the central questions facing nuclear structure studies today and in the future, especially with the advent of new generation facilities for the study of exotic nuclei. The role of the valence p-n interaction is key. Several new developments relating to these issues will be discussed. A new interpretation of the light Pt isotopes without the need for intruder states is related to the choice of single and multiple Hilbert spaces in the description of spherical-deformed transition regions. A new set of empirical proton-neutron interactions, extracted from double differences of atomic masses, shows striking bifurcations near closed shells, a generic relation to shell filling, and an empirical correlation to the different growth rates of collectivity in particle-particle and particle-hole regions. Finally, an extensive series of (p,t) experiments to locate 0^+ states in a broad range of transitional and deformed nuclei has disclosed a number of new features of these modes ranging from evidence from non-collective degrees of freedom to the statistics of nearest neighbor spaces and a new signature for critical point nuclei.

INTRODUCTION

Understanding the emergence of collectivity in nuclei and its competition with single particle degrees of freedom is one of the key challenges of modern nuclear structure physics. These issues will take on enhanced importance, and new challenges, with the growing availability of exotic nuclei and the appearance of new types of shell structure. Of particular significance in understanding how structure evolves is the valence p-n interaction which is central to the development of configuration mixing and collectivity [1, 2, 3], and to the presence of intruder states and shape coexistence [4]. In this paper we briefly address some recent work on the manifestations of collectivity and of the effects of the p-n interaction in nuclei.

INTRUDER STATES AND SHAPE COEXISTENCE

It was shown by Heyde *et al.* [4] that the monopole p-n interaction is critical to the onset of deformation and shape coexistence in nuclei. Specifically, this component of the p-n interaction can lower the energies of specific (*e.g.*, deformed) configurations sufficiently that, in some regions, they can cross the energy of the spherical ground state, leading to an abrupt change in equilibrium deformation with neutron or proton number. Before and after the crossing point, two minima can exist, spherical–like and more deformed.

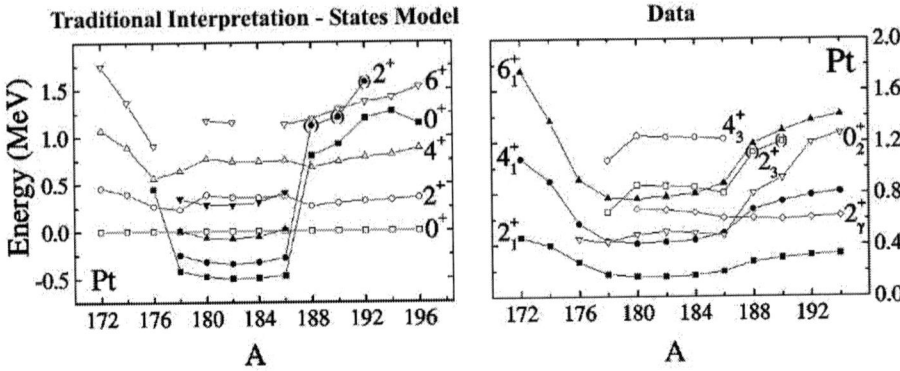

FIGURE 1. a) Traditional interpretation of the Pt isotopes. Based on ref. [7]. b) Experimental energies of low-lying states in Pt. Based on ref. [9].

This shape coexistence, as well as the discontinuous change in equilibrium deformation at the crossing point, are characteristic features of first order phase transitions.

In other regions, where the spherical and deformed configurations are initially quite separate in energy, the latter may not cross the spherical ground state and can appear, instead, in the low lying spectra as so-called "intruder states" [4]. Such states are well-established in the Hg and Pb isotopes [5, 6]. They have also been invoked to describe the low lying states of Pt isotopes [7, 8]. However, if shape coexisting states are well-mixed, a single Hilbert space description may be more appropriate. We will show that this occurs in the light Pt isotopes where there is, in fact, no need to resort to an intruder mechanism.

Figure 1a shows the traditional interpretation [7] of the Pt isotopes with intruder configurations crossing the spherical "ground" state between $^{178-186}$Pt. However, this plot is not a mere connection of data points, but a *model* of their N-dependent relationships. In contrast, a model-*independent* presentation of the raw data itself, shown in Fig. 1b, reveals a very different picture [9]. The changes in energy are much smaller and less drastic than on the left. One can ask why, given this data, is there a need for the intruder crossing interpretation on the left?

Model calculations in the intruder picture do, in fact, work quite well [8, 10]. However, they involve 6-13 parameters, stemming from the presence of three categories of terms in the Hamiltonian—those for the spherical configurations, those for the deformed, and those that mix the two.

In contrast, a simple 2-parameter Ising model of the generic type

$$H = aH_{sph} + bH_{def} \qquad (1)$$

or, in the form of the IBA [11]

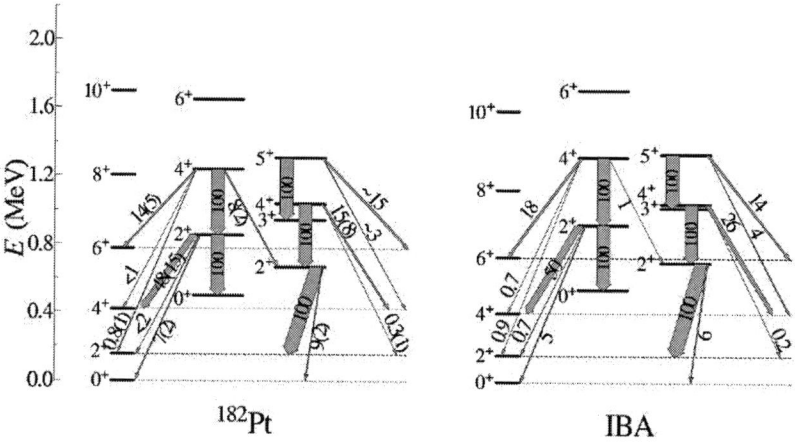

FIGURE 2. Comparison of single space 2-parameter IBA calculations with the data for ^{182}Pt. Based on the calculations in ref. [9].

$$H = \varepsilon n_d + \kappa Q \cdot Q \text{ with } Q = (s^\dagger \tilde{d} + d^\dagger s) + \chi (d^\dagger \tilde{d})^{(2)} \quad (2)$$

actually provides better fits [9]. This is illustrated in Fig. 2 for the case of ^{182}Pt. This result, and comparable ones for the other Pt isotopes from $^{178-194}$Pt, show that, at least in these nuclei, there is apparently no need to incorporate a separate Hilbert space representing cross-shell intruder states. The role of single and 2-space models of transitional nuclei was recently discussed in ref. [12]. While the intruder mechanism *per se* is well-established, its application in specific cases must be considered with care. In nuclei such as Pt, much simpler, single space 2-parameter calculations provide equal or better agreement with the data. The validity of the single space interpretation of Pt is being tested with studies of Pt isotopes underway at WNSL [13].

EMPIRICAL P-N INTERACTIONS, SHELL STRUCTURE AND GROWTH RATES OF COLLECTIVITY

As noted above, a principal mechanism for the lowering of collective states, the appearance of intruder states, and the onset of deformation is the valence p-n interaction. This interaction should be orbit- and region-dependent. Therefore, it is of high interest to determine p-n interaction strengths empirically. It is possible, in fact, to empirically isolate the interaction of the last proton with the last neutron, called δV_{pn}, by using double differences of atomic masses [14]. With the latest, 2003, mass evaluation [15], many new δV_{pn} values have become accessible and reveal fascinating features [16]. Two of these are illustrated in Fig. 3. Figure 3 (top) shows δV_{pn} values near ^{208}Pb, exhibiting a striking bifurcation below $N \leq 126$, and a sharp jump in the lower trajectory after $N = 126$. This is easily understood in terms of shell structure. Short-range proton-neutron interactions reflect the overlap of the respective orbits. The normal parity orbits of shells in heavy nuclei follow generic filling patterns from high j-low n to low j-high n. Hence, when both protons and neutrons are just below closed shells, they occupy highly overlapping orbits and δV_{pn} should be large [Region I in Fig. 3 (top)]. When protons are above and neutrons below (or vice versa but there is no data), the overlap is less and δV_{pn} should be smaller (Region II). When both are above (Region III), δV_{pn} should again be large. The behavior in Fig. 3 (top) is so striking that it can provide signatures of shell structure in exotic nuclei where masses may well be the first available data, and where shell structure may evolve in currently unknown ways.

Away from closed shells, such simple interpretations are more difficult due to precisely the complex correlations that the p-n interactions induce. Nevertheless, the generic sequences of shell model orbits suggest that δV_{pn} should, on average, be largest when protons and neutrons are at similar fractional occupations of each particles' major shell. The varying location of the unique parity orbit will modulate this simple picture but Fig. 3 (middle) shows that the effect very clearly persists. δV_{pn} values are dramatically larger near the diagonal than they are well away from it [16].

Finally, Fig. 3 (middle and bottom) show two further new results, one a newly observed relation between particle and hole regions, and the other a direct empirical link between p-n interaction strengths and collectivity never before observed [17]. Consider the $R_{4/2}$ values plotted against $N_p N_n$ in Fig. 3 (bottom): note that there is a clear difference in the growth of $R_{4/2}$ in particle-particle (p-p) and particle-hole (p-h) regions, being faster in the former. If the p-n interaction is responsible, this phenomenon should be reflected in p-n interaction strengths. This is exactly what is observed. δV_{pn} values are clearly larger in the lower left (p-p) quadrant of Fig. 3 (middle) than in the upper left (p-h). The average values in the two quadrants are 333 and 249 keV, respectively. Thus, we have observed [17] a direct link between an empirical measure of collectivity, $R_{4/2}$, and empirical values of p-n interaction strengths.

SYSTEMATICS OF 0^+ STATES

Given the venue of this conference, it is particularly fitting to present a followup to the pioneering (p,t) study of Lesher, Aprahamian, Graw and colleagues [18]. Following their (p,t) study of 0^+ states in ^{158}Gd several years ago, which revealed an unexpectedly large number of 0^+ states, we have carried out an extensive set of (p,t) experiments using the same spectrometer [19, 20, 21] at the MLL in Münich. We investigated [22, 23] a set of 17 nuclei spanning the rare earth region from spherical to deformed and γ-soft nuclei (Sm to Hg). The experiments were performed at a proton energy of 25 MeV and the outgoing tritons were analyzed with the Q3D-focal plane detector system described in refs. [19, 20, 21]. The extraordinarily high resolution (4-5 keV) and the nearly background-free spectrum enabled a large range of sensitivity. The data for 8 of the nuclei have been analyzed. 0^+ states were assigned on the basis of their angular distributions, in particular large values of the ratio of the cross sections at $5°$ and $17.5°$ scattering angles [$R(5°/17.5°) > 3$]. Of 30 previously assigned 0^+ states, 29

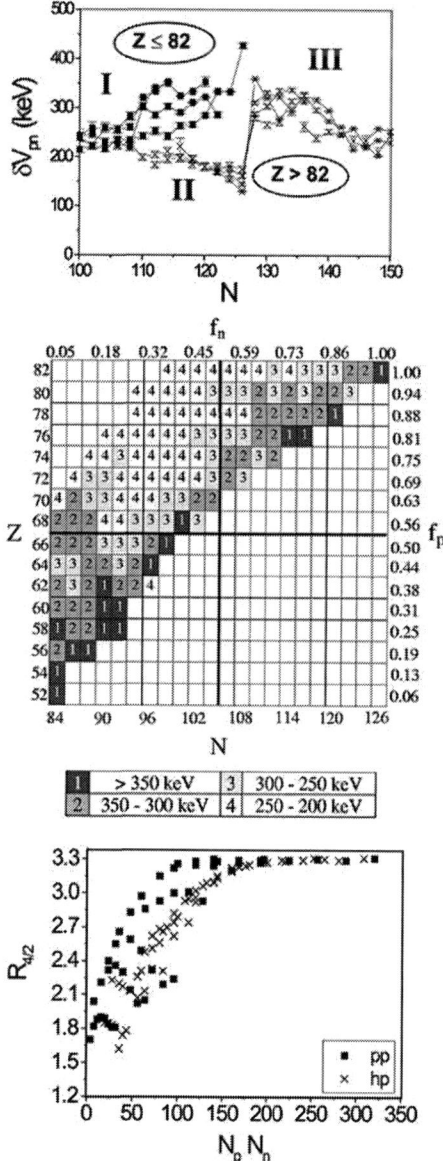

FIGURE 3. Top: δV_{pn} values near ^{208}Pb; Middle: δV_{pn} values in the major shell $Z = 50\text{-}82$, $N = 82\text{-}126$; Bottom: $R_{4/2}$ values for the $Z = 50\text{-}82$ shell for neutron numbers $N = 82\text{-}104$ (p-p region, solid square symbols) and $N = 106\text{-}126$ (p-h region, crosses). The left and middle panels are based on ref. [16], the right panel on ref. [17].

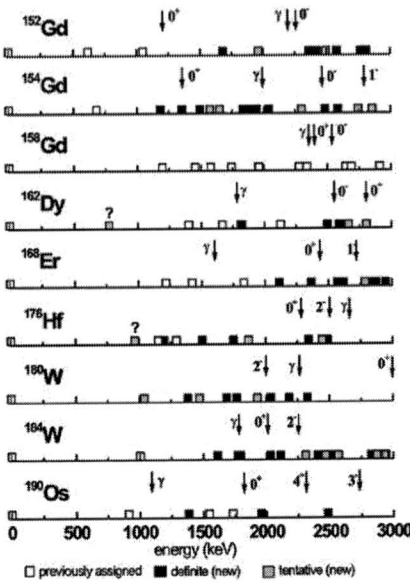

FIGURE 4. 0^+ states in the nuclei studied in refs. [22, 23]. (The ^{158}Gd results are identical to those of ref. [18].) Harmonic energies for 2-phonon states are labeled and previously known, new, and tentative 0^+ assignments are explicitly noted. Based on refs. [22, 23].

were seen and the 30th is itself inconclusively established. Although these previously known 0^+ states tend to have rather large cross sections, our sensitivity, down to $\sim \mu$b/sr, gives us confidence that nearly all 0^+ states below ~ 3.0 MeV have been identified in these experiments. In all, 75 definite 0^+ states were found, of which 46 are new. One significant finding [22] is an enhanced density of low-lying 0^+ states at the critical point nucleus (^{154}Gd) of the spherical-deformed phase transition. This will be discussed in these Proceedings by Jolie [24]. Here we will focus on other aspects [23].

Figure 4 shows the 0^+ states observed. Different symbols are used to distinguish previously known and newly found 0^+ states, and tentative assignments. The figure also indicates the harmonic estimates of 2-phonon (γ-γ, 0^+-0^+, Oct-Oct) states. While anharmonicities are expected, this gives a first order guideline. The pairing gap, 2Δ, in these nuclei is roughly at 1800 keV. Several features of Fig. 4 are striking. First are the variations in the number, density, and clustering of 0^+ states. While successful microscopic calculations exist for [25] ^{158}Gd, it will be a far more sensitive challenge to accommodate such nucleus-by-nucleus variations, even for close lying nuclides such as ^{180,184}W. Secondly, there are a larger than expected number of 0^+ states below the regime of 2-phonon modes. This suggests that many of these states are of weakly-collective or even 2-quasi-particle character. However, this raises the challenge of accounting for such a large number of 0^+ states below the pairing gap.

Finally, the mixing of these 0^+ states can be looked at through the distribution of

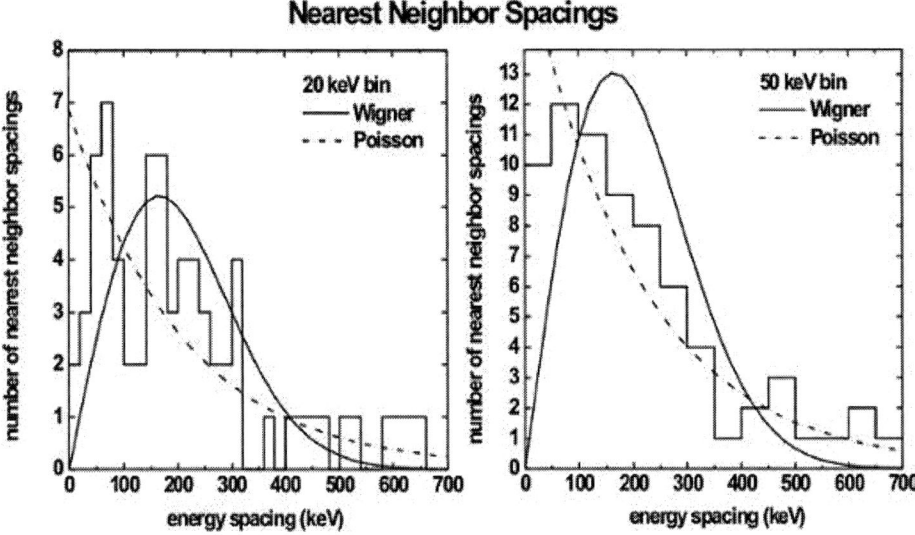

FIGURE 5. Distribution of nearest neighbor spacings of the 0^+ states studied in refs. [22, 23]. Left: 20 keV bins. Right: 50 keV bins. Based on ref. [23].

nearest neighbor spacings. Histograms of such spacings, and their implications, are highly dependent on the chosen bin size. Two examples, for a 20-keV bin and a 50-keV bin are shown, summing over all excited 0^+ states in these nuclei, in Fig. 5. The 20 keV results show a drop at the closest spacings, the 50 keV results do not. This may suggest that 2-state mixing matrix elements are greater than $\sim$10-20 keV on average but that many are less than about 50 keV.

CONCLUSION

To summarize, we have discussed a number of recent features of nuclear collectivity and structural evolution, focusing on the profound role of the p-n interaction in the development of configuration mixing and correlations. We studied this in terms of shape coexistence without intruder states in the Pt isotopes and the contrast with the evident need for the intruder state ansatz in Hg and Pb. We then turned to the study of empirical p-n interaction strengths, their relation to shell structure and the link between their magnitudes and the different growth rates of collectivity in particle-particle and particle-hole regions. Finally, we summarized a recent study using the (p,t) reaction to identify an extensive set of 0^+ states spanning a variety of nuclei in the rare earth region. We showed that many 0^+ states are below the pairing gap, and well below harmonic estimates of 2-phonon energies. We also noted striking variations in the distribution in energy of these 0^+ states as a function of N and Z, and pointed to a new signature of phase transitional behavior in the density of low-lying states near the phase transition.

ACKNOWLEDGMENTS

I am grateful to F. Iachello, W. Nazarewicz, T. Otsuka, K. Heyde, G. Dracoulis and K. Blaum for very helpful discussions and to my principal collaborators in the studies described here, including especially R.B. Cakirli, D.A. Meyer, N.V. Zam£r, E.A. McCutchan, J. Jolie, P. von Brentano, G. Graw and D. Bucurescu.

I would like to dedicate this paper to the memory of David D. Warner, who died suddenly and unexpectedly in June, 2005. His contributions to nuclear physics were many and profound. He was a wonderful collaborator, colleague, and friend for nearly 30 years. His loss is a terrible blow for nuclear physics worldwide, for U.K. nuclear physics, and for me personally. I miss him enormously.

Work supported by US DOE under Grant No. DE-FG02-91ER-40609.

REFERENCES

1. A. de Shalit and M. Goldhaber, Phys. Rev. **92**, 1211 (1953).
2. I. Talmi, Rev. Mod. Phys. **34**, 704 (1962).
3. P. Federman and S. Pittel, Phys. Lett. **69 B**, 385 (1977).
4. K. Heyde, P. Van Isacker, R.F. Casten and J.L. Wood, Phys. Lett. **155 B**, 303 (1985).
5. A.N. Andreyev et al., Nature (London) **405**, 430 (2000).
6. R. Julin, K. Helariutta and M. Muikku, J. Phys. G: Nucl. Part. Phys. **27**, 109(R) (2001).
7. J.L. Wood, K. Heyde, W. Nazarewicz, M. Huyse and P. van Duppen, Phys. Rev. **215**, 101 (1992).
8. P.M. Davidson, G.D. Dracoulis, T. Kibédi, A.P. Byrne, S.S. Anderssen, A.M. Baxter, B. Fabricius, G.J. Lane and A.E. Stuchbery, Nucl. Phys. **A 657**, 219 (1999).
9. E.A. McCutchan, R.F. Casten and N.V. Zam£r, Phys. Rev. **C 71**, 061301(R) (2005).
10. M.K. Harder, K.T. Tang and P. Van Isacker, Phys. Lett. **B 405**, 25 (1997).
11. F. Iachello and A. Arima, Phys. Lett. **53 B**, 309 (1974).
12. K. Heyde, J. Jolie, R. Fossion, S. De Baerdemacker and V. Hellemans, Phys. Rev. **C 69**, 054304 (2004).
13. R.B. Cakirli et al., to be published.
14. J-Y. Zhang, R.F. Casten and D.S. Brenner, Phys. Rev. Lett. **74**, 4607 (1995).
15. G. Audi, A.H. Wapstra and C. Thibault, Nucl. Phys. **A 729**, 337 (2003)
16. R.B. Cakirli, D.S. Brenner, R.F. Casten and E.A. Millman, Phys. Rev. Lett. **94**, 092501 (2005).
17. R.B. Cakirli and R.F. Casten, to be published.
18. S.R. Lesher et al., Phys. Rev. **C 66**, 051305 (R) (2002).
19. M. Löfþer, H.J. Scheerer and H. Vonach, Nucl. Instru. Meth. **111**, 1 (1973).
20. H.F. Wirth, Ph.D. thesis, Techn. Univ. München, (2001).
21. H.F. Wirth, et al., Beschleunigerlaboratium München Annual Report, 2000, p. 71.
22. D. Meyer et al., to be published.
23. D. Meyer et al., Phys. Rev. **C**, to be published.
24. J. Jolie, these proceedings.
25. N. Lo Iudice, A.V. Sushkov, and N.Yu. Shirikova, Phys. Rev. **C 70**, 064316 (2004).

Symmetry and Wigner Energy in Nuclear Matter

P. Van Isacker

GANIL, BP 55027, F-14076 Caen Cedex 5, France

Abstract. The nature of the symmetry and Wigner energy in the nuclear mass formula is studied. It is shown that their effects are intertwined and that one term cannot be reliably determined without knowledge of the other. This leads to considerable uncertainty in the value for the symmetry energy that should be adopted in nuclear matter.

Keywords: nuclear masses, symmetry energy, Wigner energy
PACS: 21.10.Dr, 26.60.+c

CONTEXT AND MOTIVATION

A simple, yet surprisingly accurate formula for the binding energy of an atomic nucleus is given by

$$B(N,Z) = a_{\text{vol}}A - a_{\text{sur}}A^{2/3} - a_{\text{cou}}\frac{Z(Z-1)}{A^{1/3}} - a_{\text{vsym}}\frac{(N-Z)^2}{4A} + a_{\text{pai}}\frac{\Delta(N,Z)}{A^{1/2}}, \quad (1)$$

where N is the number of neutrons, Z is the number of protons, and A is the total number of nucleons. Equation (1) is known as the von Weizsäcker or liquid drop mass formula [1, 2]. The various terms appearing in the mass formula are referred to as volume, surface, Coulomb, symmetry, and pairing. The last term represents a simple parametrization of the pairing effect in nuclei with $\Delta(N,Z) = +1, 0,$ and -1 in even–even, odd-mass, and odd–odd nuclei, respectively. We follow here the convention of positive binding energies such that the volume and pairing contributions are positive while others are negative. As a result all a coefficients in (1) are positive. Note also that a factor 4 is included in the denominator of the volume-symmetry term which leads to coefficients a_{vsym} that are 4 times bigger than what is usually quoted. This convention is adopted here for notational consistency with the quantal modification of the nuclear symmetry energy discussed below.

The nuclear symmetry energy has its origin in the Pauli principle: identical nucleons are forbidden to occupy the same quantum state and therefore symmetric nuclear matter with $N = Z$ is favoured. However, as will be discussed below, a liquid drop mass formula of the form (1) is inconsistent on fundamental grounds since a correct thermodynamic treatment requires the existence of a surface-symmetry as well as a volume-symmetry term, the former of which is lacking from the basic mass formula. Furthermore, the dependence $(N-Z)^2$ is but a semi-classical approximation to what is obtained from exact quantal theories. In part, this quantal modification introduces an additional term which depends on $|N-Z|$, known as the Wigner energy.

The basic liquid drop mass formula (1) gives only a rudimentary description of nuclear masses and much more sophisticated versions of it have been developed over the years.

(For a recent review, see Ref. [3].) Nevertheless, it is a useful formula since its simplicity allows the consecutive introduction of increasingly more sophisticated mass terms and a study of their inter-dependence.

The purpose of this contribution is to investigate modifications of the symmetry term in the liquid drop mass formula. The long-term aim of this analysis is a better understanding of the volume-symmetry coefficient $a_{\rm vsym}$ and its asymptotic behaviour for large systems, $A \to \infty$. Current uncertainties in our description of neutron stars, for example, are essentially due to a poorly-known symmetry energy [4, 5].

THE SURFACE COMPONENT OF THE SYMMETRY ENERGY

The extrapolation of the symmetry energy from finite nuclei to infinite nuclear matter is delicate because of the existence of a surface-symmetry term. Its inclusion in the liquid drop mass formula is based on two arguments.

First, a surface-symmetry term is required on fundamental grounds [6, 7]. Indeed, a consistent thermodynamic treatment of finite, asymmetric matter modifies the surface energy into

$$\frac{S_{\rm v}}{1+S_{\rm s}A^{-1/3}/S_{\rm v}} \approx -a_{\rm vsym}\frac{(N-Z)^2}{4A} + a_{\rm ssym}\frac{(N-Z)^2}{4A^{4/3}}, \qquad (2)$$

where the constants $S_{\rm v}$ and $S_{\rm s}$ on the left-hand side are related to coefficients that occur in an expansion of the neutron and proton chemical potentials as a function of the asymmetry $N-Z$ [8, 9]. The right-hand side of (2) follows from a large-A approximation and a conversion to the notation adopted in (1).

Secondly, there is empirical evidence for the existence of a surface-symmetry term. If this term is incorporated in the liquid drop mass formula, the latter is modified into

$$\begin{aligned}B(N,Z) &= a_{\rm vol}A - a_{\rm sur}A^{2/3} - a_{\rm cou}\frac{Z(Z-1)}{A^{1/3}} - a_{\rm vsym}\frac{(N-Z)^2}{4A} + a_{\rm ssym}\frac{(N-Z)^2}{4A^{4/3}}\\ &\quad + a_{\rm pai}\frac{\Delta(N,Z)}{A^{1/2}}.\end{aligned} \qquad (3)$$

The coefficients in both mass formulas (1) and (3) can be adjusted to the measured nuclear binding energies. In the following we have excluded the lightest nuclei with $N,Z < 8$ and used the mass the compilation of Ref. [10] to obtain the results shown in Table 1. Comparison of columns 2 and 3 of the table shows that a lower root-mean-square deviation σ is obtained with the modified mass formula. (The deviation σ is obtained by maximizing the likelihood function and includes experimental errors [11]. In all the fits shown in the table σ is only slightly lower than the usual root-mean-square deviation.) However, even if the inclusion of a surface-symmetry term leads to a fit which is better by about 240 keV, it appears not possible to reliably determine the two coefficients $a_{\rm vsym}$ and $a_{\rm ssym}$ separately, as will be shown below.

TABLE 1. Summary of coefficients and σ deviations (in MeV) in the various fits

	Equation (1)	Equation (3)	Equation (5)*	Equation (6)
a_{vol}	15.7	15.7	15.8	15.9
a_{sur}	17.7	18.1	18.0	17.9
a_{cou}	0.71	0.71	0.71	0.72
a_{vsym}	92.3	106.	125.	129.
a_{ssym}	—	76.0	224.	237.
a_{pai}	12.7	12.8	12.4	11.9
$10*a_f$	—	—	—	1.39
$100*a_{ff}$	—	—	—	0.49
σ	2.93	2.69	2.39	1.28

* With $r = 2.5$.

THE WIGNER CUSP

In semi-empirical mass formulas the Wigner energy usually is decomposed into two parts [12, 13]

$$B_W(N,Z) = -W(A)|N-Z| - d(A)\delta_{N,Z}\pi_{np}, \qquad (4)$$

where $W(A)$ and $d(A)$ are (smooth) functions of the nuclear mass number A. The first term on the right-hand side contributes to all $N \neq Z$ nuclei. The quantity π_{np} equals 1 for odd–odd nuclei and vanishes otherwise, and therefore $d(A)$ is relevant in odd–odd $N = Z$ nuclei only. With the convention of positive binding energies, $W(A)$ and $d(A)$ are positive and the first term in (4) is thus associated with the additional stability observed in $N = Z$ nuclei. The Wigner effect mainly stems from the first term in expression (4) which is equivalent to the replacement of $(N-Z)^2$ with $T(T+r)$, leading to the modification

$$B(N,Z) = a_{vol}A - a_{sur}A^{2/3} - a_{cou}\frac{Z(Z-1)}{A^{1/3}} - a_{vsym}\frac{T(T+r)}{A} + a_{ssym}\frac{T(T+r)}{A^{4/3}}$$
$$+ a_{pai}\frac{\Delta(N,Z)}{A^{1/2}}. \qquad (5)$$

We note that in the application of formula (5), T is to be identified with $|N-Z|/2$. A fit to all known masses with $N,Z \geq 8$ gives the lowest deviation ($\sigma = 2.39$ MeV) for $r = 2.5$ (see Table 1). Conclusions that can be drawn from these results are, however, made uncertain by the absence of shell corrections to which we now turn.

SHELL CORRECTIONS

In Fig. 1 is shown the difference between measured nuclear binding energies and those calculated with the mass formula (5). The figure clearly shows that doubly magic nuclei are more bound than what is predicted with a macroscopic droplet formula which lacks shell corrections. In fact, inspection of the figure shows that these corrections are

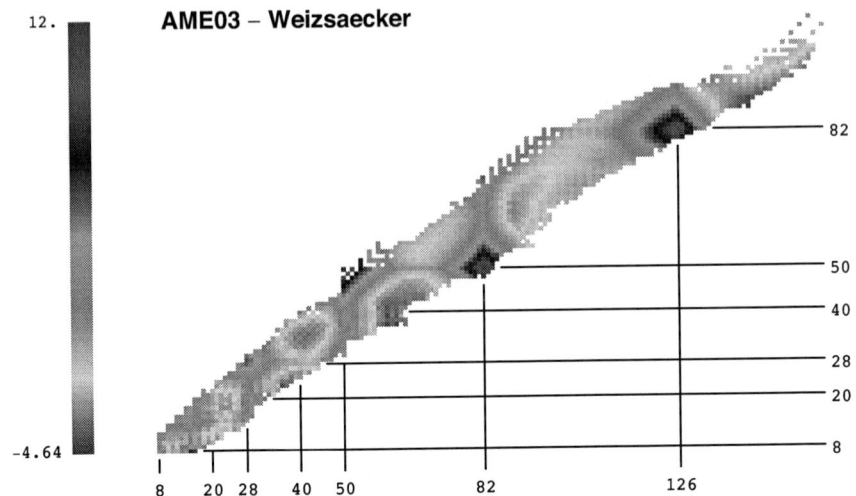

FIGURE 1. Differences between measured and calculated binding energies for nuclei with $N,Z \geq 8$. Data are taken from Ref. [10]. The binding energies are calculated with Eq. (5) which includes the surface-symmetry and Wigner contributions but not yet shell effects. The figure suggests the use of an expansion in F-spin as a shell correction.

essentially governed by the sum $n_\nu + n_\pi$, where n_ν (n_π) is the number of valence neutron (proton) particles or holes counted from the nearest closed shell. Since this is identical to the counting of bosons adopted in the neutron–proton interacting boson model [14], we refer to $(n_\nu + n_\pi)/2$ as $F_{\max}$. A simple parametrization of the shell corrections can thus be proposed based on an expansion in $F_{\max}$,

$$B(N,Z) = a_{\text{vol}}A - a_{\text{sur}}A^{2/3} - a_{\text{cou}}\frac{Z(Z-1)}{A^{1/3}} - a_{\text{vsym}}\frac{T(T+r)}{A} + a_{\text{ssym}}\frac{T(T+r)}{A^{4/3}}$$
$$+ a_{\text{pai}}\frac{\Delta(N,Z)}{A^{1/2}} - a_{\text{f}}F_{\max} + a_{\text{ff}}F_{\max}^2. \quad (6)$$

We note that Davis et al. [15] have used similar ideas to predict nuclear masses far from stability.

With this simple shell correction the σ deviation decreases to 1.28 MeV while the macroscopic a coefficients remain stable (see Table 1). This lends confidence to the results of the analysis. An important aspect is illustrated in Fig. 2 which shows the quantities a_{vsym}, a_{ssym}, and σ as a function of r. The first conclusion to be drawn from the figure is that r is not well determined from masses. While values of r below 1 can be excluded since they lead to substantially higher deviations, σ is fairly flat in the range $1 < r < 4$. Figure 2 also demonstrates that, while r is ill-determined, its value has a big impact on the coefficients of the symmetry terms, in particular the surface-symmetry term, which has an uncertainty of a factor 2. Part of the reason for this is that volume- and surface-symmetry coefficients are strongly correlated, as illustrated with Fig. 3. For fixed values of a_{vsym} and a_{ssym} other coefficients in the mass formula (6)

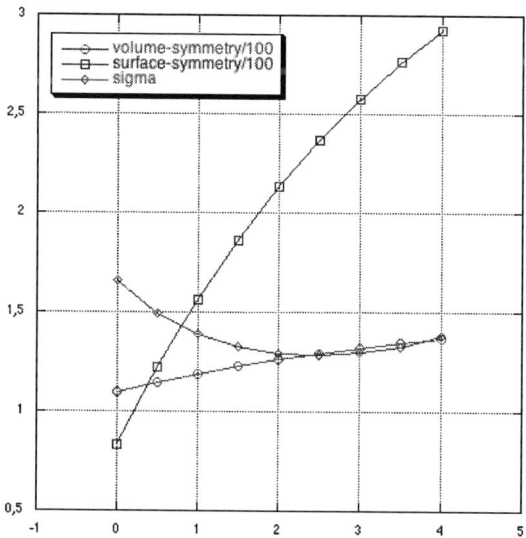

FIGURE 2. The volume-symmetry and surface-symmetry coefficients, and the σ deviation (in MeV) as a function of r, the Wigner parameter. The coefficients a_{vsym} and a_{ssym} are scaled by a factor 100.

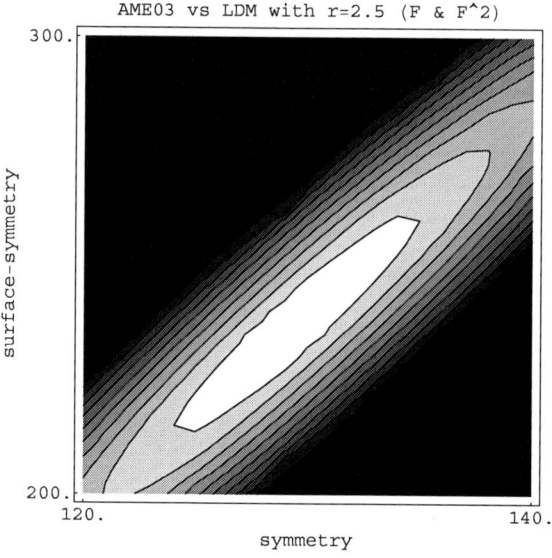

FIGURE 3. Correlation plot between the volume-symmetry and surface-symmetry coefficients in the liquid-drop mass formula (6). For a given pair (a_{vsym}, a_{ssym}), other coefficients in the mass formula are searched as to minimize σ. Each grey scale corresponds to an increase in σ by 100 keV.

are searched as to minimize σ. The minimum value thus obtained obviously is larger than the overall minimum which occurs for $a_{\text{vsym}} \approx 129$ MeV and $a_{\text{ssym}} \approx 237$ MeV. The perverse combination of a poorly-known Wigner parameter r and the existence of a strong correlation between the volume- and surface-symmetry coefficients leads us to conclude that there is considerable uncertainty in the value for the symmetry energy that should be adopted in nuclear matter.

QUANTAL SYMMETRY EFFECTS

A bothersome aspect of the previous analysis, based on the liquid drop mass formula (6), is that the entire Wigner effect is represented by a single parameter r which, as a consequence, is difficult to understand intuitively. It is, therefore, of interest to obtain an understanding of the effect from its original perspective due to Wigner [16].

Supermultiplet theory is based on two assumptions. (i) The forces between nucleons are independent of spin and isospin. Since two spin states (spin up and down) and two isospin states (neutron and proton) are available for each nucleon, this leads to a four-fold degeneracy structure and to SU(4) symmetry. (ii) The dominant component of the residual nucleon–nucleon interaction is short-range and attractive. As a consequence of (i), nuclear states can be classified according to their spatial symmetry [which is equivalent to SU(4) symmetry]. Because of (ii), the favoured (i.e., ground) state has maximal spatial symmetry since that is the configuration which provides maximal overlap of the nucleon wave functions. Of course, the spatial symmetry must be consistent with the Pauli principle and as such depends on the numbers of neutrons and protons in the nucleus.

These basic symmetry considerations lead to a recipe for the correlation energy of a nucleus in its ground state: this energy is related to the expectation value $g(\lambda,\mu,\nu)$ of the so-called quadratic Casimir operator of SU(4) where g is a quadratic polynomial in the SU(4) labels λ, μ, and ν which are uniquely determined by N and Z. This contribution of spatial correlations to the binding energy can be written as [17]

$$-K(A)g(\lambda,\mu,\nu) = -K(A)\left[(N-Z)^2 + 8|N-Z| + 8\delta_{N,Z}\pi_{\text{np}} + 6\Delta'(N,Z)\right], \quad (7)$$

where $K(A)$ is a mass-dependent, positive scale factor. The first term on the right-hand side is the semi-classical symmetry energy and suggests that $K(A) \propto A^{-1}$. The last term is due to pairing and follows here the unusual convention that $\Delta'(N,Z)$ equals 0 for even–even, 1 for odd-mass, and 2 for odd–odd nuclei. The two remaining terms in (7) correspond *exactly* to the Wigner energy (4) with the constraint $W(A) = d(A)$. We thus find that Wigner's original symmetry arguments lead to the correct terms in the binding energy.

The above derivation also implies a relation between the symmetry and Wigner energy, through $W(A) = d(A) = 8K(A)$, which overestimates both $W(A)$ and $d(A)$. Another way of formulating the same observation is to say that the Wigner parameter r obtained from masses is smaller than its SU(4) value which is $r = 4$. Note, however, that the *second* part of the Wigner energy, namely $\delta_{N,Z}\pi_{\text{np}}$, has not been included in any of the empirical mass formulas considered here.

Since the original work of Wigner many years have passed and it is now understood that SU(4) symmetry is increasingly broken as the atomic mass number A increases, mainly as a consequence of spin–orbit interactions. This gives at least a qualitative argument for understanding the deviations noted above, namely that $W(A) \neq d(A) \neq 8K(A)$. This argument can be made more quantitative by considering nuclear ground states that do not carry exact SU(4) (or spatial) symmetry but are an admixture of two SU(4) representations. This leads to binding-energy contributions of the type (7) but more complicated, and which also depend on the mixing parameter.

CONCLUSION AND UNRESOLVED QUESTIONS

The main conclusion of this contribution is that the effects of symmetry and Wigner terms in the nuclear mass formula are intertwined and that one term cannot be reliably determined without knowledge of the other. This leads to the conclusion that there is more uncertainty in the value that should be adopted for the volume-symmetry energy in infinite nuclear matter than was *hitherto* accepted.

This study reveals a number of unresolved questions and identifies approaches to be explored. Since the Wigner parameter r is not well determined from masses, other methods for its determination should be tried. A possible way is via the energies of isobaric analogue states of $T = 5/2$ multiplets. This procedure was recently suggested by Jänecke and O'Donnell [18] and should be further explored with special attention to possible shell effects. Such studies, while certainly of merit, are based on a simplified parametrization of the Wigner effect in terms of a single parameter r. A more microscopic understanding of the effect is thus called for and is, in fact, provided by Wigner's supermultiplet theory. The present analysis suggests the use of a more sophisticated Wigner term in the mass formula based on Wigner's original idea but modified to take account of SU(4) symmetry breaking. Another problem encountered in this study concerns the separate determination of the bulk and surface parts of the symmetry energy. As was pointed out in previous studies [8, 9], these terms are strongly correlated and cannot be reliably determined from masses. Mass data should thus be combined with other nuclear properties, notably differences between matter and charge radii, to constrain both terms separately. Such an analysis has been carried out [8] but only in a semi-classical approximation to the symmetry energy (i.e., with $r = 0$). The present results suggest a re-analysis of this problem with due care for the quantal character of the symmetry energy [19].

ACKNOWLEDGMENTS

I dedicate this contribution to the memory of Dave Warner in collaboration with whom my early work on the Wigner effect was carried out. However much I would have liked to investigate with him the unresolved questions raised by this study, fate has decided otherwise. The work reported here was also done in collaboration with Lex Dieperink.

REFERENCES

1. C.F. von Weizsäcker, *Z. Phys.* **96**, 431 (1935).
2. H.A. Bethe and R.F. Bacher, *Rev. Mod. Phys.* **8**, 82 (1936).
3. D. Lunney, J.M. Pearson, and C. Thibault, *Rev. Mod. Phys.* **75**, 1021 (2003).
4. J.M. Lattimer and M. Prakash, *Astrophys. J.* **550**, 426 (2001).
5. J.M. Lattimer and M. Prakash, *Science* **304**, 536 (2004).
6. W.D. Myers and W.J. Swiatecki, *Ann. Phys. (NY)* **55**, 395 (1969).
7. A. Bohr and B.R. Mottelson, *Nuclear Structure. II Nuclear Deformations*, Benjamin, New York (1975).
8. P. Danielewicz, *Nucl. Phys. A* **727**, 233 (2003).
9. A.W. Steiner, M. Prakash, J.M. Lattimer, and P.J. Ellis, *Phys. Reports* **411**, 325 (2005).
10. G. Audi, A.H. Wapstra, and C. Thibault, *Nucl. Phys. A* **729**, 337 (2003).
11. P. Möller, J.R. Nix, W.D. Myers, and W.J. Swiatecki, *At. Data Nucl. Data Tables* **59**, 185 (1995).
12. P. Möller and R. Nix, *Nucl. Phys. A* **536**, 20 (1992).
13. W. Satula, D.J. Dean, J. Gary, S. Mizutori, and W. Nazarewicz, *Phys. Lett. B* **407**, 103 (1997).
14. A. Arima, T. Otsuka, F. Iachello, and I. Talmi, *Phys. Lett. B* **66**, 205 (1977).
15. E.D. Davis, A.F. Diallo, B.R. Barrett, and A.B. Balantekin, *Phys. Rev. C* **44**, 1655 (1991).
16. E.P. Wigner, *Phys. Rev.* **51**, 106, 947 (1937).
17. D.D. Warner, M. Bentley, and P. Van Isacker, to be published.
18. J. Jänecke and T.W. O'Donnell, *Phys. Lett. B* **605**, 87 (2005).
19. A.E.L. Dieperink and P. Van Isacker, to be published.

Varying shell structure in *sd-pf* nuclei

Takaharu Otsuka

Department of Physics and Center for Nuclear Study, University of Tokyo, Hongo, Bunkyo-ku, Tokyo, 113-0033, Japan
RIKEN, Hirosawa, Wako-shi, Saitama, 351-0198, Japan

Abstract. Recent experiments on exotic nuclei with N=18 and 19 are discussed in connection to the variation of the shell gap at N=20. It is suggested that the gap is really small in such exotic nuclei. We show how the gap can be calculated from the shell-model interaction SDPF-M. Its characteristic feature can be explained as a effect of the tensor force. Thus, one of the subjects Dave Warner has been studying opens a new field of nuclear structure physics.

Keywords: N=20, shell gap, shell evolution, tensor force
PACS: 21.10.-k,21.30.-x,27.30.+t

INTRODUCTION

In the study of the structure of neutron-rich exotic nuclei, there have been many discussions on the behavior of excess neutrons as loosely bound objects and/or sources of more diffuse mean-potentials. The loose binding yields phenomena like neutron-halo, neutron-skin, *etc*. The diffuse mean potential can produce a weaker spin-orbit splitting.

One can then ask himself/herself whether there can be any other mechanisms which are equally or maybe more basic and robust ? Among those who shared this question, there was Dave Warner. I would like to quote a statement in his paper [1], as saying "Such structural changes stem principally from the dominant attractive strength of the spin-isospin part of the effective nucleon-nucleon interaction which manifests itself by shifts in the effective single-particle energies through the monopole part of the shell-model Hamiltonian." This statement is consistent with a recently proposed paradigm that the variation of the shell structure, i.e., the shell evolution, may be frequently due to some specific features of the nucleon-nucleon interaction, especially a strong coupling between spin-flip partners, $j_>$ and $j_<$ of a proton and a neutron [2].

Several very recent experiments have suggested that the shell gap at N=20 can be rather small for $Z \sim 10$. In the conventional picture of the shell structure and the magic number, the gap is rather stable respect to the variation of Z and/or N, as can be expected from the very successful model of a harmonic oscillator potential plus a spin-orbit splitting by Mayer and Jensen [3]. We will present two recent examples showing such a trend in sect. 2, and discuss, in sect. 3, how this trend can be explained naturally with a systematic shell-model calculation. In sect. 4, we shall show a microscopic explanation of for the shell evolution.

N=20 GAP, SMALL OR LARGE

The structure of an exotic nucleus ^{31}Mg has been unknown for many years, even including its ground-state spin. Very recently, Neyens et al. reported [4] that the ground state of ^{31}Mg is 1/2$^+$, and also that there are three other levels near the ground state. Figure 1 shows this experimental results compared to three different shell model calculations.

The simplest shell model calculation shown in Fig. 1 is the one at far right. This calculation was carried out in the sd shell by using the USD interaction. Although the USD interaction works well in general for stable nuclei in the sd shell, apparently ^{31}Mg is not the case. This simple fact already conveys a rather important message: in the island of inversion model of Warburton, Brown and Becker [5] (See Fig. 2), ^{31}Mg should be a normal nucleus with the normal ground state of 3/2$^+$, which corresponds to a neutron hole in the $d_{3/2}$ orbit. The second column from the right-hand side in Fig. 1 shows combined results of the shell model calculations in the 1p-1h space and those in the 2p-2h spaces. Here, np-nh space means that n particles are created on the top of the $N=20$ core, resulting in n+1 holes in the sd shell in the case of ^{31}Mg. The calculations have been performed by using the ANTOINE code [6] with the interaction presented in [7]. The negative parity ground state appears in this calculation. While the correspondence between experiment and theory has been improved, the 1/2$^+$ remains still too high. The third column from the right-hand side is the result of the Monte Carlo Shell Model with the SDPF-M interaction proposed originally in [8] and named in [9]. In this kind of shell-model calculations, there is no restriction about configurations in the given single-particle space. So all possible configurations are mixed in. The agreement with experiment is further improved, and one finds four eigenstates within 500 keV from the ground state, similarly to observed levels. However, the ground state is 7/2$^-$, differing from experiment. Thus, we see that the interaction still has to be improved. On the other hand, it is often really hard to reproduce the spectrum of an odd-mass nucleus because of many levels close to each other. Any tiny details can change the ordering of the levels. Apart from such a problem, the four lowest levels with different parities within 500 keV suggest that the gap should be really small. We shall show, in the next section, which gap corresponds to such high level density.

The nucleus ^{29}Na is discussed next. The neutron number is $N=18$ for this nucleus, implying that ^{29}Na is two units away from the island of inversion (See Fig. 2). We already know that an $N=19$ nucleus ^{30}Na has an intruder ground state [9], whereas this nucleus was misinterpreted to be normal [10]. It is therefore a very intriguing question whether the next nucleus ^{29}Na is normal or not. Figure 3 shows the level scheme of ^{29}Na [11]. Since this is a β-decay experiment, spin-parities are not known for excited states. Nevertheless, one can see easily that the sd-shell calculation with the USD interaction cannot reproduce the number of levels near 1.5 MeV (See [11] for details). On the other hand, the MCSM calculation with the SDPF-M interaction mentioned above produces four states around 1.7 $\sim$ 2.4 MeV with intruder characters. Two of them are consistent with log ft values. Thus, the intruder configurations are dominant components of low-lying states of ^{29}Na.

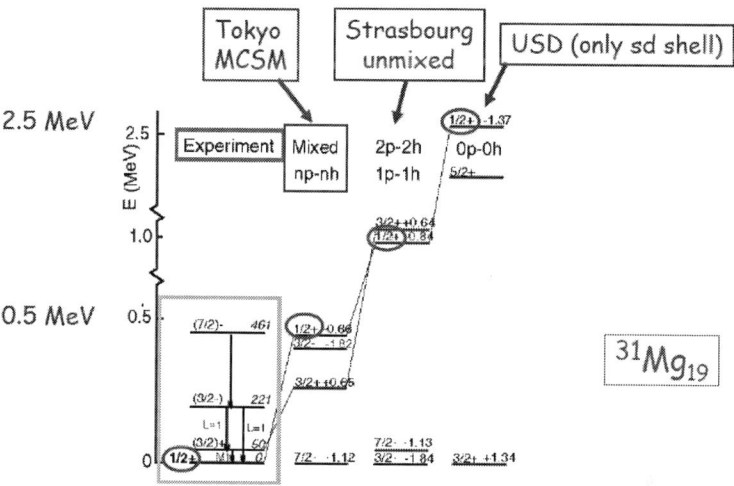

FIGURE 1. Energy levels of ^{31}Mg. The major part of this figure was taken from [4] which includes both experimental and theoretical results. For details, refer to the caption of Fig. 3 of [4]. Note that the vertical scale is changed twice.

N=20 GAP VARYING

We now discuss what gap can be obtained from the SDPF-M interaction which reproduces basic features of stable and exotic $sd - pf$ nuclei. Figure 4 shows the N=20 gap as a function of Z. The gap is about 6 MeV for Ca, but becomes smaller as Z decreases. It is as small as about 3 MeV for Ne. This small gap creates various interesting and rich properties of exotic nuclei even for $N < 20$, whereas all nuclei below N=20 are supposed to be normal in the island-of-inversion model of [5]. Such a small gap produces not only intruder positive-parity states but also low-lying negative-parity states as we have seen in the case of ^{31}Mg discussed above.

The gap in Fig. 4 increases from Z=8 up to 14. This characteristic feature can be explained partly as a result of the strong coupling between the proton $d_{5/2}$ orbit and the neutron $d_{3/2}$ orbit. Such a large change is not obtained in Skyrme calculations with reasonable parameters. After reaching Z=14, the gap remains rather stable. In the next section, we discuss the origin of this coupling.

SHELL EVOLUTION AND THE TENSOR FORCE

The tensor force has been known over decades. Its important effects have been investigated well. However, there has been one missing point. That is its first-order effects on single-particle properties, especially simple and robust nature of the tensor force. This nature can be presented in an analytic identity and its intuitive picture can be drawn. These points can be referred to in [12], and are not repeated here. The above proton-

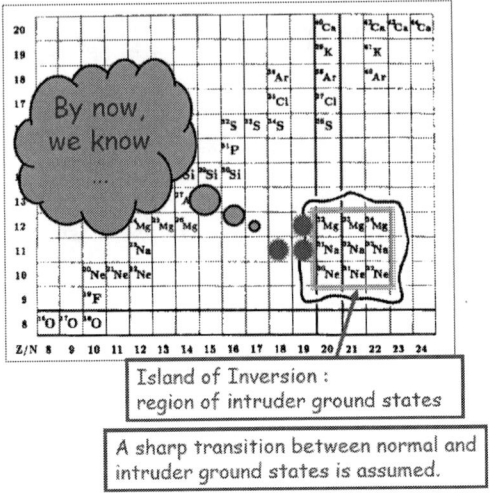

FIGURE 2. The island of inversion as suggested in [5] and three nuclei known by now to have intruder ground states, as marked by star symbols.

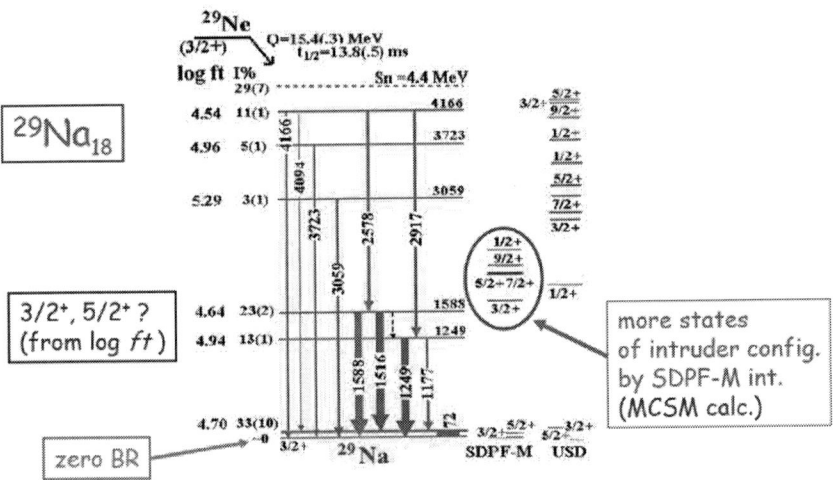

FIGURE 3. Energy levels of ^{29}Na. The major part of this figure was taken from [11] which includes both experimental and theoretical results. For details, refer to the caption of Fig. 2 of [11].

neutron $d_{5/2}$- $d_{3/2}$ coupling is a clear example.

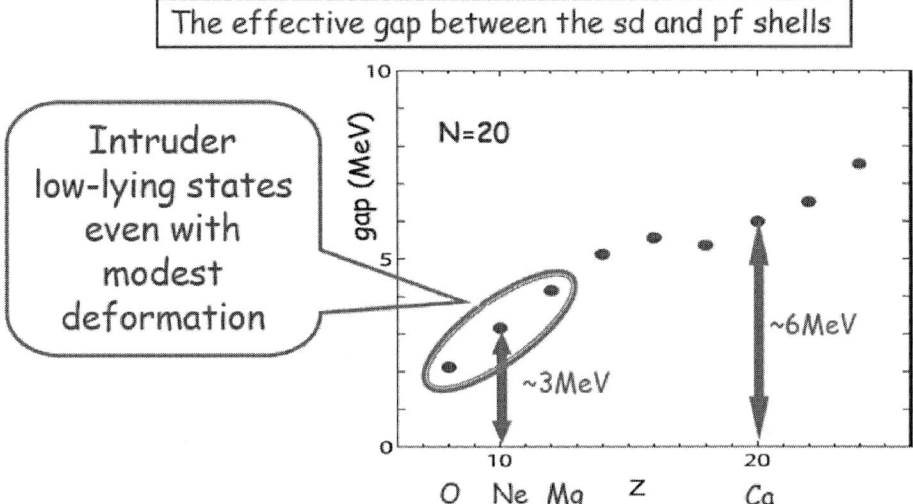

FIGURE 4. $N=20$ gap as a function of Z. The SDPF-M interaction is used.

SUMMARY

The $N=20$ gap is shown to change particularly with Z from 8 to 14. There are increasing number of experimental evidences. The origin of such a change appears to be the tensor force.

ACKNOWLEDGMENTS

This work was supported in part by a Grant-in-Aid for Specially Promoted Research (13002001) from the MEXT. This work has been a part of the RIKEN-CNS joint research project on large-scale nuclear-structure calculations. The authors acknowledge Drs. M. Serra, W.B. Walters and A. Gelberg for valuable help.

REFERENCES

1. D.E. Appelbe, C.J. Barton, M.H. Muikku, J. Simpson, D.D. Warner, C.W. Beausang, M.A. Caprio, J.R. Cooper, J.R. Novak, N.V. Zamfir, R.A.E. Austin, J.A. Cameron, C. Malcolmson, J.C. Waddington, and F.R. Xu, Phys. Rev. C **67**, 034309 (2003).
2. T. Otsuka *et al.*, Phys. Rev. Lett. **87**, 082502 (2001); T. Otsuka, Prog. Theor. Phys. Suppl. **146**, 6 (2002).

3. M.G. Mayer, Phys. Rev. **75** 1969 (1949); O. Haxel, J.H.D. Jensen and H.E. Suess, Phys. Rev. **75** 1766 (1949).
4. G. Neyens, M. Kowalska, D. Yordanov, K. Blaum, P. Himpe, P. Lievens, S. Mallion, R. Neugart, N. Vermeulen, Y. Utsuno, and T. Otsuka, Phys. Rev. Lett. **94**, 022501 (2005)
5. E.K. Warburton, J.A. Becker, and B.A. Brown, Phys. Rev. C **41**, 1147 (1990).
6. E. Caurier, F. Nowacki, A. Poves, and J. Retamosa, Phys. Rev. C **58**, 2033 (1998).
7. S. Nummela *et al.*, Phys. Rev. C **63**, 044316 (2001).
8. Y. Utsuno, T. Otsuka, T. Mizusaki, and M. Honma, Phys. Rev. C **60**, 054315 (1999).
9. Y. Utsuno, T. Otsuka, T. Glasmacher, T. Mizusaki, and M. Honma, Phys. Rev. C **70**, 044307 (2004).
10. B.V. Pritychenko, T. Glasmacher, P.D. Cottle, R.W. Ibbotson, K.W. Kemper, K.L. Miller, L.A. Riley, and H. Scheit, Phys. Rev. C **66**, 024325 (2002).
11. Vandana Tripathi, S.L. Tabor, P.F. Mantica, C.R. Hoffman, M. Wiedeking, A.D. Davies, S.N. Liddick, W.F. Mueller, T. Otsuka, A. Stolz, B.E. Tomlin, Y. Utsuno, and A. Volya, Phys. Rev. Lett. **94**, 162501 (2005)
    Phys. Rev. C **66**, 024325 (2002).
12. T. Otsuka, T. Suzuki, R. Fujimoto, H. Grawe, and Y. Akaishi, Phys. Rev. Lett. (2005).

Recent advances in studies of mirror-symmetry in the A=40-56 shell.

A.M.Bruce

School of Engineering, University of Brighton, Brighton, BN2 4GJ.

Abstract. The study of mirror symmetry has been extended to the $T_z=\pm 1/2$ nuclei ^{45}V and ^{45}Ti using a ^{24}Mg + ^{24}Mg fusion-evaporation reaction at the Vivitron accelerator at IReS Strasbourg. In order to extend the study to more-exotic nuclei, a new experimental technique using fragmentation of a secondary beam has been tried at GSI. The aim was to study excited states in ^{53}Ni using fragmentation of a ^{55}Ni beam produced in the initial fragmentation of 600 MeV/A ^{58}Ni by a ^{9}Be target. Details and results of both experiments will be presented.

Keywords: Coulomb energies, Shell model, Gamma-ray spectroscopy
PACS: 21.10.Hw, 21.10.Sf, 23.20.Lv, 27.40.+z

In the absence of the Coulomb force, charge independence and charge symmetry imply that the excitation energies and structure of analogous states in mirror nuclei should be identical. Differences in these measurable quantities can therefore yield subtle information about the Coulomb force. Nuclei in the A=40-56 shell are of particular relevance in studying this phenomenon as it is now possible to perform complete shell-model calculations in this $f_{7/2}$ space. The results of such calculations can be compared with the experimental information to give significant insights.

Dave Warner and Mike Bentley have been at the forefront of recent developments in this field, following on from work by John Cameron in the early 1990's [1] who used the Recoil Mass Separator and the POLYTESSA array at Daresbury to study A=49 nuclei. A review of the work done in this area is available in [2] so I will report on recent developments in two areas, namely 1) the extension to the A=45 mirror pair ^{45}V and ^{45}Ti and 2) the development of the "fragmentation" technique to extend the study to T=3/2, A=53 nuclei.

The A=45 mirror pair ^{45}V and ^{45}Ti.

^{45}V and ^{45}Ti were studied using the ^{24}Mg + ^{24}Mg fusion-evaporation reaction at a beam energy of 83 MeV at the Vivitron accelerator at IReS, Strasbourg. Gamma-rays were measured in the Euroball array with charged particles measured in the Euclides array and neutrons detected in the neutron wall. ^{45}Ti (2pn) was the stronger of the two reaction channels with an estimated cross-section of 300 mb, ^{45}V was estimated to be populated at the 1 mb level. Further details of the experiment can be found in [3] but figure 1 shows spectra gated on a) 2 protons and 1 neutron, b) 2 neutrons with the added condition that the two neutrons should not be in adjacent detectors and that the time of flight for each of the neutrons should be about the same. These two conditions are

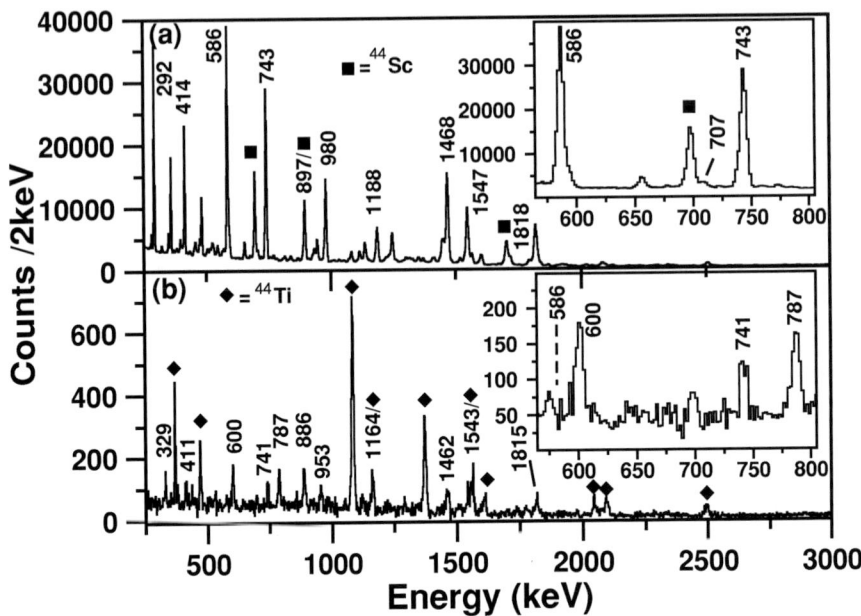

FIGURE 1. Singles gamma-ray spectra recorded in coincidence with a) two protons and one neutron and b) two "clean" neutrons (see text for details).

necessary to discriminate against events where one neutron is scattered and registers events in two detectors. In the top spectrum, the peaks labelled by energy correspond to known [4] transitions in ^{45}Ti while those labelled with a solid square belong in ^{44}Sc which is produced in the 3p1n channel of the reaction and where one of the protons escapes undetected. In the bottom spectrum, the peaks labelled by energy have been assigned to ^{45}V while those labelled with a solid diamond belong in ^{44}Ti which is produced in the 2p2n channel of the reaction. The assignment of the gamma-rays to ^{45}V has been done mainly on the basis of the similarity of the gamma-ray spectra and, by analogy, the partial level scheme for the positive-parity structure in ^{45}V, shown on the left in figure 2, has been constructed mainly by referring to the scheme for ^{45}Ti since there were insufficient statistics to combine gamma-gamma coincidence measurements with neutron detection.

In the case of the A=51 and A=53 mirror pairs, good agreement has been achieved between the observed mirror energy differences (defined as $E_x(J)[Z>]-E_x(J)[Z<]$) and shell-model calculations modified to include the J=2 anomaly [5]. The situation for the negative-parity structures in the A=45 pair is shown in figure 3. Part a) of the figure shows the empirical values and those from the shell model as a function of spin (J). The data and the calculation have been normalised to agree at spin J=7/2 but apart from the J=5/2, 15/2 and possibly 23/2 levels, the level of agreement is not as good as in previous cases. The shell-model calculation is made up of three components as indicated in part b) of the figure. The circles indicate the contribution from the Coulomb

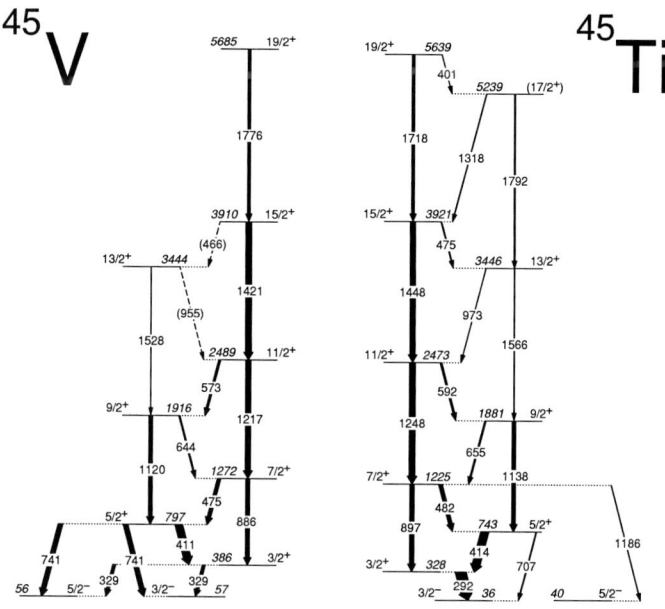

FIGURE 2. The partial level scheme for ^{45}V and ^{45}Ti from this work.

multipole term which deals with the angular momentim re-coupling of the protons. The diamonds indicate the contribution from the Coulomb monopole term which deals with the Coulomb effects associated with changes in the radii as a function of spin. The squares indicate the contribution from the additional multipole term which is added to deal with the J=2 anomaly mentioned earlier [5]. Although the calculation does reproduce the general trend of the data, the analysis of the three individual components sheds little light on the evolution of the structure. This is a different situation from that in the heavier cases and is perhaps an indication that any calculation needs to also include particles excited from the sd-shell. Further evidence to support this is presented in the next paragraph.

The A=45 nuclei are low in the A=40-56 shell and as figure 2 shows, there is the opportunity for particles to be excited from the orbits below the N/Z=20 shell gap. In both ^{45}V and ^{45}Ti there is evidence for the excitation of a particle from the $d_{3/2}$ orbit to produce a $J^\pi=3/2^+$ level at excitation energies of 386 and 328 keV respectively. A $J^\pi=5/2^+$ state is observed in each of the nuclei built on the $3/2^+$ level and the decay of this $5/2^+$ level is of particular interest. In ^{45}V this level decays not only to the $3/2^+$ level but also, quite strongly to low-lying $3/2^-$ and $5/2^-$ levels (in fact the $3/2^-$ and $5/2^-$ levels are so closely degenerate that it is not possible to separate the decays). The decay is split 56% to the negative parity states and 44% to the $3/2^+$ level. In ^{45}Ti the decay pattern is very different with 10% to the negative parity 3/2 state (the $5/2^-$ level is not populated at all) and 90% to the $3/2^+$ level. This difference in the decay pattern is

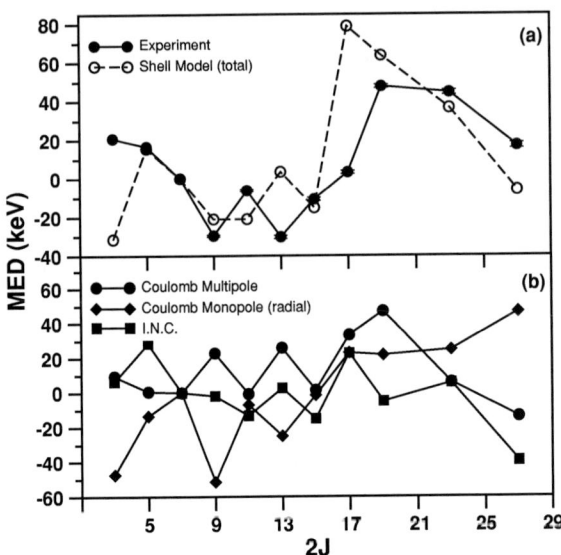

FIGURE 3. Mirror energy differences (defined as $E_x(J)[Z>]-E_x(J)[Z<]$) between states in ^{45}V and in ^{45}Ti. a) The comparison between data and the results of a shell-model calculation. b) The three components in the shell-model calculation (see text for details).

very striking since it is the only major discrepancy between the decay schemes. Similar behaviour has been observed in the A=31 mirror pair ^{31}S and ^{31}P [6, 7] and in the A=35 mirror pair ^{35}Ar and ^{35}Cl [8, 9]. However, it is perhaps of note that in the A=35 and 45 pairs, the stronger transition is observed in the nucleus which has Z>, while in the A=31 pair it is observed in the nucleus with Z<. A word of caution should however be expressed since it is only relative intensities which have been measured here and there is no indication of the absolute strength of the transitions. In general, electric and magnetic transition operators have isoscalar and isovector parts, the former of which scale with Z and the latter with T_z. By definition, the E1 operator does not have an isoscalar part and hence the transition rate should not depend on Z. E1 strengths should therefore be identical in the T=1/2 mirror pair. However, since in the A=45 pair the main component of low-lying negative parity states consists of 5 nucleons in the $f_{7/2}$ shell, while the positive parity states are 6 nucleons in $f_{7/2}$ and a hole in $d_{3/2}$, the (normal, first-order) E1 transition between those configurations is forbidden (i.e. $d_{3/2} \to f_{7/2}$ doesn't go via E1). There are two mechanisms by which this transition may therefore proceed: 1) an E1 transition through admixtures in the wave functions which might depend strongly on the details of the interaction and hence be different in the two mirror nuclei or 2) not by E1, but M2 or E3. With the exception of E1 (and to some extent M1) there is no reason that electromagnetic transition matrix elements in mirror nuclei should be the same. Hence it is very important to measure the multipolarity of these transitions before further conclusions can be made.

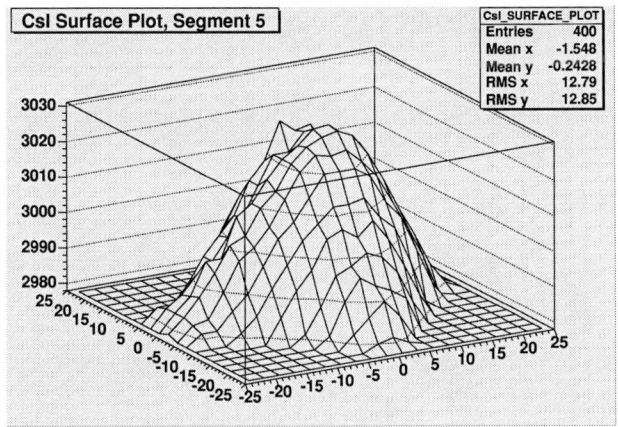

FIGURE 4. The energy response of the CsI detector as a function of position.

The development of the "fragmentation" technique to extend the study to $T_z=-3/2$ nuclei.

The limit of nuclei which can be populated by fusion-evaporation reactions is rapidly being reached so, in order to extend this study to $T_z=-3/2$ nuclei, a secondary fragmentation experiment was tested at the GSI laboratory in Darmstadt. The focus of the experiment was ^{53}Ni and the idea was to fragment a beam of ^{58}Ni "twice" such that in the first fragmentation ^{55}Ni was produced and in the second ^{53}Ni. The experiment utilised a primary beam (5 x 10^8 pps) of ^{58}Ni accelerated to 600 MeV/A. This impinged on a 4 g/cm^2 ^{9}Be target at the entrance to the **FR**ragment **S**eparator (FRS) and secondary fragments, including ^{55}Ni, were produced at energies of $\sim$ 170 MeV/A and with intensities $\sim$ 5 x 10^3 pps. These secondary fragments were identified and transported through the FRS and impinged on a second ^{9}Be target, this time of thickness 700 mg/cm^2. EPAX [10] calculations indicate that N=Z nuclei down to ^{24}Mg will be populated with about the same cross-section as ^{53}Ni and so, in addition to the focus of the work (^{53}Ni), nuclei in which excited states have previously been unobserved (e.g ^{44}V) should also be available for study. Although the cross-section for producing ^{53}Ni using ^{55}Ni as the secondary fragment is similar to that for using ^{54}Ni, it was decided to use the heavier fragment as it was considered that the removal of two particles from ^{55}Ni would introduce more angular momentum into the system. Further experimental details can be found in [11, 12, 13].

The fragments from the first reaction are identified in the FRS and the secondary fragments were detected downstream in a **CA**lorimeter **TE**lescope (CATE) [14, 15] which consisted of 9 individual position sensitive silicon detectors arranged in a square geometry with 9 caesium iodide detectors arranged behind them in the same geometry. The silicon detectors provide an energy loss signal (ΔE) which can be used for Z identification and the caesium iodide detectors give a measure of the remaining energy (E) such that (ΔE)/E can be used for mass identification. Gamma-rays emitted at the second fragmentation stage were identified using the RISING array which consisted of

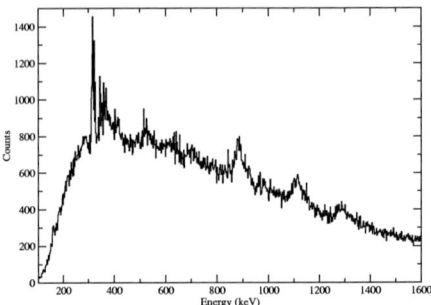

FIGURE 5. Preliminary spectrum for gamma rays associated with titanium fragments identified in the CATE detector. The spectrum clearly shows the 889, 1121 and 1289 keV yrast transitions in ^{46}Ti.

Euroball cluster and clover detectors. 8 BaF$_2$ detectors were placed at backward angles.

The positional information from the CATE detectors is required for three reasons, firstly so that the fragments can be tracked so that the correct Dopppler correction can be applied to the gamma-ray spectra (the fragments are moving at v/c $\sim$ 50%), secondly so that the distribution of reaction products as a function of angle can be investigated (there is a suggestion that the angle between the final fragment vector and the incoming fragment vector might give another degree of freedom in attempting to determine mass) and thirdly so that the position dependence of the energy response of the caesium iodide detectors can be accounted for. This latter feature arises because there was only one pin diode on each segment and it was centrally placed. Details of the position correction process can be found in [13, 14] and figure 4 displays the importance of this correction. This plot shows the energy response of the CsI detector as a function of position for the calibration mono-energetic beam. This should be a horizontal surface but it is quite clear that there is a big positional dependence and this is the correction which is the focus of current work. However, it is the spread in the velocity of the primary fragments that is the ultimate limiting factor in the achievable mass resolution. Currently the achieved resolution is 1.3% for mono-energetic beam but it is hoped that when the energy correction discussed above is applied, this can be reduced to 0.8% so that a figure of 1% can be achieved with the reaction target in place.

Figure 5 shows the gamma-ray spectrum associated with titanium fragments. The EPAX calculations indicate that the highest cross-sections should be for ^{46}Ti (1.98 x 10^{-2} b), ^{45}Ti (1.59 x 10^{-2} b) and ^{47}Ti (9.91 x 10^{-3} b) and the gamma-rays in the spectrum can be associated with the 1289,1121,889 keV yrast cascade in ^{46}Ti. The next step in the analysis is to try and distinguish transitions from the neighbouring isotopes and to improve the resolution of this spectrum. In addition, it may be possible to use the improved timing properties of the BaF$_2$ detectors to reduce the background in this spectrum which is thought to originate from bremsstrahlung.

FIGURE 6. July 1994

ACKNOWLEDGMENTS

Dave Warner was a driving force in this work and also in my whole career. He was a patient teacher who instilled in me my love for the subject. Others have focussed on Dave's brilliance as a scientist so I will concentrate on a more personal level. My first interaction with Dave was at Brookhaven National Laboratory where I was a graduate student under his tutelage. It was here that he instilled in me my love for physics and the idea that physics (and indeed life) is fun. During my time at Brookhaven, the group leader Rick Casten left to take up a one-year fellowship at the University of Koln, and Dave took over the leadership of the group for this period. Obviously a good excuse for a party, we bought a fun T-shirt for Dave and had it emblazoned with the words 'Pretend Group Leader' as shown in figure 6. However, I very quickly learnt that 'pretend' was not a word to be associated with Dave. If asked to do a job or task, he would always do it to the best of his ability. He would never pretend to be anything he was not. Moreover, there was never any falsehood in his dealings with me. Instead he would finish any statement which was not perhaps what I wanted to hear with that distinctive chuckle of his, so that I always came away thinking that the reply was not so bad after all.

Dave fought long and hard for a UK radioactive beam facility. The first proposal was the SIRIUS project which was based on using 800 MeV protons from ISIS at the Rutherford laboratory and the second was the CASIM project which was to be based at Daresbury Laboratory using 230 MeV protons from a cyclotron. Both of these were ranked very highly on the science but, mainly due to the idiosyncrasies of the UK funding system, they were not funded. He did however realise that in order to maintain the vibrancy of the field either a UK facility or a partnership in a European facility was required. To this end, he has been heavily involved in both the FAIR (Facility for Antiproton and Ion Research) project at GSI and the SPIRAL 2 developments at GANIL. He was also on the International Advisory Group for the Isospin Laboratory in the US. At GSI he was one of 2 UK representatives on the Scientific and Technical Issues working group. He was one of 3 special advisors to the French Ministry on the SPIRAL 2 project and an active member of the EURISOL development committee. The SPIRAL 2 project has now been given the go-ahead and one of Dave's final projects was to facilitate a request to the UK funding council for a UK contribution to this project in terms of a direct beam-line from the CIME cyclotron to the beam-lines in experimental areas G1 and G2. This would enable up to 5 beams to be used simultaneously at GANIL as opposed to the 1 currently, thus having a marked effect on the efficiency of the laboratory. The fruition of a significant UK involvement in the SPIRAL2, EURISOL or FAIR projects will be a fitting tribute to his memory.

I would also like to acknowledge those who have played a major part in the physics discussed here, especially (in alphabetical order) Mike Bentley, Gavin Hammond, Piet van Isacker, Rady Lozeva and Mike Taylor.

REFERENCES

1. J.A.Cameron et al, *Phys. Letts.* **235B**, 239 (1990).
2. M. A. Bentley, Proceedings of the International Conference on Frontiers in Nuclear Structure, Astrophysics and Reactions, Kos, September 2005. To be published in AIP Conference Proceedings Series.
3. M. A. Bentley et al, submitted to Phys. Rev. C (2005).
4. P. Bednarczyk et al, *Eur. Phys. J.*, **A2**, 157 (1998).
5. S. J. Williams et al *Phys. Rev.*, **C68**, 011301 (2003).
6. F. Della Vedova et al, Legnaro National Laboratory, Ann. Rep. 2003 INFN (REP) 202/2004 p.3.
7. D.G.Jenkins et al, *Phys. Rev.* ,**C72**, 031303 (2005).
8. F. Della Vedova et al, Legnaro National Laboratory, Ann. Rep. 2004 INFN (REP) 204/2004 p.7.
9. J. Ekman et al, *Phys. Rev. Letts.*, **92**, 637 (2004).
10. K. Summerer and B. Blank *Phys. Rev.*, **C61**, 034607 (2000).
11. H. J. Wollersheim et al, *Nucl. Instr. and Meth.*, **A537**, 132502 (2005).
12. G. Hammond et al, *Acta. Phys. Pol.*, **B36**, 1253 (2005).
13. M. J. Taylor et al, *Journal of Physics G.*, **31**, S1527 (2005).
14. R. Lozeva et al, *Nucl. Instr. and Meth.*, **B204**, 678 (2003).
15. R. Lozeva et al, *Acta. Phys. Pol.*, **B36**, 1245 (2005).

FUNDAMENTAL PHYSICS

Measurement of Parity-Violating Gamma-ray Asymmetry in Compound Nuclei with Cold Neutrons

P.-N. Seo[1], J.D. Bowman[2], R.D. Carlini[3]*, T.E. Chupp[4], S.D. Covrig[5], M. Dabaghyan[5], S.J. Freedman[6], T. Gentile[7], M.T. Gericke[8], R.C. Gillis[8], G.L. Greene[9,10], F.W. Hersman[5], T. Ino[11], G.L. Jones[12], M. Kandes[4], B. Lauss[6], M.B. Leuschner[13], W.R. Lozowski[13], R. Mahurin[9], M. Mason[5], Y. Masuda[11], G.S. Mitchell[2†], S. Muto[11], H. Nann[13], S.A. Page[8], S.I. Penttilä[2], W.D. Ramsay[8,14], S. Santra[13‡], E.I. Sharapov[15], T.B. Smith[16], W.M. Snow[13], W.S. Wilburn[2], V. Yuan[2], H. Zhu[5]

[1]*Department of Physics, North Carolina State University, Raleigh, NC 27695, USA*
[2]*Los Alamos National Laboratory, Los Alamos, NM 87545, USA*
[3]*Thomas Jefferson National Accelerator Facility, Newport News, VA 23606, USA*
[4]*University of Michigan, Ann Arbor, MI 48104, USA*
[5]*Department of Physics, University of New Hampshire, Durham, NH 03824, USA*
[6]*Department of Physics University of California, Berkeley, CA 94720, USA*
[7]*National Institute of Standards and Technology, Gaithersburg, MD 20899, USA*
[8]*Department of Physics, University of Manitoba, Winnipeg, Manitoba R3T2N2, Canada*
[9]*Department of Physics University of Tennessee, Knoxville, TN 37996, USA*
[10]*Oak Ridge National Laboratory, Oak Ridge, TN 37831, USA*
[11]*High Energy Accelerator Research Organization (KEK), Tukuba-shi, 305-0801, Japan*
[12]*Department of Physics, Hamilton College, Clinton, NY 13323, USA*
[13]*Department of Physics, Indiana University, Bloomington, IN 47405, USA*
[14]*TRIUMF, 4004 Wesbrook Mall, Vancouver, British Columbia V6T2A3, Canada*
[15]*Joint Institute for Nuclear Research, Dubna, Russia*
[16]*Department of Physics, University of Dayton, Dayton, OH. 45469, USA*

Abstract. The NPDGamma collaboration has constructed and commissioned an apparatus on flight path 12 at LANSCE to measure with a high precision, 5×10^{-9}, the small parity-violating gamma-ray asymmetry, A_γ, in polarized neutron capture on protons. This asymmetry can be determined unambiguously the weak pion-nucleon coupling constant. To study the hadronic weak interaction at low energy, the collaboration has used the NPDGamma apparatus to measure parity-violating gamma-ray asymmetries in compound nuclei with cold neutrons. Using the statistical model of compound nuclei and spectroscopic information of the target nuclei, we can set upper limit on the spreading width of the hadronic weak interaction for intermediate-mass nuclei. We describe the experiment and the preliminary results of measured gamma-ray asymmetries of Al, Sc, Ti, Mn, and Co.

Keywords: parity violation, hadronic weak interaction, compound state, weak spreading width
PACS: 11.30.Er; 13.75.Cs; 07.85.-m; 25.40.Lw

INTRODUCTION

Although the weak interaction between nucleons has been studied for about 50 years and many parity-violating (PV) effects have been observed in NN reactions and in nuclei,

---

*Current address: Los Alamos National Laboratory, Los Alamos, NM 87545, USA
†Current address: Department of Physics, University of California, Davis, CA 95616, USA
‡Current address: Bhabha Atomic Research Centre, Trombay, Mumbai 400085, India

we still do not have a quantitative description for these results. The standard model of weak interaction successfully explains how point-like leptons and quarks interact weakly by exchanging weak bosons. The masses of the weak bosons are large[1] and thus their interaction ranges are very short, ~10^{-3} fm. This should be compared to the characteristic distances of low-energy NN interactions that take place in the range of ~1.5 fm. Therefore, the PV weak interactions observed at low energies between nucleons, cannot be simply explained by the exchange of the weak bosons. The widely used meson-exchange model[2] describes PV effects in NN interactions as a product of two vertices: a parity-conserving strong interaction vertex and a PV weak interaction vertex. PV effects on the order of 10^{-7} have been observed in NN scattering[3] and in nuclei.[4,5] However, quantitative results are still incomplete.

PV in gamma decays of nuclei is one example where statistical methods may be employed to estimate observables. Neutron capture on elements with a large number of nucleons produces compound nuclei in highly excited states. These nuclei exhibit a huge number of possible states with different angular momentum and parity. The number of possible transitions with different amplitudes that the compound nucleus may make to its ground state is correspondingly large as well. It is impossible to calculate all the possible electromagnetic transitions in the compound nucleus. However, the calculation of the mean square matrix elements for the transition amplitudes amounts to a summation of a large number of uncorrelated random contributions. One can then use statistical arguments to estimate the root mean square (RMS) width of the PV asymmetry from materials close to a neutron separation energy.

We have done a series of experiments with low-energy neutrons to characterize the hadronic weak interaction by observing gamma-ray asymmetries arising from PV interactions. In this paper we discuss an experiment that studies the PV weak interaction in the mass region A≈50, where level densities on compound nuclei are so high that the levels exhibit statistical behavior. This allows us to apply the statistical model of the compound nuclei in order to obtain an upper limit for the spreading width of the weak interaction from measured A_γ's on Al, Sc, Ti, Mn, and Co. The physics issues probed in this work are similar to the Time Reversal Invariance and Parity Experiment (TRIPLE) measurements[6] of the weak spreading width at A≈100 mass region in neutron resonances.

THE EXPERIMENTAL SETUP

The experiments to measure A_γ in polarized cold neutron capture on nuclear targets were performed on the flight path 12 (FP12) at the Los Alamos Neutron Science Center (LANSCE). The experiment used the NPDGamma[7] setup shown in Fig. 1 and took full advantage of the pulse nature and the high intensity of the FP12 cold neutron beam. The spallation neutrons at LANSCE are generated every 50 ms, but because of their energy distribution they arrive at the experiment at different times. Through the time-of-flight (tof) the neutron energy is known accurately and can be used to control systematic uncertainties with high levels of precision. The neutron flux out of the guide has a maximum at 3-meV neutron energy.[8] The distance from the surface of the FP12 moderator to the apparatus is about 21 m. The 1.3-meV neutrons are the last neutrons in the 50-ms tof frame to be counted in the detector before the next proton pulse hits the spallation target. To prevent neutrons slower than 1.3 meV from the previous tof frame from mixing with neutrons in the present frame, the slower neutrons are removed from

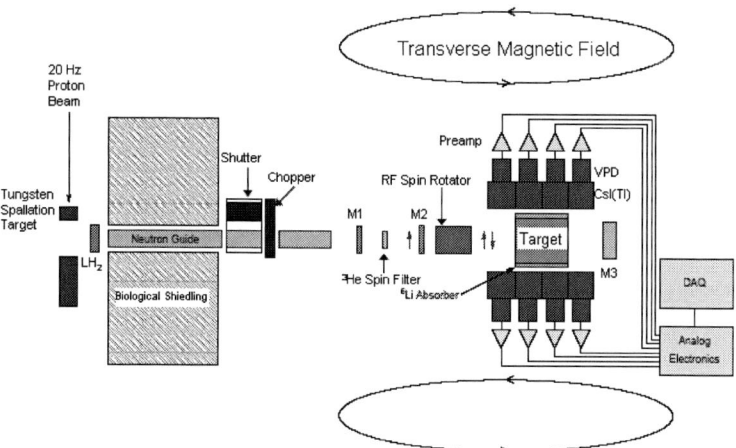

FIGURE 1. The NPDGamma experimental setup on FP12 at LANSCE. M1, M2, and M3 are ^{3}He ion chambers used for monitoring the neutron beam intensity, neutron polarization, and for the measurement of the efficiency of the RF Spin Rotator. VPD indicates vacuum photodiodes.

the beam by using a frame-definition chopper.[9]

The peak height of the proton beam pulses varies, and therefore, it is important to have a stable, accurate, and linear neutron beam monitor to normalize the neutron beam intensity for the measurement. The monitor M1, a ^{3}He ionization chamber filled with a mixture of ^{3}He, ^{4}He, and N_2 to a pressure of one atmosphere, monitors the beam flux out of the guide. The neutrons are polarized by transmitting the beam through a laser-polarized ^{3}He cell which has a large beam acceptance of ~100 cm^2. The cell is operated in a 10-G homogeneous (gradient <1mG/cm), vertical magnetic guide field which encompasses the entire setup. Using the adiabatic fast passage (AFP) technique the ^{3}He polarization and thus the beam polarization can be flipped without any changes in the magnetic fields. This spin flip is used to control systematic uncertainties of the experiment. However, it cannot be applied frequently because of polarization losses of less than 1% per spin reversal. With M2, an identical monitor to M1, the neutron beam polarization is measured.

Then the polarized beam passes through a resonant RF Spin Rotator (RFSR) used for 20-Hz pulse-by-pulse neutron spin reversal with efficiency of 99%.[10] The RFSR is a solenoid in an Al can with its axis along the beam and perpendicular to the guide field. The frequent spin reversal with the RFSR is used to control gain drifts in the detector system to second order using a specific 8-step spin reversal sequence and to make the detector array symmetric to the up-down direction.

After the RFSR the neutrons enter the target area. Gamma rays from the neutron capture reaction are detected by an array of 48 CsI(Tl) detectors around the target.[11] The array is shielded with Li-loaded plastic to prevent neutron activation of the detectors. The one-ton detector array is mounted on a movable table for the alignment of the detector with the neutron spin direction to minimize the leakage of any possible left-right gamma-ray asymmetry into the up-down asymmetry. Scintillation light from the CsI crystals is converted to currents in vacuum photodiodes (VPD), amplified by low-noise solid-state electronics[12], and then read by transient digitizers. Because of the high gamma rates, the detector system is operated in current mode. The transient digitizer samples detector signals every 20 µs. The homogeneous guide field from the polarizer to the target suppresses the up-down Stern-Gerlach steering of the neutron beam.

DISCUSSION AND THE PRELIMINARY RESULTS

The sensitivity of the NPDGamma apparatus has been thoroughly measured and results are described in detail in Ref. 13. Recent improvements for the apparatus are an accurate measurement of efficiency of the RFSR[10], and better holding field shimming and alignment.

A study of physics of the PV gamma-ray asymmetry of an integral gamma spectrum following a neutron capture in nuclei is interesting in its own right. The gamma transitions following the capture are either E_1 or M_1, and the E_1 transition rate is known to be ~10 times faster than the M_1 transition rate. The total PV gamma asymmetry after the capture is dominated by the gamma transitions with the largest energies. The PV mixing in the capture state causes the E_1 transition of gamma rays that depopulate the capture state, to interfere with the M_1 transition. Because the level spacing decreases exponentially with excitation energy, A_γ is dominated by the parity mixing in the capture state, not in states of the daughter nuclei.

It can be shown that the average A_γ depends upon the weak spreading width (Γ_W), an integral over a function that depends on the excitation energy, and the ratio of the M_1 and E_1 transition rates. The distribution of A_γs is a non-Gaussian shape which has a peak at $A_\gamma \approx 0$ and a broad distribution with a long tail. When measuring several A_γs on nuclear targets, we can expect most of them to be zero. If we are lucky, we might observe a statistically significant non-zero A_γ. The goal of this experiment was to measure A_γ on several targets in the region of $A \approx 50$ and then use the statistical model of the compound nuclei and well known spectroscopic information of the targets to set an upper limit for the weak spreading width. Accuracy of the deduced weak spreading width will depend upon the number of studied targets and the statistics-limited errors on the A_γ values. Table 1 lists the measured asymmetries on Al, Sc, Ti, Mn, and Co. No statistically-significant PV effects at the 2σ level were observed among the targets studied. The Mn and Ti targets were in a form of metal power, the Co and Sc targets were in the form of a metal oxide powder, and the Al target was in the form of thin plates. Typically, the target thickness was selected so that ~90% of the beam was absorbed by the target. The approximate number of total running days is also listed.

TABLE 1. Preliminary result of measured gamma-ray asymmetries

Target	Running time (days)	Asymmetry (10^{-7})
Al	19	-0.0±3.1
Sc	36	-7.0±2.8
Ti	72	7.1±4.0
Mn	11	5.3±7.8
Co	7	6.4±3.3

Figure 2 shows a histogram of A_γ values on a log scale with a fit to a Gaussian distribution for a 40-hour Al-target run. The quality of the fit is an indication of the low noise level of the system. All uncertainties listed in table are dominated by counting statistics. The systematic uncertainties[11] are at the level of 10^{-9}, the Mott-Schwinger left-right asymmetry has an up-down contribution of ~10^{-10}, left-right parity-allowed asymmetry contributes to up-down asymmetry at the ~10^{-10} level, and up-down asymmetry from the Stern-Gerlach beam steering is estimated to be ~10^{-10}. Other contributions to errors in A_γ come from the accuracy of beam polarization, the spin rotator efficiency, the geometry factor due to the finite size of detectors and targets, and neutron depolarization in the targets.

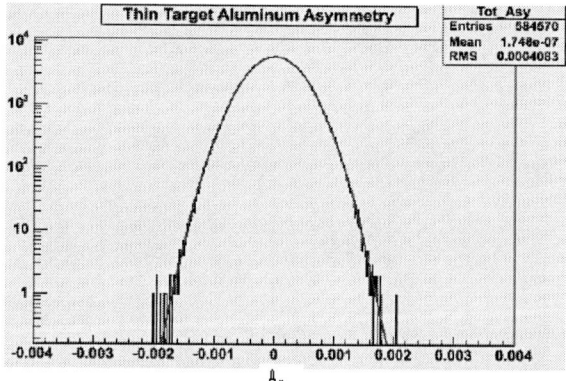

FIGURE. 2. Histogram of gamma-asymmetry values from eight-step sequences for the Al target. The mean and error in the mean are determined from a fit to a Gaussian distribution.

CONCLUSION

The NPDGamma collaboration has successfully completed commissioning of the apparatus with the exception of the liquid hydrogen target which is still under testing. The ^{3}He polarization has been increased to 55% and efficiency of the RFSR has been measured to be 99%. The NPDGamma apparatus was used to measure PV gamma-ray asymmetries on Al, Sc, Ti, Mn, and Co. Preliminary results do not show any statistically-significant PV effects, as expected. The systematic uncertainty of the preliminary asymmetries is at about the 10^{-10} level, which is below the goal of 5×10-9 for the systematic effect in the NPDGamma asymmetry.

REFERENCES

1. S. Eidelman *et al.*, *Phys. Lett.* B **592**, 1 (2004).
2. B. Desplanques, J.F. Donoghue, and B.R. Holstein, *Ann. Phys.* **124**, 449 (1980).
3. V. Yuan *et al.*, *Phys. Rev.* C **57**, 1680 (1986).
4. E.G. Adelberger and W.C. Haxton, *Ann. Rev. Nucl. Part. Sci.*, **35**, 501 (1985).
5. J.D. Bowman *et al.*, *Ann. Rev. Nucl. Part. Sci.*, **43**, 829 (1993).
6. G.E. Mitchell *et al.*, Phys. Rep. 354, 157 (2001).
7. J.D. Bowman *et al.*, "Measurement of the Parity-Violating Gamma Asymmetry A_γ in the Capture of Polarized Cold Neutrons by Para-Hydrogen", LA-UR-1999-5432; W.M. Snow *et al.*, *Nucl. Inst. and Meth.* **A 440**, 729 (2000); W.M. Snow *et al.*, *Nucl. Inst. and Meth.* **A 515**, 563 (2003).
8. P.-N. Seo *et al.*, *Nucl. Inst. and Meth* **A 517**, 285 (2004).
9. P.-N. Seo *et al.*, *J. Res. Natl. Inst. Stand. Technol.* **110**, 145 (2005).
10. P.-N. Seo *et al.*, to be submitted to *Nucl. Inst. and Method.* (2005).
11. M.T. Gericke *et al.*, *Nucl. Inst. and Meth.* **A 540**, 328 (2005).
12. W.S. Wilburn *et al.*, *Nucl.Inst. and Meth* **A 540**, 180 (2005).
13. M.T. Gericke *et al.*, in the proceedings of the 17th International Conference on Advanced Neutron Sources, Santa Fe, New Mexico, April 26-29 (2005); P.-N. Seo, *et al.*, the 2004 International Conference on Nuclear Data for Science and Technology (ND2004), ed. R.C. Haight, M.B. Chadwick, T. Kawano, and P. Talou, (AIP Conference Proceedings **796**, New York, 2005) 696 (2005).

Scattering Of KeV Neutrons And The Problem Of Quantum Entanglement

R. Moreh*¶, R.C. Block¶, Y. Danon¶ and M. Neuman¶

*Physics Department, Ben-Gurion University of the Negev, Beer-Sheva, Israel
¶Gaerttner LINAC Laboratory, Rensselaer Polytechnic Institute, Troy, New York 12180

Abstract. A scattering study of ~ 30 keV neutrons from H_2O, D_2O and H_2O/D_2O mixtures was carried out to search for a ~ 40% anomalous drop in the n-p scattering cross section. Such an anomaly was reported using 10-200eV neutrons scattered from many H-containing compounds. The anomaly was attributed to n-scattering from quantum entangled proton pairs in the sample. In the present measurement no anomaly was observed.

Keywords: Neutron scattering; keV neutrons; Entanglement; water.
PACS: 34.90. +q; 61.12.-q; 03.65.-w
http://www.aip.org/pacs/index.html

INTRODUCTION

Neutron Compton scattering (NCS) [1] is a term used when neutrons at epithermal energies (5 – 200 eV) are scattered from atomic nuclei. In a way it is similar to Compton scattering of γ-rays from free electrons. When a photon of incident energy E_0 and momentum $\mathbf{P_0}$ scatters from an electron, it transfers energy, $\varepsilon = E_0 - E_f$ and momentum $\mathbf{Q} = \mathbf{P_0} - \mathbf{P_f}$, to the recoiling electron, where E_f and $\mathbf{P_f}$ are the energy and momentum of the scattered photon. By making a detailed measurement of the energy and momentum of the scattered photons, the initial momentum distribution of electrons in solid and liquid systems can be determined. A similar process, NCS, may be applied to the scattering of epithermal neutrons from nuclei, whereby a detailed measurement of the momentum and energy of the scattered neutrons can yield information on the momentum distributions of atomic nuclei in solids and liquids. One point to note is that because the neutron energy is so much higher than the binding energy of the atom in the system, the neutron is assumed to interact with a single atom, and the NCS process can be treated theoretically using the *impulse approximation* [2]. The NCS technique, was used during the last twenty years for studying the momentum distributions and zero-point energies of atomic nuclei in many isotropic (liquid He) [3] and non-isotropic systems (Zn single crystal) [4].

The NCS technique was also applied to study H-containing systems such as water, H_2O, employing 10 – 200 eV neutrons [5]. The scattering intensity from the H-atoms was found to be anomalously lower than the conventional values by ~ 40%. The conventional n-p scattering cross section in the lab system, is deduced from the known n-p total cross section σ_t, by using the relation: $d\sigma(\theta_{lab})/d\Omega_{lab} = \sigma_t (\cos\theta_{lab})/\pi$.

This anomaly was also reported in several other samples such as formvar ($C_8H_{14}O_2$) [6], acetylene C_3OH_6 [7], and also Nb-, Pd-hydrides [8], all of which seemed to reveal more or less the same effect. In these measurements, the n-p scattering intensity was determined relative to some other heavier nuclei in the same sample. In formvar, and Pd- and Nb hydrides, the anomaly revealed an angular dependence [8], increasing with the scattering angle. Analogous samples with D partially replacing H revealed a smaller anomaly depending on amount of the H replaced. In the case of Pd and Nb hydride, the anomaly was also shown to depend on the n-p scattering time, starting at $\tau_{sc} < 6 \times 10^{-16}$s and increasing for shorter times. Note that all above measurements were made using the same instrument, VESUVIO formerly EVS, at Rutherford Lab, UK.

THEORETICAL REMARKS

The shortfall in n-p scattering intensity was attributed to the effect of short-lived quantum entanglement (QE) in which the incident neutron interacts with two or more quantum entangled protons in the liquid or solid samples at room temperature. Before proceeding, we shall explain what QE means. A pair of protons is said to be quantum entangled if its wave function can not be written as a product of two wave functions. The protons, being fermions are indistinguishable quantum objects and their wave function [9] in the initial state <i| may be represented as:

$$<i| = \frac{1}{\sqrt{2}}[\psi_1(R_\alpha)\psi_2(R_\beta) + (-1)^J \psi_1(R_\beta)\psi_2(R_\alpha)]\chi_{JM}(\alpha,\beta) \quad (1)$$

where $\psi_i(\mathbf{R}_\alpha)$ represents particle α located at position i =1,2; the same with particle β. $\chi_{JM}(\alpha,\beta)$ is the spin part of the total wave function with total spin J and projection M.

A typical example of a quantum entangled system is a H_2 molecule: where the two protons are indistinguishable fermions, either in a para state, J=0 (with a symmetric nuclear wave function) or in an ortho state, J=1 (with anti-symmetric nuclear wave function). The scattering of neutrons from this proton pair may become coherent for thermal neutrons when the wave length λ_n is of the order of molecular dimensions and when H_2 is a liquid at ~10 K. It was shown by Schwinger and Teller [10] that the calculated scattering cross section from the exchange correlated proton pair drops by about 50% relative to that calculated from two independent protons. This is caused by quantum mechanical interference between the neutron and the proton pair. The calculated n-scattering cross section from para-H_2 at ~ 10 K is 4 b, while that from ortho-H_2 is 111 b. The huge difference arises from the strong dependence of the n-p interaction upon the relative spin orientation of the particles.

It was suggested [9,11] that in order to observe the anomaly at room temperature, the neutron must interact with the two protons and that the phase relations between the proton pair must stay the same during the scattering process. This means that the n-scattering time τ_{sc} must be shorter than the decoherence time τ_{dec}, where decoherence refer to any process which can destroy the coherency or the relative phases of the two protons of the pair; it is caused by collision or by random fluctuations in the atomic environment.

EXPERIMENTAL METHOD

Experimentally, we carried out an independent test [12] of the claimed drop in the n-p scattering intensity, in which a higher final neutron energy (24.3 keV) was selected to ensure a much shorter n-interaction time $\tau_{sc} \sim 10^{-17}$s, thus eliminating any chance of decoherence. In addition, a high momentum and energy transfer, could also test if an alleged breakdown of the Born-Oppenheimer (BO) approximation [13,14] may explain the measured results. The neutron source for the scattering measurements was obtained from the Gaerttner electron linear accelerator of the Rensselaer Polytechnic Institute (RPI) operated at ~50 MeV with a repetition rate of 225/s, a pulse width of 1-2 µs, and an average current ~ 50 µA, striking ten water-cooled, Ta plates (5 cm x 5 cm, 1.6 mm to 6.4 mm thick). Water scatterers, were placed a distance of 15 cm from the Ta source (Fig. 1). The neutron energy was obtained by a time of flight (TOF) measurement where the Ta to detector distance is 25.3 m (see Ref. 12 for more details). The neutrons were passed through a 20 cm iron filter, allowing only 24.3 keV (ΔE = 2.1 keV) neutrons and other higher energies neutron lines to pass through. Those energies correspond to deep minima in the ^{56}Fe total n-cross section, caused by destructive interference between the resonance scattering of

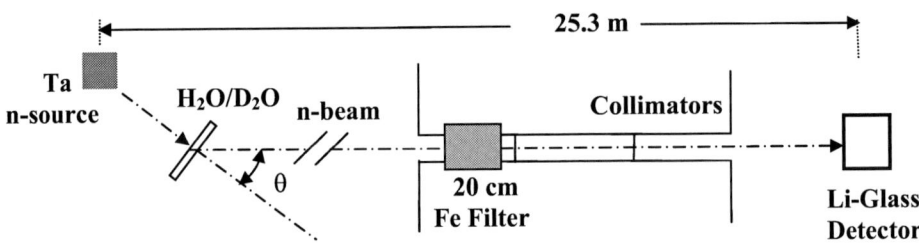

FIGURE 1. Schematic view (not to scale) of the experimental system, showing the Ta n-source, scatterers, Fe filter and Li-glass n-detector.

neutrons and potential scattering. The n-scattering intensity ratios from pure H_2O and various mixtures of H_2O+D_2O samples (contained in thin Al cans of internal dimensions: 100 x 63 x 1.8 mm), were measured.

All samples had identical geometries and contained the same number of water molecules. Typical TOF spectra measured from samples of pure H_2O, a mixture of (50%H_2O +50%D_2O), and background are shown in Fig. 2. The flat part of the peak (showing the 24.3 keV neutron group) corresponds to the long (2µs) electron pulse width which assured higher counting rates. The samples subtended scattering angles between 25° and 65° with the central angle at 43°. The variation of the incident n-flux with energy was determined and found to vary as $1/E^\delta$ with δ = 0.65. Table 1 lists the measured and calculated ratios of scattering intensities from pure H_2O to the mixed H_2O-D_2O samples after accounting for multiple scattering effects in both samples. The MS effects were calculated using the Monte Carlo MCNP general code. The scattering cross sections were deduced from the *total* n-cross section tables. Excellent agreement with the measured data was obtained, revealing no evidence for any shortfall in the n-p scattering intensity from H_2O. Note also the huge deviation between the measured points for pure H_2O/D_2O to that calculated by assuming a 40% drop in the n-p scattering intensity (last column in Table 1).

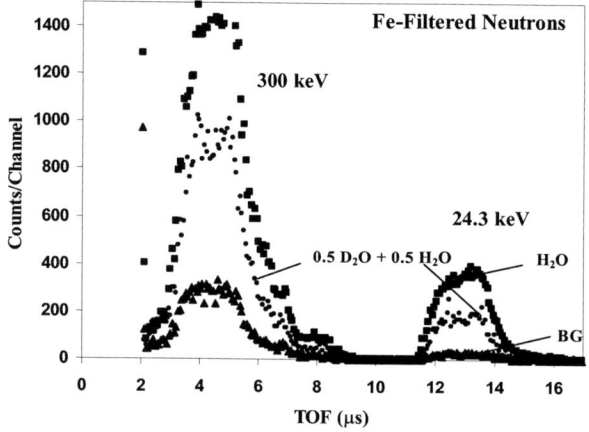

FIGURE 2. TOF of neutrons scattered from H2O, mixed 0.5 H2O+0.5 D2O and the background.

DISCUSSION AND CONCLUSION

The result of the present experiment shows beyond any doubt that there is no deficiency in the n-p scattering intensity at a final n-energy of 24.3 keV. Nevertheless, it is of interest to review the theoretical models dealing with the n-p scattering anomaly at 10 - 200 eV reported using the NCS method. As mentioned above, some attempts were made by Karlsson and Lovesey [9,11] to explain the claimed ~ 40% deficiency in the n-p scattering intensity, at room temperature, measured using the NCS technique. The basic idea relies on a similar process to that used in ref. [10] to explain the thermal neutron scattering cross section from ortho and para H_2 at ~ 10 K. It was thus suggested that the proton pair is assumed to be quantum entangled for time durations shorter than 10^{-14}s, being the decoherence time, τ_{dec}, required to destroy the phase relation between the proton pair in the sample. The scattering time [15] may be obtained from the relation: $\tau_{sc}(\theta) \sim M/[k(\theta)<p^2>^{1/2}]$, where θ, the n-scattering angle; M, scattering nuclear mass; $k(\theta)$, the momentum transfer, and $<p^2>^{1/2}$ is related to the

TABLE 1. Measured and calculated n-scattered intensity ratios of H_2O relative to H_2O+D_2O mixtures versus the D_2O concentration of the mixture, X_D. The last column shows the calculated ratio obtained after reducing the n-p scattering cross section by 40%.

D_2O concentration	Measured Ratio	Calculated Ratio	Reduced Ratio
70%	2.16 ± 0.07	2.15	1.98
84%	2.92 ± 0.10	2.96	2.52
100%	5.50 ± 0.19	5.45	3.69

square root of the mean vibrational kinetic energy of H in H_2O. At 10 eV and $\theta = 45°$, $\tau_{sc}(\theta) \sim 2 \times 10^{-15}$s, being shorter than the decoherence time τ_{dec}. This explanation was however disputed by Colognesi [16] who questioned some of the assumptions and concluded that this approach can not describe the reported anomalous behavior of n-p scattering. Further, in a very recent paper [17], a new calculation of the n-scattering intensity from a quantum entangled proton pair or a deuteron pair was made using the Heitler-London approximation. It is concluded that there can be no observable drop in the n-scattering intensity by two nuclei in entangled state when the impulse approximation is satisfied. The anomaly is said to be undetectable because of the high energy transfer and broad energy resolution of the NCS measurements.

In a different attempt to explain the anomaly, it was suggested that because of the high momentum and energy transfers to the nuclei, occurring during times of $\sim 10^{-16}$s, a breakdown of the Born-Oppenheimer approximation [13,14] may occur. It was suggested that the electronic excitation in the n-nucleus collision may diminish the energy of the scattered neutrons, causing less neutrons to reach the peak energy. However, this problem was evaluated in detail by Colognesi [18], for the specific case of n-scattering from H_2, concluding that it has a negligible effect on the measured scattering cross sections.

From the above it may be seen that there is as yet no accepted theoretical explanation of this anomaly reported using the NCS technique with 10-200eV neutrons. In the present experiment, no evidence was found for any drop in the n-p scattering intensity from water samples for neutron incident energies in the 30 keV range.

REFERENCES

1. H. Rauh, N. Watanabe, Phys. Lett. A **100**, 244-246 (1984).
2. P. C. Hohenberg and P. M. Platzman, Phys. Rev. **152**, 198-200 (1966).
3. R. Senesi et al., Phys. Rev. B **68**, 214522-1-7 (2003).
4. D. Nemirovsky, R. Moreh, K.H. Andersen and J. Mayers, J. Phys: Cond. Mat. **12**, 4293-4302 (2000).
5. C.A. Chatzidimitriou-Dreismann, T. Abdul-Redah, R.M.F. Streffer, and J. Mayers, Phys. Rev. Lett. **79**, 2839-2842 (1997).
6. C. Chatzidimitriou-Dreismann, M. Vos, C. Kleiner, T. Abdul-Redah, Phys. Rev. Lett. **91**, 057403-01-04 (2003).
7. T.Abdul-Redah and C.A.Chatzidimitriou-Dreismann, Appl. Phys. A **74** (2002) S1379-S1381.
8. E.B. Karlsson, T. Abdul-Redah, R F. Streffer, B. Hjörvarsson, J. Mayers, C.Chatzidimitriou-Dreismann, Phys. Rev. B **67**, 184108-01-13 (2003).
9. E.B. Karlsson and S.W. Lovesey, Physica Scripta **65**, 112 (2002).
10. J. Schwinger and E.Teller, Phys. Rev. **52**, 286-295 (1937).
11. E.B. Karlsson, Phys. Rev. Lett. **90**, 95301-95304 (2003).
12. R. Moreh, R.C. Block, Y. Danon, M. Neuman, Phys. Rev. Lett. **94**, 185301-1-4 (2005).
13. N.I. Gidopoulos, Phys. Rev. B **71**, 054106-1-5 (2005).
14. G.F. Reiter and P.M. Platzman, Phys. Rev. B **71**, 054107-1-6 (2005).
15. G.I. Watson, J. Phys.: Condens. Matter **8**, 5955-5976 (1996); V.F. Sears, Phys. Rev. B **30**, 44-51 (1984).
16. D. Colognesi, Physica B. **344**, 73-81 (2004).
17. H. Sugimoto, H. Yuuki, A. Okumura, Phys. Rev. Lett. **94**, 165506-1-4 (2005).
18. D. Colognesi, Physica B. **358**, 114-125 (2005).

Random Matrix Thermodynamics

G. E. Mitchell[*,†], T. N. Nogueira[**], M. P. Pato[**], J. C. Sartorelli[**] and A. P. B.Tufaile[**]

North Carolina State University, Raleigh NC 27695-8202, USA
†*Triangle Universities Nuclear Laboratory, Durham NC 27708-0308, USA*
**Instituto de Fisica, Universidade de Sao Paulo, C.P. 66318, 05315-970 Sao Paulo, S.P., Brazil*

Abstract. We have performed a study of the statistical mechanics of correlated spectra first introduced by Dyson and Mehta some 40 years ago. We have derived a modified thermodynamical statistics (a number and its variance) for linear spectra. This approach was used to analyze the statistical properties of the eigenvalues of random matrices of Gaussian ensembles and of the experimental eigenfrequencies of acoustic resonances of an aluminum plate cut in the shape of a chaotic billiard. The results obtained suggest that this statistics provides a robust tool for the investigation of spectral properties.

Keywords: Random Matrix Thermodynamics, Acoustic Resonances
PACS: 05.45.Mt, 05.20.-y

INTRODUCTION

Recently we have focused on evaluation of the reliability and uncertainties in nuclear level densities obtained from direct counting methods. This is important because level densities are key for a wide variety of applications, including the study of nuclear structure far from stability, astrophysical studies of the *r* and *p* processes, and radiochemical analyses relevant to stockpile stewardship measurements. The conventional approach to the correction for missing levels assumes that the resonance widths obey a Porter-Thomas distribution and that all levels below some cutoff are not observed (and that all levels above this cutoff are observed). With this assumption and the experimental distribution, one can estimate the fraction of missing levels. This approach is used nearly universally and works reasonably well. However, non-statistical effects lead to errors in the determination of the missing fraction of levels.

The Porter-Thomas distribution follows from the Gaussian Orthogonal Ensemble (GOE) version of Random Matrix Theory (RMT) [1]. Another prediction of GOE is that the nearest neighbor spacings for a set of states with the same symmetry should obey the Wigner distribution. Since in RMT the widths and spacings are not correlated, analysis of the spacing distribution can provide an independent determination of the missing level fraction. However, in practice this approach has been neither developed nor applied. The problem is that as soon as levels are missed, one is studying not just the nearest neighbor spacing distribution, but also the higher order distributions. This introduces a number of complications.

We have derived a general expression for the spacing distribution when a fraction of the levels is missed and have also developed a practical method to analyze such data. The probability density function for the imperfect sequence is

$$P(x)dx = \sum_{k=0}^{\infty} f(1-f)^k p(k;x)dx. \tag{1}$$

The parameter f is the observed fraction of levels. The k-th nearest neighbor probability function $p(k;x)$ gives the probability for a spacing x between two levels, given that there are k levels between the two. Although the method is general, our primary interest was for GOE. In this case $p(0;x)$ is the Wigner distribution. Details of the derivation and testing procedure are given by Agvaanluvsan et al. [2].

After first testing the method extensively on numerically generated data, we then applied the method to some of our extensive proton data sets [3] and to neutron resonance data [4]. We also have developed and tested the method for both the Gaussian Unitary Ensemble (GUE). the Gaussian Symplectic Ensemble (GSE). Following our work, Bohigas and Pato [5] approached this general problem from a somewhat different view. They confirmed our results and extended the analysis to other measures such as Δ_3 and Σ^2.

RANDOM MATRIX THERMODYNAMICS

All of these efforts addressed the issue of the quality of the entire experimental distribution – what fraction of levels were missed. A natural next question is whether one can determine whether specific levels are missed or misassigned. Whether or not a missing level can be observed depends on the character of the distribution. Clearly if a level is removed from a picket fence distribution, then this is obvious – the probability of observing the removal is one. On the other hand, if a level is removed from a Poisson distribution, then the probability of observing this is zero. All partially correlated distributions (such as the RMT ensembles) must be somewhere in between these extremes. One can phrase the issue as that of the information content of a state in an RMT ensemble.

In fact this problem was addressed and solved over 40 ago. By considering levels as positions of unidimensional charges, Dyson constructed a statistical mechanics of correlated spectra treated as a Coulomb gas [6, 7]. Thermodynamical quantities such as free energy, entropy, internal energy and specific heat were derived using Dyson's circular ensembles, which are characterized by a parameter β that plays the role of an inverse temperature. In these ensembles the levels are points on a circle.

In spite of the remarkable developments of RMT since that time, and of the many applications, the specific statistical mechanics that Dyson and Mehta developed in 1963 has to our knowledge never been pursued either theoretically or experimentally. Therefore we decided to address this issue. One practical motivation was the continuing search for 'statistics' that robustly characterize correlated spectra without requiring essentially perfect data. We present here a test of the model using high quality experimental data obtained measuring eigenfrequencies of aluminum plates. These analog systems are known to produce an excellent sequence of levels. In addition one can obtain sequence sizes of the order of 10^3 [8, 9], as opposed to the typical values of 10^2 obtained in nuclear data [10, 11].

Instead of taking directly the statistics as proposed by Dyson and Mehta [7, 6], we considered the statistical mechanics for a sequence of points on a line, i.e., for spectra. In the limit of long spectra, the values obtained with the modified statistics approach the values derived from the circular ensembles.

For linear spectra the electrostatic potential energy is given by

$$W = -\sum_{j>i} \ln|x_j - x_i|, \qquad (2)$$

where x_i with $i = 1, 2, ..., N$ are the levels of the spectrum. The key difference between the linear and circular cases is that in the circular ensembles the eigenvalues are points confined in a circle, but a linear Coulomb gas requires a confining potential to ensure a stable density. The connection between density and potential is implied by Dyson's mean field equation

$$P \int \frac{\rho(t)\,dt}{x-t} = V'(x), \qquad (3)$$

where P denotes principal value. Consider a rescaled sequence of levels in an interval $[-L,L]$, such that the density is constant and equal to one (i.e., $\rho(t) = 1$). We obtain after integrations

$$V(x) = \left\{ (L-x)\left[\frac{1}{2} + \ln\left(\frac{L-x}{2L}\right)\right] + (L+x)\left[\frac{1}{2} + \ln\left(\frac{L+x}{2L}\right)\right] \right\}, \qquad (4)$$

where a constant term has been added such that $\langle V(x) \rangle = 0$. To complete our definition we require that, as in the case of the circular ensembles, the total energy vanishes in the ground state. As the temperature decreases the levels stop fluctuating and freeze at equidistant points separated by a constant unit distance. In other words, the picket fence spectrum is the equivalent on the line of the equilateral polygon on the circle. The energy for the picket fence spectrum is

$$W_0 = -\sum_{j>i} \ln|j - i|. \qquad (5)$$

Of course a confining potential is also needed for the picket fence spectrum. We label the eigenvalues of the picket fence by ξ. Combining these four terms we can finally write the expression

$$W(L) = -\sum_{j>i} \ln\left|\frac{x_j - x_i}{j - i}\right| + \sum_i [V(x_i) - V(\xi_i)] \qquad (6)$$

for the thermodynamic energy of a linear sequence x_i with $i = 1, 2, ..., N$ of unfolded levels; $\xi_i = -L + 1/2 + i$ are the positions of the picket fence levels, where $N = 2L$.

APPLICATIONS

As a first application of the above formalism we considered eigenvalues of GOE matrices of size 800 generated numerically. For each spectrum of an ensemble of 11,000 matrices we calculated $\frac{W(L)}{N}$. The values follow a Gaussian distribution whose center and variance are respectively the internal energy and the specific heat (given by $C = N\sigma^2$). We obtain the values U = 0.362 and C = 0.28. The slight discrepancies between these numbers and the predicted values (U = 0.365 and C = 0.266) can be attributed to the finite size of the spectra.

We now apply the statistics to experimental spectra. The spectrum consists of a sequence of 985 eigenfrequencies of an aluminum plate. The plate has the shape of a Sinai billiard with a slit cut on one face to break the mirror symmetry through the middle plane of the plate. This ensures that flexural and in-plane modes are completely mixed [8]. The spectrum was first analyzed by calculating its nearest neighbor distribution (NND) and its spectral rigidity (Δ-statistics). Although the NND agrees very well with the GOE distribution, the Δ-statistics show a deviation that suggest a spectrum less correlated than a GOE one. To determine the thermodynamics quantities the spectrum was scanned with a window of length w that moves at steps $w/2$. The internal energy shows little dependence on the window size and its mean value $U_w = 0.41 \pm 0.07$ also can be obtained using all of the levels. However, the use of windows is necessary in order to extract the specific heat from the variance around the mean energy. In the range $[100, 150]$ the specific heat shows only a weak dependency on the window size. The average value is $C_w = 0.58 \pm 0.12$.

Both U and C show experimental values above the GOE ones previously determined from numerical simulations. The high value of the specific heat can in part be attributed to the smallness of the window size ~ 100 imposed in order to have a sufficient number of steps and thus sufficient statistics. Although the value of U is statistically compatible with the distribution, we note that the Δ-statistics of the data is also above the GOE prediction.

The decrease of long range correlations can be caused by missing levels [5]. However, this seems unlikely for eigenfrequency spectra obtained via this experimental method. It seems more reasonable to attribute this behavior to a real physical origin, for example as a manifestation of the fact that an aluminum plate is not truly a two-dimensional object, but instead is three-dimensional.

Additional measurements of plates with different shapes are planned. We anticipate performing measurements of acoustic resonances with quartz blocks. In addition we plan to apply the thermodynamics method to other experimental spectra, including nuclear spectra.

ACKNOWLEDGMENTS

Fruitful discussions with J. F. Shriner, Jr. are acknowledged. This work is supported in part by the U.S. Department of Energy, Office of Nuclear Physics, under grant DE-FG02-97-ER41042, by the U.S. National Science Foundation under grant No. INT-0112421, and by the Brazilian agencies CNPq and FAPESP.

REFERENCES

1. T. Guhr, A. Müller-Groeling, and H. A. Weidenmüller, Phys. Rep. **299**, 190 (1998).
2. U. Agvaanluvsan, G. E. Mitchell, J. F. Shriner, Jr., and M. P. Pato, Nucl. Instrum. Methods Res. A **498**, 459 (2003).
3. U. Agvaanluvsan, G. E. Mitchell, J. F. Shriner, Jr., and M. P. Pato, Phys. Rev. C **67**, 064608 (2003).
4. D. Dashdorj and G. E. Mitchell (private communication).
5. O. Bohigas and M. P. Pato, Phys. Lett. B **595**, 171 (2004).
6. F. J. Dyson and M. L. Mehta, J. Math. Phys. **4**, 701 (1963).
7. M. L. Mehta, *Random Matrices* (Academic Press, Boston, 1991).
8. P. Bertelsen, C. Ellegaard, and E. Hugues, Eur. Phys. J. B **15**, 87 (2000).
9. K. Schaadt, Ph.D. thesis, University of Copenhagen (2001).
10. G. E. Mitchell, E. G. Bilpuch, P. M. Endt, and J. F. Shriner, Jr., Phys. Rev. Lett. **61**, 1473 (1988).
11. J. F. Shriner, Jr., C. A. Grossmann, and G. E. Mitchell, Phys. Rev. C **62**, 054305 (2000).

Symmetry of Isoscalar Matrix Elements and Systematics in the sd and beginning of fp shells

J.N. Orce*, P. Petkov[†,**], V. Velázquez[‡], C.J. McKay*, S.R. Lesher*, S. Choudry*, M. Mynk*, A. Linnemann**, J. Jolie**, P. von Brentano**, V. Werner**, S. W. Yates[§,*] and M. T. McEllistrem*

Department of Physics and Astronomy, University of Kentucky, Lexington, Kentucky 40506-0055
[†]*Institute for Nuclear Research and Nuclear Energy, Sofia 1784, Bulgaria*
**Institut für Kernphysik, Universität zu Köln, 50937 Köln, Germany*
[‡]*Depto. de Física, Facultad de Ciencias, Universidad Nacional Autónoma de México, Apartado Postal 70-543, 04510 México, D.F., México*
[§]*Department of Chemistry, University of Kentucky, Lexington, Kentucky 40506-0055*

Abstract. A careful determination of the lifetime and measurement of the branching ratio for decay of the first $2^+_{T=1}$ state in ^{42}Sc has allowed an accurate experimental test of charge independence in the $A = 42$ isobaric triplet. A lifetime of 69(17) fs was measured at the University of Kentucky, while relative intensities for the 975 keV and 1586 keV transitions depopulating the first $2^+_{T=1}$ state have been determined at the University of Cologne as 100(1) and 8(1), respectively. Both measurements give an isoscalar matrix element, M_0, of 6.4(9) (W.u.)$^{1/2}$. This result confirms charge independence for the A=42 isobaric triplet. Shell model calculations have been carried out for understanding the global trend of M_0 values for $A = 4n+2$ isobaric triplets ranging from $A = 18$ to $A = 42$. The $2^+_{1\,(T=1)} \rightarrow 0^+_{1\,(T=1)}$ transition energies, reduced transition probabilities and M_0 values are reproduced to a high degree of accuracy. The trend of M_0 strength along the sd shell is interpreted in terms of the shell structure. Certain discrepancies arise at the extremes of the sd shell, for the $A = 18$ and $A = 38$ isobaric triplets, which might be explained in terms of the low valence space at the extremes of the sd shell.

Keywords: Charge independence, transition strengths, isoscalar matrix elements, shell model
PACS: 21.10.Tg; 21.60.Cs; 23.20.-g; 23.40.Hc; 24.80.+y

If one assumes charge symmetry, isospin symmetry can experimentally be examined in $T = 1$ isobaric triplets by the comparison of $2^+_{1\,(T=1)} \rightarrow 0^+_{1\,(T=1)}$ $E2$ transition strengths [1]. For a particular mass number, A, isospin symmetry could be examined by comparing the isoscalar matrix elements, M_0, extracted from the two nuclides with $T_z = \pm 1$ with that extracted from the $T_z = 0$ nuclide. This would test the degree to which the approximate charge independence of nuclear force leads to a single M_0 for isobaric triplets. Accordingly, in mirror pairs, $T'_z = -T_z$, the relationship between matrix elements [1] in the neutron or proton isospin representation yields,

$$M_0(T=1) = M_p(T_z) + M_p(-T_z) \qquad (1)$$

which implies that for a transition between $T = 1$ states in a $T_z = 0$ nucleus:

$$M_0(T=1) = 2M_p(T_z=0) \qquad (2)$$

where the reduced proton matrix elements, once the Wigner-Eckart theorem is applied, are given by $M_p = <J_f||\sum r_i^\lambda Y_\lambda(\Omega_i)||J_i>$, and related to the reduced transition probabilities by,

$$M_p = [(2J_i+1) \ B(E2;J_i \to J_f)]^{1/2}. \tag{3}$$

A recent series of experiments initiated by Cottle *et al.*, and supplemented by others, provide reliable experimental results for $E2$ transition rates in $T=1$ isobaric triplets through refined techniques of Coulomb excitation and lifetime measurements [2, 3, 4, 5, 6, 7]. In this work we address an earlier apparent problem [2] of broken isospin symmetry for some isobaric triplets between $A=18$ and $A=42$. Both $A=18$ and $A=42$ systems are being studied at the University of Kentucky [6, 8]. Here, we focus on the $A=42$ isobaric triplets and the intriguing picture of the systematic trend of isoscalar matrix elements through the sd shell, and into the fp shell. This systematic trend is examined within a shell model framework.

EXPERIMENT AND LINE-SHAPE ANALISIS

To populate the $2^+_{T=1}$ state at 1586 keV in ^{42}Sc, we used the ^{40}Ca(^{3}He,pγ)^{42}Sc reaction at beam energies of 4.2 and 5.1 MeV. The ^{3}He beam was provided by the 7 MV accelerator at the University of Kentucky. A detailed description of the experimental set-up can be found in Ref. [6]. The beam energy of 4.2 MeV was chosen to provide adequate statistics on relevant γ-ray yields while minimizing the side-feeding. Even this (relatively) low bombarding energy leaves the nucleus at an excitation energy above 7 MeV, although the Coulomb barrier for emitted protons would mean that very little excitation would occur above 4 MeV. Only 11% of the de-excitation intensity of the 1586 keV, $2^+_{T=1}$ level could be identified as coming from observed higher-lying levels.

Line-shape analysis [12, 13, 14] with the inclusion of the calculated side-feeding paths [15] for the (^{3}He,p) reaction has been used for determining lifetimes. This method employs a realistic description of the stopping process of the recoiling ^{42}Sc nuclei in the ^{40}Ca target and is well described in Ref. [12, 13]. The line-shape analysis strongly indicates two main feeding paths, one very fast, and the other slow. The code PACE2 [15], a MonteCarlo simulation for fusion-evaporation reactions, was used to calculate proton emission spectra, which were employed to take into account the alteration of the recoil velocity distributions. The use of PACE2 is supported by previous work [9, 10, 11], where, from experimental angular correlation and angular distribution of protons from the (^{3}He,p) reaction, it was concluded that the reaction proceeds predominantly through the compound nucleus mechanism up to about $E_{^3He}=8$ MeV. The additional data taken at 5.1 MeV bombarding energy allowed tests of the accuracy of side-feeding corrections since the feeding paths and the proton emission spectra at this energy are different from those at 4.2 MeV. Figure 1 shows fits to the line-shape of the 975 keV transition depopulating the first $2^+_{T=1}$ state at 1586 keV and observed in the 4.2 MeV experiment. This strong γ-ray is detected as a single peak and displays varying line-shapes at different angles. The results from both 4.2 and 5.1 MeV experiments are consistent and can be combined to yield the average value

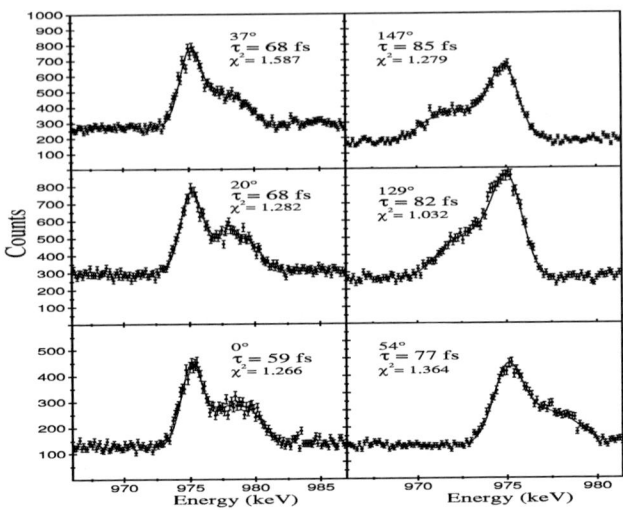

FIGURE 1. Lifetime determination by line-shape analysis of the 975 keV γ ray at E $_{3He}$=4.2 MeV.

$<\tau>$ = 69 (17) fs for the lifetime from all fits. The uncertainty of 17 fs includes the statistical errors and an estimated 10% systematic error associated with the incomplete knowledge of the feeding and the stopping powers. Together with the lifetime experiment, we carried out a γ-γ coincidence experiment at the University of Cologne using the ^{40}Ca(^{3}He,pγγ) reaction. Preliminary results for the relative intensity of the 975 keV and 1586 keV transitions depopulating the first $2^{+}_{T=1}$ state give 100(1) and 8(1), respectively, in agreement with previous work [16]. The determination of the lifetime and the branching ratio lead to a value for the isoscalar matrix element of M_0=6.4(9) (W.u.)$^{1/2}$. As noted earlier, previous results led to M_0=5.7(9) (W.u.)$^{1/2}$. A comparison of the isoscalar matrix element obtained in the current work with that given by the mirror nuclei [2] indicates that isospin symmetry is conserved for the A = 42 isobaric triplet.

SHELL MODEL CALCULATIONS AND CONCLUSIONS

Figure 2 shows the isoscalar matrix elements for isobaric triplets from $A = 18$ and up to the $A = 42$ region. In principle, the global trend could be tested if one considers that the behaviour of the isoscalar matrix elements is the result of shell effects. In view of the successful predictions for ^{42}Ca and the sd shell, we have used the m-scheme numerical code ANTOINE [20] to elucidate the behaviour of the $A = 4n+2$ isobaric triplets. As noted in Ref. [20], this code can be applied with relatively large valence spaces. In our calculations for the sd shell ($\hbar\omega_0$), which include the $1d_{5/2}$, $2s_{1/2}$ and $1d_{3/2}$ subshells in the valence space for protons and neutrons, we used the Wildenthal USD interaction [21] and ^{16}O as the inert core, following the earlier work of Brown et al. [22]. For ^{42}Ca, ^{42}Sc and ^{42}Ti, calculations without truncations have been performed using the $2s_{1/2}$, $1d_{3/2}$,

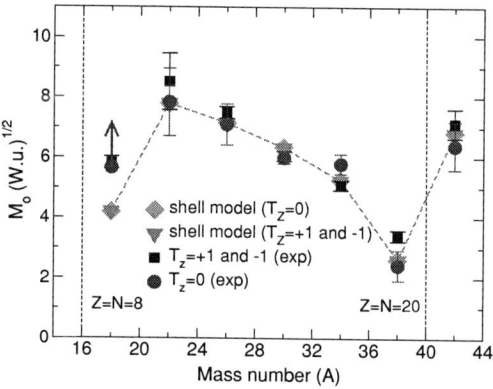

FIGURE 2. Experimental and calculated isoscalar multipole matrix elements as a function of mass number for 2_1^+ ($T=1$) → 0_1^+ ($T=1$) transitions in the $A=4n+2$ isobars ranging from $A=18$ to $A=42$. The experimental results are taken from Refs. [2, 4, 5, 6, 17, 18, 19]

$1f_{7/2}$ and $2p_{3/2}$ valence space, with ^{28}Si as the inert core. Here, we applied a version of the *sdpf.fin* interaction [23] with modified single particle energies, in agreement with the results of Caurier *et al.* in the light Ca isotopes [24]. In both regions, the calculations were performed with the usual effective charges $e_p = 1.5, e_n = 0.5$, and the harmonic oscillator parameter $b = 1.01A^{1/6}$. Generally, the agreement is excellent, and only the predictions for $A = 18$ and $A = 38$ present certain discrepancies with the experimental data; reasons for this disagreement will be suggested below. Our calculations for ^{42}Ca are in good agreement with Caurier *et al.*, and the results for the $A = 42$ isobaric triplet supports *sd* shell excitations into the *f p* shell, as proposed for ^{42}Ca and ^{44}Ca [24].

The predicted and experimental isoscalar matrix elements, M_0, for the whole chain of isobaric triplets are presented in Fig. 2. The assumption of isospin symmetry in the nucleon-nucleon interaction leads automatically to the equality of $M_{0_{th}}$ for calculations at each A-value. The results are in good agreement with the experimental data, and again, discrepancies arise in the $A = 18$ isobars, with deviations between calculated and experimental $B(E2)$ values. This is an interesting result, since these isobars are only a couple of nucleons away from the ^{16}O core. Similar results were found by Brown *et al.* [22] for $A = 18$ using shell model calculations with the Chung-Wildenthal interaction. Indications for a broken core in the $A = 18$ system were previously suggested [25, 26] and attributed to weakly coupled particle-hole excitations of the ^{16}O core. Consequently, we have also performed further calculations taking a $2p + 2n$ core, extending the valence space to the $p_{3/2}, p_{1/2}, d_{5/2}, s_{1/2}$ and $d_{3/2}$ orbits, and using the *psd.int* interaction [27]. However the $B(E2)$ strength predicted for the $2_1^+{}_{(T=1)} \rightarrow 0_1^+{}_{(T=1)}$ transition in ^{18}F is still much lower than the experimentally determined one. Similar discrepancies occur at both extremes of the *sd* shell. These discrepancies seem to be closely related to each other since for the $A = 18$ isobaric triplets there are only two particles in the sub-shell above the shell closure, whereas in $A = 38$ triplets there are two holes below the shell closure. This suggests that the low and nearly full occupancy in these orbits of the sd shell play an important role in the increased collectivity of the experimental $B(E2)$ strengths, through core polarization effects not accomodated by effective charges designed for nuclei throughout the sd shell. This might suggest that opening the fp shell is important to describe the A=38 triplet. Additional calculations using the sdpf.fi n interaction provide a better result for the $2^+_{1\,(T=1)} \to 0^+_{1\,(T=1)}$ energy difference.

Finally, the two notable results of the present calculations is fi rst to see that the progressive decrease of the M_0 values through the sd shell is well reproduced by the shell model calculations within the valence space of that shell. This ties together the results for M_0 from several different experiments studying $T = 1$ isobaric triplets. Secondly, the sharp increase of M_0 as the $Z = N = 20$ point is crossed is probably well represented through including core contributions from the sd shell into the fp valence space, just as those same core excitation contributions solved the earlier problem of g factor issues for the light Ca isotopes. Despite small deviations for $A = 18$ and $A = 38$, the recent experimental results are well reproduced by the calculated $2^+_{1\,(T=1)} \to 0^+_{1\,(T=1)}$ transition energies, reduced transition probabilities and isoscalar matrix elements. The transition rate agreements shown in Fig. 2 provide an explanation of the systematic trend noted earlier in this paper, which was the motivation for these calculations. This material is based upon work supported by the U.S. National Science Foundation under Grant No. PHY-0354656.

REFERENCES

1. Aron M. Bernstein, V. R. Brown and V. A. Madsen, Phys. Rev. Lett. **42**, (1979) 425.
2. P. D. Cottle et al., Phys. Rev. C **60**, (1999) 031301(R).
3. P. D. Cottle et al., Phys. Rev. Lett. **88**, (2002) 172502.
4. P. D. Cottle et al., Phys. Rev. C **64**, (2001) 057304.
5. L. A. Riley et al., Phys. Rev. C **68**, 044309.
6. J. N. Orce et al., Phys. Rev. C **70**, (2004) 014314.
7. T. Glasmacher, Ann. Rev. Nucl. Part. Sci. **48**, (1998) 1.
8. S. Choudry et al., to be published in this proceedings.
9. B. Heusch, A. Gallmann, Phys.Rev. C**7**, 1810 (1973).
10. J. Hazan and G. Merkel, Phys. Rev. **139**, B835 (1965).
11. D.P. Balamuth, G.P. Anastassiou, R.W. Zurmuhle, Phys. Rev. C**2**, 215 (1970).
12. G. Böhm et al., Nucl. Instrum. Methods Phys. Res. A **329**, 428 (1993).
13. P. Petkov et al., Nucl. Phys. A **640**, 293 (1998).
14. D. Tonev et al., Phys. Rev. C**65**, 034314 (2002).
15. *Angular-momentum evaporation code for fusion-evaporation reaction.* U. of Liverpool (2003).
16. S.W. Kikstra et al., Nucl. Phys. A **496**, 429 (1989).
17. D. R. Tilley, H. R. Weller, C. M. Cheves, R. M. Chasteler, Nucl. Phys. A **595**, (1995) 1.
18. P. M. Endt, Nucl. Phys. A **633**, (1998) 1.
19. B. Singh, J. A. Cameron, Nucl. Data Sheets **92**, (2001) 1.
20. E. Caurier and F. Nowacki, Acta Physica Polonica B **30**, (1999) 705.
21. B. H. Wildenthal, Prog. Part. Nucl. Phys. **11**, (1984) 5.
22. B. A. Brown et al., Phys. Rev. C **26**, (1982) 02247.
23. E. Caurier, private communications, CRN, Strasbourg, 1999.
24. E. Caurier, A. Poves and A. Zuker, Phys. Lett. B **96**, (1980) 15.
25. P. J. Ellis and T. Engeland, Nucl. Phys. A **144**, (1970) 161.
26. T. Engeland and P. J. Ellis, Nucl. Phys. A **181**, (1972) 368.
27. F. Nowacki, private communications, CRN, Strasbourg, 1999.

Symmetry and structure tests in ^{18}O and ^{18}Ne

S.N. Choudry[1], J. N. Orce[1], V. Varadarajan[1], S. Lesher[1],
D. Bandyopadhyay[1], S. Mukhopadhyay[1], S.W. Yates[1,2], M.T. McEllistrem[1]

[1]*Department of Physics and Astronomy, University of Kentucky, Lexington, Kentucky 40506-0055, USA*
[2]*Department of Chemistry, University of Kentucky, Lexington, Kentucky 40506-0055, USA*

Abstract. Isospin (T) symmetry assumes charge symmetry and charge independence are explicit in nuclear structure. Whereas charge independence implies the nn, np and pp interactions are the same, charge symmetry states that the nn and pp interactions are equal. The latter approximate symmetry is experimentally verified by the comparison of the excited levels lying at about the same energy in light mirror nuclei. Both of these symmetries are broken by the electromagnetic interaction. The neutron facility at the University of Kentucky provides a unique opportunity to examine charge independence, and in particular, charge symmetry from the reduced electromagnetic transition probabilities and neutron scattering cross sections. Here, we use the isospin formalism by Bernstein, Brown and Madsen [1], which relates the proton and neutron matrix elements, M_p and M_n respectively, for equivalent excited states in T=1 mirror nuclei (T_z = +/-1). The nucleus ^{18}O has been studied using the (n,n') reaction in order to measure the neutron cross sections of the 2^+_1, T=1 state. Using the previously determined proton matrix elements for the T_z = -1 mirror nucleus (^{18}Ne), along with the determination of the neutron matrix element in ^{18}O from neutron scattering, allows an experimental test of charge symmetry in the A=18 mirror system.

INTRODUCTION

Bernstein, Brown and Madsen [1] suggested that isospin (T) sensitive symmetry tests of nuclear structure could be obtained by comparing electromagnetic transition intensities between 0^+ and 2^+ levels in an isobaric triplet. P. D. Cottle and his graduate and postdoctoral students at Florida State, in a series of quite recent studies, completed experiments updating knowledge of these reduced E2 intensities for nuclei in several of the T = 1 triplets in the sd shell. Those tests revealed generally good agreement with isospin symmetry expectations in that shell except for deviations for multiplets at A = 34, 38, and 42. Subsequently, as noted in another contribution to this conference experiments of Orce, et al. resolved the discrepancy at A = 42 in favor of symmetry, by redetermining the E2 transition intensity in ^{42}Sc. These tests affirm the observation of symmetry associated with the charge independence of the T = 1 nucleon-nucleon force, except for the T = 1 isobars at A = 34 and 38.

A separate determination of another, even stronger symmetry, charge symmetry of forces, would be to compare target proton and neutron excitation intensities of T = 1, 2^+ levels in mirror pairs, such as ^{18}O and ^{18}Ne directly. Charge symmetry in this mirror pair would mean, for example, that the proton excitation matrix element, M_p of ^{18}O would be equal to the neutron excitation matrix element, M_n of ^{18}Ne, and vice versa. Grabmayr, et al. completed a detailed examination of neutron scattering in ^{18}O at 24 MeV incident energy. In their experiment they determined the ratio of neutron and proton matrix elements [2] for excitation of the 2^+ level, $M_n/M_p = 2.5$. That goes a long way toward determining the separate matrix elements. Raman, et al, in their extensive compilation [4] of reduced E2 transition rates for excitation of 2^+ levels (B(E2; $0^+ \to 2^+$)) reported that the B(E2) for ^{18}O is 0.00451 e^2b^2. This yields an E2 excitation matrix element, $[B(E2)]^{1/2} = 0.067$ (0.003) eb = M_p. Thus, with the ratio of Grabmayr, et al of 2.5 ± 0.6, one finds a neutron matrix element of $M_n(^{18}O) = 0.168$ (20) eb.

Just two years ago Riley, et al. made a fresh determination of the proton matrix element in ^{18}Ne in a direct lifetime measurement [3] of the first excited 2^+ level of ^{18}Ne.

They measured a mean life of $\tau = 0.77$ ps, which yields reduced transition excitation intensity of B(E2) = 222 e^2fm^4 = 0.0222 e^2b^2, with an uncertainty of 20 e^2fm^4. This in turn gives a proton matrix element $M_p(^{18}Ne) = 0.149$ (0.007). Thus, within uncertainties, the neutron matrix element of ^{18}O does equal the proton matrix element of ^{18}Ne, as charge symmetry requires. The scattering experiment of Grabmayr, et al, completed in 1980 was also focused on many other highly valuable properties of scattering from ^{18}O. The ratio cited above was an additional determination of the experiment. Since this is such a clean and potentially accurate test of the application of charge symmetry, it was deemed useful to attempt a new determination of the key ratio, M_n/M_p, cited above.

NEW NEUTRON SCATTERING EXPERIMENT

In addition to noting the comparison shown above, a new experimental effort was begun at a much lower neutron incident energy, to attempt also to learn something about the structure of other excited levels. The 24 MeV scattering experiment had to work with a neutron energy spread of about 500 keV. The next higher 2^+ level, at 3.92 MeV, is separated from the next lower level by about 290 keV, and from a reasonably strongly excited 4^+ level by 360 keV. We chose to attempt the experiment at 8.5 MeV incident, which, subsequent results have shown, is too low to learn anything valuable about the structures of excited levels higher than the 2_1^+. This report reveals the progress reached so far in neutron inelastic scattering in that nucleus.

Companion measurements are being made of the gamma rays following neutron inelastic scattering to support the neutron detection measurements, with measurements of much greater resolution in gamma-ray detection than can be achieved in neutron detection. These results, now being finalized, will be released in a later report.

Direct excitation of levels in neutron scattering yields information about the properties of the target nucleus. The neutron scattering matrix element may be written

$$M_{n,n'} = (\chi_{nn}M_n + \chi_{np}M_p)/ \chi_{nn}N + \chi_{np}Z \qquad (1)$$

where χ_{nn} and χ_{np} are strength parameters of the incident neutron-target neutrons and incident neutron-target protons respectively. Many studies of scattering properties [1] have shown that the ratio $\chi_{np}/\chi_{nn} \equiv \chi = 3$, quite closely, for incident neutrons well below 50 MeV. Thus one can alter the above ratio for $M_{n,n'}$ by dividing it first by χ_{nn} and subsequently M_p, the matrix element for target protons. With that done, the ratio now becomes:

$$M_{n,n'}/M_p = (M_n/M_p + \chi)/(N + \chi Z) \qquad (2)$$

where χ is the strength ratio noted above.

It is worth noting also that the ratio M_n/M_p is quite insensitive to the exact value of χ, since it appears in both the numerator and denominator.

The key experimental parameters are the direct excitation matrix element from neutron scattering, and the proton matrix element which is most often determined either from Coulomb excitation or lifetime determination, as was done for ^{18}Ne.

There are two mechanisms involved in the neutron scattering experiment, one the direct excitation mechanism noted, and the other a statistical model mechanism which is structure independent. Thus these mechanisms must be separated in the neutron scattering experiment. At the 24 MeV of the earlier experiment, the statistical model component is negligible for low-lying target levels. The statistical mechanism shifts to much higher excited levels. However, at 8.5 MeV, both mechanisms are present.

EXPERIMENTAL SETUP AND PRELIMINARY RESULTS

The ^{18}O(n,n') measurement was performed at the accelerator facility located at the University of Kentucky. A pulsed beam of ~1ns duration at a repetition rate of 1.875 MHz and beam currents of 1-2 μA was used. Neutrons were produced at 8.5 MeV through the D(d,n)^{3}He reaction in the deuterium cell. ^{18}O, empty vial and polyethylene samples were suspended in the beam path individually and their corresponding time-of-flight spectra recorded. Events with no sample were also recorded at various angles to determine the dynamic bias and the main detector efficiencies. The resulting neutrons from the ^{18}O(n,n') reaction were recorded by a liquid organic scintillator detector, BC501, which was shielded . Another NE218 neutron detector, placed at 42.5 degrees with respect to the beam line, and at 5.3 m from the scattering sample, was used as the neutron flux monitor. A tungsten wedge was used to shield the detector from the neutrons coming directly from the deuterium cell, or scattering from the Cu collimator wall.

The ^{18}O sample consisted of a mixture of ^{18}O and ^{16}O water in a cylindrical vial with ^{18}O water enriched to 95%. The mass of the sample and the empty vial were carefully measured before the experiment.

The time-of-flight spectrum was recorded with the electronic bias set at 1.27 MeV. The forward monitor bias was set at ~1.8 MeV. Two dynamic bias windows were employed with a 'wide' bias of 55% of the full pulse height spectrum and a 'narrow' bias of 45%. Pulse shape discrimination was also employed in time-of -flight spectra of the main detector and the monitor to get rid of the gamma events.

The events recorded by the monitor were used to normalize the detector efficiencies and the differential cross-sections. Empty vial runs were used to subtract the background from the main time-of-flight spectra. Polyethylene sample spectra at 25, 30 and 40 degrees enabled us to determine the absolute differential cross-sections using the known elastic cross-section for hydrogen. The yields of the ground state and excited states of ^{18}O were obtained by fitting the time-of flight spectra using a special neutron analysis code designed at Kentucky called SAN12. The ^{18}O angular distributions were obtained for the ground state, the first excited 2^+ (1.982 MeV) and the second excited state at 3.920 MeV (figure 1 and 2). The combined cross-sections for the 4^+ (3.555 MeV) and 0^+ (3.634 MeV) were also obtained since these states could not be individually resolved.

It is evident, of course, that the first 2^+ excitation has both direct excitation and structure independent, statistical model components. The completely forward and backward symmetric 3.92 MeV angular data suggest that this distribution has only statistical model components. This is in accord with the earlier findings of Grabmayr, *et al.* [2] who found that while the valence neutrons of ^{18}O were dominant in the first 2^+ excitation, they were quite unimportant in the 3.92 MeV results.

The distribution-data for the two levels near 3.6 MeV excitation are dominated also by statistical model components, as would be expected for 4^+ and 0^+ levels at this excitation energy. Thus we have excellent information to fix the strength of the statistical model contributions to our inelastic cross sections. The analysis and interpretation are ongoing.

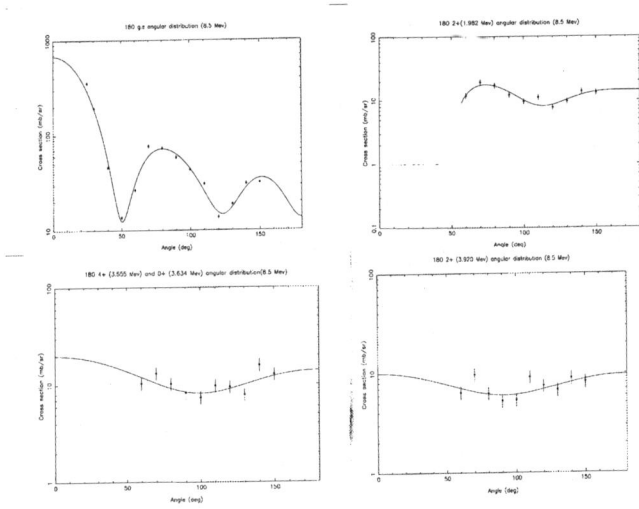

Figure 1. ground stated and excited state angular distributions for ^{18}O at 8.5 MeV

REFERENCES

[1] A. M. Bernstein, V. R. Brown and V. A. Madsen Phys. Rev. Lett. 42, 425 (1979)
[2] P. Grabmayr, J.Rapaport and R. W. Finlay Nucl. Phy. A 350, 167 (1980).
[3] L. A. Riley et al., Phys. Rev. Lett. 42, 425 (1979)
[4] S. Raman and C. W. Nestor, Jr. and K. H. Bhatt Phys. Rev.C37, 805 (1988).

Gamma-Ray Transitions In the Decay of the Superallowed Beta Emitter ^{62}Ga

B. Hyland*, C.E. Svensson*, G.C. Ball[†], J.R. Leslie**, D. Albers[†],
C. Andreoiu*, P. Bricault[†], R. Churchman[†], D. Cross[‡], M. Dombsky[†],
P.E. Garrett*, C. Geppert[§], G.F. Grinyer*, G. Hackman[†], V. Hanemaayer[†],
J. Lassen[†], J.P. Lavoie[¶], D. Melconian[∥,†], A.C. Morton[†], C.J. Pearson[†],
M. Pearson[†], A.A. Phillips*, M.A. Schumaker* and J.J. Valiente-Dobón*

*Department of Physics, University of Guelph, Guelph, Ontario N1G 2W1, Canada
[†]TRIUMF, 4004 Wesbrook Mall, Vancouver, British Columbia, V6T 2A3, Canada
**Department of Physics, Queen's University, Kingston, Ontario K7L 3N6, Canada
[‡]Department of Chemistry, Simon Fraser University, Burnaby, British Columbia V5A 1S6, Canada
[§]Johannes Gutenburg-Universitat Mainz, Staudinger Weg 7, 55099 Mainz, Germany
[¶]Department de Physique, Universite Laval, Quebec, QC, G1K 7P4, Canada
[∥]Department of Physics, Simon Fraser University, Burnaby, British Columbia V5A 1S6, Canada

Abstract. A measurement of the ground state β-decay branching ratio of ^{62}Ga has been made as part of a program of high-precision superallowed Fermi β decay studies at the ISAC radioactive beam facility. The experiment was conducted by detecting γ rays and β particles from the decay of ^{62}Ga using the 8π γ-ray spectrometer and the SCEPTAR plastic scintillator array.

Keywords: ^{62}Ga, branching ratio, CKM matrix, superallowed β decay
PACS: 23.20.Lv, 23.40.-s, 24.80.+y, 27.50.+e

INTRODUCTION

In the Standard Model of particle physics the Cabibbo-Kobayashi-Maskawa (CKM) matrix relates the weak interaction eigenstates and mass eigenstates of the quarks. Theory predicts this matrix to be unitary, but current experimental measurements [1, 2] on the first row of the matrix indicate that $|V_{ud}|^2 + |V_{us}|^2 + |V_{ub}|^2 = 0.9966 \pm 0.0014$, a 2.4$\sigma$ discrepancy. Recent measurements of V_{us} [3, 4, 5] may indicate a revised value that could push the above sum into agreement with unity; however the value for V_{us} is not resolved, as another recent measurement [6] obtains a lower value consistent with older measurements. An additional measurement [7] may also indicate a lower value for V_{ud}. Improved confidence in V_{ud}, which contributes approximately half of the uncertainty in the unitarity sum, is therefore desirable. The least well-known component of V_{ud} is δ_C, the isospin symmetry breaking correction [8] which is relatively large in the decay of ^{62}Ga. By measuring the half life, branching ratio and Q-value for the decay of ^{62}Ga, the ft value can be calculated, and then, assuming the conserved vector current (CVC) hypothesis, δ_C can be deduced and compared with theory.

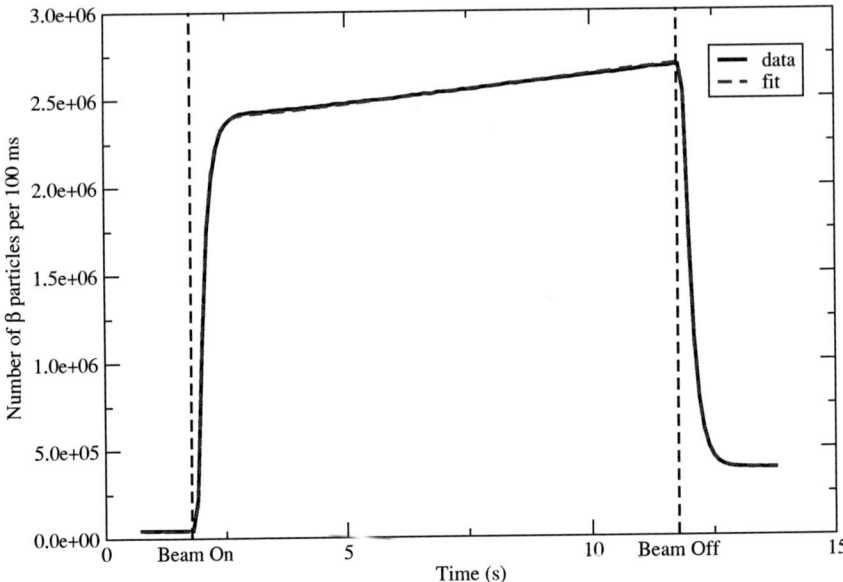

FIGURE 1. Number of β particles detected as a function of time. The first segment is when the beam is off. Then the beam is turned on, and the ^{62}Ga grows in quickly. The slope in the middle region is due to the grow-in of the longer lived ^{62}Cu. When beam is turned off, the ^{62}Ga quickly decays. The remaining background level is from ^{62}Cu.

EXPERIMENT AND ANALYSIS

The experiment was performed at the ISAC (Isotope Separator and ACcelerator) facility at TRIUMF (TRI-University Meson Facility) located in Vancouver, Canada. A beam of approximately 2000 ^{62}Ga ions/s was produced using TRILIS (TRIUMF Resonant Ionization Laser Ion Source) [9] and a ZrC production target. This was the first beam produced at ISAC using TRILIS. The 8π spectrometer array of 20 HPGe detectors was used to detect γ rays and SCEPTAR (SCintillating Electron-Positron Tagging ARray) detected the β particles. The beam was implanted into a collector tape at the centre of the 8π array. The beam was turned on for 10 s, then turned off for 2 s, and then the tape was moved to remove long lived ($T_{1/2}$ = 9.74 min) ^{62}Cu background (Figure 1).

Gamma-ray spectra from β-γ coincidence data with Compton and Bremsstrahlung suppression [10, 11] were produced for the portion of a cycle when the beam was on, and again for the period after the beam was turned off. A comparison of these spectra allowed the identification of peaks from ^{62}Ga decay, and from a long-lived background.

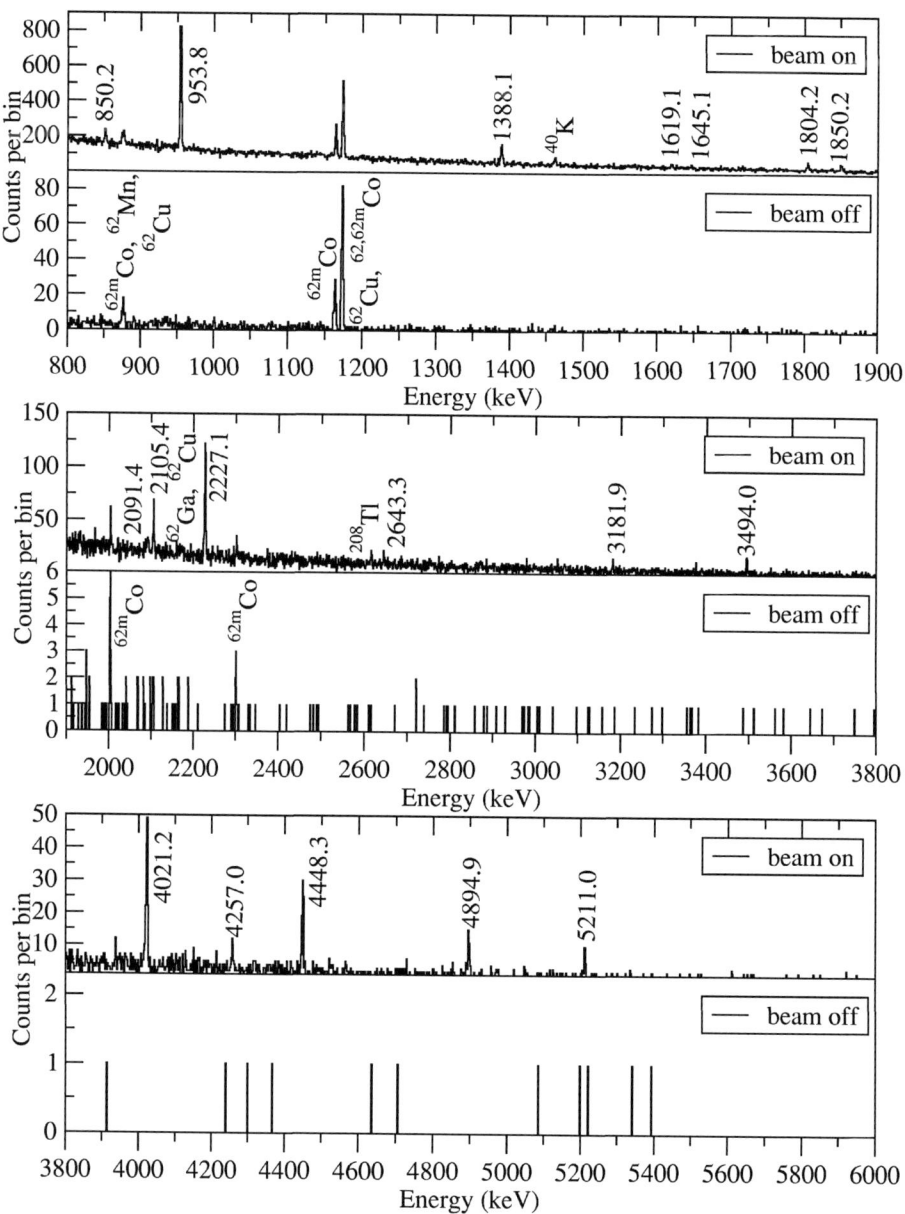

FIGURE 2. Gamma-ray spectra gated on the portion of the cycle when the beam was on (top panels) and off (bottom panels), between 800 and 1900 keV (top), 1900 and 3800 keV (middle), and 3800 and 6000 keV (bottom).

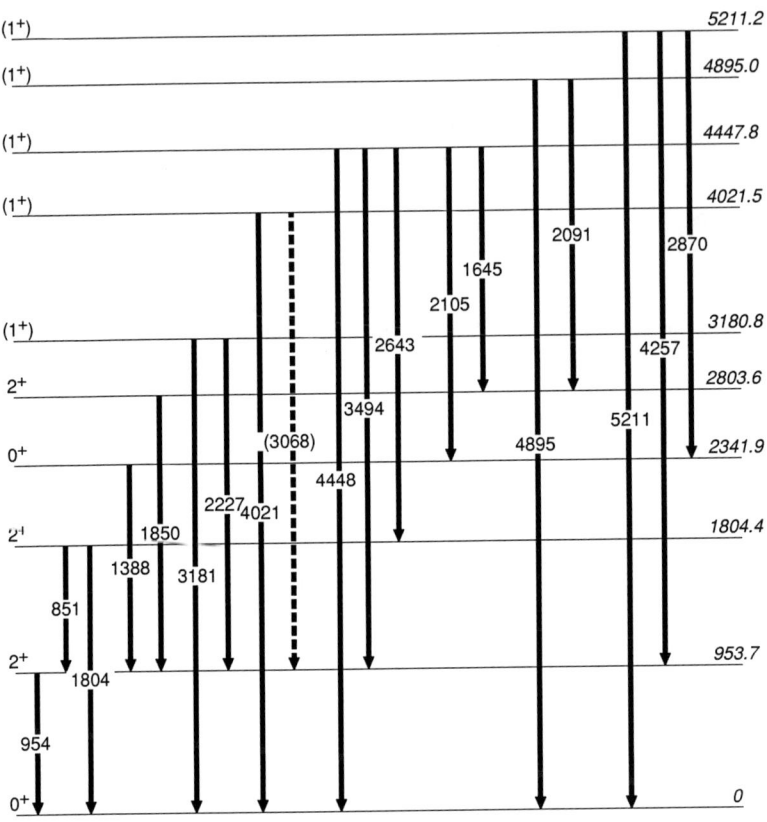

FIGURE 3. Levels in ^{62}Zn determined from the observed γ-ray spectra.

Spectra with energies between 800 keV and 6 MeV are shown in Figure 2. Levels and transitions in ^{62}Zn were deduced from these spectra, as shown in Figure 3. The level scheme was produced using the differences in energies between the γ rays. There were insufficient statistics to study γ-γ coincidences.

Previous works [12, 13] have reported the 954 keV $2^+ \rightarrow 0^+$ γ ray, and also tentatively identified the $(1^+) \rightarrow 2^+$ 2227 keV γ ray. This work identifies twenty-two γ rays, nineteen of which are placed in the level scheme. The level scheme includes five new levels at energies of 3180.8 keV, 4021.5 keV, 4447.8 keV, 4895.0 keV, and 5211.2 keV.

The number of implanted β particles was determined by fitting the decay curve (Figure 1) to a function incorporating a flat background, ^{62}Cu background, and ^{62}Ga. Analysis of the absolute γ ray efficiencies in order to establish absolute γ ray branching ratios per β decay is ongoing.

CONCLUSIONS AND OUTLOOK

Nineteen γ-ray transitions and five new levels in ^{62}Zn were observed in the decay of ^{62}Ga. Analysis is still ongoing to estimate the unobserved γ ray flux to the ground state [14, 15]. The final uncertainty of the superallowed β branching ratio is expected to be <0.02%. This measurement, when combined with high precision measurements of the half life and Q-value, will establish the superallowed ft value. Using CVC, this will test the δ_C correction. The world average half life of ^{62}Ga is 116.17 $\pm$ 0.04 ms [16, 17]. The current uncertainty in the world average is dominated by a single high-precision measurement [13]. A higher statistics measurement of similar precision is planned at TRIUMF for 2005.

ACKNOWLEDGMENTS

This work has been partially supported by the Natural Science and Engineering Research Council of Canada and the United States Department of Energy. TRIUMF receives federal funding via a contribution agreement through the National Research Council of Canada.

REFERENCES

1. Hardy, J.C. and Towner, I.S. 2005 *Phys. Rev. C* **71** 055501
2. Eidelman, S. *et al* 2004 *Phys. Lett. B* **592** 1
3. Alexopoulos, T. *et al* 2004 *Phys. Rev. Lett.* **93** 181802
4. Cabibbo, N., Swallow, E.C. and Winston, R. 2004 *Phys. Rev. Lett.* **92** 251803
5. Sher, A. *et al* 2003 *Phys. Rev. Lett.* **91** 261802
6. Lai, A. *et al* 2004 *Phys. Lett B* **602** 41
7. Savard, G. *et al* 2005 *Phys. Rev. Lett.* **95** 102501
8. Towner, I.S. and Hardy, J.C. 2002 *Phys. Rev. C* **66** 035501
9. Rauth, C *et al* 2005 *Nucl. Instr. Meth. B* **215** 268; and to be published
10. Ball, G.C. *et al* 2005 *J. Phys. G***31** No 10 S1491-S1498
11. Svensson, C.E. *et al* 2003 *Nucl. Instr. Meth. B* **204** 660-665.
12. Hyman B. C. *et al* 2003 *Phys. Rev. C* **68** 015501
13. Blank B. *et al* 2004 *Phys. Rev. C* **69** 015502.
14. Hardy, J.C. and Towner, I.S. 2002 *Phys. Rev. Lett.* **88** 252501.
15. Piechaczek, A. *et al* 2003 *Phys. Rev. C* **67** 051035.
16. Hyland, B. *et al J. Phys. G* (2005) in press.
17. Canchel, G. *et al* 2005 *Eur.Phys.J* **A23** 409-415

DATA FOR NUCLEAR AND ASTROPHYSICS APPLICATIONS

Nuclear Structure and Decay Data: Current Status and Future Perspectives

Filip G. Kondev[1] and Jagdish K. Tuli[2]

on behalf of the International Nuclear Structure and Decay Data Network [#]

[1]*Nuclear Engineering Division, Argonne National Laboratory, Argonne, IL 60439, USA*
[2]*National Nuclear Data Center, Brookhaven National Laboratory, Upton, NY 11973, USA*

Abstract. The nuclear structure databases provide physicists around the world with a useful collection of reliable and well documented datasets. The Evaluated Nuclear Structure Data File (ENSDF) database, produced by the International Nuclear Structure and Decay Data Network (NSDD) under the auspices of the International Atomic Energy Agency (IAEA), contains evaluated experimental information for all known nuclei. The bibliographical database Nuclear Science References (NSR) provides references to published data in the field of Nuclear Physics. The Experimental Unevaluated Nuclear Data List (XUNDL) provides a method for rapid access to formatted (compiled) data from recently published articles. Detailed information regarding these databases, as well as other products and services, can be found at the National Nuclear Data Center (NNDC) and IAEA web portals.

Keywords: Nuclear Structure and Decay Data, Nuclear Data Compilations and Evaluations
PACS: 21.10.-k, 23.20.-g, 23.40.-s, 23.50.+z, 23.60.+e, 29.87.+g

INTRODUCTION

The expression "nuclear structure and decay data" refers to libraries of numerical values that quantify fundamental nuclear physics information. These data are at the core of the basic nuclear science research and are implicated in many applied nuclear technologies, including nuclear energy, radioactivity production and disposal, radiation transport and shielding, and all other processes and issues that are ultimately related to a variety of nuclear reactions or the decay of radioactive nuclei. There is a growing demand for high quality, evaluated decay data in the fields of medical diagnostic and radiotherapy, health physics, geophysics and astrophysics.

The evaluated, and hence reliable, data are particularly useful to the nuclear physics research community whose major goal is to seek a deep understanding of the structure of the nucleus and, for example, its connection to the nucleosynthesis of elements.

[#] A collaboration of scientists from Australia, Argentina, Belgium, Brazil, Bulgaria, Canada, China, France, India, Japan, Kuwait, Russia and United States under the auspices of the International Atomic Energy Agency, Vienna Austria.

Various theoretical models predict the existence of more than 6000 nuclei. Although all them are expected to exist in nature, in the Cosmos for example, it is not easy to produce many of them in sufficient quantities in laboratory environment and study their properties. Until now (end of July 2005) experimental information is available for about 2923 nuclei, with more than 3000 yet to be discovered. Most of the new discoveries are sought at the neutron-rich side of the valley of stability, where the experimental information is sparse due to difficulties in production. In fact, study of neutron rich nuclei is becoming one of the challenging frontiers of the modern nuclear structure physics. This is an area where future experimental facilities, such as RIA, will have an enormous impact. It should be realized, however, that the expected new discoveries will come out gradually. Most of them will rely on systematic approaches that use the knowledge accumulated in the past 100 years of dedicated nuclear structure research. This brings the issue of availability of reliable, complete and comprehensive nuclear structure databases to the forefront. In addition, the nuclear data effort can be used as a means of making the results of basic physics research quickly and conveniently available to a broader audience, and hence having a profound effect on the socio-economical applications of the modern nuclear science.

THE INTERNATIONAL CONNECTION

Nuclear structure and decay data are compiled and evaluated by a collaboration of scientists within the International Nuclear Structure and Decay Data network (NSDD) that was established in 1974 under the auspices of International Atomic Energy Agency (IAEA). Several countries, including Belgium, Canada, China, France, Japan, Kuwait, Russia and United States, have been contributing over a long period of time to these activities, coordinated by IAEA through biennial Advisory Group meetings. Recently, new evaluation groups have emerged in other countries, such as Australia, Argentina, Brazil, Bulgaria and India. The total NSDD evaluation effort is equivalent to about 9 full-time scientists.

THE MAJOR PRODUCTS

The National Nuclear Data Center at Brookhaven National Laboratory (BNL) (www.nndc.bnl.gov) hosts and provides management controls to the major nuclear structure and decay databases on behalf of the NSDD network. Three of the most popular databases that are related to nuclear structure research are:

- **ENSDF (Evaluated Nuclear Structure Data File)**- www.nndc.bnl.gov/ensdf

This is by all means the world's most complete and comprehensive nuclear structure and decay database. It contains evaluated data for all known nuclei, compiled in over 290 isobaric chains. The information is organized in different datasets, including

   a) **Reaction Data.** The data from various reactions, such as (HI,xnγ), (p,p'γ), (n,γ), (d,p), (α,α'), (p,α), etc. are included

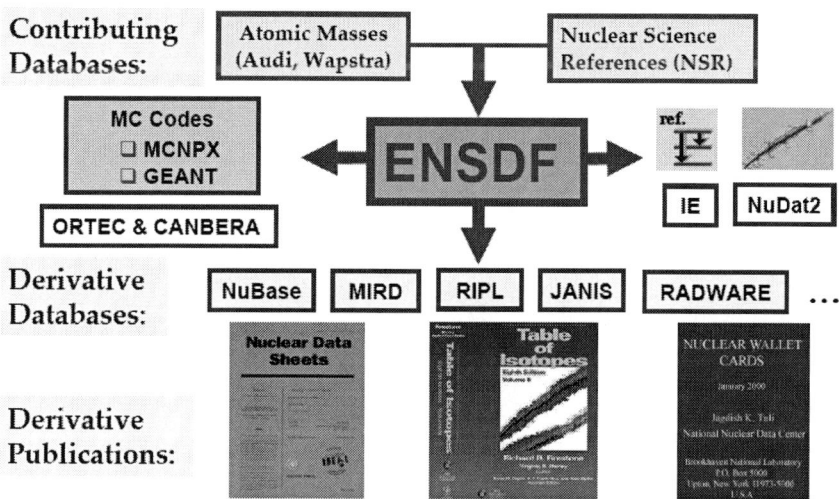

FIGURE 1. Some examples of the pivotal role the ENSDF database plays in the area of basic and applied nuclear physics.

b) **Decay Data.** The data from different decay processes, such as β^-, $\varepsilon+\beta^+$, α, β^-n, p, IT, etc. are included

c) **Adopted Data.** This is the heart of the database and it contains recommended (best) values for a range of nuclear level properties, including excitation energy, quantum numbers, half-life, static moments, configurations, gamma-ray energies, branching ratios, transition multipolarity, mixing ratios, etc. These values are derived by critical analysis of all experimental information available from the reaction and decay datasets

The ENSDF database is a treasure to the Nuclear Physics community world-wide. It is a primary source for many specialized databases and publications, some of them are indicated in Figure 1. While for many applications ENSDF data provides sufficient precision and completeness, for the specialists they can be used as a stepping stone towards more complete and comprehensive studies. The ENSDF-style evaluations can be particularly useful in identifying contradictory results that exist in the literature and hence, in stimulating needs for new, and improved, measurements that would eventually resolve these discrepancies.

- **NSR (Nuclear Science References)**-www.nndc.bnl.gov/nsr

This is a bibliographical database that contains more than 180,000 reference citations covering nearly 100 years of nuclear physics research. About 80 peer reviewed scientific journals and selected secondary references (conference proceedings articles, PhD thesis's, laboratory annual reports, private communications, etc) are scanned continuously. Approximately 4500 new entries are added into the database annually. Many of the entries are linked by doi directly to the publisher websites that allows easy access to the original papers, provided the users have the necessary journal subscription. The database has comprehensive searching capabilities based on keyword abstracts that are prepared at NNDC. It is updated on a weekly basis.

- **XUNDL (Unevaluated Nuclear Data List)**- www.nndc.bnl.gov/ensdf

This is a special-purpose database that provides a rapid access to formatted data from latest publications in the experimental nuclear structure area. It contains data for more than 1080 nuclides and includes more than 1600 data sets from 1280 journal publications. The compiled data are extracted directly from the .PDF files of relevant publications using a semi-automatic procedure developed at McMaster University, Hamilton, Canada. Several utility codes are applied on the produced ENSDF-style files in order to ensure that the data are correctly represented and that the format is strictly followed. These activities are performed by undergraduate students that are trained and supervised by an experienced staff scientist at McMaster University.

WHERE TO FIND THE DATA?

The technology advances in computing and internet development have made it possible to enhance the present databases dissemination capabilities enormously. Since the database size is no longer a major issue (the whole ENSDF database is only a "tiny" ~200 MB in size), it is now possible to accumulate a large amount of data at individual centers and disseminate them effectively through the World Wide Web to various users. The focal point of reliable nuclear structure and decay data is NNDC (www.nndc.bnl.gov) where interactive access to the major databases is available. A large amount of data can also be found at the Isotope Project, Lawrence Berkeley National Laboratory (ie.lbl.gov) and the IAEA Nuclear Data Services (www-nds.iaea.org) web sites. Specialized data for light nuclei (A=1-20) is available at the Triangle University Nuclear Laboratory (TUNL) Nuclear Data Project web site (www.tunl.duke.edu/nucldata).

FUTURE PERSPECTIVES & CHALLENGES

The value of a scientific database is determined largely by its properties. A good database should be a) comprehensive – it should include all measured quantities and their uncertainties; b) reliable – data should be correctly represented; c) complete – data produced from all measurements should be documented and considered; d) current - results from new measurements should appear promptly; e) accessible – data should be easily made available to the users in a user-friendly format. The main goal of the NSDD network is to improve the existing databases in all these aspects. In addition, it is envisioned to develop specialized modules tailored to various needs of the basic and applied nuclear physics communities. Special attention will be also given to the development of new evaluation methodologies, especially these that deal with discrepant data. Higher impact in the advanced research areas will be achieved by performing specialized evaluations and topical reviews, in many instances in collaboration with scientists from the user communities.

Unfortunately, mainly due to a limited space available for journal publications, a large amount of data is never completely moved into the public domain. A dedicated repository for such data has been suggested some time ago [1] where researchers can upload their analyzed data and provide additional details on the measured properties. Future enhancement of this capability will be made in close cooperation with the research community in order to make this effort worthwhile.

The quality of the databases and the survival of this seminal work depend crucially on the recruitment of new, high-quality evaluators. There is a recent decline in the evaluation effort that has led to delays in updating the ENSDF database. One of the major NSDD priorities is to reverse such a trend and to make sure that younger scientists are entering the evaluation network. In this respect, a fruitful activity has been initiated by IAEA through the establishment of regular evaluator's workshops to attract and encourage new nuclear structure evaluators. The IAEA also provides a modest amount of seed-funding to some of the new evaluators (mainly for those residing in developing countries) through research contracts. The recruitment of new evaluators is an area where future collaboration and support from the nuclear physics research community would be most valuable and welcome.

ACKNOWLEDGMENTS

The authors would like to express their gratitude to all colleagues within the NSDD network for their valuable effort in maintaining the quality of the existing databases. Special thanks are due to Dr. B. Singh (Mc Master University) and Dr. D. Winchell (NNDC) for providing information on XUNDL and NSR databases, respectively, and for commenting on the manuscript. Thanks are also due to Dr. C. Baglin (LBNL) for providing information prior the beginning of this conference, and Dr. J.H. Kelly (TUNL) for the critical reading of the manuscript and valuable comments. This work is supported by the U.S. Department of Energy, Office of Nuclear Physics under Contracts No. W-31-109-ENG-38 (ANL) and DE-AC02-98CH10886 (BNL).

REFERENCES

1. J.K. Tuli, *Nucl. Phys.* *A***682**, 236c-238c (2001).

An Updated Library of Reaction Rates for the Astrophysical *rp*-Process.

A.Sakharuk[*,†], T.Elliot[*,†], J.L.Fisker[**,†], S.Hemingray[*,†], A.Kruizenga[*], T.Rauscher[‡], H.Schatz[*,§], K.Smith[*,†], F.-K.Thielemann[‡] and M.Wiescher[**,†]

[*]*National Superconducting Cyclotron Laboratory, Michigan State University, East Lansing, MI 48824, USA*
[†]*Joint Institute for Nuclear Astrophysics*
[**]*Department of Physics, University of Notre Dame, Notre Dame, IN 46556, USA*
[‡]*Departement für Physik und Astronomie, Universität Basel, Klingelbergstrasse 82, CH-4056 Basel, Switzerland*
[§]*Department of Physics and Astronomy, Michigan State University, East Lansing, MI 48824, USA*

Abstract. We are developing an updated library with reaction rates important for the study of the astrophysical rp- process. It includes both experimental and theoretical rates parametrized in terms of the standard REACLIB format. The experimental data contain various compilations updated with the most recent data. The theoretical data are taken from shell-model calculations whenever available otherwise statistical Hauser-Feshbach calculations were used. All theoretical rates were recalculated using the latest reaction Q-values. All new rates include stellar enhancement factors. The data will be accessed through a convenient web interface. It is anticipated that this JINA project evolves into an evaluated public data archive for reaclib library data.

Keywords: reaction rate, REACLIB library, stellar enhancement factor
PACS: 26.30.+k

LIBRARY OF REACTION RATES IMPORTANT FOR THE *RP*-PROCESS

Nuclear reaction rates determine the energy generation and nucleosynthesis in stars and many types of stellar explosions. Model calculations therefore depend on accurate and comprehensive libaries of nuclear reaction rates. One standard set of nuclear reaction rates used in astrophysical applications is the REACLIB library [1]. It contains a vast amount of conveniently parametrized reaction rates together with their Q-value, lists of participants (to uniquely identify the reaction rate), references, and some additional information. REACLIB is used in a large number of astrophysical simulations.

Remarkable progress has been made both in experimental and theoretical nuclear physics in the last years. Several new direct measurements of reaction rates at astrophysically important energies were carried out [2, 3]. New compilations of reaction rates appeared in the literature [4, 5]. An impressive and fast development of theoretical models, primarily the shell-model, made available calculation of the structure and hence the reactions with sd- and pf- shell nuclei. And, finally, we know the nuclear masses far more accurately than ever before. All of this, together with the new demands from astrophysical problems, urgently requires a coordinated effort to continuously update and document the REACLIB library. Our work can be considered as a first step in this direction - for now we limited ourselves to updates relevant for simulations of explosive helium and hydrogen burning such as the astrophysical *rp*-process. The typical number of nuclei, involved in such calculations, is around seven hundred and the number of reactions is several thousands [6, 7, 8].

Masses are needed to calculate inverse reaction rates using detailed balance, and they are also critical for theoretical reaction rate calculations. Here we used experimental mass values from latest atomic mass evaluation [9]. For nuclei with experimentally unknown masses we used theoretical predictions from Coulomb shift calculations [10] believed to be accurate to 100 keV. These data were tested against experimentally known masses or nuclear lifetimes around the proton dripline. Note that we used the new masses for the calculation of the inverse reaction rates for all updated reactions even for the experimental ones, whenever the inverse reaction was not investigated directly.

We also calculated a new set of Hauser-Feshbach rates using the new set of masses with the NON-SMOKER code [1]. To illustrate the changes in the reaction rates, caused by the new mass data, we compared two Hauser-Feshbach rates, one with the new and another with the old Q-values. The comparison shows that the rates can differs by a few orders of magnitude in the temperature interval 0.1 - 2 GK, the most important temperature range for the X-ray burst simulations. For a particular reaction $p+^{97}In \rightarrow ^{98}Sn+\gamma$ the ratio of the new to the old reaction rates reaches 10^5 at the lowest boundary of the temperature interval. The difference in the new reaction Q-value (0.946 MeV) and the old one (1.651 MeV) for this reaction is 0.7 MeV. For the inverse reaction rates this effect looks even more pronounced, for example, for the same reaction the ratio new/old reaction rates exceeds 10^{25}. Even though for the majority of updated Hauser-Feshbach reactions this ratio is of the order of 10 or less, it demonstrates the importance of updates. The overall number of updated direct Hauser-Feshbach reaction rates is 2945 (1973 of p and α capture reactions and 972 (p,α) ones).

In another rather large update of the REACLIB library we recalculated with new masses 25 proton radiative capture rates on rather heavy nuclei in the framework of p-f shell model [11]. They substituted the reaction rates calculated in the Hauser-Feshbach statistical approach. Fig. 1 compares the previously used Hauser-Feshbach rate of the $p+^{45}Cr \rightarrow ^{46}Mn+\gamma$ reaction with a new shell model rate. The ratio of the new to the old reaction rate is of the order of 10 in 0.1-2 GK temperature range and increases steeply at lower temperatures. This difference comes both from the change in the reaction Q-value (the old Q-value is 247 keV and the new one is 836 keV) and from the detailed treatment of the structure of the compound ^{46}Mn nucleus in the shell model.

Beyond A=65 shell model calculations are not available anymore. Van Wormer at al. [12] calculated some reaction rates in the A=60-74 region using information from mirror nuclei. We also updated these calculations using new masses and level information. As an example we performed a new calculation of the $p+^{65}As \rightarrow ^{66}Se+\gamma$ reaction rate, based on estimates of the unknown level structure of ^{66}Se from the mirror nucleus ^{66}Ge, and compared the results previous estimates [12]. In our computation we used Q=2.432 MeV whereas the old value used in [12] was Q=1.909 MeV, as new experimental data on the ^{66}Ge mirror nucleus are now available. We took into account significantly more resonances - 27 (from mirror nucleus ^{66}Ge [13]) compared to [12] where only three resonances were accounted for. Note that the lowest resonance from [12] at

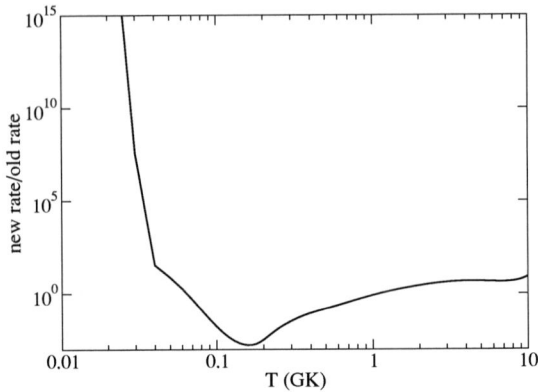

FIGURE 1. New/old rate ratio for p+^{45}Cr→^{46}Mn+γ.

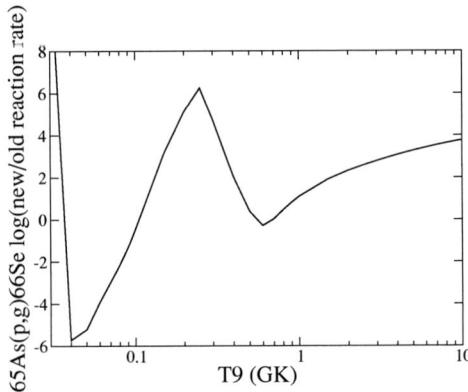

FIGURE 2. New/old rate ratio for p+^{65}As→^{66}Se+γ.

E=2.17 MeV has not been confirmed in more recent experiments and we therefore did not take it into account. Fig. 2 shows the comparison of reaction rates based on new and old data. The ratio new/old rate spans twelve orders of magnitudes in the temperature interval 0.02 - 10 GK. This rather simple calculation shows that any new experimental measurement of the nuclear structure or any new microscopic shell-modell theoretical calculation can, in principle, rather strongly change the reaction rate especially for the nuclei far from valley of stability where level densities are low.

We also have updated 44 proton capture and (p,α) reactions in REACLIB based on [5] compilation and 55 reactions based on the NACRE compilation [4]. In addition, we used recent experimental information whenever it was possible.

All reaction rates have been parametrized in the standard REACLIB format by fitting the original data. The accuracy of the fit was better then 5% for all reactions under consideration. An appropriate physical behaviour of reaction rates at low and high temperatures was established.

Another important feature of all our updated reaction rates are stellar enhancement factors (SEF). In the stellar environment the reactions proceed not only through the ground state but through an ensemble of thermally excited states which are in thermodynamic equilibrium with the ground state. The importance of SEFs was impressively revealed in [14] where for the reaction p+^{32}Cl→^{33}Ar+γ the value of stellar enhancement factor could reach 5 considerably exceeding the value calculated in Hauser-Feshbach approach. Nevertheless, for now we follow the standard way of calculating SEFs in the Hauser-Feshbach model. We will step by step replace these SEFs by the shell model ones where possible.

REACTION RATE DATABASE

In parallel with the REACLIB library updates we started to work on more general issue - the development of a reaction rate database. The our main goals in this project are

- To create database which contains both unpublished reaction rates and data from refereed literature, though each rate has to be finally "evaluated" by some group of experts.
- To implement a versioning system and to associate with each rate a discussion thread
- To allow a selection of any arbitrary set of rates and output it as a nuclear reaction rate library in formats used by broad astrophysical community.
- To allow both the analytical and the table form to represent rates
- And, finally, to create a continually updated web-based public archive, maintained by JINA in collaboration with other nuclear astrophysics data efforts.

At this preliminary stage our reaction rate database is used as a local tool at NSCL and is under extensive testing.

ACKNOWLEDGMENTS

We thank all authors of nuclear data which we used in the REACLIB update. This work was supported by NSF grant PHY 0216783 (Joint Institute for Nuclear Astrophysics).

REFERENCES

1. T. Rauscher, and F. Thieleman, *Atom. Data nucl. Tabl.*, *79*, 47 (2001).
2. D.W. Bardayan, J.C. Blackmon, C.R. Brune, *et al*, *Phys. Rev.*, *C 62*, 055804 (2000).
3. Neng-Chuan Shu, D.W. Bardayan, J.C. Blackmon, *et al*, *Chin. Phys. Lett.*, *20*, 1470 (2003)
4. C. Angulo, M. Arnould, M. Rayet, *et al*, *Nucl. Phys.*, *A 656*, 3 (1999).
5. C. Illadis, J.M. D'Auria, S. Starrfield, W.J. Thompson, M. Wiescher, *The Astrophysical Journal Supplement Series*, *134*, 151 (2001).
6. R.K. Wallace, S.E. Woosley, *Astrophys. J. Suppl. Ser.*, *45*, 389 (1981).
7. H. Schatz, *et al.*, *Phys. Rep.*, *294*, 167 (1998).
8. S.E. Woosley, *et al.*, *Astrophys. J. Suppl.*, *15*, 75 (2004).

9. G. Audi, A.H. Wapstra, C. Thibault, *Nucl. Phys.*, *A729*, 337 (2003).
10. B.A. Brown, R.R.C. Clement, H. Schatz, A. Volya, W.A. Richter, *Phys.Rev.*, *C65*, 045802 (2002).
11. J.L. Fisker, *et al.*, *Atom. Data nucl. Tabl.*, *79*, 241 (2001).
12. L. Van Wormer, J. Görres, C. Illadis, M. Wiescher, F.-K. Thielemann, *Astrophys. Journ.*, *432*, 326 (1994).
13. E.A. Stefanova, I. Stefanescu, G. de Angelis, *et al.*, *Phys. Rev.*, *C67*, 054319 (2003).
14. H. Schatz, C.A. Bertulani, B.A. Brown, *et al.*, to be published in *Phys. Rev. C*.

KADoNiS- The Karlsruhe Astrophysical Database of Nucleosynthesis in Stars

I. Dillmann[*,†], M. Heil[*], F. Käppeler[*], R. Plag[*], T. Rauscher[†] and F.-K. Thielemann[†]

[*]*Institut für Kernphysik, Forschungszentrum Karlsruhe, Postfach 3640, D-76021 Karlsruhe, Germany*
[†]*Departement Physik und Astronomie, Universität Basel, Klingelbergstrasse 82, CH-4056 Basel, Switzerland*

Abstract. The "Karlsruhe Astrophysical Database of Nucleosynthesis in Stars" (KADoNiS) project is an online database for experimental cross sections relevant to the s process and p process. It is available under http://nuclear-astrophysics.fzk.de/kadonis and consists of two parts. Part 1 is an updated sequel to the previous Bao et al. compilations from 1987 and 2000 for (n,γ) cross sections relevant to the big bang and s-process nucleosynthesis. The second part will be an experimental p-process database, which is expected to be launched in winter 2005/06. The KADoNiS project started in April 2005, and a first partial update is online since August 2005. In this paper we present a short overview of the first update of the s-process database, as well as an overview of the status of stellar (n,γ) cross sections of all 32 p isotopes.

Keywords: stellar neutron cross sections, database, compilation, s process, p process
PACS: 25.40.Lw, 26.30.+k, 27.30.+t, 27.40.+z, 27.50.+e, 27.60.+j, 27.70.+q, 97.10.Cv

STELLAR NEUTRON CAPTURE COMPILATIONS

The first collection of stellar neutron capture cross sections was published in 1971 by Allen and co-workers [1]. This paper reviewed the role of neutron capture reactions in the nucleosynthesis of heavy elements and presented also of a list of recommended (experimental or semi-empirical) Maxwellian averaged cross sections at $kT = 30$ keV (MACS30) for nuclei between carbon and plutonium.

The idea of an experimental and theoretical stellar neutron cross section database was picked up again by Bao and Käppeler [2] for s-process studies. This compilation published in 1987 included cross sections for (n,γ) reactions (between ^{12}C and ^{209}Bi), some (n,p) and (n,α) reactions (for ^{33}Se to ^{59}Ni), and also (n,γ) and (n,f) reactions for long-lived actinides. A follow-up compilation was published by Beer, Voss and Winters in 1992 [3].

In the update of 2000 this compilation [4] was extended to big bang nucleosynthesis. It now included a collection of recommended MACS30 for isotopes between ^{1}H and ^{209}Bi, and – like the original Allen paper – also semi-empirical recommended values for nuclides without experimental cross section information. These estimated values are normalized cross sections derived with the Hauser-Feshbach code NON-SMOKER [5], which account for known systematic deficiencies in the nuclear input of the calculation. Additionally, the database provided stellar enhancement factors and energy-dependent MACS for energies between $kT = 5$ keV and 100 keV.

The most recent KADoNiS version of this compilation has the aim to provide a clearly arranged and user-friendly online database, which is regularly updated and will be in later stages also extended to p-process studies.

PART 1: UPDATED BIG BANG AND S-PROCESS DATABASE

Included in the present update (status August 2005) were only cross sections, which are already published. Six semi-empirical estimates (see Table 1) were replaced by experimental data, and 20 recommended cross sections were updated by inclusion of new measurements (Table 2). A full list of measurements with references, which were (will be) included in the update(s) can be found on the KADoNiS homepage in the menu section "Logbook".

Future efforts will be focussed on the re-evaluation of semi-empirical cross sections, as well as the inclusion of theoretical results derived with the Hauser-Feshbach code MOST [6]. Another topic will be the re-calculation of cross sections for isotopes, where a recent change in physical properties (e.g. $t_{1/2}$, I_γ...) leads to changes in already measured cross sections.

TABLE 1. List of recommended semi-empirical stellar cross sections, which were now replaced by experimental values.

Isotope	Old recomm. value [mb]	New exp. value [mb]
^{128}Xe	248 ± 66	262.5 ± 3.7
^{129}Xe	472 ± 71	617 ± 12
^{130}Xe	141 ± 51	132.0 ± 2.1
^{147}Pm	1290 ± 470	709 ± 100
^{151}Sm	2710 ± 420	3031 ± 68
^{180}Ta^m	1640 ± 260	1465 ± 100

The KADoNiS homepage provides a datasheet with all necessary informations for each isotope similar to the layout in Ref. [4]. On the top of this page the recommended MACS30 for the total and all partial cross sections are shown. In the "Comment" line one can find the previous recommended values, special comments, and the date of the last review. The field "List of all available values" includes the original values as given in the respective publications, renormalized values, year of publication, type of value (theoretical, semi-empirical or experimental), a short comment about the method (accelerator, neutron and reference source), and the (linked) reference(s).

This section is followed by the tabulated MACS, reaction rates and stellar enhancement factors for energies between kT= 5 and 100 keV. A "click" on the field "Show/hide mass chain" gives a graphical plot of all available recommended total MACS30 for the isotopic mass chain of the respective element. The bottom part of each datasheet shows a chart of nuclides, which can be zoomed by selecting different sizes (S, M, L, or XL). By clicking on an isotope in this chart, one can easily jump to the respective datasheet.

TABLE 2. List of previous and new recommended stellar cross sections, which were updated by inclusion of new experimental values.

Isotope	Old recomm. value [mb]	New recomm. value [mb]
^{22}Ne	0.059 ± 0.006	0.058 ± 0.004
^{40}Ar	2.6 ± 0.2	2.6 ± 0.2
^{96}Ru	238 ± 60	207 ± 8
^{102}Ru	186 ± 11	151 ± 7
^{104}Ru	161 ± 10	156 ± 5
^{110}Cd	246 ± 10	237 ± 2
^{111}Cd	1063 ± 125	754 ± 12
^{112}Cd	235 ± 30	187.9 ± 1.7
^{113}Cd	728 ± 80	667 ± 11
^{114}Cd	127 ± 5	129.2 ± 1.3
^{116}Cd	59 ± 2	74.8 ± 0.9
^{135}Cs	198 ± 17	160 ± 10
^{139}La	38.4 ± 2.7	31.6 ± 0.8
^{175}Lu	1146 ± 44	1219 ± 10
^{176}Lu	1532 ± 69	1639 ± 14
^{176}Hf	455 ± 20	626 ± 11
^{177}Hf	1500 ± 100	1544 ± 12
^{178}Hf	314 ± 10	319 ± 3
^{179}Hf	956 ± 50	922 ± 8
^{180}Hf	179 ± 5	157 ± 2

PART 2: EXPERIMENTAL P-PROCESS DATABASE

The second part of KADoNiS will be an experimental *p*-process database and is expected to be launched in winter 2005/06. It will be a collection of the available experimental reaction rates relevant for *p*-process studies, e.g. (γ,n), (γ,α), (γ,p), (n,p), (n,α), (p,α), and their inverse rates.

The role of (n,γ) reactions in the *p* process was early recognized by Rayet et al. [7]. The (n,γ)↔(γ,n) competition hinders the photodisintegration flux towards lighter nuclei. Additionally the decrease in temperature at later stages of the *p* process leads to a freeze-out ($T_9 \simeq$ 0.3, corresponding to kT= 25 keV) via neutron captures and mainly β^+ decays, resulting in the typical *p*-process abundance pattern with maxima at ^{92}Mo (N=50) and ^{144}Sm (N=82). The influence of a variation of reaction rates on the final *p* abundances has been demonstrated repeatedly [8, 9].

Thus, it is necessary for *p*-process studies to know the neutron capture rates for both, at freeze-out energies (kT= 25 keV) and at the *p*-process energies (kT= 170-260 keV). Table 3 gives an overview of the status of neutron capture cross sections of all 32 *p* nuclei at kT= 30 keV. The Bao et al. compilation from 2000 [4] provided measured cross sections for 20 isotopes, but 9 of them (^{92,94}Mo, ^{96}Ru, ^{124,126}Xe, ^{130}Ba, ^{156}Dy, ^{180}W, and ^{190}Pt) with uncertainties ≥9%. For the remaining 12 *p* isotopes (^{74}Se, ^{84}Sr, ^{98}Ru, ^{102}Pd, ^{120}Te, ^{132}Ba, ^{138}La, ^{158}Dy, ^{168}Yb, ^{174}Hf, ^{184}Os, and ^{196}Hg) only theoretical predictions were available.

The (preliminary) results of our extended measuring program of stellar neutron capture cross sections for p nuclei are shown in Table 3. All of our measurements were carried out on natural samples at the Karlsruhe 3.7 MV Van de Graaff accelerator using the activation technique [10, 11]. Neutrons were produced via the ^{7}Li$(p,n)^7$Be reaction by bombarding 30 μm thick layers of metallic lithium on a water-cooled copper backing with protons of E_p= 1912 keV. The resulting quasi-stellar neutron spectrum approximates a Maxwellian distribution for kT= 25.0 $\pm$ 0.5 keV [12]. In all eight cases (^{74}Se, ^{84}Sr, ^{96}Ru, ^{102}Pd, ^{120}Te, ^{130,132}Ba, and ^{174}Hf) we are able to reproduce the previous recommended total cross sections from [4] within 20%. Thus, only 6 p isotopes (^{98}Ru, ^{138}La, ^{158}Dy, ^{168}Yb, ^{184}Os, and ^{196}Hg) remain without any experimental stellar neutron cross section. With exception of ^{98}Ru and ^{138}La, all isotopes can be measured with the activation technique.

However, for an inclusion into the planned p-process database, those MACS30 have to be theoretically extrapolated to p-process temperatures. Another step is then the calculation of inverse reaction rates by detailed balance.

ACKNOWLEDGMENTS

We thank E. P. Knaetsch, D. Roller and W. Seith for their help during the irradiations at the Van de Graaff accelerator. This work was supported by the Swiss National Science Foundation Grants 2024-067428.01 and 2000-105328.

REFERENCES

1. B. Allen, J. Gibbons, and R. Macklin, *Adv. Nucl. Phys.*, **4**, 205 (1971).
2. Z. Bao, and Käppeler, *ADNDT*, **36**, 411 (1987).
3. H. Beer, F. Voss, and R. Winters, *Ap. J. Suppl.*, **80**, 403 (1992).
4. Z. Bao, H. Beer, F. Käppeler, F. Voss, K. Wisshak, and T. Rauscher, *ADNDT*, **76**, 70 (2000).
5. T. Rauscher, F.-K. Thielemann, and H. Oberhummer, *Ap. J.*, **451**, L37 (1995).
6. S. Goriely, *Hauser-Feshbach rates for neutron capture reactions (version 08/26/05)*, http://www-astro.ulb.ac.be/Html/hfr.html (2005).
7. M. Rayet, N. Prantzos, and M. Arnould, *Astron. Astrophys.*, **227**, 271 (1990).
8. T. Rauscher, *Nucl. Phys. A*, **758**, 549c (2005).
9. W. Rapp, *Report FZKA 6956, Forschungszentrum Karlsruhe* (2004).
10. W. Rapp, M. Heil, D. Hentschel, F. Käppeler, R. Reifarth, H. Brede, H. Klein, and T. Rauscher, *Phys. Rev. C*, **66**, 015803 (2002).
11. I. Dillmann, M. Heil, F. Käppeler, T. Rauscher, and F.-K. Thielemann, *subm. to Phys. Rev. C* (2005).
12. W. Ratynski, and F. Käppeler, *Phys. Rev. C*, **37**, 595 (1988).
13. T. Rauscher, and F.-K. Thielemann, *ADNDT*, **79**, 47 (2001).
14. I. Dillmann, M. Heil, F. Käppeler, R. Plag, T. Rauscher, and F.-K. Thielemann, *Proceedings Nuclear Physics in Astrophysics II, Debrecen/Hungary, May 16-20, 2005, subm. to Eur. Phys. J. A* (2005).

TABLE 3. Status of MACS30 of all 32 p nuclei. Recommended cross section were taken from Ref. [4], unless another reference is given.

Isotope	Hauser-Feshbach predictions		Recommended MACS30		Comments/
	MOST [6] [mb]	NON-SMOKER [13] [mb]	previous [4] [mb]	new [mb]	Refs.
^{74}Se	247	207	$267 \pm 25^*$	276 ± 15	[11]
^{78}Kr	388	351	312 ± 26		
^{78}Kr$\rightarrow^m$	-	-	92.3 ± 6.2		
^{84}Sr	246	393	$368 \pm 125^*$	302 ± 17	[11]
^{84}Sr$\rightarrow^m$	-	-	-	190 ± 10	[11]
^{92}Mo	46	128	70 ± 10		
^{94}Mo	85	151	102 ± 20		
^{96}Ru	338	281	238 ± 60	207 ± 8	[10]
^{98}Ru	358	262	$173 \pm 36^*$		
^{102}Pd	670	374	$373 \pm 118^*$	379 ± 16	preliminary
^{106}Cd	365	451	302 ± 24		
^{108}Cd	206	373	202 ± 9		
^{113}In	316	1202	787 ± 70		
^{113}In$\rightarrow^m$	-	-	480 ± 160		
^{112}Sn	154	381	210 ± 12		
^{114}Sn	74	270	134.4 ± 1.8		
^{115}Sn	247	528	342.4 ± 8.7		
^{120}Te	309	551	$420 \pm 103^*$	451 ± 18	[14] prelim.
^{120}Te$\rightarrow^m$	-	-	-	61 ± 2	[14] prelim.
^{124}Xe	503	799	644 ± 83		
^{124}Xe$\rightarrow^m$	-	-	131 ± 17		
^{126}Xe	335	534	359 ± 51		
^{126}Xe$\rightarrow^m$	-	-	40 ± 6		
^{130}Ba	493	730	760 ± 110	694 ± 20	[14] prelim.
^{132}Ba	228	467	$379 \pm 137^*$	368 ± 25	[14] prelim.
^{132}Ba$\rightarrow^m$	-	-	-	33.6 ± 1.7	[14] prelim.
^{136}Ce	208	495	328 ± 21		
^{136}Ce$\rightarrow^m$	-	-	28.2 ± 1.6		
^{138}Ce	61	290	179 ± 5		
^{138}La	337	767	$-^\dagger$		
^{144}Sm	39	209	92 ± 6		
^{156}Dy	2138	1190	1567 ± 145		
^{158}Dy	1334	949	$1060 \pm 400^*$		
^{162}Er	1620	1042	1624 ± 124		
^{168}Yb	875	886	$1160 \pm 400^*$		
^{174}Hf	763	786	$956 \pm 283^*$	1056 ± 53	preliminary
^{180}W	751	707	536 ± 60		
^{184}Os	709	789	$657 \pm 202^*$		
^{190}Pt	634	760	677 ± 82		
^{196}Hg	469	372	$650 \pm 82^*$		

* Semi-empirical estimate.
† No recommended value available, since ^{138}La is of pure p-process origin.

Nuclear Data Resources for Capture γ-Ray Spectroscopy and Related Topics

B. Pritychenko*, M.W. Herman*, P. Obložinský*, A.A. Sonzogni* and V. Zerkin[†]

*National Nuclear Data Center, Brookhaven National Laboratory, Upton, NY 11973-5000, U.S.A.
[†]Nuclear Data Section, International Atomic Energy Agency P.O. Box 100, A-1400, Vienna, Austria

Abstract. We discuss nuclear data resources of the National Nuclear Data Center (NNDC) of relevance to nuclear structure, reactions, astrophysics as well as applied technology applications. These resources include databases, tools, publications and powerful Web service at http://www.nndc.bnl.gov.

Keywords: Nuclear Data
PACS: 29.85.1c

INTRODUCTION

Nuclear data are crucial for both basic science and applied nuclear technology communities, including nuclear structure physics, astrophysics, reactors, accelerator design, nonproliferation and safeguards, radiation protection and medicine as well as homeland security applications.

National Nuclear Data Center plays a pivotal role in the national nuclear data effort, coordinating the United States Nuclear Data Program (USNDP) and the Cross Section Evaluation Working Group (CSEWG). The NNDC is the custodian of a complex system of nuclear physics databases that represents a national nuclear data resource and a national treasure.

MAJOR NUCLEAR DATA RESOURCES

NNDC bibliographical, nuclear reaction and structure data resources were substantially improved in recent years [1, 2] to provide a better access to nuclear data evaluations and compilations. A brief description of nuclear data resources is presented below.

Nuclear Science References (NSR)

Nuclear physics research and applied technologies developments require an instant access and search of nuclear physics publications. NNDC actively contributes into development and maintenance of bibliographical nuclear data resources by keeping Nuclear Science References (NSR) database [3] content up-to-date.

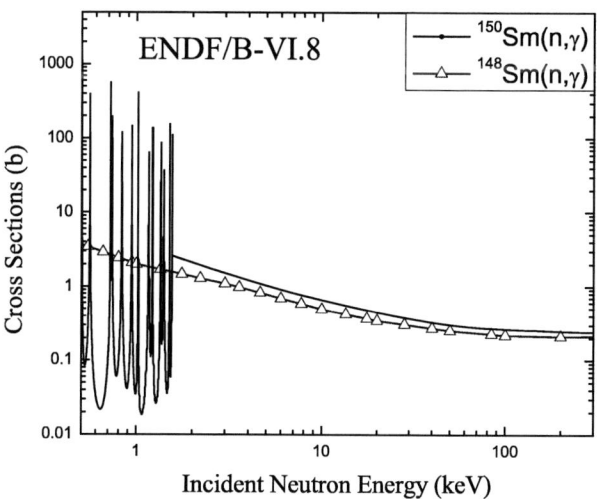

FIGURE 1. Neutron-capture cross sections from 0.5 to 300 keV are shown for astrophysically important ^{148,150}Sm nuclei. Resonance region in ^{150}Sm is clearly visible, whereas resonance data for ^{148}Sm [5] are not shown.

Nuclear Science References database (http://www.nndc.bnl.gov/nsr) is a corner stone for nuclear physics research and application developments. It covers more than 80 journals and offers some 180,000 references, providing an extremely effective way of searching scientific literature on a broad range of nuclear physics topics. In 2004, NSR usability was improved by addition of 35,000 new digital object identifier (doi) links that allow users to access actual publications directly from the Web.

Evaluated Nuclear (reaction) Data File (ENDF)

New reaction Web interfaces, developed in collaborative effort with IAEA's Nuclear Data Section [4], provide wide range of options for data retrievals and analysis using standard and interpreted formats as well as graphic tools.

Evaluated Nuclear (reaction) Data File (http://www.nndc.bnl.gov/endf) contains neutron cross sections for 340 isotopes relevant to nuclear technology, including capture in the keV region of interest to astrophysics. For example information on the strength of stellar neutron flux can be deduced from a comparison of the $<\sigma>N$ - rate values of the s-only process isotopes ^{148,150}Sm. ENDF/B-VI.8 evaluated (recommended) (n,γ) cross sections for these isotopes are shown in Figure 1.

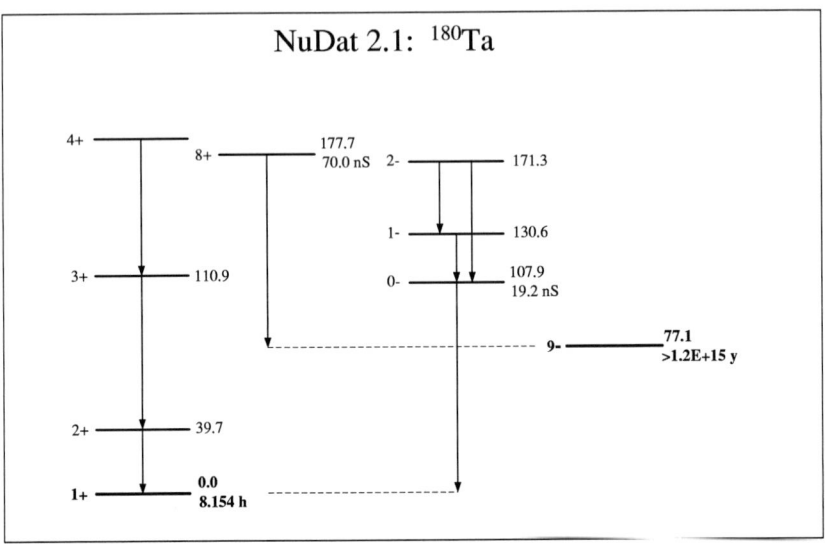

FIGURE 2. Low-energy part of level scheme for ^{180}Ta was created with NuDat 2.1 nuclear database software. Long-lived 9^- isomer of ^{180m}Ta is shown on the right.

A new version of the library, ENDF/B-VII, is under development, tentatively scheduled for release by the end of 2005.

Evaluated Nuclear Structure Data File (ENSDF)

Evaluated Nuclear Structure Data File (http://www.nndc.bnl.gov/ensdf) represents enormous resource for nuclear structure and γ-ray spectroscopy [6]. It provides information on almost all-known nuclei (2,929), including some 140,102 levels and 204,017 γ-rays. Related database NuDat 2.1 [7] and highly popular Web service (http://www.nndc.bnl.gov/nudat2) offers extensive options for simple and complex retrievals.

ENSDF data are actively used by the nuclear physics community to study nuclear isomers and their properties, including energy storage, nuclear astrophysics and structure applications. Figure 2 shows level scheme of ^{180}Ta isotope and 9^- isomer of ^{180m}Ta produced by NuDat 2.1 software. Astrophysically important [8], long-lived isomer of ^{180m}Ta with $T_{1/2} \geq 1.2 \times 10^{15}$ y is responsible for the existence of the Nature's heaviest odd neutron and proton number radioactive isotope of ^{180}Ta with $T_{1/2} = 8.154$ h.

ADDITIONAL NUCLEAR DATA RESOURCES

In addition to major nuclear data resources such as NSR, ENDF and ENSDF, the National Nuclear Data Center offers a wide variety of products that can be used to perform specific capture γ-ray spectroscopy tasks.

Experimental Nuclear Reaction Data (EXFOR/CSISRS)

Experimental nuclear reaction database (http://www.nndc.bnl.gov/exfor), covering more than 15,500 experiments (105,000 datasets) and being comprehensive resource for neutrons, contains extensive coverage for charged particles and partial coverage for photons. Direct links to journals and tabulated format (T4) that contains reference and tabulated data significantly simplify database access, making it user friendly for less experienced users.

Thermal Neutron Capture γ-Rays (CapGam)

Thermal neutron capture γ-ray database (CapGam, http://www.nndc.bnl.gov/capgam) is providing energies and intensities of γ-rays. The data for this resource are extracted from ENSDF and consists of target and γ-ray ordered tables for 256 target nuclides up to about 12 MeV in energy.

Nuclear Reaction Model Code (EMPIRE)

EMPIRE [9] is a powerful code for nuclear reaction calculations (http://www.nndc.bnl.gov/empire). It is equipped with input parameters libraries, suitable for stable and unstable targets and can be readily used for calculations of capture cross sections to be converted into reaction rates.

NNDC Publications

Other resources of interest include 7th edition of Nuclear Wallet Cards booklet [10], Nuclear Data Sheets journal and Atlas of Neutron Resonances handbook [5] to be published by Elsevier early 2006.

The latest edition of Nuclear Wallet Cards (http://www.nndc.bnl.gov/wallet) contains up-to-date information on properties of 5657 ground and isomeric states of all known nuclei.

Nuclear Data Sheets is a journal prepared for publication by the NNDC and devoted for evaluated nuclear structure and decay data.

Atlas of Neutron Resonances will represent one of the most comprehensive low-energy neutron data resources in this century. It contains evaluated neutron data for Z=1-100 (473 isotopes and 353 resolved resonance regions) and covers 30 keV cross sections for 173 out of 180 s-process nuclei. The 30 keV (n,γ) cross sections for s-process nuclei are shown in Figure 3. Neutron-capture cross section values clearly plunge near the neutron closed shells and fluctuate for odd and even nuclei.

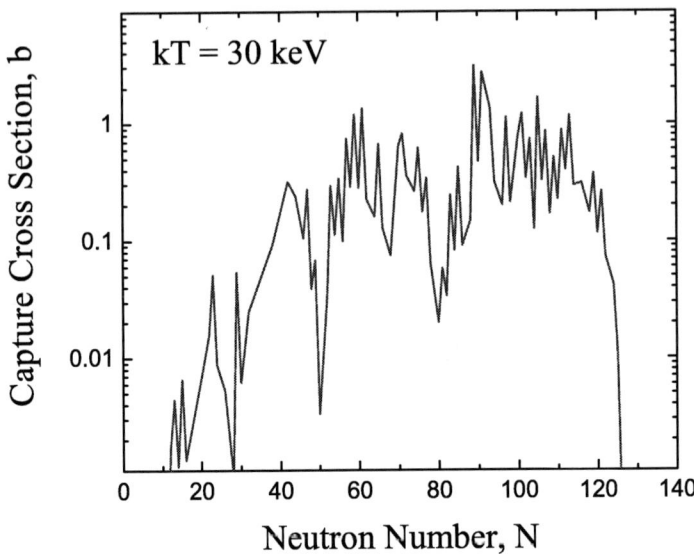

FIGURE 3. Evaluated neutron-capture cross sections at 30 keV are shown as a function of neutron number N of s-process nuclei. Cross section values plunge near the neutron closed shells at N = 20,28,50,82, and 126. Data are taken from new Atlas of Neutron Resonances [5].

ACKNOWLEDGMENTS

This work is supported by the Office of Nuclear Physics, Office of Science of the U.S. Department of Energy, under contract no. DE-AC02-98CH10886 with Brookhaven Science Associates, LLC.

REFERENCES

1. National Nuclear Data Center, *http://www.nndc.bnl.gov* (accessed August 26, 2005).
2. B. Pritychenko *et al*, Proc. Int. Conf. Nuclear Data for Sci. Tech., Santa Fe 2004, AIP **769** (2005) 132.
3. D.F. Winchell, Proc. Int. Conf. Nuclear Data for Sci. Tech., Santa Fe 2004, AIP **769** (2005) 570.
4. V.V. Zerkin *et al*, Proc. Int. Conf. Nuclear Data for Sci. Tech., Santa Fe 2004, AIP **769** (2005) 586.
5. S.F. Mughabghab, *Atlas of Neutron Resonances*, to be published by Elsevier in 2006.
6. D.F. Winchell, Proc. Int. Conf. Nuclear Data for Sci. Tech., Santa Fe 2004, AIP **769** (2005) 578.
7. A.A. Sonzogni, Proc. Int. Conf. Nuclear Data for Sci. Tech., Santa Fe 2004, AIP **769** (2005) 574.
8. D. Belic *et al*, Phys. Rev. **C 65**, 035801 (2002).
9. M. Herman *et al*, Proc. Int. Conf. Nuclear Data for Sci. Tech., Santa Fe 2004, AIP **769** (2005) 1184.
10. J.K. Tuli, *Nuclear Wallet Cards*, 7th edition, April 2005.

Measurements of Neutron Capture Cross Sections for Gd Isotopes in the Energy Region from 10 keV to 90 keV

G. N. Kim[+], W. C. Chung[**], T. I. Ro[*], T. Ohsaki[*], and M. Igashira[*]

[+]*School of Physics and Energy Science, Kyungpook National University,*
1370 Sankyok-dong, Buk-gu, Daegu 702-701, Korea
[**]*Department of Physics, Dong-A University,*
840 Hadan 2-dong, Saha-gu, Busan 604-714, Korea
[*]*Research Laboratory for Nuclear Reactors, Tokyo Institute of Technology,*
2-12-1-N1-26, O-okayama, Meguro-ku, Tokyo 152-8550, Japan

Abstract. The neutron capture cross sections of Gd isotopes (^{155}Gd, ^{156}Gd, ^{157}Gd, and ^{158}Gd) have been measured in the neutron energy range from 10 to 90 keV using the 3-MV Pelletron accelerator of the Research Laboratory for Nuclear Reactors at the Tokyo Institute of Technology. Pulsed keV neutrons were produced from the ^{7}Li(p,n)^{7}Be reaction by bombarding the lithium target with the 1.5-ns bunched proton beam from the Pelletron accelerator. The incident neutron spectrum on a capture sample was measured by means of a TOF method with a ^{6}Li-glass detector. Capture γ-rays were detected with a large anti-Compton NaI(Tl) spectrometer, employing a TOF method. A pulse-height weighting technique was applied to observed capture γ-ray pulse-height spectra to derive capture yields. The capture cross sections were obtained by using the standard capture cross sections of ^{197}Au. The present results were compared with the previous measurements and the evaluated values of ENDF/B-VI.

Keywords: neutron capture, cross section, Gd isotopes, anti-Compton NaI(Tl) spectrometer, time-of-flight method, pulse-height weighting technique.
PACS: 25.40.Lw, 27.60.+j, 29.30.Kv

INTRODUCTION

Capture cross-sections for gadolinium (Gd) isotopes in the keV-neutron energy region are important in the design of reactors as well as in the studies of the nuclear physics and astrophysics. The published experimental data are poor in both quality and quantity. One of the reasons is the difficulty of preparing pure enriched isotopes enough to perform keV-neutron capture cross-section measurement. An anti-Compton NaI(Tl) spectrometer [1] developed by the Research Laboratory for Nuclear Reactors at the Tokyo Institute of Technology (Titech) made this measurement possible with a small amount of sample.

We have measured the capture cross sections of ^{155}Gd, ^{156}Gd, ^{157}Gd, and ^{158}Gd samples in the incident neutron energy range from 10 to 90 keV with the high efficient capture γ-ray spectrometer. Pulsed keV neutrons were produced from the ^{7}Li(p,n)^{7}Be reaction by bombarding a lithium target with the 1.5-ns bunched proton beam from the Pelletron accelerator. This is the first experiment to use metal foils instead of the gadolinium oxide (Gd_2O_3). Since gadolinium oxide powder is generally hygroscopic, the effect of the water on the sample in the previous measurements was taken into account in the analysis of experimental data. We have compared the present results with the previous ones and the evaluated values of ENDF/B-VI.

EXPERIMENTAL PROCEDURE

The neutron capture cross sections of the enriched Gd isotopes have been measured in the incident neutron energy range from 10 to 90 keV using the 3-MV Pelletron accelerator of the Research Laboratory for Nuclear Reactors at the Tokyo Institute of Technology. Pulsed keV-neutrons were produced via $^7Li(p,n)^7Be$ reaction by bombarding metallic lithium target with the 1.5 ns bunched proton beam from the accelerator. The pulse repetition rate was 4 MHz and the average proton beam current was about 9 μA. The experimental arrangement, which is similar to the previous one [1, 2], is shown in Figure 1.

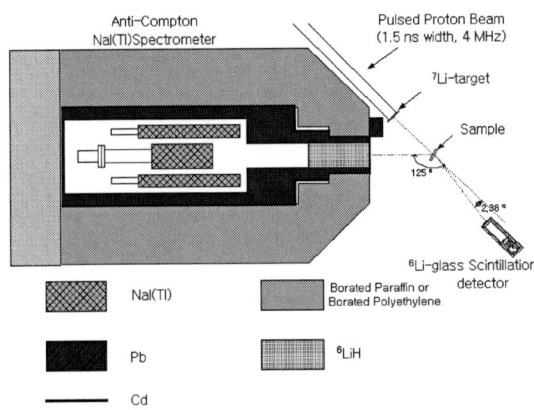

FIGURE 1. The experimental arrangement for cross section measurement

The Gd samples are enriched metallic plates with 0.1 and 0.2 mm in thickness. Two gold plates with 20 and 15 mm in diameter and 1.0 mm in thickness have been used as a standard. The characteristics of the samples are summarized in Table 1. The capture sample was aligned at zero degree with respect to the proton beam direction. The distance between the lithium target and the sample was 12 cm.

The incident neutron spectrum on the capture sample was measured by means of the TOF method with a ^{6}Li-glass scintillation detector. The γ-rays emitted from the sample were detected by a large anti-Compton NaI(Tl) spectrometer located at 86.0 cm from the sample with an angle of 125 degrees with respect to the proton beam direction.

TABLE 1. Characteristics of Samples

Sample	^{155}Gd	^{156}Gd	^{157}Gd	^{158}Gd	^{197}Au
Weight (g)	0.203	0.198	0.357	0.353	6.022
Chemical Purity (%)					99.99
Isotopic Composition (%)					
^{152}Gd	0.04	<0.01	<0.02	<0.1	
^{154}Gd	0.64	0.11	0.16	<0.1	
^{155}Gd	91.74	1.96	0.81	0.96	
^{156}Gd	5.11	93.79	2.21	1.7	
^{157}Gd	1.12	2.53	90.96	3.56	
^{158}Gd	0.94	1.20	5.08	92.0	
^{160}Gd	0.41	0.41	0.80	1.82	
Thickness (mm)	0.101	0.099	0.192	0.195	1.0
Area (cm^2)	2.55	2.55	2.35	2.29	3.14

The runs with and without sample (sample run and blank run) and the run with standard gold sample (gold run) were made cyclically in each measurement to average changes in experimental conditions such as the incident neutron spectrum. Since the measurements were carried out cyclically, systematic changes in experimental conditions could be corrected. The blank run was performed in order to determine the background, and also to monitor any changes in the incident neutron energy and the thickness of the ^{7}Li neutron-target. The three runs were normalized with the neutron counts of the ^{6}Li-glass detector. The total measuring times were about 31, 76, 32, and 67 hours for ^{155}Gd, ^{156}Gd, ^{157}Gd, and ^{158}Gd samples, respectively.

DATA ANALYSIS AND RESULTS

The incident neutron energy spectra on the sample were obtained from the ^{6}Li-glass detector by the neutron TOF method. The cut-off energy of neutrons was set to 10 keV, because the signal-to-noise ratios of the corresponding TOF spectrum were not so good.

A pulse-height weighting technique [3] with the weighting function $W(I)$ of the γ-ray spectrometer [1] was applied to each of the net capture γ-ray pulse-height spectra $S(I)$ for each channel I and the capture yields Y, i.e., the number of capture events, of the Gd isotopes and ^{197}Au samples were obtained by

$$Y_k = \frac{\sum_I W(I)S_k(I)}{B_n + \langle E_n \rangle_k} \equiv C_k N \phi_k \langle \sigma \rangle_k , \qquad (1)$$

where B_n is the neutron binding energy of target nucleus and $<E_n>_k$ the average incident neutron energy for the k-th digital window (DW) in the center of mass system. N is the number of target nuclei per cm^2, ϕ_k the number of incident neutrons in the k-th DW, and C_k the correction factor. The average capture cross-section, $<\sigma>_k$, of each sample for the k-th DW was defined as follows:

$$\langle \sigma \rangle_k = \int_{E_{k,\min}}^{E_{k,\max}} \sigma(E_n)\eta(E_n)dE_n, \quad (2)$$

where $\sigma(E_n)$ is the capture cross section, and $E_{k,min}$ and $E_{k,max}$ are low- and high-energy boundaries of the k-th DW, respectively. The normalized neutron energy spectra $\eta(E_n)$ equal to the net neutron spectra $N(E_n)$ divided by the neutron detection efficiency $\varepsilon(E_n)$. The number of incident neutrons for k-th DW of the Au run was derived by using equations (1) and (2) and the evaluated values of ^{197}Au in ENDF/B-VI [4]. Then, the number of incident neutrons for the corresponding DW of each sample was obtained by using the neutron monitor counts of the ^{6}Li-glass detector for the sample and Au runs. Finally, the average capture cross section of the sample for k-th DW was derived from the corresponding capture yield of the sample and the number of incident neutrons.

$$\langle \sigma \rangle_{k,Gd} = \frac{M_{Au}}{M_{Gd}} \cdot \frac{N_{Au}}{N_{Gd}} \cdot \frac{C_{k,Au}}{C_{k,Gd}} \cdot \frac{B_{n,Au}+\langle E_n \rangle_{k,Au}}{B_{n,Gd}+\langle E_n \rangle_{k,Gd}} \cdot \frac{\sum_I W(I) S_{k,Gd}(I)}{\sum_I W(I) S_{k,Au}(I)} \cdot \langle \sigma \rangle_{k,Au}. \quad (3)$$

The M_{Gd} and the M_{Au} are the neutron monitor counts for Gd and Au runs corrected for the attenuation by the Gd and Au sample runs, respectively. Correction factors were evaluated considering different effects coming from the neutron self-shielding and the multiple-scattering, and the γ-ray scattering and absorption in the sample, from the dependence of the γ-ray detection efficiency on the γ-ray source position, from the impurities in the sample, and from the dead time.

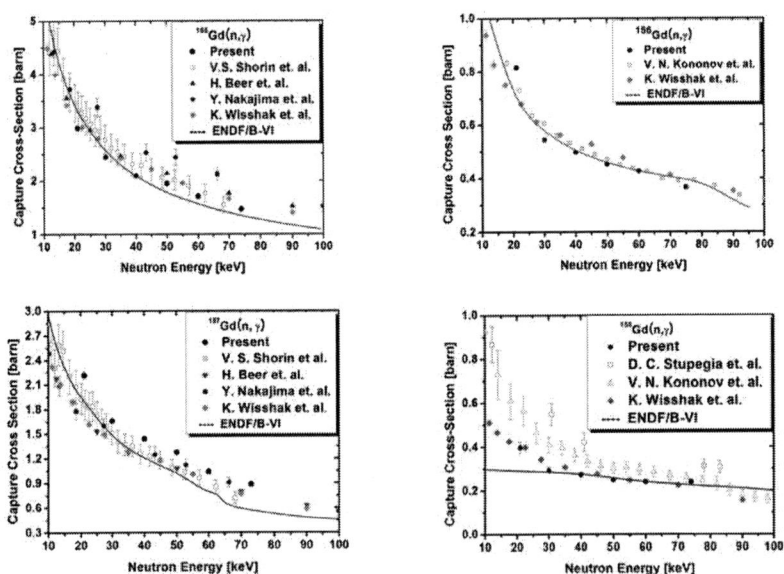

FIGURE 2. Neutron capture cross sections of ^{155,156,157,158}Gd

The present results of Gd isotopes are compared in Fig. 2 with previous experimental data [5-11] and the evaluated values in ENDF/B-VI [12]. In addition to the statistical error (about 1-10 %), the following errors were taken into account for those of capture cross sections of Gd isotopes: the errors due to the number of target nuclei (<1%), the standard capture cross sections of ^{197}Au (3 %), the weighting function of the γ-ray spectrometer (1 %), the extrapolation of net capture γ-ray pulse-height spectrum below the discrimination level (0.6 MeV) in deriving the capture yield with the pulse-height weighting technique (2 %), and the estimation of correction factor (1-2 %).

ACKNOWLEDGMENTS

The authors would like to express their sincere thanks to the staff of the Research Laboratory for Nuclear Reactors, Tokyo Institute of Technology (Titech) for the excellent accelerator operation. This research is partly supported by the research fund from the Korea Atomic Energy Research Institute (KAERI) and by the short term dispatch of overseas-research fund from the Dong-A University (2003). One of authors (G. N. Kim) is partly supported by the Center for High Energy Physics of Kyungpook National University in Daegu, Korea.

REFERENCES

1. S. Mizuno, M. Igashira, and K. Masuda, *J. Nucl. Sci. Technol.*, **36**, 493-507 (1999).
2. G. N. Kim *et al.*, *Annals of Nuclear Energy*, **28**, 1549-1562 (2001).
3. R. L. Macklin and J. H. Gibbons, *Phys. Rev.*, **159**, 1007-1012 (1967).
4. ENDF/B-VI data file for ^{197}Au (MAT=7925), evaluated by P. G. Young (1991).
5. K. Wisshak *et al.*, *Phys. Rev.* **C 52**, 2762-2779 (1995).
6. H. Beer and R. L. Macklin, *Astrophus. J.* **331**, 1047-1057 (1988).
7. Y. Nakajima *et al.*, *Annals of Nuclear Energy*, **16**, 589-597 (1989).
9. V. S. Shorin, V. N. Kononov, and E. D. Poletaev, *Yaderno-Fizicheskie Issledovaniya*, Reports No. 17, 9 (1974).
10. V. N. Kononov *et al.*, *Yadernye Konstanty*, **22**, 29 (1977).
11. D. C. Stupegia *et al.*, *J. Nuclear Energy*, **22**, 267 (1968).
12. ENDF/B-VI data file for ^{155}Gd (MAT=6434), ^{156}Gd(MAT=6437), ^{157}Gd(MAT=6440), and ^{158}Gd(MAT=6443), evaluated by B. A. Magurno et al. (1990).

The Evaluated Gamma-ray Activation File (EGAF)

R B. Firestone[*], G.L. Molnár[†], Zs. Révay[†], T. Belgya[†], D.P. McNabb[**] and B.W. Sleaford[**]

[*] *Lawrence Berkeley National Laboratory, Berkeley, CA 94720*
[†] *Institute of Isotope and Surface Chemistry, H-1525, Budapest, Hungary*
[**] *Lawrence Livermore National Laboratory, Livermore, California 94551*

Abstract. The Evaluated Gamma-ray Activation File (EGAF), a new database of prompt and delayed neutron capture γ ray cross sections, has been prepared as part of an International Atomic Energy Agency (IAEA) Coordinated Research Project to develop a "Database of Prompt Gamma-rays from Slow Neutron Capture for Elemental Analysis". Recent elemental γ-ray cross-section measurements performed with the guided neutron beam at the Budapest Reactor have been combined with data from the literature to produce the EGAF database. EGAF contains thermal cross sections for ≈35,000 prompt and delayed γ-rays from 262 isotopes. New precise total thermal radiative cross sections have been derived for many isotopes from the primary and secondary gamma-ray cross sections and additional level scheme data. An IAEA TECDOC describing the EGAF evaluation and tabulating the most prominent γ-rays will be published in 2004. The TECDOC will include a CD-ROM containing the EGAF database in both ENSDF and tabular formats with an interactive viewer for searching and displaying the data. The Isotopes Project, Lawrence Berkeley National Laboratory continues to maintain and update the EGAF file. These data are available on the Internet from both the IAEA and Isotopes Project websites.

1. INTRODUCTON

Low-energy neutron capture produces prompt γ-rays unique to each isotope that can be exploited in Prompt Gamma-ray Activation Analysis (PGAA) for nondestructive elemental analysis. Until now, PGAA has been severely limited by the lack of a reliable neutron capture γ-ray database. These data are also an important component of the Evaluated Neutron Data File (ENDF) used in neutron transport calculations. Groshev *et al* [1] published the first compilation of prompt capture γ-ray energies and intensities, and Greenwood *et al* [2] the first spectrum catalog during the era of NaI detectors. With the advent of Ge detectors in the 1960s, Rasmussen [3] and Orphan[4] measured capture γ-ray spectra for all elements. These data were compiled by Lone et al [5] who published a database of over 10,000 γ-rays in 1981. This database has been used for many years despite the inadequacies inherent to those early measurements. Prompt neutron capture γ-ray data are also compiled from the literature in the Evaluated Nuclear Structure Data File (ENSDF) [6]. These data were used primarily to extract nuclear structure information and were not evaluated for applied use. Reedy and Frankel [7] carefully reevaluated the literature for light elements from hydrogen to zinc and provided this information in ENSDF format. The Lone *et al*, ENSDF, and Reedy γ-ray intensities are normalized to units of per 100 neutron captures. In the 1990's new capture γ-ray measurements were performed for all stable elements by Molnar *et al* at the Budapest Reactor. These measurements utilized a guided neutron beam with the target station far from the reactor where both primary and secondary γ-rays could be measured under low background conditions. They measured absolute γ-ray cross sections rather than relative intensities. An International Atomic Energy Agency (IAEA) Coordinated Research Project (CRP) was organized to evaluate these data. The IAEA CRP produced a database of ≈35,000 neutron capture γ-rays [8] which has been named the Evaluated Gamma-ray Activation File (EGAF) [9]. In this paper we will discuss the Budapest measurements, evaluation of the EGAF database, and recent developments in EGAF analysis.

2. BUDAPEST MEASUREMENTS

The Budapest Research Reactor is a light-water moderated and cooled reactor operating at 10 MW thermal power. A curved neutron guide transports the thermal neutron beam to the target position where the neutron flux is $2 \times 10^6 cm^{-2} s^{-1}$. The measurements discussed in this paper were performed with the thermal beam. In 2001 a liquid-hydrogen cooled source was commissioned increasing the neutron flux to $5 \times 10^7 cm^{-2} s^{-1}$. A pneumatic beam shutter at the end of the neutron guide allows the neutrons to enter a 3-m long evacuated aluminum tube that extends across the experimental area to a beam stop at the rear wall of the guide hall. The neutron flux profile at the sample position is shown in Fig. 1.

2.1 Detectors

A 25% efficient high-purity germanium (HPGe) detector with a BGO-scintillator Compton suppression system was used in these measurements. The target to detector distance was 25 cm. Compton suppression improved the background by factors of ≈5 (1332 keV) to ≈40 (7000 keV). Energy and efficiency calibrations were determined over the range of 50 keV to 10 MeV using several multi γ-ray sources and (n,γ) reactions. Data were analyzed using the fitting code HYPERMET-PC [10].

2.2. Cross Section Standardization

Partial γ-ray cross sections were measured with internalstandards that included H, N, Cl, Au, Ti, and S. Whenever possible, measurements were made with high purity compounds of stable stoichiometry containing a standard element, e.g. NaCl. If no stoichiometric compounds were available, homogeneous mixtures, typically water solutions, were used. The advantage of this method of standardization is that the measurement requires no knowledge of the beam flux and is independent of target geometry, impurities, or neutron scattering consideration. In many cases more than one measurement was made with different compounds or mixtures to check the stoichiometry. Measurements were also made with pure target materials or oxides to record the complete elemental spectrum without interference from the internal standard. Low-energy γ-rays were corrected for attenuation when necessary.

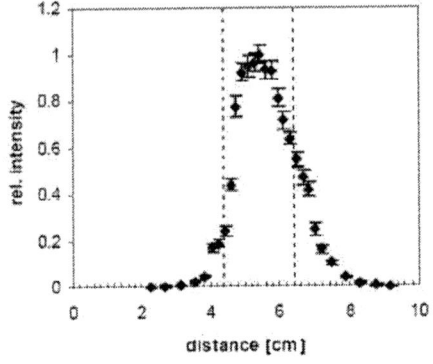

FIGURE 1. Neutron flux profile at the sample position.

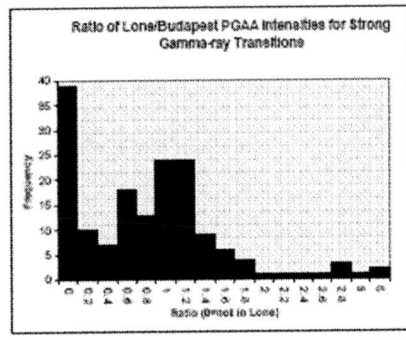

FIGURE 2. Comparison of Budapest and Lone *et al* data.

2.3. Results

Measurements were performed on 79 elements from Z=1-83,90,92 with the exception of He, Ne, Ar, Kr, Tc and Pm. The accurate new energy and intensity data were sufficient to identify 13,000 γ-rays. Figure 2 shows a comparison of the intensities of the most intense capture γ-ray transitions measured at Budapest with those reported by Lone. About 25% of these transitions were not reported by Lone, and 60% of them differed by >20%.

3. CAPTURE γ-RAY DATABASE EVALUATION

The first IAEA Research Coordination Meeting for the Development of a Database for Prompt γ-ray Neutron Activation Analysis convened November, 1999 in Vienna[11]. Representatives from seven countries gathered to plan the compilation of the Budapest capture γ-ray data into a database suitable for nuclear applications. The Isotopes Project, Lawrence Berkeley National Laboratory was asked to take the lead role in this evaluation effort, and other members reviewed the evaluation and performed benchmark tests of the data.

3.1. Isotopic Assignment

The Budapest measurements were performed on natural elemental targets. ENSDF format isotopic capture γ-ray datasets were prepared from the literature compiled in the ENSDF [6] and Reedy [7], and were updated from the Nuclear Science References [NSR][12] file. The Budapest γ-rays were then assigned to their respective level schemes by a comparison of their energies and relative intensities to the literature dataset and then entered into a second, ENSDF format Budapest dataset. Additional γ-rays were placed into the Budapest dataset by comparison with expected transitions from the *Table of Isotopes* [13].

TABLE 1. First iteration of a least squares fit of γ-ray energies to the level scheme for ^{24}Mg(n, γ). Numbers in parentheses are the number of standard deviations discrepancy in the number to the right, compared to the adopted value. The uncertainties in each dataset were increased and additional iterations were performed until $\chi^2/f=1$.

ENSDF	Budapest	Adopted	Level 1	Level 2
389.69 5	(1)389.64 3	389.685	3	2
(2)585.06 3	(2)584.936 24	584.994 16	2	1
611.8 10		611.80 9	7	6
(1)836.95 10	836.75 8	836.82 6	6	4
849.9 3	849.93 16	850.01 3	7	5
(2)863.09 5	(2)862.88 4	862.962 23	8	7
(3)974.84 5	(1)974.61 3	974.669 18	3	1
989.7 4		989.98 9	4	3
1379.7 3	1379.69 19	1379.65 9	4	2
1448.7 10		1448.61 9	7	4
1474.8 10		1474.74 9	8	6
1588.65 9	(1) 1588.40 9	1588.58 3	5	3
-	-	-	-	-
3691.07 16	3690.98 18	3691.03 3	8	2
3916.86 4	(1)3916.65 16	3916.85 3	11	7
4141.4 3	4141.38 24	4141.31 14	10	3
	4357.9 6	4357.8 5	9	1
4528.47 20	4528.66 22	4528.55 9	11	6
4766.86 23	4766.68 25	4766.71 4	11	5
6355.02 10	6354.9 3	6354.96 3	11	3
(1)6744.9 3		6744.54 3	11	2
(1)7330.6 9		7329.37 3	11	1
Level	Energy		Level	Energy
1.	0.0		7.	3413.341 23
2.	585.001 16		8.	4276.32 3
3.	974.689 18		9.	4358.2 5
4.	1964.69 9		10.	5116.36 14
5.	2563.32 3		11.	7330.52 3
6.	2801.53 9			

ENSDF $\chi^2/f=1.561$, f=25; Budapest $\chi^2/f=1.907$, f=17. $\chi^2/f=1.429$, (fit of 61 γ-ray transitions to 10 levels

3.2. Adopted γ ray Energies

Gamma-ray energies were determined by a weighted least-squares fit of both the isotopic and experimental γ-ray energies to the level energies. Since the adopted γ-ray energies are the level energy differences after correction for recoil, weak transitions could be determined to good precision. A chi-squared analysis was performed by comparing the input to the adopted data, and the uncertainties of individual outliers with $\chi^2 f>4$ and/or all data in datasets with > 1 were increased and the fit repeated until $\chi^2/f = 1$. Badly discrepant outliers were discarded, particularly when more accurate data were available. A typical fit of γ-ray energies is shown in Table 1 for ^{24}Mg(n,γ).

3.3. Adopted γ-ray Cross Sections

The Budapest experimental γ-ray intensities were reported as elemental cross sections, whereas the corresponding literature values were typically compiled in units of per 100 neutron captures of the isotope. These data were averaged by one of two methods: (1) If a well-defined γ-ray cross section existed in the literature, the γ-ray intensities in the literature dataset were renormalized to that value, converted to an elemental cross section by means of the isotopic abundance, and averaged with the experimental values. (2) If no precise normalization factor existed for mostcross sections, the intensities in the literature dataset were renormalized by a factor chosen to minimize the weighted average difference between the literature and experimental intensity data. The renormalized intensities were then averaged with the experimental data to obtain the adopted cross sections. A similar chi-squared analysis to that described for the energies was performed to handle outliers and discrepant data. The skew in the chi-squared distribution as a function of energy was used to probe systematic differences in the underlying efficiency curves, and discrepant data were adjusted or removed as necessary.

3.4. Intensity Balances

The level scheme γ-ray intensity balances were used to determine the quality and completeness of the evaluated data. The total γ-ray cross section feeding the ground state was compared with the corresponding values from Mughabghab *et al* [14-16], and the ratio of the total primary γ-ray cross section to the cross section feeding the ground state indicated the completeness of the dataset. Intensity balances through intermediary levels indicate missing or anomalous intensities, and such problems were corrected whenever possible. An example of an intensity balance analysis with no important discrepancies is shown in Table 2. Level schemes are complete for the more abundant isotopes of the light nuclei, but significant inconsistencies in the intensity balance may arise for heavier nuclei and remain unresolved in the continuum.

TABLE 2. Intensity balance for ^{24}Mg(n,γ).

ENSDFE(Level)	σ (in)	σ (out)	$\delta\sigma$
0	0.0536(14)	0.0	0
585.01(3)	0.0406(11)	0.0398(14)	0.0008(18)
974.68(3)	0.0157(4)	0.0158(4)	0.0001(6)
1964.69(10)	0.00022(2)	0.00026(3)	0.00004(4)
2563.35(4)	0.00202(10)	0.00179(7)	0.00023(12)
2801.54(9)	0.00047(4)	0.00061(5)	0.00013(6)
3413.35(3)	0.0411(14)	0.0416(11)	0.0005(18)
4358.2(5)	0.00009(2)	0.0	0.00009(2)
5116.37(15)	0.00038(4)	0.00027(3)	0.00011(5)
7330.53(4)	0.0	0.0539(14)	0.0539(14)

σ (Mughabghab)=0.0536±0.015 b, σ (this work)=0.0538±0.014 b

82-Lead-207

Isotopic Abundance(%) 22.1(1)

Isotopic Capture Cross Section (barns): 0.63(3)

Number of Gammas: 23

Westcott g-factor: 1

#	E_γ(keV)	σ_γ^Z(b)	Type	Half-life	k_0
1	1025(3)	2.2E-05(5)	p	Stable	3.2E-07(7)
2	1155.06(16)	1.10E-05(19)	p	Stable	1.6E-07(3)
3	1332.82(19)	5E-06(3)	p	Stable	7E-08(4)
4	1383(4)	3.3E-05(7)	p	Stable	4.8E-07(10)
5	1437.40(17)	8E-06(3)	p	Stable	1.2E-07(4)
6	1523.36(19)	1.1E-05(3)	p	Stable	1.6E-07(4)
7	1615.16(16)	2.2E-05(3)	p	Stable	3.2E-07(4)
8	1640.17(8)	3.6E-05(4)	p	Stable	5.3E-07(6)
9	1983.13(15)	3.8E-05(8)	p	Stable	5.6E-07(12)
10	2086.8(21)	3.8E-05(8)	p	Stable	5.6E-07(12)
11	2322.62(8)	3.0E-05(4)	p	Stable	4.4E-07(6)
12	2430.72(8)	3.0E-05(4)	p	Stable	4.4E-07(6)
13	2614.53(3)	0.00040(3)	p	Stable	5.9E-06(4)
14	2770.21(15)	1.6E-05(3)	p	Stable	2.3E-07(4)
15	3113.17(8)	3.6E-05(6)	p	Stable	5.3E-07(9)
16	3282.39(7)	8.5E-05(9)	p	Stable	1.24E-06(13)
17	3729(3)	2.2E-05(11)	p	Stable	3.2E-07(16)
18	4085.46(7)	1.04E-04(17)	p	Stable	1.52E-06(25)
19	4753.31(7)	2.33E-04(17)	p	Stable	3.41E-06(25)
20	4937.12(8)	2.5E-05(25)	p	Stable	4E-07(4)
21	5384.71(15)	8E-06(3)	p	Stable	1.2E-07(4)
22	5844.46(28)	1.4E-05(4)	p	Stable	2.0E-07(6)
23	7367.78(7)	0.137(3)	p	Stable	0.03200(4)

p - Prompt, d - Delayed

FIGURE 3. Tabular listing of ^{207}Pb data displayed by the PGAA-IAEA Viewer.

4. IAEA TECDOC AND THE EGAF DATABASE

The database of ≈35,000 neutron capture γ-rays hasbeen published as an IAEA TECDOC with accompanying CD-ROM [8]. The TECDOC includes discussions of neutron capture terminology, Westcott g-factors, characteristics of PGAA facilities, results of CRP benchmark experiments, and total radiative neutron capture cross sections and neutron separation energies. Transitions with cross sections greater than 1% of the most intense transition for each element are tabulated, and transitions >10% are listed in an energy-ordered table. The original Budapest Reactor cross section data and an extensive bibliography of measurements is given in the appendices. The complete data library is given on the CDROM in text, EXCEL, and Adobe Acrobat PDF formats. The TECDOC report and all of the data is available from the IAEA at http://www-nds.iaea.org/pgaa/.

4.1. PGAA-IAEA Database Viewer

A neutron capture γ-ray database viewer is available at http://www-nds.iaea.org/pgaa/pgaa7/index.html and on the CD-ROM. A periodic table interface provides access to the data by element and isotope. The viewer can display both tables of γ-ray cross sections and spectrum plots. Selection of ^{207}Pb data produces the γ-ray table shown in Fig. 3. The capture γ-ray database can also be searched by energy, atomic number, mass, and cross section with the PGAA-IAEA viewer.

4.2. EGAF Database

The archival capture γ-ray database has been named the Evaluated Gamma-ray Activation File (EGAF). This file is maintained in ENSDF format to preserve the nuclear structure aspects of the database. The EGAF file consists of an adopted dataset for each isotope, with supporting datasets prepared from the Lone et al[5] and Reedy [7] databases. The adopted γ-ray intensities are given as elemental cross sections. A normalization is provided to convert this intensity to isotopic cross section assuming normal abundance [17]. The EGAF database is maintained by the Isotopes Project, Lawrence Berkeley National Laboratory. It is disseminated by the IAEA on the TECDOC CD-ROM and on the Internet at http://www-nds.iaea.org/pgaa/egaf.html.

TABLE 3. History of the ^{12}C cross section measurements.

Measurement Method	σ_0(mb)	Reference
Capture	3.50 ± 0.16	Prestwich (1981)
Capture	3.53 ± 0.07	Jurney (1963)
Reactivity	3.57 ± 0.03	Nichols (1960)
Pile Oscillator	3.65 ± 0.15	Nichols (1960)
Pulsed Neutrons	3.72 ± 0.15	Sagot (1963)
Pulsed Neutrons	3.83 ± 0.06	Starr (1962)
Pile Oscillator	3.85 ± 0.15	Koechlin (1957)
Capture	3.87 ± 0.05	Molnar (2003)
Capture	4.01 ± 0.15	Yonezawa (2003)
Average of Higher Values	3.84 ± 0.03	
Current Adopted Value	3.53 ± 0.07	Mughabghab (1981)

5. APPLICATIONS OF THE EGAF DATABASE

The capture γ-ray database has been used for Prompt Gamma-ray Activation Analysis (PGAA) at the Budapest Reactor for many years. Notable examples include the analysis of deep sea vent [18] and reagent materials [19]. Total radiative cross sections were determined for most isotopes [8,20]. This work has contributed to a new evaluation of the ^{238}U total radiative cross section [22].

TABLE 4. Comparison of experimental and theoretical level feeding for ^{105}Pd(n, γ).

Level (keV)	Feeding per 100 captures Experiment	Feeding per 100 captures Theory	χ2	J$^\pi$ ENDSF	J$^\pi$ Fit
0.0	99.95	100.00		0+	0+
511.8	85.37±0.85	84.34±1.01	0.8	2+	2+
1128.0	20.32±0.23	22.27±1.40	1.4	2+	2+
1133.8	2.69±0.15	2.50±0.41	0.4	0+	0+
1229.2	16.59±0.19	16.77±1.24	0.1	4+	4+
1557.7	10.21±0.16	11.86±1.03	1.6	(3)+	3+
1562.2	8.91±0.20	7.78±0.76	1.4	2+	2+
1706.4	0.93±0.05	0.63±0.16	1.8	0+	0+
1904.3				2-, 3-	No level
1909.4	2.97±0.14	3.85±0.36	2.3	2+	2+
1932.4	2.81±0.08	3.67±0.42	2.0	4+	4+
2001.6	0.44±0.03	0.41±0.08	0.4	0+	0+
2077.1	0.56±0.04	0.44±0.15	0.8	6+	6+
2077.4	3.00±0.07	2.78±0.31	0.7	4+	4+
2084.4	4.93±0.53	5.69±0.91	0.7	3-	3-
2242.4	2.38±0.12	2.32±0.28	0.2	2+	2+
2278.5	0.27±0.04	0.29±0.07	0.3	0+	0+
2282.9	1.31±0.07	1.80±0.25	1.9	4+	4+
2306.0	2.59±0.07	2.52±0.49	0.1	4-	4-
2308.7	1.21±0.06	1.43±0.25	0.9	2+	1+
2351.0	1.45±0.08	1.60±0.25	0.6	4+	4+
2366.1	0.55±0.04	0.75±0.13	1.4	5+	5+
2397.4	1.25±0.05	0.94±0.16	1.9	(5)-	(5)-
2401.0	1.43±0.09	2.39±0.53	1.8	2-, 3-	2-
2439.1	1.39±0.10	1.73±0.20	1.5	2+	2+
2472.1	0.12±0.01	0.23±0.05	1.9	1+, 2+	0+
2484.8	1.21±0.09	1.25±0.45	0.1	(1-)	(1-)
2500.0	1.41±0.09	2.10±0.90	0.8	2-	2-

5.1. Total Radiative Cross Sections

Many neutron cross sections were determined from the EGAF database with comparable precision to those reported by Mughabghab et al [14-16], and some disagreements were notable. For example, the most accurate $_{12}$C measurements, including our current measurement, are summarized in Table 3. Our new, recommended cross section is 3.84±0.03 mb and differs significantly from the previously accepted value [14-16] of 3.53±0.07 mb.

5.2. Quasi-continuum Calculations

The EGAF database is often incomplete because continuum γ-rays can comprise up to 90% of the spectrum. We have been applying the γ-ray cascade code DICEBOX by Becvar [22] to calculate this continuum. These calculations are constrained by the EGAF γ-ray cross sections de-exciting low-lying levels in the capture nucleus. Several independent calculations were performed with DICEBOX to estimate theoretical variation. Table 3 shows preliminary results for ^{105}Pd(n,γ). An excellent fit with $\chi^2/f=1.1$ was obtained, but only after revising the ENSDF spin/parity values, indicated in bold type, and removing one level that was inconsistent with the calculations and not confirmed by other experiments. The total radiative cross section derived from this calculation is 21.2±0.5 b, consistent with the compiled value [14-16] of 21.0±1.5 b.

6. FUTURE PLANS FOR EGAF

A new series of elemental and isotopic capture γ-ray measurements are planned at the Budapest Reactor with the cold neutron beam. EGAF will be updated to include these data as well as continuum data from the quasicontinuum calculations discussed above. The data will be benchmarked with experimental spectra and provided in ENDF format. Evaluation of the continuum data will provide new nuclear structure information available for ENSDF evaluation. EGAF will also be expanded to include epithermal and fast neutron capture and reaction γ-ray data from ENSDF and other literature sources. Additional measurements are planned with LNBL 2.5 MeV D+D (10^{10} n/s) Neutron Generator Facility [23]. In addition to reaction data, we will measure fission γ-ray data. The EGAF database provides a repository for evaluated neutron-induced γ-ray data that can support a variety of other databases and applications. It offers a data bridge between the traditionally reaction based and nuclear structure based communities.

7. ACKNOWLEDGEMENTS

This paper is dedicated to the memory of Gabor Molnar. The research presented here would not have been possible without his support and encouragement. This work was supported by Director, Office of Science, Office of Basic Energy Sciences, of the U.S. Department of Energy under Contract No. DE-AC03- 76SF00098 and performed under the auspices of the U.S. Department of Energy by University of California, Lawrence Livermore National Laboratory under Contract W-7405-Eng-48.

REFERENCES

1. L.V. Groshev, A.M. Demidov, V.N. Lutsenko, and V. Pelekhov,*Atlas of the Spectra of Gamma Rays fromthe Radiative Capture of Thermal Neutrons*, Pergamon, London (1961).
2. R.C. Greenwood and J.H. Reed, *Prompt Gamma Rays from Radiative Capture of Thermal Neutrons* IIT Research Institute Report IITRI-1193-53 (1965).
3. N.C. Rasmussen, Y. Hukai, T. inouye, and V.J. Orphan, *Thermal Neutron Capture Gamma Ray Spectra of the Elements* Massachusetts Institute of Technology Report AFCRL-69-0071 (1969).
4. V.J. Orphan, N.C. Rasmussen, and T.L. Harper, *Line and Continuum Gamma-ray Yields from Thermal Neutron Capture in 75 Elements* Gulf General Atomic Report DASA 2570 (GA 10248) (1970).
5. M.A. Lone, R.A. Leavitt, and D.A. Harrison, At. Data Nucl. Data Tables **28**, 511 (1981).
6. *Evaluated Nuclear Structure Data File*, a computer file of evaluated experimental nuclear structure data maintained by the National Nuclear Data Center, Brookhaven National Laboratory.
7. R.C. Reedy and S.C. Frankel At. Data Nucl. Data Tables **80**,1 (2002).
8. R.B. Firestone, H.D. Choi, R.M. Lindstrom, G.L. Molnar, S.F. Mughabghab, R. Paviotti-Corcuera, Zs. Revay, V. Zerkin, and C.M. Zhou, *Database of Prompt Gamma Rays from Slow Neutron Capture for Elemental Analysis*

IAEA TECDOC, IAEA Nuclear Data Section, International Atomic Energy Agency P.O. Box 100 A-1400 Vienna, Austria, TECDOC, in press (2003)..
9. *Evaluated Gamma-ray Activation File*, a computer file of evalulated experimental neutron capture gamma ray data maintained by R.B. Firestone, Lawrence Berkeley National Laboratory and disseminated by the International Atomic Energy Agency.
10. Zs. Revay, T. Belgya, P.P. Ember, and G.L. Molnar, J. Radioanal. Nucl. Chem. **248**,401 (2001).
11. Development of a Database for Prompt γ-ray Neutron Activation Analysis, International Atomic Energy Agency Report INDC(NDS)-411(1999).
12. *Nuclear Science Reference File* a bibliographic computer file of nuclear science references continually updated and maintained by the National Nuclear Data Center, Brookhaven National Laboratory. Recent literature scanned by D. Winchell and A. Sonzogni.
13. R.B. Firestone, V.S. Shirley, C.M. Baglin, S.Y.F. Chu, and J. Zipkin, *Table of Isotopes*, John Wiley and Sons, New York (1996, 1998, 1999).
14. S.F. Mughabghab, *Thermal Neutron Capture Cross Sections, Resonance Integrals, and g-factors*, INDC(NDS)-440 (2003).
15. S.F. Mughabghab, M. Divadeenam, and N. HOLDEN, *Neutron Cross Sections*, Vol. 1, Part A, Z = 1-60, Academic Press, New York (1981).
16. S.F. Mughabghab, *Neutron Cross Sections*, Vol. 1, Part B, Z = 61-100, Academic Press, New York (1984).
17. K.J.R, Rosman and P.D.P. Taylor, Pure Appl. Chem.**70**,217 (1998).
18. D.L. Perry, R.B. Firestone, G. Molnar, Zs. Revay, Zs. Kasztovszky, R.C. Gatti, and P. Wilde, J. Anal. At, Spectrom. **16**, 1 (2001).
19. D.L. Perry, G. A. English, R. B. Firestone, K.-N. Leung, G. Garabedian, G. L. Molnar, and Zs. Revay, *Use of Prompt Gamma Activation Analysis (PGAA) and Related Neutron Techniques for the Analyses of Metal Oxyanion Salts*, submitted to J. Radioanal. Nucl. Chem. (2004).
20. *New Capture Gamma-Ray Library and Atlas of Spectra for All Elements*, R. B. Firestone, Zs. Revay, and G. L. Molnar, Proceedings of the Eleventh International Symposium on Capture Gamma-Ray Spectroscopy and Related Topics, Pruhonice near Prague, Czech Republic, September 2 - 6, 2002, invited talk, World Scientific, p.507 (1903).
21. A. Trkov, G.L. Molnar, Zs. revay, S.F. Mughabghab, R.B. Firestone, V.G. Pronyaev, A.L. Nichols, and M.C. Moxon, *Revisiting the U-238 Thermal Neutron Capture Cross Section and Gamma-emission Probabilities from Np-239 Decay*, submitted to Nucl. Sci. Eng.
22. F. Becvar, Nucl. Instrum. Meth. Phys. Res. **A417**, 434 (1998).
23. J. Reijonen, K.-N. Leung, R.B. Firestone, *et al*, Nuclear Instruments and Methods in Physics Research A 522, 598 (2004).
24. R.B. Firestone, G.A. English, J. Reijonen, K-N. Leung, Zs. Revay and G.L. Molnar, J. Radioanal. Nucl. Chem., in press (2004)

NUCLEAR MASSES AND SHELL MODEL

Nuclear Forecasting as Pattern Recognition: Can we predict Nuclear Masses?

A. Frank*, J.C. López Vieyra*, J. Barea*, J.G. Hirsch*, V. Velazquez† and P. Van Isacker**

*Instituto de Ciencias Nucleares, Universidad Nacional Autónoma de México, Apartado Postal 70-543, 04510 México, D.F., Mexico
†Facultad de Ciencias, Universidad Nacional Autónoma de México, 04510 México, D.F., Mexico
**GANIL, BP 55027, F-14076 Caen Cedex 5, France

Abstract. The challenging task of predicting nuclear masses is analyzed as a pattern recognition problem on the N-Z plane. A well defined pattern is built by taking the differences between measured masses and Liquid Drop Model (LDM) predictions. After removing the smooth LDM mass contributions, what remains are the microscopic components, which have proved to be extremely hard to model and predict. These contain the information related with shell closures, nuclear deformations, and the residual nuclear interactions. In the present work the more than 2000 known nuclear masses are studied as an array in the N-Z plane viewed through a mask, behind which are hidden the unknown 7000 unstable nuclei that can exist between the proton and neutron drip lines. Employing a Fourier transform deconvolution method these masses are predicted. Measured masses are reconstructed with and r.m.s. error of less than 200 keV. The existence of an island of stability around ($Z \approx 116$, $N \approx 194$) is strongly suggested. Different potential applications of the present approach are outlined.

INTRODUCTION

The importance of an accurate knowledge of nuclear masses to understand fundamental processes in nuclear physics and astrophysics is well known [1]. Though great progress has been made in the difficult task of measuring the masses of exotic nuclei, theoretical models are necessary to *predict* them in regions far from stability [2]. The simplest example is that of the liquid drop model (LDM), which incorporates the essential macroscopic terms, the nucleus being pictured as a very dense, charged liquid drop, and including other important nuclear effects, such as the pairing interactions. The finite range droplet model (FRDM) [3], which combines the macroscopic effects with microscopic shell and pairing corrections, has become the *de facto* standard for mass formulas. A microscopically inspired model has been introduced by Duflo and Zuker (DZ) [4] with surprisingly good results, although the model introduces magic numbers by hand and is not capable of predicting other observables. Finally, among other mean-field methods it is worth mentioning the Skyrme-Hartree-Fock approach [5],which leads to mass formulas that calculate and predict the masses (and often other properties) of as many as 8979 nuclides, although its predictability has not been too impressive. More troublesome is the fact that different approaches tend to diverge from each other in their predictions, so there is a permanent search for better theoretical models that reduce the

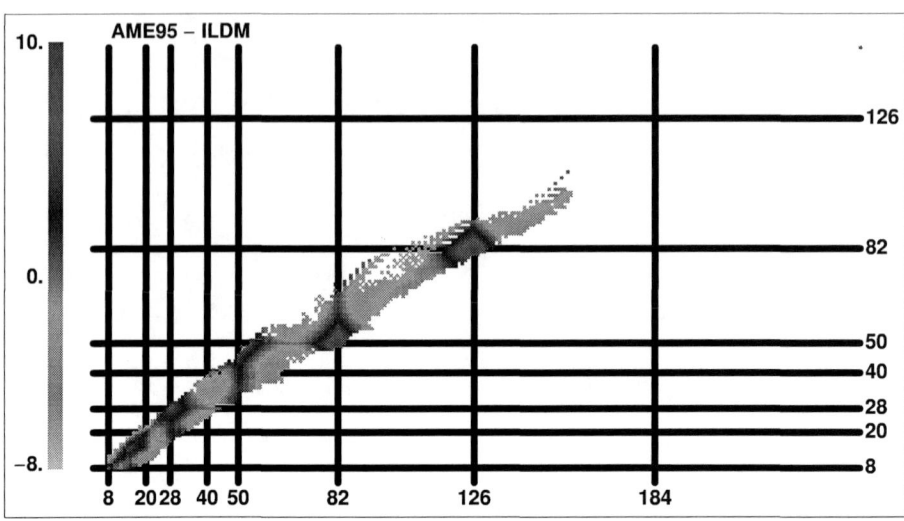

FIGURE 1. Nuclear mass differences in the nuclear landscape. The horizontal axis corresponds to *neutron number N*. The vertical axis corresponds to *atomic number Z*.

difference with the experimental masses and produce reliable predictions for unstable nuclei.

Besides the "global" formulas of which the FDRM method has become the standard, there are a number of "'local" mass formulas. These local methods are usually effective when calculating the mass of a nucleus, or a set of nuclei, which are fairly close to nuclei of known mass, taking advantage of the relative smoothness of the masses M(Z,N) as a function of proton (Z) and neutron (N) numbers to deduce systematic trends. Among these methods, there are a set of algebraic relations for nuclear neighbors known as the Garvey-Kelson (GK) relations, which can be deduced from an extreme single-particle picture of nuclei [6].

Detailed analyses of nuclear mass errors have been performed, using the FRDM [3], the shell-model-inspired mass calculations of Duflo and Zuker (DZ)[4], the Hartree-Fock-Bogoliubov mass calculations of Goriely et al. [5], as well as other methods, including the Garvey-Kelson (GK) relations [6]. The presence of strong correlations between mass errors calculated in a mean-field approach in neighboring nuclei has been exhibited, as well as the existence of a well defined chaotic signal in its power spectrum, but further analysis demonstrates that the inclusion of many-body interactions or the introduction of local information removes the chaotic signal [8, 9].

In the present work a new approach to the problem of nuclear predictions is introduced. The starting point is the striking color pattern observed on the nuclear landscape when taking the differences between measured masses [10] and Liquid Drop Model (LDM) predictions [11], as shown in Fig. 1.

After removing the smooth LDM mass contributions, what remains are the microscopic components, which have proved extremely difficult to predict by the different methods, all of which attempt to minimize the differences with measured masses. The residual pattern observed in Fig. 1, however, is quite remarkable and suggests a different approach. The "unfolding" of the data contains information related with shell closures, nuclear deformations, and the residual nuclear interaction, in a compelling graphic form. We therefore suggest that the approximately 2000 known nuclear masses can be studied as an array in the N-Z plane viewed through a window. The remaining nuclei (approximately 7000 of them) which can exist between the proton and neutron drip lines, lie hidden, covered by a "mask". So the question is how to "'open the window"' and watch and scrutinize the rest of the pattern? We show here that employing a Fourier transform deconvolution method these masses can be predicted. In the following sections the deconvolution approach is briefly described and some of the preliminary results presented.

MATHEMATICAL FORMULATION OF THE PROBLEM

The first step in our analysis consists in translating the nuclear mass table into an image. A standard two dimensional array is built in which the horizontal position corresponds to the number of neutrons and the vertical position to the number of protons. Differences between experimental mass values and those calculated using the liquid drop model for each isotope define the function

$$i(n,z) = m^{exp}(n,z) - m^{LDM}(n,z), \qquad (1)$$

which is plotted in the (n,z) plane introducing a color code associated with the mass deviations, as shown in Fig. 1. As both the proton and neutron numbers are integers, this is a discrete function defined in a restricted domain of $\mathbb{I}^2$. It is appropriate to emphasize that $i(n,z)$ is only defined in the limited region where measured masses $m^{exp}(n,z)$ are known, i.e. the colored region in Fig. 1. It can be extended to the whole N-Z rectangle of interest by assigning null values to addresses where there are no measured experimental masses.

To predict the nuclear mass differences $m(n,z)$ along the whole rectangle, a binary mask function $w(n,z)$ is introduced. This mask takes the value 1 for those positions n and z on which the experimental nuclear masses are known, and 0 on the others. Known mass differences $i(n,z)$ are related to the total mass differences $m(n,z)$ by

$$i(n,z) = m(n,z) \cdot w(n,z). \qquad (2)$$

In order to predict unknown masses, we need to extract $m(n,z)$ from Eq. (2). Formally, this is a deconvolution problem, which can be solved using discrete Fourier transforms. If $I(k_n,k_z)$, $M(k_n,k_z)$ and $W(k_n,k_z)$ are the Fourier transforms of $i(n,z)$, $m(n,z)$ and $w(n,z)$, respectively, the Fourier transform of Eq. (2) is given by

$$I(k_n,k_z) = (M * W)(k_n,k_z), \qquad (3)$$

where $(M * W)$ represents the convolution of the functions M and W. Given that both $i(n,z)$ and $w(n,z)$ are known for the whole rectangular domain, their Fourier transforms $I(k_n,k_z)$ and $W(k_n,k_z)$ can be evaluated directly. The problem is narrowed to obtaining the function $M(k_n, k_z)$ (the discrete Fourier transform of $m(n, z)$), from which $m(n, z)$ can be recovered applying the inverse Fourier transform.

We thus need a deconvolution method. Deconvolution is usually non-trivial and may lead to non-unique solutions, but there exist several linear algorithms, such as the inverse filtering and the Wiener filtering, and nonlinear algorithms, such as the CLEAN method, the maximum entropy method (MEM) and LUCY, which provide testable methodologies.

The CLEAN method is a deconvolution algorithm often used in radioastronomy. We have applied a specially adapted version of the "CLEAN" algorithm, developed by Clark et al. [12], to the mass reconstruction problem. Our version employs a deconvolution algorithm to remove the artifacts introduced by the shape of the window (or mask), thereby allowing the Fourier analysis of non-rectangular texture patches. We assume that the "'full'" mass pattern $m(x, y)$ can be modeled by a finite number of harmonic components. The deconvolution algorithm proceeds by detecting which components, when corrupted by the shape mask, provide the best explanation of the patterns observed in the Fourier spectra. Each component is then relocated to the *clean spectrum* and its footprint erased from the corrupted spectrum. The process is repeated until only noise residuals are left in the corrupted spectrum. Details of the procedure will be published elsewhere [13].

NUCLEAR FORECASTING AS IMAGE RECONSTRUCTION

The CLEAN algorithm described above was programmed in *Mathematica* and Fortran, and applied to the pattern of differences between experimental nuclear masses (AME95) [10] and the theoretically calculated masses using the Liquid Drop Model [11]. As one possible test of predictability, we compare the calculated masses using the pattern reconstruction algorithm with the data for a larger set of nuclear masses (AME03) [7]. A (consistent)calculation for the nucleon *driplines* is also included. The results are shown below.

Reconstruction of measured masses

To test the ability of the present method to reconstruct known masses, the set of 1888 nuclear masses (AME95) was employed as input data. Here we used $Z_{max} = 150, N_{max} = 250$, which allows for the exploration of possible stability islands associated with superheavy elements.

The algorithm was applied to reconstruct the "AME95 vs ILDM" data. The differences between the reconstructed masses and those obtained using the Liquid Drop Model are presented in Fig. 2. There is a clear resemblance in the plots,as seen in Fig. 1. In order to gauge the quality of the mass reconstruction, the *rms* average error between reconstructed and experimental masses was studied. After 200 iterations (the introduction of 200 frequencies) it attains the value 0.2165 MeV, a number that can be lowered by increasing the number of steps. The differences between the experimental and re-

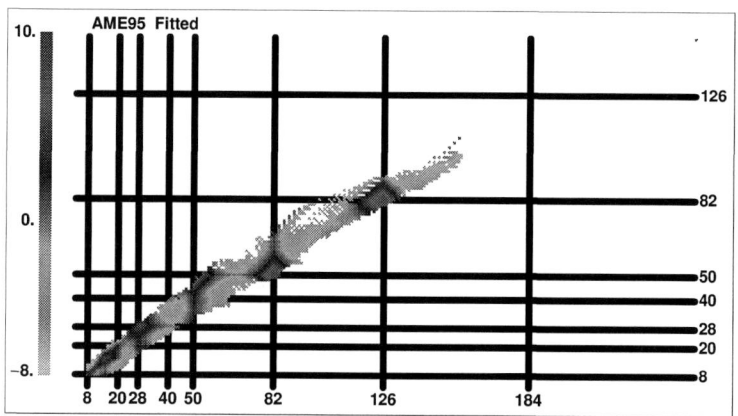

FIGURE 2. Reconstructed image of the nuclear mass differences "AME95 vs ILDM" after 200 iterations.

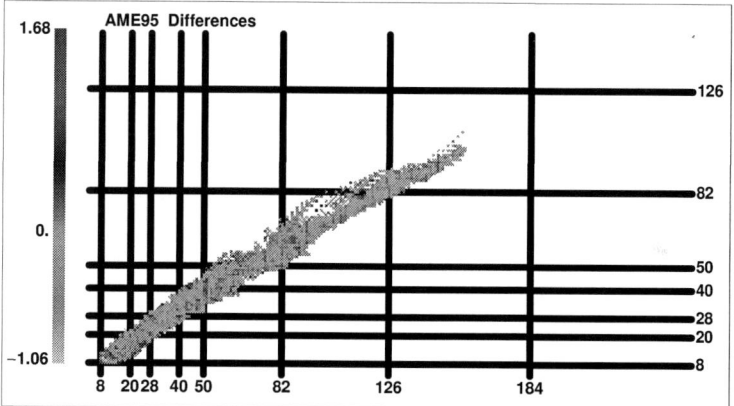

FIGURE 3. Differences Experimental vs Reconstructed "AME95 vs ILDM" after 200 iterations.

constructed masses are plotted in Fig. 3, where we cannot detect any remaining pattern.

Prediction of new masses

Besides reproducing known masses, the method predicts the masses of all possible isotopes in the area between the the proton and neutron driplines. Using the reconstructed pattern the value of 289 new nuclear masses (AME03) [7] is predicted.

The new masses have and an rms value of 0.8195 MeV, which is a good preliminary result in comparison with most of the other mass prediction formalisms [2]. In Fig. 4 we observe a remarkable result: the GK relations are almost exactly satisfied in the

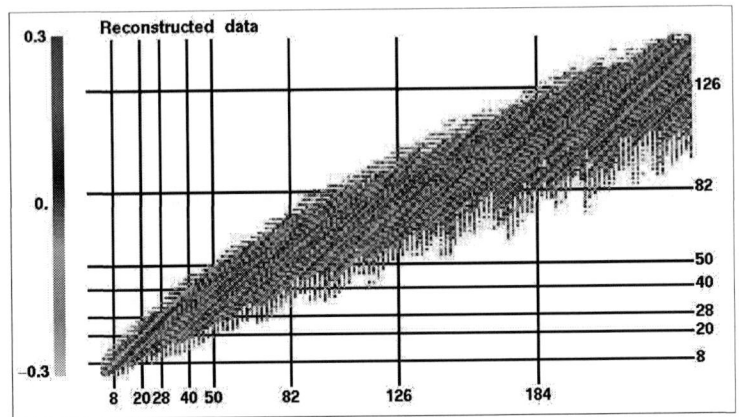

FIGURE 4. Garvey-Kelson deviations map calculated using the GK relations iteratively.

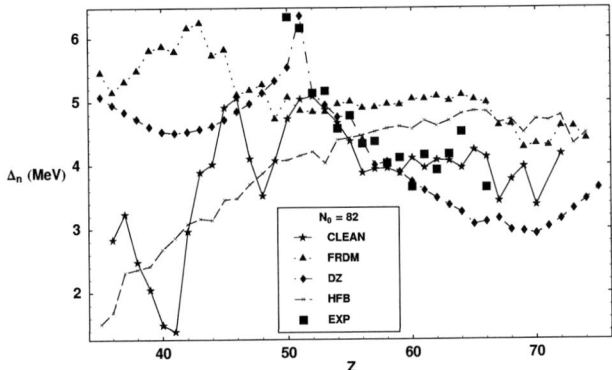

FIGURE 5. Neutron shell gaps as a function of Z for the four mass tables CLEAN, Moller and Nix (FRDM), Duflo-Zuker (DZ) and Hartree-Fock Bogoliubov (HFB).

predicted areas, which is not the case for the FDRM and HFB calculations. The DZ and LDM formulas do satisfy these relations accurately, but this is due to the simple form of the expansion functions in N and Z that they use [6, 13].

In Fig. 5 we show a comparison of the N=82 shell gaps for different proton numbers, as given by the different models discussed in this paper.

Improving mass reconstruction

The process of mass reconstruction is performed selecting the frequency with the largest squared Fourier amplitude, filtering it through the mask and reproducing the

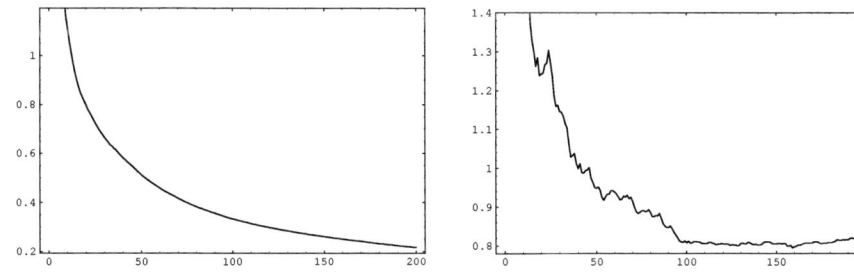

FIGURE 6. (a) RMS (input data vs reconstructed data), (b) RMS (predicted vs new AME'05 data) as function of number of iterations. Max number of iterations was 200, with constant gain $g = 1$. Rectangular grid used was $(Z_{max}, N_{max}) = (150, 250)$.

data with it. In the following step the frequency with the second largest squared Fourier amplitude is employed, next the third, etc. The process goes on until the required precision in the data reconstruction is obtained.

In Fig. 6(a) the rms value of the adjusted data is shown as a function of the iteration number, i.e. as the number of frequencies employed in the reconstruction. The *rms* value of fitted vs experimental data displays a monotonous decreasing behavior as the number of iterations increase. Eventually, the rms value can be reduced to very small values for a large number of iterations. The rms error after 200 iterations was 0.2165 MeV. On the other hand, the *rms* value of predicted vs experimental AME03 data displays a decreasing but non-monotonous behavior as function of the number of iterations. However, it seems that the *rms* value (predicted vs experimental AME03 data) reaches a minimal value (rms$\simeq$ 0.795 MeV) for certain number of iterations ($\sim$ 159) after which the prediction is essentially constant. The final rms after 200 iterations was 0.8195 MeV.

Opening the window

The reconstructed pattern allows us to enlarge the window and find the predictions all the way up to the *driplines*. The results of the 200 frequencies iteration procedure is shown in Fig. 7 . A stability island is predicted around the superheavy element $(n,z) \simeq (194, 116)$.

CONCLUSIONS

Pattern recognition techniques have been applied to the mass deviations obtained from the LDM formula smooth contributions. Measured masses can be reconstructed with a precision of 200 keV, while new masses from AME03 are predicted with an average error of 800 keV. A new stability island is predicted around $(n,z) \simeq (194, 116)$. Other observables besides nuclear masses, like nuclear deformation and beta decay properties, may be analyzed with the same techniques. We are currently carrying out diverse

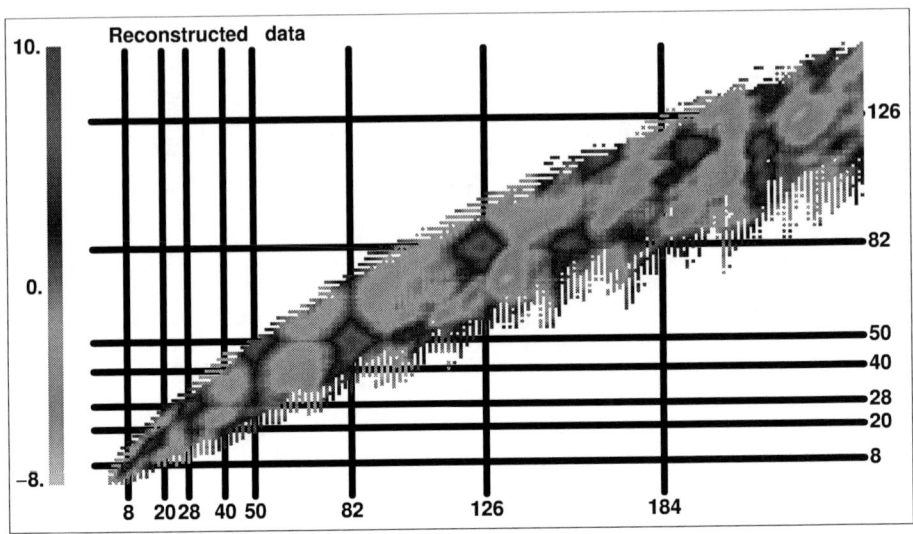

FIGURE 7. Reconstructed pattern bounded by driplines in the prediction after 200 iterations.

tests and studying further constraints imposed by nuclear physics (like the GK relations) or mathematical consistency (such as smoothness-continuity requirements) in the pattern recognition analysis, in order to attempt to achieve a more accurate and robust predictability [13].

Acknowledgments: Fruitful conversations with J. Dukelsky, A. Raga and C. Ruiz are gratefully acknowledged. This work was supported in part by PAPIIT-UNAM and Conacyt-Mexico.

REFERENCES

1. C.E. Rolfs and W.S. Rodney, *Cauldrons in the Cosmos*, University of Chicago Press (1988).
2. D. Lunney, J.M. Pearson, and C. Thibault, Rev. Mod. Phys. **75**, 1021 (2003).
3. P. Möller, J.R. Nix, W.D. Myers, W.J. Swiatecki, At. Data Nucl. Data Tables **59**, 185 (1995).
4. J. Duflo, Nucl. Phys. A **576**, 29 (1994); J. Duflo and A. P. Zuker, Phys. Rev. C **52**, R23 (1995).
5. S. Goriely, F. Tondeur, and J.M. Pearson, Atom. Data Nucl. Data Tables **77**, 311 (2001); M.V. Stoitsov, J. Dobaczewski, W. Nazarewicz, S. Pittel, and D.J. Dean, Phys. Rev. C **68**, 054312 (2003).
6. G.T. Garvey and I. Kelson, Phys. Rev. Lett. **16**, 197 (1966); G.T. Garvey, W.J. Gerace, R.L. Jaffe, I. Talmi, and I. Kelson, Rev. Mod. Phys. **41**, S1 (1969).
7. G. Audi, A.H. Wapstra, and C. Thibault, Nucl. Phys. A **729**, 337 (2003).
8. J.G. Hirsch, V. Velázquez, and A. Frank, Phys. Lett. B **595**, 231 (2004).
9. J. Barea, A. Frank, J.G. Hirsch, and P. van Isacker, Phys. Rev. Lett. **94**, 102501 (2005).
10. G. Audi and A. H. Wapstra, Nucl. Phys. A 595, 409-480 (1995).
11. S. R. Souza, et al., Phys. Rev. **C67**, 051602(R) (2003).
12. A. A. Clark and B. T. Thomas and N. W. Campbell and P. Greenway *"Texture Deconvolution for the Fourier-Based Analysis of Non-Rectangular Regions"*, in it British Machine Vision Conference, pp 193–202, British Machine Vision Association, 1999.
13. J. Barea et al., to be published.

Mass Measurements with the CSS2 and CIME cyclotrons at GANIL

M.B. Gomez Hornillos*, M. Chartier*, W. Mittig[†], B. Blank**,
L. Caballero[‡], F. Chautard[†], C.E. Demonchy*, G. Georgiev[†], A. Gillibert[§],
B. Jacquot[†], B. Jurado[†], N. Lecesne[†], A. Lépine-Szily[¶], N. Orr[‖], G. Politi[††],
M. Rousseau[†], P. Roussel-Chomaz[†], H. Savajols[†] and A.C.C. Villari[†]

*University of Liverpool, Oliver Lodge Laboratory, L69 7ZE Liverpool, UK.
[†]GANIL, Bld Henri Becquerel, BP 5027, 14076 Caen Cedex 5, FRANCE.
**CEN Bordeaux-Gradignan, Le Haut-Vigneau, 33175 Gradignan, FRANCE.
[‡]IFIC, CSIC-Universidad de Valencia, E-46071 Valencia, SPAIN.
[§]CEA/DSM/DAPNIA/SPhN, CEN Saclay, 91191 Gif-sur-Yvette, FRANCE.
[¶]IFUSP, Caixa Postal 66318, CEP 05315-970 São Paulo, BRAZIL.
[‖]LPC, 6 Bld du Maréchal Juin, 14050 Caen Cedex, FRANCE.
[††]Università di Catania, 64 Via S. Sofia, 95123 Catania, ITALY.

Abstract. This paper presents two original direct mass-measurement techniques developed at GANIL using the CSS2 and CIME cyclotrons as high-resolution mass spectrometers. The mass measurement with the CSS2 cyclotron is based on a time-of-flight method along the spiral trajectory of the ions inside the cyclotron. The atomic mass excesses of ^{68}Se and ^{80}Y recently measured with this technique are $-53.958(246)$ MeV and $-60.971(180)$ MeV, respectively. The new mass-measurement technique with the CIME cyclotron is based on the sweep of the acceleration radio-frequency of the cyclotron. Tests with stable beams have been performed in order to study the accuracy of this new mass-measurement method and to understand the systematic errors.

Keywords: Atomic Mass, Cyclotron, ^{68}Se, ^{80}Y, Rp-process.
PACS: 21.10, 29.20

INTRODUCTION

Atomic masses provide essential information about nuclear structure since they reflect the interaction of the nucleons inside the nucleus. Therefore they are one of the first quantities usually studied, as they can also be measured with very low statistics. The mass of a nucleus is a very important quantity in the development of new models of nuclear physics and astrophysics. For instance it defines the boundaries of existence of exotic nuclei far from the valley of stability. Data on masses of exotic nuclei also provide a stringent test of the prediction reliability of models built upon data from nuclei near stability.

As we intend to measure the masses of nuclei lying very far away from the line of stability, it becomes more challenging to develop new techniques, which are compatible with the production mechanisms of these exotic nuclei, which are by definition short-lived and have very low production cross sections.

In the last years two direct techniques have been developed at GANIL for mass measurements of exotic nuclei based on the use of cyclotrons as high-resolution mass spectrometers [1]. Since both techniques are direct, nuclei with well-known masses are used as references and therefore accelerated together with the unknown-mass nuclei.

TIME-OF-FLIGHT TECHNIQUE WITH THE CSS2 CYCLOTRON

The time-of-flight (ToF) technique along a linear flight path, in conjunction with the SPEG magnetic spectrometer, has been successfully used at GANIL for nearly two decades, achieving a resolution of $2\text{-}3 \times 10^{-4}$ for extremely neutron-rich nuclei around N=20 and 28, [2].

The more recent CSS2 mass-measurement technique is based on the measurement of the ToF of the ions along their spiral trajectory inside the CSS2 cyclotron. Ions with different mass/charge ratios, m/q, arrive at the detector with a delay, δt, corresponding to a phase difference, $\delta \phi$, proportional to their difference in m/q, according to:

$$\frac{\delta(m/q)}{(m/q)} = \frac{\delta t}{t} = \frac{\delta \phi}{\phi} \tag{1}$$

The mass resolution of the ToF method depends on the length of the flight path, which makes the cyclotron an excellent tool for the ToF technique since the flight path is of the order of 1 Km, which gives a resolution of $\sim 3 \times 10^{-5}$.

A mass measurement of neutron-deficient nuclei with A=64-80 near the N=Z line was recently performed at GANIL using the CSS2 technique. In this experiment a primary beam of ^{58}Ni was accelerated in the CSS1 cyclotron to an energy of 4.3 MeV/nucleon. After extraction from CSS1 the beam impinged onto a Mg target located between the CSS1 and CSS2 cyclotrons, where the secondary beam of exotic nuclei was produced by fusion-evaporation reactions. Two different Mg targets were used for the production of the secondary beam: a 2.2 mg/cm^2 thick ^{nat}Mg target and a 2.0 mg/cm^2 thick ^{24}Mg target. In both cases there was a pretarget of 0.4 mg/cm^2 of ^{12}C and some ^{16}O was also present, both of which contributed to the production of the lighter nuclei.

The ions of the secondary beam fulfilling the condition A/q=4 were injected simultaneously into the CSS2 cyclotron where they were accelerated until intercepted by a Si-detector telescope mounted on a radial probe inside the cyclotron and located at a radius close to extraction. The Si-detector telescope consisted of two 2cm $\times$ 2cm Si detectors: a 30 μm thick detector used for the identification of ions by energy loss and a 300 μm thick detector where the ions deposited all their remaining energy. The energy loss, residual energy and ToF of the ions were measured on an event-by-event basis. The ToF of the ions was measured with respect to the cyclotron radio-frequency.

One of the main difficulties of this technique used in past experiments has been the reliability of the identification of the ions and the separation of isobars. In the present experiment the use of the two different Mg targets, ^{24}Mg and ^{nat}Mg, was essential to confirm the identification thanks to the production of ^{80}Rb and ^{80}Kr when the ^{nat}Mg target was used. More importantly, the high intensity of the primary beam made possible the use of a bunch suppressor, which allowed the event-by-event measurement of the number of turns intercepted and hence the unambiguous separation of the isobars.

During the experiment, nine independent measurements were performed, with the two different Mg targets and different values of the cyclotron magnetic field, B, and of the radial position of the detector inside the cyclotron.

The reference ions used for the mass calibration were $^{64}Zn^{16+}$, $^{64}Ga^{16+}$, $^{68}Ge^{17+}$, $^{68}As^{17+}$, $^{72}Se^{18+}$, $^{76}Kr^{19+}$, $^{76}Rb^{19+}$ and $^{80}Sr^{20+}$. Their relative m/q differences with respect to $^{80}Sr^{20+}$, $\delta(m/q)/(m/q)$, were plotted versus their measured phase differences and fitted to a second order polynomial for each measurement. In order to check the validity of the data analysis and to study the systematic errors, the masses of the reference ions were extracted using the calibration fits. The final measured mass excesses result from the average of the nine independent measurements and the errors were obtained from their dispersion. When compared to their corresponding tabulated values [3], a systematic error of 42 keV was found, which was subsequently added quadratically to the dispersion errors.

In the case of $^{68}Se^{17+}$ and $^{80}Y^{20+}$, their $\delta(m/q)/(m/q)$ with respect to the reference ion $^{80}Sr^{20+}$ is rather large. Hence instead of performing a long extrapolation of the calibration fits, we performed a more complex analysis whereby the local slope of the fits in the region of phase of these two ions was calculated and used for the determination of their masses. The atomic mass excesses of ^{68}Se and ^{80}Y obtained from seven independent measurements are -53.958(246) MeV and -60.971(180) MeV, corresponding to an accuracy of 4×10^{-6} and 2×10^{-6} respectively (including the 42 keV systematic error).

The experimental atomic mass excesses obtained in this work [4] are presented on figure 1 and compared to the values of the most recent Atomic Mass Evaluation (AME) [3] and to values from other authors [6, 7, 8, 9, 10, 11, 12, 13]. All the experimental values for ^{68}Se shown in the figure are in good agreement with the table and indicate that this nucleus is a waiting point of the rp-process. Regarding ^{80}Y, there are two groups of experimental values, one of them in good agreement with the table and an older set of measurements based on β^- decay which gives a more bound mass. The value from this work is in agreement with the tabulated value and the situation regarding this mass seems clarified.

RADIO-FREQUENCY SWEEP TECHNIQUE WITH THE CIME CYCLOTRON

One limitation of the CSS2 technique is the fact that the CSS2 cyclotron is coupled in frequency to CSS1. This makes the change of frequency during the experiment a time consuming process and limits the range of masses that can be simultaneously accelerated to 4×10^{-4}. This drawback can be overcome with the CIME cyclotron since its frequency can be efficiently changed during the experiment. CIME thus offers the possibility to accelerate a wider range of masses than the CSS2 cyclotron.

The basis of the new CIME mass-measurement technique is the sweeping of the acceleration radio-frequency of the cyclotron. According to the cyclotron equation, $m/q = B/\omega$, ions with a given m/q are accelerated at a corresponding mean frequency, $f = \omega/2\pi$. Therefore, as the frequency of the cyclotron is being changed different ions, injected directly from the SPIRAL ion source, are successively accelerated. This makes possible the acceleration of a wider range of masses in the CIME cyclotron (up

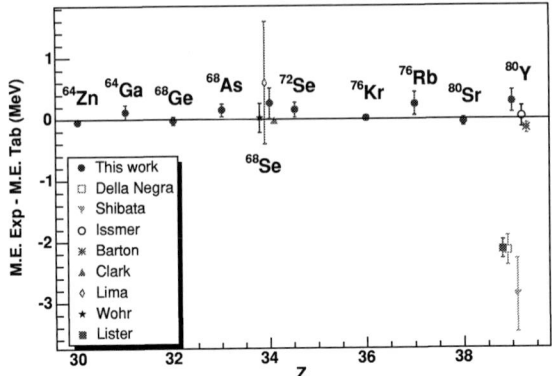

FIGURE 1. Experimental atomic mass excesses, measured with the CSS2 cyclotron, compared to the AME values [3] and to experimental results from other authors, [6, 7, 8, 9, 10, 11, 12, 13].

to $\delta(m/q)/(m/q) \approx 10^{-3}$) than in the CSS2 cyclotron, which is very important since reference masses are scarce in exotic regions of the chart of nuclides. In this technique the difference in m/q between two ions is proportional to the difference in their mean acceleration frequency, f, according to:

$$\frac{\delta(m/q)}{(m/q)} = -\gamma^2 \frac{\delta f}{f} \quad (2)$$

Tests have been performed with stable beams (injected in the SPIRAL ion source) in order to study the feasibility of the radio-frequency sweep and to understand its associated systematic errors. The frequency sweeps were done in short time intervals to ensure the stability of the magnetic field of the cyclotron. The initial frequency was ~12.970 MHz and the range of frequency swept was ~30 KHz in steps of 0.1 KHz and 2 s/step.

A cocktail beam of stable ions composed of $^{12}C^{3+}$, $^{16}O^{4+}$, $^{20}Ne^{5+}$, $^{32}S^{8+}$, $^{36}Ar^{9+}$ and $^{40}Ar^{10+}$ was injected into the CIME cyclotron with the condition A/q=4. The ions were detected in a 2cm×2cm, 300μm-thick Si detector mounted on a radial probe inside CIME. Their total energy, phase and frequency were measured for each event.

The identification of the ions was unambiguous since the acceleration in a cyclotron is done on a per-nucleon basis and there were no isobars present. The phase-versus-frequency spectra were plotted, gating on the total energy peaks corresponding to each ion species. These spectra were rotated until horizontal, with the purpose of reducing the width of the distributions when projected onto the frequency axis [5].

The mean value of the frequency distribution for each ion species was then plotted versus its corresponding relative m/q difference, $\delta(m/q)/(m/q)$, calculated with respect to $^{12}C^{3+}$. The ions used as reference masses were $^{12}C^{3+}$, $^{16}O^{4+}$, $^{20}Ne^{5+}$, $^{32}S^{8+}$, $^{36}Ar^{9+}$, which were fitted to a first order polynomial, from which the $\delta(m/q)/(m/q)$ value of

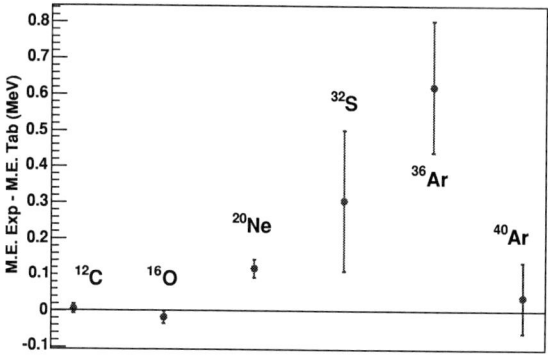

FIGURE 2. Experimental atomic mass excesses obtained with the new CIME cyclotron technique compared to AME values [3].

$^{40}\text{Ar}^{10+}$ was extrapolated. The masses of the reference nuclei were also determined using this fit in order to study the systematic errors.

Nineteen independent measurements were taken at different radial positions of the detector. The final atomic mass excesses result from the weighted average of these nineteen measurements and the error from the mean of the individual errors. The measured atomic mass excess for ^{40}Ar is -35.004(98) MeV, corresponding to a 36 keV difference with the AME [3].

The results of this work [5] are compared on figure 2 with the AME values [3]. The large error bars for ^{32}S and ^{36}Ar are due to low statistics. A quadratic trend, increasing with A, can be appreciated in the deviation from the tabulated values. This could be due to the walk in the electronics and is still to be fully understood.

An experiment has been performed this summer using this new technique in order to measure the mass of ^{31}Ar. The data analysis is in progress.

REFERENCES

1. M. Chartier, *et al, Hyp. Inter.*, **132**, 275–281 (2001).
2. H. Savajols, *Hyp. Inter.*, **132**, 245–254 (2001).
3. G. Audi, *et al., Nucl. Phys.*, **A729**, 1–676 (2003).
4. M. Chartier, *et al. J. Phys. G* **31** 1771–1774 (2005).
5. M.B. Gomez Hornillos, *et al. J. Phys. G* **31** 1869–1872 (2005).
6. S. Della Negra *et al., Z. Phys.* **A307**, 305 (1982).
7. M. Shibata *et al., J. Phys. Soc. Jap.* **65**, 3172 (1996).
8. S. Issmer *et al., Eur. Phys. J.* **A2**, 173–177 (1998).
9. C.J. Barton *et al.*, *Phys. Rev.* **C67**, 034310 (2003).
10. J.A. Clark *et al., Phys. Rev. Lett.* **92**, 192501 (2004).
11. G.F. Lima, *Phys. Rev. C*, **65**, 044618 (2002).
12. A. Wöhr *et al., Nucl. Phys.* **A742**, 349–362 (2004).
13. C.J. Lister *et al., Phys. Rev.* **C24**, 260–278 (1981).

Recent Achievements and Future of Mass Measurements with Fast Radioactive Beams

Milan Matoš [1]

National Superconducting Cyclotron Laboratory & Joint Institute for Nuclear Astrophysics, Michigan State University, East Lansing, Michigan 48824, USA

Abstract. Fast radioactive beams offer mass measurements of exotic short-lived nuclides. Primary beams accelerated to relativistic energies can produce in a production target a large variety of ions. GSI in Germany and NSCL at MSU in the USA are facilities that have devices and projects for mass measurement with fast radioactive beams. Two complementary methods at the storage ring at GSI are already implemented and are producing results. Two experiments that used the fast mass spectrometry of individual ions with the storage ring in isochronous mode produced new mass values of exotic neutron-rich nuclides. A new TOF-Bρ mass measurement method project has been started at the NSCL facility.

Keywords: Mass measurements; Radioactive beams
PACS: 21.10.Dr, 29.20.-c, 26.50.+x

INTRODUCTION

Mass measurements have always played an important role in exploring nuclear properties, so it is surprising that a systematic exploration that reached exotic regions and moved towards the drip-lines began only a decade ago. At that time only a few techniques allowed mass measurements in the large unexplored areas: the pioneering techniques of the Penning-trap facility ISOLTRAP at CERN [1], the Schottky Mass Spectrometry in the storage ring at GSI [2], and indirect Q_α mass measurements [3]. Mass measurements focused on the neutron-deficient areas about a decade ago, but very short half-lives and difficulty in production did not allow measurements of masses of more exotic neutron-rich nuclides, which are required for astrophysical r–process path calculations. Only recently the neutron-rich area has become a region of interest for new investigations. One of the techniques that has a capability to improve the knowledge about masses of neutron-rich nuclides is mass measurements with fast radioactive beams.

MASS MEASUREMENTS WITH FAST RADIOACTIVE BEAMS

Radioactive Beams

Fast radioactive beams are produced in peripheral reactions of primary beam ions with target nuclei. Relativistic energies exceeding 50 MeV/nucleon are reached

---

[1] E-mail address: matos@nscl.msu.edu

by heavy-ion accelerators; at NSCL, MSU it is a coupled superconducting cyclotron facility that provides energies from 10 to 200 MeV/nucleon [4], while the synchrotron SIS at GSI reaches energies up to 1 GeV/nucleon for all ions including uranium [5].

There are two reaction processes that produce exotic nuclei: projectile fragmentation and projectile fission [6]. Most of the exotic nuclides lighter than the projectile are produced in a fragmentation reaction with a large range of cross-sections that decrease rapidly away from the valley of stability: less on the proton-rich side, more steeply on the neutron-rich side. A lack of high yields from fragmentation in very neutron-rich regions could be balanced by fission reactions. At the present facilities the ^{238}U primary beam is exclusively used to produce fission fragments. Two processes, reactions with overlap of target and projectile nuclei and Coulomb fission without overlap, produce neutron-rich ions with highest intensity for nuclides around N=50 and N=82.

Mass Measurements

Mass measurements of fast exotic ions are in general based on the equation between their mass-to-charge ratio m/q, magnetic rigidity $B\rho$ and time-of-flight t over a distance l,

$$\frac{m_0}{q}\gamma = B\rho \frac{t}{l}. \qquad (1)$$

Facilities offering fast radioactive beams are suitable for mass measurements of very exotic nuclides. They are designed to produce species far away from stability and require only a very short time for a measurement. Trajectories of relativistic ions are well controlled by magnetic optics, only the time-of-flight and magnetic rigidity are required for a mass measurement. In reality, an absolute $B\rho$ measurement would be very complicated. The problem is solved by fixing the $B\rho$ for all ions (which is, within its acceptance, fulfilled in the magnetic optical system). The mass value of a nuclide is determined from its time-of-flight with respect to the time-of-flight of a nuclide with well-known mass, the so-called reference mass (or reference nuclide).

In general the magnetic optics does not fix the magnetic rigidity well enough for precise mass measurement. Two methods are used to solve this: **1. Relative measurement of magnetic rigidity** (by position measurement at the dispersive optical plane), which is applied at the SPEG mass measurements at GANIL [7], and TOF-$B\rho$ technique at NSCL, MSU. Time-of-flight values are consequently corrected with respect to the central beam line based on the relative magnetic rigidities. **2. A constraint in some parameter** is used in several techniques. Isochronous setting of magnetic optics requires

$$\gamma \frac{l}{B\rho} = const., \qquad (2)$$

so that the time-of-flight of an ion depends on the mass-to-charge ratio only and is independent of its velocity (or magnetic rigidity). This technique was used at TOFI in Los Alamos [8]. The method that uses isochronous mode in the storage ring at GSI [9, 10] is described in this work. However, it was shown in the experiment [11] that the restriction of magnetic rigidity by scrapers in the beam line improves measurement precision. It is a complementary technique to the Schottky Mass Spectrometry [2] that uses an electron cooler to the fix velocity of all ions. For a specific nuclide and charge state, magnetic rigidity and trajectory are fixed.

One should mention other methods of mass determination that use radioactive beams, but these are out of the scope of this paper. Two of them are running at NSCL; The High Resolution Array Detector (HIRA) will be used for mass measurements of the rp-process nuclides via Q-value measurements of (p, d) transfer reactions in inverse kinematics [12]. The low-energy beam and ion trap LEBIT [13] which slows down fast exotic ions has measured masses of fragmentation products for the first time in a Penning trap.

FAST MASS SPECTROMETRY OF INDIVIDUAL IONS AT THE STORAGE RING

At GSI the fast radioactive beam is injected into the storage ring ESR that is used for multiple time-of-flight (or frequency) measurements [5]. The first method developed for mass measurements was Schottky Mass Spectrometry (SMS). The ion stored in the ring are cooled to momentum spreads of the order of 10^{-6} [14] and the frequencies are obtained from the Fourier-transformed noise signal in the Schottky detector. The electron cooler and the Schottky detector require few seconds for successful operation which limits the mass measurements to long-lived species, and is not suitable for most of the exotic neutron-rich nuclides.

Fast mass spectrometry of individual ions [15] at the storage ring operated in isochronous optical mode with time-of-flight measured by micro-channel plate detectors was developed to reach nuclides with half-lives as short as few tens of microseconds [9, 10]. Pilot experiments were performed in neutron-deficient regions with many reference masses and new mass values of rp-process nuclides were obtained [16, 17], see left panel of Fig. 1. In 2002 two experiments of ^{70}Zn projectile fragments and ^{238}U fission fragments were performed [18, 19] to determine masses

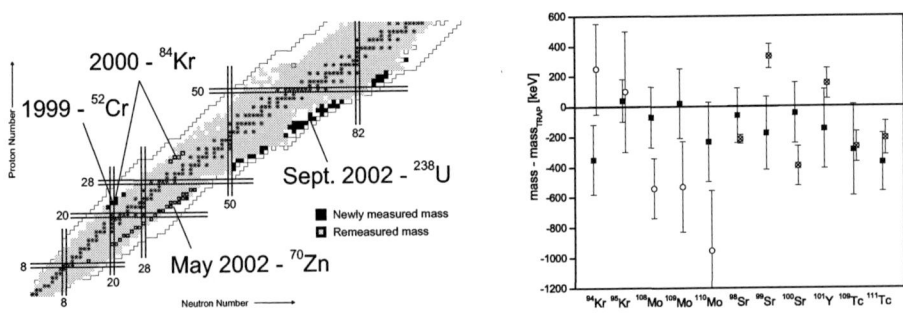

FIGURE 1. Left: Chart of mass measurements with ESR-TOF techniques. Neutron-deficient nuclides were investigated in first two pilot experiments [10, 17], a large areas of neutron-rich nuclides were covered in two 2002 experiments [18, 19]. **Right:** A comparison of mass values obtained from ESR-TOF measurements [11] and AME-2003 extrapolated (empty circles) or evaluated (crossed circles) values [20] with recent Penning-trap results (Kr isotopes from the ISOLTRAP [23], rest from the JYFLTRAP [24]).

in the large regions of exotic neutron-rich nuclides. Details of the data analysis and final new mass values for more than 40 nuclides with accuracies in the range 100–300 keV can be found in [11]. A comparison of previously measured mass values with evaluated ones [20] shows for the first time the tendency of Q_β mass measurements to carry systematic errors, as it was also later observed by other techniques [21, 22]. New measurements provided by Penning-trap facilities [23, 24] confirmed the correctness of the results for 5 new mass values and 6 remeasured ones. The comparison is shown in the right panel of Fig. 1. Large uncertainties are caused by the lack of reliable reference masses and peculiarities of the isochronicity condition that is only good enough for part of the time-of-flight spectrum. The series of the mentioned new Penning-trap measurements could solve the first problem; the uncertainties will be reduced by 11 new reference masses shown in right panel of Fig. 1. The second difficulty could be solved only in the future; a second TOF detector (both to be in the linear part of the storage ring) would allow velocity measurement with relative uncertainty of 10^{-4} and decrease the isochronicity requirement by a factor of 100.

TOF-Bρ MASS MEASUREMENT AT NSCL, MSU

At the NSCL a fast radioactive beam is produced in the A1900 fragment separator [25]. The 58 m long time-of-flight line starts at the A1900 extended focal plane and ends at the focal plane of the spectrograph S800 [26], see Fig. 2. Fast scintillator detectors provide the timing resolution of about 20 ps, the relative magnetic rigidity will be measured at the momentum dispersive plane by position-sensitive micro-channel plates detector [27]. First mass measurement experiment with ^{86}Kr primary beam will be aiming for the neutron-rich nuclides around ^{68}Fe, whose mass values are important for astrophysical r-process calculations.

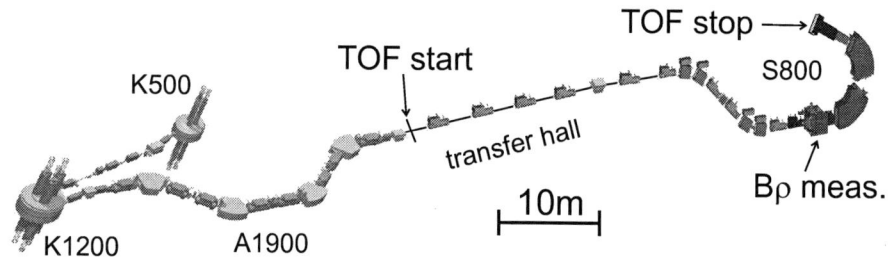

FIGURE 2. An overview of the facility. The primary relativistic beam is produced in coupled cyclotron facility K500-K1200. Radioactive ions are obtained in fragment separator A1900, where TOF measurements starts, magnetic rigidity measurement and TOF stop detector are placed in the S800 spectrograph that is rotated by 90°, thus the bending plane of its dipoles is vertical.

CONCLUSIONS

Fast radioactive beams provide a production of exotic short-lived nuclides and their mass measurements. These measured mass values are of a high importance for many aspect of physics, e.g. for astrophysical r-process calculations that have to use mass values calculated by nuclear mass models such as FRDM [28] fitted over measured mass values. However, discrepancies with previous mass measurements in neutron-rich area were observed recently by several authors [18, 21, 22]. It is important to point out that these disagreements are exclusively with values from β-decay spectroscopy that tend to carry systematic errors. These uncertainties could have a strong effect on r–process path calculations [29]. To minimize this effect, new mass measurements in neutron-rich areas are required. Mass measurements with fast radioactive beams are especially suitable for the most exotic nuclides. Their accuracy of 100 keV or more is much larger than the accuracy of Penning-trap mass measurements that is better than few keV. However, time-of-flight mass measurements of fast radioactive ions are always a step further and provide mass values of nuclides that are a few isotope more exotic than nuclides with masses measured by Penning traps. New experiments and also new facilities should offer new exciting information about masses and other properties of very exotic neutron-rich nuclides.

REFERENCES

1. K. Blaum, et al. *Nucl.Phys.*, **A752**, 317c–320c (2005).
2. T. Radon, et al., *Nucl.Phys.* **A677** 25–99 (2000).
3. Yu. Novikov, et al. *Nucl.Phys.*, **A697**, 92–106 (2002).
4. C. K. Gelbke, *Prog. Part. Nucl. Phys.*, **53**, 363–372 (2004).
5. G. Münzenberg, *Prog. Part. Nucl. Phys.*, **53**, 351–362 (2004).
6. T. Enquist, et al., *Nucl. Phys.* **A658**, 47–66 (1999).
7. A. Gillibert, et al., *Phys.Lett.* **B176** 39–43 (1986).
8. J. M. Wouters, et al., *Nucl. Instrum. Methods Phys. Res.* **B26** 286–293 (1987).
9. H. Wollnik, et al., *Nucl.Phys.* **A626**, 327c–331c (1997).
10. M. Hausmann, et al., *Nucl. Instrum. Methods Phys. Res.* **A446** 569–580 (2000).
11. M. Matoš, Ph.D. Thesis, Justus Liebig Universität Giessen, Germany 2005, http://geb.uni-giessen.de/geb/volltexte/2004/1582.
12. M. Wallace, private comm. (2005).
13. G. Bollen, et al., *Nucl. Phys.* **A746**, 597c–603c (2004).
14. M. Steck, et al., *Nucl.Phys.* **A626** 495c–498c (1997).
15. Yu. Novikov, *Proc. 6th Int. Conf. on Nuclear Physics at Storage Rings, May 2005, Bonn, Germany (STORI-2005)*, in press.
16. M. Hausmann, et al., *Hyperfine Interact.* **13** 289–295 (2001)
17. J. Stadlmann, et al., *Phys.Lett.* **B586** 27–33 (2004).
18. M. Matoš, et al., *Proc. International Symposium on Exotic Nuclei, July 2004, Peterhof, Russia (EXON2004)*, World Scientific Publishing, Singapore, 2005, pp. 90–95.
19. M. Matoš, et al., *Proc. International Symposium on Exotic Nuclei, July 2004, Peterhof, Russia (EXON2004)*, World Scientific Publishing, Singapore, 2005, pp. 96–99.
20. G. Audi, et al., *Nucl. Phys.* **A729**, 1–676 (2003).
21. J. Äystö, this proceedings.
22. J. A. Clark, et al., *Nucl.Phys.*, **A746**, 342c–346c (2004).
23. D. Lunney, *Eur. Phys. J.* **A25** 3–8 (2005).
24. J. Äystö, private comm. (2005). U. Hager, et al., *Phys. Rev. Lett.*, in press.
25. A. Stolz, et al., *Nucl. Instrum. Methods Phys. Res.* **B**, in press (2005).
26. D. Bazin, et al., *Nucl. Instrum. Methods Phys. Res.* **B204** 629–633 (2000).
27. D. Shapira, et al., *Nucl. Instrum. Methods Phys. Res.* **A454** 409–420 (2000).
28. P. Möller, et al., *At. Data Nucl. Data Tables*, **59**, 185–381 (1995).
29. K.-L. Kratz and B. Pfeiffer, private comm. (2005).

NUCLEAR ASTROPHYSICS

How Accurately Do We Know the Cross Section of the $^7Be(p,\gamma)^8B$ Reaction?

Moshe Gai

Laboratory for Nuclear Science at Avery Point, University of Connecticut, 1084 Shennecossett Rd., Groton, CT 06340-6097, USA
e-mail: moshe.gai@yale.edu, URL: http://astro.uconn.edu

Abstract.
The "world average" of the astrophysical cross section factor, $S_{17}(0)$, is driven by the Seattle result due to the very small quoted uncertainty, which we however demonstrate it to be an overestimated accuracy. We propose more realistic error bars for the Seattle results based on the published Seattle data. This leads to a an uncertainty of the measured slope of S_{17} and thus an uncertainty due to extrapolation that can be reasonably estimated to be $^{+0.0}_{-3.0}$ eV-b.

Keywords: Solar Fusion, Solar Neutrinos, Astrophysical Cross Section factor
PACS: 25.20.Dc, 25.70.De, 95.30.-K, 26.30.+K, 26.65.+t

INTRODUCTION

The value of $S_{17}(0) = 21.4$ eV-b and the (impressive) accuracy of $\pm 4\%$ has been adopted from the Seattle group [1]. It should be compared to the previously quoted value of $S_{17}(0) = 19 +4 -2$ eV-b [2]. The new quoted value of $S_{17}(0)$ leads to an increase of 12.6% in the predicted ^8B solar neutrino flux. Since the claimed high precision of the Seattle measured data points and quoted extracted $S_{17}(0)$ [1] drive the value and the error of so called "world average" it seems reasonable to examine it in detail, especially at low energies.

HOW ACCURATE ARE THE SEATTLE DATA

In Fig. 1 we show the results of the target-beam calibration data shown by the Seattle group using the $^7Be(\alpha,\gamma)^{11}C$ reaction. As can be seen in Fig. 1 both calibration spectra published in the Phys. Rev. Lett. and the Journal Phys. Rev. C (which where measured with different ^7Be targets) exhibit a resonance energy which is off by 9 keV from the known energy in ^{11}C. This 9 keV shift is intolerable and the authors as yet did not explain this discrepancy in a published erratum. The repeated publication of the same mistake first in the PRL paper and a year later in the PRC long paper does not lend credence to the claimed high precision experiment.

---

[1] Work Supported by USDOE Grant No. DE-FG02-94ER40870.

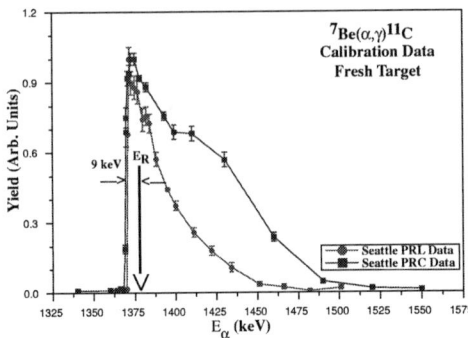

FIGURE 1. The measured beam-target calibration spectra [1] using the $^7Be(\alpha,\gamma)^{11}C$ reaction. Both spectra published in PRL and PRC exhibit resonance energy which is off by 9 keV from the well known resonance energy in ^{11}C as indicated.

Furthermore, in their paper they show the ^7Be target profile measured in the middle (PF2) and at the end (PF3) of the experiment [1]. These profiles are superimposed on top of each other in Fig. 2a, from which it becomes clear that we can not support the claim [1] that "they are similar". In fact there is evidence that the ^7Be moved further into the target in the intervening period between these two measurements. The unstable nature of the ^7Be during the experiment must be reflected in a systematic uncertainty due to variation of the expected yield, as we discuss below. Another systematic uncertainty that is not considered here is associated with the evaluation of the effective center of mass energy.

These profiles are measured with alpha-particle beams with energies of approximately 1.4 MeV in steps of $\Delta E = E_\alpha - E_R$ which can be translated to steps of $\Delta E = (E_\alpha - E_R) \times \frac{dE}{dX}_{p+(Mo+^7Be)} / \frac{dE}{dX}_{\alpha+(Mo+^7Be)}$ for proton beams. Note that the ratio $\frac{dE}{dX}_{p+(Mo+^7Be)} / \frac{dE}{dX}_{\alpha+(Mo+^7Be)}$ is almost independent of the exact $Mo + ^7Be$ mixture of the target [1], as well as variation among the various tabulated individual energy loss. Hence for an incoming proton beam (E_p) the ^7Be target profiles during the $^7Be(p,\gamma)^8B$ reaction measurement is given by:

$$\frac{dN_{tgt}}{dE}((E_p - (E_\alpha - 1370) \times \frac{\frac{dE}{dX}_{p+(Mo+^7Be)}}{\frac{dE}{dX}_{\alpha+(Mo+^7Be)}}) \propto \text{Profile}(E_\alpha) \quad \text{(equ. 2)}$$

The so obtained profile is shown in Fig. 2b, it leads to very large target thickness, for example of the order of 35 keV at a beam energy of approximately 130 keV. Over such a thick target the cross section varies by almost a factor of five, as shown in Fig. 2c. A convolution of the so obtained target profile with the expected variation of the cross section leads to the expected yields:

$$\text{Yield} \propto \Sigma_{E_i} \frac{dN_{tgt}}{dE_p}(E_i) \times \sigma(E_i) \quad \text{(equ. 3)}$$

The yields evaluated for the two shown profiles measured in the middle (PF2) and at the end (PF3) of the experiments differ by 7.5% at the lowest measured energy, almost

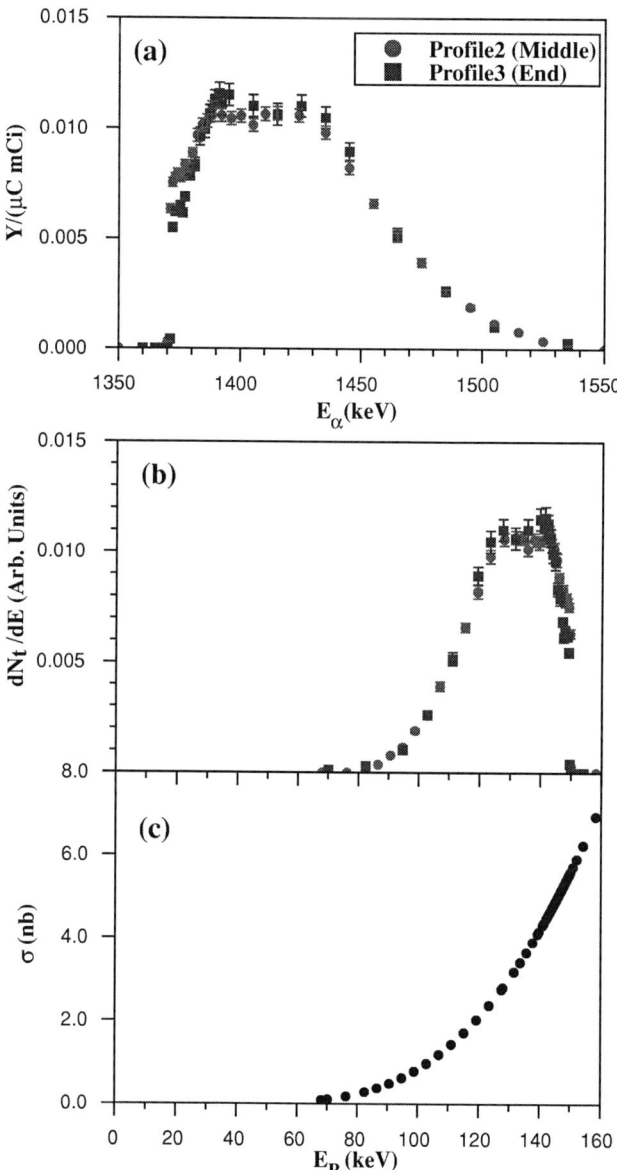

FIGURE 2. (a) Target profiles measured using the $^7Be(\alpha,\gamma)^{11}C$ reaction in the middle (PF2) and at the end (PF3) of the experiment [1], (b) the target profile for a 150 keV proton beam, (c) the predicted cross section variation across the target.

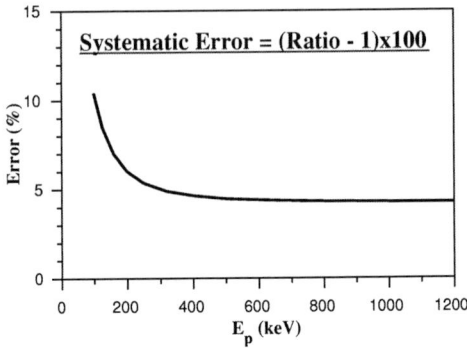

FIGURE 3. One systematic error of the Seattle experiment [1] due to variation in ^7Be target profile shown in Fig. 2. The two profiles measured in the middle (PF2) and at the end (PF3) of the experiment are used.

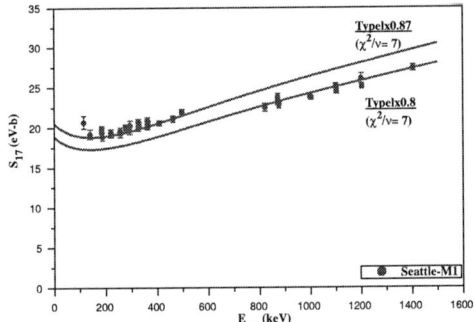

FIGURE 4. A comparison of the Typel potential model [3] with the Seattle data [1].

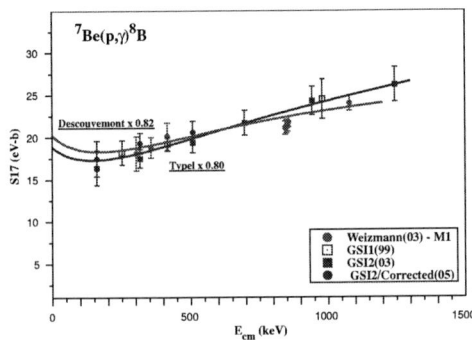

FIGURE 5. A comparison of the Descouvemont cluster model [4] and Typel potential model [3] with the GSI [3] and Weizmann data [5].

a factor of 3 larger than the quoted systematic uncertainty [1]. In Fig. 3 we show the differential yields = Y(PF2)/Y(PF3) - 1 evaluated for all incoming proton beam energies. This differential yield must be considered a lower limit of the systematical uncertainty. This uncertainty is a factor of 3 larger than the quoted varying systematical uncertainty [1]. Clearly the most troubling uncertainty is at energies below 400 keV. This energy range was emphasized in the Seattle paper as most useful for extracting $S_{17}(0)$, but it is measured with the largest systematical uncertainty.

The large systematical uncertainty of Seattle data at low energies is particularly bothersome vis-a-vis the exclusion of theoretical models such as the potential model of Typel [3] that exhibit a considerably different slope of the cross section factor S_{17}. The different slope is due to smaller s-wave and thus larger d/s ratio. In Fig. 4 we show a comparison of that model to the Seattle data where we find that the model reproduce well the high energy data but not the low energy data (or vice versa). The large systematical error at low energies shown in Fig. 3 cast doubt on the claim that the Typel model is "rejected" by the data. This model fits well all other data including the larger slopes exhibited by the early GSI data [3] and Weizmann data [5]. The potential model consistently yield extrapolated values of $S_{17}(0)$ which are approximately 3 eV-b smaller, as shown in Fig. 5.

We conclude that in the absence of an accurate measurement of the slope we must add an extrapolation error to the quoted $S_{17}(0)$. And in view of the fact that the Typel model is consistent with all existing data except the Seattle data, and in view of the large systematical error of the Seattle data at low energies, a reasonable estimate of an extrapolation error is of +0.0 -3.0 eV-b. Such a large error must be alleviated by future experiments that will determine the slope of S_{17} with high precision.

REFERENCES

1. A.R. Junghans *et al.*; Phys. Rev. Lett **88**, 041101 (2002), ibid Phys. Rev. **C68**, 065803 (2003).
2. E.G. Adelberger *et al.*; Rev. of Modern Phys. **70**, 1265 (1998).
3. F. Schumann *et al.*; Phys. Rev. Lett. **90**, 232501 (2003).
4. P. Descouvemont; Phys. Rev. **C70**, 065802 (2004).
5. L.T. Baby *et al.*, Phys. Rev. Lett. **90**, 022501 (2003), ibid Phys. Rev. **C67**, 065805 (2003), ER **C69**, 019902(E) (2004).

Nuclear Astrophysics Experiments at CIAE

Weiping Liu[*], Zhihong Li, Xixiang Bai, Gang Lian, Bing Guo, Sheng Zeng, Shengquan Yan, Baoxiang Wang, Nengchuan Shu, and Yongshou Chen

China Institute of Atomic Energy, P. O. Box 275(1), Beijing 102413, P. R. China

Abstract. This paper describes the nuclear astrophysical studies using the unstable ion beam facility, GIRAFFE. We measured the angular distributions for some low energy reactions, such as $^7Be(d,n)^8B$, $^{11}C(d,n)^{12}N$, $^8Li(d,n)^9Be$, $^8Li(d,p)^9Li$ and $^{17}F(d,n)^{18}Ne$ in inverse kinematics, and derived the astrophysical S-factors or reaction rates of $^7Be(p,\gamma)^8B$, $^{11}C(p,\gamma)^{12}N$, $^8Li(n,\gamma)^9Li$, $^{17}F(p,\gamma)^{18}Ne$ by asymptotic normalization coefficient approach at astrophysically relevant energies.

Keywords: Nuclear strophysics experiment, (d,n) reactions, (p,γ) reactions
PACS: 25.60.-t, 27.20.+n

DESCRIPTION OF THE UNSTABLE ION BEAM FACILITY GIRAFFE

Aiming at the studies of nuclear astrophysics, the secondary beam facility (GIRAFFE) [1,2] for producing and utilizing low energy beams of unstable nuclei has been constructed at the HI-13 tandem laboratory in 1993. The facility made use of the transfer and charge exchange reactions in inverse kinematics to yield some beams of unstable ions (A < 20) near the β-stability line with the acceptable intensities (10^4 - 10^6 pps). It comprises a primary reaction chamber, a dipole-quadrupole doublet (D-Q-Q) magnetic separation and focusing system, as well as a secondary reaction chamber, as shown in Fig. 1. Up to now, the ion beams of ^6He, ^7Be, ^8Li, ^{11}C, ^{13}N, ^{15}O, ^{17}F and ^{10}C have been delivered. They are summarized in Table 1. A velocity filter is installed between quadrupole doublet and focal plane by the end of 2004, which greatly improved the secondary beam purity.

EXPERIMENTS AND THEORETICAL ANALYSIS

The astrophysical S-factor for the $^7Be(p,\gamma)^8B$ reaction at solar energies is a crucial nuclear physics input for the *"solar neutrino problem"*. The S-factor can be indirectly determined through the asymptotic normalization coefficient (ANC) [3] extracted from the proton pickup reaction of ^7Be, with accuracy comparable to that from direct radiative capture or Coulomb Dissociation reaction, and thus can provide a significant cross examination. We measured the $^7Be(d,n)^8B$ angular distribution in inverse kinematics at E_{cm} = 5.8 MeV and extracted the ANC for the virtual decay $^8B \rightarrow {}^7Be + p$ based on

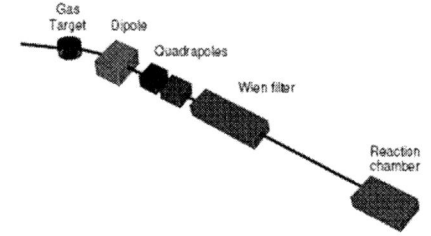
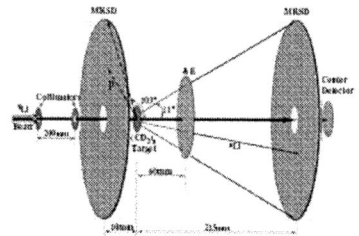

FIGURE 1. Sketch of GIRAFEE   **FIGURE 2.** Experimental setup

TABLE 1. Summary of the produced unstable ion beams at GIRAFFE

RNB	Reaction	Energy (MeV)	FWHM (MeV)	Purity (%)	Beam intensity (pps)[a]
^{6}He	^{2}H(^{7}Li, ^{6}He)^{3}He	35.3	0.5	90	500
^{7}Be	^{1}H(^{7}Li, ^{7}Be)n	23.0	1.3	99	1000
^{8}Li	^{2}H(^{7}Li, ^{8}Li)^{1}H	39.0	0.5	90	500
^{11}C	^{1}H(^{11}B, ^{11}C)n	38.2	2.7	85	1000
^{13}N	^{2}H(^{12}C, ^{13}N)n	57.8	2.1	86	500
^{15}O	^{2}H(^{14}N, ^{15}O)n	66.0	3.6	91	300
^{17}F	^{2}H(^{16}O, ^{17}F)n	76.1	3.7	90	300
^{10}C	^{1}H(^{10}B, ^{10}C)n	55.9	3.5	96	200

[a]With 2 mm diameter collimator and primary beam intensity 100-700 enA.

DWBA [4] analysis. The astrophysical S-factor for the ^{7}Be(p,γ)^{8}B reaction at zero energy was found to be $S_{17}(0)$ = 27.4 ± 4.4 eVb [5]. Our experimental data were re-analyzed by other groups, as shown in Fig. 5.

One of the key reactions in the hot pp chains is the ^{11}C(p,γ)^{12}N which is believed to play an important role in the evolution of Population III stars. As a result of the low Q-value, its cross section at astrophysically relevant energies is likely dominated by the direct capture into the 1$^+$ ground state of ^{12}N, and the resonant captures into the first and second excited states of ^{12}N at 2$^+$ 0.960 MeV and 2$^-$ 1.191 MeV, respectively. Angular distribution of the ^{11}C(d,n)^{12}N reaction at E_{cm} = 9.8 MeV was measured with the secondary ^{11}C beam. The experimental data were analyzed with DWBA calculations and thereby the (ANC)^2 was extracted to be 2.86 ± 0.91 fm^{-1} for the virtual decay ^{12}N → ^{11}C + p. The zero energy astrophysical S-factor for the direct capture ^{11}C(p,γ)^{12}N reaction was then derived to be 157 ± 50 eV b. We have also estimated the contributions from resonant captures into the first and second excited states of ^{12}N and the interference between direct capture into the ground state and resonant capture into the second excited state. The astrophysical S-factor of ^{11}C(p,γ)^{12}N in the astrophysically relevant energies are illustrated in Fig. 3. The temperature dependence of the direct capture, resonant capture and total reaction rates for ^{11}C(p,γ)^{12}N were derived [6]. This work shows that the direct capture dominates the ^{11}C(p,γ)^{12}N in the wide energy range of astrophysical interest except the ranges corresponding to two resonances.

In the baryon inhomogeneous big-bang models for primordial nucleosynthesis, (IBBNs) [7], many nuclear reactions of unstable nuclei are involved, which can bridge the stability

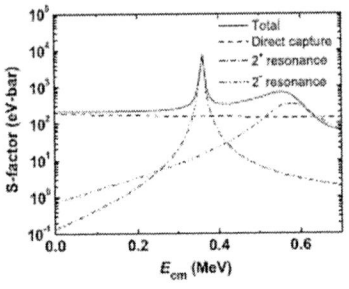

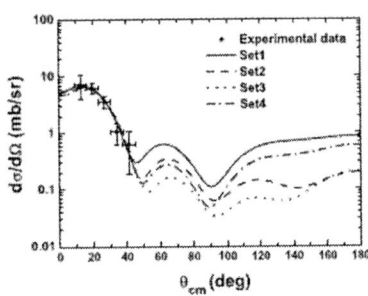

FIGURE 3. Deduced $^{11}C(p,\gamma)^{12}N$ astrophysical S-factors.

FIGURE 4. Angular distribution of $^8Li(d,p)^9Li$ at $E_{cm} = 7.8$ MeV

gap at mass number A = 8, and predict a higher production of elements beyond 7Li and a larger universal mass-density parameter of baryons Ω_B. The reaction chains involving unstable nuclei ^8Li, ^9Li, ^8B, etc. are found to play a pivotal role in IBBNs. The production of succeeding heavier elements scales with the abundances of these unstable isotopes during primordial nucleosynthesis and thus all the reactions for generating or destroying them are of importance. We have measured the angular distribution of $^8Li(d,p)^9Li$ reaction at $E_{cm} = 7.8$ MeV, through coincidence detection of ^9Li and recoil proton, and obtained the cross section and astrophysical S-factor. By using spectroscopic factor deduced from the $^8Li(d,p)^9Li_{g.s}$ angular distribution, we have successfully derived the $^8Li(n,\gamma)^9Li$ direct capture cross section and astrophysical reaction rate for the first time [11].

The typical experimental setup for the $^8Li(d,p)^9Li$ reaction is shown in Fig. 2, the setup of $^7Be(d,n)^8B$ reaction and that of $^{11}C(d,n)^{12}N$ were described elsewhere [5,6], respectively. Two Multi-Ring Semiconductor Detectors (MRSDs) with center hole were used in this experiment. The upstream one aimed at detection of the recoil protons, and the down-stream one served as a residue energy (E_r) detector which composed a $\Delta E - E_r$ silicon counter telescope. This setup enabled the ^9Li-recoil proton coincidence measurement. We applied the similar experimental setup to other reactions except upstream MRSD. Such a detector configuration covered the full laboratory angular region. This setup also facilitated to precisely determine the accumulated quantity of incident unstable beams because the beams themselves were recorded by the counter telescope simultaneously. As examples, Fig. 4 demonstrates the angular distribution of $^8Li(d,p)^9Li$ reaction, where set 1 to set 4 refer to four sets of optical potential parameters; Fig. 5 shows the comparison of $^7Be(p,\gamma)^8B$ $S_{17}(0)$ factor with other measurements described in Ref. [8] and references therein; Fig. 6 displays the reaction rate of $^8Li(n,\gamma)^9Li$ derived through transfer reaction approach and those of theoretical calculations and Coulomb dissociation measurements presented in Ref. [9] and references therein. This data was also used to extract the ANC of mirror system, by assuming the identical nuclear spectroscopic factor as a result of mirror symmetry [12].

Recently the measurement of $^{17}F(d,n)^{18}Ne$ reaction angular distribution at center of mass energy of 7.0 MeV was finished. The data analysis is underway, this data will be very important to determine the contribution of direct capture component of $^{17}F(p,\gamma)^{18}Ne$ reaction. All astrophysical reactions and their deduced parameters are summarized in Table 2.

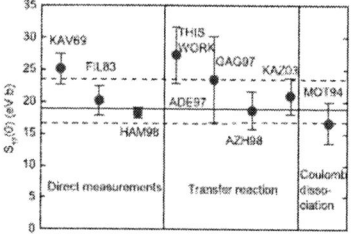

FIGURE 5. The $^7Be(p,\gamma)^8B$ S-factor by different approach.

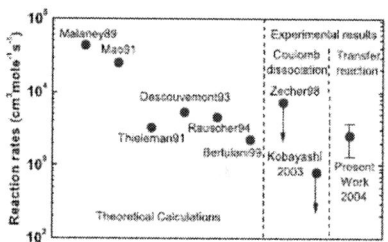

FIGURE 6. Comparison for reaction rates of $^8Li(n,\gamma)^9Li$.

TABLE 2. Summary of astrophysics experiment results

Reaction	E_{cm} (MeV)	σ_{tot} (mb)	(ANC)^2 (fm^{-1})	Indirect Reaction	S-factor or reaction rate	reference
$^7Be(d,n)^8B$	5.8	58 ± 8	0.711 ± 0.090	(p,γ)	27 ± 4 eV b	[5]
$^7Be(d,n)^8B$	8.3	28 ± 3	0.62 ± 0.12	(p,γ)	24 ± 5 eV b	[10]
$^{11}C(d,n)^{12}N$	9.8	23 ± 5	2.86 ± 0.91	(p,γ)	157 ± 50 eV b	[6]
$^8Li(d,p)^9Li$	7.8	7.9 ± 2.0	1.25 ± 0.25	(n,γ)	3970 ± 950 cm^{3}mole^{-1}s^{-1}	[11]
$^8Li(d,p)^9Li$	7.8	7.9 ± 2.0	1.10 ± 0.23^a	(p,γ)^a	42 ± 9 eV b	[12]
$^{17}F(d,n)^{18}Ne$	7.0	tbd	tbd	(p,γ)	tbd	[13]

^{a}For $^8B(p,\gamma)^9C$ mirror system.

SUMMARY

In summary, GIRAFFE, a tandem based one stage unstable beam facility proved to be effective to produce beams suitable for the study of nuclear astrophysics reactions. Angular distribution measurements of transfer reaction in inverse kinematics, together with DWBA/ANC theoretical approach have been used to study the astrophysical reactions indirectly. The astrophysical S-factors and/or reaction rates for $^7Be(p,\gamma)^8B$, $^{11}C(p,\gamma)^{12}N$, $^8Li(n,\gamma)^9Li$, $^{17}F(p,\gamma)^{18}Ne$ were deduced by using the measurements of $^7Be(d,n)^8B$, $^{11}C(d,n)^{12}N$, $^8Li(d,p)^9Li$ and $^{17}F(d,n)^{18}Ne$ reactions at the energies of astrophysical interest.

ACKNOWLEDGEMENT

The above research programs were supported by the Major State Basic Research Development Program under Grant Nos. G200077400 and 2003CB716704, the National Natural Science Foundation of China under Grant Nos. 19935030, 10025524 and 10375096.

REFERENCES

1. X. Bai, W. Liu, J. Qin et al., Nucl. Phys. A 588 (1995) 273c.
2. W. Liu, Z. Li, X. Bai et al., Nucl. Instr. Meth. B 204 (2003) 62.
3. H. Xu, C. Gagliardi, R. Tribble, Phys. Rev. Lett. 73 (1994) 2027.
4. P. Kunz, computer code DWUCK4.
5. W. Liu, X. Bai, S. Zhou et al., Phys. Rev. Lett. 77 (1996) 611.
6. W. Liu, Z. Li, X. Bai et al., Nucl. Phys. A 728 (2003) 275.
7. T. Kajino and R. N. Boyd, Astrophys. J. 359 (1990) 267.
8. K. Ogata, M. Yahiro, Y. IseriH et al., Phys. Rev. C 67 (2003) 011602(R).
9. H. Kobayashi, K. Ieki, A. Horvath et al., Phys. Rev. C 67 (2003) 015806.
10. Y. Wang, W. Liu, X. Bai et al., Chin. Phys. Lett. 16 (1999) 873.
11. Z. H. Li, W. P. Liu, Z. H. Li et al., Phys. Rev. C 71(2005)052801(R)
12. G. Guo, W. P. Liu, Z. H. Li et al., Nucl. Phys. (2005) (in press)
13. G. Lian, W. P. Liu, Z. H. Li et al., (unpublished).

An independent measurement of the $^{12}C(\alpha,\gamma)^{16}O$ cross section with the Karlsruhe 4π BaF_2 detector

R. Plag*, M. Heil*, F. Käppeler* and K. Wisshak*

Institut für Kernphysik, Forschungszentrum Karlsruhe, Postfach 3640, D-76021 Karlsruhe, Germany

Abstract. Many measurements of the $^{12}C(\alpha,\gamma)^{16}O$ cross section have been performed up to now but only a few could measure at energies as low as 1 MeV. All of these measurements at low energy have been carried out using Germanium detectors. The low gamma-ray efficiency of these detectors was compensated by very intense beam currents, which may cause considerable uncertainties due to severe target degradation. A verification of these data with a completely different approach, which could reveal systematic uncertainties, was still missing.

The realization of such an independent measurement was performed with the Karlsruhe 4π BaF_2 array. Due to its high gamma-ray efficiency the beam current could be substantially reduced thus minimizing the thermal load of the target and avoiding sputtering effects.

Keywords: carbon, stellar helium burning, alpha capture, $^{12}C(\alpha,\gamma)^{16}O$
PACS: 27.20.1n, 25.55.2e, 26.201f

INTRODUCTION

The $^{12}C(\alpha,\gamma)^{16}O$ reaction is considered as one of the most important processes in nuclear astrophysics since it determines the ratio of ^{12}C to ^{16}O during stellar Helium burning. Together with the triple-alpha-process this reaction is crucial for the further evolution of a star.

The direct measurement at stellar energies, where the cross section is in the sub-femto barn range, appears impossible. Therefore, the precise determination of the $^{12}C(\alpha,\gamma)^{16}O$ reaction rate remains difficult. Although the cross section could be measured down to $E_{c.m.}$=891 keV [1], it still needs to be extrapolated to astrophysical energies around $E_{c.m.}$=300 keV. So far, all relevant direct measurements have been performed with Germanium detectors [1, 2, 3, 4], which provide a high energy resolution. However, to compensate for the low efficiency, high beam currents of up to 700 μA [4] had to be used in spite of possible target problems.

A different approach is the use of detectors with high efficiency such as NaI [5] or BGO [6]. Although in principle promising, these setups suffered from a small solid angle and relatively high backgrounds. Hence, a verification of these data with a completely different approach, which could reveal systematic uncertainties was still missing.

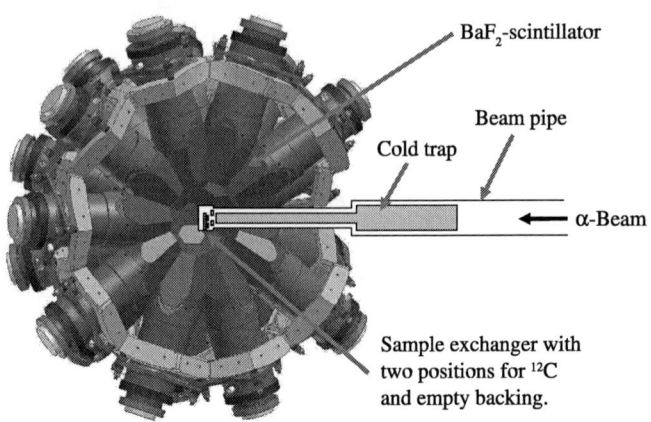

FIGURE 1. One half of the BaF$_2$ detector including sample exchanger, cold trap and beam pipe.

SETUP

The Karlsruhe 4π BaF$_2$ Detector [7] offers a high gamma-ray efficiency of more than 90% for single gamma-rays and practically 100% for the detection of gamma-ray cascades. The detector consists of 41 independent BaF$_2$ detector modules assembling a shell of BaF$_2$ with a thickness of 15 cm. As shown in Fig. 1 the detector array provides a simultaneous measurement of 12 different angles with respect to the beam axis. These angular distributions allow one to disentangle the E1 and E2 contributions of the cross section. The separation of the E1 and E2 part improves the precision of the subsequent cross section extrapolation to stellar energies significantly.

The high gamma-ray efficiency, the low neutron sensitivity and several techniques for background supression allowed to decrease the beam current by more than a factor of 100 compared to previous experiments. Hence, no target degradation occured as was verfied several times (see below).

The He$^+$ beam was delivered by the 3.7 MV Van de Graaff accelerator of Forschungszentrum Karlsruhe. In order to discriminate neutron-induced background from ^{13}C$(\alpha,n)^{16}$O, the α beam was pulsed with a frequency of 1 MHz and a pulse width of less than 2 ns. Thus, background from cosmic rays is suppressed by a factor of 500, while neutrons from ^{13}C(α,n) reactions are entering the BaF$_2$ crystals approx. 4 ns after the prompt capture gamma-rays, and are, therfore, almost completely removed. Since the minimal pulse width provided by the accelerator is only 10 ns, an external Mobley-Bunching-System had to be used for further compression. This resulted, however, in an increase of the pulse width in horizontal direction, and hence hampered the beam transport to the target through 5 quadrupole magnets and the 80 cm long cold trap. Therefore, beam currents of only approx. 5 μA have been achieved. The setup of an optimized beam line for further experiments is currently in progress.

Plastic scintillator panels as active shielding against cosmic rays were mounted on top of the detector but were only needed for the run at lowest energy.

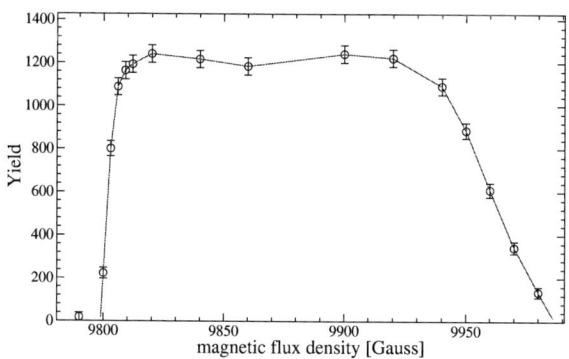

FIGURE 2. Typical thickness profile of a 120 $\mu g/cm^2$ sample.

A sample exchanger with two sample positions for ^{12}C and an empty backing for background subtraction was installed in the center of the BaF_2 detector. The housing of the sample exchanger was used as a Faraday cup for the measurement of the number of He^+ particles hitting the sample. Secondary electrons were suppressed by an electrical field between the sample and a collimator in front of the sample. Both samples and the empty backing were irradiated alternately, until a predefined charge was accumulated, typically after 15 minutes. In this way, fluctuations in the beam related background and in the pulse width were properly considered in the subtraction of the background spectra.

Samples have been produced by the isotope separator SIDONIE at CSNSM, Orsay, France. The ^{12}C ions have been deposited with a kinetic energy of only 200 eV on either a gold coated copper backing or directly on an 99.9999% pure copper backing. Two different sample thicknesses of 30 and 120 $\mu g/cm^2$ ^{12}C have been used. The count rate could be heavily increased using the 120 $\mu g/cm^2$ samples while only a small correction factor for the energy loss in the target had to be applied.

The thickness of each sample was determined before and after the actual measurement of the $^{12}C(\alpha,\gamma)^{16}O$ cross section by scanning the narrow 2^+ resonance of the $^{12}C(\alpha,\gamma)^{16}O$ cross section at 2.68 MeV. The difference in thickness of the sample before and after the measurement was always well below the uncertainty of the thickness measurement of 4-6%. Nevertheless, carbon build-up could never be completely prevented and even though the amount of ^{12}C did not significantly change, the amount of ^{13}C (1.1% in natural carbon) did. A representative thickness profile is shown in Fig. 2.

RESULTS

The determination of the E1 and E2 S-factors requires several corrections of the measured spectra with respect to the detector efficiency, the cross-talk between neighboring detectors affecting angular distributions and multiplicities, and the effect of the housing of the sample exchanger, which may scatter gamma-rays and hence blur the angular distributions. All corrections have been obtained by means of detailed Geant3 [8] simulations, which have been shown [9] to agree very well with measured data. Hence the

TABLE 1. Measured S-factors for the E1 and E2 component as well as cross section ratio σ_{E2}/σ_{E1} and phase difference ϕ.

$E_{c.m.}$ (keV)	$S_{E1}(keVb)$	$S_{E2}(keVb)$	σ_{E2}/σ_{E1}	$\phi(°)$
1002 keV	29 ± 15	$10 \pm 7,7$	$0,35 \pm 0,3$	67 ± 13
1308 keV	$14 \pm 3,7$	$6,2 \pm 2,4$	$0,44 \pm 0,2$	62 ± 8
1416 keV	$14 \pm 3,3$	$5,7 \pm 2,8$	$0,41 \pm 0,3$	55 ± 8
1510 keV	$12 \pm 2,2$	$4,7 \pm 1,5$	$0,40 \pm 0,2$	58 ± 6

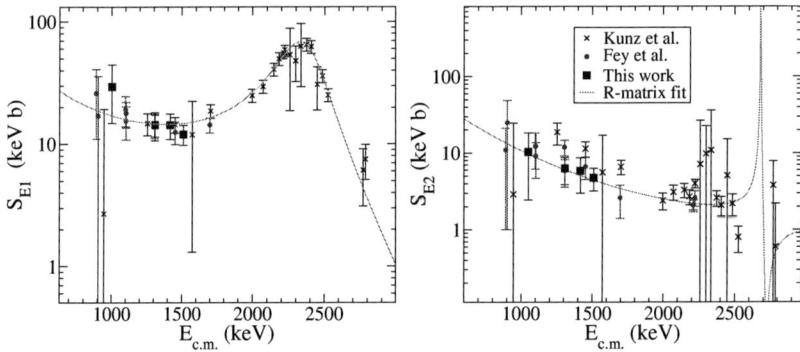

FIGURE 3. Final results of the measurement in comparison to the most recent measurements at low energies [1, 2].

application of these corrections is justified. In addition, the simulations of the current setup were tested with several calibration sources confirming the simulated efficiency and distortion of angular distributions.

Due to the limited time of four months, only four energies could be measured (see Table 1 and Fig. 3). The results are in good agreement with Refs. [1, 2] thus confirming the latest measurements with Germanium detectors [1]. Even though the uncertainties of all measurements are in the same order of magnitude, the uncertainties of our measurements are dominated by beam related background and not by counting statistics of $^{12}C(\alpha,\gamma)^{16}O$ events. Nevertheless, the main uncertainties in the E1 and E2 S-factors are introduced by the σ_{E2}/σ_{E1} ratio which are determined by a fit of the angular distributions.

The continuation of this experiment is scheduled for 2006. With a significantly improved beam line the beam current can be increased, while the beam related background, especially ^{13}C build-up, can be strongly reduced. Hence, measurements at energies around 800 keV seem to be possible.

REFERENCES

1. M. Fey et al., to be published (2004).

2. R. Kunz, M. Jaeger, A. Mayer, J. W. Hammer, G. Staudt, S. Harissopulos, and T. Paradellis, *Phys. Rev. Lett.*, **86**, 3244 (2001).
3. J. M. L. Ouellet, T. K. Alexander, M. N. Butler, H. C. Evans, H. W. Lee, J. R. Leslie, J. D. MacArthur, W. McLatchie, H.ÜB. Mak, P. Skensved, J. L. Whitton, and X. Zhao, *Phys. Rev. C*, **54**, 1982 (1996).
4. A. Redder, H. W. Becker, T. R. Donoghue, J. W. Hammer, T. C. Rinckel, C. Rolfs and H. P. Trautvetter, *Nucl. Phys. A*, **462**, 385 (1987).
5. P. Dyer and C.A. Barnes, *Nucl. Phys. A*, **233**, 495 (1974).
6. G. Roters, C. Rolfs, F. Strieder and H. P. Trautvetter, *Eur. Phys. J. A*, **6**, 451 (1999).
7. K. Wisshak, K. Guber, F. Käppeler, J. Krisch, H. Müller, G. Rupp and F. Voss, *Nucl. Instrum. Methods A*, **292**, 595 (1990).
8. Application software group, GEANT, Detector description and simulation tool, Computing and Networks Division, CERN, Geneva, Switzerland.
9. M. Heil, R. Reifarth, M. M. Fowler, R. C. Haight, F. Käppeler, R. S. Rundberg, E. H. Seabury, J. L. Ullmann, J. B. Wilhelmy and K. Wisshak, *Nucl. Instrum. Methods A*, **459**, 229 (2001).

$^{19}F(p,\gamma)^{20}Ne$: Putting a Lid on the CNO Cycle

A. Couture*, M. Beard*, M. Couder*, J. Görres*, L. Lamm*, P. LeBlanc*, H.-Y. Lee*, S. O'Brien*, A. Palumbo*, E. Stech*, E. Strandberg*, W. Tan*, E. Uberseder[†], C. Ugalde** and M. Wiescher*

* *University of Notre Dame, Department of Physics, Notre Dame, IN 46556, USA*
[†] *Forschungszentrum Karlsruhe, IK, P.O. Box 3640 D-76021 Karlsruhe, Germany*
** *University of North Carlina, Chapel Hill, NC 27599, USA*

Abstract. A new measurement of the $^{19}F(p,\gamma)^{20}Ne$ reaction has been made at the University of Notre Dame, covering proton energies from E_p=200-800 keV. Upper limits are placed on the strength of the E_p=224 keV and E_p=597 keV resonances.

Keywords: CNO Cycle, ^{19}F, ^{20}Ne
PACS: 25.40.Lw, 25.40.Ny, 26.20.+f, 27.30.+t

INTRODUCTION

The CNO cycle, illustrated in figure 1, is a hydrogen burning cycle active in core hydrogen burning in main-sequences stars ranging in mass from slightly heavier than the sun to the most massive stars in the galaxy. In addition, the hydrogen burning mechanism is found in hydrogen shell burning and some novae scenarios. Because it is often found early in the burning sequence, the nucleosynthesis taking place is limited to the conversion of hydrogen into helium. As a result, any material leaks from the cycle would impact both the burning cycle and the ash composition at the completion of the main burning stage.

The $^{19}F(p,\gamma)$ reaction represents the only possible breakout from the CNO cycle at temperatures attainable in steady-state burning. While measurements of the reaction of protons on fluorine have been studied since the earliest days of nuclear physics [1, 2], it has been a reaction that has been difficult to isolate due to the competing $^{19}F(p,\alpha)$ reactions. A level scheme for ^{20}Ne shown in figure 2 illustrates part of the difficulty in measuring the reaction. Because of the 6-8 MeV gamma-rays from the populated excited states of ^{16}O, the gamma signature is largely washed out. While this does indicate that the branching between (p,γ) and (p,α) is strongly dominated by the α-branch, this reaction participates in a cyclic process with the same seeds being reused many times; a branching of even 0.5% could contribute to a significant Ne production over time and many cycles.

Past measurements [3, 4, 5] have always looked for the primary to the first excited state, making the measurements particularly sensitive to pile-up from the 6-8 MeV gamma-rays from ^{16}O. Furthermore, with the exception of the measurement of Subotić *et al.*, all of the measurements used low resolution detectors, further challenging the clean determination of a (p,γ) event. No observations or limits are reported on the lowest energy resonance at 224 keV, and the strength of the 597 keV resonance is disputed. Fi-

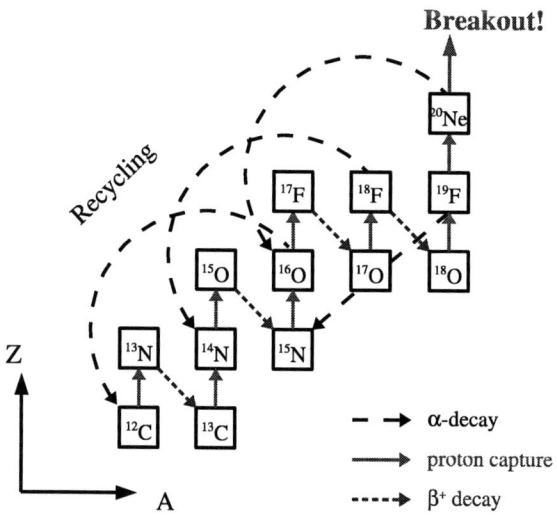

FIGURE 1. The CNO cycle operating a temperatures below $T_6=100$

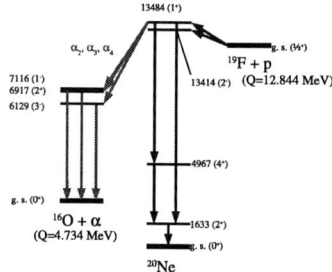

FIGURE 2. Level schematic for ^{20}Ne. Only selected levels are shown for clarity.

nally, reaction rate compilations recommend a 50% uncertainty on the ^{19}F(p,γ) reaction rate due to the undetermined signs of the interference effects between the multiple 1^+ resonances at low energies. This dominant uncertainty required a new series of measurements to both clarify the contribution of the 224 keV and 597 keV resonances as well as determine the interference contributions of all of the resonances at these energies.

EXPERIMENTAL PROCEDURE

Beams and Targets

All of the experiments were performed at the University of Notre Dame, using the low energy accelerators at the Nuclear Structure Laboratory. The JN model Van de Graaff accelerator was used for the runs between 200 keV and 700 keV. The KN accelerator was used for the region from 700 keV to 800 keV. Runs down to 640 keV were made with the KN to include a significant overlap between the two datasets. The two accelerators service the same beamlines, so the detection station was unchanged between the runs. Beams between 500 nA and 20 μA were delivered to the target for the experiment.

All of the targets used were prepared by the evaporation of CaF_2 powder onto 0.010" Ta backings. The target stability was checked throughout the experiment by scanning the yield profile of the resonance at 484 keV in the (p,α_2) channel. This allowed accounting of target degradation effects in the analysis. Because solid targets were used, it was possible to integrate the total charge on target in order to determine the total number of protons delivered to target. A copper cold finger was biased to -300 V prevent the escape of electrons liberated by the impingent beam. Furthermore, the cold finger was cooled with liquid N_2 in order to prevent the buildup of carbon on the target. Carbon was not found by either visual inspection nor by a sign of a dead-layer in the data.

γ-ray Detection and Data Acquisition

In order to isolate the (p,γ) reaction channel from the Compton background of the $(p,\alpha-\gamma)$ background, both the difference in multiplicity and Q-value between the two reactions were exploited. A HPGe clover detector was placed at 0° directly in front of the target. Four NaI(Tl) cylindrical detectors (8" diameter, 6" thickness) were placed at backward angles, rotated off-axis by 45° in both θ and ϕ in order to optimize both solid angle and photopeak efficiency. The yield was determined by looking for the 1.63 MeV ground-state transition from the first excited state in ^{20}Ne in the clover detector. A coincidence between the NaI(Tl) and clover was required for all gated events in addition to a sum energy of $>$ 10 MeV. Background suppression achieved by this technique can be seen in figure 3.

RESULTS AND CONCLUSIONS

The dataset was fit using a series of Breit-Wigner resonance shapes integrated over the target in order to reproduce a yield curve. The upper limits for the resonances were obtained by using the literature values for the 340, 484, and 669 resonances, minimizing the deviation by an appropriate choice of interference parameters and then increasing the resonance strength of the 224 and 597 resonances until the fit was seen to be significantly worse. In both cases, the best fit is obtained with a value of the strength at least on order of magnitude smaller than the upper limit. The values can be seen in table 1

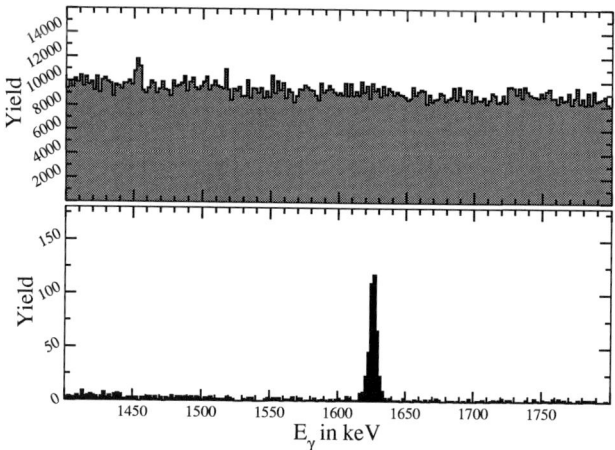

FIGURE 3. The upper plot shows an HPGe singles spectrum near 1630 keV gamma-ray energy at E_p. The lower plot is the same spectrum gated on total gamma energy and multiplicity as discussed above.

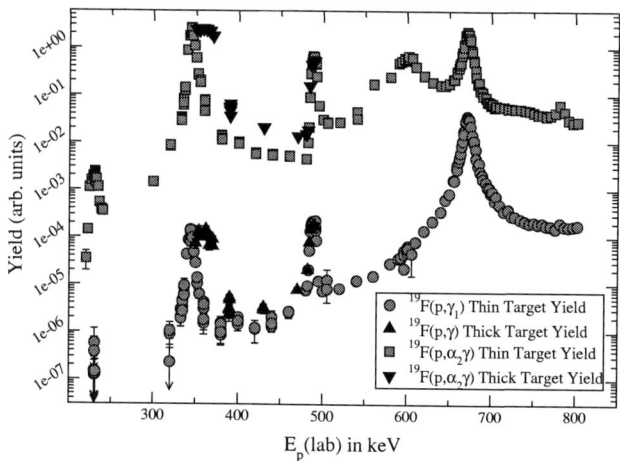

FIGURE 4. Capture yield curves for both the $(p,\gamma$ and $(p,\alpha_2\gamma)$ reaction channels normalized to charge. Statistical error bars are included, though they are generally smaller than the data point and not visible.

TABLE 1. Resonance parameters for low-lying $J^\pi = 2^-$ states in ^{20}Ne populated via ^{19}F(p,γ)

Resonance Energy (keV)	This Experiment (meV)	Subotić et al. * (meV)	Berkes et al. † (meV)	TUNL Evaluation ** (meV)
224	< 0.036	not observed	not observed	none given
597	< 2.0	5.6	20.	22.

* see reference [3]
† see reference [4]
** see reference [6]

The datasets are of sufficient quality that improved resonance parameters are likely to be obtained. In addition, it will be possible to obtain the interference terms for the 1^- resonances, moving the largest source of uncertainty in the reaction rate to the actual resonance strengths.

ACKNOWLEDGMENTS

The authors would like to thank Mr. Bradley Mulder, Mr. James Kaiser, and Mr. Jerry Lingle for their technical assistance with the maintenance and operation of the accelerator at the Nuclear Structure Laboratory at the University of Notre Dame.

REFERENCES

1. J. D. Cockroft, and T. S. Walton, *Proceedings of the Royal Society of London* **137**, 229–242 (1932).
2. H. R. Crane, L. A. Delsasso, W. A. Fowler, and C. C. Lauritsen, *Physical Review* **46**, 531–533 (1934).
3. K. M. Subotić, R. Ostojić, and B. Z. Stepančić, *Nuclear Physics A* **331**, 491–501 (1979).
4. I. Berkes, I. Dézsi, I. Fodor, and L. Keszthelyi, *Nuclear Physics* **43**, 103–109 (1963).
5. G. K. Farney, H. H. Given, B. D. Kern, and H. T. M., *Physical Review* **97**, 720–725 (1955).
6. D. R. Tilley, C. M. Cheves, J. H. Kelley, S. Raman, and H. R. Weller, *Nuclear Physics A* **636**, 249–364 (1998).

Threshold states in ^{19}Ne and the CNO breakout reaction ^{15}O$(\alpha,\gamma)^{19}$Ne

W. P. Tan*, J. Görres*, J. Daly*, M. Couder*, A. Couture*, H.Y. Lee*, E. Stech*, E. Strandberg*, C. Ugalde* and M. Wiescher*

Department of Physics, University of Notre Dame, Notre Dame 46556, USA

Abstract. The ^{15}O$(\alpha,\gamma)^{19}$Ne reaction is one of the most important breakout reactions for the hot CNO cycles. However, the relevant states in ^{19}Ne at excitation energies of 4-5 MeV have not been well studied. The lifetimes of these states are not known and are only constrained by experimental upper/lower limits. In particular, the accurate knowledge of the γ- and α- decay widths of the 4.03 MeV state of ^{19}Ne is important, since the resonance strength of this level dominates the reaction rate for the astrophysically relevant temperatures $T_9 < 0.6$. In this work, we employed an improved DSAM approach to obtain lifetime values of this and other states via ^{17}O(^{3}He,$n-\gamma)^{19}$Ne. For the 4.03 MeV state, the measured excitation energy is 4034.5 ± 0.8 keV and the mean lifetime, measured here for the first time, is 13^{+9}_{-6} fs at the confidence level of 1σ and 13^{+16}_{-9} fs at the confidence level of 2σ. This result is in excellent agreement with the 9 fs prediction by Langanke, Wiescher, Fowler, and Görres.

Keywords: CNO breakout, Reaction rate, Doppler-shift attenuation method
PACS: 26.50.+x, 21.10.Tg

INTRODUCTION

The hot CNO cycles and the ensuing rp-process after breakout play a principal role in the energy production and nucleosynthesis of explosive hydrogen burning processes occurring in astrophysical sites such as novae and accreting neutron stars [1, 2, 3, 4]. The reaction ^{15}O$(\alpha,\gamma)^{19}$Ne provides the initial break-out route from the hot CNO cycles, triggering the rp-process [1]. Calculations have shown that the luminosity of X-ray bursters is dramatically increased by the breakout from the hot CNO cycles due to the subsequent energy produced by the rp-process [1]. Multi-mass zone model simulations of the evolution of X-ray bursters indicate that the ^{15}O$(\alpha,\gamma)^{19}$Ne reaction rate actually provides the trigger conditions for periodic X-ray bursters [5].

This reaction rate is dominated by the resonance contributions at temperatures $T_9 > 0.1$ [6]. However, the partial widths Γ_α and Γ_γ of the relevant states with excitation energies of 4-5 MeV are not sufficiently well known for reliably calculating the reaction rate. Of particular interest is the resonance at an excitation energy of 4.03 MeV in ^{19}Ne, just above the ^{15}O+α threshold. This resonance dominates the reaction rate at temperatures below 0.6 GK [6] and determines therefore the temperature conditions for the breakout from the hot CNO cycles. The relative α widths $\Gamma_\alpha/\Gamma_{tot}$ of the resonance states in ^{19}Ne have been successfully measured [7] except for the critical level at 4.03 MeV. The main obstacle is the extremely small branching ratio $\Gamma_\alpha/\Gamma_\gamma$ which has been predicted to be of the order of 10^{-4} [6]. Its measurement has been attempted by various groups [8, 9, 10, 11] but only upper limits could be established in these studies.

The critical parameters for deriving the resonance strengths from the relative α-widths are the total widths or lifetimes of the resonance states. This information is not known for any of the resonance states in question. Previous attempts to determine the lifetimes have provided only upper limits for the lifetimes of the bound states near the α threshold (see the TUNL compilation of Ref. [12]). Lifetime measurements for the unbound 4.03 MeV and higher energy excited states with the Doppler-shift attenuation method (DSAM) [13] also yielded only upper limits. A recent approach of a direct measurement of the lifetimes has been made using the the Coulomb excitation technique resulting only in an lower limit of 1.5 fs for the lifetime of the 4.03 MeV level [14]. In this work, we successfully employed an improved DSAM approach to obtain lifetime values for this state and other threshold states via $^{17}O(^{3}He,n-\gamma)^{19}Ne$.

EXPERIMENTAL DETAILS

The experiment was conducted using the KN Van de Graaff accelerator at the University of Notre Dame. A beam of 3.0 MeV ^{3}He with average intensity of 12 μA was directed upon an implanted ^{17}O target with Ta backing. The target was prepared by implanting a low dose of ^{17}O ions ($1.25 \times 10^{17}/cm^2$) with an implantation energy of 70 keV. An ORTEC HPGe detector with active volume of 225 cm^3 was used to measure γ rays from ^{19}Ne decay in coincidence with neutrons detected with a liquid scintillation detector. Three geometries were used to obtain the Doppler shifted γ spectra. The unshifted gamma energy is obtained by averaging two γ spectra measured by the same Ge detector at 45° and 135°, respectively, in coincidence with neutrons detected at 0°. In the third geometry, γ rays were measured at 28.5^o and neutrons at 90° to maximize the Doppler shift as the angle between emitting γ-ray and recoiling ^{19}Ne is 0° (denoted as 0° setup in the rest of the paper).

In Fig. 1, the raw gamma spectra in coincidence with neutrons are shown for the portions relevant to the states in ^{19}Ne. Spectra shown in the top and middle panels were measured in the 45° and 135° setups, respectively, for the determination of unshifted gamma energies. The spectrum shown in the bottom panel has been obtained with the maximized Doppler-shift (0°) setup. In this setup, the peaks are not only significantly shifted, but broadened and deformed as well. The simple DSAM approach [13] widely used in the lifetime measurements takes into account only the Doppler shift of peak centroids and, therefore, is insufficient to study these cases.

The simulations of the γ-line Doppler broadening have been done with Geant4 [15] for the 0° spectrum adopting the unshifted γ energy obtained in this experiment and taking into account the calibrated energy response of the Ge detector and the geometry of the setup. The low energy extensions to the Geant4 electromagnetic processes are applied [16] and the stopping power model for the energy loss calculation has been adopted from the Ziegler parametrization [17]. Details of the analysis will be presented in a forthcoming publication[18] and the relevant aspects are summarized in the following. The sensitivity of the target profile and composition has been studied and found negligible. Because of the low implantation dose the stopping powers are dominated by the interaction with the Ta atoms [18]. Dependence on the stopping powers is also found insignificant [18]. Corrections for beam spread and beam direction have been applied

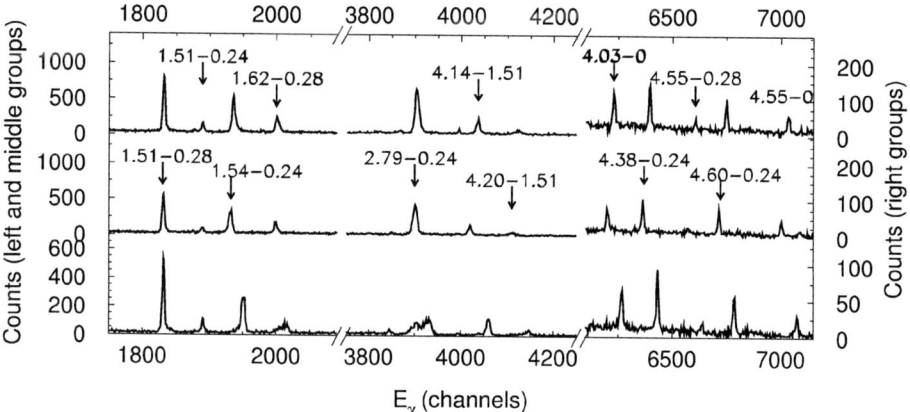

FIGURE 1. Three gamma spectra are shown from top to bottom panels for the 45°, 135°, and 0° setups, respectively. Three groups of peaks from the significant transitions of the states in ^{19}Ne are shown in each panel. See text for detailed discussion.

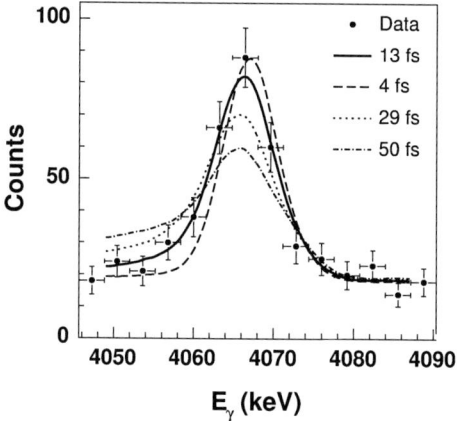

FIGURE 2. Black dots are the experimental data of the transition 4.03→0 for the 0° setup. The solid line is the best fit at $\tau_m = 13$ fs. The dashed and dotted lines represent the 2σ uncertainty limits of the here presented analysis. The dot-dashed line simulates the peak shape for a 50 fs lifetime corresponding to the upper limit from previous measurements[13] and very close to the suggested value in Ref. [14].

according to the bombarding spot on the target after the experiment.

The solid line in Fig. 2 shows the best fit obtained for a lifetime of $\tau_m = 13$ fs. The error of the lifetime is estimated by examining the χ^2 values of the fit as the lifetime is

TABLE 1. Values of excitation energies and lifetimes of the states in ^{19}Ne are listed in the right columns from present work. For reference, the adopted values from Ref. [12] are shown in the left columns.

Compilation [12]			Present work	
E_X(keV)	J^π	τ_m (fs)	E_X(keV)	τ_m (fs)
1507.56±0.3	$\frac{5}{2}^-$	$1.4^{+0.5}_{-0.6}\times 10^3$	1507.51±0.35	$1.7\pm 0.3\times 10^3$
1536.0±0.4	$\frac{3}{2}^+$	28±11	1536.05±0.36	16±4
1615.6±0.5	$\frac{3}{2}^-$	143±31	1615.4±0.4	80±15
2794.7±0.6	$\frac{9}{2}^+$	140±35	2794.2±0.4	100±12
4032.9±2.4	$\frac{3}{2}^+$	<50	4034.5±0.8	13^{+9}_{-6}
4140±4	$(\frac{9}{2})^-$	<300	4143.5±0.6	18^{+2}_{-3}
4197.1±2.4	$(\frac{7}{2})^-$	<350	4200.3±1.1	43^{+12}_{-9}
4379.1±2.2	$\frac{7}{2}^+$	<120	4377.8±0.6	5^{+3}_{-2}
4549±4	$(\frac{1}{2},\frac{3}{2})^-$	<80	4547.7±1.0	15^{+11}_{-5}
4600±4	$(\frac{5}{2}^+)$	<160	4601.8±0.8	7^{+5}_{-4}
4635±4	$\frac{13}{2}^+$	$>1\times 10^3$	4634.0±0.9	$>1\times 10^3$

varied. At the 1σ confidence level the lifetime is 13^{+9}_{-6} fs and at the 2σ confidence level the value is 13^{+16}_{-9} fs. The dashed and dotted lines in Fig. 2 depict the 2σ limits with the lifetimes of 4 and 29 fs, respectively. The dot-dashed line represents the previous upper limit of 50 fs set by Ref. [13], which is too broad and unable to account for both the peak height and the low energy tail. In Ref. [14], a gamma decay width of 12^{+9}_{-5} meV was suggested (very close to the previous upper limit of 50 fs lifetime) based on the 1.5 fs lower limit for the lifetime set by the Coulomb excitation measurements and a theoretical calculation of the E2/M1 ratio. This value is more than a factor of four smaller than the one resulting from our new measurement.

Similar procedures have been carried out for other relevant states at excitation energies of 4-5 MeV. The resulting values for excitation energies and lifetimes are listed in Table 1. The errors of the lifetime values are evaluated at the confidence level of 1σ in the χ^2 analysis. For comparison the currently adopted values from Ref. [12] are listed in the left columns.

SUMMARY

In summary, an improved DSAM approach with Geant4 simulation has been applied to the γ spectra data from the reaction ^{17}O(^{3}He,$n-\gamma$)^{19}Ne. The excitation energy values of all threshold states are improved and lifetimes for all states up to 4.6 MeV have been extracted. The lifetime of the important 4.03 MeV state has been determined for the first time to $\tau=13^{+9}_{-6}$ fs at the confidence level of 1σ and 13^{+16}_{-9} fs at the confidence level of 2σ.

The present results represent significant improvements in the accuracy of the excitation energies and in the final experimental determination of the lifetimes. Both results reduce significantly the uncertainties in the reaction rate of ^{15}O(α,γ)^{19}Ne. Recent estimates of the reaction rate are based on a variety of estimates for these γ widths, which were either taken directly from lifetime measurements of ^{19}F mirror states [12] or were based on upper/lower limit arguments or shell model calculations [6, 10, 11, 5]. These selected best values for gamma widths, in particular for the two critical resonance states at 4.03 MeV and at 4.38 MeV, differ substantially from our work and within themselves by a factor of three. Our new results remove these uncertainties, particularly reduce the uncertainty of the gamma width of the 4.03 MeV state by a factor of three, and will significantly improve our knowledge of the reaction rate of ^{15}O$(\alpha,\gamma)^{19}$Ne.

ACKNOWLEDGMENTS

This work is supported by the National Science Foundation under grant No. PHY01-40324 and the Joint Institute for Nuclear Astrophysics, NSF-PFC under grant No. PHY02-16783.

REFERENCES

1. M. Wiescher, J. Görres, and H. Schatz, *J. Phys. G* **25**, R133–R161 (1999).
2. R. Wallace, and S. Woosley, *Astrophys. J. Suppl.* **45**, 389 (1981).
3. A. Champagne, and M. Wiescher, *Annu. Rev. Nucl. Part. Sci.* **42**, 39 (1992).
4. H. Schatz, A. Aprahamian, J. Görres, M. Wiescher, T. Rauscher, J. Rembges, F.-K. Thielemann, B. Pfeiffer, P. Möller, K.-L. Kratz, H. Herndl, B. Brown, and R. Rebel, *Phys. Rep.* **294**, 167–263 (1998).
5. J. L. Fisker, J. Görres, M. Wiescher, and B. Davids, URL http://arxiv.org/PS_cache/astro-ph/pdf/0410/0410561.pdf, Astrophys. J. (2005) submitted.
6. K. Langanke, M. Wiescher, W. Fowler, and J. Görres, *Astrophys. J.* **301**, 629 (1986).
7. P. V. Magnus, M. S. Smith, A. J. Howard, P. D. Parker, and A. E. Champagne, *Nucl. Phys. A* **506**, 332 (1990).
8. M. Kurokawa, et al, in *International Symposium on Origin of Matter and Evolution of Galaxies 97 : Atami, Japan, 5-7 November 1997.*, edited by S. Kubono, T. Kajino, K. I. Nomoto, and I. Tanihata, World Scientific, Singapore ; New Jersey, 1998, p. 245.
9. A. M. Laird, et al, *Phys. Rev. C* **66**, 048801 (2002).
10. K. E. Rehm, et al, *Phys. Rev. C* **67**, 065809 (2003).
11. B. Davids, et al, *Phys. Rev. C* **67**, 012801(R) (2003); ibid **67**, 065808 (2003).
12. D. Tilley, H. Weller, C. Cheves, and R. Chasteler, *Nucl. Phys. A* **595**, 1 (1995), URL http://www.tunl.duke.edu/nucldata/.
13. J. Davidson, and M. Roush, *Nucl. Phys. A* **213**, 332 (1973).
14. G. Hackman, et al, *Phys. Rev. C* **61**, 052801(R) (2000).
15. S. Agostinelli, et al, *Nucl. Instr. Meth. Phys. Res. A* **506**, 250–303 (2003), URL http://geant4.web.cern.ch/geant4/, Geant4 Collaboration.
16. S. Chauvie, G. Depaola, V. Ivanchenko, F. Longo, P. Nieminen, and M. Pia, "Geant4 Low Energy Electromagnetick Physics," in *Proceedings of Computing In High Energy Physics*, Beijing, 2001, URL http://www.ge.infn.it/geant4/lowE/index.html.
17. H. Andersen, and J. Ziegler, *The Stopping and Ranges of Ions in Matter*, vol. 3, Pergamon Press, London, 1977; J. F. Ziegler, J. P. Biersack, and U. Littmark, *The Stopping and Ranges of Ions in Solids*, vol. 1, Pergamon Press, London, 1985; Srim-2003 (2003), URL http://www.srim.org.
18. W. P. Tan, et al, in preparation.

Nuclear isomers: structures and applications

Yang Sun*, Michael Wiescher*, Ani Aprahamian* and Jacob Fisker*

Department of Physics and Joint Institute for Nuclear Astrophysics, University of Notre Dame, Notre Dame, Indiana 46545, USA

Abstract. Isomeric states in the nuclei along the rapid proton capture process path are studied by the projected shell model. Emphasis is given to two waiting point nuclei ^{68}Se and ^{72}Kr that are characterized by shape coexistence. Energy surface calculations indicate that the ground state of these nuclei corresponds to an oblate-deformed minimum, while the lowest state at the prolate-deformed minimum can be considered as a shape isomer. Due to occupation of the orbitals with large K-components, states built upon two-quasiparticle excitations at the oblate-deformed minimum may form high K-isomers. The impact of the isomer states on isotopic abundance in X-ray bursts is studied in a multi-mass-zone X-ray burst model by assuming an upper-lower limit approach.

Keywords: Nuclear isomers, rp-process, projected shell model, multi-mass-zone X-ray burst model
PACS: 21.10.-k, 21.60.Cs, 26.30.+k

INTRODUCTION

It has been suggested that in X-ray binaries, nuclei are synthesized via the rapid proton capture process (rp-process) [1, 2], a sequence of proton captures and β decays responsible for the burning of hydrogen into heavier elements. Recent reaction network calculations [3] have shown that the rp-process can extend up to the heavy Sn-Te mass region. The rp-process proceeds through an exotic mass region with $N \approx Z$, where the nuclei exhibit unusual structure properties. Since the detailed reaction rates depend on the nuclear structure, information on the low-lying levels of relevant nuclei is thus valuable for the isotopic abundance study.

Depending on the shell filling, some nuclei along the rp-process path can have excited metastable states, or isomers [4], by analogy with chemical isomers. Of particular interest are two kinds of isomers, as illustrated in Fig. 1. It is difficult for an isomeric state either to change its shape to match the states to which it is decaying, or to change its spin orientation relative to an axis of symmetry. Therefore, isomer half-lives can be very long. If such states exist in nuclei along the rp-process path, the astrophysical significance could be that the proton-capture on long-lived isomers may increase the reaction flow, thus reducing the timescale for the rp-process nucleosynthesis during the cooling phase.

Coexistence of two or more stable shapes in a nucleus at comparable excitation energies has been known in nuclei with $A \approx 70 - 80$. The expected nuclear shapes include, among others, prolate and oblate deformations. In an even-even nucleus, the lowest state with a prolate or an oblate shape has quantum numbers $K^{\pi} = 0^+$. An excited 0^+ state may decay to the ground 0^+ state via an electric monopole (E0) transition. For lower excitation energies, the E0 transition is usually slow, and thus the excited 0^+

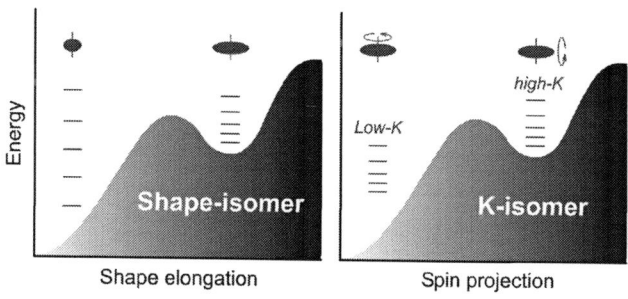

FIGURE 1. Schematic illustration for nuclear excitation as functions of various nuclear variables. The secondary energy minima are responsible for the different kinds of isomers.

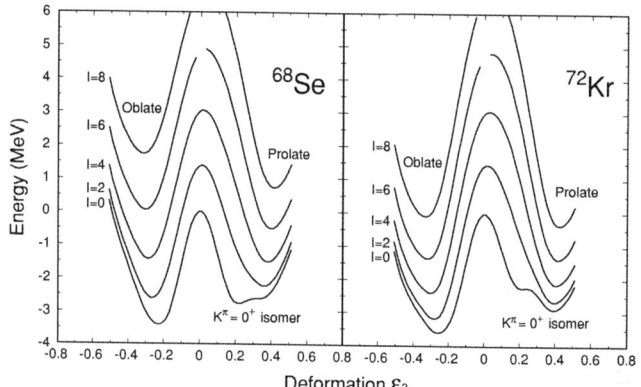

FIGURE 2. Energy surfaces for various spin states in ^{68}Se and ^{72}Kr as functions of deformation variable ε_2.

state becomes a "shape isomer". There are also excited states based on two-quasiparticle (qp) excitation. If the two quasiparticles occupy the orbitals having large K (where K is the quantum number representing the projection of the total nuclear spin along the symmetry axis), the decay path to lower energy states having a small or zero K requires a large change in K quantum number, and therefore the emission of radiation with high multipolarity is required to match the change. Such emissions are usually strongly hindered, and thus the excited high-K state becomes a "K-isomer" (see the schematic illustration in Fig. 1).

SHAPE-ISOMERS AND K-ISOMERS IN ^{68}SE AND ^{72}KR

Calculations on the structure are performed by the projected shell model [5]. Fig. 2 shows calculated total energies as a function of deformation variable ε_2 for different spin states in ^{68}Se and ^{72}Kr. The configuration space and the interaction strengths in the Hamiltonian are the same as those employed in the previous calculations for the

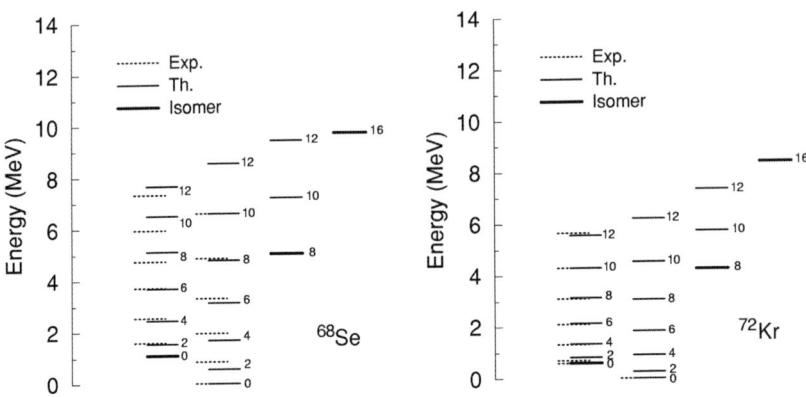

FIGURE 3. Calculated energy levels for ^{68}Se and ^{72}Kr by the projected shell model. The theoretical results are compared with available data.

same mass region [6]. Under these calculation conditions, it is found that in both nuclei, the ground state takes an oblate shape with $\varepsilon_2 \approx -0.25$. As spin increases, the oblate minimum moves gradually to $\varepsilon_2 \approx -0.3$. Another local minimum with a prolate shape ($\varepsilon_2 \approx 0.4$) is found to be 1.1 MeV (^{68}Se) and 0.7 MeV (^{72}Kr) high in excitation. Bouchez et al. [7] observed the 671 keV shape-isomer in ^{72}Kr with half-life $\tau = 38 \pm 3$ ns. The one in ^{68}Se is our prediction, awaiting experimental confirmation. Similar isomer states have also been calculated by Kaneko et al. [8].

Most nuclei near the $N = Z$ line with $A \sim 70 - 80$ are well-deformed. At the deformed potential minimum, the high-j $g_{9/2}$ orbit intrudes into the pf-shell. With an oblate shape, one finds the largest K components ($K = \frac{7}{2}$ and $\frac{9}{2}$) of this j-orbit near the Fermi levels of ^{68}Se and ^{72}Kr. Thus, a 2-qp state can have $K = \frac{7}{2} + \frac{9}{2} = 8$, and a 4-qp state $K = 16$ which is built from a neutron 2-qp and a proton 2-qp state. If K is approximately a conserved quantum number, the K value in these 2- and 4-qp states is much larger than that of the ground state band ($K = 0$). Once having been populated, this makes it rather difficult for such 2- or 4-qp states to decay back to the ground state.

In Figs. 3, we present the energy levels calculated by the projected shell model, and compare them with available experimental data [9]. For ^{72}Kr, with the newly confirmed 0^+ isomer [7] which should be the bandhead of the prolate band, the rotational band at the prolate minimum is now known. However, there have been no experimental data to compare with the predicted oblate band. In contrast, an oblate band in ^{68}Se was observed and a prolate one was also established [9], except for the missing bandhead which we predict as a shape isomer. For both nuclei, we predict low-lying high-K isomers, indicated by bold lines. In particular, the spin-16 states are so low in excitation (much lower than the spin-16 state in the ground band) that one may consider them as a spin trap [4].

IMPACT ON ISOTOPIC ABUNDANCE IN X-RAY BURSTS

The recent observation of a low energy 0^+ shape isomer in ^{72}Kr [7] has opened new possibilities for the rp-process reaction path. A similar shape isomer has been predicted for ^{68}Se in this paper. Since the ground states of ^{73}Rb and ^{69}Br are bound with respect to these isomers, proton capture on these isomers may lead to additional strong feeding of the ^{73}Rb$(p,\gamma)^{74}$Sr and ^{69}Br$(p,\gamma)^{70}$Kr reactions. However, whether these branches have any significance depends on the associated nuclear structure parameters, such as

- how strong is the feeding of the isomer states?
- what is the lifetime of the isomer with respect to γ-decay and also to β-decay?
- what are the lifetimes of the proton unbound ^{69}Br and ^{73}Rb isotopes in comparison to the proton capture on these states?

Two processes can be envisioned to populate the isomeric states in appreciable abundance, through thermal excitation of the ground state at high temperatures, or through proton capture induced γ-feeding. Thermal excitation is very efficient for feeding levels at low excitation energy since the population probability scales with $e^{-E_{is}/kT}$. Contributions of low energy states ($E_x \leq Q$) are negligible since proton capture on those states is balanced by inverse proton decay [2]. This is not the case for proton capture on the isomeric states. The peak temperature in the here used X-ray burst model is around 1.1 GK, the isomer states in ^{68}Se at 1.1 MeV and in ^{72}Kr at 0.67 MeV are therefore only very weakly populated with $\leq 0.02\%$ and $\leq 0.5\%$, respectively. Feeding through ^{67}As$(p,\gamma)^{68}$Se* ($Q \approx 3.19$ MeV) and ^{71}Br$(p,\gamma)^{72}$Kr* ($Q \approx 4.1$ MeV) is a more likely population mechanism. A quantitative prediction of the feeding probability requires a more detailed study of the γ-decay pattern of low spin ($J \leq 3$) states above the proton threshold in ^{68}Se and ^{72}Kr, respectively.

The lifetime of the isomeric states must be sufficiently long to allow proton capture to take place. No information is available about the lifetime of the ^{68}Se* isomer while the 55 ns lifetime of the isomer in ^{72}Kr is rather short [7]. Based on Hauser Feshbach estimates [2] the lifetime against proton capture is in the range of ≈ 100 ns to 10 μs depending on the density in the environment. Considering the uncertainties in the present estimates a fair fraction may be leaking out of the ^{68}Se, ^{72}Kr equilibrium abundances towards higher masses.

This however also depends on the actual proton decay lifetimes of ^{69}Br and ^{73}Rb. Based on model dependent fragmentation cross section predictions for these isotopes lifetimes have been estimated to be less than 24 ns and 30 ns respectively [10]. Again, within the present systematic uncertainties this is in the possible lifetime range of proton capture processes in high density environments.

While it is likely that equilibrium is ensued between all these configurations within the presently given experimental limits a considerable flow towards higher masses through the isomer branch cannot be excluded. Fig. 4 shows the comparison between the two extreme possibilities for the reaction sequence calculated in the framework of a multi-mass-zone X-ray burst model [11]. The left-hand figure shows the mass fractions of ^{64}Ge, ^{68}Se, and ^{72}Kr as a function of time neglecting any possible isomer contribution to the flow. The right-hand figure shows the results from the same model assuming

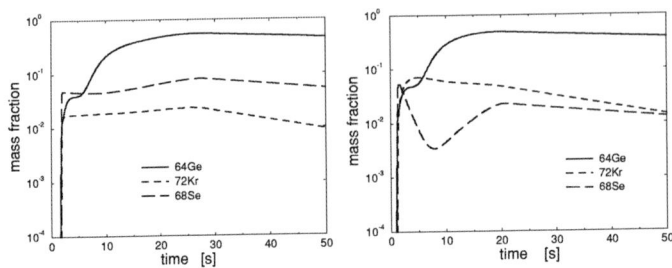

FIGURE 4. Mass fractions in the X-ray burst model with two extreme cases. Left: no isomer contribution. Right: full isomer contribution.

full reaction flow through the isomeric states in ^{68}Se and ^{72}Kr rather than through the respective ground states. The K-isomers predicted in this paper have not been considered. The main differences in ^{68}Se and ^{72}Kr mass fractions are due to rapid initial depletion in the early cooling phase of the burst. This initial decline is compensated subsequently by decay feeding from the long lived ^{64}Ge abundance. The results of our model calculations shown in Fig. 4 are based on upper and lower limit assumptions about the role of the shape isomer states. The possible impact on the general nucleosynthesis of ^{68}Se and ^{72}Kr turns out to be relatively modest. These assumptions are grossly simplified. Improved calculations would require better nuclear structure data to identify more stringent limits on the associated reaction and decay rate predictions.

We are just beginning to look at the impact that isomers may have on various nucleosynthesis processes such as the rp-process. We look for cases in which an isomer of sufficiently long lifetime (probably longer than microseconds) can change the paths of reactions taking place and lead to a different set of elemental abundances. This aspect of nuclear isomers is very much in its infancy [12].

ACKNOWLEDGMENTS

This work is partly supported by NSF under contract PHY-0140324.

REFERENCES

1. L. Van Wormer et al., *Astrophys. J.* **432**, 326 (1994).
2. H. Schatz et. al., *Phys. Rep.* **294**, 167 (1998).
3. H. Schatz et. al., *Phys. Rev. Lett.* **86**, 3471 (2001).
4. P. Walker and G. Dracoulis, *Nature* **399**, 35 (1999).
5. K. Hara and Y. Sun, *Int. J. Mod. Phys. E* **4**, 637 (1995).
6. Y. Sun, *Eur. Phys. J. A* **20**, 133 (2004).
7. E. Bouchez et al., *Phys. Rev. Lett.* **90**, 082502 (2003).
8. K. Kaneko, M. Hasegawa, and T. Mizusaki, *Phys. Rev. C* **70**, 051301(R) (2004).
9. S.M. Fischer et al., *Phys. Rev. Lett.* **84**, 4064 (2000); **87**, 132501 (2001).
10. R. Pfaff et al., *Phys. Rev. Lett.* **53**, 1753 (1996).
11. J. Fisker et al., *Nucl. Phys. A* **758**, 447 (2005).
12. A. Aprahamian and Y. Sun, *Nature Physics* **1**, 81 (2005).

Radiative capture reactions and α-elastic scattering on ^{106}Cd for the astrophysical p-process

Gy. Gyürky*, Z. Elekes*, G.G. Kiss*, Zs. Fülöp*, E. Somorjai*, Z. Máté*,
J. Görres[†], A. Palumbo[†], M. Wiescher[†], H.-Y. Lee[†], N. Özkan**,
R.T. Güray**, G. Efe**, D. Galaviz[‡], A. Kretschmer[‡], K. Sonnabend[‡],
A. Zilges[‡] and T. Rauscher[§]

*Institute of Nuclear Research (ATOMKI), P.O.Box 51 H-4001 Debrecen, Hungary
[†]University of Notre Dame, Notre Dame, Indiana 46556, USA
**Kocaeli University, Department of Physics, 41380 Umuttepe, Kocaeli, Turkey
[‡]Technische Universität Darmstadt, D-64289 Darmstadt, Germany
[§]Universität Basel, CH-4056 Basel, Switzerland

Abstract.

In the present work both the (α,γ) and (p,γ) cross sections on the p-nucleus ^{106}Cd have been measured in the energy range relevant to the astrophysical p-process. The results are compared with the predictions of the statistical model calculations implemented with the NON-SMOKER code using different input parameters. The proton capture cross section has also been measured for ^{108}Cd.

The α + ^{106}Cd optical potential, an important input parameter for the ^{106}Cd(α,γ) reaction rate determination, can be determined directly by measuring the deviation from the Rutherford scattering in the ^{106}Cd(α,α)^{106}Cd elastic scattering experiment. This experiment has also been performed in a wide angular range and the results are compared with different global optical potentials.

Keywords: Astrophysical p-process, Capture reactions, Statistical model calculations, Optical potential
PACS: 26.30.+k, 26.50.+x, 27.60.+j

INTRODUCTION

In spite of the experimental and theoretical efforts of recent years, the astrophysical p-process is still one of the least understood mechanisms of stellar nucleosynthesis. This process is responsible for the production of the heavy (Z $\geq$ 34) proton rich isotopes which cannot be synthesized by the slow and rapid neutron capture processes since they are shielded by stable isobars.

It is generally accepted that the astrophysical p-process mainly involves γ-induced reactions on abundant seed nuclei produced by the s- or r-processes. The dominant (γ,n) reactions drive the material towards the neutron-deficient side of the isotopic chart. Within an isotopic chain this photodisintegration with neutron emission will proceed until a nucleus is reached at which (γ,α), (γ,p) reactions or β^+ decays become faster than (γ,n), feeding into another chain. For the γ-induced reactions high energy γ-photons are necessary. This puts a constraint on the possible astrophysical site of the p-process. The most favored site is the O-Ne rich layer of type II supernovae, however, other explosive stellar environments are also under consideration. If there is high enough hydrogen abundance in the high temperature environment, proton capture reactions may also contribute to the production of the p-nuclei. A recent, exhaustive overview of the astrophysical p-process can be found in Ref. [1].

The aim of the p-process modeling is to reproduce the abundances of p-isotopes found in the Solar System. The modeling involves a huge reaction network where thousands of nuclei are linked together by tens of thousands of reactions. The reaction rates of all these reactions are inputs to such a network calculation. Due to lack of experimental data, these reaction rates are calculated by the Hauser-Feshbach statistical model. Generally, the p-process model calculations are able to reproduce the Solar System p-isotope abundances with an accuracy not better than about one order of magnitude. Even larger deviations can be found for light p-nuclei near the Mo-Ru mass region. One possible reason for these deviations can be found in nuclear physics, i.e. the reaction rates calculated by statistical models may not be reliable. The calculated cross sections (from which the reaction rates can be derived) depend strongly on some input parameters such as optical model potentials, nuclear level densities, γ-ray strength functions, ground states properties, etc. Since experimental data is scarce, it is not possible to select those input parameters which give the best results, leaving the nuclear physics input of p-process models uncertain.

Thus, it is highly necessary to determine experimental cross sections in the mass and energy range relevant to the p-process and compare the results with the predictions of statistical model calculations. This need has been realized some years ago and several groups have started a systematic study of reactions relevant to the p-process. The most important γ-induced reactions, however, are very difficult to measure directly. There are a few cases where (γ,n) cross section have successfully been measured (see e.g. [2, 3]), and fast development is expected on the γ-induced cross section measurements in the near future. The reaction rate of charged particle emitting (γ,α) and (γ,p) can, however, still be calculated from the inverse capture cross sections using the detailed balance theorem. The capture cross sections must be known in the astrophysically relevant energy range. The typical temperature for p-process stellar environments is $T = 1.5 - 3.5 \times 10^9$ K. For these temperatures the astrophysically relevant energy region (the Gamow window) for (p,γ) reactions lies between 1 and 6 MeV (for lower mass nuclei the lower part and for higher mass nuclei the higher part of this energy region applies). The same energy region for (α,γ) reactions is between 4 and 13 MeV. Due to the Coulomb barrier at the entrance channel, the α and proton capture cross sections are very low at these low energies making the cross section measurements difficult. In those cases where the capture reaction leads to a radioactive isotope, the cross section can be determined using the activation technique. In the case of stable reaction products, however, the technically more difficult on-line γ-detection method must be used.

In Table 1 those nuclei are listed where experimental (p,γ) and (α,γ) cross sections are available relevant for the p-process. The method (activation or on-line γ-detection) is also shown. Compared with the huge number of reactions playing a role in p-process networks there is still very limited experimental information for charged particle induced reactions (especially for α-captures). It is therefore very important to continue the systematic study and determine further (p,γ) and (α,γ) cross sections and compare the results with the statistical model predictions.

TABLE 1. List of nuclei for which experimental (p,γ) or (α,γ) cross sections are available relevant to the p-process. For isotopes marked with asterisk (*) no tabular cross section results are available. Several other (p,γ) and (α,γ) cross sections have been measured recently [5]; however, the still unpublished results are not included in the Table.

Isotope	Method	Ref.	Isotope	Method	Ref.	Isotope	Method	Ref.
			(α,γ) reactions					
^{70}Ge	on-line	[4]	^{91}Zr*	on-line	[5]	^{96}Ru*	activation	[6]
^{112}Sn	activation	[7]	^{118}Sn*	on-line	[5]	^{144}Sm	activation	[8]
			(p,γ) reactions					
^{74}Se	activation	[9]	^{76}Se	activation	[9]	^{84}Sr	activation	[10]
^{86}Sr	activation	[10]	^{87}Sr	activation	[10]	^{88}Sr	on-line	[11]
^{89}Y	on-line	[12]	^{90}Zr	on-line	[13]	^{96}Zr	activation	[14]
^{93}Nb	on-line	[15]	^{92}Mo	activation	[16]	^{94}Mo	activation	[16]
^{95}Mo	activation	[16]	^{98}Mo	activation	[16]	^{96}Ru	activation	[17]
^{98}Ru	activation	[17]	^{99}Ru	activation	[17]	^{104}Ru	activation	[17]
^{102}Pd	activation	[7]	^{112}Sn	activation	[14, 18]	^{116}Sn	activation	[7]
^{119}Sn	activation	[14]						

INVESTIGATED REACTIONS

In the present work the ^{106}Cd(α,γ)^{110}Sn, ^{106}Cd(p,γ)^{107}In and ^{108}Cd(p,γ)^{109}In reaction cross sections have been measured. Since the reaction products in all three reactions are radioactive, the cross sections have been measured with the activation technique. The proton capture cross sections have been measured in the proton energy range between E_p = 2.4 and 4.8 MeV which is the relevant energy range for the p-process. The α capture cross section has been determined between E_α = 8.0 and 12.5 MeV, in the upper half and above the relevant energy range. The (p,γ) cross sections have been measured at the ATOMKI, Debrecen, Hungary, while (α,γ) experiments have been carried out both in ATOMKI and at the University of Notre Dame, USA. Details of the experimental procedure are described elsewhere [19, 20, 21]. The derived experimental cross sections are compared with the prediction of the statistical model calculations implemented with the NON-SMOKER code [22].

RESULTS

Fig. 1 shows the results for both the ^{106}Cd(p,γ)^{107}In and ^{108}Cd(p,γ)^{109}In reactions in the form of the astrophysical S-factor calculated from the cross sections. The results of the NON-SMOKER calculations using the standard input parameter set [22] are also plotted. For these two proton capture cross sections the model is able to fairly reproduce the measured data. The final experimental results in tabular form will be published in a forthcoming paper. The dependence of the model predictions on the input parameters will also be investigated in details. This work is still in progress.

Fig. 2 shows the results for the ^{106}Cd(α,γ)^{110}Sn reaction. The results of the two independent measurements at the ATOMKI and at the Notre Dame University are in

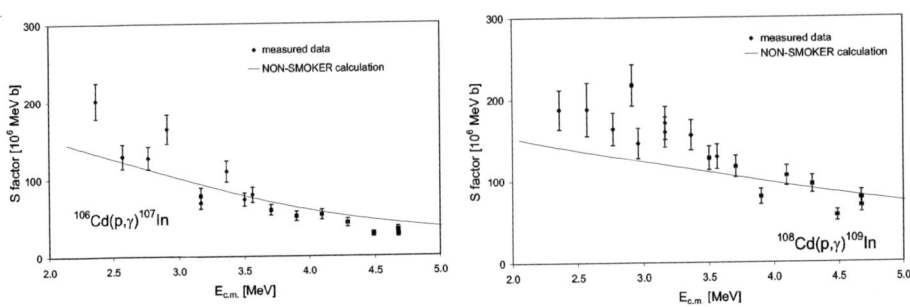

FIGURE 1. Astrophysical S-factor for the ^{106}Cd(p,γ)^{107}In (left panel) and ^{108}Cd(p,γ)^{109}In (right panel) reactions. Results of the NON-SMOKER calculations are also plotted.

good agreement. The standard NON-SMOKER calculation using the α-nucleus optical model potential by McFadden and Satchler [23] (thick solid line) overestimates the measured cross section by about a factor of two; therefore the model calculations were carried out using different optical potentials. It is clearly seen in Fig. 2 that the cross section predicted by the statistical model depends very strongly on the optical potential. The potential by Avrigeanu et al. [24] (dashed line) gives a very similar result as the potential by McFadden and Satchler [23]. The recently developed potential by Fröhlich and Rauscher [25] (dash-dotted line) reproduces very well the experimental results.

Knowing the importance of the optical model potentials in statistical model calculations, elastic α scattering cross section of ^{106}Cd has been measured at energies $E_\alpha \approx$ 15.5, 17, and 19 MeV in a wide angular range. Details of this experiment can be found in [26]. Elastic scattering cross sections can also be calculated with different global optical model potentials and the calculations can be compared with the experimental results. The comparison shows that potential by Fröhlich and Rauscher [25] (which gives the best description of the capture cross section) is not able to reproduce the scattering cross section. In contrast, the potential by McFadden and Satchler [23] (which overestimated the scattering cross section) reproduces reasonably well the scattering data. To resolve this apparent discrepancy, further theoretical considerations are needed and are in progress.

It is also possible to derive local optical potentials for the ^{106}Cd + α system by fitting the experimental cross section with a given function. In the present work, double-folding potential in the real part and a combination of surface and volume Woods Saxon potential in the imaginary part has been fitted to the scattering data. This potential can also be used in the statistical model to calculate the capture cross section. The preliminary result of this calculation is also shown in Fig. 2 (dotted line). Final results for both the capture and scattering cross sections will be published in details in forthcoming papers where the astrophysical conclusions will also be drawn.

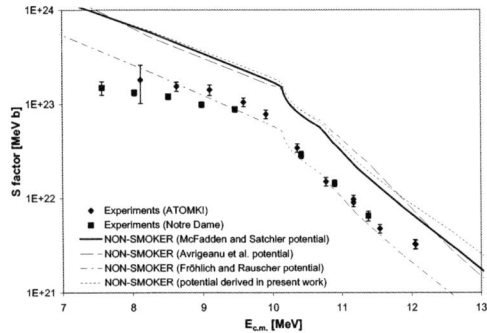

FIGURE 2. Astrophysical S-factor for the ^{106}Cd$(\alpha,\gamma)^{110}$Sn. Results of the NON-SMOKER calculations using different optical model potentials are also plotted. See text for details.

ACKNOWLEDGMENTS

This work was supported by OTKA (T42733, T49245, F43408, D48283), the DFG (SFB634), TUBITAK - Grant TBAG-U/111 (104T2467), the NSF-Grant PHY01-40324 and through the Joint Institute of Nuclear Astrophysics (www@JINAweb.org), NSF-PFC grant PHY02-16783. Zs. Fülöp is a Bolyai fellow.

REFERENCES

1. M. Arnould, and S. Goriely, *Phys. Rep.* **384**, 1. (2003)
2. P. Mohr et al., *Phys. Lett. B* **488**, 127. (2000)
3. H. Utsunomiya et al., *Phys. Rev. C* **67**, 015807. (2003)
4. Zs. Fülöp et al., *Z. Phys.* **A355**, 203. (1996)
5. S. Harissopulos et al., *Nucl. Phys. A* **758**, 505. (2005)
6. W. Rapp et al., *Nucl. Phys. A* **688**, 427. (2001)
7. N. Özkan, et al., *Nucl. Phys. A* **710**, 469. (2002)
8. E. Somorjai, et al., *Astron. Astrophys.* **333**, 1112. (1998)
9. Gy. Gyürky et al., *Phys. Rev. C* **68**, 055803. (2003)
10. Gy. Gyürky et al., *Phys. Rev. C* **64**, 065803. (2001)
11. S. Galanopoulos et al., *Phys. Rev. C* **67**, 015801. (2003)
12. P. Tsagari et al., *Phys. Rev. C* **70**, 015802. (2004)
13. C.E. Laird, D. Flynn, R.L. Hershberger, and F. Gabbard, *Phys. Rev. C* **35**, 1265. (1987)
14. F. R. Chloupek et al., *Nucl. Phys. A* **652**, 391. (1999)
15. S. Harissopulos et al., *Phys. Rev. C* **64**, 055804. (2001)
16. T. Sauter and F. Käppeler, *Phys. Rev. C* **55**, 3127. (1997)
17. J. Bork, H. Schatz, F. Käppeler, and T. Rauscher, *Phys. Rev. C* **58**, 524. (1998)
18. N. Özkan, et al., *Eur. Phys. J. A.* in press.
19. Gy. Gyürky et al., *Nucl. Phys. A* **758**, 517. (2005)
20. Gy. Gyürky et al., *Eur. Phys. J. A.* in press.
21. A. Palumbo et al., these proceedings
22. T. Rauscher, and F. K. Thielemann, *At. Data Nucl. Data Tables* **79** 47. (2001)
23. L. McFadden, and G.R. Satchler, *Nucl. Phys.* **84**, 177. (1966)
24. M. Avrigeanu, W. von Oertzen, A.J.M. Plompen, and V. Avrigeanu, *Nucl. Phys. A* **723**, 104. (2003)
25. C. Fröhlich, diploma thesis, University of Basel, Switzerland, 2002.; T. Rauscher, *Nucl. Phys. A* **719** 73c (2003); T. Rauscher, *Nucl. Phys. A* **725** 295 (2003).
26. G.G. Kiss et al., *Eur. Phys. J. A.* in press.

NUCLEAR REACTIONS
(REACTIONS WITH NEUTRON-RICH NUCLEI)

Neutron-Transfer Reactions with Exotic Neutron-Rich Beams: Surrogates for Neutron-Capture Reactions

J.A. Cizewski, K.L. Jones, S.D. Pain, J.S. Thomas

Department of Physics and Astronomy, Rutgers University, New Brunswick, NJ 08901

D.W. Bardayan, J.C. Blackmon, M.S. Smith

Physics Division, Oak Ridge National Laboratory, Oak Ridge, TN 37831

R.L. Kozub

Physics Department, Tennessee Technological University, Cookeville, TN 38505

M.S. Johnson

Oak Ridge Associated Universities, Oak Ridge, TN 37831

Abstract. A new program to measure (d,p) reactions on rare isotopes of fission fragments has been established at Oak Ridge National Laboratory. Initial measurements on N=50 isotones and prospects for Z=50 experiments are reported.

Keywords: ^{83}Ge, ^{85}Se deduced Q-values and spectroscopic strengths. Rare isotope beam (d,p) reaction studies.
PACS: 25.60.Je, 27.50.+e, 21.20.Dr, 21.20.Jx

INTRODUCTION

In nuclei near stability the shell structure has been well established with magic numbers at 50 and 82 for both neutrons and protons. However, as we go further from stability on the neutron-rich side and the nuclear surface becomes more diffuse, the nuclear structure is expected to change, evolving to a situation where the shell structure is quenched (e.g., [1]). These very neutron-rich nuclei also lie along the path of rapid neutron-capture nucleosynthesis, the r process. Therefore, it is not only important to map out the single-particle excitations in these exotic nuclei, but it is important to provide data needed for nucleosynthesis calculations, such as neutron-separation energies and neutron-capture cross sections.

However, these exotic nuclei have short half lives, so it will not be possible to make targets for (n,γ) reaction measurements. Alternatively, we can develop *beams* of these rare isotopes and measure neutron-transfer, (d,p), reactions in inverse kinematics to determine masses, excitation energies and single-neutron strengths, spectroscopic information important for nuclear structure and astrophysics.

At Oak Ridge National Laboratory we are able to measure neutron-transfer reactions on rare isotopes at energies above the Coulomb barrier[2]. The present focus is on beams of fragments following proton-induced fission of ^{238}U that are ionized, mass separated and accelerated in the 25-MV electrostatic accelerator for delivery to the target areas. Deuterated CD_2 targets are used and the reaction protons are measured, often in coincidence with heavy reaction products.

(d,p) REACTIONS ON NEUTRON-RICH RARE ISOTOPES: RESULTS AND PROSPECTS

Measurements of N=51 isotones

The first measurement performed was a study of the (d,p) reaction on the very neutron-rich nucleus ^{82}Ge to probe the structure of neutron-rich N=51 isotones and the location of the path of r-process nucleosynthesis near N≈50. For this measurement a mixed A=82 beam impinged on a ≈400 μg/cm^2 CD_2 target. The beam and beam-like recoils were detected in a segmented ionization chamber at zero degrees. Good, $\Delta Z=1$, elemental identification was obtained; the beam was dominated by stable ^{82}Se, with about 10^4 particles/s of ^{82}Ge, and minimal ^{82}As. The reaction protons were detected at back angles in the segmented silicon-strip detector array SIDAR[3], in a lampshade configuration covering 105°-150° in the laboratory, corresponding to forward angles in the center-of-mass frame. True coincidences between heavy recoils and protons in SIDAR were required. The calibration of the energies and angles of the reaction protons was facilitated by concurrent measurement of the ^{82}Se(d,p) reaction, which had been previously reported by Montestruque, et al.[4]. The Q-value spectrum for reaction protons associated with Ge recoils is displayed in Fig.1a. The highest-energy group is considerably broader than the instrumental resolution of ≈300 keV determined from the ^{83}Se data, implying the existence of an unresolved doublet. Deconvoluting the highest energy ^{83}Ge peak using the measured instrumental resolution resulted in ground and first-excited states separated by 280 keV. The ground state Q-value was determined to be 1.47(2) MeV, which implies a neutron-separation energy of S_n=3.69 MeV. With such a low neutron-separation energy and relatively long half life ($t_{1/2}$ = 1.85s [5]), ^{83}Ge is, therefore, an r-process waiting point. The angular distributions of the reaction protons associated with the ground and 280-keV states in ^{83}Ge are displayed in Fig. 1b. These distributions have been fitted with distorted wave Born approximation (DWBA) calculations using global optical model parameters of Lohr-Haeberli (deuterons, ref. [6]) and Varner (protons, ref. [7]). They are consistent with $\ell=2$, presumably $d_{5/2}$ transfer to the ground state (with spectroscopic factor S=0.48) and $\ell=0$, $s_{1/2}$ transfer to the 280-keV state (with S=0.50). These results have been published in Ref. 8.

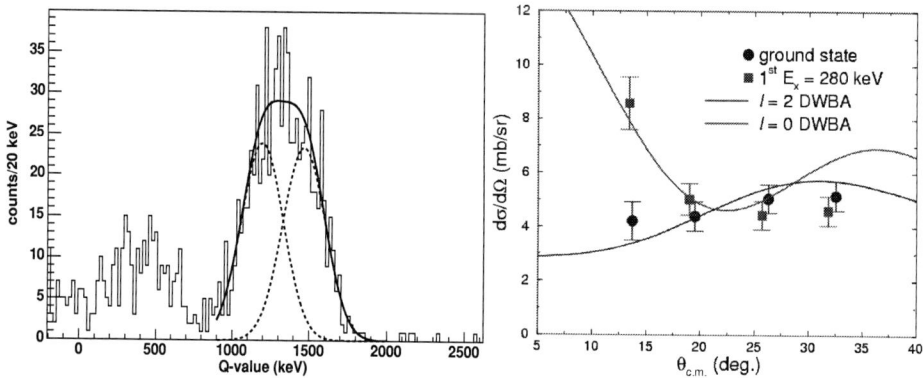

FIGURE 1. ^{82}Ge(d,p) Q-value spectrum (left) and angular distributions of protons (right) for the ground and first excited states in ^{83}Ge.

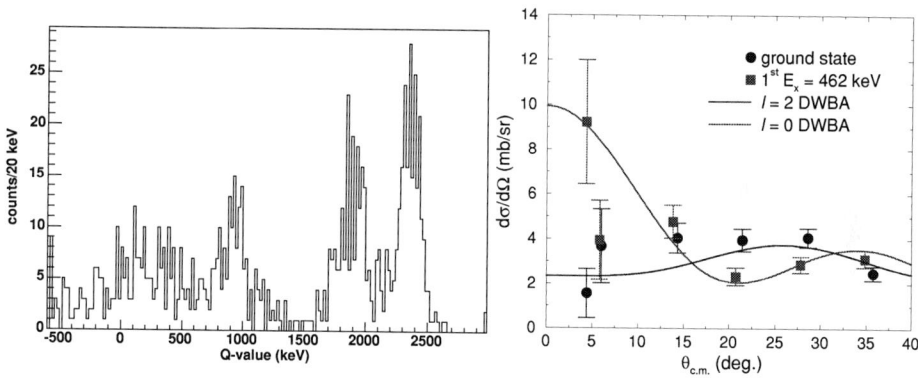

FIGURE 2. ^{84}Se(d,p) Q-value spectrum (left) and angular distributions of protons (right) for the ground and first excited states in ^{85}Se.

Subsequently, we measured the (d,p) reaction on the neutron-rich N=50 isotope ^{84}Se. A similar experimental setup was used, supplemented with an annular "CD-type" segmented silicon strip detector at back angles of 160°-170°. Compared to the initial ^{82}Ge measurement, a somewhat higher beam energy and thinner target were used. The calibrations of the proton Q-values and angles were facilitated by the previously measured mass and excitation energies of ^{85}Se [9]. The Q-value spectrum and angular distributions of the protons associated with the ground and first excited states in ^{85}Se are displayed in Fig. 2. Again, the angular distributions of the reaction protons are consistent with ℓ=2, presumably $d_{5/2}$ transfer to the ground state (with S=0.33) and ℓ=0, $s_{1/2}$ transfer to the 462-keV state (with S=0.30). The results from the ^{82}Ge and ^{84}Se (d,p) measurements are summarized in Table 1 and compared with the systematics of the N=51 isotones in Fig. 3. In all of these N=51 isotones the ground state is consistent with 5/2$^+$, with a first excited 1/2$^+$ state that comes down in energy from above 1 MeV near stability in ^{91}Zr and ^{89}Sr to only 280 keV in ^{83}Ge. In addition, the single-particle strength in the lightest isotopes is significantly less than near stability. The lowering of the excitation energy of the 1/2$^+$ state and fragmentation of the single-particle strengths are reproduced by preliminary shell model calculations of B.A. Brown, reported in Table 1, based on a ^{78}Ni core[12].

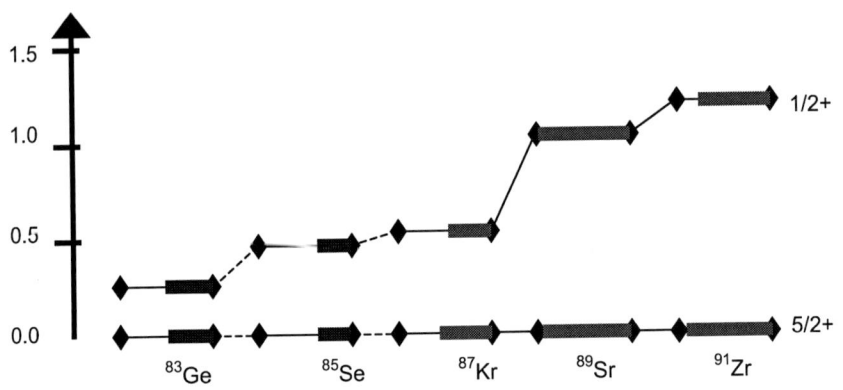

FIGURE 3. Systematics of ground and first excited states in N=51 isotopes. The lengths of the thick lines are proportional to the measured spectroscopic factors. Data taken from refs. 8, 10, and 11.

TABLE 1. Summary of experimental results from the ^{82}Ge(d,p) and ^{84}Se(d,p) measurements.

	E_x(MeV)	ℓ	J^π	$S_{\ell j}$	E_x(MeV) Ref. 12	$S_{\ell j}$ Ref. 12
^{83}Ge	0.00	2	(5/2)$^+$	0.48(12)	0.00	0.85
	0.28(7)	0	1/2$^+$	0.50(13)	0.47(30)	0.51
^{85}Se	0.00	2	(5/2)$^+$	0.33(8)		
	0.462	0	1/2$^+$	0.30(8)		

Prospects for Z=50 studies

To benchmark future measurements on heavy fission fragments, such as ^{132}Sn, we used the ^{124}Sn(d,p) reaction in inverse kinematics to develop the experimental configuration for such measurements and confirm that spectroscopic factors deduced in inverse kinematics are the same as those determined in traditional kinematics. Because measurements at HRIBF are restricted to beam energies of <5 MeV-A, the reaction protons for $\ell \geq 2$ transfers are focused at angles forward of 120° in the laboratory frame (corresponding to backward of 30° in the center of mass). Therefore, it was necessary to place ΔE-E telescopes of position-sensitive silicon strip detectors near 90°. In the test experiment reported in Ref. 13, ΔE position-sensitive silicon strip detectors of 65 (backward angles) and 140 micron (forward angles) and E silicon detectors of 1000 micron thicknesses were used. The energy loss in the ΔE detectors relative to the energy deposited in the E detectors enabled the particle identification necessary to separate reaction protons from the dominant elastically scattered deuterons, carbon and protons from the target. The spectroscopic factors for the low-spin doublet at 28+215 keV ($3/2^+$ and $1/2^+$, respectively) and the $7/2^-$ state at 2.77 MeV deduced from our measurement in inverse kinematics are consistent with values in the literature[10].

We are poised to measure the ^{130,132}Sn(d,p) reactions with 6-9 ΔE-E telescopes of position-sensitive silicon strip detectors. Measurements are anticipated in Fall 2005.

New experimental developments

To measure angular distributions of protons, especially with heavier fragments, will require considerable solid angle coverage near 90°, to complement the backward angle detection with SIDAR. Therefore, we are developing the Oak Ridge Rutgers University Barrel Array (ORRUBA) of position-sensitive silicon strip detectors. ORRUBA will consist of two-rings of position-sensitive silicon strip detectors, including ΔE-E telescopes at forward angles to realize proton particle selection. We anticipate commissioning this array in 2006.

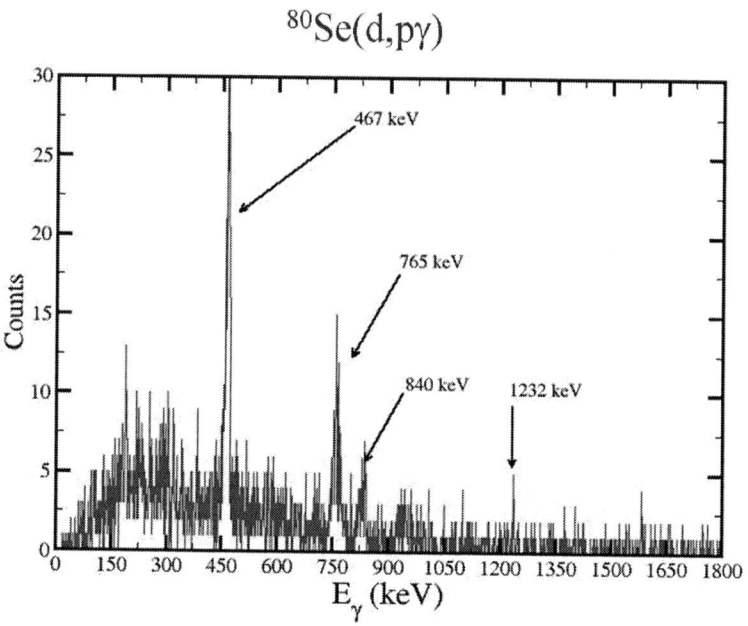

FIGURE 4. Preliminary spectrum of gamma rays gated on protons from the ^{80}Se(d,p) reaction measured in inverse kinematics.

In many cases it is difficult to resolve by charged-particle detection alone the states populated in reactions with rare isotopes beams. The detection of gamma rays, e.g., with high-purity germanium detectors, would provide significantly improved resolution, as well as help to identify levels populated in γ-ray de-excitation, but not directly populated in transfer, and provide γ-ray transition strengths and branching ratios. Initial tests of a set-up to measure γ rays in coincidence with (d,p) reaction protons have been completed with stable beams of ^{90}Zr and ^{80}Se. Reaction protons were detected in SIDAR at back angles and in 4 ΔE-E telescopes of position-sensitive silicon strip detectors surrounding the target in a box-like configuration near 90°. Four segmented Ge clover detectors at 90° in a closed-packed geometry were used to detect photons. A preliminary spectrum of γ rays detected in coincidence with protons from the ^{80}Se(d,p) reaction is displayed in Fig. 4. The dominant line at 467 keV is the known [10] 5/2⁻ to ground-state transition in ^{81}Se.

SUMMARY

We have successfully completed the initial studies of neutron-transfer (d,p) reactions with rare isotope beams of the N=50 isotones ^{82}Ge and ^{84}Se and have demonstrated the feasibility to extend these studies to heavier rare isotopes, including ^{132}Sn. These first results have enabled the extraction of ground-state Q values, energies of excited states, and spectroscopic strengths of the low-lying single-particle states in the neutron-rich N=51 isotones ^{83}Ge and ^{85}Se. These observables are important in calculating direct neutron-capture cross sections, and have allowed us to identify ^{83}Ge as an r-process waiting point. We have also developed the techniques to measure gamma rays in prompt coincidence with (d,p) reaction protons. We are in the process of developing a new barrel array of charged-particle detectors to enhance the capabilities of measurements of reaction protons near and forward of 90°.

The present efforts are only the initial studies of neutron-rich nuclei far from stability, studies that will allow us to explore nuclear structure in very neutron-rich nuclei, including those along the r-process path of nucleosynthesis. While the measurement of neutron-capture on these rare isotopes may never be feasible, the present efforts help to provide the data necessary to calculate capture cross sections.

ACKNOWLEDGEMENTS

This work is supported in part by the U.S. Department of Energy under contract numbers DE-FC03-03NA00143 (Rutgers), DE-AC05-00OR22725 (ORNL), DE-FG02-96-ER40955 (TTU), the National Science Foundation, and the LDRD program of ORNL.

REFERENCES

1. J. Dobaczewski et al., *Phys. Rev.* **C 53**, 2809 (1996).
2. D. W. Stracener, *Nucl. Instrum. Methods Phys. Res.,* **B 204**, 42 (2003).
3. D. W. Bardayan et al., *Phys. Rev.* **C 62**, 055804 (2000).
4. L.A.Montestruque et al., *Nucl.Phys.* **A305**, 29 (1978).
5. J.A. Winger et al., *Phys. Rev.* **C 38**, 285 (1988).
6. J.M. Lohr and W. Haeberli, *Nucl. Phys.* **A232,** 381 (1974).
7. R.L. Varner et al., *Phys. Rep.* **201**, 57 (1991).
8. J.S. Thomas, et al., *Phys. Rev.* **C 71**, 021302R (2005).
9. J. P. Omtvedt, B. Fogelberg, and P. Hoff, *Z. Phys.* **A 339**, 349 (1991).
10. Evaluated Nuclear Structure Data Files, http://www.nndc.bnl.gov.
11. J.S. Thomas, Ph.D. Dissertation, Rutgers University (unpublished, 2005).
12. B.A. Brown, private communication.
13. K.L. Jones, et al., *Phys. Rev.* **C 70**, 067602 (2004).

Do halos exist on the dripline of deformed nuclei?

F.M. Nunes, B. Avez, T. Duguet

National Superconducting Cyclotron Laboratory and Department of Physics and Astronomy, Michigan State University, East-Lansing MI 48824-1321

Abstract. A study of the effect of deformation and pairing on the development of halo nuclei is presented. Exploratory three-body $core + n + n$ calculations show that both the NN interaction and the deformation/excitation of the core hinder the formation of the halo. Preliminary self-consistent mean-field calculations are used to search for regions in the nuclear chart where halos could potentially develop. These are also briefly discussed.

Keywords: Halo Nuclei, three-body models, HFB, deformation, pairing
PACS: 21.10.-k,21.10.Dr,21.45.+v,21.60.-n

MOTIVATION

During the early years of Radioactive Nuclear Beam physics, while the dripline for light nuclei was being explored, *nuclear halos* became a very hot topic [1]. Typical examples of halo nuclei include ^{6}He, ^{11}Li, ^{14}Be and ^{19}C, all in the low mass region of the nuclear chart. The big open question, when moving toward heavier systems, is whether halos can develop when for instance $A > 40$.

In order to successfully describe a halo nucleus, the structure model needs to take into account [3]: i) the very low density region in which the halo nucleons move, subject to an interaction that is closer to the free NN interaction than the realistic in-medium nuclear interaction; ii) the long tails of the wavefunctions and correct asymptotics of these tails, which contribute decisively to many nuclear properties; iii) the few-body dynamics of the valence nucleons relative to the core and between themselves. For these reasons, it is acceptable to decouple the halo degrees of freedom from the core's, simplifying the standard microscopic treatment: this is the basis for applying few-body models to light nuclear halos.

In heavier systems, the decoupling of core and valence nucleons may not be as straightforward. Mean-field studies in the past decades have shown that pairing is important, and in some regions of the nuclear chart (namely the deformed region), deformation is also necessary. As matter densities in Hartree Fock Bogolyubov (HFB) have a faster radial decay when pairing is included, one expects the pairing force to act against the halo formation. The pairing anti-halo effect was shown in Ref. [5] through self-consistent Hartree Fock Bogolyubov calculations for Carbon isotopes. In opposition to [5], the results of Hamamoto et al. [6, 7] indicate that when an $s_{1/2}$ neutron approaches the dripline, it becomes decoupled from the mean field, which means they are less paired and consequently may give rise to a halo. One can then claim that, if the neutron becomes decoupled from the mean field, the HFB theory should not be used. Unfortunately, the calculations in [6, 7] are not self-consistent; the mean field is reduced arbitrarily, from a well bounded situation, to force the system toward the dripline.

If the situation with pairing is controversial, the effect of deformation appears to be well settled. Nilsson model calculations performed by Hamamoto [8] suggest that $J_z = 1/2^+$ states (J_z being the projection of the angular momentum on the deformation axis), which contain s-wave components of the valence particle as well as other components with higher partial waves, become pure s-waves when the system it forced toward threshold. Thus, it is concluded that deformation does not hinder the halo formation [8].

This exotic feature is not solely associated to nuclear physics: halo manifestations appear also in atomic, molecular or condensed matter physics. A detailed review [2] compiles the results across the nuclear border, and develops a halo signature that removes the specific scale (nuclear, atomic, etc).

RESULTS WITH THREE-BODY MODEL

In Ref. [4] we study the structure of the ground state of a nucleus with two valence nucleons as the system approaches the two particle threshold. A three-body model of $core + n + n$ is used where the core is deformed and allowed to excite. The Faddeev Equations are solved within the hyperspherical method. Our starting point is the ^{12}Be model [9]. An important ingredient of the model [9] is the core+n interaction fitted to reproduce the properties of ^{11}Be. We artificially decrease this $core - n$ interaction,

$$V_{n-core}(\vec{r}) = \lambda\, V_{n-core}^{be12}(\vec{r}) . \tag{1}$$

with $\lambda \to 0$ to simulate the approach to the neutron dripline [4].

The correspondence from a few-body language to the mean-field terminology appropriate for heavy systems, is far from trivial. Nevertheless, in some way, pairing and deformation are included in this three-body model. In the limit of a very heavy core, the core-n interaction plays the role of the mean field in the microscopic description. The NN interaction V_{nn} included in the three-body model would then be related to the pairing associated with the valence pair only, in the mean-field language. However the pairing force and V_{nn} are not identical. Here V_{nn} is a sum of gaussians fitted to the low energy NN phases shifts, containing central s, p and d terms, as well as a spin orbit and a tensor force, while pairing in a typical mean-field calculation is zero range and s-wave only (see [4, 10] for a further discussion).

The hyperspherical expansion introduces a new quantum number K related to the sum of the angular momenta associated with the Jacobi coordinates (x,y) [9]: $K = (l_x + l_y)/2$. The resulting coupled channel equations contain a centrifugal barrier of the form $(K+3/2)(K+5/2)/\rho^2$, where $\rho = \sqrt{x^2 + y^2}$ is the size of the halo. Even for the lowest hyper momentum $K = 0$, there is a barrier reducing the halo effect. For a halo to appear, the K=0 component needs to be the dominant component (>50%) [2]. Thus, we look at the probability of K=0 in the ground state wavefunction as a function of the three-body binding energy (Fig. 1). The probabilities for the lowest hyperspherical harmonic component (K=0) as a function of the two-neutron separation energy are shown for four different cases: i) the case where a realistic NN interaction is included, as well as a quadrupole deformation $\beta_2 = 0.67$ for the ^{10}Be core that reproduces the experimental

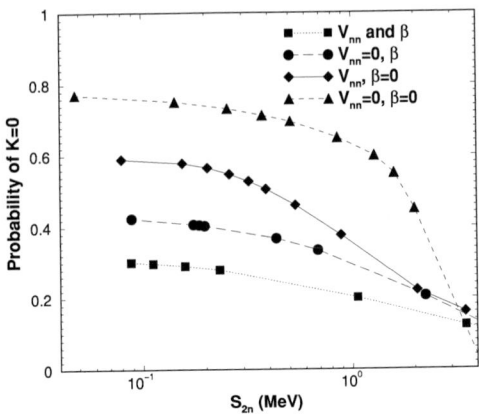

FIGURE 1. Probabilities for the K=0 component in the ground state wavefunction of a ^{12}Be-like system as a function of the two valence neutron separation energy.

$B(E2;0^+ \to 2^+)$; ii) the case where only the NN interaction is switched off; iii) the case where deformation/excitation is neglected but the NN interaction is included; iv) the case where both the NN interaction and core deformation/excitation are neglected.

The first thing that should be noted is that, even for no deformation/excitation and for no NN interaction, the system never develops a pure K=0 state in the limit of zero binding. This result is associated with the fact that we are treating the system as a three-body system and it is in contrast with what would be obtained in the two body case [2]. Secondly, both the NN interaction and collective core degrees of freedom decrease considerably the limiting value for Prob(K=0) when $S_{2n} \to 0$. Ultimately, when both effects are included, the probability is well below 50 %, suggesting that no halo will appear.

Note that the physical input for this exploratory study was provided by ^{12}Be. One can repeat the calculations reducing the deformation parameter and increasing the mass of the core, to simulate a heavier system more realistically [4]. However the model would still be inconsistent, in the sense that the core-n interaction would not be determined from the physical subsystem, and the approach to the dripline would remain artificial.

PRELIMINARY HFB RESULTS

The appropriate way to address the problem for medium mass to heavy nuclei is through a self-consistent model that contains both pairing and deformation [11]. Let us first concentrate on pairing only. A systematic HFB study was performed, searching for nuclei with the $3s_{1/2}$ neutron orbital intercepting the Fermi level in the limit of stability [12]. All calculations were performed with the HFB code written in a 3D mesh [13]. We found that the Cr isotopes satisfied this condition. Using SLy4 [14] for the particle-hole channel and ULB zero-range density-dependent pairing [15] for the particle-particle channel, the neutron dripline is reached for ^{80}Cr. We plot the proton and neutron Helm radii [16] and geometrical radii (Fig. 2) as a function of isotopic mass. The large difference between the neutron and proton Helm radii indicate a neutron skin. A neutron halo exists when there is a large difference between the neutron Helm and geometrical radii. We see that both are present on the Cr dripline. A detailed analysis of our results do not confirm those from Ref. [6] concerning the decoupling of the s-wave orbital from the core, when reaching the dripline [12].

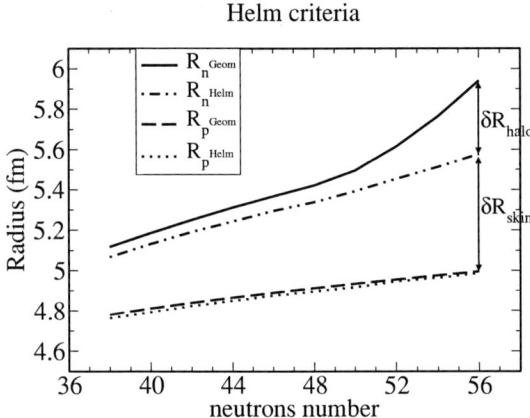

FIGURE 2. Helm criteria for halos and skins: geometrical and Helm radii for both neutrons and protons in Cr isotopes.

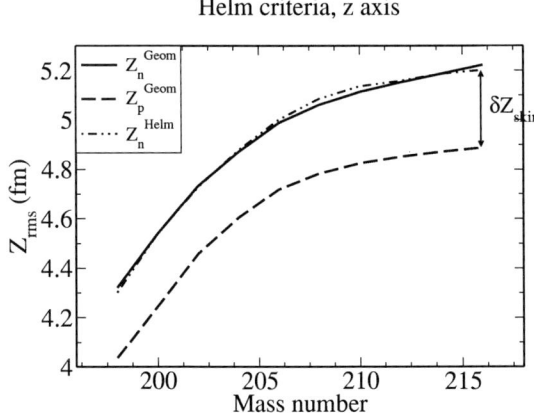

FIGURE 3. Same as Fig.(2) for Dy isotopes along the deformation axis.

Similar searches where performed in the region of deformed nuclei using HFB with even multipole deformation [17]. For the deformed region the orbital we need to concentrate on is the harmonic oscillator $1/2^+$ Nilsson orbital coming from the $4s_{1/2}$ shell in the limit of sphericity. Only for the prolate Dy isotopes did we find this valence neutron orbital to be close to the Fermi level at the dripline. The analysis based on the Helm and geometrical radii was performed on the Dy isotopes (see Fig. 3), similarly to the Cr case. For that purpose, an extension of the Helm criterion to deformed systems was proposed [12]. As there is no significant difference between the neutron's Helm and geometrical radii, we expect no halo to develop. Note that Dy isotopes demonstrate another minimum for oblate shape, which means that the correct ground state should have configuration admixture. Such a calculation was not performed due to its complexity, yet we do not expect it would reverse the result concerning halos.

From these preliminary self-consistent HFB studies we can draw the following conclusions:

- In a realistic situation, pairing does indeed oppose the development of the halo. This confirms the results of [5] and the three-body model [4]. The effect is not sufficiently strong to hinder completely a halo in ^{80}Cr.
- The self-consistent HFB results do not show the decoupling of the neutron from the mean field seen by [6].
- Deformation in heavy nuclei seem to strongly hinder the appearance of a halo, as suggested by the three-body model [4] and in disagreement with [8]. No halos were found in the deformed region of the nuclear chart. In particular, the fully self-consistent calculations do not confirm the purification of the $J_z = 1/2^+$ orbitals at the Fermi level into s-waves when approaching the dripline.

An article containing all details of these HFB studies will become available soon. As part of our near future plans, we will study the dependence on the choice of the microscopic pairing force, namely a better low-density behaviour and finite-range effect [10].

* This work has been partially supported by National Superconducting Cyclotron Laboratory at Michigan State University and the National Science Foundation under grants PHY-0456656 and PHY-0456903.

REFERENCES

1. P.G. Hansen, New Scientist, pg 38 October 1993.
2. A.S. Jensen, K. Riisager, D.V. Fedorov and E. Garrido, Rev. Mod. Phys. 76 (2004) 215.
3. F.M. Nunes, C.R. Physique 4 (2003) 489.
4. F.M. Nunes, Nucl. Phys. A 757 (2005) 349.
5. K. Bennaceur, J. Dobaczewski, M. Ploszajczak, Phys. Lett. B 496 (2000) 154.
6. I. Hamamoto and B.R. Mottelson, Phys. Rev. C 68 (2003) 034312.
7. I. Hamamoto and B.R. Mottelson, Phys. Rev. C 69 (2004) 064302.
8. I. Hamamoto, Phys. Rev. C 69 (2004) 041306.
9. F.M. Nunes et al., Nucl. Phys. A 609 (1996) 43.
10. T. Duguet, Phys. Rev. C 69 (2004) 054317.
11. M. Bender, P.-H. Heenen, and P.-G. Reinhard, Rev. Mod. Phys. 75 (2003) 121.
12. Benoit Avez, Report of Summer Training on *Halo Effect on Heavy nuclei* NSCL, August 2005.
13. B. Gall et al., Z. Phys. A 348 (1994) 183.
14. E. Chabanat et al., Nucl. Phys. A 627 (1997) 710.
15. C. Rigollet and P. Bonche and H. Flocard and P.-H. Heenen, Phys. Rev. C (1999) 3120.
16. R. Helm, Phys. Rev. 104 (1956) 1466.
17. M. Bender, P. Bonche, T. Duguet, and P.-H. Heenen, Phys. Rev. C 69 (2004) 064303.

Doppler shift as a tool for studies of resonant (p,n) reactions with RIBs: Spectroscopy of ^{7}He.

P. Boutachkov*, G.V. Rogachev[†], V.Z. Goldberg**, A. Aprahamian*, F.D. Becchetti[‡], J.P. Bychowski[§,*], Y. Chen[‡], G. Chubarian**, P.A. DeYoung[§], J.J. Kolata*, L.O. Lamm*, G.F. Peaslee[¶], M. Quinn*, B.B. Skorodumov* and A. Wöhr*

Physics Department, University of Notre Dame, Notre Dame, IN 46556
[†]*Physics Department, Florida State University, Tallahassee, FL 32306*
**Texas A&M University, College Station, TX 77843*
[‡]*Physics Department, University of Michigan, Ann Arbor, MI 48109*
[§]*Physics Department, Hope College, Holland, MI 49422*
[¶]*Chemistry Department, Hope College, Holland, MI 49422*

Abstract. We report on a new methods for studies of neutron rich systems through resonant (p,n) reaction with radioactive ion beams. A specific example of the spectroscopy of ^{7}He and future application of the proposed methods are discussed.

Keywords: resonance reactivation, RIBs, thick target, 7He
PACS: 25.60.-t, 25.40.Kv, 27.20.+n

Radioactive Ion Beams (RIB) alow us to study nuclei close and beyond the proton and neutron drip lines. One of the experimental challenges in performing experiments with RIBs is the low beam intensity. This leads to the necessity for new experimental techniques and improving the current ones. Recently three new techniques [1, 2, 3] have been developed for study of neutron rich nuclei close to the drip line. These techniques use resonance scattering in a thick proton target [4]. In ref. [1] elastic resonant scattering was used to study ^{9}He while in refs. [2, 3] we used inelastic resonance scattering to obtain information on the spectroscopy of ^{7}He.

The above methods have several characteristics in common. They use resonance reaction to populate states of high isospin in the compound system: RIB+proton. These state allow us to probe indirectly the more neutron rich systems using the isospin symmetry of the nuclear force. For example, in ^{6}He+p scattering one can populate states with isospin $T = 3/2$ in ^{7}Li. These states are isobaric analogs of levels in ^{7}He. There are two isospin allowed channels for the populated ^{7}Li($T = 3/2$) states: proton decay back to ^{6}He and a proton or neutron decay to $T = 1$ states in ^{6}Li, see Fig. 1a. As it follows from the wave function of the populated $T = 3/2$ states:

$$\frac{1}{\sqrt{3}}\Psi(^6He)\Psi(p) + \sqrt{\frac{2}{3}}\Psi(^6Li, T = 1)\Psi(n) \qquad (1)$$

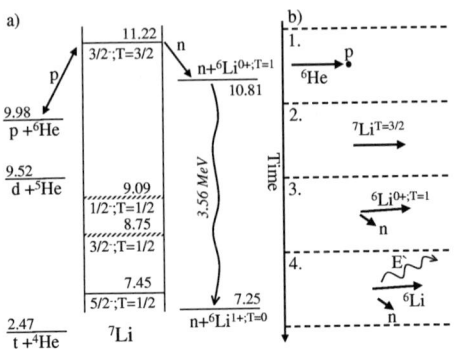

FIGURE 1. a) Decay pathways for the T=3/2 resonance in ^{7}Li, and b) the successive kinematics stages of the studied reaction.

the decay to n+^{6}Li is dominant and the reduced widths for the two decay channels are related as:

$$(\gamma_n/\gamma_p)^2 = (\sqrt{\frac{2}{3}}/\frac{1}{\sqrt{3}})^2 = 2 \qquad (2)$$

The $T = 1$ state of ^{6}Li populated via the (p,n) reaction can decay only through isospin forbidden channels. Therefore, the probability for γ transition is enhanced with respect to the particle decays. For the specific case of the first $T = 1$ state in ^{6}Li, particle decay also violates the parity conservation law and therefore only γ decay is allowed. Thus, measuring the characteristic 3.56 MeV γ-ray from the decay of the first $T = 1$ state in ^{6}Li is a clear signature that a $T = 3/2$ state in ^{7}Li was populated. This chain of transmutations is typical for the isobaric analog states of neutron rich nuclei close to the line of stability. What is special for the described example is the 100% probability for γ decay from the first $T = 1$ state of ^{6}Li.

Suppose that a thick proton target $((CH_2)_n)$ is used to stop the ^{6}He beam. Interaction of ^{6}He with protons can take place at any energy from the maximum (beam energy) to zero populating $T = 1/2$ and $T = 3/2$ resonances in ^{7}Li. When a $T = 3/2$ resonance is populated, it will decay with the highest probability to neutron and ^{6}Li(0$^+$,$T = 1$) state (see Fig.1). The velocity of ^{6}Li will depend on the velocity of ^{7}Li and the angle at which the neutron was emitted. The excited ^{6}Li nucleus decays by γ emission before it loses any energy in the target (the width of the 0$^+$,$T = 1$ resonance is 8 eV) and information on the velocity of ^{6}Li is preserved in the Doppler shift of the γ ray. Therefore one can place a γ detector at 90° relative to the beam direction and detect the Doppler broadened 3.56 MeV γ-rays in confidence with a neutrons [2] or place a γ detector at the beam axis and measure the doppler profile of the γ-ray in question [3]. In the proof of principle experiment [2], we showed that by neutrons-γ coincidences one can extract the $T = 3/2$ resonance energies and widths with accuracy limited only to the neutron detector resolution. In a second experiment, we showed that by combining the results of the neutron-γ coincidence and the Doppler profile measurement, one can determine the spins and parities of the states in question [3].

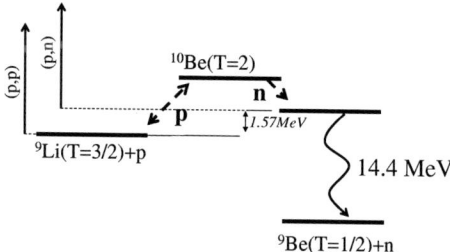

FIGURE 2. Decay pathways for the T=2 isobaric analog states of ^{10}Li in ^{10}Be. The arrow labelled (p,p) shows the energy from which states can be studied through elastic resonance scattering while the arrow labelled (p,n) shows excitation energy above which inelastic resonance scattering can be used.

With the measurements mentioned here, we obtained data on the ^{7}Li(T=3/2) states. These data provides information about ^{7}He. The techniques discussed here should work for any neutron rich exotic nucleus. What is specific for the ^{7}He case is that the populated T=1 state in ^{6}Li can decay only by a γ-ray. In Figure 2, we show schematically show how the discussed techniques can be applied to study the isobaric analog states of ^{10}Li in ^{10}Be. In this case, the branching ratio for γ-ray decay from the 14.4 MeV ^{9}Be(T=1/2) state is 10% [5].

We chose ^{7}He nucleus as first case to study with the this techniques since the existence of a state in ^{7}He with very low excitation energy (E*=0.56(10) MeV) was claimed in two recent publications [6, 7]. It was identified as the 1/2$^-$ spin-orbit partner of the 3/2$^-$ ^{7}He ground state. If confirmed, this would be a very important result since the low energy of the state indicates a dramatic breakdown of the shell model at large neutron excess. The width of the state was reported to be Γ=0.75(8) MeV, close to the single-particle limit, implying a nearly pure $p_{1/2}$ neutron configuration coupled to the ^{6}He ground state. Hence, this state is expected to decay predominantly to the ground state of ^{6}He.

In refs. [2, 3] the mass excess of ^{7}He is obtained from that of its analog in ^{7}Li by adding the neutron-proton mass difference and subtracting the Coulomb energy difference of the ^{6}He-^{6}Li*(T=1) pair. From the assumption that there are only two decay channels for the ^{7}Li(T=3/2) states, follows that total width is:

$$\Gamma(^{7}Li) = 2P_{n+^{6}Li(T=1)} \times \gamma_n^2 + 2P_{p+^{6}He} \times \gamma_p^2 \qquad (3)$$

where $P_{n+^{6}Li(T=1)}$ and $P_{p+^{6}He}$ are the corresponding penetrability factors for the $n+^{6}Li(T=1)$ and $p+^{6}He$ channels. To first order from the isospin symmetry of the nuclear forces follows that the reduced widths for the $^{7}Li(T=3/2) \longrightarrow n+^{6}Li(T=1)$ and the $^{7}He(T=3/2) \longrightarrow n+^{6}He(T=1)$ channels are the same. The $^{7}He(T=3/2) \longrightarrow n+^{6}He(T=1)$ decay channel dominates for the studied ^{7}He states. Therefore their total width is:

$$\Gamma(^{7}He) = 2P_{n+^{6}He} \times \gamma_n^2 \qquad (4)$$

Combining equations 2, 3 and 4 we obtain that the widths of the states of interest in ^{7}He are related to the measured ^{7}Li(T=3/2) widths by:

$$\frac{\Gamma(^7He)}{\Gamma(^7Li(T=3/2))} = \frac{2P_{n+^6He}}{P_{n+^6Li(T=1)} + 0.5P_{p+^6He}} \quad (5)$$

The width of the ^{7}He ground state is measured to be: $\Gamma(^7He) = 150 \pm 20$ keV [8]. Using equation 5, we obtain the width of the corresponding isobaric analog state in ^{7}Li: $\Gamma(^7Li(T=3/2)) = 306 \pm 40$ keV. We have measured a width of $\Gamma(^7Li(T=3/2)) = 265 \pm 40$ keV [2]. These numbers agree within the error bars (One can take higher order effects into account by using the potential model [9]).

We used the same approach to find the parameters of the isobaric analog state in ^{7}Li of the (E*=0.56(10) MeV, 1/2$^-$) state reported in refs. [6, 7]. We excluded the existence of this state conclusively [2, 3]. Instead, we found evidence for the population of a broad (T=3/2, 1/2$^-$) resonance in ^{7}Li instead.

The current experimental status on ^{7}He and ^{7}Li(T=3/2) can be summarized as follows:

- The ground state of ^{7}He was observed by P.H.Stokes, et al. [10]. In this study the ^{7}Li(n,p)^{7}He reaction was used. The g.s. has a spin and parity of J^π =3/2$^-$ and width Γ=0.150 MeV.
- The isobaric analog state of the g.s. of ^{7}He in ^{7}Li was found by C. Detraz, et al.[11] in the reaction ^{9}Be(p,^{3}He)^{7}Li. E_{ex} = 11.28 MeV, Γ=0.260 MeV
- The first observation of a ^{7}He excited state was reported in the reaction p(^{8}He,d)^{7}He by A.A.Korsheninnikov, et al. [12]. E_{ex} =3.3 MeV, Γ=2.2 MeV and J^π =5/2$^-$.
- H.G.Bohlen, et al. [13] reported on an excited state at E_{ex} = 2.95 MeV and Γ=1.9 MeV observed in the reaction ^{9}Be(^{15}N,^{17}F)^{7}He. Considering the low density of states in the studied loosely bound systems this is probably the same state as the one observed in ref. [12].
- M.Meister, et al. [7, 6] reported an excited state at E_{ex}=0.56(10) MeV, Γ=0.75(8) MeV and J^π =1/2$^-$ populated in the breakup of ^{8}He.
- G.V. Rogachev, et al. [2] observation of T=3/2 states in ^{7}Li using p(^{6}He,^{6}Li(T=1))n reaction. This experiment excludes the existence of the state reported in refs. [7, 6] and indicates the existence of a higher lying T=3/2 resonance.
- P. Boutachkov, et al. [3], observe of T=3/2 states in ^{7}Li using p(^{6}He,^{6}Li(T=1))n reaction. The existence of the state reported in refs. [7, 6] is excluded beyond the 90% confidence level. Indications for higher lying (T=3/2,J^π=1/2$^-$) resonance is shown.

More experimental data on ^{7}He will become available in near future. There are number of results reported in 2004-2005 including:

- A. Wuosmaa, et al. [14], using the d(^{6}He,p)^{7}He reaction.
- W. Mittig, et al., [15], using the p(^{8}He,d)^{7}He reaction.
- P. von Neumann-Cosel, et al. [16], using the ^{7}Li(d,^{2}He)^{7}He reaction.

The techniques described in refs. [2, 3] and summarized in this work can be broadly applied to the study of other exotic neutron rich systems. It is possible to find a nucleus where the (p,n) channel is energetically closed for some of the high isospin states. This is the case for example in some of the isobaric analog states of ^{10}Li in ^{10}Be (see Figure 2). In this case, the states in question can be studied by elastic resonance scattering. It has been shown in a prove of principle experiment on the isobaric analog states of ^{9}He in ^{9}Li [1] that such an approach works. Therefore in the general case the combination of the techniques described in refs. [1],[2] and [3] will provide access to the properties of the isobaric analog states of interest in a comprehensive way.

In conclusion, two new approaches to the spectroscopy of neutron-rich nuclei were tested for the case of p(^{6}He,n)^{6}Li(0^{+},T=1), and found to give excellent results. The first application of these new techniques helped to resolve a dilemma related to a low lying narrow $1/2^{-}$ state in ^{7}He. We showed that the state in question does not exist. Instead, we found evidence for the analog of a very broad ($1/2^{-}$, T=3/2) resonance lying at an excitation energy above 2.2 MeV in ^{7}He.

We believe that the combination of the methods described in [1],[2] and [3] provide a compleat new set of tools to study the spectroscopy of neutron rich nuclei close to the border of stability.

This work was supported by the NSF under Grant Numbers PHY99-01133, PHY00-98061, PHY01-00102, PHY02-03099, and PHY02-44989, and DOE Grant DE-FG03-93ER40773.

REFERENCES

1. G. V. Rogachev, et al., *Phys. Rev. C* **67**, 041603 (2003).
2. G. V. Rogachev, et al., *Phys. Rev. Lett.* **92**, 232502 (2004).
3. P. Boutachkov, et al., *Phys. Rev. Lett.* **95**, 132502 (2005).
4. V. Z. Goldberg, "Resonance Scattering to Study Nuclei at the Borders of Nuclei Stability," in *Proceedings of Exotic Nuclei and Atomic Masses (ENAM98) International Conference*, edited by B. M. Sherrill, D. J. Morrisey, and C. N. Davids, Springer, New York, 1998, p. 319.
5. D. R. Tilley, et al., *Nuclear Physics A* **745**, 155 (2004).
6. K. Markenroth, et al., *Nucl. Phys. A* **679**, 462 (2001).
7. M. Meister, et al., *Phys. Rev. Lett.* **88**, 102501 (2002).
8. D. R. Tilley, et al., *Nuclear Physics A* **708**, 3 (2002).
9. L. Axelsson, et al., *Phys. Rev. C* **54**, R1511 (1996).
10. R. H. Stokes, and P. G. Young, *Phys. Rev. Lett.* **18**, 611 (1967).
11. C. Detraz, J. Cerny, and R. H. Pehl, *Phys. Rev. Lett.* **14**, 708 (1965).
12. A. A. Korsheninnikov, et al., *Phys. Rev. Lett.* **82**, 3581 (1999).
13. H. G. Bohlen, et al., *Phys. Rev. C* **64**, 024312 (2001).
14. A. H. Wuosmaa, et al. (2005), submitted to Phys. Rev. C.
15. W. Mittig, et al. (2005), reported on DREB 2005 workshop.
16. P. Neumann-Cosel, et al. (2005), submitted to Phys. Lett. B.

Model calculations of radiative capture of nucleons in MeV region

E. Běták

Institute of Physics, Slovak Acad. Sciences, 84511 Bratislava, Slovakia
Faculty of Philosophy and Sciences, Silesian Univ., 74601 Opava, Czech Rep.

Abstract.
We address calculations of the neutron and the proton radiative capture at incident energies up to 20 MeV on medium and heavy nuclei. The main formalism used is the pre-equilibrium (exciton) model of γ emission. A link to the Consistent Direct-Semidirect model is noticed as well. The resulting pre-equilibrium (plus equilibrium) calculations of the radiative capture excitation functions are compared to experimental data and also some cross section trends important for possible production of therapeutic radioisotopes are extracted.

Keywords: gamma emission, cross sections, pre-equilibrium decay, computer codes
PACS: 25.40.Lw, 26.60.-k, 23.20.-g, 24.50.+g, 24.30.Cz

INTRODUCTION

Reactions of radiative capture of nucleons — even though very difficult to be measured due to their low cross sections — serve as a challenge for different γ emission mechanisms already for decades. The pre-equilibrium model (see, e.g., [1]) appeared to be rather successful to describe the γ energy spectra in the continuum region in 14 MeV neutron-induced reactions [2, 3, 4]. The study of excitation functions extended the applicability of the model to energies starting from few MeV to about 30 MeV. Significant improvement was the incorporation of spin into the formalism of the pre-equilibrium exciton model [5, 6], which enabled also pre-equilibrium calculations leading to discrete states and the comparison to the direct-semidirect model calculations.

Both models have some common features, even though they are strictly complementary in their underlying physics. Whereas the direct-semidirect model deals with wavefunctions and specific interactions of nucleons, while completely ignoring competing processes, the pre-equilibrium models are of statistical nature and they deeply involve quantities like the level densities etc. The competition — e.g. of the nucleon emission — is naturally contained there. What is common to both models is that they both are capable to reproduce (more-or-less) the data corresponding to hard γ emission [7] observed in radiative nucleon capture at energies above few MeV.

The available experimental data are not frequent due to low cross sections and — at the same time — low detector efficiency for these hard γ's.

EMPIRE AND TALYS COMPUTER CODES

Whereas one can use simple exciton-model based codes at nucleon energies exceeding 10 MeV and sufficiently far from the closed shells, like non-spin code PEQAG [8] or its spin-dependent successor DEGAS [9], it is advisable to incorporate sophisticated nuclear reaction codes which include whole range of approaches and are also coupled to extensive libraries of parameters, if one needs some estimate for a reaction not just tailored to use of simple statistical pre-equilibrium code only. Two codes of this family have been recently released, namely EMPIRE-II (version 2.18 in 2002 [10] and version 2.19 early this year [11] [1]) and TALYS [12, 13].

They are very close as for their underlying physics at the pre-equilibrium stage (e.g., the same single-nucleon radiative mechanism formula for the γ emission is used both in EMPIRE and in TALYS), and similarly both of them use very extensive tables of various recommended parameters.

The main differences important for the pre-equilibrium stage of the reaction may be summarized as follows: *i)* The basic approach to the pre-equilibrium stage is the two-component one (i.e. distinguishing between the neutrons and the protons) in TALYS, whereas one-component formulation with a charge factor is used in EMPIRE; *ii)* One-particle radiation mechanism for the γ emission is used in EMPIRE, but TALYS adds the quasideuteron (two-particle) [2], what may cause some differences (however, very small ones) at excitation energies above about 30 MeV; *iii)* Though the level densities (using the default option) are the same in both codes (with parameters taken from RIPL [14]), different (semi-)microscopic approaches are available for the advanced user; *iv)* Classical optical model is used to calculate the particle transmission coefficients T_l in EMPIRE with parameters from libraries, and the local and global parameterization of [15] is employed in TALYS. (This difference influences the γ emission only via the competition with that of the particles.)

REACTIONS (N, γ) LEADING TO THERAPEUTIC ISOTOPES

The need to produce isotopes for diagnostic and therapeutic purposes stimulated also calls for further measurements and evaluations of the (n,γ) reactions at energies below 20 MeV. Within the IAEA Coordinated Research Program, some very desirable isotopes for therapeutic needs have been identified and are studied [16]. With opening two excellent codes to community earlier this year, one has got a chance to predict the excitation curves with much better reliability than before. Generally, there are not many data on such (n,γ) reactions [17] in the continuum region. The exception is the reaction on ^{165}Ho with remarkable number of data points [3]. We present the data together with the calculations

---

[1] The main differences between two versions of EMPIRE-II may be characterized as replacing data libraries by their more recent versions, adding of further subroutines and also replacing some minor bugs.
[2] The quasideuteron mechanism is also included in EMPIRE-II v. 2.19, but it is considered for the photonuclear reactions only, and not for the γ emission.
[3] Data of Menlove et al. [18] and one point from Thesis of Csikai are not here, as they differ significantly from all the others.

of TALYS [12, 13] and two versions of EMPIRE-II (v. 2.18 [10] and v. 2.19 [11]) in Fig. 1. Essentially, we kept the default parameters in EMPIRE, just with allowance for full inclusion of pre-equilibrium emission and γ cascades. Details of the form of the Giant Dipole Resonance (which enters the calculations of the γ emission via the detailed balance principle) and of other parameters did not show much influence on the resulting excitation functions calculated using EMPIRE-II v. 2.18 [16], and we therefore applied this approach also to version 2.19 and TALYS.

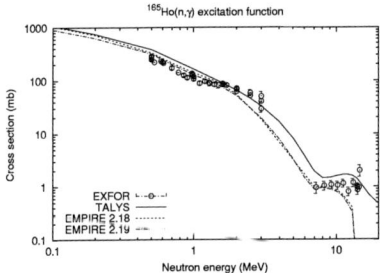 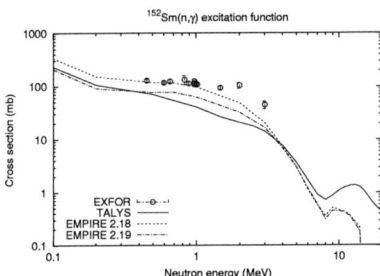

FIGURE 1. *Excitation functions of (n,γ) reactions on ^{165}Ho (left) and ^{152}Sm (right). Circles denote the experimental data [17], full curves the calculations by TALYS, and the dotted and dotted-dashed ones those by EMPIRE-II v. 2.18 and 2.19, respectively.*

Similar situation, only with less data points, is for ^{152}Sm(n,γ). Much more complicated is the situation where there are simply no data. Here, we had to rely on the experience and assume that if a reasonable description is achieved for reactions comparable to experimental data, one can expect not bad prediction using the same approach also in the absence of data. In some cases, TALYS and EMPIRE calculations practically coincide (e.g. ^{124}Xe(n,γ)), but there are also reactions (with no available data), where these two predictions differ nearly by an order of magnitude (e.g. ^{102}Pd(n,γ)).

REACTION ^{208}(P, γ)

Recent measurements of the (p,γ) reactions on ^{208}Pb leading to discrete states [19] is another challenge to check the validity of pre-equilibrium formulation of the γ emission. Unfortunately, the target nucleus is a double magic one, with extremely enhanced individual features and far off typical statistical behaviour of nuclei in its vicinity, what is — in somewhat weakened sense — true also for the final nucleus of the radiative capture, ^{209}Bi. One has to be therefore *extremely careful* with the choice of proper parameters used in this calculation [4]. As already shown previously [20], the systematics of the level density parameters fails for reactions on ^{208}Pb, and one has to apply carefully chosen individual ones. Anyway, Refs. [20] and [21] presented calculations of *neutron*-induced reactions on Pb (integrated and activation cross sections and also continuous γ spectra in [20], and cross sections to discrete states in [21]), whereas the *proton*-induced one has been measured now, and due to the presence of Coulomb barrier, the competing *pro-*

---

[4] The single-particle level density of ^{208}Pb is *three* times less than the average around.

ton channel is more influenced by not optimal choice of parameters than it was in the *neutron* reactions.

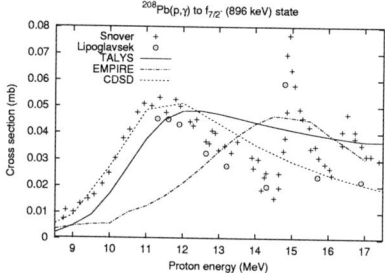

FIGURE 2. Excitation curve of $^{208}Pb(p,\gamma)$ leading to the $f_{7/2^-}$ (896 keV) discrete state. Crosses are the data of Snover [22], circles the recent ones by Lipoglavšek et al. [19], and the calculations by TALYS, EMPIRE-II v. 2.19 and within the CDSD model are drawn by a full, dotted-dashed, and dotted curves, respectively.

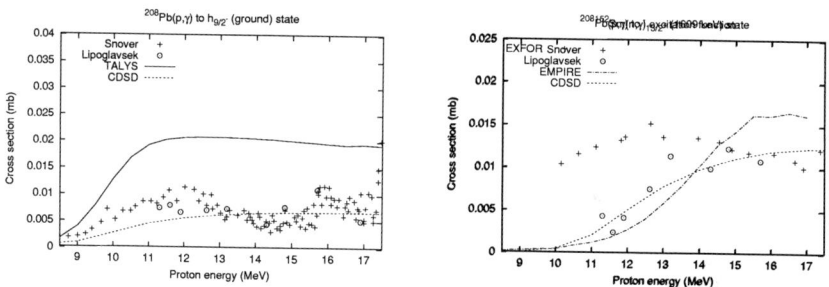

FIGURE 3. As in Fig. 2, but for the $h_{9/2^-}$ (ground state) (left) and $i_{13/2^+}$ (1609 keV) (right). Only one set of pre-equilibrium calculations is depicted for each of he sates, as the other one significantly overpredicts the measured data.

In Figs. 2 and 3, we present the excitation functions leading to discrete states calculated using two pre-equilibrium codes together with the data [22, 19] and the Consistent Direct-Semidirect Model (CDSD) [23] calculations [19] [5]. The sharp experimental peaks are due to the analogue states.

CONCLUSIONS

The pre-equilibrium model of γ emission depicts the same reality which is contained in the direct-semidirect model, but using statistical approach. Two excellent computer codes, EMPIRE v. 2.19 and TALYS, yield similar results far off the closed shells. In their vicinity, their predictive power weakens and discrepancies become large.

---

[5] The original paper by Snover [22] brings data which differ by a factor of 2 from those used now [19, 24].

ACKNOWLEDGMENTS

The author thanks M. Herman, A. Likar, P. Obložinský and T. Vidmar for valuable discussions; A. Likar also for availability of their (partially still unpublished) results on the ^{208}Pb(p,γ) reaction. The work has been supported in part by the IAEA Res. Contract No. 12425/R2 and by VEGA grant No. 2/4102/24.

REFERENCES

1. E. Gadioli, and P. Hodgson, *Pre-Equilibrium Nuclear Reactions*, Clarendon Press, Oxford, 1992.
2. E. Běták, and J. Dobeš, *Phys. Lett. B*, **84**, 368 (1979).
3. J. M. Akkermans, and H. Gruppelaar, *Phys. Lett. B*, **157**, 95 (1985).
4. F. Cvelbar, and E. Běták, *Z. Phys. A*, **332**, 163 (1989).
5. P. Obložinský, *Phys. Rev. C*, **35**, 407 (1987).
6. P. Obložinský, and M. B. Chadwick, *Phys. Rev. C*, **41**, 1652 (1990).
7. E. Běták, F. Cvelbar, A. Likar, and T. Vidmar, in *Proc. Eleventh Internat. Symp. Capture Gamma-Ray Spectroscopy and Related Topics, Pruhonice near Prague, 2–6 Sept. 202*, edited by J. Kvasil, P. Cejnar, and M. Krtička, World Scientific, Singapore, 2003, p. 212.
8. E. Běták, PEQAG: A PC version of fully pre-equilibrium computer code with gamma emission, Rep. INDC(CSR)-016/LJ, IAEA, Vienna (1989).
9. E. Běták, and P. Obložinský, PEGAS: Pre-Equilibrium–Equilibrium Gamma-And-Spin Code (PC Version), Rep. INDC(SLK)-001, IAEA, Vienna (1993).
10. M. Herman, EMPIRE-II statistical model code for nuclear reaction calculations (version 2.18 Mondovi), Report, IAEA, Vienna (2002).
11. M. Herman, P. Obložinský, R. Capote, A. Trkov, V. Zerkin, M. Sin, and B. Carlson, EMPIRE modular system for nuclear reaction calculations (version 2.19 Lodi), Report, NNDC, Brookhaven National Lab., Upton, USA (2005).
12. A. Koning, and S. H. abd M.C. Duijvestijn, in *International Conf. on Nuclear Data for Sci. and Technol. Santa Fe, NM, 26 Sept. – 1 Oct. 2004*, edited by R. Haight, M. Chadwick, T. Kawano, and P. Talou, American Inst. Phys., AIP Conf. Proc., vol. 769, Part II, Melville and New York, 2005, p. 1154.
13. A. Koning, S. Hilaire, and M. Duijvestijn, TALYS: A nuclear reaction program, Rep. 21297/04.62741/P FAI/AK/AK, NRG, Petten (2004).
14. Handbook for calculations of nuclear reaction data. reference input parameter library. http://www-nds.iaea.or.at/ripl-2/ bla bla, Tech. rep., IAEA, Vienna (2002).
15. A. Koning, and J. Delaroche, *Nucl. Phys. A*, **713**, 231 (2003).
16. Nuclear data for production of therapeutic radionuclides. Summary report of Second Research Coord. Meeting, prepared by J.-C. Sublet, and R. C. Noy, Rep. INDC(NDS)-465, IAEA, Vienna (2004).
17. EXFOR-CINDA for applications. database and rerieval systems. version 1.63i (CD-ROM), IAEA, Vienna (2004).
18. H. Menlove, K. Coop, H. Grench, and R. Sher, *Phys. Rev.*, **163**, 1299 (1967).
19. M. Lipoglavšek, R. Bark, M. Benatar, E. Gueorguieva, J. Kau, F. Komati, P. Kwinana, J. Lawrie, G. Mabala, P. Maine, S. Mukhrejee, S. Mullins, S. Murray, N. Ncapayi, R. Newman, P. Vymers, A. Likar, M. Vencelj, and T. Vidmar, *Phys. Lett. B*, **593**, 61 (2004).
20. E. Běták, J. Kopecky, and F. Cvelbar, *Phys. Rev. C*, **46**, 945 (1992).
21. E. Běták, F. Cvelbar, A. Likar, and T. Vidmar, *Nucl. Phys. A*, **686**, 204 (2001).
22. K. Snover, J. Amann, W. Hering, and P. Paul, *Phys. Lett. B*, **37**, 29 (1971).
23. A. Likar, and T. Vidmar, *Nucl. Phys. A*, **591**, 458 (1995).
24. S.S.Hanna, in *Proc. Int. Conf. Nucl. Phys. with Electromagntic Interacts., Mainz 1979*, edited by H. Arenhövel, and D. Drechsel, Springer, Berlin, 1979, p. 288.

NUCLEAR STRUCTURE (0^+ STATES)

Study of 0^+ States in Deformed Nuclei

S.R. Lesher[*,†], Z. Ammar[*], M. Merrick[*], C.D. Hannant[*], N. Warr[*,**], T.B. Brown[*,‡], N. Boukharouba[*], C. Fransen[*,**], M.T. McEllistrem[*] and S.W. Yates[*]

[*]*University of Kentucky, Lexington, KY 40506-0055 USA*
[†]*K.U. Leuven, Instituut voor Kern-en Stralingsfysica, Celestijnenlaan 200D, B-3001 Leuven, Belgium*
[**]*Institut für Kernphysik, Universität zu Köln, D-50937 Köln, Germany*
[‡]*SRTC, Westinghouse, Savannah River Co., Aiken, SC 29808, USA*

Abstract. In recent ^{160}Gd(p,t) reaction studies the existence of more than ten 0^+ states in ^{158}Gd below 3.0 MeV was revealed. We have examined ^{158}Gd with the (n,n'γ) reaction at neutron energies up to 3.5 MeV to confirm the identification of these states and to determine their lifetimes through DSAM measurements. Gamma-ray excitation function and angular distribution measurements have been performed and $\gamma - \gamma$ coincidences have been measured with the KEGS array of detectors. Moderately strong decays are observed from some of these 0^+ states.

Keywords: 0^+ states, β vibration, DSAM
PACS: 43.35.Ei, 78.60.Mq

INTRODUCTION

Since the work of Bohr and Mottelson [1], the lowest excited 0^+ state (0_2^+) and second excited 2^+ state (2_2^+) in deformed nuclei have been referred to as β- and γ-vibrational states, and rotational bands are expected to be built with these states. For deformed nuclei quadrupole distortions in shape are these vibrations. The γ vibration is a shape change against the axis of symmetry and is identified by a collective $K^\pi = 2^+$ state decaying to the ground state. This vibration exhibits a systematic behavior across the region of deformed nuclei and is the first $K^\pi = 2^+$ band. This vibration is so well characterized that two-phonon $\gamma\gamma$ vibration have been observed [2, 3, 4].

The β vibration is a shape change along the axis of symmetry and is identified as a collective $K^\pi = 0^+$ state. The lowest 0^+ states in deformed nuclei fail to display the systematic trends exhibited by the γ vibration. Although there are not many lifetime measurements available for these decays, systematics do not appear to surface. There is current debate on the collectivity and even the existence of the β type of vibration in deformed nuclei [5, 6, 7, 8, 9, 10]. The argument is two-fold. Does the band decay to the γ band or to the ground state, and is this decay collective, i.e., does it involve several nucleons or just a pair of nucleons?

The best candidate for a β band has been identified in ^{166}Er where the fourth 0^+ state exhibits an enhanced E2 transition to the first excited state [11] but the authors are uncertain in characterizing it as a β vibration. Recently, A. Aprahamian *et al.* have identified a β band candidate and have also shown evidence for states of a higher-lying 0^+ band in ^{178}Hf decaying to a lower-lying 0^+ state, suggesting a two-phonon

$\beta\beta$ vibration [12, 13]. There is some controversy surrounding this assertion, since the character of the β band has yet to be resolved. Another two-phonon vibration would be the $\beta\gamma$ vibration, where a higher lying 2^+ state should decay to the β and γ bands. To date, there is no evidence of this $\beta\gamma$ two-phonon vibration. The debate over identification of the β band has led P.E. Garrett [14] to question the general occurrence of the β vibrations and to emphasize the role of pairing in describing low-lying 0^+ states. This is one interpretation of the β band and he outlines what he believes would be a convincing argument for a β vibration:

- B(E2; $0_\beta^+ \to 2_{gs}^+$) $\sim$ 12 - 33 W.u.
- B(E2; $2_\beta^+ \to 0_{gs}^+$) $\sim$ 2.5 - 6 W.u.
- ρ^2(E0) x $10^3 \sim$ 85 - 230
- "small" two-nucleon transfer strengths
- small single-nucleon transfer strengths (for some situations)

Where the B(E2) values are the transition probabilities between the 0^+ and 2^+ states of interest. The value, ρ^2(E0) is the accepted way of denoting the E0 value of transitions between two 0^+ states and is the squared value of the monopole transition strength.

There is a paucity of lifetime information available for 0^+ states in deformed nuclei, which is an essential piece of the puzzle. Also, in searching for the β band, one must look at all the 0^+ states and not just the lowest excited 0^+ bands, as has been the focus of previous measurements. Therefore, we have initiated a study of 0^+ states in a unique case, ^{158}Gd.

In high-resolution (p,t) two-neutron transfer reactions, it has been shown that ^{158}Gd, a well-deformed, prolate nucleus, exhibits at least ten 0^+ states below 3 MeV [15]. The identification of these states has provided us with a rare opportunity to garner an abundance of information about spin-0 states in a single deformed nucleus. This nucleus was studied with the (n,n'γ) reaction, a reaction that should permit the population and detailed characterization of the decays of these states. The primary goal, after assuring that these 0^+ assignments are correct, is to determine lifetimes of as many of these states as possible and to assess how many of them exhibit collective characteristics.

Previous work on ^{158}Gd includes a detailed study by R.C. Greenwood et al. [16] through radiative neutron capture. The lowest five 0^+ levels, with the exception of a possible state at 1577 keV, were identified along with their corresponding γ rays, and a detailed level scheme for levels up to around 2.2 MeV was presented. In a recent study, with the same reaction and the GRID technique, H.G. Börner et al. [17] measured lifetimes of a few of the lowest-lying levels, including the lowest two 0^+ states, and calculated their respective transition probabilities. Spawning the present work was the ^{160}Gd(p,t)^{158}Gd experiment by S.R. Lesher et al. [15] in which excited 0^+ levels were identified at 1196.1, 1452.3, 1577.0, 1743.1, 1952.3, 1957.8, 2277.3, 2338.0, 2643.4, 2687.1, 2911.2, 3076.7, and 3109.9 keV.

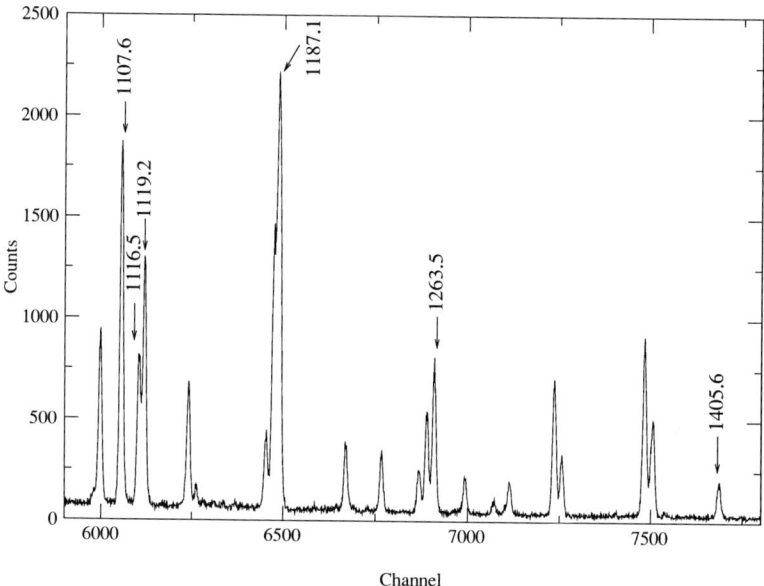

FIGURE 1. Portion of the prompt γ-ray spectrum from the ^{158}Gd(n,n'γ) reaction with 2.2-MeV neutrons, recorded at 90° with respect to the direction of the incident neutrons. Gamma rays are labeled with energies in keV. Time-of-flight gating has been employed to reduce the contributions from background events.

EXPERIMENTAL PROCEDURES

Low-lying, low-spin levels of ^{158}Gd were studied in experiments performed at the University of Kentucky 7 MV Van de Graaff accelerator facility. Neutrons were produced using the ^{3}H(p,n) reaction. For the inelastic neutron scattering measurements, the scattering sample was 54.73 g of Gd_2O_3 enriched to 95.81% in ^{158}Gd.

Gamma-ray singles measurements, including angular distributions and excitation functions, were performed with HPGe detectors having relative efficiencies of $\geq 50\%$ and energy resolutions (FWHM) of approximately 2.0 keV at 1332 keV. Time-of-flight gating and Compton suppression with an annular BGO shield were used to reduce extraneous background events. Detailed descriptions of the methods employed are given by P.E. Garrett, N. Warr, and S.W. Yates [18]. A portion of a typical spectrum is illustrated in Fig. 1. An excitation experiment was performed in neutron energy steps of ≈ 80 keV from $E_n = 1.4 - 3.27$ MeV. Gamma-gamma coincidence measurements were performed with four $\geq 50\%$ HPGe detectors in a close geometrical arrangement [19].

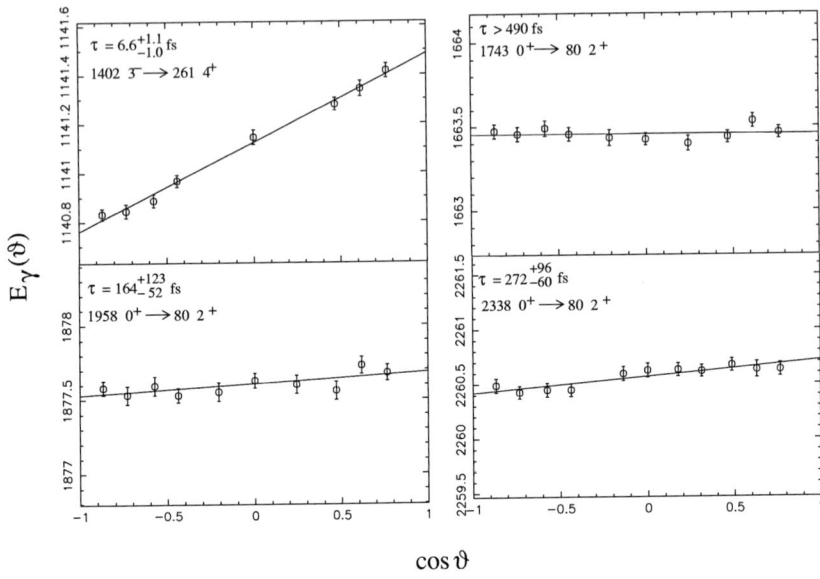

FIGURE 2. Examples of Doppler shift plots for γ rays from states in ^{158}Gd.

In all measurements, calibrations for efficiency, energy, and nonlinearity were performed using ^{226}Ra and ^{152}Eu standard sources. The placements of γ rays in the level scheme were confirmed from excitation function threshold information and from coincidence relationships; spin-parity assignments were obtained from the cross section and angular distribution data. The observed Doppler shifts of the γ-ray energies as a function of angle provided a means for the determination of level lifetimes shorter than about 1 ps [20].

0^+ STATE RESULTS

The initial challenge in this study was the observation of the γ rays de-exciting the 0^+ states assigned by S.R. Lesher *et al.* [15] and the confirmation that the (n,n'γ) results were consistent with the prior assignments. While the energies of the states obtained from the (p,t) reaction data [15] were good, the level density in a well-deformed nucleus such as ^{158}Gd increases rapidly above 1 MeV and associating the observed γ rays with specific states observed in the (p,t) reaction study is not always trivial. It was expected that the decay of the excited 0^+ states to the first excited 2^+ state by E2 transitions would be the dominant γ-ray decay mode, although transitions to other low-lying 2^+

TABLE 1. Properties of 0^+ states in ^{158}Gd. Parentheses denote a tentative assignment. The lifetimes of the 0_2^+ and 0_3^+ states are taken from Ref. [17].

E_γ keV	E_x keV	J_x^π	E_f keV	J_f^π	B(E2) (W.u.)
1116.40	1195.92	0_2^+	80	2^+	1.1
1372.66	1452.19	0_3^+	80	2^+	2.09
1663.52	1743.05	0_5^+	80	2^+	< 1.2
1877.52	1957.05	0_7^+	80	2^+	$4.2^{+0.9}_{-1.0}$
2197.08	2276.62	0_8^+	80	2^+	$4.4^{+0.8}_{-0.7}$
(2260.54)	2340.01	0_9^+	80	2^+	$1.1^{+0.1}_{-0.2}$
2564.73	2644.27	0_{10}^+	80	2^+	$7.6^{+1.5}_{-0.7}$
2832.02	2911.36	0_{12}^+	80	2^+	$1.9^{+2.3}_{-1.1}$

and 1^- states were also sought. In addition, it was required that the γ-ray excitation functions, both cross sections and thresholds, were consistent with 0^+ assignments. It was further required that the angular distributions of γ rays from these states be isotropic. Finally, from the angular distribution data, it was sought to determine the lifetimes of these states by the Doppler-shift attenuation method (DSAM). Examples of DSAM data are presented in Fig. 2. A search for 2^+ members of these presumed 0^+ bands was conducted. Although there are a few candidates, without in-band γ rays between the 2^+ and 0^+ states, such assignments remain uncertain. Unfortunately, because of absorption in our sample, we are unable to observe these ≈ 80 keV low-intensity γ rays.

Table 1 provides a listing of confirmed 0^+ states for which lifetimes or limits were obtained along with the B(E2) values measured. As internal conversion electron data are not available, the branchings to the ground state were assumed to be negligible. On the other hand, if other γ-ray branches were observed to excited states, these were considered in determining the reduced transition rates. The results for these 0^+ states require further discussion.

The $0^+ \to 2_1^+$ transition from the state at 2644.27 keV has the largest observed $B(E2; 0_{10}^+ \to 2_1^+)$ value of $7.6^{+1.5}_{-0.7}$ W.u. There are two others, from states at 1957.05 and 2276.62 keV, which have transition probabilities around 4 W.u. Although these levels do show some collectivity, none show the strong collectivity needed to characterize a β vibration according to the interpretation of Ref. [14]. Further experimental information, including a conversion electron measurements would be helpful.

There are also three previously identified 0^+ levels at 1577.0, 1952.3 and 2689 keV, which do not exhibit γ-ray decays; in addition, the 2340.01-keV level has an uncertain placement. It is expected that γ-ray emission would be the preferred mode of decay. If no γ-rays are found, then how do they decay? Further investigation is warranted to fully understand the nature and decay of the multiple 0^+ states observed in ^{158}Gd.

ACKNOWLEDGMENTS

We wish to thank A. Aprahamian and P.E. Garrett for insightful discussions and their interest in this study. This work was supported by the U.S. National Science Foundation under Grant Nos. PHY-9803784 and PHY-0098813.

REFERENCES

1. A. Bohr, and B. Mottelson, *Nuclear Structure Vol. II*, W.A. Benjamin, Reading, Massachusetts, 1975.
2. P. Garrett, M. Kadi, M. Li, C. McGrath, V. Sorokin, M. Yeh, and S. Yates, *Phys. Rev. Lett.* **78**, 4545 (1997).
3. C. Fahlander, A. Axelsson, M. Heinebrodt, T. Härtlein, and D. Schwalm, *Phys. Letts. B*. **388**, 475 (1996).
4. F. Corminboeuf, J. Jolie, H. Lehmann, K. Föhl, F. Hoyler, H. Börner, C. Doll, and P. Garrett, *Phys. Rev. C* **56**, R1201 (1997).
5. R. Casten, and P. von Brentano, *Phys. Rev. C* **50**, R1280 (1994).
6. K. Kumar, *Phys. Rev. C* **50**, 3524 (1995).
7. D. Burke, and P. Sood, *Phys. Rev. C* **50**, 3525 (1995).
8. R. Casten, and P. von Brentano, *Phys. Rev. C* **51**, 3528 (1995).
9. C. Günther, S. Boehmsdorff, K. Freitag, J. Manns, and U. Müller, *Phys. Rev. C* **54**, 679 (1996).
10. H. Lehmann, J. Jolie, F. Corminboeuf, H. Börner, C. Doll, M. Jentschel, R. Casten, and N. Zamfir, *Phys. Rev. C* **57**, 569 (1998).
11. P. Garrett, M. Kadi, C. McGrath, V. Sorokin, M. Li, M. Yeh, and S. Yates, *Phys. Lett. B* **400**, 250 (1997).
12. A. Aprahamian, R. deHaan, S. Lesher, J. Döring, A. Bruce, H. Börner, M. Jentschel, and H. Lehmann, *J. Phys. G* **25**, 685 (1999).
13. R. de Haan, A. Aprahamian, H. Börner, C. Doll, M. Jentschel, A. Bruce, and S. Lesher, *J. Res. Natl. Inst. Stand. Technol.* **105**, 125 (2000).
14. P. Garrett, *J. Phys. G* **27**, R1 (2001).
15. S. Lesher, A. Aprahamian, L. Trache, S. Deyliz, A. Gollwitzer, R. Hertenberger, B. Valnion, G. Graw, and A. Oros, *Phys. Rev. C* **66**, R0513051 (2002).
16. R. Greenwood, C. Reich, H. Baader, H. Koch, D. Breitig, O. Schult, B. Fogelberg, A. Bäcklin, W. Mampe, T. von Egidy, and E. Schreckenbach, *Nucl. Phys.* **A304**, 327 (1978).
17. H. Börner, M. Jentschel, N. Zamfir, R. Casten, M. Krticka, and W. Andrejtscheff, *Phys. Rev. C* **59**, 2432 (1999).
18. P. Garrett, N. Warr, and S.W.Yates, *J. Res. Natl. Inst. Stand. Technol.* **105**, 141 (2000).
19. C. McGrath, P. Garrett, M. Villani, and S. Yates, *Nucl. Instrum. Methods Phys. Res.* **A421**, 458 (1999).
20. T. Belgya, G. Molnár, and S. Yates, *Nucl. Phys.* **A607**, 43 (1996).

Pairing versus quadrupole collectivity of low-lying 0^+ states in deformed nuclei

N. Lo Iudice*, A. V. Sushkov[†] and N. Yu. Shirikova[†]

Università di Napoli Federico II, Dipartimento di Scienze Fisiche, Monte S. Angelo, via Cintia, I-80126 Napoli
[†]*Bogoliubov Laboratory of Theoretical Physics, Joint Institute for Nuclear Research, 141980 Dubna, Russia*

Abstract. The low-lying 0^+ states observed in large abundance in several deformed nuclei are investigated systematically within the microscopic quasiparticle-phonon model. Attention is paid at their quadrupole and pairing collective properties.

Keywords: 0^+ states, deformed nuclei, E0 E2 (p,t) transfer strengths
PACS: 21.10.Jx,21.10.Ky,21.10.Re,21.60.Jz,27.70.+q

INTRODUCTION

The large abundance of 0^+ levels, established first by a (p,t) experiment on ^{158}Gd [1], has been confirmed also in some actinide nuclei [2] and ^{168}Er [3]. Especially large is the number, about 22, of 0^+ levels detected in ^{168}Er up 4 MeV [4]. In all nuclei, only one among these 0^+ states is strongly populated in the (p,t) transfer reaction. An exception is represented by ^{168}Er, where the strength is fragmented into several small peaks.

The theoretical studies have been concentrated mostly on ^{158}Gd. The first investigation was carried out within the extended $(sdpf)$ interacting boson model (IBM) [5] and has pointed out the importance of the octupole degrees of freedom. A projected shell model (PSM) calculation [6] within a restricted space spanned by two and four quasiparticle states has reproduced well all levels and yielded small $E2$ decay strengths for the corresponding states.

A more exhaustive study, carried out within the quasiparticle-phonon model (QPM) [7], has confirmed the IBM and PSM results and has provided additional information on the collective properties of these 0^+ states. Very recently, the same QPM approach was adopted to carry out a systematic study of the low-lying 0^+ states covering ^{168}Er and some Th and U isotope [8]. All these nuclei were explored in the latest (p,t) experiments [2, 3, 4]. The goal of this systematic is to offer a complete consistent picture of 0^+ states in deformed nuclei, by disentangling the properties common to all systems from the ones related to their peculiar shell structure.

CALCULATION AND RESULTS

Calculation procedure

We have adopted a Hamiltonian composed of a deformed axially symmetric Woods-Saxon potential plus a two-body interaction consisting of a sum of multipole, including monopole and quadrupole, pairing potentials plus a sum of separable particle-hole interactions of different multipolarity λ. We have diagonalized the above Hamiltonian following the QPM procedure [9] in a truncated space spanned by one- plus two-phonon states built out of twenty quadrupole phonons $\lambda = 2, \mu = 0$ and ten phonons for each of the other multipolarities up to $\lambda = 5$. The parameters were fixed by an independent fit of low-lying energies and transition strengths in the nearby nuclei.

Using the eigenstates so obtained, we have computed the reduced strengths of the $E2$ decay transitions from the 0_n^+ to the $I = 2$ ground rotational state, as well as the normalized monopole strengths

$$\rho^2(E0; 0_n^+ \to 0_0^+) = \frac{4\pi}{e^2 R_0^4} B(E0; 0_n^+ \to 0_0^+). \tag{1}$$

These transitions provide information on the quadrupole collectivity of the 0^+ levels. To test pairing collectivity, we have computed the (p,t) normalized spectroscopic factors

$$\mathscr{S}_n(p,t) = \left(\frac{\Gamma_n(p,t)}{\Gamma_0(p,t)}\right)^2, \tag{2}$$

where $\Gamma_n(p,t) = \langle 0_n^+, N-2 \mid \sum_q a_q a_{\bar{q}} \mid 0_0^+, N \rangle$ are the QPM (p,t) transfer amplitudes.

Results

Let us make a comparative analysis between ^{158}Gd and ^{168}Er (Fig. 1). The QPM calculation accounts for all levels observed in both nuclei. In particular, it yields for ^{168}Er about 40 0^+ levels up to 4 MeV. Several of them, however, carry a negligible (p,t) strength and, therefore, cannot be detected. This explains why the number of observed levels is about half (~ 22). In ^{158}Gd, the (p,t) reaction populates mainly a single 0^+ level. In ^{168}Er, instead, the (p,t) strength is distributed among several states. This fragmentation is a pure anharmonic effect being induced by the phonon coupling, accounted for in QPM. In both nuclei, the calculation reproduces fairly well the magnitude and the energy distribution of the strengths. In ^{168}Er, all levels above 3 MeV are weakly excited. This reflects the two-phonon nature of most of the states describing the 0^+ excitations above 3 MeV. In ^{158}Gd, several 0^+ levels are associated to states with dominant two-phonon components, some of octupole nature in agreement with the IBM findings [5]. The octupole phonons do not play any special role in the states below 3 MeV of ^{168}Er.

In both nuclei, the RPA one-phonon states are built out of pairing correlated $q\bar{q}$ quasiparticle components. Pairing, however, acts coherently only into one 0^+ state, which,

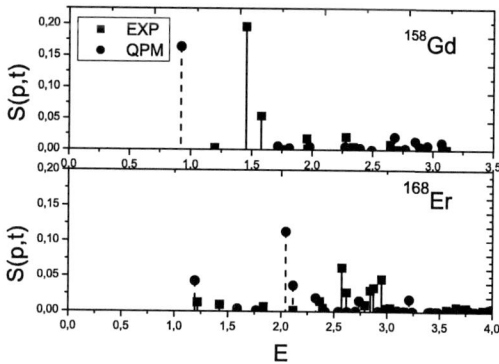

FIGURE 1. QPM versus experimental spectroscopic factors in ^{158}Gd

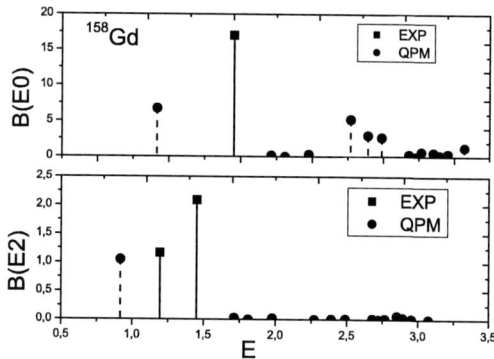

FIGURE 2. E0 and E2 strength of 0^+ states in ^{158}Gd

thereby, collects most of the two-nucleon transfer strength $\mathscr{S}(p,t)$. This coherence is not spoiled by the coupling with other phonon configurations in ^{158}Gd. In ^{168}Er, instead, the phonon coupling reduces the strength of the strongly populated RPA 0^+ state yielding a more fragmented spectrum.

Being dominated by pairing correlations, all the states carry little if at all quadrupole collectivity. Indeed, all the $E2$ decay strengths are very small in ^{158}Gd (Fig. 2) and even smaller in ^{168}Er, where the largest QPM strength amounts to $B^{(QPM)}(E2;0_1^+ \to 2_0^+) = 0.46$ W.u. in agreement with the experimental value $B^{(exp)}(E2;0_1^+ \to 2_0^+) = 0.08(1)$ W.u. [10]. Equally weak are the normalized monopole transitions in ^{158}Gd, as shown in Fig. 2, as well as in ^{168}Er, where the strength collected by the bandhead 0_1^+ is $10^3 \rho_{exp}^2(E0;0_1^+ \to 0^+) = 1.71$, close to the experimental value $10^3 \rho_{exp}^2(E0;0_1^+ \to 0^+) = 0.8(8)$ [11, 12, 13].

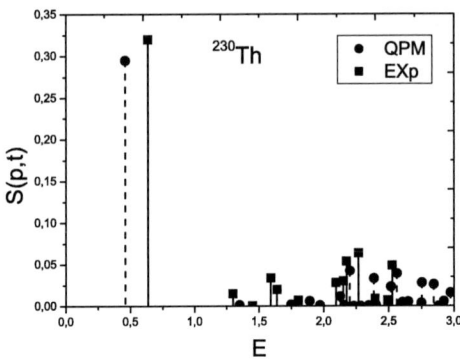

FIGURE 3. QPM versus experimental spectroscopic factors in ^{230}Th

We find therefore an almost complete lack of quadrupole collectivity in all 0^+ states of ^{158}Gd and, especially, of ^{168}Er.

As for the actinides, we will discuss ^{230}Th as an illustrative example. Like in the case of ^{158}Gd, the (p,t) reaction populates mainly one 0^+ state (Fig. 3). The QPM calculation reproduces fairly well the magnitude and distribution of the measured normalized spectroscopic factors $\mathscr{S}(p,t)$. The $E2$ strengths in the actinides are generally larger than in ^{158}Gd and ^{168}Er. The lowest excited state, in particular, is dominated by a quadrupole phonon and collects an appreciable $E2$ strength. In ^{230}Th, this is $B^{(QPM)}(E2; 0_1^+ \to 2_0^+) = 1.71$ (W.u.), close to the measured value $B^{(exp)}(E2; 0_1^+ \to 2_0^+) = 1.10$ W.u [14]. The monopole transition strength collected by the first 0^+ is estimated to be $10^3 \rho_{exp}^2(E0; 0_\beta^+ \to 0_g^+) = 50(20)$ [13], few times larger than the QPM value and orders of magnitude larger than in ^{168}Er. Still, neither the $E2$ or the $E0$ transitions are sufficiently strong to qualify the 0^+ of the actinides explored here as beta vibrational levels [15].

With respect to ^{168}Er, the two-phonon content is more pronounced in the 0^+ states of the actinides. On the other hand, the octupole phonons have a relatively modest impact even on the 0^+ states of these nuclei, at variance with the IBM prediction [2] and contrary to expectations. The actinides, indeed, are nuclei of strong octupole collectivity, displaying collective octupole levels at very low energy. A closer look at the QPM procedure shows that the octupole coherence is spoiled once Pauli principle is enforced on the two-phonon octupole components.

CONCLUDING REMARKS

The first conclusion to be drawn is that, with the exception of ^{158}Gd, where several two-phonon states intrude into the 2-3 MeV region, practically all 0^+ levels observed below 3 MeV in deformed nuclei are described by states with dominant one-phonon components.

Nonetheless, the (p,t) spectra are appreciably affected by the coupling with other phonon configurations. In ^{168}Er, the states with dominant two-phonon components are above 3 MeV and carry very little (p,t) strength, in agreement with experiments.

The role octupole phonons play is important in ^{158}Gd, negligible in ^{168}Er below 3 MeV and marginal in the actinides. The suppression of the octupole coherence in the actinides is due to the repulsive effect of the Pauli principle which redistributes the strength of the collective octupole phonons among several, closely lying, 0^+ states. The QPM results suggest that the octupole correlation is not a common feature of the 0^+ states in all nuclei, but is to be associated to the peculiar shell structure of some nuclei, in our case ^{158}Gd.

The QPM analysis of the $E2$ and $E0$ transitions leads to the conclusion that quadrupole collectivity is lacking in all 0^+ states of ^{168}Er and ^{158}Gd and is more appreciable, but still not strong, in the actinides. Pairing, instead, is dominant in all one-phonon 0^+ states of all nuclei, but acts coherently only in one 0^+ state, which thereby collects most of the (p,t) strength. Pairing correlations are not spoiled in general by the coupling to complex configurations. Only in ^{168}Er the phonon coupling has a damping effect. Ground state pairing fluctuations are large in all nuclei considered, except for ^{168}Er. We may then conclude that the 0^+ state collecting the (p,t) strength describes a pairing vibrational mode. Such a mode is damped and fragmented in ^{168}Er.

ACKNOWLEDGMENTS

This work was partly supported by the Italian Ministero dell'Istruzione, Universitá and Ricerca (MIUR) and by a RFBR grant no. 05-02-17606. It is a pleasure to thank Dr. D. Bucurescu for providing the ^{168}Er data prior publication and the Munich-Köln-Yale group for stimulating discussions.

REFERENCES

1. S. R. Lesher et al., *Phys. Rev. C* **66**, 051305-(1-4)(R) (2002).
2. H.-F. Wirth et al., *Phys. Rev. C* **69**, 044310-(1-13) (2004).
3. D. Bucurescu et al., in *Annual Report of the Meier-Leibnitz Laboratory, University and Technical University of Munich*, and to be submitted for publication.
4. D. Bucurescu, private communication.
5. N. V. Zamfir, Jing-ye Zhang, and R. F. Casten, *Phys. Rev. C* **66**, 057303-(1-3) (2002).
6. Y. Sun, A. Aprahamian, J. Zhang, C. Lee, *Phys. Rev. C* **68**, 061301-(1-4)(R) (2003).
7. N. Lo Iudice, A. V. Sushkov, and N. Yu. Shirikova, *Phys. Rev. C* **70**, 064316-(1-9) (2004).
8. N. Lo Iudice, A. V. Sushkov, and N. Yu. Shirikova, *Phys. Rev. C* **72**, 034303-(1-8) (2005).
9. V. G. Soloviev, *Theory of Atomic Nuclei: Quasiparticles and Phonons*, Institute of Physics Publishing, Bristol, 1992.
10. T. Härtlein, M. Heinebrodt, D. Schwalm, C. Fahlander, *Eur. Phys. J. A* **2**, 253-261 (1998).
11. W. F. Davidson et al., *J. Phys. G* **7**, 455-538 (1981).
12. H. Lehmann et al., *Phys. Rev. C* **57**, 569-576 (1998).
13. J. L. Wood, E. F. Zganjar, C. De Coster, K. Heyde, *Nucl. Phys. A* **651**, 323-368 (1999).
14. Y. A. Akovali, *Nuclear Data Sheets* **69**, 155-208 (1993).
15. P. E. Garrett, *J. Phys. G* **27**, R1-R22 (2001).

Recent Progress On The 0^+ Dominance

N. Yoshinaga*, A. Arima** and Y. M. Zhao***

*Department of Physics, Saitama University, Saitama, 338-8570, Japan
**Science Museum, Japan Science Foundation, 2-1 Kitanomaru-koen,
Chiyoda-ku, Tokyo, 102-0091, Japan
***Department of Physics, Shanghai Jiao Tong, University, Shanghai 200030, China

Abstract. We propose a new formula to evaluate the ground state energy of the two-body random Hamiltonian in terms of the average energy and the width of eigen-energies. For fermions in a single orbital, we obtain an excellent agreement between the spin I ground state probabilities predicted by using our formula and those obtained by diagonalizing the random Hamiltonian.

Keywords: spin-zero dominance, width-prediction method
PACS: 21.60.Cs, 21.10.Hw, 24.60.Ky, 24.60.Lz

INTRODUCTION

The ground states of all even-even nuclei have angular momentum equal to zero, $I = 0$, and positive parity, $\pi = +$. This feature was believed to be a consequence of the attractive short-range interaction between nucleons. However, a predominance of $I^\pi = 0^+$ ground states was discovered by Johnson, Bertsch and Dean in 1998 by using the two-body random ensemble (TBRE) [1]. See Refs. [2] for recent reviews. In this paper we propose a novel method to study the spin-zero ground state (0 g.s.) dominance in the presence of the TBRE. First we study the problem of evaluating the ground-state energy in terms of the average energy, the width of eigen-energies, and the number of spin I states. This expression for the ground state turns out to achieve a remarkable agreement between the predicted I g.s. probabilities and those obtained by diagonalizing the random Hamiltonian.

AVERAGE ENEGIES AND VARIANCES OF EIGEN-ENEGIES

In order to simplify our argument, we take single j^n configuration for fermions, but the method is easily generalized to fermions in many-j shells. The matrix elements of $\hat{H}$ for spin I states can be expressed as

$$H_{I\beta\gamma} = < j^n I\beta | \hat{H} | j^n I\gamma > = \sum_{J=0}^{2j-1} \alpha_{I\beta\gamma}^J G_J, \tag{1}$$

where G_J's are two-body interactions between fermions, and are assumed to follow the TBRE, i.e., they are a set of random numbers with a distribution function

$$\rho(G_J) = \frac{1}{\sqrt{2\pi}} \exp(-G_J^2/2), \quad J = 0, 2, \ldots, 2j-1. \tag{2}$$

For convenience, we define $(\alpha_I^J)_{\beta\gamma} \equiv \alpha_{I\beta\gamma}^J$, $(\beta, \gamma = 1, \ldots, d_I)$, where d_I is the dimension of spin I states. By diagonalizing the Hamiltonian for each I with a specific set of interactions $\{G_J\}$, we obtain eigen-energies $E_{I\beta}$ ($\beta = 1, \ldots, d_I$) of the matrix $H_{I\beta\gamma}$. For a specific set of interactions, the squared average energies, $(\bar{E}_I)^2$, are written as $(\bar{E}_I)^2 = \left(\sum_J^{2j-1} \bar{\alpha}_I^J G_J\right)^2$, where $\bar{\alpha}_I^J \equiv \frac{1}{d_I} \sum_\beta^{d_I} \alpha_{I\beta\beta}^J = \frac{1}{d_I} \text{Tr}(\alpha_I^J)$. The average of the squared energies is expressed as $\overline{(E_I)^2} = \frac{1}{d_I} \sum_{\beta=1}^{d_I} (E_{I\beta})^2 = \frac{1}{d_I} \text{Tr}(\hat{H}^2)$. Using these $(\bar{E}_I)^2$ and $\overline{(E_I)^2}$, the variance of $E_{I\beta}$ around $\bar{E}_I$ is denoted as

$$(\sigma_I\{G_J\})^2 = \frac{1}{d_I} \sum_{J,K}^{2j-1} \text{Tr}\left[(\alpha_I^J - \bar{\alpha}_I^J \mathbf{I})(\alpha_I^K - \bar{\alpha}_I^K \mathbf{I})\right] G_J G_K, \tag{3}$$

where $\mathbf{I}$ is the unit matrix with dimension d_I.

ESTIMATATION OF THE LOWEST EIGENVALUE

In order to evaluate the lowest eigen-energy $E_I^{(\min)}$ of the matrix $H_{I\beta\gamma}$, we assume that eigen-energies $E_{I\beta}$ ($\beta = 1, \ldots, d_I$) follow a Gaussian distribution with width σ_I,

$$\rho(E_I) = \frac{d_I}{\sqrt{2\pi}\sigma_I} \exp\left[-\frac{(E_I - \bar{E}_I)^2}{2(\sigma_I)^2}\right]. \tag{4}$$

To estimate $E_I^{(\min)}$, we need to solve the following equation

$$\int_{E_I^{(\min)}}^{\bar{E}_I} \rho(E_I) \, dE_I = \frac{d_I}{2} - 1, \tag{5}$$

which means that one half of all eigen-energies, except $E_I^{(\min)}$, can be found above $E_I^{(\min)}$ and below $\bar{E}_I$. Eq. (5) is simplified as $\text{Erfc}(t^M) = \sqrt{\pi}/d_I$, where the error function is defined as $\text{Erfc}(x) = \int_x^\infty \exp(-t^2) \, dt$. This equation is not analytically solvable. For large d_I, however, we obtain $t^M \approx -\sqrt{\ln d_I - \frac{1}{2}\ln(4\pi \ln d_I)}$ by using the asymptotic expansion for the error function for its large argument. Then we obtain $E_I^{(\min)} \approx \bar{E}_I - \sqrt{2\ln d_I - \ln(4\pi \ln d_I)} \, \sigma_I\{G_J\}$. Note that this assumption is only valid when $d_I \gg 1$. In fact $E_I^{(\min)} = \bar{E}_I$ for $d_I = 1$ and $E_I^{(\min)} = \bar{E}_I - \sigma_I\{G_J\}$ for $d_I = 2$. Our result is similar to that in Ref. [3] where they obtained a formula

$E_I^{(\min)} \approx \bar{E}_I - \sqrt{\ln d_I / \ln 2}\ \sigma_I\{G_J\}$, although our derivation is very different from theirs.

DETERMINATION OF THE DIMENSIONAL FACTORS

Taking into account the preceding discussion for large dimensional matrices, let us estimate the ground-state energy of the Hamiltonian $\hat{H}$ with a relatively small dimension by assuming that it can be expressed in terms of the average energy (centroid), the standard deviation (width) and the number of states as,

$$E_I^{(\min)} = \bar{E}_I - \Phi(d_I)\ \sigma_I\{G_J\}, \qquad (6)$$

where $\Phi(d_I)$ is assumed to depend only on the dimension of the Hamiltonian matrix and to take the following form:

$$\Phi(d_I) = \sqrt{a \ln d_I + b}. \qquad (7)$$

To determine the parameters a and b, we study the system of $n=4$ and $j=31/2$. Our procedure is as follows: First for one case of I with d_I, e.g., $I=0$ with $d_I=5$, the factors $\left[\bar{E}_I - E_I^{(\min)}\right]/\sigma_I\{G_J\}$ are calculated for 1000 TBRE runs. By taking the ensemble average of these factors, we have one data point shown in Fig. 1.

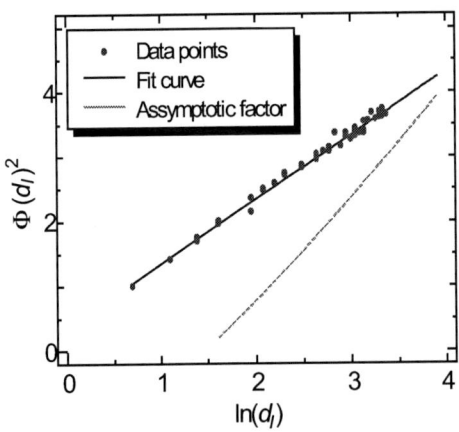

FIGURE 1. Phenomenological factors $\Phi(d_I)^2$ determined numerically for the case of the shell $j=31/2$ and $n=4$ fermions (dots) and their linear fit curve as a function of $\ln d_I$ for $2 \le d_I \le 29$ (solid line). Dotted line indicates the asymptotic factor valid for large d_I [$2\ln d_I - \ln(4\pi \ln d_I)$].

Then for states with a different I we repeat the above procedure. Then, for all the data points a χ-square fit to Eq. (7) is carried out, as shown in Fig. 1 where $a = 0.99$ and $b = 0.36$ is obtained. We should note that a similar method was suggested and applied to estimate $P(I)$ by Papenbrock and Weidenmüller, who assumed the minimum energy $E_I^{(\min)}$ in Ref. [5] as, $E_I^{(\min)} = -r_I \sqrt{\text{Tr}(\hat{H}^2)/d}$, although they obtained the numerical factors r_I phenomenologically for each case of j and n without analytical forms.

APPLICATION TO FERMION IN SIGLE ORBITALS

Using Eq. (6) with (7) to estimate $E_I^{(\min)}$, we can calculate $P(I)$, the probability that the ground state has spin I (called the "width-prediction method"). The results for fermions with $n = 4$ and $j = 11/2$ are shown in Fig. 2 and are compared with $P(I)$ obtained by diagonalizing $\hat{H}$ with TBRE, and also with the empirical prediction $P_{\text{emp}}(I)$ suggested in Ref. [6] (called the "emp-prediction method"). The two predictions agree reasonably well with $P(I)$ obtained by diagonalizing with TBRE, but the empirical rule gives large disagreements at $I = 2$ and $I = 8$ in addition to a small disagreement at $I = 0$. On the other hand the present prediction (width-prediction) agrees very well with $P(I)$ by TBRE.

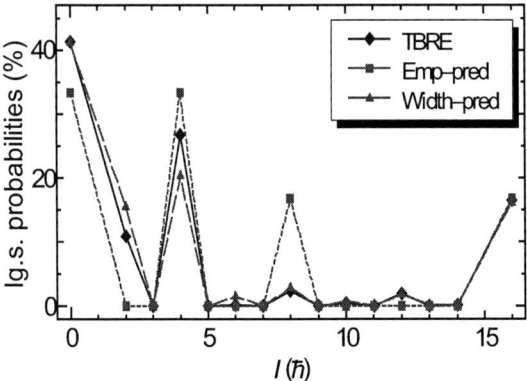

FIGURE 2. TBRE indicates $P(I)$ for 1000 random interactions with $j = 11/2$ and $n = 4$, Emp-pred indicates the empirical method of Ref. [6] and Width-pred indicates the method proposed in this paper.

The $P(0)$'s of four fermions in single-j shells are shown in Fig. 3(a). The diamonds are obtained by 1000 runs of the TBRE Hamiltonian. The squares are the $P_{\text{emp}}(0)$'s, and the triangles are the results of the width-prediction method. It is seen that the both predictions well describe the staggering behavior of $P(0)$'s, although a slight disagreement in case of the emp-prediction is seen for $j = 27/2$ and $j = 29/2$. In Fig. 3(b) $P(I^{\max})$'s are shown for three cases. It is seen that a slight staggering of the TBRE results is better reproduced by the width-prediction method proposed in this paper. We have numerically confirmed that the present approach, i.e., Eq. (6) with (7), well describes $P(I)$ for odd-n systems such as $n = 5$ from $j = 9/2$ up to $j = 19/2$, as well as $n = 6$ systems from $j = 11/2$ up to $j = 19/2$, *without any change* of a and b, and also for boson systems [4].

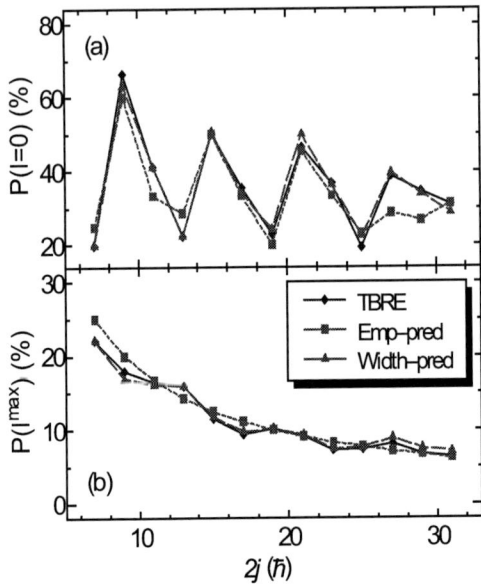

FIGURE 3. (a) $P(0)$'s of four fermions in a single-j shell. The diamonds are obtained by 1000 runs of the TBRE Hamiltonian. The squares are by using the emp-prediction method and the triangles, by the width-prediction method. (b) The same figure as in (a), but for $P(I^{max})$.

CONCLUSIONS

To summarize, we proposed a new analytical expression to estimate the ground state energy of spin I states in terms of the energy centroid, the width, and the number of states for the two-body random ensemble. As an application this simple formula presents a remarkably successful description of the spin I ground state probabilities in the presence of the two-body random ensemble, for fermion systems in a single orbital.

REFERENCES

1. C. W. Johnson, G. F. Bertsch, and D. J. Dean, *Phys. Rev. Lett.*, **80**, 2749 (1998).
2. Y. M. Zhao, A. Arima, and N. Yoshinaga, *Phys. Rept.* **400**, 1 (2004); V. Zelevinsky and A.Volya, *Phys. Rept.* **391**, 311 (2004).
3. V. Velazquez and A.P. Zuker, *Phys. Rev. Lett.*, **88**, 027502 (2002).
4. N. Yoshinaga, A. Arima and Y. M. Zhao, *to be published*.
5. T. Papenbrock and H. A. Weidenmüller, *Phys. Rev. Lett.*, **93**, 132503 (2004).
6. Y. M. Zhao, A. Arima, and N. Yoshinaga, *Phys. Rev. C*, **66**, 034302 (2002).

Gamma-Ray Spectroscopy at TRIUMF-ISAC

P.E. Garrett[1], C.E. Svensson[1], G.C. Ball[2], G. Hackman[2], E.F. Zganjar[3],
C. Andreoiu[1], A. Andreyev[2], S.F. Ashley[4], R.A.E. Austin[5],
D. Bandyopadhyay[1], J.A. Becker[6], S. Chan[2], H. Coombes[2],
R. Churchman[2], R.S. Chakrawarthy[2,10], P. Finlay[1], G.F. Grinyer[1],
B. Hyland[1], E. Illes[1], G.A. Jones[4], W.D. Kulp[7], J.R. Leslie[8], C. Mattoon[9],
A.C. Morton[2], C.J. Pearson[2], A.A. Phillips[1], P.H. Regan[4], J.J. Ressler[10],
F. Sarazin[9], M.A. Schumaker[1], J. Schwarzenberg[11], M.B. Smith[2],
J.J. Valiente-Dobón[1], P.M. Walker[4], S.J. Williams[4], J.C. Waddington[12],
L.M. Watters[12], J. Wong[1], J.L. Wood[7]

1) Department of Physics, University of Guelph, Guelph, Ontario, N1G2W1, Canada
2) TRIUMF, 4004 Wesbrook Mall, Vancouver, British Columbia, V6T 2A3, Canada
3) Department of Physics and Astronomy, Louisiana State University, Baton Rouge, Louisiana, 70803-4001, United States
4) Department of Physics, University of Surrey, Guildford, GU2 7XH, United Kingdom
5) Department of Astronomy and Physics, Saint Mary's University, Halifax, Nova Scotia, B3H3C3, Canada
6) Lawrence Livermore National Laboratory, PO Box 808, Livermore, California, 94551, United States
7) School of Physics, Georgia Institute of Technology, Atlanta, Georgia, 30332-0430, United States
8) Department of Physics, Queen's University, Kingston, Ontario, K7L3N6, Canada
9) Department of Physics, Colorado School of Mines, Golden, Colorado, 80401, United States
10) Department of Chemistry, Simon Fraser University, Burnaby, British Columbia, V5A1S6, Canada
11) Institut für Isotopenforschung und Kernphysik, Währinger Strasse 17, 1090 Wein, Austria
12) Department of Physics, McMaster University, Hamilton, Ontario, L8S4M11, Canada

Abstract. The 8π spectrometer at TRIUMF-ISAC consists of 20 Compton-suppressed germanium detectors and various auxiliary devices. The Ge array, once used for studies of nuclei at high angular momentum, has been transformed into the world's most powerful device dedicated to radioactive-decay studies. Many improvements in the spectrometer have been made, including a high-throughput data acquisition system, installation of a moving tape collector, incorporation of an array of 20 plastic scintillators for β-particle tagging, 5 Si(Li) detectors for conversion electrons, and 10 BaF_2 detectors for fast-lifetime measurements. Experiments can be performed where data from all detectors are collected simultaneously, resulting in a very detailed view of the nucleus through radioactive decay. A number of experimental programmes have been launched that take advantage of the versatility of the spectrometer, and the intense beams available at TRIUMF-ISAC.

Keywords: gamma-ray spectrometer, conversion electrons, lifetime measurements.
PACS: 23.20.-g, 23.20.Lv, 23.20.Nx, 23.40.-s, 23.40.Bw, 23.40.Hc

INTRODUCTION

With a new generation of radioactive beam facilities coming on-line, many researchers are re-visiting techniques pioneered decades ago in order to address the most pressing issues in nuclear physics research today. Radioactive-decay studies, eclipsed in the 1980's and 1990's by in-beam γ-ray spectroscopy and heavy-ion fusion-evaporation reactions, are enjoying a revitalization as one of the most important techniques to learn about nuclear structure, address questions in nuclear astrophysics, and explore physics beyond the current Standard Model. The advantages of modern β-decay studies over those performed even a decade ago are the tremendous increases in instrumental resolving powers, sensitivities, and data analysis with modern computers. In the present paper, work involving β-decay spectroscopy at the Isotope Separator and Accelerator (ISAC) facility of the TRI-University Meson Facility (TRIUMF) is reported. Much effort has been devoted in the past five years to build a highly sensitive and versatile device for decay studies centered on the 8π γ-ray spectrometer and its auxiliary detectors. A wide and varied programme of nuclear structure, nuclear astrophysics, and weak-interaction studies is being pursued.

INSTRUMENTAL CAPABILITIES

The 8π spectrometer and its associated auxiliary detectors currently comprises 4 different detector systems: Compton-suppressed Ge detectors (the 8π spectrometer) for γ-ray detection, plastic scintillators (named the SCintillating Electron Positron Tagging ARray – SCEPTAR) for detection of β particles, BaF_2 detectors (named the Dipentagonal Array for Nuclear Timing Experiments – DANTE) for γ-ray detection with fast-timing measurements, and Si(Li) detectors (named the Pentagonal Array for Conversion Electron Spectroscopy – PACES) for conversion electron studies. The ISAC low-energy beam is focused on a segment of ½" wide tape at the center of these arrays. The Moving Tape Collector (MTC) can then be used to transport the deposited sample into a shielded tape box away from the detectors at predetermined time intervals using a computer-controlled stepping motor. The beam ON/OFF times, and tape movement intervals, frequencies, and dwell times are all variable limited only by the beam-pulsing response time («10 ms), and tape movement speed. The MTC system allows separation of parent/daughter/granddaughter/etc. decays of differing half-lives, an indispensable feature at an online isotope separator like the ISAC facility.

The 8π Spectrometer

The 8π spectrometer is based on a geometrical arrangement of 20 hexagonal and 12 pentagonal shapes. The 20 Ge detectors, Ortec HPGe with a nominal relative efficiency of ~25%, occupy the hexagonal positions, with 4 rings of 5 detectors at angles of ±37° and ±79° with respect to the beam direction. The target-to-front-face distance for the BGO suppression shields is 13 cm, and that for the Ge detectors is 14 cm. The absolute photopeak efficiency of the reconfigured array has been measured for the 1332-keV ^{60}Co line to be 1.5%, with a peak-to-total ratio of 0.41. To demonstrate the sensitivity of the reconfigured 8π array, shown in Fig. 1 are results from an experiment [1] using a ^{26}Na beam of 10^6 ions/s of 10-hour duration. A sensitivity on the relative β-decay branching on the order of 10^{-6} was achieved.

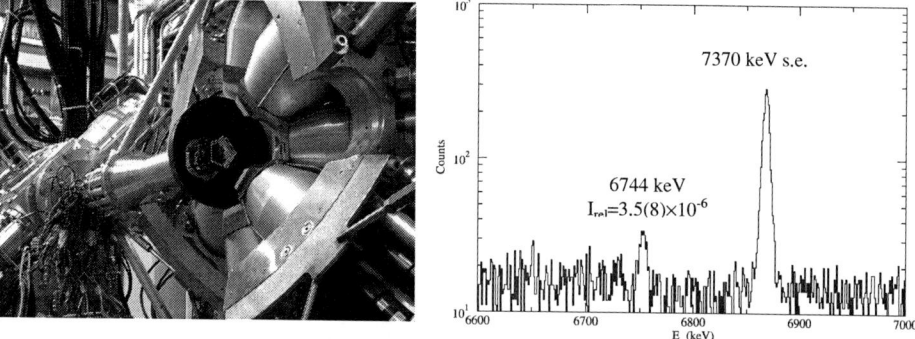

FIGURE 1. Photograph (left) of the upstream portion of the SCEPTAR plastic scintillator array (the pentagon shaped object in the center of the target chamber) and one hemisphere of the 8π array. The SCEPTAR array covers ~80% of the solid angle, and each plastic scintillator shadows a particular Ge detector. The partial spectrum (right) was obtained from a 10hr run with a beam of 10^6 ^{26}Na s^{-1} and displays the sensitivity of 10^{-6} for weak β branches achieved.

SCEPTAR

Complementing the Ge detectors are 20 plastic scintillators of SCEPTAR. Arranged into 2 rings of 5 trapezoidal pieces and 2 rings of 5 rectangular pieces, the positioning is such that one plastic scintillator overlaps the solid angle of one Ge detector. This permits the rejection of bremsstrahlung events in a Ge detector due to the stopping of the very-high-energy β particles often encountered in far-from-stability radioactive decays. The total solid angle coverage of SCEPTAR is approximately 80% of 4π. The plastic scintillators are BC410 of thickness 1.6 mm. Light is collected from the edge of the scintillators and transported via ~25 cm long light-guides to the phototubes located outside of the main frame of the array. Figure 1 contains a photo showing portions of the 8π and SCEPTAR arrays.

PACES

The most recent addition to the arsenal of detectors is PACES. An array of 5 Si(Li) detectors, PACES makes available both conversion-electron and internal-pair spectroscopy, the latter of which is advantageous far from stability where the Q-value is large. Inclusion of conversion-electron data provides not only multipolarity information, but also reveals electric monopole (E0) transitions indicative of shape coexistence. The Si(Li) detectors are approximately 5 mm in thickness and have a typical resolution of 2.5 keV at 1 MeV. A close-up view of the PACES array is shown in Fig. 2. A recent experiment [2] using PACES investigated the decay of a newly discovered isomer in ^{174}Tm. The 2.29(1) s isomer was first observed with the 8π spectrometer only [3]; based on X-ray yields, it was suggested that the 100-keV and 152-keV transitions observed were *M1* in nature. However, it was not known if the full decay intensity had been observed, making a spin assignment for the isomeric level uncertain. This experiment was recently repeated with the inclusion of PACES, where an *E3* multipolarity for the 152-keV transition was determined based on conversion-electron sub-shell ratios as shown in Fig. 2. The spin-parity for the isomeric level is now suggested to be 0$^+$ [2].

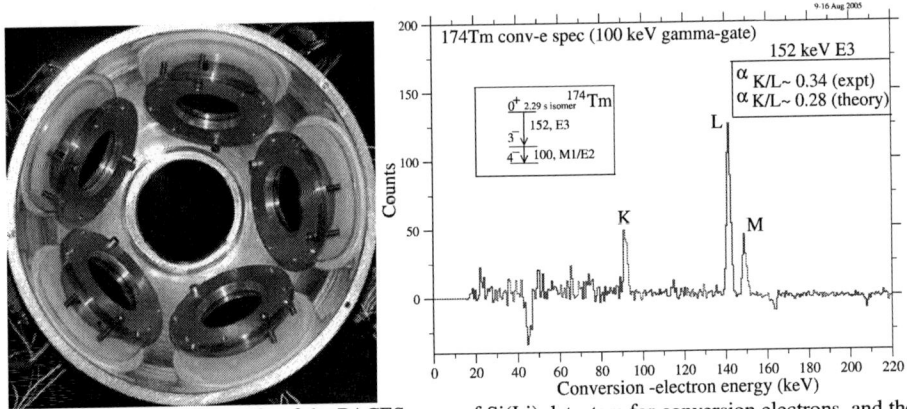

FIGURE 2. Photograph (left) of the PACES array of Si(Li) detectors for conversion electrons, and the low-energy portion of spectrum (right) collected with PACES gated on the 100-keV transition following the isomer decay in ^{174}Tm.

DANTE

The ten BaF$_2$ detectors of DANTE are being installed in the available open pentagonal positions of the spectrometer for fast timing measurements. The BaF$_2$ scintillator is the fastest known; one of its two scintillating-light components has a decay constant of 0.6 ns and emits light in the UV region. The detectors are coupled to very fast photomultiplier tubes, the Photonis XP2020/URQ, having a quartz entrance window for maximum transmission of the fast-component UV light. An initial test experiment [4] to accurately map the time response of the SCEPTAR plastic scintillators and test the feasibility of the DANTE array used 4 BaF$_2$ detectors from the University of Surrey. Using a ^{26}Na radioactive beam, it was found that the individual SCEPTAR detectors had timing resolutions (FWHM) in the range of 1–1.5 ns. With BaF$_2$-BaF$_2$ coincidences, on the other hand, the timing FWHM of ≈200 ps was ultimately achieved. In addition to determining the timing resolution, a novel Compton-rejection scheme for the BaF$_2$ detectors was also investigated. Since five BGO shields of neighbouring Ge detectors form a pentagonal ring around each BaF$_2$ detector, the BGO signals were used to form a BaF$_2$ Compton veto. This reduced the background significantly, even though the geometry was not optimum. This test may be the first time that such timing BaF$_2$ detectors have been Compton suppressed [4]. This will lead to a significant reduction in the systematic uncertainty assigned to the short level lifetimes, since one of the dominating contributions arises from the Compton distribution under the peak of interest.

EXPERIMENTAL PROGRAMMES

As the 8π spectrometer is such a versatile array, a widely varied set of experiments have been proposed addressing nuclear structure, nuclear astrophysics, and Standard Model tests. Studies to date have included: 1) the β decay of ^{11}Li, to study the possibility of neutron-halo "survival" during the β decay process [5], 2) ^{18}Ne [6], 3) ^{35}Ar [7], and 4) ^{62}Ga [8], as part of the super-allowed Fermi β-decay programme, 5) ^{26}Na [1], (our workhorse test beam), 6) ^{32}Na to investigate the structure of ^{32}Mg and the island of inversion, 7) ^{156}Ho and 8) ^{158}Tm decay as part of the programme to study shape-phase transitions in the $N=90$ isotones, and isomeric states in 9) ^{174}Tm [2–3], 10) ^{178}Hf [9], and 11) ^{178,179}Lu [10]. There are a number of other accepted experiments: search for seniority isomers in the mass 80 and 100 regions, structure of neutron-rich Cd and Pd isotopes in connection with r-process nucleosynthesis in the mass 120 region, additional experiments in the $N=90$ region, studies of ^{14}Be, a study of the possible s-process branch point ^{115}Cd, etc. This list will continue to grow as ISAC develops more production targets and ion sources to provide a richer variety of radioactive beams.

ACKNOWLEDGEMENTS

This work has been supported by; the Natural Sciences and Research Council of Canada, the US Department of Energy, the Brookhaven Technology Group, the Engineering and Physical Sciences Research Council, and TRIUMF through the National Research Council. C.A. acknowledges the support offered by the Swedish Foundation for Higher Education and Research and the Swedish Research Council. Work by J.A.B. performed under the auspices of the US Department of Energy by the University of California, Lawrence Livermore National Laboratory under contract no. W-7405-ENG-48.

REFERENCES

1. G.F. Grinyer, *et al.*, Phys. Rev. C **71**, 044309 (2005).
2. R.S. Chakrawarthy, *et al.*, to be published.
3. R.S. Chakrawarthy, *et al.*, *The Fourth International Conference on Exotic Nuclei and Atomic Masses, ENAM'04*, Pine Mountain, Georgia, Sept. 2004, European Physical Journal **A25 S01**, 125 (2005).
4. S.J. Williams, *et al.*, J. Phys. G **31**, S1979 (2005).
5. F. Sarazin, Phys. Rev. C **70**, 031302(R) (2004).
6. M.B. Smith, G.F. Grinyer, *et al.*, to be published.
7. G.F. Grinyer, *et al.*, to be published.
8. B. Hyland, *et al.*, J. Phys. G, **31**, S1885 (2005), and these proceedings.
9. M.B. Smith, Phys. Rev. C **68**, 031302(R) (2005).
10. M.B. Smith *et. al.*, Nucl. Phys. **A746**, 617c (2004).

NUCLEAR ASTROPHYSICS
(S-PROCESS NUCLEOSYNTHESIS)

Neutron capture reaction rates for stellar nucleosynthesis

A. Mengoni

Department of Nuclear Sciences and Applications, Division of Physical and Chemical Sciences
Nuclear Data Section
International Atomic Energy Agency, A-1400 Wagramer Strasse 5, Vienna, Austria

Abstract. A short summary of the three fundamental neutron capture reaction processes of interest in stellar nucleosynthesis is presented. Three examples of actual situations in which the capture reaction proceeds through one of these processes are identified and an overview of the experimental data as well as of the model calculations used in the determination of the capture rates are presented.

PACS: 25.40.Lw, 25.40.Ny, 24.50.+g, 21.10.Jx

INTRODUCTION

The neutron capture process represents the basis for the formation of heavy elements in stars. Neutrons are produced in stars and captured by seed material (typically the Iron-group elements) in stellar environment at energies in the keV range. Under different conditions and in different environments the capture reaction mechanisms proceeds according to specific nuclear structure and neutron-nucleus interaction mechanisms. There are essentially three capture reaction processes of interest for the energy range in consideration for stellar nucleosynthesis. These are the direct radiative capture (DRC) process, the compound nucleus reaction mechanism and a statistical mechanism, taking place over many unresolved compound states. Three examples in which each of these fundamental reaction mechanisms plays the key role are shown in the following. The examples have been chosen also on the basis of the presence of discrepancies between the calculations and the experimental information in an attempt to point out rationales for possible developments, both from the modeling point of view as well as for possible additional experiments.

The $^{14}C(n,\gamma)$ cross section

The neutron capture on ^{14}C presents some interesting characteristics which we will briefly reviewed here. For neutron energies up to a few hundred keV above the neutron separation threshold, there are only two states in ^{15}C which can be populated by neutron capture (Figure 1). These are the ground state with $J^\pi = 1/2^+$, and the first excited $J^\pi = 5/2^+$ state at 0.740 MeV. The only possible capture process leading to these final states is a direct capture process dominated by *p*-wave neutrons in the continuum and E1 transitions. The neutron capture cross section for a DRC process is given by

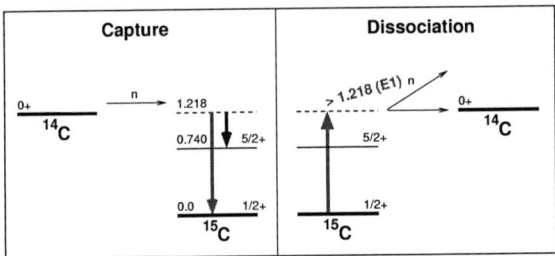

FIGURE 1. Neutron capture process in ^{14}C (*left panel*). There are only two bound states in ^{15}C which can be populated in the capture process. The dissociation of ^{15}C (*right panel*) can be induced by γ-rays of energy larger than 1.218 MeV. A correspondence between the neutron capture and the dissociation processes can be established only for the transition to (or from) the ground state of ^{15}C.

$$\sigma_{n,\gamma}(E_n) = \frac{16\pi}{9\hbar v} \bar{e}^2 k_\gamma^3 \mid Q^{E1}_{c \to b} \mid^2$$

where $k_\gamma = \varepsilon_\gamma/\hbar c$ is the wave number corresponding to the emitted γ-ray energy ε_γ, v is the incident neutron velocity relative to the target nucleus, and $\bar{e} = -Ze/A$ is the neutron effective charge. Only electric dipole transitions are considered here (a more general expression for different multipolarities can be found in the reference [1]). Clearly, the essential ingredient in the calculation is represented by the matrix elements

$$Q^{E1}_{c \to b} = <\Psi_b \mid \hat{T}^{E1} \mid \Psi_c>$$

where with Ψ_c and Ψ_b we have indicated the wave functions in which the neutron is in the continuum and in the final bound state, respectively, and $\hat{T}^{E1}$ is the electromagnetic transition dipole operator. In its simplest approximation, the wave functions can be replaced by single-particle wave functions in a given mean-field potential. Plane-waves for the continuum, and bound-state wave functions from square-well potentials have been used in the past in order to derive analytical expressions for the capture cross section[2]. However, present computational capabilities allow for a more realistic representation of the neutron-nucleus interaction potential for the continuum and for bound states. A Woods-Saxon form of the mean-field potential can be used as a suitable approximation. The mean-field potential can be used to calculate the radial part of the single-particle wave functions and the related overlap integrals. In the present calculation, standard parameters $r_0 = 1.25$ fm, $d = 0.65$ fm, and $V_{so} = 6.5$ MeV have been used as radius, diffuseness and spin-orbit strength, respectively. The well-depth has been adjusted to reproduce the binding energy of the ^{15}C ground state at -1.218 MeV, which resulted in $V_0 = 50.3$ MeV. It should be noticed here that this choice of the mean-field potential ensures that the continuum (above the neutron emission threshold) does not include $l = 1$ single-particle quasi-bound states (pole of the S-matrix imbedded in the continuum). In fact, these potential parameters place the p−shell below the $2s_{1/2}$ orbit, which is already bound by definition. This is an important aspect of the calculation which would, otherwise, be affected by possible perturbation of the continuum. Also to be noted here is the fact that there is a very low sensitivity with respect to the potential used, when the transition concerned is a $p \to s$ transition. In the present case, varying the potential in the

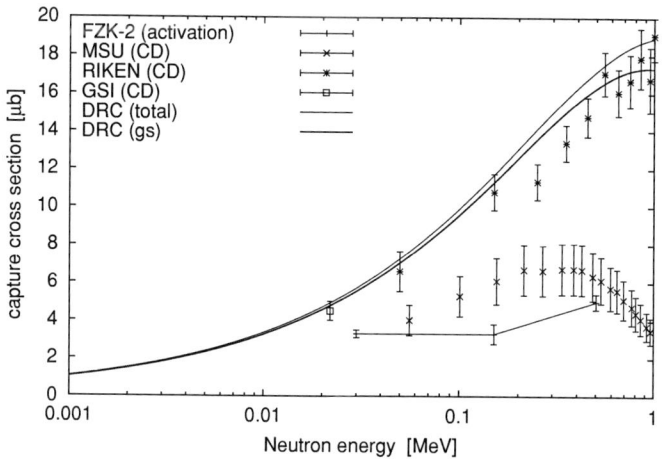

FIGURE 2. The $^{14}C(n,\gamma)^{15}C$ cross section in the relevant energy range. The experimental data are labeled as FZK-2 [5] (supersedes [4]), MSU [8], RIKEN [11], and GSI [9]. Only the FZK data are from a direct measurement of the capture cross section. The other values are derived from Coulumb dissociation measurements (see text for details).

range $V_0 = 45 - 55$ MeV produces a difference in the MACS-30 (maxwellian averaged cross section at $kT = 30$ keV) of less than 10%.

In order to account for the effect of the residual interaction in the description of the final state, the bound state wave function is normalized by its fractional parentage amplitude. In practice, this is given by the spectroscopic factor for a given single-particle configuration. In the present case, the two bound states of ^{15}C are represented as

$$B_{1/2} |^{14}C(0^+) \times \nu(2s_{1/2}); 1/2^+ > \quad \text{and} \quad B_{5/2} |^{14}C(0^+) \times \nu(1d_{5/2}); 5/2^+ >.$$

The coefficients $B_{1/2}$ and $B_{5/2}$ can be derived from the measurement of the spectroscopic factor in a one-neutron transfer (d,p) reaction. In the present case, the spectroscopic factors are $S_{1/2} = 0.88$ and $S_{5/2} = 0.62$ for the ground and first excited state in ^{15}C, respectively[3]. Alternatively, shell model calculations can be performed for the determination of these coefficients.

The result of the DRC cross section calculation for the $^{14}C(n,\gamma)^{15}C$ is shown in Figure 2. A comparison is made with different experimental information. A direct neutron capture cross section measurement for the $^{14}C(n,\gamma)^{15}C$ reaction is possible by the neutron activation method. A first measurement of this type has been performed by Beer et al.[4] for a quasi-stellar incident neutron spectrum corresponding to $kT = 23.3$ keV. The measurement resulted in a MACS of 1.72 ± 0.43 μb. An additional measurement performed with the same technique resulted in a capture cross section roughly a factor of two higher [5], a discrepancy due to a wrong determination of the sample mass in the first measurement. In this case, a quasi-monoenergetic neutron spectrum was used and the measurement has been performed for three different neutron energies.

TABLE 1. Maxwellian averaged capture cross section of ^{14}C in a comparison between measured and calculated values.

Source	Method	$<\sigma_{n,\gamma}>$ in μb at $kT = 23.3$ keV	Reference
FZK-2	(n,γ), activation	2.7 ± 0.2	[5]
MSU	Coulomb dissociation	2.6 ± 0.9	[8]
RIKEN	Coulomb dissociation	7.6 ± 0.7	[11]
DRC	theory	7.6	present

A second class of experimental data can be obtained from Coulomb dissociation measurements. This process represents the time-reversal invariant of the neutron capture process (see Figure 1). Therefore, a simple correspondence between the neutron capture and the photo-dissociation cross sections can be established from the detailed-balance relation, which in the present case reads

$$\sigma_{\gamma,n} = \frac{k_n^2}{k_\gamma^2} \frac{1}{2} \sigma_{n,\gamma}$$

where k_n is the wave number for the neutron motion relative to the target nucleus. However, the relation between the photo-dissociation and the Coulomb dissociation process is not immediate. One can use the equivalent photon method [6], which is in fact based on a first-order perturbation theory. Following this method, the Coulomb dissociation cross section is related to the photo-dissociation cross section by

$$\frac{d\sigma_{CD}}{d\varepsilon_\gamma} = \frac{1}{\varepsilon_\gamma} N_{E1}(\varepsilon_\gamma) \sigma_{\gamma,n}$$

where, for a given excitation energy ε_γ, the equivalent photon number N_{E1} depends on the kinetic energy of the incident ion and on the nuclear charge of the nucleus used as target in the dissociation experiment. According to recent studies, the validity of the first-order perturbation theory should be justified in the case of a neutron dissociation process, while corrections need to be applied in the case a dissociation into the proton emission channel, and with both $E1$ and $E2$ components coming into play [7]. In any case, even for the extreme case in which the incident ion is very loosely bound (such as in the ^{8}B dissociation where the proton is bound by 0.137 MeV), the correction to be applied is at most of the order of 20%. Three different Coulomb dissociation experiments of ^{15}C have been performed so far. Using the equivalent photon method mentioned above, the Coulomb dissociation data can be compared to the direct neutron capture cross section results. This comparison is shown in Figure 2.

There appears to be a strong discrepancy between the direct measurement (performed by activation) and the Coulomb dissociation results. In addition, two of the Coulomb dissociation measurements show a strong disagreement amongst each other. The DRC calculation performed as outlined above, seems to be in a fair agreement with the RIKEN and GSI experimental results, both in magnitude as well as in terms of energy dependence. An earlier DRC calculation [10], based on a cruder approximation of the interaction potentials, produced values in complete agreement with the present

TABLE 2. The ^{62}Ni(n,γ) cross sections. The calculation for the JENDL-3.3 library has been performed with a Γ_γ of 2.38 eV for the 4.54 keV s-wave resonance [14], while Mughabghab [13] includes a subthreshold resonance at -0.077 keV.

Quantity		Experimental	Ref.	JENDL-3.3	Mughabghab	p-wave contribution
$\sigma_{n,\gamma}^{th}$	(b)	14.5±0.5	[13]	14.6	14.9	0
MACS-25	(μb)	28.4±2.8	[15]	23.1	13.3	8.3
MACS-25	(μb)	49.5±4.4	[16]			

calculation. At present it is therefore impossible to disentangle the origin of these discrepancies.

The ^{62}Ni(n,γ) cross section

The neutron capture rate of ^{62}Ni plays an important role as bottleneck in the reaction flow of the weak s process, responsible for the synthesis of nuclei with masses between 60 and 90, in massive stars. This has been originally pointed out by Rauscher et al. [12]. The rate is dominated by a strong s-wave resonance at 4.5 keV and, in general, it is determined completely by isolated resonances in the energy range of interest for stellar nucleosynthesis. In this situation, the theoretical approach for the calculation of the capture rate can only rely on experimental information derived from direct cross section measurements. The cross section can be expressed in terms of the multi-level Breit-Wigner formalism which, in the vicinity of an isolated resonance is simply given by

$$\sigma_{n,\gamma}(E_n) = \frac{\pi}{k_n^2} \frac{g_\lambda \Gamma_{\lambda n} \Gamma_{\lambda \gamma}}{(E_n - E_\lambda)^2 + \frac{1}{4}\Gamma_\lambda^2}$$

where g_λ is the statistical factor for angular momentum coupling and the parameters E_λ, $\Gamma_{\lambda n}$, $\Gamma_{\lambda \gamma}$ (energy, neutron width and γ width of the resonance state) are usually derived from an R-matrix fit of the experimental data.

The present information concerning the resonance parameters of ^{62}Ni is not consistent with the thermal neutron capture cross section, which is $\sigma_{n,\gamma}^{th} = 14.5 \pm 0.5$b. The parameters of the first s-wave resonance, $E = 4.54$ keV, $g\Gamma_n = 1.88 \pm 0.20$ keV, and $\Gamma_\gamma = 0.76 \pm 0.12$ eV [13], do not reproduce the thermal value with the low-energy tail, typical of s-wave resonances. In order to reproduce the thermal value, either a subthreshold state with arbitrary parameters has to be introduced, or a much higher value of Γ_γ (roughly a factor of 3) needs to be adopted (see for example [14]). While for the thermal value one could live with either assumptions, the resulting MACS for the keV energy range will be very different in the two cases. This can be seen in Table 2, where the MACS-25 values are reported as derived from calculations performed with these two different assumptions.

The MACS-25 values recently derived from two different experiments [15, 16] are in disagreement which each other. The capture cross section is shown as a function of

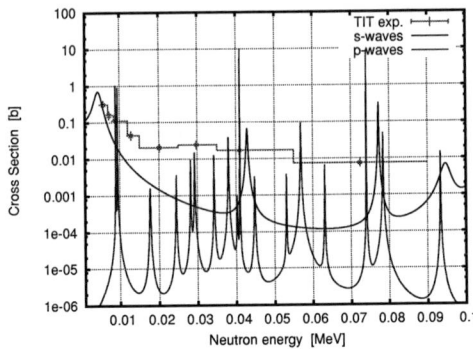

FIGURE 3. The ^{62}Ni(n,γ) cross section in the energy range up to 100 keV. Experimental data are from reference [16]. The contributions of s-wave and p-wave resonances are shown separately.

the neutron energy in the keV region in Figure 3. Considering the large uncertainties on the parameters used for the calculation of the cross section, in particular those of p-wave resonances which can contribute up to 30% of the total capture strength assuming average parameters for the γ-widths, it is clear that only a high resolution measurement can contribute to solve the situation.

The ^{186,187}Os(n, γ) stellar cross sections

For nuclei with higher mass the neutron capture process is dominated by overlapping compound states at excitation energies around the neutron separation energy and the cross section is well represented by the Hauser-Feshbach formalism [17] (Hauser-Feshbach statistical model theory, HFSM). This theory consists in averaging over a large number of compound states the Breit-Wigner resonant cross section relation shown above. The final expression for the capture cross section is given by

$$\sigma_{n,\gamma}(E_n) = \frac{\pi}{k_n^2} \sum_{J\pi} g_J \frac{\sum_{ls} T_{n,ls} T_{\gamma,J}}{\sum_{ls} T_{n,ls} + \sum_{n',ls} T_{n',ls} + T_\gamma} W_{\gamma,J}$$

where the neutron transmission coefficients, T_n and $T_{n'}$ respectively for the elastic and inelastic channels, can be calculated by solving the two-body Schrödinger equation for an optical potential. The γ-ray transmission coefficients for single transitions are calculated, usually, from the isovector giant dipole response to nuclear excitations (giant dipole resonance, GDR) and the nuclear level density is utilized to obtain the summed value over low-lying nuclear states, T_γ. The factor W_γ, defined by the relation

$$\left\langle \frac{\Gamma_{\lambda n}\Gamma_{\lambda \gamma}}{\Gamma_\lambda} \right\rangle_\lambda \approx \frac{\langle \Gamma_{\lambda n}\rangle \langle \Gamma_{\lambda \gamma}\rangle}{\langle \Gamma_\lambda \rangle} W_\gamma$$

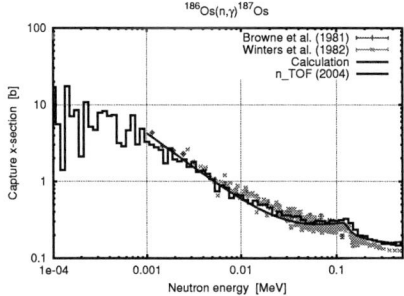 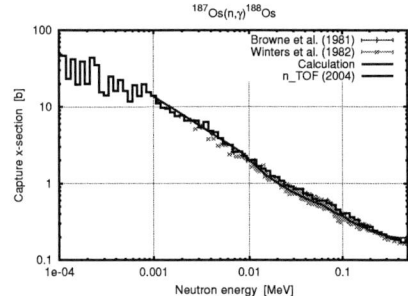

FIGURE 4. The ^{186}Os$((n,\gamma)$ (*left* panel), and ^{187}Os(n,γ) (*right* panel) cross sections in a comparison between available experimental data and HFSM calculations.

takes into account the different statistical fluctuation properties of the neutron, γ-ray, and other particle channels' widths. $\langle\rangle_\lambda$ indicate averaging over the ensemble of λ compound states.

The HFSM theory has been around since long time and has proved to be reliable [18]. The key issue here, is the determination of the model parameters. Traditionally, these have been derived from global systematics such as those derived from neutron scattering cross section data for the optical model potential. Much on the same line, the parameters of the fermi-gas expression with pairing and shell corrections for the level densities can be derived from discrete spectra and matched to the experimental data on average spacings of neutron resonances at higher excitation energies. Finally, the parameters for the calculation of the γ-ray transmission functions can be deduced from (γ,n) measurements in the GDR region. This approach allows for model calculation of the neutron capture rates for a large sample of nuclei of interest in nucleosynthesis network calculations [19].

The example we have chosen here is the case of the capture cross sections of ^{186,187}Os. These are crucial for the determination of the s process component of the ^{187}Os abundance which in turn, is essential for the Re/Os nuclear cosmochronometry [20]. A detailed analysis of this case goes beyond the scope of the present work. What we would like to notice here is that an accurate calculation of these two capture cross sections is the essential basis for the determination of their stellar rates. For these, one needs to rely on the calculation of the capture cross sections from excited states of the target nuclei which will be all the more realistic once the cross sections for the ground states are accurate. Therefore, all the parameters used for the calculation need to be verified and the result of the calculation compared to as much experimental information as possible. Here we only show (see Figure 4) the results of the ^{186,187}Os capture cross section calculation in comparison with the preliminary results of a recent measurement performed at CERN n_TOF [21].

Recent developments of the HFSM approach concern the determination of model parameters from microscopic and/or semi-microscopic nuclear models. In particular, the developments of realistic microscopic description of the mean-field potential such as those based on the Brückner-Hartree-Fock theory [22], or those based on quasi-particle random-phase-approximation (QRPA) for GDR excitations have been used in the determination of the HFSM model parameters [23]. These approaches can be used in network calculations where a large number of neutron capture cross section rates are involved. In particular, for reactions taking place in nuclei far from the β stability, microscopic approaches should provide a better description of the nuclear properties just mentioned because of their more fundamental basis. These studies represent the present frontier line for the accurate determination of theoretical neutron capture reaction rates.

ACKNOWLEDGMENTS

I would like to thank T. Nakamura, Y. Nagai, M. Wiescher, R. Reifarth, M. Heil, and F. Käppeler, for fruitful discussions on the experimental aspects of the reaction rates of interest in this work.

REFERENCES

1. A. Mengoni, and T. Otsuka, Proceeding of the 10th International Conference *Capture Gamma-Ray Spectroscopy and Related Topics*, Santa Fe, New Mexico (USA), September 1999. S. Wender Ed., AIP Conference Proceedings, New York, **529**, pp. 119-125 (2000).
2. A.M. Lane, and J.E. Lynn, *Nucl. Phys.*, **17**, p. 563 (1960); *ibid.*, p. 586.
3. F. Ajzenberg-Selove, *Nucl. Phys.*, **A523**, p. 1 (1991).
4. H. Beer, *et al.*, *Ap. J.*, **363**, p. 387 (1992).
5. R. Reifarth, *et al.*, *Nucl. Phys.*, **A758**, p. 787c (2005).
6. C. Bertulani, and G. Baur, *Phys. Rep.*, **163**, p. 299 (1988).
7. H. Esbensen, these proceedings. Also, H. Esbensen, *et al.*, *Phys. Rev. Lett.*, **94**, p. 042502 (2005).
8. Á. Horváth, *et al.*, *Ap. J.*, **570**, p. 926 (2002).
9. U. Datta Pramanik, *et al.*, *Phys. Lett. B*, **551**, p. 63 (2003).
10. M. Wiescher, *et al.*, *Ap. J.*, **363**, p. 340 (1990).
11. T. Nakamura, private communication (2004).
12. T. Rauscher, *et al.*, *Ap. J.*, **576**, p. 323 (2002).
13. S.F. Mughabghab, *et al.*, *Neutron Cross Sections*, Academic Press, New York, 1981, Vol. 1 (Part A).
14. K. Shibata, *et al.*, *Japanese Evaluated Nuclear Data Library Version 3 Revision-3: JENDL-3.3*, J. Nucl. Sci. Technol. **39**, p. 1125 (2002).
15. H. Nassar, *et al.*, *Phys. Rev. Lett.*, **94**, p. 092504 (2005).
16. A. Tomyo, *et al.*, *Ap. J.*, **623**, L153 (2005).
17. W. Hauser, and H. Feshbach, *Phys. Rev.*, **87**, p. 366 (1952).
18. P.A. Moldauer, *Phys. Rev. C*, **11**, p. 426 (1975).
19. T. Rauscher, and F-K. Thielemann, *Atomic Data Nucl. Data Tables*, **75**, p. 1 (2001).
20. D.D. Clayton, *Ap. J.*, **139**, p. 5190 (1964).
21. M. Mosconi, *et al.*, Proceeding of the International Conference on *Nuclear Data for Science and Technology*, Santa Fe, New Mexico (USA), September 2004. M. Chadwick *et al.* Eds., AIP Conference Proceedings, New York, **769**, pp. 1335 (2005).
22. E. Bauge, *et al.*, *Phys. Rev. C*, **63**, p. 024607 (2001).
23. S. Goriely, *et al.*, *Nucl. Phys*, **A739**, p. 331 (2004).

The weak s-process and its relation to explosive nucleosynthesis

M. Heil* and M. Pignatari[†]

*Forschungszentrum Karlsruhe, Postfach 3640, 76021 Karlsruhe, Germany
[†]Dept. of General Physics, University of Torino, Via P. Giuria 1, 10125 Torino, Italy

Abstract. Considerable effort in experimental nuclear astrophysics, stellar modeling, and observations led to an improved understanding of various nucleosynthesis scenarios in the last decades. This is particularly true for the main s-process in low-mass AGB stars, which is largely responsible for the production of about half of the elemental abundances in the mass range $90 \leq A \leq 209$. The weak s-process, which produces elements with $A \leq 90$, however, is much less understood. Since this process operates in massive stars it is ultimately linked with the abundance contributions of explosive nucleosynthesis in supernovae (SN II). In this field more accurate neutron capture cross sections in the mass range $56 \leq A \leq 90$ are indispensable for a meaningful comparison of model predictions with observational data. Also, the abundant light elements with $A < 56$ play an important role, since they act as neutron poisons and affect the stellar neutron balance. In this context, the impact of new results for neutron capture cross sections on light and medium mass nuclei will be discussed.

Keywords: nuclear reactions, nucleosynthesis, abundances, s-process
PACS: 25.40.Lw, 26.20.+f, 26.30.+k, 27.50.+e, 97.10.Tk

INTRODUCTION

The elements heavier than iron are mainly produced by two neutron capture processes, the s(slow)- and the r(rapid)-process, both contributing about half of the observed solar abundances. A third process, the so called p(photodissociation)-process is responsible for the origin of about 30 rare, proton rich nuclei, but does not contribute significantly to the synthesis of the elements in general (<1%). The s-process is characterized by low neutron densities (10^6-10^{11} cm^{-3}) and, as a consequence, the average neutron capture time is of the order of years. Therefore, unstable nuclei have time to undergo β-decay, and the s-process proceeds along the valley of stability by a sequence of neutron capture reactions and β-decays until the termination point at ^{209}Bi is reached (Fig. 1). The r-process, in contrast, operates at much higher temperatures and neutron densities (>10^{20} cm^{-3}) and many neutrons are captured instantaneously by the seed nuclei until $(n,\gamma) \Leftrightarrow (\gamma,n)$ equilibrium is established at the very neutron rich side of an isotopic chain. Only after β-decay, further neutrons can be captured, and the r-process works its way on the neutron rich side up to the heaviest elements like thorium and uranium. After the neutron source deceases the neutron-rich isotopes fall back to the valley of stability by a series of β-decays. As can be seen in Fig. 1 there are nuclei, which are produced by the s- or r-process exclusively. These nuclei are called s-only and r-only, respectively.

Since 1957 [1] most progress was made in the field of the s-process. It became apparent that a single s-process was not sufficient to explain the observed solar abundances. At least two components, the main and the weak s-process, are necessary and could

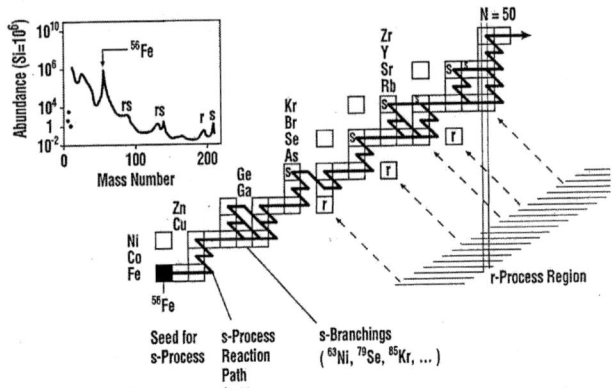

FIGURE 1. Section of the chart of nuclei showing the s-process reaction path (black line). s-only and r-only isotopes are marked with "s" and "r" respectively.

be connected with the corresponding stellar objects and sites. A third component, the strong s-process, which was considered to be responsible for the production of ^{208}Pb is superseded since it was found that the main s-process in low metallicity stars can account for the observed ^{208}Pb [2]. The main s-process, by far the most studied process, occurs in the He-rich intershell of thermally pulsing AGB stars and produces predominantly nuclei with mass numbers A>90. These evolved Red Giants have burnt already all the H and He in the core to carbon and oxygen. The energy generation occurs by alternate H- and He-burning episodes, which are separated by a thin He-rich intershell. The main s-process takes place in the He-rich intershell where neutrons are liberated by the ^{13}C$(\alpha,n)^{16}$O reaction at temperatures of $T_8 \sim 1$ and at neutron densities of about 10^7 cm^{-3}. Since there are not many seed nuclei in this thin shell, the neutron/seed ratio is high and the s-process operates very efficiently over a long period of time. During the subsequent convective He-flashes, the freshly synthesized material is mixed and diluted with the He-intershell and again exposed to neutrons liberated by the ^{22}Ne$(\alpha,n)^{25}$Mg reaction at temperatures $T_8 \geq 2.5$. The second neutron exposure is rather weak and not sufficient to produce s-isotopes on a grand scale but strong enough to alter the isotope ratios of s-process branchings. After the He-flash, where peak neutron densities of 10^{10} cm^{-3} are reached, part of the freshly synthesized material is mixed with the envelope and brought to the surface of the star, where it is detectable by spectroscopy.

The weak component, which is responsible for the production of nuclei between iron and yttrium (56<A<90), takes place during convective core He burning in massive stars ($M_\odot$>8), where for a short time temperatures of $(2.2-3.5) \cdot 10^8$ K are reached and neutrons are liberated by the activation of the ^{22}Ne$(\alpha,n)^{25}$Mg reaction. Since the neutron exposure is small, the s-process flow can not overcome the bottle neck at the closed neutron shell N=50. Recent stellar models discuss also the possibility of a second neutron exposure during convective carbon shell burning of massive stars [3]. There, neutrons are produced by various reactions, e.g. ^{13}C$(\alpha,n)^{16}$O, ^{22}Ne$(\alpha,n)^{25}$Mg, and ^{17}O$(\alpha,n)^{20}$Ne.

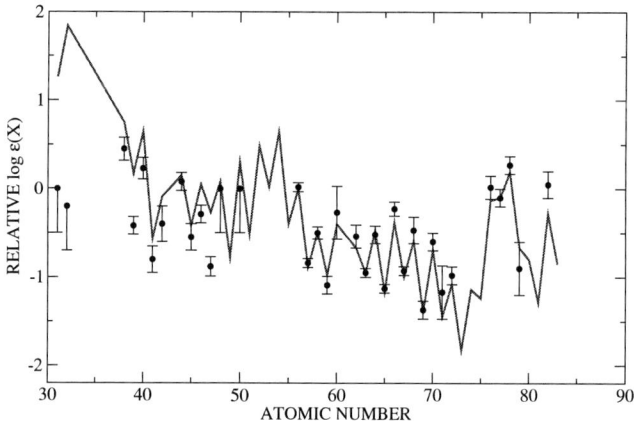

FIGURE 2. Comparison of the scaled solar r-abundances (solid line) with the observed abundances (symbols) of the metal-poor halo star CS22892-052 [5] (log $\varepsilon(X) \equiv \log_{10}(N_X/N_H)+12$).

The high temperatures during carbon burning of $T_9 \sim 1$ cause high neutron densities, which start at $\sim 10^{12}$ cm^{-3} and then decrease exponentially.

The nucleosynthesis yields of the weak component in massive stars are also important for the r-process, since they determine the composition of a star before the supernova explosion. Since the s-process abundances can be determined reliably on the basis of experimental (n,γ) cross sections, the r-abundances are commonly inferred by the r-residual method, that is by subtracting the s-abundances from solar values: $N_r = N_\odot - N_s$. The r-abundances obtained in this way are then used for testing r-process models.

This contribution focuses on the weak s-process. In the second chapter the consequences of recent observations of halo stars for the s-process will be discussed and new neutron capture cross section measurements for the weak s-process will be presented in chapter 3. Finally, further experimental quests will be addressed in chapter 4.

OBSERVATIONS

The quantitative understanding of nucleosynthesis requires the close collaboration between various fields in physics. Models, for describing stellar evolution and nucleosynthesis occurring in the deep stellar interior are the starting point. The task of nuclear physics is to provide the necessary nuclear data input complemented by the important theoretical contribution. Finally, the models have to be tested and validated with the information obtained by astronomers. There are several observations, which help to constrain stellar models. First, there is the solar abundance distribution [4], and it is clear that every theory on the formation of the elements must explain the solar abundance pattern and all its features. Although, s-, r-, and p-only nuclei help to disentangle the various contributions, one has to keep in mind that the solar abundance distribution is a mixture of many generations of stars with different evolution times, masses, metallicities, mass loss rates, etc. In contrast, stellar spectroscopy delivers information on single stars and

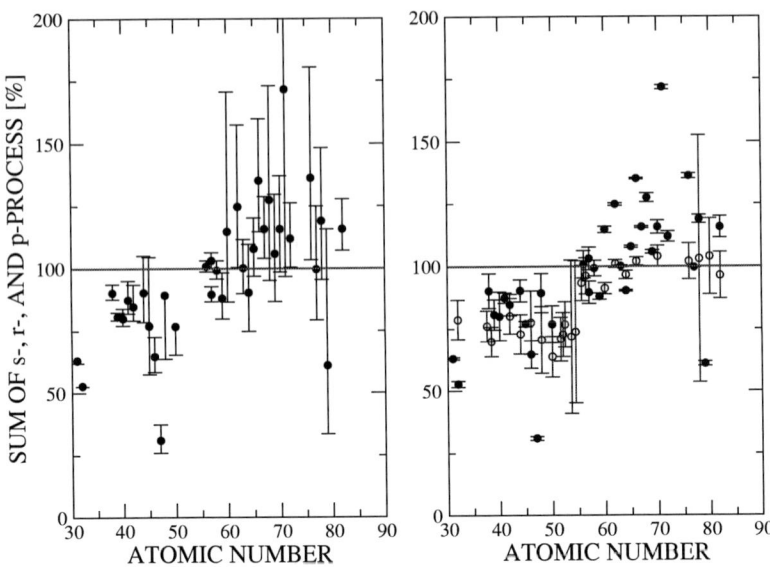

FIGURE 3. Sum of s-, r-, and p-process abundances with error bars from the observation of the r-arbundances (left) and uncertainties caused by neutron capture cross sections (right). In the right figure also the abundance of s-only isotopes (open symbols) are plotted.

by looking at stellar objects with different ages and metallicities valuable information on galactic chemical evolution is obtained.

An excellent example are recent spectroscopic observations of ultra-metal-poor halo stars. Since these stars are very old they show no s-contribution since the s-process is a secondary process and needs seed nuclei, which were not present at early times. Correspondingly, these stars are expected to exhibit a pure r-process abundance distribution. Indeed, comparison of the observed abundances and the scaled solar r-abundances, which were derived by the r-residual method, are in excellent agreement for elements with atomic numbers $Z \geq 56$, suggesting a very robust and unique r-process [5]. This observation is meanwhile confirmed for a few metal-poor halo stars like CS22892-052 [5], HD 115444 [6], BD +17°3248 [7], and CS 31082-001 [8]. Fig. 2 shows the observed elemental abundances of the metal-poor halo star CS22892-052 in comparison with the solar r-distribution, normalized to reproduce the Eu r-abundance. Below $Z=56$, however, the observed abundances fall systematically below the solar r-curve. This might be interpreted as a missing r-process contribution. However, such an argumentation relies on the known s-process abundances, since the solar r-abundances are derived by the r-residual method. This is an assumption, which might not be true for lighter nuclei, where the weak s-process contributes significantly to the s-abundances.

The observations in metal-poor halo stars have not only consequences for the r-process, but can also be used to test the s-process models. Since for the first time observed r-abundances are available, they can be added to the predicted s-process abundances in order to test stellar s-process models via the sum rule. Fig. 3 shows the sum of the abundances for the weak s-process [3] and the main s-process [9], including the chemical evolution of the Galaxy [10]. The p-process contributions of 24 % for Mo and 7 % for Ru are also taken into account. In the left part of the figure the uncertainties of the astronomical observations are indicated, while the uncertainties of the main s-component, which are introduces by the uncertainties of the (n,γ) cross sections, are shown in the right panel. Not plotted are the uncertainties from the weak s-component, which should contribute significantly below Z=40. At higher atomic numbers beyond Ba the uncertainties of the observations dominate the picture, but at lower atomic numbers also the uncertainties from the cross sections contribute significantly. Nevertheless, beyond Ba the elemental abundances can be fully accounted for by the sum of the s-, p-, and r-contributions, but below Ba a contribution of about 10 to 20% is missing. The fact that also s-only isotopes (open symbols in Fig. 3) are underproduced by the same amount points to a missing s-like process as already mentioned in Ref. [10]. In this paper we want to investigate to what extent the weak s-process, which is still not fully understood, could contribute to the missing part.

NEUTRON CAPTURE MEASUREMENTS

Most important input for stellar models of the s-process are Maxwellian averaged neutron capture cross sections and β-decay rates, but also stellar enhancement factors (SEF) have to be known. The weak s-process is responsible for the production of elements in the mass range $56 \leq A \leq 90$. In this mass range the (n,γ) cross sections need substantial improvement for deriving a reliable description of the abundance contributions from massive stars. This is especially important since the local approximation ($<\sigma>$N=const) is not valid during the weak s-process. Therefore, any change in the cross section of a light isotope, e.g. ^{62}Ni [11], can affect the abundances of all the heavier isotopes up to zirconium and maybe even higher. This underlines that neutron capture cross sections in the mass range A=50-90 have to be measured with higher accuracy.

Therefore, a measuring campaign was launched at Forschungszentrum Karlsruhe with the aim to improve the neutron capture cross sections in this region. A reliable and accurate approach to derive Maxwellian averaged cross section at kT=25 keV is the activation method [12], where the ^{7}Li(p,n)^{7}Be reaction is used to produce a quasi-stellar neutron spectrum as sketched in Fig. 4. After irradiation in that spectrum the induced sample activity is counted in a low background environment. The proton beam with an energy of E_p=1912 keV and typical intensities of 100 μA was delivered by the Karlsruhe 3.7 MV Van de Graaff accelerator. The neutron production target consists of a metallic Li layer, which is evaporated onto a water cooled copper backing. The sample is placed inside the resulting neutron cone, which has an opening angle of 120 degrees. The neutron flux is monitored throughout the irradiation by means of a ^{6}Li-glass detector, positioned at a distance of 1 m from the target. After the irradiation the total number of activated nuclei A is given by

$$A = \Phi \cdot N \cdot \sigma \cdot f_b \qquad (1)$$

where Φ is the time integrated neutron flux, N the number of sample atoms per cm^2,

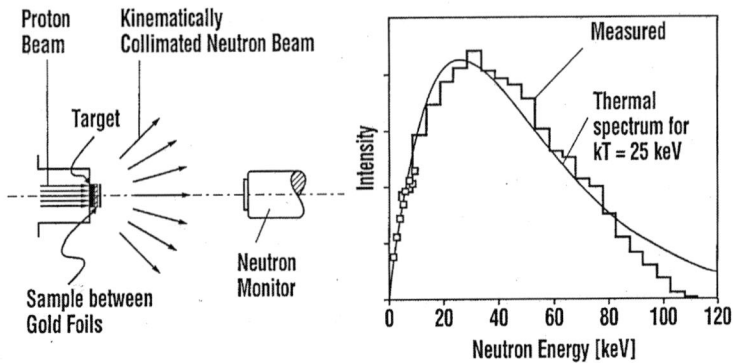

FIGURE 4. The experimental setup for activation measurements (left) and a comparison of the produced neutron spectrum and a thermal neutron spectrum at kT=25 keV (right).

TABLE 1. Measured MACS at kT=30 keV (in mbarn).

Isotope	This work	Ref [13]
^{45}Sc	57±2	69±5
^{59}Co	41±2	38±4
^{63}Cu	53±2	94±10
^{65}Cu	29±2	41±5
^{79}Br	626±19	627±42
^{81}Br	241±9	313±16
^{87}Rb	16.1±2.0	15.5±1.5

and σ the spectrum averaged neutron capture cross section. In order to determine the neutron flux, the sample is sandwiched between gold foils. Since the gold cross section is well known, the total number of neutrons can be obtained by measuring the 412 keV line from the decay of ^{198}Au with HPGe detectors. The factor f_b accounts for the variation of the neutron flux and for the decay during activation. The cross section can then be calculated from the number of counts in a characteristic γ-ray line

$$C_\gamma = A \cdot K_\gamma \cdot \varepsilon_\gamma \cdot I_\gamma \cdot (1 - exp(-\lambda t_m)) \cdot exp(-\lambda t_w) \qquad (2)$$

where K_γ is a correction factor for γ-ray self-absorption, ε_γ the efficiency of the Ge-detector, I_γ the line intensity, t_w the waiting time between irradiation and counting, and t_m the duration of the activity measurement. In this way, we have measured the Maxwellian averaged capture cross section of seven isotopes. The results are shown in Table 1 together with the previously recommended values [13].

For some isotopes the measured cross sections differ significantly from previous recommendations. To explore the effect for the weak s-process, stellar model calculations

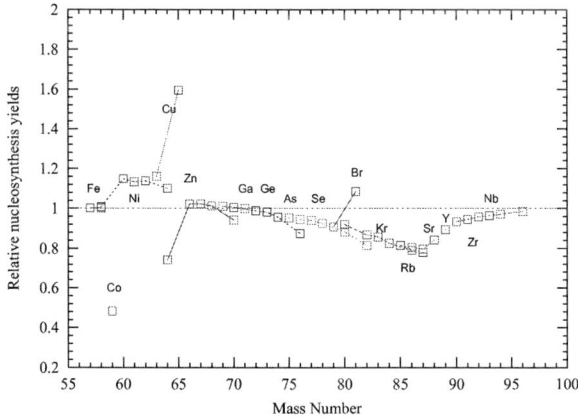

FIGURE 5. Abundance yields obtained for a 25M$_\odot$ star using the neutron capture cross sections of this work relative to the standard case.

for a 25 M$_\odot$ star were performed with an updated post-processing code of Refs. [14, 15]. Fig. 5 shows the combined effect of all new cross section measurements on the abundance distribution of the weak s-process. Plotted are the nucleosynthesis yields at the end of carbon burning relative to the yields obtained with the previous cross sections.

Again, just like in the case of ^{62}Ni [11] one can see a strong influence of the measured cross sections. Therefore, one has to conclude that a reliable abundance prediction of the weak s-process is only possible if all neutron capture cross sections of the involved isotopes are known with high accuracy.

SUMMARY AND OUTLOOK

The present measurements have shown that the nucleosynthesis yields for the weak s-process, calculated with stellar models for massive stars, still show large variations due to the uncertainties of the involved neutron capture cross sections. Unlike the main s-process, flow equilibrium is not reached during the weak s-process. Therefore, improved neutron capture cross sections do not only influence the yield of the respective isotope, but also the production of all heavier nuclei on the weak s-process path. Observations of metal-poor halo stars show that the abundances of the elements below Ba can not be fully accounted for by the present nucleosynthesis processes. There seems to be a need for an addional s- and/or r-process, which is expected to add 10 to 20% of the abundances for some elements between Fe and Ba. Before this problem can be settled, more accurate neutron capture cross sections are needed to see whether the weak s-process itself may provide the missing fraction. It is planned to further pursue activation measurements of isotopes relevant for the weak s-process at Karlsruhe, but for isotopes, which are not accessible by this method, time-of-flight experiments have to be performed. Especially important would be neutron capture measurements of ^{58}Fe, all Ge isotopes, all Ga isotopes, and all Zn isotopes. In spite of their small uncertainties some of the elemental abundances in Fig. 3, e.g. Ag, are particularly low. This could point to a problem with the neutron capture cross sections of the related isotopes. In case of Ag, the cross sections of the two stable isotopes ^{107}Ag and ^{109}Ag are known with uncertainties of 3%. However, a large fraction of elemental Ag is produced by the radiogenic contribution of ^{107}Pd, which has a half-life of 6.5·10^6 years. So far, there is only one measurement of the neutron capture cross section of ^{107}Pd [16], which was performed with a fission product Pd sample with a very low (16%) enrichment in ^{107}Pd.

Also the abundant light isotopes below Fe are important, since they may constitute crucial neutron poisons for the s-process. The respective neutron capture cross sections show large uncertainties, in particular for ^{12}C, ^{16}O, ^{22}Ne, and ^{23}Na.

ACKNOWLEDGMENTS

We would like to thank R. Gallino and F. Käppeler for many stimulating discussions. We are also thankful to D. Roller, E.-P. Knaetsch, and W. Seith for their support during the measurements.

REFERENCES

1. E. M. Burbidge, G. R. Burbidge, W. A. Fowler, and F. Hoyle, *Rev. Mod. Phys.* **29**, 547 (1957).
2. R. Gallino, C. Arlandini, M. Busso, M. Lugaro, C. Travaglio, O. Straniero, A. Chieffi and M. Limongi, *Ap. J.*, **497**, 388 (1998).
3. C. M. Raiteri, R. Gallino, M. Busso, D. Neuberger, and F. Käppeler, *Ap. J.*, **419**, 207–223 (1993).
4. E. Anders and N. Grevesse, *Geochim. Cosmochim. Acta*, **53**, 197 (1989).
5. C. Sneden, J. J. Cowan, J. E. Lawler, I. I.Ivans, S. Burles, T. C. Beers, F. Primas, V. Hill, J. W. Truran, G. M. Fuller, B. Pfeiffer, and K.-L. Kratz, *Ap. J.*, **591**, 936–953 (2003).
6. J. Westin, C. Sneden, B. Gustafsson, and J. J. Cowan, *Ap. J.*, **530**, 783–799 (2000).
7. J. J. Cowan, C. Sneden, S. Burles, I. I. Ivans, T. C. Beers, J. W. Truran, J. E. Lawler, F. Primas, G. M. Fuller, B. Pfeiffer, and K.-L. Kratz, *Ap. J.*, **572**, 861–879 (2002).
8. V. Hill, B. Plez, R. Cayrel, T. C. Beers, B. Nordström, J. Andersen, M. Spite, F. Spite, B. Barbuy, P. Bonifacio, E. Depagne, P. Francois, and F. Primas, *Astron. Astrophys.*, **387**, 560–579 (2002).
9. C. Arlandini, F. Käppeler, K. Wisshak, R. Gallino, M. Lugaro, M. Busso, and O. Straniero, *Ap. J.*, **525**, 886 (1999).
10. C. Travaglio, R. Gallino, E. Arnone, J. Cowan, F. Jordan, and C. Sneden, *Ap. J.*, **601**, 864–884 (2004).
11. H. Nassar, M. Paul, I. Ahmad, D. Berkovits, M. Bettan, P. Collon, S. Dababneh, S. Ghelberg, J.P. Greene, A. Heger, M. Heil, D.J. Henderson, C. L. Jiang, F. Käppeler, H. Koivisto, S. O'Brien, R. C. Pardo, N. Patronis, T. Pennington, R. Plag, K. E. Rehm, R. Reifarth, R. Scott, S. Sinha, X. Tang, and R. Vondrasek, *Phys. Rev. Lett.* **94**, 092504 (2005).
12. H. Beer and F. Käppeler, *Phys. Rev. C*, **21**, 534–544 (1980).
13. Z. Y. Bao, H. Beer, F. Käppeler, F. Voss, K. Wisshak, and T. Rauscher, *Atomic Data Nucl. Data Tables*, **76**, 70 (2000).
14. A. Chieffi and O. Straniero, *Ap. J. Suppl.*, **71**, 74 (1989).
15. O. Straniero, *Ap. J. Suppl.*, **76**, 157 (1989).
16. R. L. Macklin, *Nuc. Sci. Eng.*, **89**, 79–86 (1985).

Measurement of Neutron Capture Cross Section of ^{62}Ni in the keV-Region

A.M. Alpizar-Vicente[1,2], T.A. Bredeweg[2], E.-I. Esch[2], U. Greife[1], R.C. Haight[2], R. Hatarik[1,2], J.M. O'Donnell[2], R. Reifarth[2], R.S. Rundberg[2], J.L. Ullmann[2], D.J. Vieira[2], J.B. Wilhelmy[2], J.M. Wouters[2]

[1]*Colorado School of Mines, Golden, CO 80401, USA*
[2]*Los Alamos National Laboratory, Los Alamos, NM 87545, USA*

Abstract. The neutron capture cross section of ^{62}Ni, relative to gold as a standard, was determined in the energy range from 250 eV to 100 keV. This energy range covers the region between 5 keV to 20 keV, which is not available in ENDF. Capture events are detected with the 160-fold 4π BaF$_2$ Detector for Advanced Neutron Capture Experiments (DANCE) at the Los Alamos Neutron Science Center. One of the challenges was to process the high count rate of 4 MHz, which required an optimization of the data acquisition software. The neutron energy was determined by the time-of-flight technique using a flight path of 20.25 m. The sample mass of the 96% enriched ^{62}Ni target was 210 mg and it was mounted in a 1.5 μm thick Mylar foil.

Keywords: keV neutron capture, solid state detectors, neutrons sources, nuclear reactions.
PACS: 28.20.Fc, 29.40.Vj, 29.25.Dz, 29.40.Wk

INTRODUCTION

The science of nucleosynthesis attempts to interpret the measured abundances of the nuclear species in terms of their nuclear properties and a set of environments in which nuclei can be synthesized by nuclear reactions. There are a number of astrophysical processes, which are believed to be responsible for nucleosynthesis in the universe. The ^{62}Ni(n,γ) cross section is related to the weak s-process component during helium burning in massive stars of 10 to 25 solar masses [1]. During the weak s-process mainly isotopes with masses between 60 and 90 are produced.

The Maxwellian-averaged neutron-capture cross section (MACS) of ^{62}Ni was only poorly known until this experiment; there are time-of-flight measurements available but the determination of the total capture cross section is affected by large uncertainties due to the unknown direct capture (DC) component. The estimated values [2,3] for the MACS at kT = 30 keV vary from 12.5 ± 4 mb to 35 ± 5 mb. This wide range suggests the need for an independent experiment.

EXPERIMENT

The measurements were performed on flight path 14 (FP14) of the Manuel Lujan Jr. Neutron Scattering Center at the Los Alamos Neutron Science Center (LANSCE) at Los Alamos National Laboratory (LANL). The neutrons are produced via spallation reactions by an 800 MeV pulsed proton beam hitting a tungsten target with typical beam currents of 100 µA. The neutron energies were measured by time of flight over a 20.25 m flight path. The DANCE detector is a 4π array that consist of 162 barium fluoride (BaF_2) crystals in four different shapes. Two of the 162 crystals are left out in order to leave space at the entrance and the exit of the neutron beam pipe, which reduces the solid angle coverage to about 3.6π. The neutron capture events are detected via the prompt γ-rays following the capture. DANCE is a high efficiency, highly segmented detector to measure neutron capture cross sections of targets with masses in the order of a few mg from stable and unstable isotopes.

A 96% enriched ^{62}Ni sample with a mass of 210 mg and the isotopic composition shown in Table 1 was mounted in a 1.5 µm Mylar bag. The bag containing the target was pasted to an aluminum ring and placed inside the beam pipe. The target was positioned in the center of the DANCE array and irradiated. The Mylar bag also helped to hold the sample in the center of the neutron beam, which has a diameter of 1 cm. The Mylar foil was chosen after an analysis made by Hatarik [4] with different plastics (Kapton and Mylar of different thickness) because of its low (mainly scattering) background contribution.

TABLE 1. Isotopic composition of the ^{62}Ni enriched target.

Abundance %	Isotope	Q-Value (n,γ) MeV
1.95	^{58}Ni	8.99951
1.31	^{60}Ni	7.82005
0.15	^{61}Ni	10.59733
96.45	^{62}Ni	6.83791
0.14	^{64}Ni	6.09806

The ^{62}Ni target was interchanged with a carbon disc of 44 mg/cm^2 and a thin foil of gold of 4 mg/cm^2 for background and neutron flux determination respectively. Also empty container runs and empty Mylar bag runs were performed for background subtraction purposes. A 6 cm thick ^{6}LiH sphere with an inner radius of 10.5 cm was used to surround all samples (Ni, Au, C) to reduce the scattered neutron background. The beam time required for the ^{62}Ni irradiation was three days.

NEUTRON CAPTURE ANALYSIS

DANCE measures the total energy of the γ-ray cascade that follows the capture, this summed energy is approximately equal to the Q-value of the reaction and can be used to identify the nuclide on which the capture occurred. Figure 1 shows the number of events as a function of the energy deposition in the detector for the energies above

1 keV. The peak due to the neutron capture on ^{62}Ni slightly below the Q-value of 6.837 MeV is clearly visible.

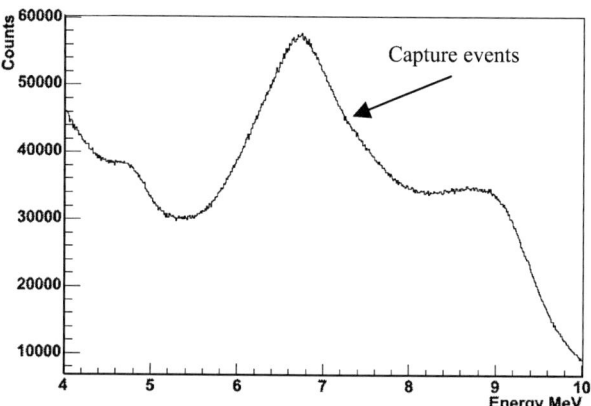

FIGURE 1. Summed γ-energy spectrum for the ^{62}Ni(n,γ) reaction, the peak around 6.8 MeV are the neutron capture events.

The neutron energies are calculated by time-of-flight, Fig. 2 shows the very well known resonance at around 4.5 keV and others at ~ 9.6, 17.9, 24.8 and 29.1 keV, that are not found in the ENDF-VI library, but consistent with the ones shown in JENDL 3.3. The resonances at 9.6 and 29.1 keV are two not resolved resonances according to JENDL (these two resonances are broader than the others in Fig. 2), so the DANCE array shows 5 of the 7 shown in JENDL. There is a resonance at about 6.2 keV that appears to belong to ^{62}Ni but further analysis is in progress.

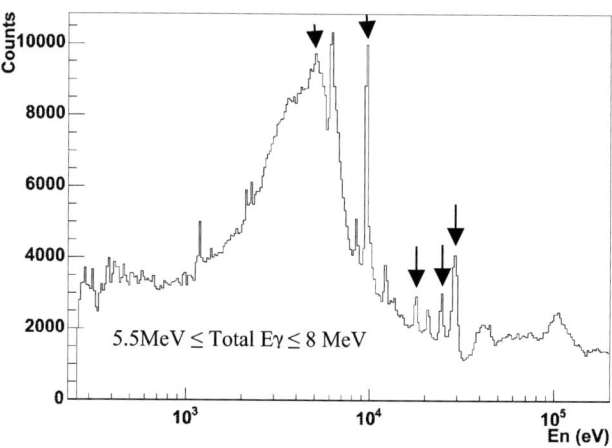

FIGURE 2. Neutron energies with a γ-energy cut around the Q-value peak of ^{62}Ni.

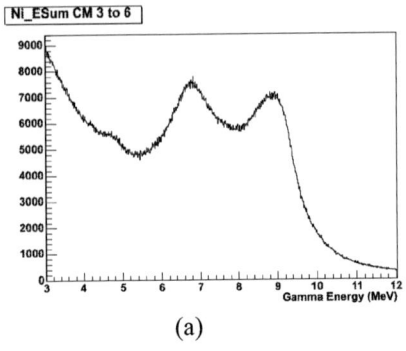 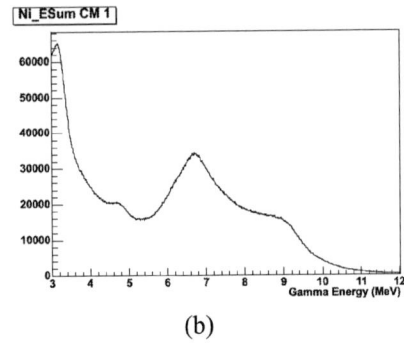

FIGURE 3. (a) Summed γ–energy for ^{62}Ni with a cluster multiplicity from 3 to 6. (b) Summed γ–energy for ^{62}Ni including only cluster multiplicity one.

The DANCE array can separate events based on their multiplicity. Fig. 3 shows a cluster multiplicity (CM) analysis. Neighboring crystals, which have fired during an event, are combined to one single cluster in order to correct for cross talking. As a result, the cluster multiplicity is almost exactly the same as the cascade multiplicity (number of γ-rays).

The signal to background ratio in Fig. 3(a) shows that by considering cluster multiplicities 3 to 6, most of the real events from ^{62}Ni are being eliminated. Fig. 3(b) has only cluster multiplicity one and the increase of the signal over the background is evident. A similar analysis (in this particular case) allows distinguishing between s and p-wave resonances, by observing the average multiplicity of the resonances. For instance the resonance at 4.5 keV (s-wave) decrease by increasing the cluster multiplicity while the p-wave resonances increase by increasing the cluster multiplicity.

The neutrons scattered in the target make (n, γ) reactions with the barium inside the crystals, this scattering contribution is suppressed by the ^{6}LiH shell [5] and the remaining can be estimated using a C target, since the scattering cross section of the carbon is much greater than its capture cross section [6]. Fig. 4 shows in black the events of the ^{62}Ni target and the red curve is the contribution of the scattered neutrons captured inside the detector, which needs to be subtracted. This curve is normalized to the high energy region, because ^{135}Ba has a neutron capture Q-value of 9.1 MeV, which is clearly above the 6.8 MeV of ^{62}Ni(n,γ).

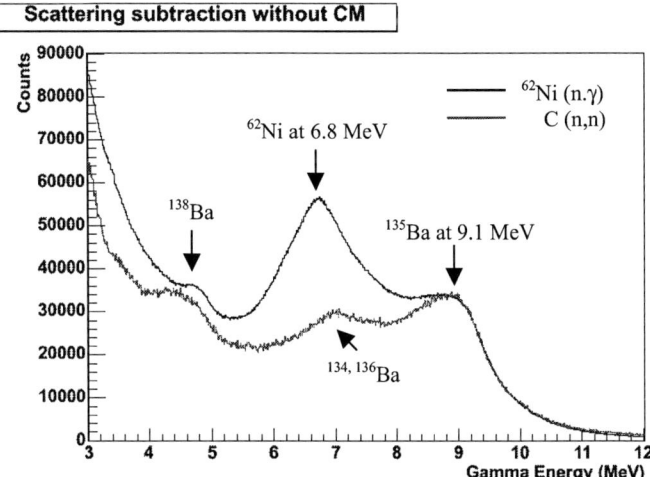

FIGURE 4. Scattering background subtraction

SUMMARY

The preliminary results of the ^{62}Ni(n,γ) measurement are in agreement with the JENDL 3.3 evaluation. The analysis to date shows that all background contributions can be subtracted using the Q-value information and therefore the direct capture component can be determined experimentally. Final cross sections calculations are in progress.

ACKNOWLEDGMENTS

This work has benefited from the use of the Los Alamos Neutron Science Center at the Los Alamos National Laboratory. This facility is funded by the US Department of Energy and operated by the University of California under Contract W-7405-ENG-36. The Colorado School of Mines group is funded via DOE Grant: DE-FG02-93ER40789.

REFERENCES

1. R. Reifarth, et al., "Nuclear Astrophysics at DANCE", International Conference on Nuclear Data for Science and Technology 2004, AIP Conference Proceedings (Santa Fe, NM, 2005).
2. Z.Y. Bao, et al., Nuclear Physics **A 621** (1997) 295c.
3. T. Rauscher and K. Guber, Phys. Rev. C **66** (2002) 028802.
4. R. Hatarik, et al., this proceedings
5. R. Reifarth, et al., Nucl. Instru. Meth. **A 531**, 530-543 (2004).
6. J.L. Ullmann, et al., this proceedings

^{102}Pd(n, γ) Cross Section Measurement Using DANCE

R. Hatarik[*,†], A.M. Alpizar-Vicente[*,†], T.A. Bredeweg[†], E.-I. Esch[†],
U. Greife[*], R.C. Haight[†], J.M. O'Donnell[†], R. Reifarth[†], R.S. Rundberg[†],
J.L. Ullmann[†], D.J. Vieira[†] and J.M. Wouters[†]

Colorado School of Mines, Golden, CO 80401
†*Los Alamos National Laboratory, Los Alamos, NM 87545*

Abstract. The neutron capture cross section of the proton rich nucleus ^{102}Pd was measured with the Detector for Advanced Neutron Capture Experiments (DANCE) at the Los Alamos Neutron Science Center. The target was a 2 mg Pd foil with 78% enriched ^{102}Pd. It was held by a 0.9 μm thick Mylar bag which was selected after comparing different thicknesses of Kapton and Mylar for their scattering background. To identify the contribution of the other Pd isotopes the data of a natural Pd sample was compared to the data of the ^{102}Pd enriched sample. A ^{12}C sample was used to determine the scattering background. The ^{102}Pd(n, γ) rate is of importance for the p-process nucleosynthesis.

Keywords: Nucleosynthesis, p-process, keV neutron capture
PACS: 28.20.Fc, 29.40.Vj, 29.25.Dz, 29.40.Wk

INTRODUCTION

The stable nuclei heavier than Fe are mainly produced via neutron capture reactions followed by β-decay sequences in either the s-process, which follows the valley of stability or in supernovae explosions with high neutron flux as a rapid sequence of neutron captures and β-decays close to the neutron drip line. The few isotopes on the proton rich side of the valley of stability cannot be produced in either s or r-process. Those isotopes have very low abundances and represent only 0.1% to 1% of the nuclides with $Z \geq 34$. These nuclides are mainly produced in the so called p-process were neutron rich nuclei produced in s or r-process are undergoing (γ, n) and (γ, α) reactions during a supernovae explosion. The p-process contribution to the nucleosynthesis is very small compared to s and r-process, which explains the low abundance of the proton rich elements that can only be produced by the p-process and are therefore called p only nuclei.

NEUTRON CAPTURE CROSS SECTIONS

The free neutrons produced via the (γ, n) reactions during the p-process will be captured by other isotopes. If the neutron separation energy is high enough, capture reactions are found to compete with photo induced reactions and can even dominate, which applies specially to even-even nuclei with masses $A < 130$. Therefore neutron capture cross sections are needed to better understand p-process nucleosynthesis.

The Detector for Advanced Neutron Capture Experiments (DANCE)

DANCE is a high-efficiency 4π array of 160 BaF$_2$ scintillation crystals. It is designed to calorimetrically detect γ-rays emitted after neutron capture on mg sized samples. DANCE is located on flight path 14 in the Manual Lujan Jr. Neutron Scattering Center at Los Alamos Neutron Science Center (LANSCE) [4]. In order to cover the full solid angle 162 crystals in four different shapes would be needed; two of them were left out to have space for the beam pipe, which reduces the covered solid angle to 3.52π [2].

Neutrons are produced via spallation reactions by an 800 MeV proton beam hitting a tungsten target with typical beam currents of 100 μA. DANCE is located 20.25 m from the tungsten target and the neutron energy is determined via time-of-flight. Because the proton beam has a low repetition rate of 20 Hz, neutron energies from 10 meV up to 200 keV can be measured.

Neutron capture on ^{102}Pd

As explained above proton rich isotopes have very low natural abundances (1% for ^{102}Pd). Highly enriched samples in larger quantities of these isotopes are therefore fairly expensive and measurements of the ^{102}Pd(n,γ) cross sections using enriched material have not been performed before. The high efficiency of the DANCE detector makes it possible to measure energy dependent cross sections on samples with masses of only milligrams.

A 2 mg ^{102}Pd foil with 78% enrichment was obtained as a loan from Indiana University Cyclotron Facility (IUCF). The isotopic composition of the target, the capture Q-values and Maxwellian averaged cross sections of each isotope are shown in Table 1.

TABLE 1. Isotopic composition of the 2 mg Palladium foil

%	Isotope	Q-value (n,γ)	$\sigma(kT=30\,\text{keV})$ *
77.89	^{102}Pd	7624.7 keV	375±118 mbarn [†]
5.49	^{104}Pd	7094.1 keV	289±29 mbarn
6.41	^{105}Pd	9562.6 keV	1200±60 mbarn
5.41	^{106}Pd	6538.7 keV	252±25 mbarn
3.51	^{108}Pd	6153.3 keV	203±20 mbarn
1.28	^{110}Pd	5750.4 keV	146±20 mbarn

* Maxwellian averaged cross section [1]
[†] estimated

CALIBRATIONS

Energy calibration

DANCE has 160 BaF$_2$ scintillation detectors and each of them needs to be energy calibrated. In order to do that, it was necessary to have an automated routine that goes through all 160 histograms of the individual crystals and fits the corresponding peaks of a source run. There are two data acquisition (DAQ) modes: segmented and continuous. Since the continuous mode uses only the integral over the slow component of the BaF$_2$ light output a separate calibration needed to be made for each DAQ mode. Table 2 shows the average resolutions obtained at the different source lines.

TABLE 2. Energy resolution (average) for continuous and segmented mode

Energies [keV] (SOURCE)	511 (^{22}Na)	898 (^{88}Y)	1173.2 (^{60}Co)	1274.5 (^{22}Na)	1332.5 (^{60}Co)	1836.1 (^{88}Y)
Avg. res. segm.	19.1%	12.4%	11.0%	10.2%	10.5%	8.4%
Avg. res. cont.	18.6%	13.5%	12.3%	11.6%	11.5%	11.2%

Time synchronization

During the offline analysis an event is defined as all crystals that detect a γ-ray within a coincidence window of 10 to 50 ns. By selecting all events that contain crystal 0 and then plotting the time difference of each crystal to crystal 0 vs crystal number, it is possible to obtain a plot like the ones shown in Figure 1. By fitting a Gaussian to the time distribution of each crystal, one can use the centroid of the Gaussian as static offset and subtract this offset from the corresponding time of flight value for each crystal. Figure 1B shows the result of this synchronization. The comparison of Figure 1A and 1B shows that as a result the coincidence window can be significantly reduced, which is specially useful at higher count rates because it reduces the probability of random coincidences.

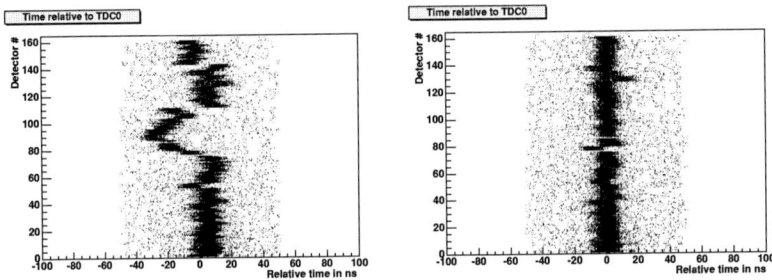

FIGURE 1. Time differences between detectors (A) before and (B) after synchronization

DATA ANALYSIS

The sample was held in a bag of thin plastic, which was glued to the aluminum target holder. Plastic foils have the advantage over metal foils (e.g. titanium) that they have very low neutron capture cross sections, i.e. the only background comes from scattered neutrons. To determine the scattering background of the backing, blank runs of target holders made with $0.9\,\mu$m and $1.5\,\mu$m thick Mylar as well as $8\,\mu$m thick Kapton foil were taken and compared with an empty run. The results showed that the Kapton foil had a significant higher scattering background than the Mylar foils and therefore the $0.9\,\mu$m thick mylar foil was chosen as backing for the Pd target.

Since the target consists of different isotopes, it is necessary to identify the corresponding isotope for each resonance (see Figure 2). This is possible because DANCE measures calorimetrically the emitted γ-energy after neutron capture which allows identifying the capturing isotope using the Q-value of the neutron capture reaction [3]. This is done by plotting the summed energy of each event in the neutron energy region of the resonance.

Neutrons scattered into the detector are seen via neutron capture on Ba isotopes. To reproduce the influence of scattering events a ^{12}C run was used. Above a total energy deposition of 8 MeV there are two components: scattering and ^{105}Pd(n,γ). To determine the contribution of these two components an energy spectrum following ^{105}Pd(n,γ) reactions was obtained by gating on neutron energies corresponding to the well known ^{105}Pd resonances. By making a fit in the region from 8.2 MeV to 9.6 MeV of the form $y = a \cdot C + b \cdot {}^{105}$Pd, where C stands for the energy spectrum gained from carbon and ^{105}Pd for the one gained from the ^{105}Pd resonances, the scattering contribution a and the ^{105}Pd contribution b can be determined. Figure 3 shows the result of such a fit.

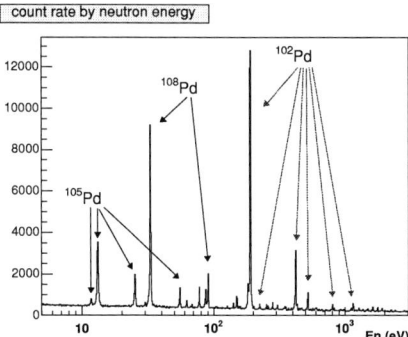

FIGURE 2. Count rate by neutron energy spectrum of the target. Every resonance has been identified using the Q-value information.

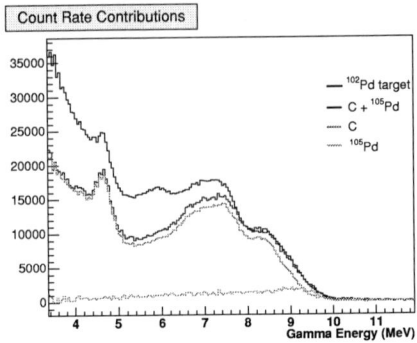

FIGURE 3. Subtracting scattering and ^{105}Pd contribution

SUMMARY

Before this experiment the only known resonance of ^{102}Pd was the one at 190 eV. As shown in Figure 2 there are at least 5 more resonances in the energy range from 200 eV to 2 keV. Figure 3 shows all contributions that need to be subtracted in order to get the capture cross section and it has been shown that all of them can be separated from the ^{102}Pd(n,γ) events.

ACKNOWLEDGMENTS

This work has benefited from the use of the Los Alamos Neutron Science Center at the Los Alamos National Laboratory. This facility is funded by the US Department of Energy and operated by the University of California under Contract W-7405-ENG-36. The Colorado School of Mines group is funded via DOE Grant: DE-FG02-93ER40789.

REFERENCES

1. Z.Y. Bao et al., Atomic Data and Nucl Data Tables **76** (2000)
2. J.L. Ullmann et al., this proceedings
3. R. Reifarth et al., "Nuclear Astrophysics at DANCE", International Conference on Nuclear Data for Science and Technology 2004, AIP Conference Proceedings (Santa Fe, NM 2005)
4. P.W. Lisowski, C.D. Bowman, G.J. Russel, S.A. Wender, Nucl. Sci. Eng. 106 (1990) 208.
5. D.A. Smith et al., Phys. Rev. C **65**, 024607 (2002).

Measurement of ^{139}La(n,γ) Cross Section

R. Terlizzi[19], U. Abbondanno[20], G. Aerts[7], H. Álvarez[35], F. Alvarez-Velarde[31], S. Andriamonje[7], J. Andrzejewski[26], P. Assimakopoulos[16], L. Audouin[12], G. Badurek[1], P. Baumann[10], F. Bečvář[6], E. Berthoumieux[7], F. Calviño[34], D. Cano-Ott[31], R. Capote[3,36], A. Carrillo de Albornoz[27], P. Cennini[37], V. Chepel[28], E. Chiaveri[37], N. Colonna[19], G. Cortes[33], A. Couture[41], J. Cox[41], M. Dahlfors[37], S. David[9], I. Dillmann[12], R. Dolfini[23], C. Domingo-Pardo[32], W. Dridi[7], I. Duran[35], C. Eleftheriadis[13], M. Embid-Segura[31], L. Ferrant[9], A. Ferrari[37], R. Ferreira-Marques[28], L. Fitzpatrick[37], H. Frais-Koelbl[3], K. Fujii[20], W. Furman[30], R. Gallino[22], I. Goncalves[28], E. Gonzalez-Romero[31], A. Goverdovski[29], F. Gramegna[18], E. Griesmayer[3], C. Guerrero[31], F. Gunsing[7], B. Haas[8], R. Haight[39], M. Heil[12], A. Herrera-Martinez[37], M. Igashira[25], S. Isaev[9], E. Jericha[1], Y. Kadi[37], F. Käppeler[12], D. Karamanis[16], D. Karadimos[16], M. Kerveno[10], V. Ketlerov[29,37], P. Koehler[40], V. Konovalov[30,37], E. Kossionides[15], M. Krtička[6], C. Lamboudis[13], H. Leeb[1], A. Lindote[28], I. Lopes[28], M. Lozano[36], S. Lukic[10], J. Marganiec[26], L. Marques[27], S. Marrone[19 a], P. Mastinu[18], A. Mengoni[3,37], P.M. Milazzo[20], C. Moreau[20], M. Mosconi[12], F. Neves[28], H. Oberhummer[1], S. O'Brien[41], M. Oshima[24], J. Pancin[7], C. Papachristodoulou[16], C. Papadopoulos[14], C. Paradela[35], N. Patronis[16], A. Pavlik[2], P. Pavlopoulos[11], L. Perrot[7], R. Plag[12], A. Plompen[5], A. Plukis[7], A. Poch[33], C. Pretel[33], J. Quesada[36], T. Rauscher[38], R. Reifarth[39], M. Rosetti[17], C. Rubbia[23], G. Rudolf[10], P. Rullhusen[5], J. Salgado[27], L. Sarchiapone[37], I. Savvidis[13], C. Stephan[9], G. Tagliente[19], J.L. Tain[32], L. Tassan-Got[9], L. Tavora[27], G. Vannini[21], P. Vaz[27], A. Ventura[17], D. Villamarin[31], M.C. Vincente[31], V. Vlachoudis[37], R. Vlastou[14], F. Voss[12], S. Walter[12], H. Wendler[37], M. Wiescher[41], K.Wisshak[12]

[1]*Atominstitut der Österreichischen Universitäten,Technische Universität Wien, Austria,* [2]*Institut für Isotopenforschung und Kernphysik, Universität Wien, Austria,* [3]*International Atomic Energy Agency (IAEA), Nuclear Data Section, Vienna, Austria* [4]*Fachhochschule Wiener Neustadt, Wiener Neustadt, Austria,* [5]*CEC-JRC-IRMM, Geel, Belgium,* [6]*Charles University, Prague, Czech Republic,* [7]*CEA/Saclay - DSM, Gif-sur-Yvette, France,* [8]*Centre National de la Recherche Scientifique/IN2P3 - CENBG, Bordeaux, France,* [9]*Centre National de la echerche Scientifique/IN2P3 - IPN, Orsay, France,* [10]*Centre National de la Recherche Scientifique/IN2P3 - IReS, Strasbourg, France,* [11]*Pôle Universitaire Léonard de Vinci, Paris La Défense, France,* [12]*Forschungszentrum Karlsruhe GmbH (FZK), Institut für Kernphysik, Germany,* [13]*Aristotle University of Thessaloniki, Greece,* [14]*National Technical University of Athens, Greece,* [15]*NCSR, Athens, Greece,* [16]*University of Ioannina, Greece,* [17]*ENEA, Bologna, Italy,* [18]*Istituto Nazionale di Fisica Nucleare(INFN), Laboratori Nazionali di Legnaro, Italy,* [19]*Dipartimento di Fisica and Istituto Nazionale di Fisica Nucleare, Bari, Italy,* [20]*Istituto Nazionale di Fisica Nucleare, Trieste, Italy,* [21]*Dipartimento di Fisica, Università di Bologna, and Sezione INFN di Bologna, Italy,* [22]*Dipartimento di Fisica, Università di Torino and Sezione INFN di Torino, Italy,* [23]*Università degli Studi Pavia, Pavia, Italy,* [24]*Japan Atomic Energy Research Institute, Tokai-mura, Japan,* [25]*Tokyo Institute of Technology, Tokyo, Japan,* [26]*University of Lodz, Lodz, Poland,* [27]*Instituto Tecnológico e Nuclear(ITN), Lisbon, Portugal,* [28]*LIP - Coimbra & Departamento de Fisica da Universidade de Coimbra, Portugal,* [29]*Institute of Physics and Power Engineering, Kaluga region, Obninsk, Russia,* [30]*Joint Institute for Nuclear Research, Frank Laboratory of Neutron Physics, Dubna, Russia,* [31]*Centro de Investigaciones Energeticas Medioambientales y Technologicas, Madrid, Spain,* [32]*Istituto de Fisica Corpuscolar, CSIC-Universidad de Valencia, Spain,* [33]*Universitat Politecnica de Catalunya, Barcelona, Spain,* [34]*Universidad Politecnica de Madrid, Spain,* [35]*Universidade de Santiago de Compostela, Spain,* [36]*Universidad de Sevilla, Spain,* [37]*CERN, Geneva, Switzerland,* [38]*Department of Physics and Astronomy - University of Basel, Basel, Switzerland,* [39]*Los Alamos National Laboratory, New Mexico, USA,* [40]*Oak Ridge National Laboratory, Physics Division, Oak Ridge, USA,* [41]*University of Notre Dame, Notre Dame, USA,*

[a] Corresponding author. Address: via Orabona 4, I-70125 Bari (Italy), Tel. +390805442511, Fax +390805442470, e-mail: stefano.marrone@ba.infn.it .

Abstract. We measured the neutron capture cross section of ^{139}La relative to ^{197}Au in the energy range of 0.6 eV to 9 keV at n_TOF, the neutron time-of-flight facility at CERN. After a description of the experimental apparatus, we discuss data analysis procedures. The data were fitted using R-matrix formalism to extract resonance parameters which, in turn, were used to calculate average level spacings $D_0 = 268 \pm 22$ eV and $D_1 < 250$ eV, and neutron strength functions $S_0 = (0.79 \pm 0.03) \times 10^{-4}$ and $S_1 = (0.73 \pm 0.05) \times 10^{-4}$ for s- and p-wave resonances. The data also were used to determine Maxwellian-averaged neutron capture cross sections which, in turn, were used to calculate the ^{139}La abundance synthesized in a stellar model of the *main* component of the s process.

Keywords: ^{139}La, neutron capture, level density, neutron strength function, s process.
PACS: 25.40.Lw, 26.20.+f, 27.60.+j, 97.10.Cv .

INTRODUCTION

Isotopes with closed neutron shells such as ^{139}La (N=82) are known as neutron magic isotopes, and are of special importance in nuclear physics. Their nuclear structures are characterized by low level densities and small neutron capture cross sections. For these reasons, experimental information on their neutron induced reaction cross sections is very useful for both basic physics (microscopic calculation of quantum many-body systems and nuclear astrophysics) as well as nuclear technology applications.

Solar lanthanum predominantly (~ 74%) was produced by the *main* component of the slow neutron capture process (s process) whereas the remainder originated in the rapid neutron capture process (r process). The nuclide ^{139}La belongs to the second s-process peak (Ba to Nd); consequently, it is particularly suited for monitoring the heavy s-process abundances from Ba up to Pb. Moreover, this element is essentially monoisotopic (99.9% of solar lanthanum is ^{139}La) and it is easy to observe in stellar spectra. To fully exploit the peculiarities of these features, accurate measurements of the Maxwellian-averaged neutron capture cross sections (MACS) are needed at temperatures typical (kT = 8-26 keV) of the astrophysical site where the *main s* process takes place.

Previous determinations of the ^{139}La(n,γ) MACS [1-3] are in substantial disagreement (up to 40%). These large discrepancies motivated us to perform a new measurement at the innovative neutron time-of-flight facility (n_TOF). The long flight path, high neutron flux, low duty factor, and low background, together with a neutron-capture apparatus optimized for this facility, allow accurate measurements over a wide energy range for both radioactive samples as well as nuclides having low capture cross sections such as ^{139}La.

EXPERIMENTAL APPARATUS

At n_TOF, neutrons in the energy range from 1 eV to 250 MeV are generated by spallation when a 20 GeV/c proton beam from the CERN PS accelerator strikes a massive ^{nat}Pb target. The detector area is located at the end of the TOF tunnel, at 187.5 m from the spallation target. Massive concrete and iron shields, a sweeping magnet, and two collimators provide effective background suppression and a suitably shaped neutron beam [4].

The γ-rays from neutron-capture reactions were detected using two C_6D_6 scintillators. Each detector had an active volume of ~ 1000 cm^3 and the scintillator was housed in a carbon fiber container directly coupled to an EMI 9823 QKA phototube [5]. The detectors were symmetrically positioned on either side of the carbon fiber sample changer, 9.8 cm upstream with respect to the sample and 3 cm from the beam axis. Five samples (^{nat}La, Al-can, ^{nat}C, ^{208}Pb and ^{197}Au) were remotely cycled through the beam to perform sample-in as well as background and reference measurements.

The neutron flux was monitored using Si surface-barrier detectors (SiMon) placed outside the neutron beam, which viewed a ^{6}Li-loaded foil placed in the beam ~3 m upstream of the main sample position. The SiMon detected tritons and α-particles emitted from ^{6}Li$(n,\alpha)^3$H reaction and was used to determine the relative neutron flux between the various sample-in, background, and reference measurements [6].

Data were acquired using an innovative system based on fast analogue-to-digital converter (FADC) modules capable of 1 GSample/s sampling rate. The FADC system recorded the entire waveform of each detector for the duration of each neutron pulse. The raw data, stored in a disk repository, were analyzed off-line to extract time-of-flight, pulse-height, and particle identification parameters for each event [7].

DATA ANALYSIS

Because of the low efficiency of the C_6D_6 detectors, the probability of detecting a capture event depends on the multiplicity of the γ-cascade as well as on the energy of the emitted γ-rays. Therefore, the Pulse Height Weighting Technique (PHWT) was applied to extract accurate absolute neutron-capture cross sections. This method consists of modifying by software the detectors' responses so that their efficiencies do not depend on the details of the cascade but only on the total capture energy (which is known). The weighting functions, *WF*, necessary for such modification were calculated via Monte Carlo simulations. Corrections for dead-time and coincidence events also were applied. A detailed description of the *WF* determination at n_TOF and their use to extract capture yields is reported in the Ref. [8].

Three background components affect capture measurements at n_TOF: 1) The "environmental" background (sample independent for stable samples), 2) The contribution of in-beam γ-rays (sample related), and 3) A component due to scattered neutrons (sample related). Each background contribution is quantitatively estimated by means of dedicated measurements, for example, empty Al-can for environmental background, ^{208}Pb for in-beam γ-rays, and ^{nat}C for scattered neutrons. Background-subtracted yields for the main samples (^{nat}La and ^{197}Au) were calculated by subtracting measured background yields weighted with relative normalization factors extracted from Monte Carlo simulations of the raw spectra. More details on the background subtraction procedures are given in the reference [9]. The lanthanum raw capture yield together with the total background is shown in the left panel of Figure 1.

Several different techniques have been used to obtain an accurate determination of the total neutron flux at n_TOF. Measurements were performed with fission chambers, SiMon, C_6D_6, and Au activation. However, in actual capture cross section measurements, the target diameter is smaller than the neutron beam size and therefore only a fraction of the beam impinges on the capture sample. To estimate this fraction, the flux is normalized according to the standard ^{197}Au(n,γ) cross section. Applying this procedure, the extracted cross section for the sample under study is independent of the total flux determination and relies only on the accuracy of the gold capture cross section, which is effectively used as standard. This cross section is well known in the whole energy range and is a standard for the resonance at 4.9 eV and in the neutron energy range from 10 - 100 keV.

RESULTS AND IMPLICATIONS

The measured capture cross section of ^{139}La was fitted using the R-matrix code SAMMY [10] to extract resonance parameters. Effects such as Doppler broadening, intrinsic time-of-flight resolution, self-shielding, and multiple scattering are included in the SAMMY calculations. Capture strengths extracted from our data using SAMMY are lower, in general, than previously reported [1-2] by amounts ranging from 3% to 12%. In addition, we have identified three new resonances that were not reported in previous works.

A preliminary statistical analysis of our resonance parameters indicates that virtually all s-wave resonances were detected up to 9 keV and that the sequence of p-wave resonances is complete up to 2 keV (see right panel of Fig. 1). Our parameters yield average level spacings of $D_0 = 268 \pm 22$ eV and $D_1 < 250$ eV, and neutron strength functions $S_0 = (0.79 \pm 0.03) \times 10^{-4}$ and $S_1 = (0.73 \pm 0.05) \times 10^{-4}$ for s- and p-wave resonances, respectively.

We have used our data calculate the ^{139}La(n,γ) MACS for various stellar temperatures. The value obtained at $kT = 25$ keV is consistent to within a few percent with the recent activation measurement reported by O'Brien et al. [3]. Using our new MACS in a model of the s process in thermally pulsing low-mass Asymptotic Giant Branch stars [11] indicates that the s-process *main* component can account for 74.2% of the lanthanum solar abundance. This result, together with a more accurate observation of lanthanum in stellar spectra opens a new perspective for studying the chemical evolution of the Galaxy.

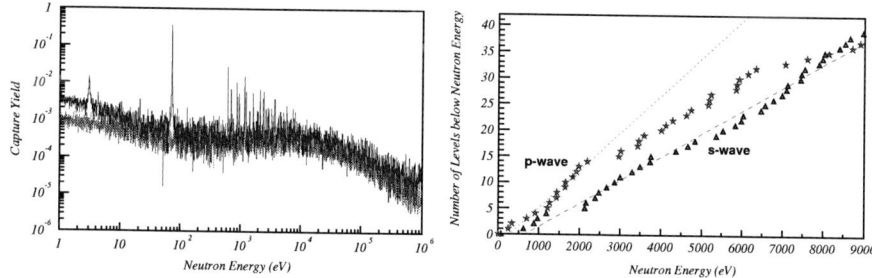

FIGURE 1. In left panel, the raw capture yield of the lanthanum sample (black line) is drawn together with the total background (red line). In right panel, the cumulative number of s-wave (blue triangles) and p-wave (red stars) levels is illustrated. Dashed and dotted lines represent the best fit with a straight line of the experimental points. Note that p-wave level sequence seems to be complete up to 2 keV.

ACKNOWLEDGMENTS

This work was supported partly by the EC under contract FIKW-CT-2000-00107 and by the funding agencies of the participating institutes.

REFERENCES

1. Musgrove, B. J. Allen, and R. L. Macklin, *Australian Journal of Physics*, **30**, 599 (1977).
2. Y. Nakajima et al., *Journal of Nuclear Science and Technology*, **20**, 183 (1983).
3. S. O'Brien et al., *Physical Review C*, **68**, 35801 (2003).
4. C. Borcea et al., *Nuclear Instruments and Methods A*, **513**, 524 (2003).
5. R. Plag et al., *Nuclear Instruments and Methods A*, **496**, 425 (2003).
6. S. Marrone et al., *Nuclear Instruments and Methods A*, **517**, 389 (2004).
7. U. Abbondanno et al., *Nuclear Instruments and Methods A*, **538**, 692 (2005).
8. U. Abbondanno et al., *Nuclear Instruments and Methods A*, **521**, 454 (2004).
9. R. Terlizzi, *n_TOF Internal Note Ba 07/2004*, CERN (2004); available on the website www.cern.ch/n_TOF .
10. N. M. Larson, *ORNL/TM-2000/252*, Oak Ridge National Laboratory (2000).
11. M. Busso, R. Gallino, and J. Wasserburg, *Annual Review of Astronomy and Astrophysics*, **37**, 239 (1999).

Measurement of the resonance capture cross section of ^{204,206}Pb and termination of the *s*-process

C. Domingo-Pardo[32(1], S. O'Brien[41], U. Abbondanno[20], G. Aerts[7], H. Álvarez[35], F. Alvarez-Velarde[31], S. Andriamonje[7], J. Andrzejewski[26], P. Assimakopoulos[16], L. Audouin[12], G. Badurek[1], P. Baumann[10], F. Bečvář[6], E. Berthoumieux[7], F. Calviño[34], D. Cano-Ott[31], R. Capote[3,36], A. Carrillo de Albornoz[27], P. Cennini[37], V. Chepel[28], E. Chiaveri[37], N. Colonna[19], G. Cortes[33], A. Couture[41], J. Cox[41], M. Dahlfors[37], S. David[9], I. Dillmann[12], R. Dolfini[23], W. Dridi[7], I. Duran[35], C. Eleftheriadis[13], M. Embid-Segura[31], L. Ferrant[9], A. Ferrari[37], R. Ferreira-Marques[28], L. Fitzpatrick[37], H. Frais-Koelbl[3], K. Fujii[20], W. Furman[30], C. Guerrero[31], I. Goncalves[28], R. Gallino[22], E. Gonzalez-Romero[31], A. Goverdovski[29], F. Gramegna[18], E. Griesmayer[3], F. Gunsing[7], B. Haas[8], R. Haight[39], M. Heil[12], A. Herrera-Martinez[37], M. Igashira[25], S. Isaev[9], E. Jericha[1], Y. Kadi[37], F. Käppeler[12], D. Karamanis[16], D. Karadimos[16], M. Kerveno[10], V. Ketlerov[29,37], P. Koehler[40], V. Konovalov[30,37], E. Kossionides[15], M. Krtička[6], C. Lamboudis[16], H. Leeb[1], A. Lindote[28], I. Lopes[28], M. Lozano[36], S. Lukic[10], J. Marganiec[26], L. Marques[27], S. Marrone[19], P. Mastinu[18], A. Mengoni[3,37], P.M. Milazzo[20], C. Moreau[20], M. Mosconi[12], F. Neves[28], H. Oberhummer[1], M. Oshima[24], J. Pancin[7], C. Papachristodoulou[16], C. Papadopoulos[14], C. Paradela[35], N. Patronis[16], A. Pavlik[2], P. Pavlopoulos[11], L. Perrot[7], R. Plag[12], A. Plompen[5], A. Plukis[7], A. Poch[33], C. Pretel[33], J. Quesada[36], T. Rauscher[38], R. Reifarth[39], M. Rosetti[17], C. Rubbia[23], G. Rudolf[10], P. Rullhusen[5], J. Salgado[27], L. Sarchiapone[37], I. Savvidis[13], C. Stephan[9], G. Tagliente[19], J.L. Tain[32], L. Tassan-Got[9], L. Tavora[27], R. Terlizzi[19], G. Vannini[21], P. Vaz[27], A. Ventura[17], D. Villamarin[31], M.C. Vicente[31], V. Vlachoudis[37], R. Vlastou[14], F. Voss[12], S. Walter[12], H. Wendler[37], M. Wiescher[41], K. Wisshak[12]

The n_TOF Collaboration

[1]*Atominstitut der Österreichischen Universitäten,Technische Universität Wien, Austria* – [2]*Institut für Isotopenforschung und Kernphysik, Universität Wien, Austria* – [3]*International Atomic Energy Agency (IAEA), Nuclear Data Section, Vienna, Austria* – [4]*Fachhochschule Wiener Neustadt, Wiener Neustadt, Austria* – [5]*CEC-JRC-IRMM, Geel, Belgium* – [6]*Charles University, Prague, Czech Republic* – [7]*CEA/Saclay - DSM, Gif-sur-Yvette, France* – [8]*Centre National de la Recherche Scientifique/IN2P3 - CENBG, Bordeaux, France* – [9]*Centre National de la Recherche Scientifique/IN2P3 - IPN, Orsay, France* – [10]*Centre National de la Recherche Scientifique/IN2P3 - IReS, Strasbourg, France* – [11]*Pôle Universitaire Léonard de Vinci, Paris La Défense, France* – [12]*Forschungszentrum Karlsruhe GmbH (FZK), Institut für Kernphysik, Germany* – [13]*Aristotle University of Thessaloniki, Greece* – [14]*National Technical University of Athens, Greece* – [15]*NCSR, Athens, Greece* – [16]*University of Ioannina, Greece* – [17]*ENEA, Bologna, Italy* – [18]*Istituto Nazionale di Fisica Nucleare(INFN), Laboratori Nazionali di Legnaro, Italy* – [19]*Istituto Nazionale di Fisica Nucleare, Bari, Italy* – [20]*Istituto Nazionale di Fisica Nucleare, Trieste, Italy* – [21]*Dipartimento di Fisica, Università di Bologna, and Sezione INFN di Bologna, Italy* – [22]*Dipartimento di Fisica, Università di Torino and Sezione INFN di Torino, Italy* – [23]*Università degli Studi Pavia, Pavia, Italy* – [24]*Japan Atomic Energy Research Institute, Tokai-mura, Japan* – [25]*Tokyo Institute of Technology, Tokyo, Japan* – [26]*University of Lodz, Lodz, Poland* – [27]*Instituto Tecnológico e Nuclear(ITN), Lisbon, Portugal* – [28]*LIP - Coimbra & Departamento de Fisica da Universidade de Coimbra, Portugal* – [29]*Institute of Physics and Power Engineering, Kaluga region, Obninsk, Russia* – [30]*Joint Institute for Nuclear Research, Frank Laboratory of Neutron Physics, Dubna, Russia* – [31]*Centro de Investigaciones Energeticas Mediomabientales y Technologicas, Madrid, Spain* – [32]*Instituto de Física Corpuscular, CSIC-Universidad de Valencia, Spain* – [33]*Universitat Politecnica de Catalunya, Barcelona, Spain* – [34]*Universidad Politecnica de Madrid, Spain* – [35]*Universidade de Santiago de Compostela, Spain* – [36]*Universidad de Sevilla, Spain* – [37]*CERN, Geneva, Switzerland* – [38]*Department of Physics and Astronomy - University of Basel, Basel, Switzerland* – [39]*Los Alamos National Laboratory, New Mexico, USA* – [40]*Oak Ridge National Laboratory, Physics Division, Oak Ridge, USA* – [41]*University of Notre Dame, Notre Dame, USA*

Abstract. The neutron capture cross sections of ^{204}Pb and ^{206}Pb have been measured at the CERN n_TOF installation using the time of flight method with the pulse height weighting technique. In a preliminary analysis of ^{204}Pb we have determined by first time the capture cross sections for two strong s-wave resonances below 2.5 keV, which seem to enhance the Maxwellian Averaged Cross Section at $kT = 5$ keV by about a factor of two with respect to the value assumed so far [1, 2]. In ^{206}Pb we have determined capture cross sections for a large number (97) of resonances, which were not reported in the two previous capture measurements [3, 4]. We discuss preliminary implications of the new cross sections in the stellar nucleosynthesis of the Pb isotopes.

INTRODUCTION

The neutron capture cross sections of ^{204,206}Pb have been measured with high resolution at the n_TOF facility of CERN (Geneva) [5].

The neutron capture cross section of ^{204}Pb is of particular relevance in astrophysics, since this s-only isotope constrains the branching of the reaction path at $A = 204$. The predicted abundance of this isotope is very sensitive to the details of the stellar model. Therefore, an accurate value of its capture cross section provides an important test and constraint for astrophysical models.

^{206}Pb is representative of the entire reaction flow, thus its abundance scales with the inverse of the cross section, becoming practically independent of the specific stellar model employed. Consequently, its r-process abundance is well determined as the subtraction of the s-abundance from the total observed abundance. The r-process abundance is important as a constraint for the U/Th-cosmochronometer.

Because of the occurrence (26%) of ^{206}Pb in natural lead, its capture cross section is also important for the design of hybrid reactors based on Pb/Bi eutectic spallation sources.

EXPERIMENT

At n_TOF, the capture cross section of rare materials and radioactive samples have been measured by using a total absorption calorimeter made of 40 BaF$_2$ crystals (see e.g. Ref. [6]). For the measurement of the Pb-isotopes however, given their almost neutron magic shell, radiative capture is hindered and the elastic scattering channel can be orders of magnitude higher. In this situation a simpler experimental setup of C$_6$D$_6$ detectors becomes a much better approach in order to avoid backgrounds related with sample-scattered neutrons. The n_TOF C$_6$D$_6$ based detection system has been optimized in terms of neutron sensitivity by developing a very light detector [7]. Further improvements in this experiment are related with i) the 125° setup used to minimize angular distribution effects, ii) a more accurate determination of the weighting functions (WF) and iii) the small and highly enriched samples used at n_TOF (99.7% and 99.8%

---

[1] Corresponding Author: address IFIC, Apdo. Correos 22085, E-46071 Valencia, Spain. Tel. +34-963543502, Fax. +34-963543488, e-mail: cesar.domingo.pardo@cern.ch

respectively for ^{204}Pb and ^{206}Pb). These samples had cylindrical shape with thicknesses of 1.18 mm and 2.27 mm respectively, both of them with a diameter of 20 mm.

ANALYSIS AND RESULTS

The Pulse Height Weighting Technique (PHWT) [8] is applied to the C_6D_6 capture data in order to achieve a cascade detection probability independent of the particular deexcitation path. It has been shown in previous works [9, 10], that detailed MC-simulations of the experimental setup are necessary to obtain accurate weighting functions. We used the GEANT4 simulation toolkit [11] to model the n_TOF experimental setup. The geometric details of the MC simulation can be appreciated in Fig. 1 of Ref. [12].

Thanks to the improvements in the experimental setup (see prev. section), we did not need to apply any correction due to scattered neutron induced background.

The effect of the electronic threshold, as well as some other smaller effects like γ-ray summing, have been accounted for by performing MC-simulations of the capture cascades, in the same manner as in previous works [10, 12]. By means of these MC simulations, we also checked that the systematic uncertainty of the calculated WF was absolutely negligible.

The largest yield corrections were due to the strong angular distribution effects in ^{206}Pb. The MC-calculated angular distribution correction factors vary between 1% and 10% for resonances of same spin and parity. However, for most of the resonances analysed in ^{206}Pb the decay schemes are unknown and, despite the 125° setup used at n_TOF, the uncertainty in the capture areas raises up to 10-20% due to this effect.

An R-matrix analysis of the experimental yield, corrected by the effects mentioned above, was performed with the SAMMY code [13]. We followed basically the same procedure as in a previous work [12].

In the case of ^{204}Pb, we were able to analyse 137 resonances in the energy range from 480 eV up to 86 keV. From these capture data we have determined a Maxwellian averaged capture cross section (MACS) at $kT = 5$ keV which is about 80% larger than the $<\sigma>_{5keV} = 168$ mb provided by Horen et al. from a previous measurement (see Fig.5 in Ref. [1]). It appears that in the later reference the first two s-wave resonances (measured in transmission, but not in capture) were not taken into account. For these two s-wave resonances at 1.686 keV and 2.48 keV, we obtain also remarkably higher capture cross sections (50% and 30% respectively) than those derived from the resonance parameters determined by Horen et al. [1].

In ^{206}Pb (see Fig. 1), resolved resonances could be observed up to 400 keV, determining capture cross sections for 97 resonances, for which no capture data has been reported in Refs. [3, 4].

In our preliminary analysis, for both ^{204}Pb and ^{206}Pb, we obtain capture areas which differ remarkably with respect to the previous capture data [1, 3, 4] (see Fig. 1). Relative differences scatter up to 35% and sometimes substantially more. Several of the experimental effects mentioned above can contribute to such discrepancies. Work is in progress in order to understand these differences.

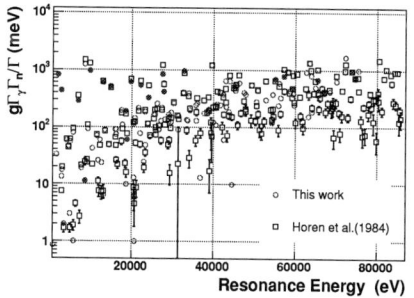 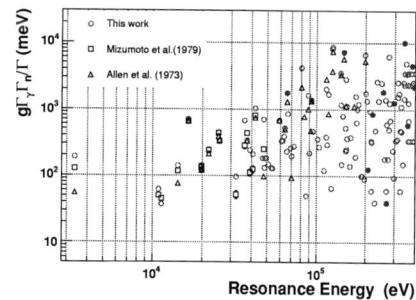

FIGURE 1. (Left) Comparison of the ^{204}Pb radiative kernels measured at n_TOF versus the previous experiment [1]. (Right) Comparison of the ^{206}Pb radiative kernels measured at n_TOF versus the two previous experiments [3, 4]. Filled symbols designate s-wave resonances.

TERMINATION OF THE S-PROCESS

In Fig. 2 the ^{204}Pb MACS calculated from the resolved resonances measured here is compared versus the MACS given by Horen et al. [1]. The recommended values for nucleosynthesis calculations [2] are based in the later experiment.

At the lower stellar temperature of $kT = 5$ keV our value is almost a factor of two larger (see prev. section). Note that above 25 keV both curves start to deviate, because in Ref. [1] also average capture cross section values beyond 85 keV were taken into account (not included yet in our case).

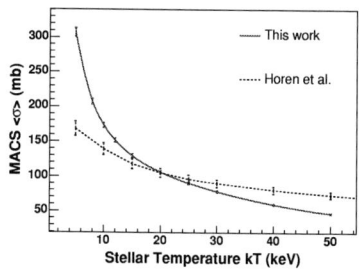

FIGURE 2. MACS calculated from the resolved resonances of ^{204}Pb measured in this work, compared versus the previous measurement due to Horen et al. [1].

Qualitatively, a higher cross section would reduce the abundance of ^{204}Pb, consequently enhancing the abundance of the next isotopes in the s-process path. However, ^{204}Pb is supposed to be synthesized almost exclusively in the recurrent He-shell flashes of thermally pulsing asymptotic giant branch (AGB) stars [14]. During these flashes, the ^{22}Ne(α,n) neutron source is activated at the higher stellar temperature of $kT \sim 23$ keV. At this temperature the MACS obtained in the present work for ^{204}Pb is in agreement with the value derived in Ref. [1]. Therefore, despite the large discrepancies found with respect to the previous experiment, our new capture cross section, according to this stellar model, will not modify substantially the calculated abundance of the s-only isotope ^{204}Pb.

The s-process abundance of ^{206}Pb is produced mostly between He-shell flashes [14], when the ^{13}C(α,n) source operates at the lower stellar temperature of $kT \sim 8$ keV. At this temperature, the MACS obtained in this preliminary analysis of the n_TOF capture data seems to verify the recommended value $< \sigma >_{5keV} = 25.5 \pm 1.3$ mb. Therefore, the synthesis calculations of ^{206}Pb might not be strongly affected either.

CONCLUSIONS

The neutron capture cross sections of ^{204,206}Pb have been measured at the CERN n_TOF installation with high resolution and with improved features in the experimental setup and in the data reduction process. Despite the notable differences found with respect to the previous experiments [1, 3, 4], the MACS in the relevant kT ranges for the synthesis of ^{204,206}Pb (23 keV and 8 keV respectively) agree surprisingly well. Therefore, important changes in the calculated s-process abundances of these isotopes are not expected. Nevertheless, detailed s-process calculations are now needed for a quantitative study of these new results.

ACKNOWLEDGMENTS

This work was supported partly by the EC under contract FIKW-CT-2000-00107, by the Spanish Ministry of Science and Technology (FPA2001-0144-C05) and by the funding agencies of the participant institutes.

REFERENCES

1. D.J. Horen et al., *Phys. Rev. C* **29** No.6 (1984) 29.
2. Z.Y. Bao et al., *At. Data and Nuc. Data Tab.* **76** No. 1 (2000) 70.
3. B.J. Allen et al., *Phys. Rev. C*, **8** No.4 (1973) 1504.
4. M. Mizumoto et al., *Phys. Rev. C* **19** No.2 (1979) 335.
5. The n_TOF Collaboration, CERN/SPSC 99-8, SPSC/P 310.
6. This Conference Proceedings, D. Cano-Ott et al.
7. R. Plag et al., *Nuc. Instr. and Meth. A* **496** (2003) 425.
8. R. Macklin et al., *Phys. Rev.*, **159** No.4 (1967) 1007.
9. J.L. Tain et al.,*Jour. of Nuc. Sci. and Tech.*, Sup. 2 (2002) 689.
10. U. Abbondanno et al., *Nuc. Instr. and Meth. A*, **521** (2004) 454.
11. S.Agostinelli et al.,*Nuc. Instr. and Meth. A* 506 (2003) 250.
12. C. Domingo-Pardo et al.,*Proc. Int. Conf. Nuc. Data Sci. Tech.*, Part 2 (2004) 1521.
13. N.M. Larson, ORNL/TM-9179, Oak Ridge NL, 2000.
14. U. Ratzel et al., *Phys. Rev. C* **70** 065803 (2004).

NEUTRON CAPTURE REACTIONS

The Radiative Strength Function Using the Neutron-Capture Reaction on ^{151,153}Eu

U. Agvaanluvsan,* A. Alpizar-Vicente,† J.A. Becker,* F. Bečvář,‡
T.A. Bredeweg,** R. Clement,* E. Esch,** C.M. Folden, III¶ R. Hatarik,†
R.C. Haight,** D.C. Hoffman,¶ M. Krtička,‡ R.A. Macri,*
G.E. Mitchell,§ H. Nitsche,¶ J.M. O'Donnell,** W. Parker,* R. Reifarth,**
R.S. Rundberg,** J.M. Schwantes,** S.A. Sheets,**§ J.L. Ullmann,**
D.J. Vieira,** J.B. Wilhelmy,** P. Wilk,* J.M. Wouters,** C.Y. Wu*

*Lawrence Livermore National Laboratory, Livermore, CA 94551
†Colorado School of Mines, Golden, CO 80401
**Los Alamos National Laboratory, Los Alamos, NM 87545
‡Charles University in Prague, CZ-180 00 Prague 8, Czech Republic
§North Carolina State University, Raleigh, NC 27695 and
Triangle University Nuclear Laboratory, Durham, NC 27708
¶University of California Berkeley and Lawrence Berkeley National Laboratory, Berkeley, CA 94720

Abstract. Radiative strength functions in ^{152,154}Eu nuclei for γ-ray energies below 6 MeV have been investigated. Neutron capture for incident neutron energies <1eV up to 100 keV has been measured for ^{151,153}Eu targets. Properties of γ decay of neutron resonances in ^{152,154}Eu nuclei are examined. The results of measurements are compared to outcome of simulation of γ cascades based on various models for the radiative strength function. Comparison between experimental data and simulation suggests existence of the low-energy resonance in these two nuclei.

Keywords: neutron capture, radiative strength function, resonances.
PACS: 24.30.-v, 24.30.Gz, 24.30.Gd

INTRODUCTION

Nuclear radiative strength functions are known best around the maximum of the giant electric dipole resonance, 10-20 MeV γ-ray energy. Extrapolation of the radiative strength functions from the giant dipole resonance region to lower γ-ray energy region is problematic at best. The radiative strength function, also known as the γ-ray or photon strength function is defined [1] as

$$f_{XL}(E_\gamma) = \frac{\langle \Gamma_{\gamma if} \rangle}{E_\gamma^{2L+1} D_i}, \quad (1)$$

where $\langle \Gamma_{\gamma if} \rangle$ is the average partial radiative width for transition from an initial state i to a final state f, E_γ is the energy of transition, and D_i is the level spacing of the initial states. Most experimental data for the RSF is from the study of photoabsorption cross sections [2, 3]. Other methods involving radiative neutron capture such as the spectrum fitting method [4] and the two-step cascade method [5, 6] provide additional information about the RSF for high-energy transitions. The investigation of primary γ-rays of different multipolarities [7, 8] and the sequential extraction method [9] are also used for obtaining experimental information for the RSF.

One of the most commonly used theoretical models is the Lorentzian function for the Giant Electric Dipole Resonance (GEDR). For deformed nuclei, experimental data can be fit as a superposition of two Lorentzians. The Giant Magnetic Dipole (or Spin-flip) Resonance (GMDR) is also described by the Lorentzian [7, 10]. Although the Lorentzian model describes data at higher transition energies, it does not adequately describe the data for low energy γ-rays [11]. To explain the non-zero limit of the RSF for $E_\gamma \to 0$, models based on the Fermi liquid theory were developed [12, 13] that give an energy and temperature dependent damping width. The strength function model developed by Kadmenskiĭ, Markushev, and Furman (KMF) is given by

$$f_{E1}(E_\gamma) = \frac{1}{3\pi^2\hbar^2 c^2} \frac{\lambda \sigma_{E1} \Gamma_{E1}^2 (E_\gamma^2 + 4\pi^2 T^2)}{E_{E1}(E_\gamma^2 - E_{E1}^2)^2}, \qquad (2)$$

where λ is the Landau-Migdal parameter, and Γ_{E1} is energy and temperature dependent. In addition to the GEDR and GMDR, the low energy ($E_\gamma \sim 3$ MeV) mode with the Lorentzian shape is considered for the total RSF.

The experimental indication of the low energy mode has been observed in several types of measurements. From a set of nuclear resonance fluorescence (NRF) [14] experiments, the existence of the so-called M1 scissors mode of deformed nuclei is clearly established. In another set of experiments using the so-called Oslo method, the level density and radiative strength function are obtained simultaneously [15]. In deformed rare earth nuclei studied with the Oslo-method, a resonance near $E_\gamma \sim 3$ MeV is observed and is identified as a pygmy resonance [16, 17]. The A-dependences of the resonance energy from above two sets of experiments do not agree. In addition, the total strengths of the pygmy resonance observed in the Oslo-type experiments are larger than the SM strength observed by the NRF. Therefore, it is unclear whether the same physics phenomenon explains both effects fully. In the so called two-step cascade (TSC) method, de-excitation of nucleus following the thermal neutron capture is studied [18] where the SM resonance is also observed. However the strength is again greater than that of observed in the NRF experiments. Recently it has been established from a TSC experiment that SM resonances are built on *all* excited levels in ^{163}Dy, including the levels in the quasicontinuum [19]. DANCE data reveal similar phenomenon in the statistical γ-ray decay cascade of Eu nuclei.

EXPERIMENTAL DESCRIPTION

The experiment was performed using the DANCE array located at the flight path 14 at Lujan Center at the Los Alamos Neutron Science Center (LANSCE). The DANCE array is a 4π γ-ray calorimeter that consists of 160 barium fluoride crystals. The neutron beam with E_n = 10 meV - 100 keV with the repetition rate 20 Hz was provided by the spallation neutron source at LANSCE. The flight path length is 20 meter. The neutron energy is determined by the time-of-flight technique. The stable ^{151,153}Eu targets with thicknesses 0.836 ± 0.040 mg/cm^2 and 1.06 ± 0.05 mg/cm^2 and enrichment 96.83 % and 98.76 %, respectively, were used. Both targets were mounted on a Be backing. The DANCE data acquisition system relies on waveform digitization. The description of the DAQ is given in reference [20]. The details of various background and methods of suppression are described in reference [21]. Event by event data analysis was performed offline. Taking advantage of the high granularity of the DANCE detector, events for each multiplicity can be separated as function of neutron energy. Various neutron energy regions and spectrum of events for each multiplicity can be selected as shown in Fig. 1 for the two resonances in ^{152}Eu.

FIGURE 1. The first two resonances in the ^{151}Eu(n,γ) reaction. Events with different γ-ray multiplicities can be separated for each resonance. The most dominant are multiplicities 3 and 4.

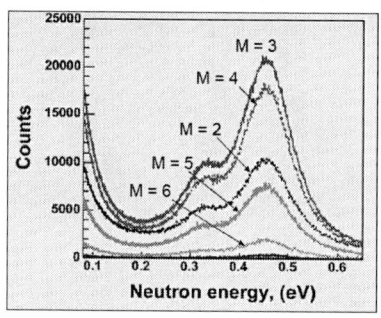

IV. SIMULATION AND MODELING

Experimental data obtained from the DANCE detector can be compared with an outcome from simulations performed by adjacent DICEBOX/GEANT simulations. The Monte Carlo code DICEBOX [18] generates γ-ray cascades initiating at the neutron capturing state and terminating at the ground state following the rules of the extreme statistical model. The level system of the nucleus and the associated decay scheme are artificially generated according to adapted level-density and radiative strength function models. Each set of the generated level structure and the decay scheme are called a nuclear realization. The level structure below a critical energy (E_{crit} ~ 400 keV in odd-odd Eu) is taken from a known level scheme [22] and kept fixed. Above E_{crit}, the level density and decay scheme are assumed to follow statistical rules. Many nuclear realizations are simulated. Introducing the technique of precursors, as described in [18], the code DICEBOX offers the unique feature of simulation of the residual Porter-Thomas fluctuations. Another important feature of the code is the treatment of conversion electrons which is in the case of odd-odd deformed nuclei vitally important for realistic simulation of the γ-cascade process.

Cascades produced by the DICEBOX code in the list mode serve as an input for GEANT simulation of the detector response to these cascades. Various types of simulated spectra are produced. Simulation and data for sum-energy spectra and γ-ray spectra for events for various multiplicities with deposited sum-energy close to Q-value of the (n,γ) reaction are compared. Comparisons between data from the Eu experiments at DANCE and simulation are shown in Fig. 2. Multiplicity 2 (top graphs) and multiplicity 3 (bottom graphs) spectra are shown as a function of the γ-ray energy. Experimental data are shown in black histograms and simulation is shown in red. In the experimental multiplicity 2 spectrum cascades consisting of two-step transitions are selected, thus the spectrum is symmetric around the center as expected. The peak on the right hand side is slightly broader than the peak on the left due to the poorer detector resolution for higher energy γ-rays. The bump structure in the center of the multiplicity 2 spectrum, and near 2.5 MeV in the multiplicity 3 spectrum is interpreted as a manifestation of the presence of SM resonance.

FIGURE 2. Comparison between data and simulation. The black histograms represent data from the neutron energy $E_n = 0.24 - 0.65$ eV gated around the first two resonances in the ^{151}Eu(n,γ) reaction. The red histograms represent simulations a) without postulating any resonance and 2) with postulating an M1 resonance.

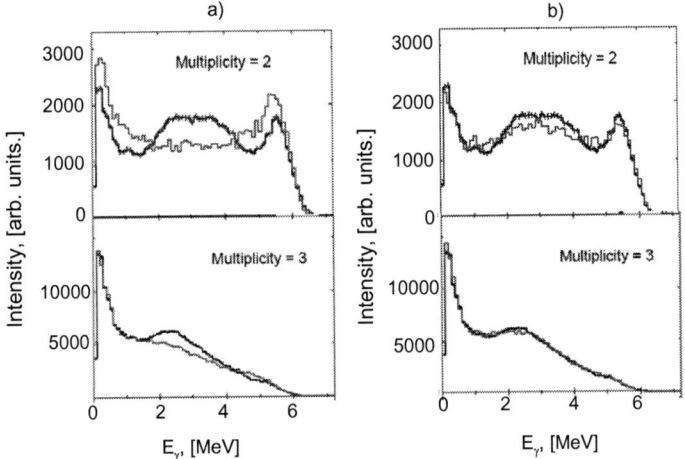

In both a) and b), the Back-Shifted Fermi gas model for the level density is employed. For the radiative strength function, a combination of the KMF (for lower E_γ) and Lorentzian (for higher E_γ) models for the E1 GEDR and the M1 spin-flip GMDR given by the Lorentzian are chosen. Without introducing any low energy resonance near 3 MeV, an agreement between data and simulation for all multiplicities is poor. As an example of poor fit, multiplicity 2 and 3 spectra are shown in the two graphs on the left in Fig. 2. Assuming an E1 resonance near 3 MeV does not improve the fit. Assuming an M1 resonance improves the fit significantly. Variation of the parameters of the M1 resonance yields better fit at $E_\gamma = 2.6$ MeV with width $\Gamma_\gamma = 1.6$ MeV, with a good agreement in the multiplicity 3 spectrum, an improved but not satisfactory agreement in the multiplicity 2 spectrum as shown in the right two graphs in Fig. 2. Similar resonance is also observed in the case of ^{153}Eu. The complicated level structure of odd-odd compound systems ^{152,154}Eu requires further refinement in the consideration of effect of isomers and varying deformation as a function of excitation energy.

V. CONCLUSION AND OUTLOOK

The 4π γ-ray calorimeter DANCE is utilized for the study of statistical decay cascade following the neutron capture on stable Eu targets. The decay cascade is simulated using the statistical code DICEBOX taking into account the detector response function with usage of the GEANT code. The outcome of the simulation is compared with the experimental spectra. A concentration of γ-ray strength near $E_\gamma \sim 2.5$ MeV is identified at present as manifestation of the scissors-mode like mechanism in odd-odd Eu nuclei. Our estimates indicate that these vibrations may be responsible for up to a 30 percent fraction of the value of the total radiative widths of neutron resonances. The impact of this assessment on systemization of nuclear data is straightforward, as well as on data for nuclear astrophysics. Further developments in the analysis and simulation are underway.

Acknowledgement. Work supported in part by the U.S. DOE by the UC under contract Nos. W-7405-ENG-48 (LLNL) and W-7405-ENG-36 (LANL), and by the NNSA Academic Alliance through DOE Research Grant Nos. DE-FG03-03-NA00076 (NCSU) and DE-FG03-03-NA00075 (UCB). Support by U.S. Department of Energy Grant No. DE-FG02-97-ER41042 (NCSU/TUNL) is acknowledged.

References:
[1] J. M. Blatt and V. F. Weisskopf, Theoretical Nuclear Physics, Wiley, New York , (1952).
[2] B. L. Berman and S. C. Fultz, Rev. Mod. Phys. **47**, 713 (1975).
[3] Samuel S. Dietrich and Barry L. Berman, At. Data Nucl. Data Tables **38**, 199 (1988).
[4] E. D. Earle et al., Can. J. Phys. **52**, 989 (1974).
[5] F. Bečvář, P. Cejnar, R. E. Chrien, and J. Kopecky, Phys. Rev. C **46**, 1276 (1992).
[6] F. Bečvář, P. Cejnar, J. Honzátko, K. Konečný, I. Tomandl, and R. E. Chrien, Phys. Rev. C **52**, 1278 (1995).
[7] J. Kopecky and R. E. Chrien, Nucl. Phys. **A468**, 285 (1987).
[8] W. V. Prestwich et al., Z. Phys. **A315**, 103 (1984).
[9] G. A. Bartholomew et al., Can. J. Phys. **48**, 687 (1970).
[10] A. Richter, Prog. Part. Nucl. Phys. **34**, 261 (1995).
[11] Y. P. Popov, Fiz. Elem. Chastits. At. Yadra **13**, 1165 (1982), [Sov. J. Part. Nucl. **13**, 483 (1982)].
[12] S. G. Kadmensiĭ, V. P. Markushev, and V. I. Furman, Yad. Fiz. **37**, 277 (1983), [Sov. J. Nucl. Phys. **37**, 165 (1983)].
[13] V. K. Sirotkin, Yad. Fiz. **43**, 570 (1986), [Sov. J. Nucl. Phys. **43**, 362 (1986)].
[14] A. Zilges et al., Nucl. Phys. **A519**, 848 (1990).
[15] M. Guttormsen et al., Phys. Scr. **T32**, 54 (1990).
[16] M. Guttormsen et al., Phys. Rev. C **68**, 064306 (2003).
[17] U. Agvaanluvsan et al., Phys. Rev. C **70**, 054611 (2004).
[18] F. Bečvář, Nucl. Instr. Methods A **417**, 434 (1998).
[19] M. Krtička et al., Phys. Rev. Lett. **92**, 172501 (2004).
[20] J. M. Wouters et al., to be published in the Proceedings of the 14th IEEE-NPSS Real-Time Conference, (2005).
[21] R. Reifarth et al., Nucl. Inst. Methods Phys. Res. A **531**, 530 (2004).
[22] T. Von Egidy et al., Z. Phys. A **286**, 241 (1978).

Determination of Thermal Neutron Capture Cross Sections Using Cold Neutron Beams at the Budapest PGAA and NIPS Facilities

T. Belgya, Zs. Révay, L. Szentmiklósi

Institute of Isotopes, Chemical Research Centre, HAS H-1525 Budapest, Hungary

Abstract. We report about our methodology developed for the determination of the thermal capture cross section of various target isotopes at our PGAA and NIPS facilities, which both use a guided cold neutron beam produced by the 10 MW Budapest Research Reactor. The two facilities provide an excellent means for determining partial gamma ray cross sections for products produced in the sample by neutron capture reactions. Both stations are equipped with HPGe detectors to detect the gamma rays coming from the excited nuclei of the samples. We present examples for the determination of thermal capture cross section of various target isotopes including the radioactive ^{99}Tc, ^{129}I nuclei and of the ^{204,206,207}Pb isotopes. The chopped beam option provides a good opportunity to study short-lived products.

INTRODUCTION

In 1997 we reported for the first time about our new neutron capture gamma-ray facility[1] located at the Budapest Research Reactor. Since then the guide system was substantially upgraded with super-mirrors, which offer their full advantage with the liquid-hydrogen neutron source completed in 2001. Due to these modifications, the neutron flux has increased more than an order of magnitude[2]. These instruments are partly operated within the framework of the Budapest Neutron Centre, which can be accessed by external users with the support of the EU FP6 (NMI3 - Integrated Infrastructure Initiative for Neutron Scattering and Muon Spectroscopy).

We have already reported most of the features of our experimental PGAA-NIPS facilities[3] and most recently in our new Handbook[4]. Cross sections for many isotopes were determined by combining the ENSDF data with our PGAA data library. The result of this comparison is also included in the Handbook[5]. Since the library data were measured mainly with a thermal beam, many cross sections can be improved by new measurements with the cold beam. In this article we report briefly about our upgraded experimental facilities and the ways we use them to determine thermal neutron capture cross sections assuming a $1/v$ dependence.

Neutron cross sections of the components of nuclear fuels, long-lived nuclear waste, and structural materials used in reactor and accelerator based nuclear energy production as well as in incineration or transmutation systems are of special significance. The thermal neutron capture cross section σ_{th} is also an important normalization point used in the evaluation of neutron resonance data[6], and thus it has significance in nuclear astrophysics as well. Its accurate value is sometimes difficult to determine by integral measurements, such as by the neutron activation method, due to the improper knowledge of the neutron energy distribution and the inaccuracy of radioactive decay data. Some of the σ_{th} values determined in the past have rather large uncertainties or, comparing with recent results, are ambiguous, so they have to be re-measured in order to use them in model calculations relevant to incineration or transmutation. Intense guided beams of cold neutrons provide the means to determine accurately the cross section for capture (and fission) at the thermal energy, without the disturbing effect of resonances. The σ_{th} values can be inferred from the partial production cross sections for the primary or ground-state transitions measured relative to an internal standard or from the beta decay lines. Using a beam chopper and a digital analyzer enables us to perform accurate activation measurements of ground state and isomer cross sections on very short-lived reaction products[7] and to reduce the intensity uncertainties of decay schemes.

INSTRUMENTATION

The neutron beam, guided by a super-mirror neutron guide from the liquid hydrogen cooled neutron source of the Budapest Research Reactor, was split at the end of our guide and now serves two measuring stations. The new facility is dual-purpose: neutron induced prompt gamma ray spectroscopy or NIPS[8], and its applications, namely prompt gamma activation analysis or PGAA.

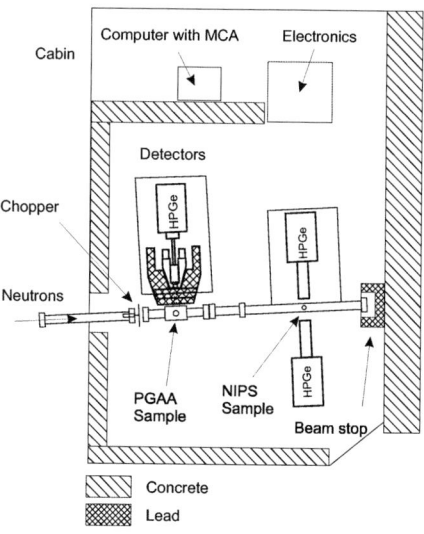

FIGURE 1. Experimental arrangement of the PGAA-NIPS facilities at the end of neutron guide No. 1.

The NIPS facility has been designed for a large variety of experiments, namely for studying nuclear reaction-induced prompt and delayed gamma radiation, including γ-γ-coincidences[8]. A recently upgraded beam chopper was also installed, which allows simultaneous measurements of prompt reaction and short-lived decay γ-rays. Figure 1 shows the simplified drawing of the experimental area located at the end of the guide No. 1, about 35 m away from the reactor core. The thermal equivalent neutron fluxes are 5×10^7 n·cm^{-2}·s^{-1} and 3×10^7 n·cm^{-2}·s^{-1} at the PGAA and NIPS sample positions respectively. The 23% efficiency PGAA main detector has an active BGO shield, which is used in Compton-suppression mode.

EXPERIMENTAL METHODS AND RECENT RESULTS

The idea of standardization with stoichiometric compounds used in PGAA[9] can be used in general for determination of partial gamma ray cross sections of isotopes. The basic equation is

$$\frac{\sigma_{\gamma,x}}{\sigma_{\gamma,c}} = \frac{n_c}{n_x} \frac{A_{\gamma,x}/\varepsilon(E_{\gamma,x})/f(E_{\gamma,x})}{A_{\gamma,c}/\varepsilon(E_{\gamma,c})/f(E_{\gamma,c})} \quad (1)$$

where x and c indexes refer to the unknown and comparator isotopes. The σ_γ is the partial gamma ray cross section, n is the stoichiometric coefficient of the compound, A_γ is the measured peak area, $\varepsilon(E_\gamma)$ is the detector efficiency, $f(E_\gamma)$ is the sample self absorption at gamma energy E_γ. The formula is complete if a homogenous sample is used. In this case the neutron absorption and inhomogenity of the neutron beam cancel out. Special treatment must be used if any of the isotopes do not follow the 1/v cross section rule[10].

The partial gamma ray cross section σ_γ can be used to determine the σ_{th} of an isotope with an abundance of θ if the production probability P_γ of the gamma ray is known, using the following equation

$$\sigma_\gamma = \theta \sigma_{th} P_\gamma \quad (2)$$

To determine P_γ is not easy. One way to avoid this problem is to sum up the partial cross sections, corrected for the internal conversion C_γ, for all of the primary gamma rays. In this case $\Sigma C_\gamma P_\gamma = 1$, if only electromagnetic decay is involved. Thus, we can easily obtain the σ_{th} from Eq. 2. The summing can also be done for all of the ground state transitions with the same result and in general it can also be calculated from the crossing intensity sum[11]. There is however a limiting factor in this procedure, namely the complete knowledge of all of the primary or ground state transitions. Even for a relatively simple decay scheme, like that of ^{210}Bi, we run into this problem[6]. Thus, this approach yields only a lower limit[7] for the σ_{th}. In some cases with the help of an estimate of completeness[12] or of model calculations, we may get definite results[13]. To improve knowledge of the decay scheme we can perform (n,γγ) coincidence experiments[11]. This is however a hard way to obtain more information. Nevertheless, for simple nuclei this approach works rather well. In this later case even repetition of earlier experiments with higher statistics can help as has happened in our work performed in co-operation with P. Schillebeeckx (EC JRC IRMM) on the ^{206}Pb(n,γ) reaction. Using a 10 g and 2×2 cm^2, highly enriched ^{206}Pb sample we were able to improve the decay scheme of the ^{207}Pb nucleus significantly. The ^{206}Pb sample was placed in the cold neutron beam at the PGAA experimental station and measured for about 100 000 s. The resulting preliminary decay scheme is shown in Fig. 2.

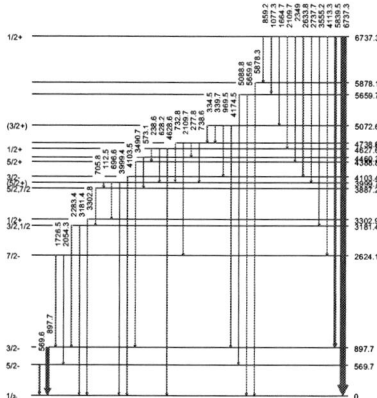

FIGURE 2. Decay scheme of the ^{207}Pb nucleus from cold neutron capture. The arrow width is proportional to the gamma ray intensity, except for the primary transition to the ground state. It should be multiplied by 50 to give the correct intensity.

This is a significant improvement, because only 4 gamma-rays were known from this reaction in the literature[14]. In the decay-scheme construction we have made use of the new crossing intensity balance method, which yielded the following crossing intensity balance figure (Fig. 3).

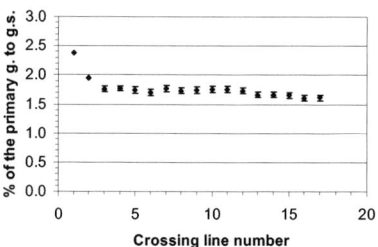

FIGURE 3. The intensities at the two lowest crossing lines are higher than the average of the rest. A possible reason is the (n,n'γ) reaction induced gamma rays in the lead shielding on the ^{207}Pb isotope from the source of the 2.4 MeV neutrons produced in the Li absorber.

To avoid the need for complete knowledge of the decay scheme, we can make use of the total energy detector concept, which is widely used in the energy differential cross section experiments. This is nothing else than the reverse of the Q-value test:

$$\sigma_{th} = \sum_i E_i \sigma_{\gamma i} c_i / B_n \qquad (3)$$

where E_i is the transition energy, c_i is the correction for conversion electrons and B^n is the binding energy. However in this case we have to identify all of the gamma rays belonging to the nucleus. Another way to obtain information on P_γ is to use the natural process of radioactive decay, which sums all of the yields of the parent nucleus. In this case, however, we have the problem of the determination of decay probabilities for the daughter nucleus. If these are known, then the σ_{th} can be obtained by determining the σ_γ value for a suitable decay gamma ray. The method of determination of σ_γ depends on the lifetime of the decaying nucleus. For short lifetime nuclei ($\tau \ll$ irradiation time) saturation during the activation process happens in a short time and the σ_γ value can be determined easily with the internal comparator method. The ^{99}Tc(n,γ)^{100}Tc(β^-)^{100}Ru reaction is a good example for this[7]. To determine the decay scheme of a short-lived decay product of the capture, the chopped neutron beam provides an excellent means; again the ^{100}Tc(β^-)^{100}Ru is a good example.

In the case of daughter nuclei with $\tau > 10$ min the luck of saturation during the irradiation must be taken into account and the decay scheme can be measured off-line. The correction formula for the peak area A for an in-beam experiment lasting time t, if the starting time of the irradiation and the counting coincide, is[8]

$$A' = A\left(1 - \frac{1-e^{-\lambda t}}{\lambda t}\right)^{-1} \qquad (4)$$

where λ represents the decay constant. Utilising this formula, a partial gamma ray cross section from the ^{128}I decay was re-determined using hydrogen as a comparator in triiodo-methane (CHI$_3$) compound. We obtained σ_γ=0.678(11) b for the 443 keV intense decay line. If we use the most recent absolute decay probability P_γ = 0.12613(77), which was determined by Miyahara et al.[15] than we obtain σ_{th}=5.4(1) b from Eq. 2. This is about 20% lower than the adopted[16] and the most recent experimental[17] values. Until now, the σ_γ values for the decay gamma rays were determined from the prompt gamma spectra, which made the evaluation difficult due to the large number of prompt gamma peaks and uncertain due to the possible interferences.

Therefore we decided to develop a new technique based on our beam chopper. Recently, electronics were set up to control the beam chopper and to make possible the parallel acquisition of two separate spectra, during the open and the closed-beam phases, respectively. The live-time normalization of the two spectra could be achieved with peaks from a radioactive source, which were measured together with the gamma rays arising from the neutron capture activated sample. The closed-beam decay spectrum contains only a few peaks on a substantially lowered background, thereby offers ideal conditions to determine the partial gamma-ray production cross-sections (σ_γ) of short- and medium-lived nuclides. A series of measurements have just begun to determine new cross-section data with this technique. Using the same compound we got σ_γ=0.710(9) b for the 443-keV peak of ^{128}I, which is slightly higher than our continuous beam PGAA value. Besides that, measurements on nuclides ^{20}F, ^{24}Na, ^{28}Al, ^{56}Mn are also in progress. The σ_γ cross sections determined this way are typically 1–2% accurate, and are in good general agreement with the literature data. For some of these nuclides P_γ is practically unity, or is accurately determined, i.e. σ_{th} data can be also readily generated.

The importance of the σ_γ value of 443 keV as a comparator is large in the determination of the σ_{th} value of the ^{129}I(n,γ)^{130}I reaction. We have recently measured the gamma transitions from this reaction on a sample received from the EC Joint Research Centre Institute for Reference Materials and Measurements (IRMM). The sample was packed in an aluminum container and contained a powder consisting of PbI$_2$ and Pb compounds[18]. The composition is important for the analysis of the recent neutron capture cross section data taken at the GELINA facilities of the IRMM[19]. The ^{129}I(n,γ) experiments were performed in collaboration with O. Bouland from CEA Cadarache at our PGAA-NIPS facilities. Since the target contains both ^{127}I and ^{129}I the 443 keV decay line of ^{128}I is the best comparator. The isotopic mass ratio (^{129}I/^{127}I) has been measured with three ICP-MS instruments (one at PSI and two at IRMM) and yielded a value of 4.9 with less than 1% uncertainty. This ratio and the lower ^{127}I neutron capture cross section however means a relatively low yield for the 443 keV internal comparator line, which in addition is a member of a complex multiplet. Furthermore the saturation correction is much more elaborate. Figure 3 shows the simplified decay scheme of the ^{129}I(n,γ)^{130}I(β⁻) reaction, which serves as a basis for the saturation model.

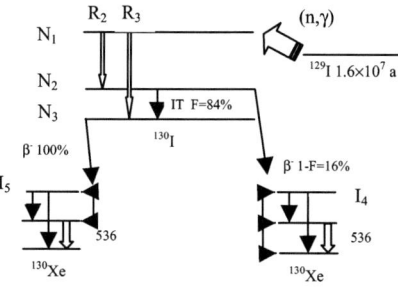

FIGURE 4. Simplified decay scheme of ^{130}I compound nucleus produced in thermal neutron capture reaction.

o confirm the half-lives for the ground state (12.4 h) and the isomer state (8.8 min) we performed decay and in-beam activation experiments. The agreement with the literature values was found to be excellent. These experiments made it also possible to calculate the isomer and ground state feeding ratios (R_2, R_3), as well as the β-decay and isomer transition branching (F). A final thermal cross section value will be given after the experiments have been repeated with the new chopper technique, which offers much better background conditions and more confidence in the data reduction process.

CONCLUSION

The upgraded Budapest PGAA-NIPS facilities are well suited to measure partial gamma ray cross sections, which can be used for non-destructive interrogation or transmutation monitoring or to infer thermal neutron capture cross sections. We are continuously developing the method to achieve the best performance of the facilities. We have already re-determined a number of cross section values for various isotopes (Te, Bi,

Pb, ^{99}Tc, ^{129}I, ^{238}U,...) which may help to clarify confusion about cross sections measured in the past and to motivate the urgent re-measurement and re-calculation of energy dependent neutron capture values. These data can be used to perform more reliable Monte Carlo calculations for the design of the planned demonstrator Accelerator Driven Systems for incineration and transmutation of nuclear waste.

ACKONWLEDGEMENTS

Two of the authors Zs. Révay and T. Belgya are grateful for the support of the Bolyai scholarship and for the support EU FP6 NMI3 project. Thorough reading of Jesse Weil is greatly acknowledged.

REFERENCES

[1] T Belgya, Zs. Révay, B. Fazekas, I. Héjja, L. Dabolczi, G. L. Molnár, Z. Kis, J. Östör, and Gy. Kaszás, presented at the 9th Int. Symp. on Capture Gamma-Ray Spectroscopy and Related Topics, Budapest, 1997, pp. 826.

[2] L. Rosta, L. Cser, and Z. Revay, Applied Physics a-Materials Science & Processing **74**, S292-S294 (2002).

[3] Z. Révay, T. Belgya, Z. Kasztovszky, J. L. Weil, and G. L. Molnár, Nucl. Instrum. & Methods **B 213**, 385-388 (2004); T. Belgya, Zs. Révay, P. P. Ember, J. Weil, G. L. Molnár, and S. M. Qaim, presented at the 11th Int. Symp. on Capture Gamma-Ray Spectroscopy and Related Topics, Pruhonice near Prague, Czech republic, 2003, pp. 562-568; T. Belgya, J. Weil, G. L. Molnár, and H. F. Wirth, presented at the workshop on Neutron Measurements and Evaluations for Applications 5-8 November 2003, Budapest, Hungary, 2004, pp. 159-163.

[4] R. M. Lindstrom and Zs. Révay, in *Handbook of Prompt Gamma Activation Analysis with Neutron Beams*, edited by G. L. Molnár (Kluwer Academic Publishers, Dordrecht, Boston, London, 2004), pp. 31-58; T. Belgya and Zs. Révay, in *Handbook of Prompt Gamma Activation Analysis with Neutron Beams*, edited by G. L. Molnár (Kluwer Academic Publishers, Dordrecht, Boston, London, 2004), pp. 71-111.

[5] Zs. Révay, R. B. Firestone, T. Belgya, and G. L. Molnár, in *Handbook of Prompt Gamma Activation Analysis with Neutron Beams*, edited by G. L. Molnár (Kluwer Academic Publishers, Dordrecht, Boston, London, 2004), pp. 173-366.

[6] A. Borella, P. Schillebeeckx, G. L. Molnár, T. Belgya, Zs. Révay, L. Szentmiklosi, E. Berthoumieux, X. Gunsing, A. Letourneau, and F. Marie, presented at the 10th Int. Conf. on Nuclear Data for Science and Technology, September 26 - October 1, 2004, Santa Fe, New Mexico, 2005, pp. 648-651.

[7] G. L. Molnár, T. Belgya, Zs. Révay, and S. M. Qaim, Radiochimica Acta **90** (8), 479-482 (2002).

[8] P. P. Ember, T. Belgya, J. L. Weil, and G. L. Molnar, Applied Radiation and Isotopes **57** (4), 573-577 (2002).

[9] Z. Revay and G. L. Molnar, Radiochimica Acta **91** (6), 361-369 (2003).

[10] Zs. Révay and T Belgya, in *Handbook of Prompt Gamma Activation Analysis with Neutron Beams*, edited by G. L. Molnár (Kluwer Academic Publishers, Dordrecht, Boston, London, 2004), pp. 1-30.

[11] T Belgya, G. L. Molnár, Zs. Révay, and J. Weil, presented at the 10th Int. Conf. on Nuclear Data for Science and Technology, September 26 - October 1, 2004, Santa Fe, New Mexico, 2005, pp. 744-747.

[12] I. Tomandl, J. Honzatko, T. von Egidy, H. F. Wirth, T. Belgya, M. Lakatos, L. Szentmiklosi, Z. Revay, G. L. Molnár, R. B. Firestone, and V. Bondarenko, Physical Review C **68** (6), 067602 (2003).

[13] R. B. Firestone, M. Kritcka, D. P. McNabb, B.W. Sleaford, U. Agvaanluvsan, T. Belgya, and Z. Révay, presented at the 12th Int. Conf. on Capture Gamma-Ray Spectroscopy and Related Topics, September 4-9, 2005 University of Notre Dame, Indiana, USA, 2005, this Conf..

[14] P. Hugenford, T. von Egidy, H. H. Schmidt, S. A. Kerr, H. G. Boerner, and E. Monnand, Z. Phys. A **313**, 349-350 (1983).

[15] H. Miyahara, H. Matumoto, G. Wurdiyanto, K. Yanagida, Y. Takenaka, A. Yoshida, and C. Mori, Nucl. Instrum. and Methods **A 353** (1-3), 229-233 (1994).

[16] S F Mughabghab, Report No. INDC(NDS)-440, 2003.

[17] T. Katoh, S. Nakamura, H. Harada, and Y. Ogata, Journal of Nuclear Science and Technology **36** (3), 223-231 (1999).

[18] C. Ingelbrecht, J. Lupo, K. Raptis, T. Altzitzoglou, and G. Noguere, Nucl. Instrum. & Methods **A 480**, 204-208 (2002).

[19] G. Noguere, A. Brusegan, A. Lepretre, N. Herault, O. Bouland, and G Rudolf, in: 10th Int. Conf. on Nuclear Data for Science and Technology, September 26 - October 1, Santa Fe, New Mexico, 2005, pp. 1462-1465.

Method for (n,γ) cross section measurements on unstable isotopes

Stephan Walter*, Michael Heil*, Franz Käppeler*, Ralf Plag* and René Reifarth[†]

*Institut für Kernphysik, Forschungszentrum Karlsruhe, Postfach 3640, D-76021 Karlsruhe, Germany
[†]Los Alamos National Laboratory, Los Alamos, New Mexiko, 87545, USA

Abstract. Branching points in the s process provide information of the stellar interior and are considered as stringent tests of stellar models. Current measurements with the time of flight (TOF) technique are limited by the available neutron fluxes which require - even under ideal circumstances - sample masses of at least several milligram. For short lived nuclei, this would imply severe backgrounds due to the activity of the sample. We propose to increase the sensitivity of the TOF method such that target masses in the microgram region can be used. This is achieved by shortening the flight path down to a few centimeters, setting up the neutron production target inside a 4π BaF$_2$ calorimeter. Some first results of background test measurements as well as of computer simulations with GEANT are shown.

Keywords: s process, unstable isotopes, branch point isotopes, 4π BaF$_2$ calorimeter, time of flight method
PACS: 29.25.Dz,29.40.Vj

INTRODUCTION

About 50% of the isotopic abundances between Fe and Bi are produced in slow neutron capture reactions in the so called s process. Since s process neutron densities are in the order of about 10^8 n/s/cm^3, the time between two neutron captures (1-10 years) is usually much longer than the time scale for β^- decay. Accordingly, the reaction path of the s process follows the valley of stability.

Some isotopes in the s process path have half-lives comparable to the time between two neutron capture reactions, leading to a branch point in the reaction path. The knowledge of (n,γ) cross sections of such radioactive branch point isotopes are important to deduce information on the physical conditions during the s process (temperature, density and neutron flux) [1, 2, 3]. Such data are also needed for explosive nucleosynthesis in supernoave in the r and p processses [4].

The (n,γ) cross sections can for example be determined in activation experiments. For TOF experiments, 4π BaF$_2$ calorimeter arrays are preferable for their advantageous combination of high efficiency and good timing properties.

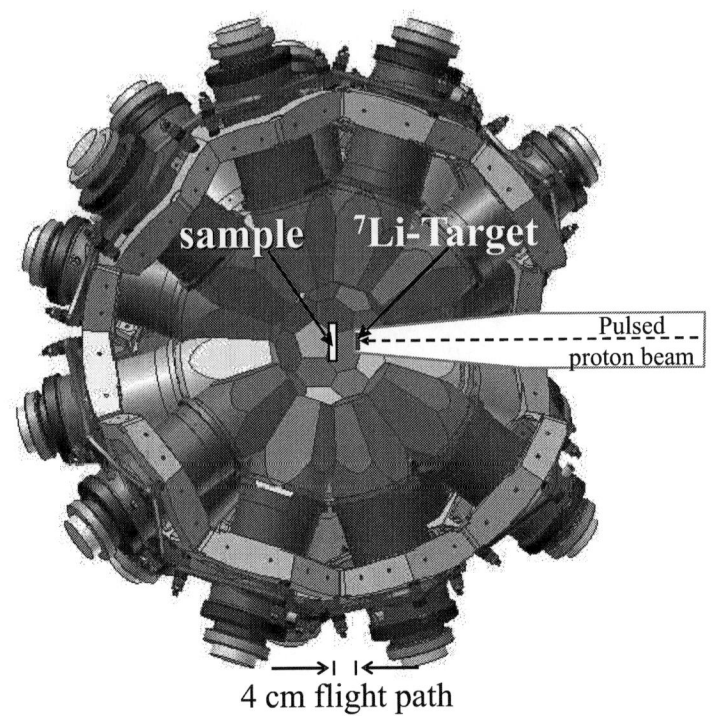

FIGURE 1. New setup with the ^{7}Li target for neutron production mounted inside the 4π detector thus increasing the neutron flux by two orders of magnitude.

THE KARLSRUHE 4π BAF$_2$ DETECTOR

In the standard setup of the Karlsruhe detector neutrons are produced via the ^{7}Li(p,n)^{7}Be reaction using a pulsed proton beam with an energy of about 2 MeV. Background from the lithium target is strongly reduced by a lead enforced collimator, which defines a neutron beam 20 mm in diameter at the sample in the center of the 42-fold segmented detector. The flight path between the neutron target and the sample is 77 cm [5]. For the investigation of radioactive samples, which can only be used in very small quantities, the sensitivity of this setup needs to be improved.

MODIFIED SETUP

One possible solution is shown in Fig. 1. The ^{7}Li target is now mounted inside the 4π detector and the flight path is reduced from 77 cm to 4 cm. In this way, the neutron flux is increased by two orders of magnitude, an improvement that allows one to use sample masses in the microgram region.

In an idealized picture, the expected signals are the following: a prompt γ-flash is

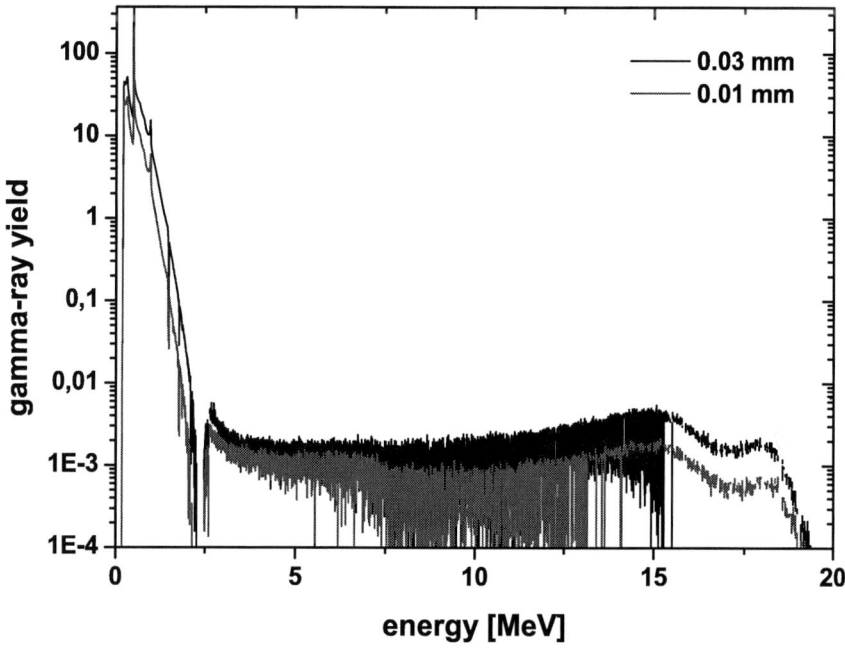

FIGURE 2. Gamma-ray yield of the γ-flash for different Li layers. The spectra were measured with a HPGe detetector, but are not corrected for detector efficiency.

released, when the proton beam hits the target. Assuming a maximum neutron energy of 100 keV, it takes 9 ns for the fastest neutrons to reach the sample. After 26 ns, background signals due to sample scattered neutrons are detected. In between, there is a time window with nearly undisturbed signals from capture events.

The main difficulty of the new setup is due to the prompt γ-flash. In the standard configuration with 77 cm flight path, this background is reduced by a factor of 10^8 by the shiedling around the neutron target. In the modified arrangement, shielding possibilities are limited, especially since the signals from capture events in the sample should not be disturbed.

TEST MEASUREMENTS

The scintillation light of BaF_2 crystals consists of two components with different decay time constants and wavelengths: a fast component with a peak emission at 220 nm and $\tau=0.6$ ns, and a slow component with a peak emission at 310 nm and $\tau=620$ ns [6].

While the fast component from signals induced by the prompt γ-flash decays in the

9 ns interval before the arrival of the fastest neutrons at the sample, the slow component is practically unchanged.

For high count rate applications, only the fast component is of interest. In order to improve the fast slow ratio, measurements with a small BaF$_2$ crystal were performed at different temperatures. Though the fast slow ratio could be increased by a factor of 1.6 by rising the temperature from 20°C to 65°C, such high crystal temperatures are not suited for the actual measurements.

The use of wavelength shifters as well as UV-filters for reducing the slow component was considered as well. However wavelength shifters such as p-terphenyl shift both components or would reduce the signal strength too much. This is also true for UV-filters affecting the slow component more than the fast one. Accordingly, both possibilities had to be abandoned.

Since the material used for the backing of the Li target has an influence on the intensity of the γ-flash, measurements with a HPGe-detector have been performed with different backing materials such as Cu, Ag, Au and Ta. While copper is used as standard backing because of the high heat conductivity, it produces by far more γ-rays compared to the other tested materials. The γ-ray ratios are Cu:Ag:Au:Ta=400:60:4:1. Therefore, tantalum was chosen to be the future backing material.

Most of the background, however, stems from the lithium itself. Figure 2 shows spectra taken with lithium targets 0.03 and 0.01 mm in thickness. Since this γ-ray yield dominates over that from the backing material, the lithium layer should be as thin as possible. Measurements with 5 μm thick Li targets showed that a count rate of 1–2 γ-rays per pulse and crystal can be expected if the beam current is limited to 1.5 μA.

SIMULATIONS

Since the main cause for the γ-flash cannot be avoided, methods to reduce this background further have to be applied. Although shielding possibilities inside the 4π calorimeter are limited, a small lead shield around the Li target was considered. The influence on γ-rays both from the target and the sample was estimated by means of simulations with the computer code GEANT [7, 8].

Figure 3 compares three cases for γ-rays of different energies up to 6 MeV, originating either from the sample or the target. The upper curve (squares) shows the result without the Pb absorber, while the middle (circles) and the lower curve (triangles) illustrate the effect of the lead shield. The lead shield consists of a massive sphere 66 mm in diameter, with a cylindrical opening for the beamline and a cone with 120° opening angle (opposite to each other) for the neutron beam. For typical capture γ-ray energies of 2 MeV, γ-rays from the sample are reduced by 7%, while γ-rays from the target are reduced by 30%. Simulations of the neutron background in the BaF$_2$ crystals show that capture events can be clearly separated from this background (which appears 26 ns after the γ-flash) in the relevant TOF interval corresponding to neutron energies between 30 keV and 70 keV.

In order to verify the simulations, test measurements with a gold sample are planned. Next, ^{147}Pm will be investigated since this is one of the important radioactive branch point isotopes in the s-process reaction path. So far, this cross section has been measured

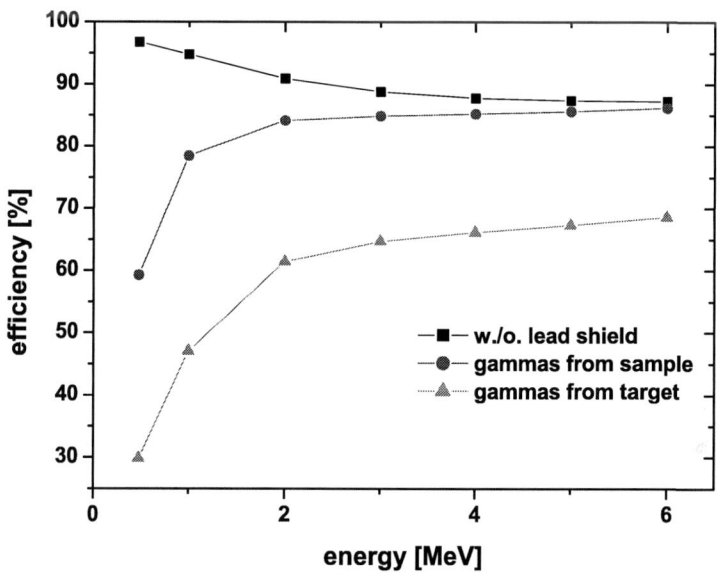

FIGURE 3. Influence of lead shield on γ-rays from the sample and from the target, simulated with GEANT.

via the activation technique [9]. With the new setup, the TOF measurement of the (n,γ)-cross section is planned with a sample mass of about 1 mg.

REFERENCES

1. F. Käppeler, *Nuclear Physics A* **752**, 500c–509c (2005).
2. F. Käppeler, M. Wiescher, U. Giesen, J. Goerres, I. Baraffe, M. E. Eid, C. M. Raiteri, M. Busso, R. Gallino, M. Limongi, and A. Chieffi, *Ap. J.* **437**, 396–409 (1994).
3. A. Jorissen, A. Smith, and A. Lambert, *Astron. Astrophys.* **261**, 164–187 (1992).
4. T. Weaver, and S. Woosley, *Phys. Reports* **227**, 65–96 (1993).
5. K. Wisshak, K. Guber, F. Käppeler, J. Krisch, H. Müller, G. Rupp, and F. Voss, *Nucl. Instr. Meth. A* pp. 595–618 (1990).
6. M. Laval, M. Moszyński, R. Allemand, E. Cormoreche, P. Guinet, R. Odru, and J. Vacher, *Nucl. Instr. Methods* **206**, 169–176 (1983).
7. S. Agostinelli, J. Allison, K. Amako, J. Apostolakis, H. Araujo, P. Arce, M. Asai, D. Axen, S. Banerjee, G. Barrand, F. Behner, L. Bellagamba *et. al*, *Nucl. Insr. Methods A* **506**, 250–303 (2003).
8. R. Brun, and F. Carminati, *CERN Program Library Long Writeup W5013 edition*, CERN, Geneva (1993)
9. R. Reifarth, C. Arlandini, M. Heil, F. Käppeler, P. V. Sedyshev, A. Mengoni, M. Herman, T. Rauscher, R. Gallino, and C. Travaglio, *The Astrophysical Journal* **582**, 1251–1262 (2003).

Neutron Capture Cross Sections of ^{236}U and ^{234}U

R.S. Rundberg*, T.A. Bredeweg*, E.M. Bond*, R.C. Haight*, L.F. Hunt*,
A. Kronenberg†, J.M. O'Donnell*, J.M. Schwantes*, J.L. Ullmann*, D.J. Vieira*, J.B. Wilhelmy* and J.M. Wouters*

*Los Alamos National Laboratory
†Oak Ridge National Laboratory

Abstract. Accurate neutron capture cross sections of the actinide elements at neutron energies up to 1 MeV are needed to better interpret archived nuclear test data, for post-detonation nuclear attribution, and the Advanced Fuel Cycle Initiative. The Detector for Advance Neutron Capture Experiments, DANCE, has unique capabilities that allow the differentiation of capture gamma rays from fission gamma rays and background gamma rays from scattered neutrons captured by barium isotopes in the barium fluoride scintillators. The DANCE array has a high granularity, 160 scintillators, high efficiency, and nearly 4-π solid angle. Through the use of cuts in cluster multiplicity and calorimetric energy the capture gamma-rays are differentiated from other sources of gamma rays. The preliminary results for the capture cross sections of ^{236}U are in agreement with the ENDF/B-VI evaluation. The preliminary results for ^{234}U lower are than ENDF/B-VI evaluation and are closer to older evaluations.

Keywords: Neutron capture cross sections, DANCE
PACS: 25.40.Lw,25.40.Ny,27.90.+b,28.20.Fc,29.30.Hs,29.40.Vj

INTRODUCTION

The neutron capture cross sections of uranium isotopes are important for the accurate prediction of uranium fission reactor performance and for the use of uranium isotopes as a diagnostic for nuclear explosions. The isotopic distribution of uranium following a nuclear explosion can be used to infer the neutron flux distribution that the uranium component was exposed to. Likewise, given an over-determined system the ingoing isotopic distribution may be inferred. In spite of the importance of these cross sections, few measurements have been made, particularly at neutron energies around 10 keV [1][2]. This is in part due to the difficulty in making these measurements. There are only a few measurements of the ^{236}U neutron capture cross section since the measurement of the resonance region by Carlson [3]. These more recent measurements differ by up to a factor of two[4], as shown in Fig 1. There are even fewer measurements of the ^{234}U neutron capture cross section. Recent measurements suggest a larger cross section in the keV neutron energy range[5], see Fig. 2.

The Detector for Advanced Neutron Capture Experiments (DANCE) is designed as a high efficiency, highly segmented 4 BaF$_2$ detector for calorimetrically detecting gamma-rays following a neutron capture. DANCE is located on the 20 m neutron flight path 14 (FP14) at the Manuel Lujan Jr. Neutron Scattering Center at the Los Alamos Neutron Science Center (LANSCE) [6]. The neutron flux at the sample position of Flight Path 14 is 3x10^5 n/s/cm^2/decade. The initial design work is described in [7]. For practical

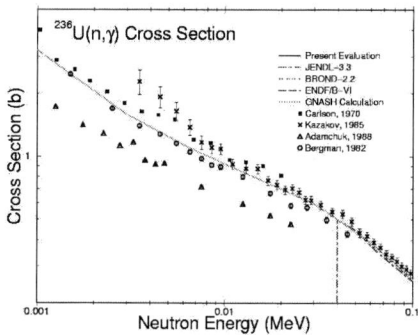

FIGURE 1. Previously measured ^{236}U capture cross sections.

reasons the detector modules do not quite cover the entire solid angle. The design of the detector is such that a full 4π array would consist of 162 crystals of four different shapes, each shape covering the same solid angle [8]. Two of the 162 crystals are left out in order to leave space for the neutron beam pipe. Thus the full array hosts 160 out of 162 possible BaF$_2$ crystals. The dimensions of the bare crystals are designed to form a BaF$_2$ shell with an inner radius of 17 cm and a thickness of 15 cm. Thanks to the fairly low repetition rate of 20 Hz, measurements can be carried out over the whole energy range from 10 meV to 500 keV. This combination of a strong neutron source and a high efficiency gamma-ray detector allows to measure (n,γ) cross section of radioactive isotopes down to a few hundred days half-life. Further details on the overall performance of the array can be found in [9],[10].The DANCE time of flight spectrometer makes these measurements possible with the ability to discriminate against backgrounds, such as, the capture of scattered neutrons in the scintillator. Backgrounds are suppressed by discriminating on the basis of gamma-ray multiplicity and calorimetric energy.

EXPERIMENTAL

Targets of ^{236}U and ^{234}U weighing 0.49 milligrams and 1.08 milligrams, repectively, were prepared by electrodeposition on 2.5 micrometer thick titanium foils [11],[12]. The thickness of both targets were about 1 milligram/cm^2. Titanium is used as a substrate because it has a low capture cross section, can be manufactured extremely thin, and is resistant to attack by mineral acids.

The data acquisition system can be operated in one two modes called segmented and continuous, respectively. The segmented mode stores the waveform from each event in internal memory by transferring the waveform from circular buffer. The waveforms collected in internal memory are compressed and dumped to the central computer every 50 milliseconds. The advantage of the segmented mode is that a wide range of flight times is recorded, ranging from less than a microsecond to 14 milliseconds. This allows

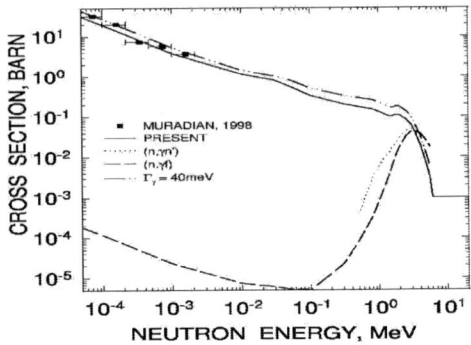

FIGURE 2. ^{234}U capture cross section evaluations.

the measurement of cross section at thermal energy, 0.0253 eV, simultaneously with energies up to hundreds of keV. The disadvantage of the segmented mode is the need for a master trigger and inherent dead time. The master trigger sets a lower limit on detector multiplicity, and a threshold of about 200 keV in pulse height, due to the minimum threshold in the constant fraction discriminators. The continuous mode stores a continuous waveform for up to 400 microseconds, thus limiting the neutron energy range from 1 MeV to about 10 eV. The advantages of the continous mode are the neglible deadtime and a very low gamma-ray threshold, 7 mV, corresponding to about 50 keV.

The cross sections were normalized to both the thermal neutron cross section and the area under the first resonance. The latter method utilizes a technique for measuring absolute disintegration rates in large calorimeters[13]. The principle is that as the efficiency of the 4π detector increases the probability that a gamma-ray or a compton scatter from one of the gamma-rays in the cascade approaches unity. The probability of detecting a decay depends on the efficency to power of the cascade multiplicity. The DANCE detector was calibrated using the events in the ^{236}U 5.45 eV resonance. The background was determined using the off-resonance events. The efficiency of the detector was varied by omitting crystals in replay. The full array has only 84 percent coverage of 4π due to insulating materials and irregularities in the crystals. The curve shown in Fig 3. was best fit by a cubic. A cubic is appropriate for a three gamma-ray cascade. The peak of the gamma-ray multiplicity distribution is at 3.

RESULTS AND CONCLUSIONS

The data were analyzed by setting gates in calorimetric energy. One for energy above 4.0 MeV and below 5.5 Mev. The lower limit is to minimize the background from scattered neutrons captured on the even barium isotopes and to avoid the natural background in the radium chain. The upper limit is slightly above the neutron binding energy for ^{234}U

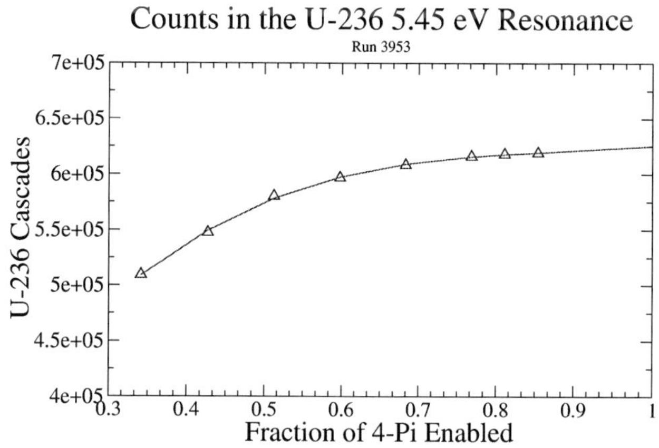

Counts in the U-236 5.45 eV Resonance

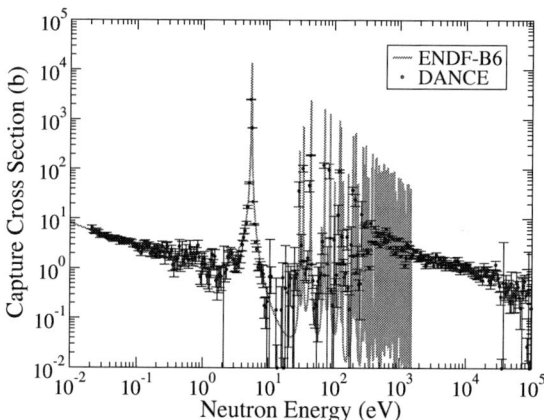

FIGURE 3. ^{236}U capture rate as function of detector efficiency.

FIGURE 4. ^{236}U capture cross section measured at DANCE.

and ^{236}U, 5.29 MeV and 5.13 MeV, respectively. A second gate is used to characterize the magnitude of the background. The limits for the second gate is from 5.51 MeV to 8.5 MeV. This range is intended to cover the range of gamma-ray cascades from neutron capture on the odd isotopes of barium. The gamma-ray cluster multiplicity is set at 3 and higher to minimize the contribution from neutron captures on the even isotopes of barium because it has been observed the peak in the multiplicity is near 2. A third gate is used for the actinides to correct for fission. This gate is set to multiplicities at and above 7 with the upper calorimetric energy range. No gamma-ray cascades have been observed in this range without fission. Cosmic-ray events generally have multiplicities greater than 15. The distribution of fission gamma-rays were determined using the fission tagging chamber [14].

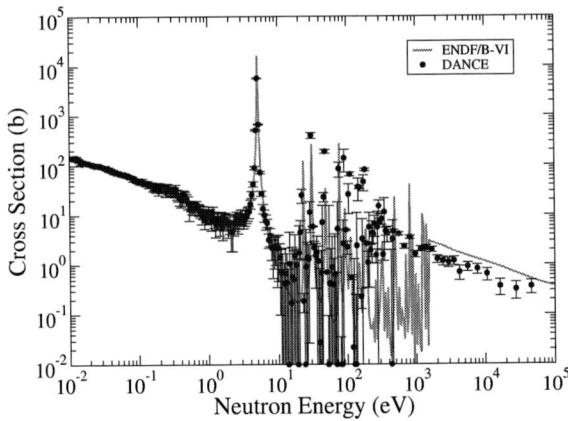

FIGURE 5. ^{234}U capture cross section measured at DANCE. The data is normalized to the thermal neutron capture cross section.

The results are shown Figs, 4 and 5. The data were normalized to the thermal neutron cross sections. The area under the 5.45 eV resonance in ^{236}U was also analyzed on an absolute basis using the Barr method. Both methods gave the same area within the error in the neutron flux. The area under the first resonance is in excellent agreement with the ENDF/B-VI value. The ^{234}U cross sections are also in good agreement between the two methods. The Barr method gives a an area of 1770 barn-eV for 5.16 eV resonance. The cross sections normalized to the thermal cross section 100 b give an area of 1740 barn-eV for 5.16 eV resonance. The ENDF/B-VI resonance parameters give an area of of 2830 barn-eV for 5.16 eV resonance. The small bump at 0.3 eV is probably due error in the (n,γ) cross section for ^{235}U, a minor contaminant in the ^{234}U target. Both methods give lower cross sections in the 1 to 10 keV range than the current ENDF evaluation.

In conclusion, DANCE is a valuable new tool that can provide a fresh new look a low energy neutron cross sections. The ablility to differentiate capture gamma-rays from backgrounds is superior, particulary for radioactive isotopes. Fission gamma-rays were shown to be a significant interference in the measurement of neutron capture by the actinides. DANCE was successful in eliminating fission gamma-rays through the use of multiplicity and calorimetric energy cuts.

ACKNOWLEDGMENTS

Work performed under the auspices of the U.S. Department of Energy by the University of California, Los Alamos National Laboratory (W-7405-ENG-48).

REFERENCES

1. J. Halperin, and R. W. Stoughton, *Proceedings of the 2nd U. N. Conf. on the peaceful uses of atomic energy, Vol. 16*, United Nations, New York, 1958, pp. 64–69.
2. T.A Eastwood, A.P. Baerg, C.B. Bigham, F. Brown, M.J. Cabell, W.E. Grummitt, J.C. Roy, and R.P. Schuman, *Proceedings of the 2nd U. N. Conf. on the peaceful uses of atomic energy, Vol. 16*, United Nations, New York, 1958, pp. 54–63.
3. A. D. Carlson, S. J. Friesenhan, W. M. Lopez and M. P. Fricke, *Nucl. Phys.*, **A141**, 577–594 (1970).
4. P.G. Young, personal communication.
5. M.B. Chadwick, personal communication.
6. P. Lisowski, C. Bowman, G. Russell, and S. Wender *Nucl. Sci. Engineering*, **106** 1990, pp. 208–213.
7. M. Heil, R. Reifarth, M. Fowler, R. Haight, F. Käppler, R. S. Rundberg, E. Seabury, J. Ullmann, J. Wilhelmy, K, Wisshak, et al., *Nucl. Instr. Meth. A*, **459**, 229–237 (2001).
8. D. Dabs, F. Stephens, and R. Diamond, Tech. Rep., Report LBL-8945, Lawrence Berkeley Laboratory (1979).
9. R. Reifarth, T. Bredeweg, A. Alpizar-Vicente, J. Browne, E.-I. Esch, U. Greife, R. Haight, R. Hatarik, A. Kronenberg, J. O'Donnell, et al., *Nucl. Instr. Meth. A*, **531**, 528 (2004).
10. J.U. Ullmann, U. Agvaanluvsan, A. Alpizar-Vicente, E. Bond, T. Bredeweg, E.-I. Esch, C. Folden, U. Greife, R. Hatarik, R. Haight, D. Hoffman, et al., in International Conference on Nuclear Data for Science and Technology, Santa Fe, NM, 2004 (AIP, 2004).
11. G.G. Miller, P.Z. Rogers, P. Palmer, D. Dry, R. Rundberg, M. Fowler, J. Wilhelmy, *J. Radioanal. and Nucl. Chem.*, **263**(2), 527–530 (2005).
12. M.M Fowler, J.C Gursky, J.B. Wilhelmy, *Nucl. Instr. Meth. A*, **303** 99–101 (1991).
13. D.W. Barr, personal communication.
14. T.A. Bredeweg, et al., This Conference.

Neutron Capture Cross Section Measurements at n_TOF of ^{237}Np, ^{240}Pu and ^{243}Am for the Transmutation of Nuclear Waste

D. Cano-Ott[31(1)], U. Abbondanno[20], G. Aerts[7], H. Álvarez[35], F. Alvarez-Velarde[31], S. Andriamonje[7], J. Andrzejewski[26], P. Assimakopoulos[16], L. Audouin[12], G. Badurek[1], P. Baumann[10], F. Bečvář[6], E. Berthoumieux[7], F. Calviño[34], R. Capote[3,36], A. Carrillo de Albornoz[27], P. Cennini[37], V. Chepel[28], E. Chiaveri[37], N. Colonna[19], G. Cortes[33], A. Couture[41], J. Cox[41], M. Dahlfors[37], S. David[9], I. Dillman[12], R. Dolfini[23], C. Domingo-Pardo[32] W. Dridi[7], I. Duran[35], C. Eleftheriadis[13], M. Embid-Segura[31], L. Ferrant[9], A. Ferrari[37], R. Ferreira-Marques[28], L. Fitzpatrick[37], H. Frais-Koelbl[3], K. Fujii[20], W. Furman[30], R. Gallino[22], I. Goncalves[28], E. Gonzalez-Romero[31], A. Goverdovski[29], F. Gramegna[18], E. Griesmayer[3], C. Guerrero[31], F. Gunsing[7], B. Haas[8], R. Haight[39], M. Heil[12], A. Herrera-Martinez[37], M. Igashira[25], S. Isaev[9], E. Jericha[1], Y. Kadi[37], F. Käppeler[12], D. Karamanis[16], D. Karadimos[16], M. Kerveno,[10] V. Ketlerov[29,37], P. Koehler[40], V. Konovalov[30,37], E. Kossionides[15], M. Krtička[6], C. Lamboudis[16], H. Leeb[1], A. Lindote[28], I. Lopes[28], M. Lozano[36], S. Lukic[10], J. Marganiec[26], L. Marques[27], S. Marrone[19], T. Martinez[31], P. Mastinu[18], A. Mengoni[3,37], P.M. Milazzo[20], C. Moreau[20], M. Mosconi[12], F. Neves[28], H. Oberhummer[1], S. O'Brien[41], M. Oshima[24], J. Pancin[7], C. Papachristodoulou[16], C. Papadopoulos[14], C. Paradela[35], N. Patronis[16], A. Pavlik[2], P. Pavlopoulos[11], L. Perrot[7], R. Plag[12], A. Plompen[5], A. Plukis[7], A. Poch[33], C. Pretel[33], J. Quesada[36], T. Rauscher[38], R. Reifarth[39], M. Rosetti[17], C. Rubbia[23], G. Rudolf[10], P. Rullhusen[5], J. Salgado[27], L. Sarchiapone[37], I. Savvidis[13], C. Stephan[9], G. Tagliente[19], J.L. Tain[32], L. Tassan-Got[9], L. Tavora[27], R. Terlizzi[19], G. Vannini[21], P. Vaz[27], A. Ventura[17], D. Villamarin[31], M. C. Vincente[31], V. Vlachoudis[37], R. Vlastou[14], F. Voss[12], S. Walter[12], H. Wendler[37], M. Wiescher[41], K. Wisshak[12]

The n_TOF Collaboration

[1]*Atominstitut der Österreichischen Universitäten,Technische Universität Wien, Austria* — [2]*Institut für Isotopenforschung und Kernphysik, Universität Wien, Austria* — [3]*International Atomic Energy Agency (IAEA), Nuclear Data Section, Vienna, Austria* — [4]*Fachhochschule Wiener Neustadt, Wiener Neustadt, Austria* — [5]*CEC-JRC-IRMM, Geel, Belgium* — [6]*Charles University, Prague, Czech Republic* — [7]*CEA/Saclay - DSM, Gif-sur-Yvette, France* — [8]*Centre National de la Recherche Scientifique/IN2P3 - CENBG, Bordeaux, France* — [9]*Centre National de la Recherche Scientifique/IN2P3 - IPN, Orsay, France* — [10]*Centre National de la Recherche Scientifique/IN2P3 - IReS, Strasbourg, France* — [11]*Pôle Universitaire Léonard de Vinci, Paris La Défense, France* — [12]*Forschungszentrum Karlsruhe GmbH (FZK), Institut für Kernphysik, Germany* — [13]*Aristotle University of Thessaloniki, Greece* — [14]*National Technical University of Athens, Greece* — [15]*NCSR, Athens, Greece* — [16]*University of Ioannina, Greece* — [17]*ENEA, Bologna, Italy* — [18]*Istituto Nazionale di Fisica Nucleare(INFN), Laboratori Nazionali di Legnaro, Italy* — [19]*Istituto Nazionale di Fisica Nucleare, Bari, Italy* — [20]*Istituto Nazionale di Fisica Nucleare, Trieste, Italy* — [21]*Dipartimento di Fisica, Università di Bologna, and Sezione INFN di Bologna, Italy* — [22]*Dipartimento di Fisica, Università di Torino and Sezione INFN di Torino, Italy* — [23]*Università degli Studi Pavia, Pavia, Italy* — [24]*Japan Atomic Energy Research Institute, Tokai-mura, Japan* — [25]*Tokyo Institute of Technology, Tokyo, Japan* — [26]*University of Lodz, Lodz, Poland* — [27]*Instituto Tecnológico e Nuclear(ITN), Lisbon, Portugal* — [28]*LIP - Coimbra & Departamento de Fisica da Universidade de Coimbra, Portugal* — [29]*Institute of Physics and Power Engineering, Kaluga region, Obninsk, Russia* — [30]*Joint Institute for Nuclear Research, Frank Laboratory of Neutron Physics, Dubna, Russia* — [31]*Centro de Investigaciones Energeticas Medioambientales y Technologicas, Madrid, Spain* — [32]*Instituto de Física Corpuscular, CSIC-Universidad de Valencia, Spain* — [33]*Universitat Politecnica de Catalunya, Barcelona, Spain* — [34]*Universidad Politecnica de Madrid, Spain* — [35]*Universidade de Santiago de Compostela, Spain* — [36]*Universidade de Sevilla, Spain* — [37]*CERN, Geneva, Switzerland* — [38]*Department of Physics and Astronomy - University of Basel, Basel, Switzerland* — [39]*Los Alamos National Laboratory, New Mexico, USA* — [40]*Oak Ridge National Laboratory, Physics Division, Oak Ridge, USA* — [41]*University of Notre Dame, Notre Dame, USA*

Abstract. Accurate and reliable neutron capture cross section data for actinides are necessary for the proper design, safety regulation and precise performance assessment of transmutation devices such as Fast Critical Reactors or Accelerator Driven Systems. In particular, the neutron capture cross sections of ^{237}Np, ^{240}Pu and ^{243}Am play a key role in the design and optimization of strategies for the Transmutation of Nuclear Waste. The listed cross sections have been measured in 2004 at n_TOF [1] with a high accuracy due to a combination of features unique in the world: high instantaneous neutron fluence and excellent energy resolution of the n_TOF facility, innovative Data Acquisition System based on flash ADCs and the use of a high performance BaF$_2$ Total Absorption Calorimeter as a detection device.

Keywords: ^{237}Np, ^{240}Pu, ^{243}Am, neutron capture cross section, nuclear waste, transmutation, total absorption calorimeter
PACS: 25.40.Lw,27.80.+w,97.10.Cv

INTRODUCTION

Nuclear waste transmutation has been proposed as a way to reduce substantially (in a factor of 1/100 or more) the radiotoxicity inventory of the long lived component of the nuclear waste, mainly the trans-uranium actinides. Actinide transmutation is proposed to take place by fission in nuclear systems like critical reactors or subcritical Accelerator Driven Systems (ADS). The detailed engineering designs, safety evaluations and the detailed performance assessment of dedicated transmutation ADS and critical reactors (i.e. with fuels highly enriched in transuranic isotopes) require more precise and complete basic nuclear data [2].

EXPERIMENT

Isotopically enriched targets of minor actinides are available only in small amounts, typically of the order of 1 - 100 mg. Even then, the activities involved present major difficulties related both to the experimental techniques and the radioprotection aspects. The targets of ^{237}Np (43.3 mg, 1.29 MBq), ^{240}Pu (51.2 mg, 458 MBq) and ^{243}Am (10 mg, 75 MBq) measured at n_TOF [1] were all assembled in the same way: the radioactive material was sandwiched between two thin Al layers (total mass < 75 mg) and canned inside a 0.35 mm thick Ti canning with ISO 2919 certification (requested by the safety regulations at CERN). Their isotopic purity was determined by γ-ray spectrometry and is >98% in all cases.

In capture measurements at n_TOF [3] with a 4 cm diameter neutron beam spot, the instantaneous fluence amounts to 10^5 neutrons/cm^2/pulse for neutron energies between 0.1 eV and 20 GeV. This is one of the key features for measuring the (n,γ) cross section of low mass and highly radioactive isotopes with a good signal to noise ratio. During the measurements, the repetition rate of the PS was on average 3 - 4 proton pulses for a PS supercycle of 16.4 s. In addition, the n_TOF facility provides an excellent energy

---

[1] Corresponding Author: address CIEMAT, Avda Complutense 22, E-28040 Madrid, Spain. Tel. +34-913466116, Fax. +34-913466576, e-mail: daniel.cano@ciemat.es

resolution of 10^{-3} to 10^{-4} at a 185 m flight path between the Pb spallation target and the counting station.

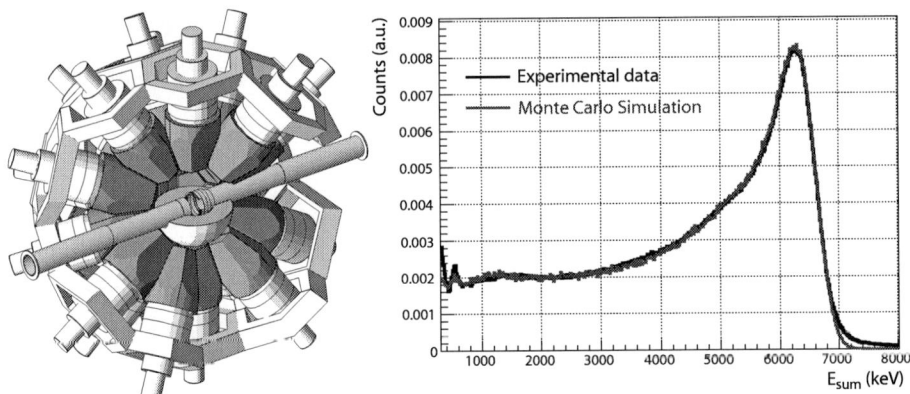

FIGURE 1. Left: view of the Total Absorption Calorimeter as it is implemented in the code GEANT4[9] used for the MC simulations. Right: experimental (black) and Monte Carlo simulated (red) energy deposition spectra in the TAC for capture events in the ^{197}Au(n,γ) reaction at 4.9 eV.

The capture cross section measurements at n_TOF have been done relative to the standard capture cross section ^{197}Au(n,γ). For this reason, several independent monitors were used permanently for a proper normalization between the ^{197}Au(n,γ) and main measurements.

The Data Acquisition System (DAQ)[4] used in the measurements consists in 54 channels of high performance flash ADCs [5]. Each channel has 8 Mbytes memory and was operated at a sampling rate of 500 Msamples/s, thus allowing to record the full detector history for neutron energy ranges between 0.3 eV and 20 GeV. After zero suppression and data formatting, the raw data are sent to CERN's massive storage facility CASTOR [6] via several Gigabit links. In parallel, especially designed pulse shape analysis routines run on a PC farm and extract from the digitized detector signals the necessary information for the data analysis. The n_TOF DAQ offers unique features such as an extremely low dead time (< 10 ns), good signal analysis and pileup discrimination among others, resulting in an excellent mechanism for controlling all kind of systematic uncertainties associated to the detector's behavior.

The neutron capture detection system consists in a segmented Total Absorption Calorimeter made of 40 BaF$_2$ crystals with ^{10}B loaded carbon fibre capsules and placed at 185 m flight path from the spallation source. The TAC has a nearly 100% detection efficiency for electromagnetic cascades (i.e. capture events) and a good energy resolution (14% at 662 keV and 6% at 6.1 MeV). The radioactive targets are placed at the geometric center of the TAC and surrounded by a $C_{12}H_{20}O_4(^6Li)_2$ neutron absorber placed inside the inner hole of the TAC. The neutron absorber and the ^{10}B loaded carbon fibre capsules reduce the sensitivity of the detector to the scattered neutrons, do not reduce the capture detection efficiency (even though they lower the total absorption efficiency) and also help in attenuating the low energy component (10 - 100 keV) of the sample

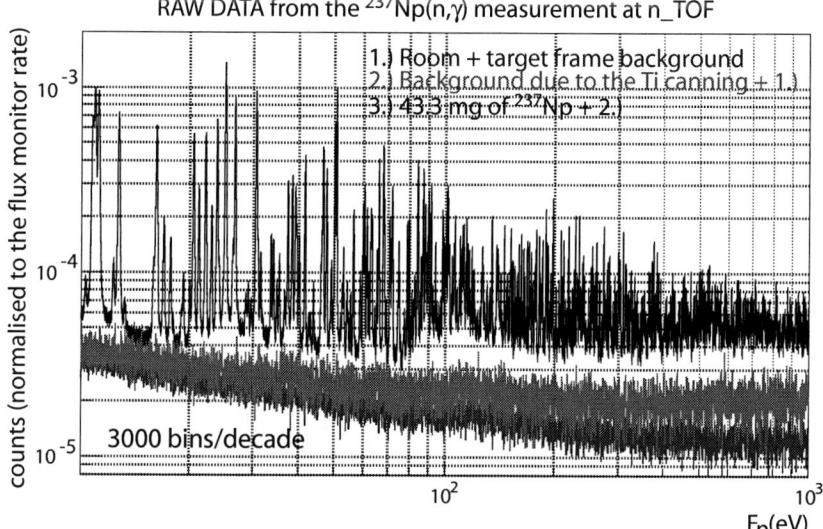

FIGURE 2. Raw data from the ^{237}Np (n,γ) measurement in the resolved resonance region between 10 eV and 1 keV. Below, the background due to the Ti canning and the empty TAC (without any sample in the beam).

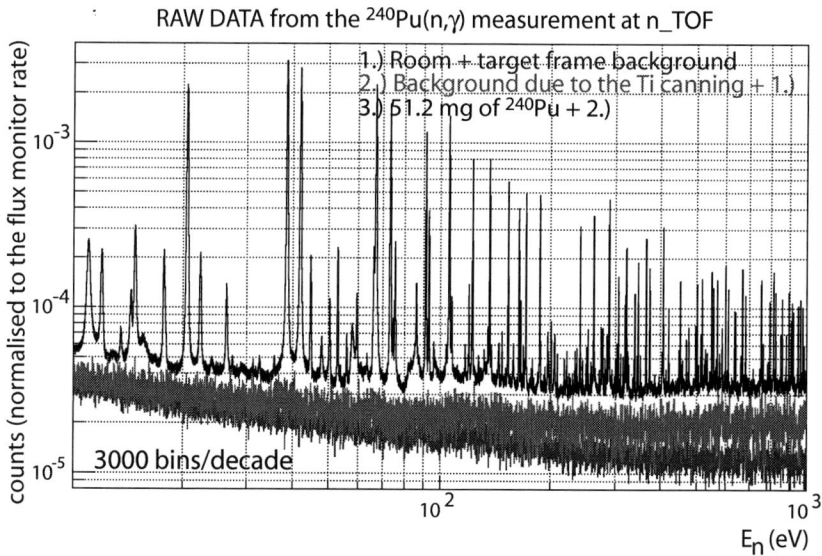

FIGURE 3. Raw data from the ^{240}Pu (n,γ) measurement in the resolved resonance region between 10 eV and 1 keV. Below, the background due to the Ti canning and the empty TAC (without any sample in the beam).

radioactivity. For the measurement of ^{243}Am, a cylindrical Pb shielding of 1 mm thickness around the target and outside the beam was necessary for suppressing the strong γ-activity with energies about 100 - 200 keV.

Fig. 1 shows a view of the experimental setup, as it is implemented in the code GEANT4[9] used for the detailed MC simulations of the TAC. The performance of the TAC has been investigated both experimentally (with standard calibration sources and the reference ^{197}Au(n,γ) cross section) and by Monte Carlo simulations [7] [8]. Furthermore, all sources of background have been measured and are being simulated for performing the background corrections necessary for an accurate capture cross section analysis.

RESULTS

The capture data for ^{237}Np, ^{240}Pu and ^{243}Am have been taken in summer 2004. A large number of resonances have been observed in the resolved resonance region with good statistics in both the raw data for ^{237}Np and ^{240}Pu, as it is shown in fig. 2 and 3, respectively. Furthermore, the level of the background there is fairly low, thus allowing to conclude already that the extraction of accurate resonance parameters is possible. The cross section analysis in the unresolved resonance region will require a more detailed study and subtraction of the background, dominated mainly by the effect of the Ti canning of the samples. However, clear capture signals are visible in the TAC energy deposition spectra up to a few tens of keV.

ACKNOWLEDGMENTS

This work is part of the European Commission 5th Framework Programme project NTOF-ND-ADS under contract FIKW-CT-2000-00107.

REFERENCES

1. D. Cano-Ott et al. (The n_TOF Collaboration), Proceedings of the Nuclear Data for Science and Technology, Santa Fe - 2004, AIP Conf. Proc. 769 (2004) 1442.
2. Accelerator Driven Systems and Fast Reactors in Advanced Nuclear Fuel Cycles, ISBN 92-64-18482-1 ENEA/OECD Report, 2002
3. U. Abbondano et al., "n_TOF Performance Report", CERN/INTC-O-011, INTC-2002-037
4. U. Abbondanno, et al. (The n_TOF Collaboration), Nucl. Instr. Meth. A 538 (2005) 692.
5. http://www.acqiris.com
6. http://castor.web.cern.ch/castor
7. M. Heil et al. *Nucl. Instr. Meth. A* 459(2001) 229
8. D. Cano-Ott et al., "Monte Carlo Simulation of the 4π Total Absorption Calorimeter at n_TOF", n_TOF internal report (2003)
9. S.Agostinelli et al.,*Nuc. Instr. and Meth. A*, 506, (2003), p. 250-303.

Gamma-Ray Cross Section Standards in the MeV Energy Range and ^{56}Fe Inelastic Scattering

R. O. Nelson*, M. Devlin*, N. Fotiades*, J. A. Becker†, P. E. Garrett†, W. Younes†, D. Dashdorj†, T. Ethvignot‡ and T. Granier‡

*LANSCE-NS, Los Alamos National Laboratory, Los Alamos, NM 87545
†Lawrence Livermore National Laboratory, Livermore, CA 94551
‡CEA, Bruyères-le-Châtel, France

Abstract. An accurate knowledge of neutron-induced γ–ray production cross sections is important in enabling the determination of absolute cross sections from relative measurements of γ–ray yields. Such data have a variety of uses, for example, in determining reaction cross sections for activation of materials used as integral neutron flux detectors, and in measuring photon production necessary to evaluate γ–ray heating in nuclear systems. Such data also find application in neutron-induced γ–ray detection and assay systems and in planetary exploration where cosmic-ray-neutron-induced photon production is used to determine elemental composition of soils. In our work using the GEANIE γ–ray spectrometer at the WNR facility of LANSCE, we observed discrepancies outside of the error estimates for evaluations of the ^{nat}Fe(n,n$_1$'γ) cross section for the 847-keV γ ray and for other reactions that are taken as standards. We have performed measurements relative to the ^{nat}Cr(n,n$_1$'γ) 1434-keV γ ray absolute cross section measurements using the ^{235}U(n,f) and ^{238}U(n,f) reactions to determine the neutron fluence.

Keywords: MeV gamma–ray cross sections, standards, neutron inelastic scattering, Iron, Chromium, Fe, Cr, HPGe detector
PACS: 25.40Fq 27.40.+z, 29.30.Kv,

INTRODUCTION

For thermal neutron capture there exist a number of well determined standard cross sections [1]. These cross sections can be used to calibrate new measurements by measuring the relative intensity of γ rays, thus circumventing many of the difficulties associated with an absolute measurement. True standard cross sections are generally accurate to a few percent and have passed a rigorous documentation process. Here we refer to convenient standards that may be accurate to better than 10% as secondary standards.

For neutron inelastic scattering in the MeV energy range there are several γ–ray cross sections that have been suggested as secondary standards, however, the situation is not very satisfactory. In particular the ^{56}Fe(n,n'γ) E$_\gamma$ = 847 keV cross section has been used as a standard in a number of measurements, using Fe of natural isotopic abundance. Other suggested cross section standards include ^{27}Al(n,n'γ) - 3000 keV, ^{28}Si(n,n'γ) - 1778 keV, and ^{52}Cr(n,n'γ) - 1434 keV [2].

Some desirable properties of a standard cross section for neutron-induced reactions producing γ–rays are: (1) Large cross section – for good signal to background ratio and short measuring times. (2) Good physical properties: readily available, relatively inexpensive, chemically stable, easily fabricated, uniform in density, etc. (3) Constant with incident neutron energy – no large fluctuations with small changes in energy. (4) Isotropic – independent of measurement angle. (5) No backgrounds from common materials or activation at the chosen incident neutron energy.

While there is no perfect cross section standard for every experiment, some cross sections appear to be better known than others, and the development of a set of standards covering a wide incident neutron energy range and a wide γ-ray energy range can provide useful tools for future data measurements and for applications. Examples of applications include: use of differential cross sections for calculation and interpretation of integral radiochemical diagnostics, calculation of activation, radiation damage, and heating in nuclear power systems, interpretation of γ–ray data from planetary exploration into elemental composition of planetary surfaces such as for the Moon and Mars, use of cross sections for dose determination in medical radiotherapy, and determination of sensitivities of nuclear-based detection systems, such as for explosives, nuclear materials, or other items of interest.

Inelastic scattering on Fe is very important in many applications. Radiation damage in steel pressure vessels of nuclear power reactors determines the working life of the vessel. The damage is calculated using nuclear reaction cross sections, of which inelastic scattering is an important component. The 2^+ to 0^+ decay in ^{56}Fe includes more than 94% of the inelastic cross section, and thus provides a means to more accurately determine this important cross section. Differences in evaluated data libraries are shown in Fig. 1.

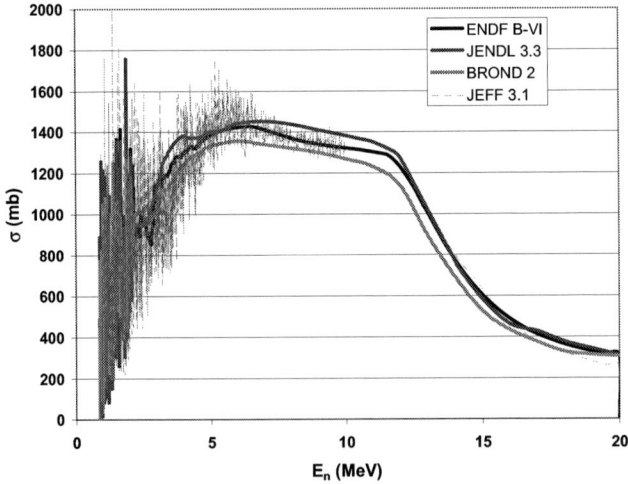

FIGURE 1. Four evaluated data libraries for the ^{56}Fe inelastic scattering cross section are shown. The JEFF 3.1 evaluation uses the very high resolution data of Dupont, *et al.* [6] (normalized to other data) up to E_n = 10 MeV. The JENDL evaluation has the highest average values, while the BROND evaluation has the lowest values.

There are 29 known γ–rays that decay to the ground state of ^{56}Fe (bypassing the 2^+ first excited level) with measured branching ratios or cross sections. The maximum cross section for γ-decays to the ground state, determined from the measured branching ratios and cross sections for the stronger γ rays, is 6% near 5 MeV. This decreases rapidly with increasing neutron energy. Thus, the 2^+ to 0^+ decay gives a good measure of the total inelastic scattering cross section.

In this paper we discuss mainly two secondary standards for neutron-induced γ–ray cross sections, ^{56}Fe and ^{52}Cr. These are typically obtained from material of natural isotopic abundance. This is followed by a discussion of other potential cross section standards that we are working to improve or develop. Because Fe is an important structural material, there are many applications where an accurate knowledge of the Fe inelastic scattering cross section is important. We show that the present data may be used to refine and extend our knowledge of this cross section.

Fe(n,n'γ) CROSS SECTION MEASUREMENTS

Measurements of inelastic neutron scattering on iron have a long history. Understanding of nuclear inelastic scattering began less than 3 years after the discovery of the neutron. Aoki [3] and Lea [4] reported measurements in 1935 using d-d neutron sources, and a PoBe source that produces neutrons by the (α,n) reaction on Be. In 1938 Grahame and Seaborg [5] published a paper on inelastic scattering of neutrons that further clarified the work of Aoki and Lea. There have been more than 40 measurements of the Fe(n,n'γ) cross sections reported since 1968 using modern techniques. Current measurements, in addition to the present work include the very high resolution data of Dupont *et al.* [6] at GEEL (used in the recent JEFF 3.1 evaluation) and ongoing work of Plompen *et al.* [7] also at GEEL.

An examination of the Fe data shows a very wide range in the measured values. Indeed, two recent evaluations differ by 25% [2,8]. This scatter is much more than is to be expected from the usual variation among different experiments that measure the same quantity. The reasons for these discrepancies appear to be many:

Backgrounds are difficult to determine well for the Fe 846.8-keV γ ray – backgrounds from ^{27}Al (844 keV) and from ^{72}Ge 835 keV + recoil energy if Ge detectors are used.

^{56}Mn activation from the ^{56}Fe(n,p) reaction beta decays to ^{56}Fe producing the 847-keV γ ray can give a large contribution – in particular in experiments using continuous rather than pulsed beams.

The angular distributions of the 847-keV γ ray vary with incident neutron energy and can cause the measured cross section to vary from the average by up to 6 % at 14 MeV and by as much as 10 to 15% at 4 MeV incident neutron energies. Note that the measured angular data, as for the cross sections, are in poor agreement, and there are many fewer angular distribution measurements than for the cross sections.

Measurement of neutron fluxes can be difficult. It appears that errors in accurately determining neutron fluxes may have been a problem in some of the measurements.

Inaccuracy in determining multiple scattering and reaction contributions for thicker samples may also contribute to the spread in the measured values for the cross section.

Fe(n,n'γ) 2^+ to 0^+ does have a large cross section and samples are easily obtained and fabricated, but it is inadequate in terms of constancy of the cross section with energy and angle, and in being well separated from common background sources.

To improve the data we have performed measurements with GEANIE [9] (Germanium Array for Neutron-Induced Excitations) consisting of 26 HPGe Detectors with BGO escape suppression shields, located at the WNR spallation neutron source – driven by the 800 MeV LANSCE proton linac [10]. The neutron energy is determined by time-of-flight on a 20 meter flight path. The typical neutron energy range is $1 < E_n < 200$ MeV. Both 25% efficient coaxial HPGe detectors and low-energy planar HPGe detectors are used. The typical γ–ray energy range observed is 15 keV $< E_\gamma <$ 4 MeV. GEANIE was built using components of the former HERA array from Lawrence Berkeley National Laboratory. The neutron flux is measured using a fission ionization chamber [11]. This enables both absolute and relative cross section measurements. Accuracy of better than 2% is expected for relative cross sections, and absolute cross sections with an accuracy of better than 4% are expected. Final analysis of a relative measurement between Fe and Cr, and absolute measurements of both is underway.

Cr(n,n'γ) - A BETTER STANDARD

In looking for a better secondary standard than Fe, Cr was suggested by Herbert Vonach [12]. For measurements of the ^{52}Cr(n,n'γ) 1434-keV γ ray near 14 MeV, 6 of 7 measured values agree within errors. The evaluation of Simakov, *et al.* at E_n = 14.5 MeV gives a 3.8% error in the evaluated cross section for the Cr with the natural abundance of ^{52}Cr being 83.789 % [13].

As a test of our absolute cross section measurements, and due to the advantages of relative measurements we undertook a measurement of the ^{56}Fe 847-keV γ–ray cross section relative to the ^{52}Cr 1434-keV γ ray. In a relative measurement many quantities are the same for both samples: neutron fluence, electronic dead times, and absolute detector acceptance. Perhaps the most important quantities to determine well in a relative measurement are the sample thickness and relative detector efficiency.

An accurate determination of relative areal densities was accomplished using x-ray radiography. The Cr plate used exhibited variations in density of 20% from edge to edge. Even the rolled Fe plate showed variations of 5 % in density – larger than expected.

One other area of concern is that of backgrounds. Determining the background from Fe in the experimental setup and from other reactions is important. This was studied using data from a Cr + V relative measurement conducted immediately after the Fe + Cr relative measurement.

EXPANDING THE SELECTION OF GAMMA-RAY CROSS SECTION STANDARDS

To improve the usefulness of secondary cross section standards, we have undertaken a series of measurements on other materials to extend both the incident-neutron and γ–ray energy ranges available. The nuclides currently being investigated, associated γ–ray energies, and cross sections are shown in Table 1.

Both absolute cross section measurements and relative measurements using ^{52}Cr(n,n'γ) as a standard have been performed or are planned in the energy range, $1 < E_n < 200$ MeV. The goal is to provide secondary cross section standards spanning the γ-ray energy range, $0.226 < E_\gamma < 6.13$ MeV. More accurate Fe cross section data may find use in many applied areas. Final corrections are in progress to refine the accuracy of the cross sections.

TABLE 1. Some possible γ–ray cross section secondary standards

Active nuclide (Abundance) [13]	Primary γ-ray energy (keV) (approximate 14.5 MeV cross section in mb) [2]	Other useful energies (keV) (approximate 14.5 MeV cross section in mb) [2]
^{51}V (99.750%)	320 (313)	226 (368), 1609 (214)
^{56}Fe (91.754%)	847 (785)	1238 (393)
^{52}Cr (83.789%)	1434 (695)	935 (210)
^{28}Si (92.230 %)	1778 (403)	2839 (59), 5100 (37), 6879(36)
^{27}Al (100%)	3004 (111)	1809 (184)
^{12}C (98.89 %)	4439 (187)	
^{16}O (99.762%)	6129 (183)	

ACKNOWLEDGMENTS

Portions of this work were performed under the auspices of the U.S. Department of Energy by the University of California, Los Alamos National Laboratory under Contract No. W-7405-ENG-36, and the Lawrence Livermore National Laboratory under Contract No. W-7405-ENG-48. This work has benefited from the use of the Los Alamos Neutron Science Center at LANL that is also funded under Contract No. W-7405-ENG-36.

REFERENCES

1. E. T. Jurney, J. W. Starner, J. E. Lynn, and S. Raman, Phys. Rev. **C56** 118 (1997) and references therein.
2. S. P. Simakov, et al., IAEA report INDC(CCP)-413, 1 (1998).
3. Aoki, Proc. Phys. Math . Soc. Japan **19**, 369 (1935).
4. D. E. Lea, Proc Roy. Soc. (London) **A150**, 637 (1935).
5. D. C. Grahame and G. T. Seaborg, Phys. Rev. **53**, 795 (1938).
6. E. Dupont, et al. in the JEFF 3.1 evaluation.
7. A. Plompen, personal communication.
8. M. V. Savin, A. V. Livke and A. G. Zvenigorodskij, IAEA report INDC(CCP)-426 95 (2000).
9. J. A. Becker and R. O. Nelson, Nucl. Phys. News Int. **7**, 11 (1997).
10. P. W. Lisowski, C. D. Bowman, G. J. Russell, and S. A. Wender, Nucl. Sci. Eng. **106**, 208 (1990).
11. S. A. Wender, et al., Nucl. Instrum. Meth. A **336**, 226 (1993).
12. H. Vonach, personal communication.
13. Nuclear Wallet Cards, J. Tuli, NNDC, Brookhaven National Laboratory, 2005.

NUCLEAR STRUCTURE

Level Dynamics of non-yrast States

J. Jolie and S. Heinze

Institute of Nuclear Physics, University of Cologne, Zülpicher Strasse 77, 50937 Cologne, Germany,

Abstract. We present a systematic study of the dynamics of non-yrast states using a simple but physically relevant hamiltonian. The emphasis is put on the behavior of 0^+ states. When the transition takes place between spherical and gamma-soft nuclei level bunching occurs which is related to the conserved seniority. Similar behavior occurs in the regular region which is not directly connected with the known symmetries. In between spherical and prolate nuclei a sharp increase in the level density of 0^+ states is found. We report on recent (p,t) transfer reactions related to these topics which were performed at the Q3D spectrometer in Garching with the aim to measure the 0^+ states in a complete way up to high energy.

INTRODUCTION

One of the most successful of recent models used in nuclear structure is the Interacting Boson Model (IBM) [1] of which we will study here a very simple, but physically relevant, hamiltonian. It consists of only two interactions: one which gives rise to harmonic vibrations and one which causes nuclear deformation. The structures this very simple hamiltonian generates are extremely rich as we will show. This parameterisation has been used in the past to study in the context of the IBM chaotic and regular behavior [2], catastrophe theory [3], and quantum phase transitions [4]. It gained recently a lot of new attention [5, 6, 7, 8, 9, 10, 11].

Most of the recent work done with this simple Hamiltonian is either related to experimental observables connected with the lowest states or concerned the groundstate. Here we will take another approach and study the behavior of all states with a given spin as a function of the changing structure. In order to enhance the visibility of these effects we will use very high boson numbers. Finally, we will discuss that effects obtained from the schematic calculations have partially been observed in recent (p,t) experiments using the Q3D spectrometer at Garching.

PHASE TRANSITIONS IN THE INTERACTING BOSON MODEL

Phase transitions involving nuclear shapes became again to the forefront of nuclear structure physics when F. Iachello develloped new symmetries which can describe atomic nuclei at the critical points [13, 14]. The new symmetries, called X(5) and E(5), are obtained within the framework of the collective model [15] under some simplifying approximations. Remarkably, the parameter free predictions provided by these symmetries are closely realised in some atomic nuclei [16, 17]. The phase transitions considered in these references are those of the ground state deformation, which can conveniently be described by the three Euler angles defining the orientation of the deformed nucleus in space and the deformation parameters β and γ defining the form of the ellipsoid. These quantum phase transitions take place at zero temperature and depend on the number of neutrons and protons. While these new symmetries provide only an analytic solution at the phase transition, they do not allow the study what happens close to the phase transition.

One of the best nuclear models to study shape phase transitions is the Interacting Boson Model [1] which is able to describe all common shapes of atomic nuclei in an unified way. We will use the following very simple Hamiltonian:

$$\hat{H} = \alpha[\eta \hat{n}_d - \frac{(\eta-1)}{N} \hat{Q}_\chi \hat{Q}_\chi] \quad (1)$$

with

$$\hat{Q}_\chi = (s^\dagger \tilde{d} + d^\dagger s)^{(2)} + \chi (d^\dagger \tilde{d})^{(2)}, \quad (2)$$

Here we have used η and χ to play the role of the control parameters and pulled out an overall energy scaling factor α. The parameter η governs the transition between spherical and deformed, and χ the transition between prolate and oblate deformation. The Hamiltonian (1) contains the three dynamical symmetries of the IBM: $U(5)$, $O(6)$ and $SU(3)$ as well as the $\overline{SU(3)}$ limit. These correspond to vibrational nuclei with a spherical form ($U(5)$), an axially symmetric rotor which is prolate ($SU(3)$) or oblate ($\overline{SU(3)}$), and an axially asymmetric rotor with a flat potential in γ ($O(6)$).

Equation 1 is thus able to generate the full range of structure of the IBM. The first term leads to a spherical solution while the second term, which is a quadrupole interaction between bosons, yields deformed solutions. The equilibrium shape is spherical for η values ranging down from unity, crosses from spherical to deformed at some value of $\eta \equiv \eta_{crit}$, and is deformed for $\eta < \eta_{crit}$. The quadrupole operator introduces another parameter χ which is related to axial asymmetry, and which notably increases the richness of the structures possible with eq. 1. The parameter $\chi \in [-\sqrt{7}/2, \sqrt{7}/2]$:negative (positive) χ corresponds (if $\eta < \eta_{crit}$) to a prolate (oblate) axially symmetric rotor and $\chi=0$ to a completely γ-soft rotor. Figure 1 illustrates the such obtained phase diagram and indicates the position of the first and second order phase transitions.

It should be mentioned that strictly speaking phase transitions happen when the number of bosons N goes to infinity, which doesn't happen in actual nuclei. Instead, we are dealing with a mesoscopic system where we can study how the finite number influences the phase transition. Indeed an extension of the concept to systems having a limited number of constituents is of importance today. Finite N effects are also important since atomic nuclei contain only an integer number of nucleons, nature does not allow us to vary the control parameters continuously in the region where the phase transition occurs. While the experimental limitation to integer nucleon number leads to discrete changes in the properties of atomic nuclei around the critical point, theoretical models do allow one to continuously vary the appropriate control parameter freely. Interestingly, we have recently showed how the general characteristics of phase transitional character in collective models or at N going to infinity can be described using the Landau theory of continuous phase transitions [7, 8].

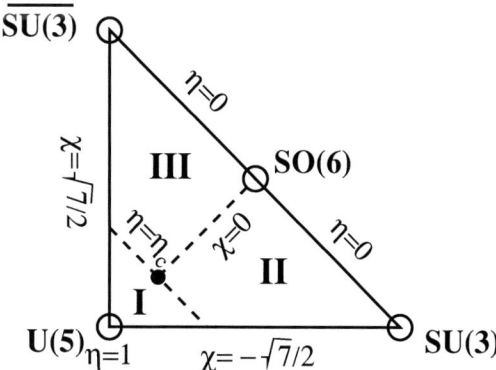

FIGURE 1. The extended casten triangle obtained using the Hamiltonian (1). Dashed lines indicate the first order phase transitions and the filled dot the isolated second order transition [5]. The open circles indicate the location of the different dynamical symmetries. The three different phases are spherical (I), prolate deformed (II) and oblate deformed (III).

LEVEL DYNAMICS OF YRARE STATES

In order to study the level dynamics we have performed theoretical calculations using the Hamiltonian (1) for 30 bosons and varying either η or χ using a fixed value of 100 keV for α. The calculations were performed keeping one control parameter fixed and varying the other one. The results for all 0^+ are given in Figure 2.

One notices striking differences between the three transitions. The transition between $U(5)$ and $SU(3)$ shows many avoided crossings and a maximal level density at the phase transition where the kink in the ground state energy is situated. The $\overline{SU(3)}$ to $SU(3)$ has a similar behavior except at the phase transition where real crossings take place. The behavior is also symmetric around $\chi = 0$. Finally, the $U(5)$ to $O(6)$ transition exhibits many real crossings everywhere. The explanation of the different crossing behavior are the occurence of either a dynamical symmetry or a partial dynamical symmetry. In the upper part there exist no additional quantum number except at the symmetries and such levels with the same spin and parity repel each other. In the middle part at $\chi = 0$ the $O(6)$ symmetry forms the phase transition and all states obtain the additional $O(6)$ and $O(5)$ quantum numbers there. Finally in the lower part everywhere the $O(5)$ quantum number is conserved leading to the many crossings. When one select the levels also by their $O(5)$ quantum number a smooth behavior is obtained showing the typical characteristics of a second order phase transition [18]. Of interest in the lower part is also the level bunching observed for excited states at energies close to zero. This phenomenon can be related to the classical orbit phase-space that becomes available at this energy and is related to monodromy [19]. It will be discussed in another contribution at this conference [20].

The existence of regular (integrable) and chaotic behavior in many-body systems forms also a major theme in many branches of physics. Atomic nuclei provide an important testing ground for such behavior because of the variety of structures they

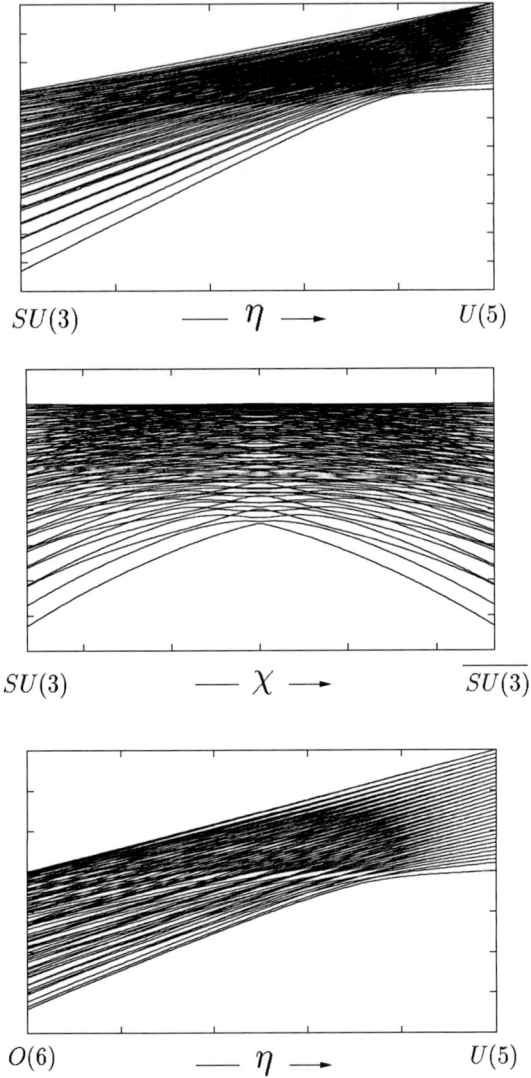

FIGURE 2. All 0^+ states as a function of χ or η for 30 bosons keeping. The upper part has $\chi = -\sqrt{(7)}/2$, the middle $\eta = 0$ and the lower part $\chi = 0$. Indicated are also the different dynamical symmetries.

exhibit and the dependence of these structures on the number of constituent nucleons.

Chaotic behavior in nuclei is usually thought of in terms of high temperature systems where shell structure melts. However, a decade ago Alhassid and Whelan carried out an important study of the chaotic behavior in nuclei using the Hamiltonian of eq.(1). As

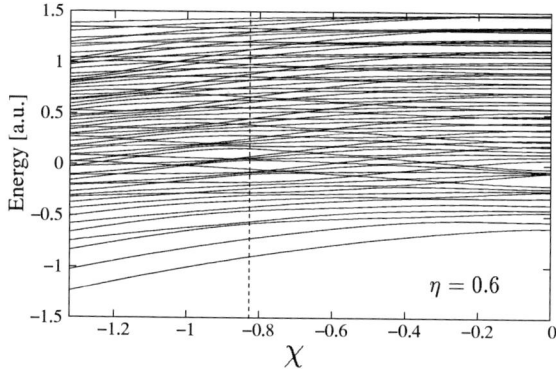

FIGURE 3. Energies of 0^+ states calculated with the Hamiltonian of eq. 2 with $N = 25$ for $\eta=0.6$ as a function of χ. The regular region, according to eq. 7, is close to $\chi=-0.8$

expected, they found that the level spectra are regular at and near the three dynamical symmetries U(5), SU(3) and O(6) of the model and that chaotic behavior develops rapidly as one moves away from the symmetries. There were two exceptions to this behavior. The first happens between U(5) and O(6) and is connected to the conserved O(5) symmetry mentioned before. The most fascinating result, however, was that there is an unexpected region of nearly regular behavior connecting SU(3) to U(5) along an interior arc in the triangle which can be parameterized by the relation [2, 21]:

$$\chi = \frac{\sqrt{7}}{2}(\eta - 1) - \frac{\eta}{2}. \tag{3}$$

In the (η,χ) plane the location of the regular region is independent of the angular momentum L, at low to moderate L values, and of the number of valence nucleons (*i.e.*, the boson number N).

Regular and chaotic behavior were distinguishable in the Alhassid–Whelan analysis by the distribution of nearest neighbor spacings of states in the same spin. Chaotic behavior is associated with the loss of good quantum numbers (except angular momentum) and hence of level mixing and avoided crossings. This leads to level repulsion and a Wigner distribution of level spacings. In regular systems, one or more quantum numbers remain valid (*e.g.*, the quantum number of O(5) along the U(5)–O(6) leg of the triangle). Hence there can be level crossings and Poisson distributions of spacings result (high probability of close lying levels). This is illustrated in Fig. 3, where the spectra of 0^+ states are plotted for a typical value of $\eta=0.6$, as a function of χ for $N=25$. For this situation, the regular region, where $\sim$90% of all trajectories are regular, is centered at $\chi \sim -0.8$. It is clear that, rather suddenly, the avoided crossings that appear for $\chi \neq 1$ values are replaced by nearly real crossings near $\chi \sim -0.8$.

When following the path given by Eq.(4) through the triangle a behavior that is very similar to the one in the $U(5)$-$O(6)$ transition is observed as shown in Fig. 4. Again one

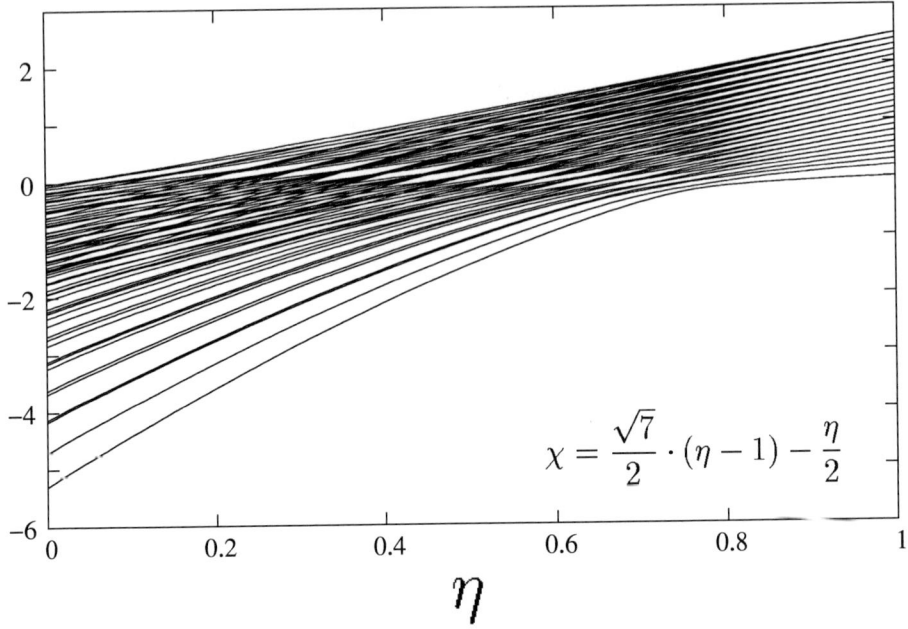

FIGURE 4. Energies of 0^+ states calculated for $N = 25$ along the path given by eq. 4.

has level bunching at the energy $E = 0$ indicating that also here monodromy might play a role.

At the time of their original work, unfortunately, there were no identified nuclei corresponding to this new regular region and this finding therefore remained merely a fascinating theoretical curiosity. However, a recent fit of nuclei in the rare-earth region using the Hamiltonian (1) which put more emphasis on describing also excited 0^+ states [10] radically changed this conclusion [11]. In fact, twelve rare earth nuclei were found to fall very near the regular region. They are ^{156}Gd, ^{158}Gd, ^{156}Dy, ^{156}Er, ^{158}Er, ^{170}Yb, ^{170}Hf, ^{172}Hf, ^{176}W, ^{178}W, ^{178}Os, and ^{180}Os. Note that they span a wide variety of structures ranging from anharmonic vibrator to near perfect rotors.

With regard to the particular distributions at the regular region shown in Figure 3, a $^{172}Yb(p,t)^{170}Yb$ was performed at the Q3D spectrometer operated by the LMU and TU Munchen in Garching. The use of the (p,t) reaction with the Q3D spectrometer recently gained a lot of attention when a collaboration led by the Notre-Dame group showed that very complete spectroscopy of 0^+ states in ^{158}Gd could be obtained up to high energies [22]. Motivated by this study we wanted to study whether the distribution in ^{170}Yb showed particular bunching. ^{170}Yb is a good case for such a study as it has a relatively high number of bosons with $N = 15$, making it one of the nuclei with the highest boson number that can be reached using the (p,t), the highest one being $N = 17$

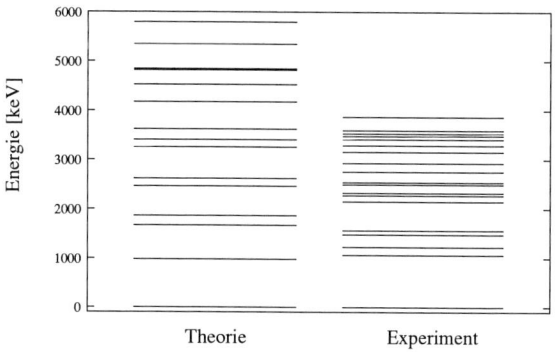

FIGURE 5. Comparison between experimental 0^+ states in ^{170}Yb and the theoretical spectrum.

for ^{174}Yb.

Up to an excitation energy of about 4 MeV, 19 0^+ states were observed of which 14 for the first time. One proposed state at 1228.84 keV could be excluded by our data. In Figure 4 the results of this experiment are compared to the theoretical fit from [10]. The first observation to be made is that the theoretical level density is much lower than the experimental one. This is especially the case once the pairing gap is reached at about 2 MeV. Clearly the Hamiltonian (1) is much too restricted to describe the complete complexity at high energies. Indeed, in a similar study for nuclei in the actinide region by Wirth et al., a comparison with states up to 2.7 MeV has shown already that a spdf model instead of a sd boson model is needed to describe the number of states [23]. Also particle-hole excitations can generate much more states than our Hamiltonian. While the data due to these extra excitations do not allow to make any valid conclusions at present, they certainly form a very interesting data set for future theoretical work.

The experiment on ^{170}Yb formed part of a much more extended search for 0^+ states that covered many even-even nuclei in the rare earth region [24, 25]. Generally most nuclei where only measured up to an energy below 3 MeV. However the study confirmed one of the properties of the level dynamics of the Hamiltonian (1). In Figure 2 (top) one notices a strong increase of the level density at the phase transition (slightly below $\eta = 0.8$). Figure 6 illustrates this observation by experimental data. Here we have plotted the number of 0^+ states below 2.5 MeV for those nuclei studied in [22, 25] where the parameters of the simple Hamiltonian have been fitted in ref [10] as a function of the fitted parameter η. In the region studied nuclei are close to the $U(5)$-$SU(3)$ transition and indeed an increase at low-energy was observed exactly were expected at $\eta = 0.8$. Note that here no corrections were applied for the changing boson number and the fitted χ parameter.

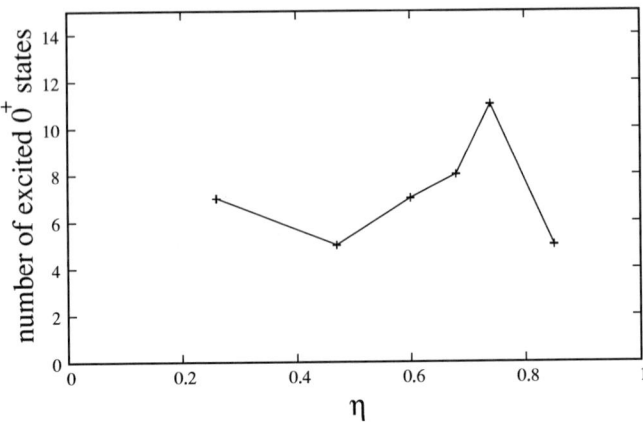

FIGURE 6. Number of observed 0^+ states below 2.5 MeV as a function of η (see text). The corresponding nuclei are from left to right: ^{162}Dy, ^{168}Er, ^{158}Gd, ^{176}Hf, ^{154}Gd and ^{152}Gd.

CONCLUSION

We have studied the level dynamics of 0^+ states with a schematic but realistic IBM Hamiltonian. In order to enhance the identification of particular effects, the calculation were performed using very high boson numbers. Effects due to underlying symmetries were identified and some of them lead to a typical level bunching. Surprisingly they also occur in the regular region. In order to experimentally study the level dynamics we have studied the near regular nucleus ^{170}Yb using the (p,t) reaction. Here 19 0^+ states were observed which could not yet be compared with the too simple calculations. On the otherhand in a more elaborated study of other rare-earth isotopes a predicted increase of the level density at the phase transition could be observed.

The authors acknowledge financial support by the DFG under grant JO 391/2-3 and would like to thank P. Cejnar, M. Macek, R.F. Casten, D.A. Meyer, H.F. Wirth, D. Bucurescu, P. von Brentano, E.A. McCutchan and N.V. Zamfir with whom part of this work was done.

REFERENCES

1. F. Iachello and A. Arima, The Interacting Boson Model (Cambridge University Press, Cambridge, 1987).
2. N. Whelan, and Y. Alhassid, Nucl. Phys. **A556**, 42 (1993)
3. E. Lopez-Moreno and O. Castanos, Phys. Rev. C**54**, 2374 (1996).
4. P. Cejnar and J. Jolie, Phys. Rev. E **61**, 6237 (2000).
5. J. Jolie, R.F. Casten, P. von Brentano, V. Werner Phys. Rev. Lett. **87**, 162501 (2001).
6. V. Werner, P. von Brentano, R.F. Casten, and J. Jolie, Phys. Lett. **B527** 55 (2002).
7. J. Jolie, P. Cejnar, R.F. Casten, S. Heinze, A. Linnemann, V. Werner, Phys. Rev. Lett. **89**, 182502 (2002).
8. P. Cejnar, S. Heinze, J. Jolie, Phys. Rev. C **68**, 034326 (2003).

9. J. Jolie, A. Linnemann, Phys. Rev. C**70**, 031301(R) (2003)
10. E.A. McCutchan, N.V. Zamfir and R.F. Casten, Phys. Rev. C**69**, 064306 (2004); E.A. McCutchan, N.V. Zamfir, Phys. Rev. C**71**, 054306 (2005).
11. J. Jolie, R.F. Casten, P. Cejnar, S. Heinze, E.A. McCutchan, N.V. Zamfir *et al.*, Phys. Rev. Lett. **93**, 132501 (2004)
12. R.F. Casten in *Interacting Bose-Fermi Systems in Nuclei*, ed. F. Iachello Plenum, p 1 (1981).
13. F. Iachello, Phys. Rev. Lett. **85**, 3580 (2000).
14. F. Iachello, Phys. Rev. Lett. **87**, 052502 (2001).
15. A. Bohr and B.R. Mottelson, Nuclear Structure (Benjamin, New York, 1975).
16. R.F. Casten and N.V. Zamfir, Phys. Rev. Lett. **85**, 3584 (2000).
17. R.F. Casten and N.V. Zamfir, Phys. Rev. Lett. **87**, 052503 (2001).
18. S. Heinze, P. Cejnar, J. Jolie, M. Macek, subm. to Phys. Rev. C nucl-th/0504016
19. M. Macek, P. Cejnar, J. Jolie, S. Heinze, subm. to Phys. Rev. C nucl-th/0504017
20. P. Cejnar, M. Macek, P. Stransky, these Proceedings.
21. P. Cejnar and J. Jolie, Phys. Rev. **E58**, 387 (1998).
22. S.R. Lesher, A. Aprahamian, L. Trache, A. Oros-Puesquens, S. Deyliz, A. Gollwitzer, R. Hertenberger, B.D. Valnion, G. Graw, Phys. Rev. C **66**, 051305(R) (2002).
23. H.-F. Wirth, G. Graw, S. Christen, D. Cutoiu, Y. Eisermann, C. Günther, R.Hertenberger, J. Jolie, A.I. Levon, O. Möller, G.Thiamova, P. Thirolf, D. Tonev, N.V. Zamfir, Phys. Rev. C **69**, 044310 (2004).
24. R.F. Casten, these proceedings.
25. D.A. Meyer et al., to be subm.
26. P. Cejnar and J. Jolie. Phys. Lett **B420**, 241 (1998).

New findings for mixed-symmetry states

V. Werner*, N. Pietralla†, P. von Brentano**, U. Kneissl‡, H.H. Pitz‡, A. Tonchev§, M.W. Ahmed§, C. Fransen**, H. von Garrel‡, C. Kohstall‡, J. Li§, A. Linnemann**, S. Müller¶, I.V. Pinayev§, D. Savran¶, M. Scheck‡, F. Stedile‡, W. Tornow§, S. Walter‡, H.R. Weller§ and Y.K. Wu§

*WNSL, Yale University, 272 Whitney Ave., New Haven, CT 06520-8124, USA
†Department of Physics and Astronomy, SUNY, Stony Brook, NY 11794-3800, USA
**Institut für Kernphysik, Universität zu Köln, Zülpicher Str. 77, D-50937 Köln, Germany
‡Institut für Strahlenphysik, Universität Stuttgart, Allmandring 3, D-70569 Stuttgart, Germany
§Triangle Universities Nuclear Laboratory, Duke University, Durham, NC 27708, USA
¶Institut für Kernphysik, TU Darmstadt, Schlossgartenstr. 9, D-64289 Darmstadt, Germany

Abstract. This report summarizes experiments performed on ^{164}Dy using photon scattering techniques. The scissors mode in ^{164}Dy has been reinvestigated using unpolarized photons from bremsstrahlung and polarized photons from a free electron laser. The current experiments lead to the observation of a new decay mode of the scissors mode in well-deformed rotors.

Keywords: scissors mode gamma vibration nuclear resonance fluorescence polarized
PACS: 24.30.Gd, 25.20.Dc, 21.60.Fw, 23.20.Lv

INTRODUCTION

Coupling of isovector resonances to γ-vibrations

The coupling of the giant dipole resonance (GDR) to γ-vibrations was investigated years ago in photon scattering experiments by Nathan and Moreh [1, 2] in ^{166}Er, a well-deformed nucleus in the rare earth region, leading to the observation of decays of the GDR to the γ-band. The experimental results were discussed in relation to the splitting of the GDR due to triaxiality, and successfully described within the dynamic collective model (DCM) [3], a geometric approach, and the interacting boson model (IBM) [4], including a p boson in order to account for the GDR.

The IBM is formulated as a valence space model, so its description of an excitation which is a relative to the GDR in the valence space would present a stringent test for its capability of describing the coupling of isovector resonances to γ-vibrations. Such a valence excitation exists and is well-known as the scissors mode [5, 6, 4, 7]. The scissors mode corresponds to a scissors-like counter-oscillation of the deformed proton versus the deformed neutron body, which only involves valence particles, lowering its energy to about 3 MeV, compared to 10-15 MeV for the GDR. In well-deformed rotors, the scissors mode is the head of a $K = 1$ band of mixed-symmetry character - that means its wave function is non-symmetric under the exchange of protons and neutrons, or $F = F_{max} - 1$ in terms of the IBM. The next member of this band is a 2^+ state which corresponds to the one-phonon mixed-symmetry $2^+_{ms,1}$ state found in more spherical nuclei (see, e.g., [8]), in which also the decay of the scissors mode to the γ-band has been observed (e.g. in [9]), while both observations are missing for well-deformed rotors.

The experiments described in the following focused on revealing the decay of the scissors mode to the γ-band in a well-deformed nucleus, and reveal a candidate for a possible mixed-symmetry 2^+ state.

^{164}Dy

From the known systematics of M1 excitation strengths [10, 11], for this experiment the nucleus ^{164}Dy was chosen, where earlier experiments using (γ, γ') [12, 13] and $(n, n'\gamma)$ [14] reactions left ambiguities in the placement of levels and transitions. The right hand side (solid lines) of Figure 1 represents how the levels of interest for this discussion are placed nowadays. A main fragment of the scissors mode is located at 3173 keV, and decays to the ground state and the first excited 2^+ state have been observed. From a similar decay pattern another state was identified at 2411 keV, which was assigned negative parity from Alaga rules [15]. Another state at 3027 keV was placed due to the observation of a γ-ray transition at that energy.

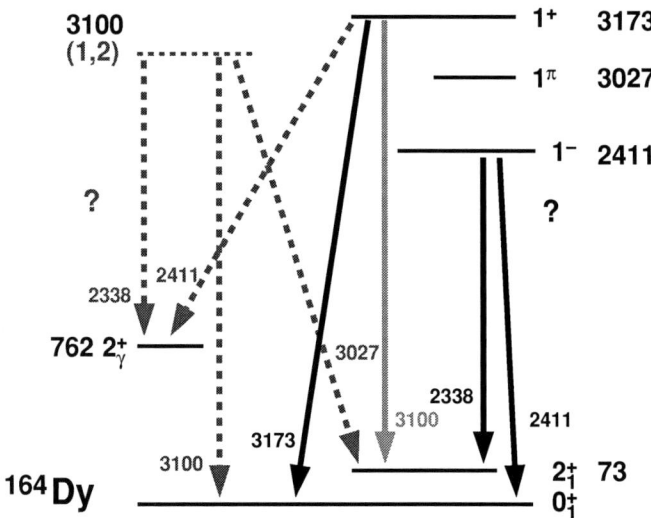

FIGURE 1. Levels and transitions in ^{164}Dy as placed so far are shown by solid lines, dashed lines denote the proposed alternative placement of levels and transitions.

However, an alternative placement is possible by assumption of a state at 3100 keV excitation energy, which was confirmed in [14], but the decay pattern remained unclear. In previous photon scattering experiments, a ground state transition of such state may have been masked by the strong transition depopulating the scissors mode fragment to the 2_1^+ state. As suggested on the left hand side of Figure 1 (dashed lines), such state might decay through a 3027 keV transition to the 2_1^+ state and through a 2338 keV transition to the 2^+_γ state, the head of the γ-band. In that case, the 2411 keV transition, so far attributed to the ground state decay of a state at that energy, might represent the searched for transition from the scissors mode to the 2^+_γ state.

EXPERIMENT

Unpolarized continuous spectrum photon beam

New high-sensitivity photon scattering experiments using continuous bremsstrahlung with endpoint energies of 2.9 and 3.6 MeV have been performed at the DYNAMITRON accelerator facility at Stuttgart University, amending 9 new states to the known level scheme. The target (1.486 g ^{164}Dy$_2$O$_3$ enriched to 96.0%, 1.100 g ^{164}Dy$^{met.}$ enriched to 95.6%, and appropriate amounts of Al and LiF for photon flux calibration) was irradiated for 6 days at 2.9 MeV endpoint energy, and another 4 days at 3.6 MeV endpoint energy. Target nuclei were excited resonantly from their 0^+ ground states, and γ-decays were detected by three HPGe detectors surrounding the target at different angles allowing spin assignments. A review of the setup and the technique of nuclear resonance fluorescence (NRF) can be found in [16].

The different maximum photon energies were chosen in order to strongly excite the scissors mode at higher energy, while avoiding its excitation at lower energy. In the first case all known decays which are discussed above have been observed. From the known photon flux integrated cross sections of each decay were determined. In the experiment at lower energy some decays were not present, most important those at 2411 keV and 2338 keV. Knowing the sensitivity limit of the measurement, they should have been present if they would stem from the decays of a dipole excited state at 2411 keV excitation energy, thus this placement can be excluded, and they have to be placed as transitions from states above 2.9 MeV excitation energy.

The existence of an excited state at 3100 keV could, however, not directly be proved, because the γ-ray deexciting it to the ground state would have had the same energy as the strong scissors mode to 2^+_1 transition. Another method was applied in order to search for this state as described below.

Polarized mono-energetic photon beam

We carried out an experiment using polarized photons produced at the Duke Free Electron Laser Laboratory (DFELL) at TUNL. In the cavity of the electron storage ring an almost mono-energetic and completely polarized photon beam can be produced by Compton backscattering [17] (width about 60 keV at 3 MeV beam energy). This allowed to investigate excitations at 3100 keV and 3173 keV separately, without the decays of one interfering with the decays of the other. Figure 2 shows the beam profile at 3100 keV measured in beam direction with reduced intensity and without target, thus giving an approximate idea of the beam profile during the experiment. The NRF setup [18] at the High Intensity γ-ray Source (HIγS) was used, consisting of four HPGe detectors at 90° relative to the beam axis in the directions of the electric and magnetic vectors of the

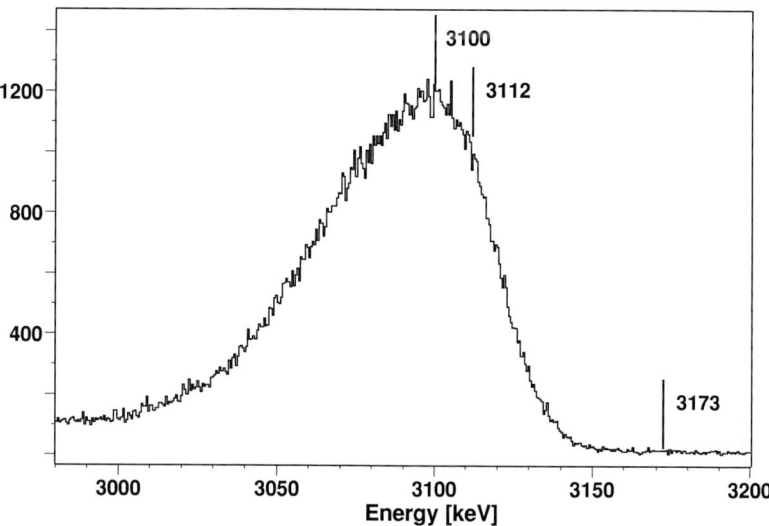

FIGURE 2. Spectrum of the polarized γ-ray beam, taken with a HPGe detector in beam direction at a beam energy of 3100 keV. 3112 keV is close to the beam maximum, while 3173 keV is not covered.

incoming photons, enabling us to confirm parities of dipole excited states. Due to the by orders of magnitude lower beam intensity compared to bremsstrahlung, in addition to the enriched target 78.7 g of natural Dy_2O_3 were used.

Tuning the beam to 3100 keV, only states at or close to this energy have been excited, as well from ^{164}Dy, as also from its neighboring isotopes, as the target contained large amounts of natural material. A peak at 2338 keV was observed, which is only known from ^{164}Dy, and thus can be concluded to indeed stem from a state at 3100 keV excitation energy, deexciting it to the 2^+_γ state. At 3100 keV also a decay from another Dy isotope is located, but we can correct for this contamination using well known cross sections [13] of decays around 3100 keV, i.e. at 3112 keV, resulting in a large error for its excitation cross section. From the information of both experiments, Stuttgart and HIγS, the lifetime of the state must be in the range of few femtoseconds, and the decay branches to the lowest two 2^+ states are roughly an order of magnitude stronger than the ground state decay. The strengths of the decays to the 2^+ states are of the order of 1-2 μ_N^2 in case of pure M1 radiation or ten's of W.u. in the case of pure E2 radiation, which are the two cases favored from angular correlations for the decay to the γ-band, which can be analyzed due to the polarization of the beam at HIγS.

DISCUSSION

The spin assignment to the at 3100 keV remains unknown. Due to the excitation by real photons, it must be $J = 1, 2$, while its decay behavior seems more typical for a 2^+ state.

This, together with the strong transitions to the lowest 2^+ states, makes it a candidate for the mixed-symmetry 2^+ member of the rotational band build on the scissors mode, and a future measurement of the spin and parity of this state is desirable.

The main result of the new photon scattering experiments is the discovery of a decay from a main fragment of the scissors mode to the γ-band, corresponding to the earlier findings for the GDR. A 1^+ state about the energy 3.14 MeV had been discussed in [19] to show large spin contributions, while the scissors mode is of orbital nature. However, the relative decay strengths of the 3173 keV 1^+ state are in well agreement with predictions of the IBM-2 in the SU(3) limit, while other 1^+ states in the energy region, some of them also found to decay to the 2_γ^+ state in this work, have a completely different decay pattern. Therefore we identify the 3173 keV state with a main fragment of the scissors mode. The IBM-2 predicts the M1 decay strengths of the scissors mode relative to the ground state decay to be 55% to the 2_1^+ state and 11% to the 2_γ^+ state, to be compared with the now measured 43(4)% and 11(4)% from experiment, respectively. Final results are in preparation.

ACKNOWLEDGMENTS

We thank F. Iachello, N. LoIudice and S. Yates for helpful discussions. This work was supported by the USDOE under grant numbers DE-FG02-91ER40609, DE-FG02-97ER41033, and DE-FG02-04ER41334 (OJI program), the DFG under contracts Br 799/11-1, Pi 393/1-2, Kn 154/31-2, and SFB 634, and the DAAD under contract number D/0247211.

REFERENCES

1. A.M. Nathan and R. Moreh, Phys. Lett. B **91**, 38 (1980).
2. A.M. Nathan, Phys. Rev. C **38**, 92 (1988).
3. M. Danos and W. Greiner, Phys. Rev. **134**, B234 (1964).
4. F. Iachello and A. Arima, *"The Interacting Boson Model"* (Cambridge University Press, Cambridge, 1987).
5. N. Lo Iudice and F. Palumbo, Phys. Rev. Lett. **41**, 1532 (1978).
6. N. Lo Iudice and F. Palumbo, Nucl. Phys. **A326**, 193 (1979).
7. A. Richter, *Proc. of the Int. Conf. on Nuclear Physics*, Florence 1983, ed. P. Blasi and R. A. Ricci, (Tipografi ca Compositori Bologna, Vol. 2), p. 189.
8. N. Pietralla *et al.*, Phys. Rev. Lett. **83** (1999) 1303.
9. P. von Brentano *et al.*, Phys. Rev. Lett. **76**, 2029 (1996).
10. N. Pietralla, Phys. Rev. C **59**, 2941 (1999).
11. J. Enders *et al.*, Phys. Rev. C **71**, 014306 (2005).
12. C. Wesselborg *et al.*, Phys. Lett. B **207**, 22 (1988).
13. J. Margraf *et al*, Phys. Rev. C **52**, 2429 (1995).
14. E.L. Johnson *et al.*, Phys. Rev. C **52**, 2382 (1995).
15. G. Alaga *et al.*, Dan. Mat. Fys. Medd. **29**, no. 9 (1955).
16. U. Kneißl, H. H. Pitz, and A. Zilges, Prog. Part. Nucl. Phys. **37**, 349 (1996).
17. V.N. Litvinenko *et al.*, Phys. Rev. Lett. **78**, 4569 (1997).
18. N. Pietralla *et al.*, Phys. Rev. Lett. **88**, 012502 (2002).
19. D. Frekers *et al.*, Phys. Lett. B **218**, 439 (1989).

Independent evidence for $M1$ scissors resonances built on the levels in the quasicontinuum of ^{163}Dy

M. Krtička [*], F. Bečvář[*], M. Heil[†], F. Käppeler[†], R. Reifarth[†,**],
I. Tomandl[‡], F. Voss[†] and K. Wisshak[†]

[*]*Charles University, Faculty of Mathematics and Physics, 180 00 Prague, Czech Republic*
[†]*Forschungszentrum Karlsruhe, Postfach 3640, D-76021 Karlsruhe, Germany*
[**]*Los Alamos National Laboratory, Los Alamos, New Mexico 87545, USA*
[‡]*Nuclear Physics Institute of the Czech Academy of Sciences, 250 68 Řež, Czech Republic*

Abstract. The spectra of γ-rays following the capture of neutrons in ^{162}Dy at neutron energy region of 90-100 keV, where p-wave neutron resonances dominate, were measured for individual γ multiplicities. These data support independently the main assessment from our previous study of the two-step γ cascades, following the thermal s-wave neutron capture in ^{162}Dy, that the $M1$ scissors-mode resonances are built on all levels of ^{163}Dy, including the levels in the quasicontinuum.

Keywords: 4π BaF$_2$ γ detector, neutron capture, γ cascades, $M1$ scissors mode, photon strength function, summed $B(M1)\uparrow$ strength
PACS: 23.20.Lv, 24.30.Gd, 25.40.Lw, 27.70.+q

INTRODUCTION

The low-energy collective isovector $M1$ mode in deformed nuclei was predicted by Hilton [1] and independently by Lo Iudice and Palumbo [2]. Since its discovery by Bohle *et al.* [3] this mode, referred to as a scissors mode (SM), is a subject of continuing theoretical end experimental studies. Rich experimental data on this interesting phenomenon have been accumulated from numerous nuclear resonance fluorescence (NRF) experiments by the Cologne-Darmstadt-Stuttgart group. In particular, reliable data were obtained for 32 even-even nuclei, see Refs. [4, 5]. The set of these data made it possible to map behavior of the $M1$ vibrational response of deformed nuclei in their *ground states* to the incident photons with energies $E_\gamma = 2.5 - 4.0$ MeV.

The first, albeit not firm claim for observation of SM of *excited* deformed nuclei was made in Ref. [6]. It was based on analysis of γ-ray spectra belonging to *all* two-step cascades (TSCs) initiating at the $J^\pi = 1/2^+$ thermal s-wave neutron capturing state of the ^{162}Dy+n system and ending at a preselected set of low-lying levels in ^{163}Dy. As neutron binding energy of ^{163}Dy is 6271 keV, the TSCs of $M1$-$M1$ type, populating the ^{163}Dy low-lying level at 251 keV with $J^\pi = 5/2^+$, turn out to be very sensitive to the SM strength of the excited ^{163}Dy nuclei. This is because the total energy carried by these TSCs equals almost exactly to $2E_{\rm SM}$, so that their steps may be enhanced simultaneously. The thermal ^{162}Dy(n,γ)^{163}Dy reaction is thus of a crucial importance for studying the $M1$ SM. The claim for observation of SM of excited nuclei was corroborated by additional TSC data for compound nuclei ^{150}Sm, ^{156}Gd and ^{158}Gd [7], and later by the data on spectra of γ rays emitted from the (^{3}He,^{3}He$'\gamma$) and (^{3}He,$\alpha\gamma$) reactions on numerous deformed targets, see, e.g., Ref. [8]. Recently, using the sum-coincidence setup at the Řež reactor, we accumulated more precise data on TSCs for the same ^{162}Dy(n,γ)^{163}Dy reaction [9]. From these data we made the following conclusions: (i) the SM, as such, displays a distinct and consistent resonance behavior, which justifies us to refer this mode as a scissors resonance (SR); (ii) without postulating SRs in the quasicontinuum of ^{163}Dy, specifically at levels with energy above 2.5 MeV, the predictions of the statistical model lead to a sharp disagreement with the data; (iii) the estimated summed $B(M1)\uparrow$ strength of SRs is unusually large, $\sum B(M1)\uparrow \approx 6.2\ \mu_N^2$.

The aim of this Contribution is to verify justifiability of somewhat striking conclusion (ii) from an independent experiment. For this purpose we studied behavior of the multi-step γ cascades following the capture of keV neutrons in ^{162}Dy when initial neutron capturing states are formed mainly by the p-wave resonances with $J^\pi = 1/2^-, 3/2^-$.

THE EXPERIMENT AND RESULTS

The measurements were undertaken using the 4π BaF$_2$ detector array [10] installed at the pulsed beam of fast neutrons produced from the ^{7}Li(p,n)^{7}Be reaction at the Karlsruhe Van De Graaff accelerator, see Fig. 1. The energy of protons was adjusted at 4.7 MeV, the duration of the proton bursts being less than 1 ns. The sample was formed by 2.14 g of Dy$_2$O$_3$ enriched in ^{162}Dy to 92.5 %. The *singles* and *sum-energy* spectra of γ rays following the capture of neutrons in ^{162}Dy at neutron energies of 90-100 keV were accumulated for individual γ multiplicities m. For the selection of the neutron energy region the neutron time-of-flight method has been used. Examples of the obtained singles γ-ray spectra for various γ multiplicities are illustrated repeatedly in Figs. 2 and 3, while an example of the sum-energy spectrum is seen in Fig. 4.

All spectra were compared with the outcome of the Monte Carlo simulations based on the validity of the statistical model. The γ cascades were generated using the DICEBOX

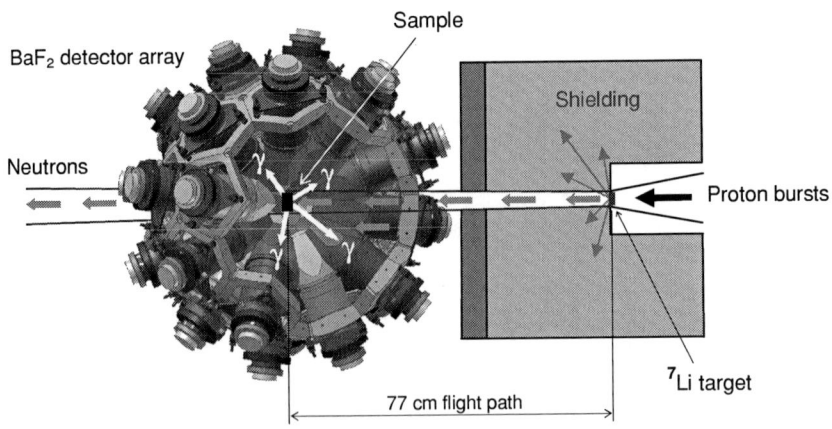

FIGURE 1. Experimental setup for capture γ-ray measurements in keV neutron energy region

algorithm [11], while the response of the 4π γ detector to these artificial cascades was simulated with the aid of the GEANT code [12].

In the same way as in our previous work [9], three separate series of simulations, based on various assumption about the makeup of the $M1$ photon strength function $S_\gamma^{(M1)}(E_\gamma)$, were performed. In the first series the role of the SRs was entirely neglected. In the second series, the SRs built on the full set of levels below 2.5 MeV were considered to contribute to $S_\gamma^{(M1)}(E_\gamma)$. In the third series we postulated that SRs reside on *all* ^{163}Dy levels. We used the following expression for the SM term contributing to $S_\gamma^{(M1)}(E_\gamma)$:

$$S_\gamma^{(M1,SM)}(E_\gamma) = \frac{16\pi}{27(\hbar c)^3} \frac{E_0}{\arctan(2E_0/\Gamma_{SM})} \frac{\Gamma_{SM} E_\gamma}{(E_\gamma^2 - E_{SM}^2)^2 + E_\gamma^2 \Gamma_{SM}^2} \sum B(M1)\uparrow$$

with $E_0 = (E_{SM}^2 - \Gamma_{SM}^2/4)^{1/2}$. Here, E_{SM} is the SM energy, Γ_{SM} – the SM damping width and $\sum B(M1)\uparrow$ – the reduced SM strength. In accordance with what had been deduced from the data in Ref. [9], we used the following values of these parameters: $E_{SM} = 3.0$ MeV, $\Gamma_{SM} = 0.6$ MeV and $\sum B(M1)\uparrow = 6.2$ μ_N^2. As far as the remaining entities responsible for the emission of γ cascades are concerned, the same assumptions as those in Ref. [9] were adopted while generating γ cascades. Specifically, for the level density the well-known constant-temperature level-density formula (CTF) is valid, the $E1$ photon strength function $S_\gamma^{(E1)}$ is given by the semi-empirical extension [13] of the Kadmenskij-Markushev-Furman model [14], and the low-energy wing of the $M1$ spin-flip resonance doublet, observed in Ref. [15], contributes to $S_\gamma^{(M1)}$. The values of parameters of the CTF, $S_\gamma^{(E1)}$ and $S_\gamma^{(M1)}$, adopted in Ref. [9], were kept unchanged.

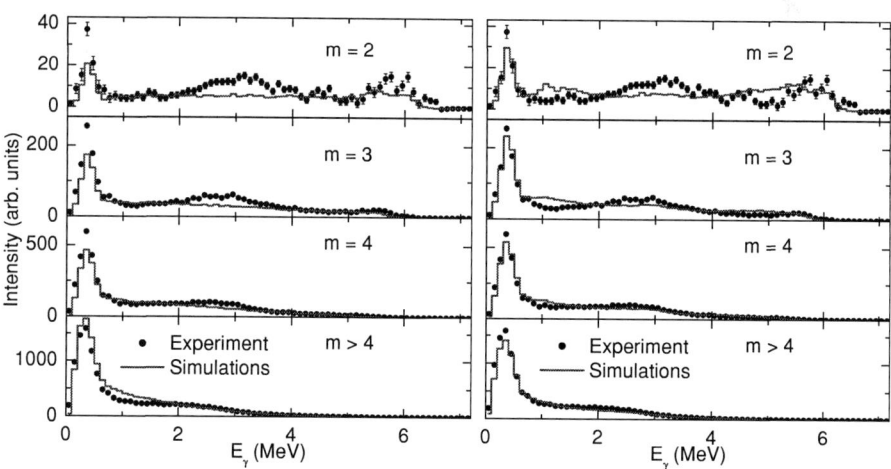

FIGURE 2. Comparison between singles γ-ray spectra for individual multiplicities m and the outcome of the DICEBOX/GEANT simulations. Left part: while generating γ-cascades only the spin-flip resonance doublet was assumed to be responsible for the $M1$ photon strength function. Right part: SRs are assumed to be built on all levels in ^{163}Dy within a restricted excitation region 0-2.5 MeV

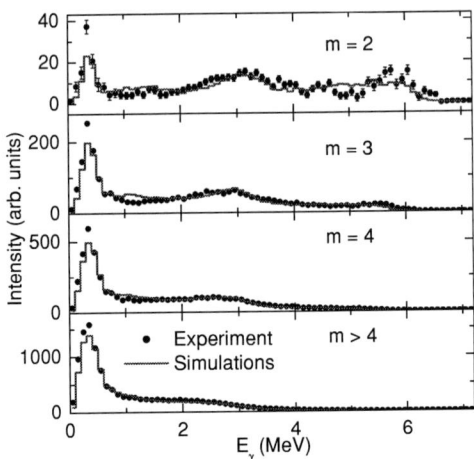

FIGURE 3. Comparison similar to that shown in Fig. 2. SRs are postulated on *all* levels in ^{163}Dy.

DISCUSSION

As is apparent from the results of simulations displayed in Figs. 2-4, our data support strongly the main conclusion made in Ref. [9] that SRs reside on all ^{163}Dy levels, including the levels in the quasicontinuum. The degree of agreement between the data and the outcome of the simulations visualized in Fig. 3 and part c) of Fig. 4 is almost ideal, in spite of the fact that the only free parameter for reaching the best fits was a common factor for normalizing the experimental spectra.

The previous estimate of the SM strength, $\sum B(M1)\uparrow \approx 6.2$ μ_N^2, is implicitly compatible with the present data, falling into the range of 5.7-6.8 μ_N^2 predicted in Ref. [16] for the well-deformed nuclei. The same estimate agrees well with what we have drawn from the available (^{3}He,$\alpha\gamma$) data in Ref. [17] for ^{161,162}Dy and ^{171,172}Yb nuclei. In this case, the deduced values of $\sum B(M1)\uparrow$ cover the range of 5.5-7.7 μ_N^2.

In contrast to this, the value $\sum B(M1)\uparrow \approx 6.2$ μ_N^2 disagrees with the estimate $\sum B(M1)\uparrow \approx 3.0$ μ_N^2 deduced for the ground state of ^{163}Dy from the SU(3) limit of the proton-neutron *sdg* interacting boson-fermion model [18]. It disagrees also with the upper limit of 2.03 μ_N^2 for ^{163}Dy determined from NRF measurements [19]. However, the NRF data for odd nuclei are to be taken with care. As can be guessed from Ref. [19], in view of a high level density of these nuclei at excitation energies near 3 MeV, even the present state-of-the-art of the NRF technique does not yield the estimates of $\sum B(M1)\uparrow$ free of a bias toward low values. Nevertheless, this objection cannot be raised against the NRF data for deformed even-even nuclei. From the latest survey of these data [5] it is apparent that the highest values of the observed SM strength $\sum B(M1)\uparrow$ are equal to 3.71 ± 0.59 and 3.68 ± 0.48 μ_N^2, belonging to ^{158}Gd and ^{168}Eu, respectively. Even these strengths are lower, by factor of 0.6, compared to our estimate $\sum B(M1)\uparrow \approx 6.2$ μ_N^2. This disagreement can be only in part accounted for by the escape of the $B(M1)\uparrow$ strength from the limited γ-ray energy region, typically 2.5-4.0 MeV, from which the data on

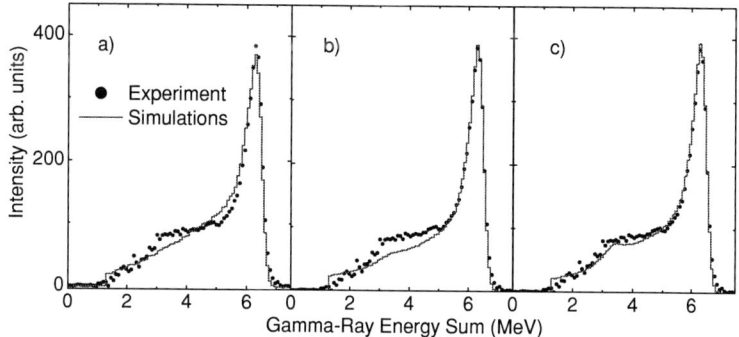

FIGURE 4. The experimental spectrum of γ-ray energy sums for multiplicities $m = 1 - 15$ compared with its simulated counterparts: a) complete absence of SR term in $M1$ photon strength function is assumed; b) SRs built only on all ^{163}Dy levels below 2.5 MeV; c) SRs built on *all* levels in ^{163}Dy.

SM strength are drawn in NRF experiments. More specifically, if the damping width of the SRs built on the ground states of even-even nuclei is as low as in the case of ^{163}Dy, i.e. $\Gamma_{SM} \approx 0.6$ MeV, about 22 % of the SM strength will escape from that region.

The most distinct features of the SRs built on excited levels are their strength and robust behavior. The phenomenon of $M1$ vibrations of excited nuclei deserves definitely deeper studies. No theory has been yet devoted to understand this striking phenomenon.

This work has been partly supported by the Grant Agency of the Czech Republic under contract No. 202/03/P136.

REFERENCES

1. Hilton, R. R., in *Proceedings of the International Conference on Nuclear Structure*, JINR, Dubna, 1976 (unpublished).
2. Lo Iudice, N. and Palumbo, F., *Phys. Rev. Lett.* **41**, 1532 (1978).
3. Bohle D. *et al.*, *Phys. Lett.* **137B**, 59 (1984).
4. Pietralla N. *et al.*, *Phys. Rev. C* **58**, 184 (1998).
5. Enders J. *et al.*, *Phys. Rev. C* **71**, 014306 (2005).
6. Bečvář, F. *et al.*, *Phys. Rev. C* **52**, 1278 (1995).
7. Bečvář, F., in *Capture Gamma-Ray Spectroscopy and Related Topics*, edited by S. Wender, *AIP Conf. Proc.* **579**, AIP, New York, 2000, pp.504-511; Bečvář, F. *et al.*, *ibid*, pp. 657-659.
8. Melby, E. *et al.*, *Phys. Rev. C* **63**, 044309 (2001).
9. Krtička, M. *et al.*, *Phys. Rev. Lett.* **92**, 172501 (2004).
10. Wisshak, K. *et al.*, *Nucl. Instr. Methods A* **292**, 595 (1990).
11. Bečvář, F., *Nucl. Instr. Methods A* **417**, 434 (1998).
12. Heil, M. *et al.*, *Nucl. Instr. Methods A* **459**, 229 (2001).
13. Kopecky, J., in *Handbook for Calculations of Nuclear Reaction Data*, IAEA, Vienna, 1998, p. 97.
14. Kadmenskij, S. G., Markushev, V. P. and Furman, V. I., *Sov. J. Nucl. Phys.* **37**, 165 (1983).
15. Richter, A., *Prog. Part. Nucl. Phys.* **34**, 261 (1995).
16. Lipparini, E. and Stringari, S., *Phys. Rep.* **175**, 103 (1989).
17. Voinov, A. *et al.*, *Phys. Rev. C*, **63**, 044313 (2001).
18. Devi, Y. D. and Kota, V. K. B., *Nucl. Phys.* **A600**, 20 (1996).
19. Nord, J. *et al.*, *Phys. Rev. C* **67**, 034307 (2003).

Low-Energy Dipole Modes of Excitation Below the Neutron Separation Energy

A.P. Tonchev[1], C. Angell[2], M. Boswell[2], C.R. Howell[1], H.J. Karwowski[2],

J.H. Kelley[3], W. Tornow[1], N. Tsoneva[4]

[1] *Duke University and TUNL, Department of Physics, Box 90308, Durham, NC 27708-0308, USA*
[2] *University of North Carolina, Department of Physics and Astronomy, Chapel Hill, NC 27599-3255, USA*
[3] *North Carolina State University and TUNL, Department of Physics, Box 8202, Raleigh, NC 27695-8202, USA*
[4] *Institute of Theoretical Physics, University of Giessen, Heinrich-Buff-Ring 16, D-35392, Giessen, Germany*

Abstract. The nuclear resonance fluorescence experiments have been performed at the High Intensity Gamma Source (HIγS) on ^{138}Ba nuclei using four 60% efficient HPGe detectors. Excitation energies, spin, parities, and decay branching ratios were measured for the low-energy dipole modes of excitations. Experimental results on the parity measurement below the neutron separation energy shows that all dipole states in this energy region exhibit E1 excitation. These results are consistent with theoretical prediction of the collective isoscalar nature of this low-energy mode of excitation.

Keywords: pygmy dipole resonance; free electron laser; monoenergetic and polarized γ-beams; nuclear resonance fluorescence.
PACS: 21.10.Hw ; 25.20.-x; 29.30.Kv

INTRODUCTION

A wide variety of nuclear structure phenomena have been investigated with the nuclear resonance fluorescence technique during the last decade [1-5]. Spectroscopic experiments on two-phonon excitations of even-even nuclei near closed shells revealed large magnetic dipole strengths in heavy deformed nuclei. The corresponding excitations have been associated with scissors-like oscillations of the deformed proton density distribution against the neutron distribution, and this excitation mode was, accordingly, called the "scissors mode". Large electric dipole transitions to the ground states have been observed in spherical nuclei near Z = 50 and N = 82. They are assumed to arise from the coupling of quadrupole and octupole vibrational modes of the nucleus.

Recent experimental activities at the HIγS have focused on investigation of the so-called Pygmy Dipole Resonance (PDR), which is observed as a clustering of states close to the neutron threshold at excitation energies E_x ~ 5.5 - 8 MeV. Although carrying only a small fraction of the full dipole strength, these states are of particular interest because they reflect the motion of the neutron skin against the isotropic symmetric core. Some theoretical calculations indicate a correlation between the observed total B(E1) strength of the PDR and the neutron-to-proton ratio N/Z [6]. The observation of this collective dipole mode near the neutron threshold has important astrophysical implications. For example, the conditions governing thermal equilibrium of (γ,n) and (n,γ) reactions in the explosive nucleosynthesis of certain neutron deficient heavy nuclei may be significantly modified.

EXPERIMENTAL THECHNIQUE

The High Intensity γ-Ray Source

The HIγS facility is used to produce high-intensity and nearly monoenergetic γ-ray beams by intracavity Compton backscattering. In γ-ray production mode two electron bunches stored in the Duke storage ring are synchronized so that the lased photons from one electron bunch are reflected by the downstream mirror, and then collide head on with the second electron bunch at the scattering point producing γ-rays [7]. The backscattered photons are collimated with "on-axis" cylindrical lead collimator, located 60 m downstream of the collision point. In addition, the γ-rays are highly polarized resulting from the Compton scattering process of nearly completely polarized FEL photons. The flux of γ-rays can be increased by increasing the current in the bunches in the storage ring, presently up to 30 mA, and/or by operating with more electrons bunches. The average γ-flux on the target position using a 1.27 cm diameter lead collimator exceeds 2×10^6 γ/sec.

Using this approach, the quality of the HIγS γ-ray beam provides tremendous advantages over bremsstrahlung beams, for example, the monoenergetic γ-beam allows excitation of only desired levels of interest. In addition, the near 100% beam polarization provides deeper insight into the dynamics of photonuclear reactions.

Nuclear Resonance Spectroscopy

The excitation of nuclear states by linearly polarized photons in connection with the measurement of the azimuthal asymmetry of the scattered photons is powerful method to distinguish between electric and magnetic dipole radiation and, hence, to determine the parity of nuclear states. The Nuclear Resonance Fluorescence (NRF) method is used at HIγS to the study low-multipolarity ground state transitions (i.e. E1, M1, E2) with large partial widths. In this process the photon has a small probability to transfer angular momentum to the atomic nucleus. Due to the low detection limit this technique represents an outstanding tool for measuring of dipole transitions. The main advantage of this method is that both the excitation and the de-excitation processes proceed via the electromagnetic interaction - the best understood interaction in physics. In Fig. 1 the angular correlation functions for the E1 and M1 transitions are shown.

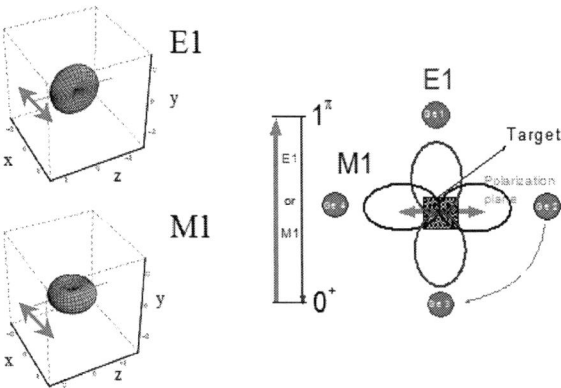

FIGURE 1. Angular distributions of E1 and M1 excitations are shown in the left panel. The NRF set-up at HIγS consists of four 60% HPGe detectors mounted perpendicular to the beam axis.

The photon beam moves in the z-direction and the electric vector of the incoming photons is parallel to the x-axis, as is the case at the HIγS facility. Therefore, even-even nuclei with a $J^\pi=0^+$ ground state, excited via E1 transition to a $J^\pi=1^-$ state by the polarized γ-rays can emit light only perpendicular to the x-axis when decaying to the ground state by dipole transition. For the M1 transition the situation is the same, but turned by 90^0 around the z-axis, as the M1 vector is perpendicular to the E1 vector. Therefore, comparing spectra in detectors at azimuthal scattering angles of $\phi=0^0$ and 90^0, one can easily distinguish between E1 and M1 excitations. In the case of photodisintegration (above the particle threshold) protons and neutrons will be predominantly emitted in direction of the electric field vector of the linearly polarized beam.

EXPERIMENTAL RESULTS

From the recent photon scattering experiments at the Darmstadt TU bremsstrahlung facility, a large data set on E1 excitations in N=82 nuclei has been compiled [8]. For example, in ^{138}Ba alone more than 70 excitations have been observed in the energy range from 4 to 8.5 MeV. For the vast majority of the observed transitions only one decay branch to the ground state has been measured. However, this data set is still incomplete since the polarization sensitivity is very low above 4 MeV making the parity assignment impossible. To obtain the parity information of the low-lying dipole states in the PDR region, NRF measurements have been performed on ^{138}Ba using the linearly polarized and monoenergetic γ-beam at HIγS.

Table 1 lists the parity assignments of 13 dipole states in ^{138}Ba in the energy range from 7.5 to 8.5 MeV. All the dipole states deexcite via E1 transitions ($J^\pi=1^-$). Earlier NRF measurements at HIγS unambiguously showed that all the dipole states in the energy region from 5.5 to 6.5 MeV are also $J^\pi=1^-$ states [2]. Combined data from both measurements are shown in Fig. 2.

TABLE 1. Parity assignments to the observed dipole excitations in ^{138}Ba. All parity quantum numbers here are assigned for the first time.

E_x (keV)	J^π Previous [8]	J^π Present
7545.9	$1^{(-)}$	1^-
7754.8	$1^{(-)}$	1^-
7774.5	$1^{(-)}$	1^-
7805.6	$1^{(-)}$	1^-
7819.9	$1^{(-)}$	1^-
7871.5	$1^{(-)}$	1^-
7913.6	$1^{(-)}$	1^-
8383.0	---	1^-
8408.0	---	1^-
8434.0	$1^{(-)}$	1^-
8456.0	---	1^-
8472.0	---	1^-
8496.0	---	1^-

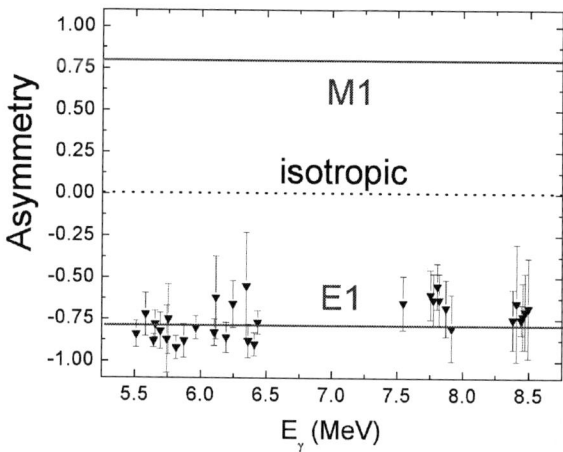

FIGURE 2. Asymmetry of the low-lying dipole states in ^{138}Ba using the 100% linearly polarized γ-beam at HIγS.

Fig. 3 shows the high (top panel) and low-energy part (bottom panel) of the gamma spectra in the vertical (left) and horizontal (right) detectors relative to the polarization plane at E_γ = 8.5±0.17 MeV. There are only a few dipole states decaying to the ground state (top-left panel). These states are observed only from the vertical detectors located on the plane perpendicular to the polarization plane. According to the azimuthal distribution from Fig. 1 these states exhibit E1 character. Strong E2 transitions from the lowest (J^π=2$^+$) excited states in ^{138}Ba at 1435.8, 2217.9, and 2639.5 keV to the ground state have been observed with the advantage of the pulsed γ-beam at the HIγS facility (bottom panel). This experimental information is usually difficult to obtain in bremsstrahlung experiments due to the increasing non-resonant scattering at low-energies and the DC beam from the electron linear accelerator [8]. As can be seen from Fig. 3, the decay from the first three J^π=2$^+$ states to the ground state is observed in both the horizontal and vertical detectors, which means they lost their initial polarization due to the multiple gamma cascade. At the same time, no primary inelastic transitions were observed from the maximum resonance energy to the J^π=2$^+$ states.

This measurement was followed with lower beam energy at E_γ = 8.2±0.16 MeV, where no resonance transitions were observed at the Darmstadt facility. No direct transition was measured at the HIγS facility ether to the ground state or the J^π=2$^+$ states. However, the low-energy J^π=2$^+$ state to ground state transitions still remain in both horizontal and vertical detectors. From this picture one can conclude that at E_γ = 8.5±0.17 MeV, there are only a couple of dipole states strongly bound with the ground state. However, according to the statistical model there should be ~300 dipole states (J^π=1$^+$ or 1$^-$) in the 170 keV energy range. Hence, there are a lot of missing dipole transitions and respectively γ-strength, which are hidden in the "background" of the previous nuclear resonance experiments. The advent of the pulsed and monoenergetic HIγS beam allows for accounting of the hidden γ-strength in ^{138}Ba at this energy. The ratio of the non-resonance versus the resonance transitions is 5.7±0.7. A microscopic study of the 1$^-$ states close to the particle threshold shows that their stucture is mostly an admixture of complex configurations. As a result the direct transitions from these states to the ground state are strongly hindered. Only a few of them decay by strong E1 transitions to the ground state due to the PDR or GDR component (for the higher lying) in the structure of the state vector. These states prefer to decay by fast E1 transitions which "rain" down to the low-energy 2$^+$ states. Single phonon component in these dipole states are only few percent which is responsible for direct transitions to the ground state.

It should be noted that the resonance peak at 8496.0 keV is the highest dipole excited state in ^{138}Ba that has been observed. Our experiments at excitation energies above the neutron threshold at 8.8±0.13, 9.1±0.14, 9.5±0.14, and 10.0±0.15 MeV, for example, did not identify any dipole transitions to the ground state or to any of the higher excited states in ^{138}Ba [9].

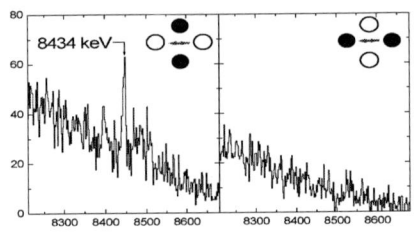
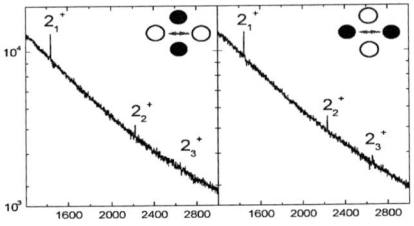

FIGURE 3. High (top panel) and low-energy part (bottom panel) of the gamma spectra in the vertical (left) and horizontal (right) detectors at $E_\gamma = 8.5 \pm 0.17$ MeV.

CONCLUSION

On the basis of the present parity measurements in ^{138}Ba it is fairly safe to assume that the observed dipole srength below the particle threshold is predominantly electric in nature. Our findings are in agreement with other experiments performed in this nucleus in the same energy region [8]. Hence, it will be very interesting to confirm the PDR structure in the other N=82 nuclei (^{140}Ce, ^{142}Nd, and ^{144}Sm). This intention is motivated by the very strong M1 excitation observed recently in ^{124}Sn in the PDR region [10]. Another important question that remains to be answered is how this low-energy collective PDR mode evolves to the GDR mode.

The present measurements show also that the monoenergetic and pulsed beam from HIγS facility opens up a new opportunity for direct measurements of the level density and the radiative γ-ray strength functions by measuring not just the direct gamma transitions but also the non-resonance part of the decay process.

ACKNOWLEDGMENTS

We thank the technical staff at the DFELL for providing excellent γ-ray beams. We are also grateful to Norbert Pietralla for fruitful discussion. This work was partially supported by the Grant Nos. DE-FG02-97ER41033, DE-FG02-97ER41042, and DE-FG02-97ER41041.

REFERENCES

1. U. Kneissl et al., Prog. Part. Nucl. Phys. **37**, 349-433 (1996).
2. N. Pietralla et al., Phys. Rev. Lett. **88**, 012502 (2002).
3. N. Pietralla et al., Phys. Rev. **C 65**, 047305 (2002).
4. C. Fransen et al., Phys. Rev. **C 70**, 044317 (2004).
5. D. Savran et al., Phys. Rev. **C 71**, 034304 (2005).
6. N. Tsoneva et al., Phys. Lett. **B 586**, 213 (2004).
7. V. Litvinenko et al., Phys. Rev. Lett. **78**, 4569 (1997).
8. A. Zilges at al., Phys. Lett. B **542**, 43 (2002).
9. A. Tonchev et al., Nucl. Ins. Meth. A (2005) in press.
10. M. Boswell, Bull. Am. Phys. Soc. **50**, 177 (2005).

Search For Enhanced Alpha Preformation in the N=Z+1 Nuclei ^{113}Ba, ^{109}Xe, ^{105}Te

A.A. Hecht[1,2], C.J. Lister[1], C.N. Davids[1], A. Heinz[3], N. Hoteling[2],
C. Mazzocchi[5], J. Palombo[2], D. Seweryniak[1], J. Shergur[1,2], M. Stoyer[6],
W.B. Walters[2], P.J. Woods[4], S. Zhu[1]

[1] Physics Division, Argonne National Laboratory, Argonne, IL 60439
[2] Department of Chemistry, University of Maryland, College Park, MD 20742
[3] WNSL, Yale University, New Haven, CT 06520
[4] Department of Physics, University of Edinburgh, Edinburgh EH9 3JZ, UK
[5] Oak Ridge National Laboratory, Oak Ridge, TN 37831
[6] Lawrence Livermore National Laboratory, Livermore, CA 94550

Abstract. α–decay is known in a few cases just above ^{100}Sn. The lowest isospin nuclei in which it has been observed are the T_z=1 nuclei ^{114}Ba, ^{110}Xe, and ^{106}Te. For the next lower isospin with T_z=$^1/_2$, N=Z+1, we used the reactions ^{58}Ni + ^{58}Ni and ^{58}Ni + ^{54}Fe to search for the ^{113}Ba→^{109}Xe→^{105}Te→^{101}Sn α–decay chain. We found no candidates in our study. We set upper limits on the production of ^{113}Ba via the ^{58}Ni + ^{58}Ni reaction and upper limits on the production of ^{109}Xe via the ^{58}Ni + ^{54}Fe reaction.

INTRODUCTION

Since the initial discovery of the α–decay branches for ^{107,108}Te [1], the prospect of very fast α–decay for light Te, Xe, and Ba nuclides has stimulated the study of other possible α emitting nuclides in the mass region beyond ^{100}Sn [2,3]. α–decay rates are usually modeled as the probability of preformation of the α particle multiplied by the tunneling probability. (This preformation, generalized to other particles, has also led to predictions of enhanced cluster emission of ^{12}C and other similar mass ejecta [4].) In the light Te, Xe, and Ba nuclides α–decay rate enhancement is expected relative to ^{212}Po (the standard candle for α–decay) because their protons and neutrons occupy the same orbitals beyond doubly magic ^{100}Sn, a feature which has been predicted to enhance preformation. This is in contrast to ^{212}Po, whose protons and neutrons lie in different shells.

Recently, α–decay of ^{114}Ba was reported at GSI [2]. ^{113}Ba was also investigated at the GSI offline mass separator but was not observed, probably due to the time needed for the separation process, on the order of 100 milliseconds. In contrast, the time needed for in-flight mass separation is on the order of a microsecond, more appropriate for a short lived ^{113}Ba and its decay products. We report here on a search for the N=Z+1 α–decay chain, ^{113}Ba → ^{109}Xe → ^{105}Te → ^{101}Sn, performed at the Fragment Mass Analyzer (FMA) using beams from the Argonne Tandem Linear Accelerator System (ATLAS) at Argonne National Laboratory. This decay chain would provide an entrance channel to ^{101}Sn, provide Q(α) values, and define the mass surface around ^{113}Ba, ^{109}Xe, and ^{105}Te, which lie very close to the proton dripline and have never before been observed.

Previous estimates for the Q(α) value for ^{113}Ba decay ranged from 3.4 to 4.0 MeV [2,5]. Mazzocchi et al. [2] note that the measured Q(α) value for ^{114}Ba of 3.54 MeV was well predicted by Möller, Nix, and Kratz: 3.55 MeV [6]. Using their predicted value of 4.02 MeV for the Q(α) from ^{113}Ba, calculations following the method of Rasmussen [7] give an estimated α-decay partial half-life ($t_{1/2\alpha}$) of under 100 ms for ^{113}Ba. As the half-life for EC/β decay in this region is usually greater than 100 ms, the α-decay branch (b_α) derived from this calculated Q(α) value should compete favorably. Extrapolations from isotopic trends suggest Q(α) = 4.2 MeV for ^{109}Xe, giving $t_{1/2\alpha} \sim 10$ ms, and Q(α) = 4.6 MeV for ^{105}Te, giving $t_{1/2\alpha} \sim 10$ µs. This indicates that b_α is 100% for these two nuclei.

EXPERIMENTS

Two experiments were performed to study these $T_Z = 1/2$ nuclei: The first experiment was designed to populate ^{113}Ba via the ^{58}Ni(^{58}Ni,3n) reaction at beam energies of 250 and 260 MeV. The second was designed to populate ^{109}Xe, the immediate α-decay daughter of ^{113}Ba, directly via the ^{54}Fe(^{58}Ni,3n) reaction at a beam energy of 240 MeV. The thickness of the ^{58}Ni targets ranged from 0.6 to 1.0 mg/cm^2 and the ^{54}Fe target was 0.7 mg/cm^2 thick. A 10 µg/cm^2 carbon charge reset foil was in place behind the target. The beam energies were chosen to maximize the 3n production channel according to HIVAP calculations [8]. HIVAP calculations predicted a maximum production cross section (σ) of 50 nb for both ^{113}Ba and ^{109}Xe, via their respective 3n channels. Based on a comparison of our calculations for ^{114}Ba (~ 2 µb) with the experimental value ($0.20^{+0.13}_{-0.09}$ µb) [2], a value of 5 nb for ^{113}Ba and ^{109}Xe was used in the rate estimates.

For both experiments the ^{58}Ni beam was delivered to the target by the ATLAS and the production recoils were separated using the FMA. The selected recoils passed through a position sensitive parallel grid avalanche counter (PGAC) and a transmission ionization chamber, and implanted into an 80x80 double-sided silicon strip detector (DSSD). Slits were put in place in front of the PGAC to select one mass/charge (A/Q) ratio.

RESULTS AND DISCUSSION

A total 113 hours of data were recorded, with 400x10^6 DSSD implant events and 3x10^6 decay events. Use of the PGAC and DSSD allowed time of flight vs. energy spectra with recoils and scattered beam. The recoil peak on which the data were gated accounted for about 90% of all time of flight measurements, demonstrating a very clean experiment with very little scattered beam at the back of the FMA.

The recoils were focused by A/Q at the PGAC, and spread out more uniformly on the DSSD. The DSSD was run at an average of 1 kHz with the main intensity of the recoils spread over about a quarter of the silicon detector, averaging ~ 1 Hz per pixel in this area.

^{113}Ba

For the ^{58}Ni + ^{58}Ni experiment to populate ^{113}Ba, the beam energy was set to maximize the 3n production channel as suggested by HIVAP calculations. This had the disadvantage of increasing the cross section for all A/Q = 112/25 products, which showed a recoil production of > 10 times that of the total 113/25 recoils (Fig. 1). Slits were put in place in front of the PGAC to select for the A/Q = 113/25 recoils. Data were gathered for 89 hours of beam time, with an average beam current of 8.8 pnA on target. Using the expected $\sigma = 5$ nb, with an FMA efficiency estimated at 10%, a DSSD full energy efficiency of 50% (25% for correlations), and assuming a maximal $b_\alpha \sim 100\%$, we expected to observe approximately 18 ^{113}Ba α–decay events.

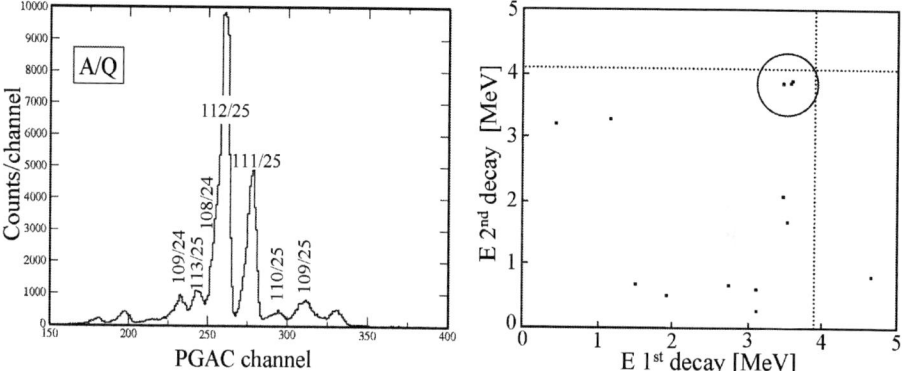

FIGURE 1(left). Intensity profile on the PGAC with no slits. A/Q ratios are listed for several peaks.

FIGURE 2(right). Correlation spectrum of 1st and 2nd decays within the same pixel in the DSSD. Time gates were set so the 1st decay occurs within 1 sec of recoil implantation and the 2nd decay within 100 msec of the 1st decay. The cluster of interest is circled (see text). E(α) from the calculated Q(α) [6] is shown for ^{113}Ba (vertical line) and E(α) from the extrapolated Q(α) is shown for ^{109}Xe (horizontal line).

Correlations between first and second decays were sorted and showed evidence of known decay chains. Background was reduced by setting time gates of 1 second from implant for the first decay energy, and 100 ms from that first decay event for the second decay. A small cluster of correlated α energies was found around $E_{1st} = 3.5$ MeV and $E_{2nd} = 3.9$ MeV (Fig. 2), near the energies expected for the ^{113}Ba $\rightarrow$ ^{109}Xe $\rightarrow$ ^{105}Te decay chain. To verify the identification of this cluster, the slit positions were changed. When the slits were set for A/Q = 111/25, within 90 minutes 5 more correlated events were gathered in that region, suggesting instead that these came from the decay chain ^{111}Xe $\rightarrow$ ^{107}Te $\rightarrow$ ^{103}Sn. The data were further analyzed by using an expanded time window for the correlation spectrum, 2.5 s for the first decay and 250 ms for the second decay. No events were found above these correlations in the place where ^{113}Ba $\rightarrow$ ^{109}Xe $\rightarrow$ ^{105}Te decays were anticipated. Assuming $b_\alpha = 100\%$ for the groundstate, our non-observation of correlated events allows us to set a limit of $\sigma < 0.25$ nb for a mean center-of-target beam energy $E_{mean\ tgt} = 246$ MeV.

There are two reasons which may explain why ^{113}Ba was not observed. The predicted β–decay partial $t_{1/2}$ for ^{113}Ba, based on odd barium systematics, is 110 ms. With the predicted $t_{1/2\alpha} \sim 378$ ms (using Q(α) = 4.0 MeV), b_α would be 23%. $t_{1/2\alpha}$ changes as a logarithm of Q(α), and a Q(α) only 200 keV lower would bring $t_{1/2\alpha}$ to 1.15 seconds, reducing b_α to 9%. For a 5 nb cross section, $b_\alpha \sim 5\%$ alone would account for the non-observation. In addition to a lower branching ratio, the production cross section at the chosen beam energies may be lower than 5 nb.

Although not consistent with the HIVAP predictions, a beam energy nearer the Q value of the 3n reaction may be more ideal, improving the fraction of nuclei in the 3n channel at ~ 230 MeV. Also, since both the 2n and α2n channels were used successfully in this region (e.g., ^{102}Sn [9], and ^{114}Ba [2]), the α3n channel may be an alternative way of populating the ^{113}Ba daughter nucleus ^{109}Xe.

^{109}Xe

To avoid the difficulties associated with a possibly small b_α from ^{113}Ba the ^{58}Ni + ^{54}Fe reaction was used to try to populate ^{109}Xe directly. Data were gathered for 24 hours, with an average beam current of 6.3 pnA on target. The beam energy was set to 240 MeV to maximize production of the 3n channel, according to HIVAP calculations. This had the disadvantage of increasing the cross section for all A/Q = 108/25 products, which showed a recoil production of about twice that of A/Q = 109/25. The FMA was optimized for A/Q = 109/25. For the ^{58}Ni + ^{54}Fe experiment to produce ^{109}Xe directly, about 6 full energy α–decay events and as many partial energy events were expected for σ = 5 nb.

The decay time for the α–decay daughter of ^{109}Xe, ^{105}Te, is expected to be on the order of microseconds. Though a primary decay with this $t_{1/2}$ would be detectable, due to electronics considerations this is too fast for ^{109}Xe → ^{105}Te → ^{101}Sn α–decay correlations. Instead, spectra showing decay energy vs. decay time from implant were analyzed. These spectra showed strong α–decays for both ^{108}Te and ^{109}Te and about 3,000 ^{109}I decay protons. With the extrapolated value of Q(α) = 4.2 MeV for ^{109}Xe, full energy decay events should be observed at an α energy of E = 4.05 MeV. No clear signal of α–decay was observed near that energy. In the case of a very rapid decay of ^{105}Te following ^{109}Xe, sum peaks should be observed up to ~ 9 MeV, but nothing at all was observed above 4 MeV during the first few milliseconds after implantation.

Reducing the background - with a tighter software gate on the PGAC than the slits, using a SiLi detector behind the DSSD for a veto, and requiring the front and back panels of the DSSD to have the same energy for an event - allowed a study of the decay energy vs. time spectra to longer times, but no new decay lines were discernible near the expected energy. While the non-observation of ^{113}Ba may be due to both a small b_α and cross section, since b_α ~ 100% for ^{109}Xe, the non-observation of ^{109}Xe events near the expected α–decay energy then means a lower cross section at a beam energy of 240 MeV than was expected. As 6 full energy ^{109}Xe α–decay events were expected for 5 nb of cross section, and additional partial energy events should be present, our non-observation sets an upper limit of $\sigma(E_{\text{mean tgt}} = 233 \text{ MeV}) < 0.8$ nb.

^{109}I

In the A = 109 data, the 3,000 ^{109}I decay protons observed can address another physics question. The Q(α) for ^{109}I → ^{105}Sb is given as 3.782(16) MeV [10], determined through the mass excesses of ^{109}I and ^{105}Sb, since the ^{109}I α–decay has not actually been observed. The mass excess for ^{105}Sb though was determined through a single measurement of p-decay out of ^{105}Sb [11] which found Q(p) = 478(15) keV. No lifetime measurements were presented in that reference, though a proton branch of 1% was reported. Though several attempts have been made, this experiment has not since been duplicated [12]. Using $Q(p) = 478(15)$ keV from ^{105}Sb, the calculated proton $t_{1/2}$ is ~10 s. Compared with the total half life, 1.3 sec, the proton branch is ~ 8 %. This discrepancy, in addition to the experiment not being reproduced, calls for a reexamination.

Following Geiger-Nuttal systematics and using the quoted $Q(\alpha)$, $t_{1/2\alpha}$ is expected to be ~ 480 ms for ^{109}I. For the neighboring isotope ^{110}I, the α-decay relative width [2] (a measure of preformation) is $W_\alpha = 1.24$. Using a similar width for ^{109}I the calculated $t_{1/2\alpha} = 400$ ms, or since the total measured $t_{1/2} = 103$ μs, $b_\alpha \sim 1/4000$.

In the current experiment, no α-decays were observed within 4 $t_{1/2\alpha}$ around 1 σ of the relevant energy, 3.643(15) MeV. This shows at least that the branch is < 1/3000, which confirms that $t_{1/2\alpha} > 310$ ms and implies that $Q(\alpha) < 3.845$ MeV. This also sets a limit on ^{105}Sb of $Q(p) > 420$ keV. A larger data set from a previous ^{109}I experiment [13] was reanalyzed which, due to a higher background, set a similar branching limit. While this work is not a remeasurement of $Q(p)$, it sets a limit on its minimum value with which the previously measured $Q(p)$ value is consistent.

SUMMARY AND CONCLUSION

Following a search for the N=Z+1 α-decay chain ^{113}Ba → ^{109}Xe → ^{105}Te → ^{101}Sn we report an upper limit on production cross sections of ^{113}Ba and ^{109}Xe. With the reaction ^{58}Ni (^{54}Fe,3n) ^{109}Xe, $\sigma(E_{mean\ tgt} = 233$ MeV$) < 0.8$ nb. The detection limit for ^{113}Ba on the other hand is influenced by both σ and b_α. An upper limit is set on the product of these, $\sigma(E_{mean\ tgt} = 246$ MeV$) < 0.25$ nb for $b_\alpha = 100\%$ via the ^{58}Ni(^{58}Ni,3n)^{113}Ba reaction.

An additional outcome of the experiment is the production of a large number of protons from ^{109}I decay. An upper limit was set on the alpha branch from ^{109}I of $b_\alpha < 1/3000$, which implies limits for ^{109}I of $Q(\alpha) < 3845$ keV and for ^{105}Sb of $Q(p) > 420$ keV. Though several attempts had been made, the previous measurement of $Q(p) = 483(15)$ has not been reproduced. These results set a limit on $Q(\alpha)$ with which the previously determined value is consistent.

This work was supported in part by the U.S. Department of Energy under Contract Numbers W-31-109-ENG-38 and DE-FG02-94-ER40834.

REFERENCES

1. R. D. Macfarlane and A. Siivola, *Phys. Rev. Lett.* **14**, 114 (1965).
2. C. Mazzocchi et al., *Phys. Lett.* **B 532**, 29(2002).
3. Z. Janas et al., *Eur. Phys. J.* **A 23**, 197 (2005).
4. S. Misicu and W. Greiner, *J. Phys. G: Nucl. Part. Phys.* **29**, L67 (2003).
5. Z. Janas et al, *Nucl. Phys.* **A 627**, 119 (1997).
6. P. Möller, J. R. Nix, and K.-L. Kratz, *At. Data Nucl. Data Tables* **66**, 131 (1997).
7. J.O. Rasmussen, *Phys. Rev.* **113**, 1593 (1959).
8. W. Reisdorf, *Z. Phys.* **A 300**, 227 (1981).
9. M. Lipoglavsek et al., *Phys. Lett.* **B 440**, 246 (1998).
10. G. Audi et al., *Nucl. Phys.* **A 729**, 337 (2003).
11. R.J. Tighe et al., *Phys. Rev.* **C 49**, R2871 (1994).
12. Z. Liu et al., *GSI Scientific Report*, 85 (2004).
13. Data courtesy of C.-H. Yu. See C.-H. Yu et al., *Phys. Rev.* **C 59**, R1834 (1999).

NEUTRON CAPTURE REACTIONS II
AND NUCLEAR STRUCTURE

Photo-Induced Population of the $h_{11/2}$ Isomeric States in (γ, n) Reactions

C. T. Angell*, H. J. Karwowski*, J. H. Kelley[†], A. P. Tonchev** and W. Tornow**

University of North Carolina, Department of Physics and Astronomy, Chapel Hill, NC, USA
[†]*North Carolina State University, Department of Physics, Raleigh, NC, USA*
**Duke University, Department of Physics, Durham, NC, USA*

Abstract. The mechanism of excitation of isomeric $h_{11/2}$ states in nuclei around the closed shell at N=82 have been studied at the High-Intensity Gamma Source (HIγS). We have taken advantage of the monoenergetic (ΔE/E=1.5%) and pulsed γ-ray-beam from HIγS to perform in-beam spectroscopy measurements with an improved level of precision and sensitivity. The giant dipole resonances at 15 MeV in the N=82 target isotopes ^{138}Ba, ^{140}Ce, and ^{142}Nd have been excited, and following neutron emission the γ-ray cascades leading to the isomeric $h_{11/2}$ state and the ground state were observed in the N=81 isotopes. For all three nuclei a very similar de-excitation scheme was found. The only level observed from which the isomeric state was populated was found to be $J_\pi = 7/2^-$. The ground state is principally populated from the states with spin and parities $J^\pi = 1/2^+$, $5/2^+$, and $7/2^+$. The structure of the N=81 isotopes and the role of the gateway states in isomer population will be discussed. The results of the measurements will be compared with statistical model calculations.

Keywords: Isomers, Photonuclear Reactions
PACS: 25.20.-x

INTRODUCTION

The population of nuclear isomers via photonuclear reactions have attracted considerable interest lately, especially the isomers in the $h_{11/2}$ shell, which is the largest island of isomerism. There are two items of interest: the isomeric ratio (IR), and the gateway states through which the isomer is populated. The IR is the ratio of the partial cross sections of producing the residual nucleus in the isomeric and ground states.

The IR's of the $h_{11/2}$ isomeric states have been measured before using an activation technique with a (γ, n) reaction to populate the residual nucleus in an excited state [1]. We aim to improve on this data by making in-beam measurements, with the primary purpose being to identify the gateway states that populate the isomeric state, and to calculate the relative transition strengths. The present work is focused on the N=81 isotopes ^{137}Ba, ^{139}Ce, and ^{141}Nd.

The population of the $h_{11/2}$ isomeric state was studied at the High Intensity Gamma Source (HIγS) using the (γ, n) reaction[2]. The Giant Dipole Resonance (GDR) was excited at 15 ± 0.3 MeV in these three nuclei, and following the emission of a neutron, the subsequent γ-ray cascades were observed, leading to the ground ($J^\pi_{gs} = 3/2^+$) or isomeric ($J^\pi_{iso} = 11/2^-$) states (see Fig. 1). These three targets were chosen since at each

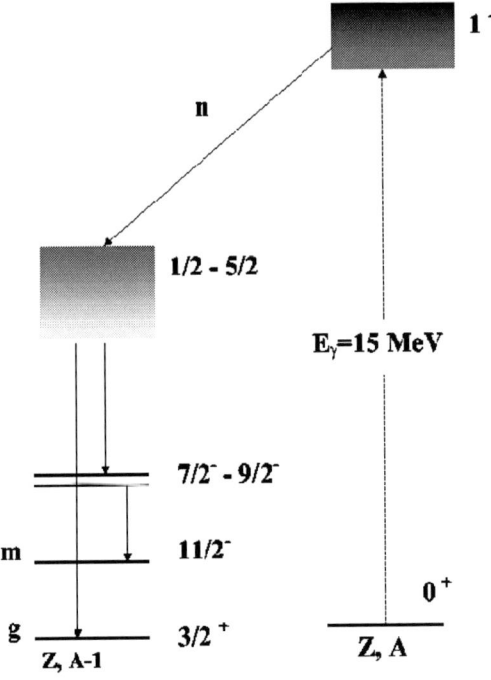

FIGURE 1. Population of the $h_{11/2}$ isomeric states in (γ,n) reaction on ^{138}Ba, ^{140}Ce, and ^{142}Nd.

step in the reaction process, the three nuclei have similar excitation energies and angular momenta. For example, starting in the target N=82 nuclei, they each have an initial spin and parity of $J_i^\pi = 0^-$. The GDR in each nucleus have similar cross sections with maximum at E_γ=15 MeV, and similar widths. In the residual nucleus the spin and parity of the final states are also the same, with similar excitation energies for the isomeric states (662 keV for ^{137}Ba, and about 755 keV for ^{139}Ce and ^{141}Nd), and the same J^π for both the isomeric and ground states. The primary uncertainty in the initial excitation energy of the residual nucleus comes from the spread in the kinetic energy of the emitted neutron. The neutron is ejected with a mean energy of 0.8 MeV, but its spectrum has pronounced right tailing, and a typical width of about 1.5 MeV.

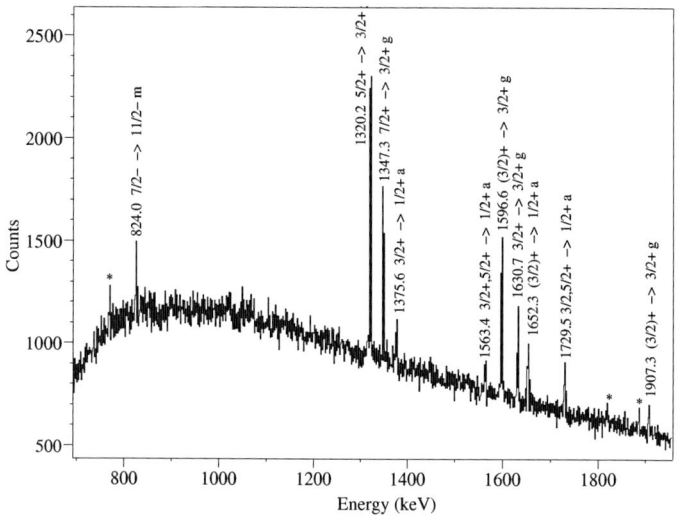

FIGURE 2. Spectrum for ^{139}Ce after taking time-of-flight cut. Energies are in keV.

EXPERIMENTAL METHOD

To perform this experiment we used the pulsed, mono-energetic, and linearly polarized γ-ray beam at HIγS. The beam has an energy resolution of approximately 2%. Four 60% HPGe detectors were positioned perpendicular to the beam axis at the azimuthal angles of 0°, 90°, 180°, and 270° to observe emitted γ-ray's energy and time-of-flight.

For targets we used either the carbonate ($BaCO_3$) or the oxide (CeO_2, and Nd_2O_3) form having a natural isotopic abundance. The samples weighed about 22 g in the case of Ba and Ce, and 14 g for Nd. For each sample, data was taken for between 2.5 and 3 hours. For Ba, we identified 19 γ-ray transitions, for Ce, 12 transitions, and for Nd, 9 transitions, all using the known γ-ray transitions summarized in the Nuclear Data Sheets[3, 4, 5]. The data analysis was done using the program Tv[6].

EXPERIMENTAL RESULTS

The observed spectrum for one of the irradiated targets (^{139}Ce) is shown in Fig. 2, and the decay diagram is shown in Fig. 3. The final stages of the decay for ^{137}Ba and ^{141}Nd are quite similar to that of ^{139}Ce, with minor variations in the energies of particular states. As can be seen from the diagram, the states principally responsible for populating the ground state are the $J^\pi = 1/2^+$, $5/2^+$, and $7/2^+$ states. The transition from the $J^\pi = 1/2^+$ to the ground state was not directly seen, but it can be inferred from the data since transitions which populate it are observed. In Ba and Ce the only state seen which populates the isomeric state is the $J^\pi = 7/2^-$ state. In Nd, it was the state at 2018.8 keV which has an unassigned J^π label in the NuDat tables, but owing to the very similar decay schemes between the N=81 nuclei, it can tentatively be assigned $J^\pi = 7/2^-$.

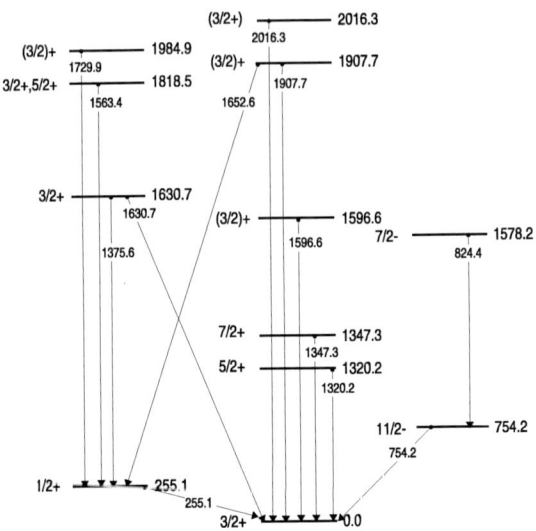

FIGURE 3. The decay diagram for ^{139}Ce. All energies are in keV.

TABLE 1. Experimental and theoretical relative cross sections. The cross sections in each case have been normalized to the $5/2^+ \to 3/2^+$ transition.

Transition	^{137}Ba		^{139}Ce		^{141}Nd	
	σ_{exp}	σ_{th}	σ_{exp}	σ_{th}	σ_{exp}	σ_{th}
$7/2^- \to 11/2^-$	0.30(3)	0.39	0.28(2)	0.56	0.16(4)	0.00
$11/2^- \to 3/2^+$	2.19(8)	1.02	1.33(3)	1.36	0.73(5)	0.46
$7/2^+ \to 3/2^+$	0.92(4)	0.92	0.60(2)	0.81	0.43(4)	0.58

The γ-ray yields were corrected for detector efficiency, internal conversion, and attenuation within the target. In the data shown in Table 1 the yields were normalized to the strongest transition in all three nuclei, that of $5/2^+ \to 3/2^+$. By comparing the intensity of the populating transition ($7/2^- \to 11/2^-$) to the intensity of the depopulating transition ($11/2^- \to 3/2^+$) we can infer that there are other weak transitions populating the isomer (see Table 1). The $7/2^- \to 11/2^-$ transition only contributes about 20% to the total isomer population for all three nuclei. The transition strengths would then have to be scattered between several states. The current detection limit allows us to measure transitions with a strength of 5 to 15% relative to the $5/2^+$ transition strength depending on energy. None were observed. The data are compared with statistical model calculations which were done using the EMPIRE code[7] with default systematics for level densities. The results are listed as σ_{th} in Table 1. Varying the parameters of the level densities within the Gilbert-Cameron model cannot account for this unique set of relative strengths.

CONCLUSIONS

The γ-ray transition strengths relative to the $5/2^+ \to 3/2^+$ transition in the even Z, N=81 isotones were obtained following a (γ, n) reaction. The primary state populating the isomeric state was identified to be the $J^\pi = 7/2^-$ state. The difference in strength between the observed transition populating the isomer and the depopulating transition however points to other states contributing to the total population of the isomer. We intend to make additional measurements to determine the location of these states as well as their J^π assignments.

The relative transition strengths were calculated with EMPRIRE. Variations within known systematics of level densities in the statistical model calculations cannot account for the differences between calculated and measured relative strengths of the transitions. The inability of the statistical model to account for the relative transition strengths suggests the possibility of the presence of further key "gateway" states higher up in energy, which would be an effect not describable within the framework of the statistical model.

REFERENCES

1. A. G. Belov, Y. P. Gangrsky, A. P. Tonchev, and N. P. Balabanov, *Phys. Atomic Nuclei* **59**, 553–559 (1996).
2. V. N. Litvinenko, et al., *Phys. Rev. Lett.* **78**, 4569–4572 (1997).
3. J. K. Tuli, *Nuclear Data Sheets* **81**, 579 (1997).
4. T. W. Burrows, *Nuclear Data Sheets* **92**, 623 (2001).
5. J. K. Tuli, and D. F. Winchell, *Nuclear Data Sheets* **92**, 277 (2001).
6. J. Theuerkauf, S. Esser, S. Krink, M. Luig, N. Nicolay, O. Stuch, and H. Wolters, *Program Tv*, Institute for Nuclear Physics, Cologne, Germany (2000).
7. M. Herman, et al., *EMPIRE*, NNDC, Brookhaven National Laboratory, Upton, USA (2005).

General Approach To Materials Classification Using Neutron Analysis Techniques

Vladimir G. Solovyev[*] and David S. Koltick[†]

[*]*Saint Gobain Crystals and Detectors, 12345 Kinsman Rd, Newbury, OH 44124, USA*
[†]*Department of Physics, Purdue University, 525 Northwestern Avenue,
West Lafayette, IN 47907, USA*

Abstract. The 'neutron in, gamma out' method of elemental analysis has been known and used in many applications as an elemental analysis tool. This method is non-intrusive, non-destructive, fast and precise. This set of advantages makes neutron analysis attractive for even wider variety of uses beyond simple elemental analysis. The question that is addressed within this study is under what conditions neutron analysis can be used to differentiate materials of interest from a group or class of materials in the face of knowing that what is truly of interest is the molecular content of any sample under interrogation. Purpose of the study was to develop a neutron-based scanner for rapid differentiation of classes of materials sealed in small bottles. Developed scanner employs D-T neutron generator as a neutron source and HPGe gamma detectors. Materials can be placed into classes by many different properties. However, neutron analysis method can be used only few of them, such as elemental content, stoichiometric ratios and density of the scanned material. Set of parameters obtainable through neutron analysis serves as a basis for a hyperspace, where each point corresponds to a certain scanned material. Sub-volumes of the hyperspace correspond to different classes of materials. One of the most important properties of the materials are stoichiometric ratios of the elements comprising the materials. Constructing an algorithm for converting the observed gamma ray counts into quantities of the elements in the scanned sample is a crucial part of the analysis. Gamma rays produced in both fast inelastic scatterings and neutron captures are considered. Presence of certain elements in materials, such as hydrogen and chlorine can significantly change neutron dynamics within the sample, and, in turn, characteristic gamma lines development. These effects have been studied and corresponding algorithms have been developed to account for them.

Keywords: elemental analysis, neutron capture, gamma ray spectroscopy.
PACS: 25.20.Lj, 25.40.Fq, 25.40.Lw, 81.70.-q

INTRODUCTION

The detection, classification and identification of materials is a common problem in a variety of applications, among which are quality control in industrial applications, nondestructive testing and the detection of hazardous materials in order to secure entry points, to name a few [1-3]. The requirements for such systems are as follows:

1. Non-intrusive;
2. Non-destructive (We cannot allow alteration of the interrogated material);
3. Quick (We cannot limit flow of people or commerce);
4. Precise (False-positive and false-negative rates are regulated by the scan time);
5. Safe (To the operator, to the interrogated substance, and to the general public).

One of the great advantages of the neutron interrogation technique is that it allows the analysis of materials without the opening of storage containers. Because neutrons are penetrating, they can enter into both metal and non-metal containers easily. Likewise, the return gamma rays are also penetrating and can easily exit a metal or non-metal container to reveal the elemental contents of the container. Unlike chemical analysis techniques, the method does not rely on sample collection, sample preparation and finally, analysis.

MATERIALS CLASSIFICATION

To explore the utility of neutron analysis, it is assumed that we need not differentiate the materials of interest as an element of a limited set, from all possible materials that might exist. Let U be the complete universe of all the materials,

$$U = \{M_i\}, \quad \forall i \qquad (1)$$

then we wish to differentiate materials within only a sub-world or subspace u containing a subset of materials Mi that will be presented for interrogation:

$$u \subset U \qquad (2)$$

This sub-world u may have a fuzzy delineation, meaning that the materials in the fuzzy areas are not to be presented for differentiation frequently and can be dealt with on a case-by-case basis.

There is great advantage in approaching the problem of materials classification in this way. For example, if we take the case where a set of materials consists of the elements that are simply bonded then by taking elemental ratios, for example

$$\left[\frac{H}{O}\right] = 2 \qquad (3)$$

it can be assumed that water is observed as opposed to hydrogen peroxide. While this is not universal or a general conclusion, in the context of the sub-world being searched this is a valid conclusion. Within another sub-world in which pure gas mixtures are being searched, it would be concluded that an explosive gas mixture is present.

What is being illustrated in these examples is that within a context or sub-world, the stoichiometric parameters can be found. However the conclusion, or whether or not a conclusion can be reached, depends on the sub-world being searched. The sub-world is a set of the materials or classes of materials to be differentiated. The elemental makeup of each material is the major input to the differentiation process. Each material can be represented as a set of elemental quantities:

$$M_i = \{\chi_1, \ldots, \chi_{j_i}\} \qquad (4)$$

where χ_j is the jth element, and j_i is the number of elements that are used to identify the ith material.

Each element in turn is defined by the following set of interrogation parameters:

$$\chi_i = \{(L_i^1, \ldots, L_i^{k_i}), (S_i^1, \ldots, S_i^{k_i}), (D_i^1, \ldots, D_i^{k_i})\}, \qquad (5)$$

where S_i^m is the strength of the mth peak in the gamma ray energy spectrum associated with ith element, L_i^m is the energy of the mth peak in the gamma ray spectrum associated with ith element, and D_i^m is the detection modality for the mth peak in the gamma spectrum associated with the ith element.

In order to proceed further, the parameters by which the sub-world can be characterized need to be detailed:
1. All the groups of the materials or individual materials of interest have to be defined. Groups have to be formed on the basis of similar elemental content.
2. For each of the groups of materials defined, a decision has to be made on whether or not identification of individual materials within the group is necessary or if group identification is sufficient.
3. All the elements necessary for materials identification have to be analyzed for the presence of neutron-induced spectral lines and their production mechanisms. Not all the elements present in the materials have to be included in the analysis. Only those required for differentiation within the sub-world will be used.
4. The expected quantity of material has to be defined in order to establish the scan time required for adequate differentiation.

Elemental Analysis

One of the necessary steps in developing the material differentiation scanner is constructing an algorithm for converting the observed gamma ray counts into quantities of the elements in the scanned sample. The first step in this process is to search the gathered gamma ray spectrum for the lines corresponding to the elements of interest, which are fitted with Gaussian-like functions to obtain the number of counts in each peak.

Most of the characteristic gamma ray spectral peaks considered for elemental detection are produced in fast neutron inelastic scattering reactions. In this case the quantity of the ith element X_i is linearly proportional to the number of counts A_i in the spectral peak(s) corresponding to the ith element:

$$X_i = \alpha_i A_i \tag{6}$$

The response linearity is due to the following reasoning. Inelastic scattering reactions as a group, have a threshold incident neutron energy of ~1 MeV. The characteristic reaction length for inelastic scattering reactions for all the materials is at least an order of magnitude larger than the mean free path of neutrons in the materials. The system is designed for scanning of relatively thin objects (up to 10 cm in diameter), which is approximately equal to the fast neutron mean free path in most materials. Therefore, the probability for a fast neutron to encounter more than one inelastic interaction in the sample before exiting the sample is negligibly small, and the system response is expected to be linear.

A correction has to be introduced for elements who's characteristic gamma ray signatures originate from neutron capture reactions. Neutron dynamics in the material containing elements with large neutron capture cross section differs from the case of inelastic scattering elements considered above. The neutron capture reaction does not have an energy threshold as does inelastic neutron scattering. On the contrary, the cross section for neutron capture for room temperature thermal neutrons with energies of kT ~0.02 eV can be up to 7 orders of magnitude larger than the neutron capture cross section for fast neutrons with energies greater than 1 MeV. For this reason, the neutron mean free path for thermal neutrons is a few millimeters in most hydrogenated materials. For our interrogation system, a scanned sample can no longer be considered thin for thermal neutrons penetration.

However, in a case when a uniform sample contains more than one element with signature gamma lines due to neutron capture reactions, the ratio of the amounts of these elements is proportional to the ratio of the number of counts in the corresponding lines:

$$\frac{X_i}{X_j} \propto \frac{A_i}{A_j} \qquad (7)$$

In most cases, elemental ratios provide sufficient information for successful material classification. If one of the elements in the material is characterized by the neutron capture gamma signature, while another is characterized by the inelastic neutron scattering gamma signature, the corresponding stoichiometric ratio $\frac{X_{NC}}{X_{NIS}}$ in general cannot be simply measured.

Eq. 7 is valid under the assumption that the sample itself does not significantly change the neutron energy distribution of the flux passing through the interrogated sample. This is a valid assumption for all but the lightest elements that might make up the sample, because elastic scattering, which is the dominating moderating reaction, can decrease neutron energy by only a small fraction per scattering event due to kinematic restrictions. A material containing both hydrogen and chlorine is a good example for this case. Hydrogen and chlorine lines are generated by neutron radiative capture. Hydrogen in this case serves as a neutron thermalizer, while chlorine, which has a much larger cross section then most elements [4], captures thermalized neutrons, producing excessive chlorine gamma signature, which would not happen for if hydrogen is not present. In this case, the following correction has to be introduced [5]:

$$A_{Cl}^{adj} = A_{Cl} - \gamma_{HCl} A_H. \qquad (8)$$

Then the elemental ratio of hydrogen and chlorine in the material can be established:

$$\frac{X_H}{X_{Cl}} = \frac{A_H + \beta_{HCl} A_{Cl}}{A_{Cl} - \gamma_{HCl} A_H}. \qquad (9)$$

RESULTS AND CONCLUSIONS

Using this reasoning, algorithms can be developed to convert the number of observed counts in characteristic elemental spectral gamma peaks to the elemental quantities or their stoichiometric ratios within the material. These algorithms work under two assumptions. First, the material is assumed to be uniform, and second, the shape of the scanned items must not differ significantly from each other. The latter requirement arises because some elemental signatures are due to the reactions occurring almost uniformly throughout the entire sample volume, while others occur in a relatively thin layer of sample material at the outer surface. Combining these types of spectral lines into single analysis is impossible if contributions from surface and volume terms change uncontrollably.

Using the methods described, a system for differentiating hazardous organophosphates from innocuous materials has been developed and tested [5]. System was able to distinguish hazardous from non-hazardous materials in 30 seconds or less in 99.9% level of correct identifications.

REFERENCES

1. J. Rynes, J. Bendahan, T. Gozani, R. Loveman, J. Stevenson, and C. Bell.. *Nuclear Instruments and Methods in Physics Research A*, 422:895–899, February 1999.
2. P. C. Womble, F. J. Schultz, and G. Vourvopoulus. *Nuclear Instruments and Methods in Physics Research B*, 99:757–760, 1995.
3. L. Zhang and R.C. Lanza. In *Nuclear Science Symposium*,1998. Conference Record., Vol.1, pp. 353–355, 1998.
4. P. F. Rose, editor. ENDF-201, ENDF/B-VI Summary Documentation. Brookhaven National Laboratory, October 1991. BNL-NCS-17541.
5. V.Solovyev "Differentiating Of Classes Of Materials Using Neutron Interrogation Techniques", PhD Thesis, Purdue University, 2005.

Measurement of Gamma Rays from keV-Neutron Capture Reaction by Zr-90, 94

Kazuya Ohgama*, Masayuki Igashira and Toshiro Ohsaki

*Research Laboratory for Nuclear Reactors, Tokyo Institute of Technology,
2-12-1-N1-26 O-okayama, Meguro-ku, Tokyo 152-8550, Japan
*Present Address: O-arai Engineering Center, Japan Nuclear Cycle Development Institute,
O-arai-machi, Higashi-Ibaraki-gun, Ibaraki 311-1393, Japan*

Abstract. The γ rays from the neutron capture reaction by ^{90,94}Zr were measured in an incident neutron energy region from 15 to 100 keV and at 550 keV. A neutron time-of-flight method was adopted with a 1.5-ns pulsed neutron source by the ^{7}Li(p,n)^{7}Be reaction and with a large anti-Compton NaI(Tl) spectrometer. A pulse-height weighting technique was applied to observed capture γ-ray pulse-height spectra to derive capture yields. Using the standard capture cross sections of ^{197}Au in ENDF/B-VI, the capture cross sections of ^{90,94}Zr were obtained with the errors from 6 to 8%. The present results were compared with previous measurements and the evaluations of JENDL-3.3 and ENDF/B-VI. The capture γ-ray spectra of ^{90,94}Zr were obtained by unfolding the observed capture γ-ray pulse-height spectra. The multiplicities of observed γ rays were derived from the γ-ray spectra.

Keywords: kev-neutron capture reaction, zirconium 90, zirconium 94, capture γ-ray spectrum, capture cross section, anti-Compton NaI(Tl) spectrometer.
PACS: 25.40.Lw, 27.60.+j, 29.30.Kv

INTRODUCTION

Currently, a great interest has been taken in the study on the nuclear transmutation of Long-Lived Fission Products (LLFPs: ^{79}Se, ^{93}Zr, ^{99}Tc, ^{107}Pd, ^{126}Sn, ^{129}I, ^{135}Cs) generated in fission reactors. In ordinary design of transmutation system, neutron capture reaction is utilized to transmute LLFPs into stable or short-lived nuclides.

As for the stable isotopes of LLFPs, they are also produced in fission reactors and accompany LLFPs to a transmutation system unless the isotopic separation of LLFPs is performed. Therefore, the neutron capture cross sections of both LLFPs and their stable isotopes are indispensable for the transmutation study.

Moreover, the systematic data analysis of both keV-neutron capture cross sections and capture γ-ray spectra of the stable isotopes of a LLFP is important to accurately predict the capture cross sections of the LLFP. However, the present status of those experimental data is quite inadequate both in quality and in quantity.

In this situation, we started a systematic measurement of the keV-neutron capture cross sections and capture γ-ray spectra of stable isotopes of an important LLFP nuclide ^{93}Zr [1]. In this paper, we present the experimental results of ^{90,94}Zr.

EXPERIMENTAL PROCEDURE

The experimental procedure has been described in detail elsewhere [1, 2], so it is summarized here. The capture cross sections and capture γ-ray spectra of $^{90, 94}$Zr were measured in an incident neutron energy region from 15 to 100 keV and at 550 keV, using the 3-MV Pelletron accelerator of the Research Laboratory for Nuclear Reactors at the Tokyo Institute of Technology. An experimental arrangement is shown in Fig.1. Pulsed neutrons were produced by the ^{7}Li(p,n)^{7}Be reaction with a pulsed proton beam (1.5-ns width, repetition rate of 4 MHz) from the Pelletron accelerator. The incident neutron energy spectrum on a capture sample was measured by means of a Time of Flight (TOF) method with a ^{6}Li-glass scintillation detector. The characteristics of the $^{90, 94}$Zr samples and the standard Au samples are given in Table 1.

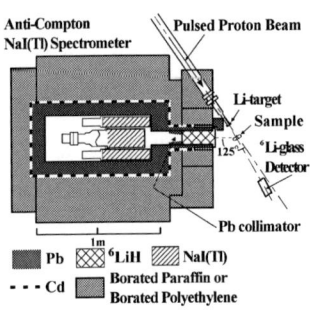

FIGURE 1. Experimental arrangement in the 15-100 keV measurement.

TABLE 1. Characteristics of Samples

Sample	^{90}Zr[*1]	^{90}Zr[*1]	^{94}Zr	^{197}Au
Chemical form	ZrO_2	ZrO_2	ZrO_2	Au
Chemical purity (%)	99.90	99.63	99.94	99.99
Weight of sample (g)	0.991 (^{90}Zr)	1.174 (^{90}Zr)	0.993 (^{94}Zr)	6.023 /12.046[*2]
Isotopic composition (%)				^{197}Au:100
^{90}Zr	95.6	97.7	3.7	-
^{91}Zr	1.7	0.9	1.1	-
^{92}Zr	1.2	0.6	2.1	-
^{94}Zr	1.3	0.7	92.6	-
^{96}Zr	0.2	0.2	0.5	-
Thickness of sample (mm)	2.7	2.9	2.8	1.0 /2.0[*2]
Diameter of sample (mm)	20	20	20	20

[*1] Both samples were used for the measurement of ^{90}Zr.
[*2] 15-100 keV measurement / 550 keV measurement.

The capture γ rays emitted from the sample were measured with a large anti-Compton NaI(Tl) spectrometer by means of a TOF method. The spectrometer was set in a heavy shield consisting of borated paraffin, borated polyethylene, Cd, ^{6}LiH and potassium-free lead [3]. The signals from the spectrometer were recorded in a personal computer as two-dimensional data of Pulse Height (PH) and TOF.

The measurements with the $^{90, 94}$Zr and ^{197}Au samples and one without any samples (Blank) were performed cyclically to average out changes in experimental conditions such as the incident neutron spectrum. The individual measurements were connected by the neutron counts of the ^{6}Li-glass detector.

DATA PROCESSING

The data processing method has been also given in detail elsewhere [1, 2], so it is summarized here.

The incident neutron spectra on the sample were derived from the TOF spectra measured with the ^{6}Li-glass detector for the blank run. The normalized spectra, $\eta(E_n)$, are shown in Fig.2, where E_n is the incident neutron energy in the laboratory system.

The TOF spectra obtained with the γ-ray spectrometer for the (a) ^{90}Zr, (b) Au and (c) Blank runs in the 15-100 keV measurement are shown in Fig.3. The broad peak around 500 ch. in Fig. 3(b) is due to the neutron capture γ rays from the Au sample. The corresponding broad peak is also observed in Fig. 3(a), but it is very weak. The sharp peak around 620 ch. is due to the γ rays from the ^{7}Li(p,γ)^{8}Be reaction in the neutron-production Li target. Gates were set in the foreground and background regions to obtain foreground and background PH spectra, as shown in Fig. 3. Then, net capture γ-ray PH spectra were obtained by subtracting the background PH spectra normalized with the gate-width ratios from the foreground PH spectra.

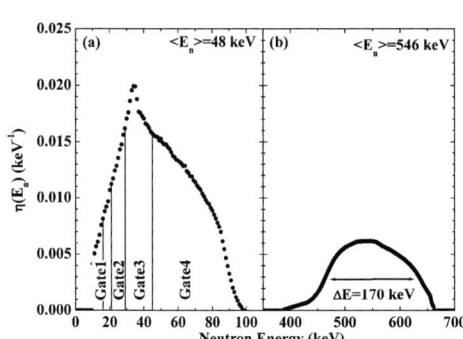

FIGURE 2. Normalized incident neutron spectra in the (a) 15-100 keV and (b) 550 keV measurements of ^{90}Zr.

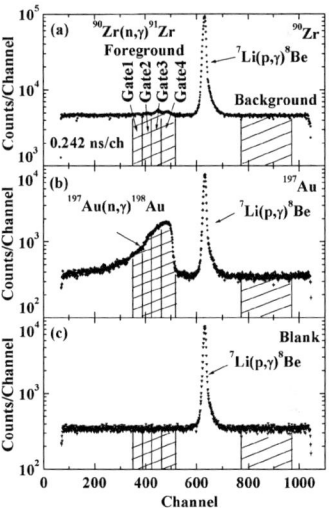

FIGURE 3. TOF spectra observed with the anti-Compton NaI(Tl) spectrometer for the (a) ^{90}Zr, (b) Au and (c) Blank runs in the 15-100 keV measurement.

In order to obtain the capture yields of the $^{90, 94}$Zr and ^{197}Au samples, a PH weighting technique [4] was applied to the net capture γ-ray PH spectra. The number of incident neutrons for each gate in the ^{197}Au run was determined by the corresponding capture yield of ^{197}Au and the averaged capture cross section of ^{197}Au, which was obtained from the capture cross sections of ENDF/B-VI [5] and the neutron energy spectrum. The number of incident neutrons for each gate in the $^{90, 94}$Zr runs was derived from that in the Au run and the neutron monitor counts of the ^{6}Li-glass detector. The averaged capture cross section for each gate of $^{90, 94}$Zr was derived from the number of incident neutrons and the capture yield for each gate. The same procedure was applied to the data processing in the 550 keV measurement.

The correction for the neutron self-shielding and multiple-scattering in the sample was made by a Mote-Carlo method [6], taking account of the graphite case of sample, the oxygen and impurities in the sample. The nuclear data used for the correction were taken from JENDL-3.3 [7] and ENDF/B-VI [5]. Other corrections described in ref. 1 were also made.

The capture γ-ray spectra were obtained by unfolding the net capture γ-ray PH spectra with the computer code FERDOR [8] and the response matrix of the γ-ray spectrometer [9].

RESULTS AND DISCUSSION

The capture cross sections of ^{90,94}Zr were derived with the errors from 6 to 8% in the neutron energy region of 15-100 keV and at 550 keV. Both present results at 550 keV were the first experimental ones above 200 keV in the keV region. The present results of ^{90}Zr are compared in Fig. 4 with previous measurements [10-12] and the evaluations of JENDL-3.3 [7] and ENDF/B-VI [13]. The evaluated values in Fig. 4 are averaged ones with appropriate energy widths for easy comparison with the experimental data.

The keV-neutron capture γ-ray spectra of ^{90,94}Zr were derived for the first time, and the spectra of ^{94}Zr are shown in Fig. 5. Strong primary transitions to low-lying states of ^{95}Zr are observed.

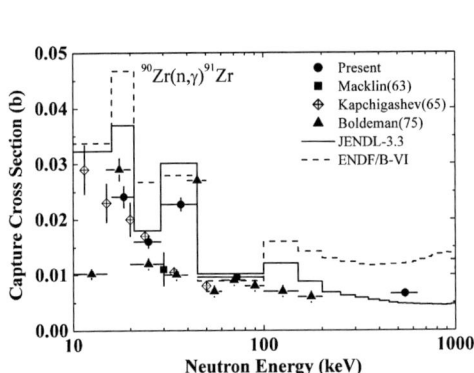

FIGURE 4. Neutron capture cross sections of ^{90}Zr in the keV region.

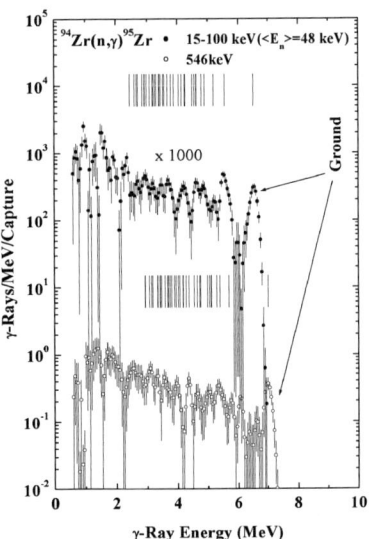

FIGURE 5. Capture γ-ray spectra of ^{94}Zr in the neutron energy region from 15 to 100 keV and at 550 keV

The multiplicities of observed capture γ rays (E≥0.6 MeV) were derived by integrating the γ-ray spectra with respect to the γ-ray energy, and are given in Table 2. The reduced multiplicities [2], which were obtained by dividing the multiplicities by the corresponding excitation energy of neutron capture states, are also given in Table 2.

TABLE 2. Multiplicities of observed capture γ rays (E≥0.6 MeV) per capture reaction of $^{90, 94}$Zr.

Isotope [Neutron binding energy (MeV)]	^{90}Zr[7.195]		^{94}Zr[6.463]	
Incident neutron energy (MeV)	0.048	0.550	0.048	0.550
Multiplicity*¹(E≥0.6MeV)	2.26±0.09	2.27±0.21	2.49±0.11	2.38±0.14
Reduced multiplicity*²	0.283±0.011	0.284±0.027	0.312±0.014	0.298±0.017

*¹ Unit: γ rays/capture
*² Defined by (multiplicity)/(B_n+<E_n>), where B_n is the neutron binding energy and <E_n> is the averaged incident neutron energy. The unit is γ rays/1-MeV excitation energy/capture.

CONCLUSION

The neutron capture γ rays of $^{90, 94}$Zr and ^{197}Au were measured in the incident neutron energy region from 15 to 100 keV and at 550 keV with the 1.5-ns pulsed neutron source by the ^{7}Li(p,n)^{7}Be reaction and the large anti-Compton NaI(Tl) spectrometer. The capture cross sections of $^{90, 94}$Zr were derived with the errors from 6 to 8%, adopting the PH weighting technique and the standard capture cross sections of ^{197}Au. The capture γ-ray spectra of $^{90, 94}$Zr were obtained by unfolding the net capture gamma-ray PH spectra. The multiplicities of observed γ rays were derived from the γ-ray spectra.

ACKNOWLEDGMENTS

The present study was supported by a Grant-in-Aid of the Japan Ministry of Education, Culture, Sports, Science and Technology and by Japan Nuclear Cycle Development Institute.

REFERENCES

1. K. Ohgama, M. Igashira and T. Ohsaki, *J. Nucl. Sci. Technol.*, **42**, 333(2005).
2. S. Mizuno, M. Igashira and K. Masuda, *J. Nucl. Sci. Technol.*, **36**, 493(1999).
3. M. Igashira, H. Kitazawa and N. Yamamuro, *Nucl. Instrum. Methods*, **A245**, 432 (1986).
4. R. L. Macklin and J. H. Gibbons, *Phys. Rev.*, **159**, 1007 (1967).
5. ENDF/B-VI data file for ^{197}Au (MAT=7925), evaluated by P. G. Young (1991).
6. K. Senoo, Y. Nagai, T. Shima, T. Ohsaki and M. Igashira., *Nucl. Instrum. Methods*, **A339**, 556 (1994).
7. K. Sibata, T. Kawano, T. Nakagawa, et al., *J. Nucl. Sci. Technol.*, **39**, 1125 (2002).
8. H. Kendrick and S. M. Sperling, *GA-9882*, (1970).
9. K. Ohgama, *"Systematic study of keV-neutron capture cross sections and capture gamma-ray spectra of Zr isotopes"*, Doctor Thesis, Tokyo Institute of Technology, 2005, [in Japanese].
10. R. L. Macklin, T. Inada and J. H. Gibbons, *Bull. Am. Phys. Soc.*, **8**, 80 (1963).
11. S. P. Kapchigashev, *Sov. At. Energy*, **19**, 1212 (1965); EXFOR40034.002, (2003).
12. J. W. Boldeman, B. J. Allen, A. R. de L. Musgrove and R. L. Macklin, *Nucl. Phys.*, **A246**, 1 (1975); EXFOR30329.004, (1982).
13. ENDF/B-VI data file for ^{90}Zr (MAT=4025), evaluated by M. Drake *et al.*, (1976).

Measurement of Capture Gamma Rays from the 46- and 84-keV Neutron Resonances of ^{24}Mg

Toshiro Ohsaki, Daisuke Ikenaga* and Masayuki Igashira

Research Laboratory for Nuclear Reactors, Tokyo Institute of Technology
N1-26, 2-12-1, O-okayama, Meguro-ku, Tokyo 152-8550, Japan

Abstract. The capture γ rays from the 46-keV d-wave and 84-keV p-wave neutron resonances of ^{24}Mg were measured by using a large anti-Compton HPGe γ-ray spectrometer. A neutron time-of-flight method was adopted with a 1.5-ns pulsed neutron source by the ^{7}Li(p,n)^{7}Be reaction. The standard capture cross sections of ^{197}Au were used to determine the incident neutron flux. Twenty-one and twenty-six primary γ-ray transitions were identified for the 46- and 84-keV resonances, respectively. The corresponding capture kernels were derived for both the resonances, and the radiative widths were obtained for the 84-keV broad resonance.

Keywords: magnesium-24, neutron resonance, capture gamma ray, capture kernel, radiative width
PACS: 23.30.Lv, 25.40.Ny, 27.30.+t, 29.30.Kv

INTRODUCTION

In general, the measurement of radiative widths of broad neutron resonances is difficult, because the neutrons strongly scattered by the broad resonances cause a huge background in an ordinary experiment. The background reduction and/or the separation of true signals from the background are very important to correctly measure the radiative widths of broad resonances. One way to separate the true signals from the background is the observation of distinguished discrete γ-ray transitions from the resonances.

For the measurement of discrete γ-ray transitions from keV-neutron resonances, NaI(Tl) or other scintillators are usually used. In the present study, we employed a large anti-Compton HPGe detector with good energy resolution, and measured capture γ rays from the narrow 46-keV d-wave and broad 84-keV p-wave neutron resonances of ^{24}Mg.

EXPERIMENT

The experimental procedure and data processing have been described in detail elsewhere[1, 2], so these are briefly described below.

The measurement was carried out by using pulsed neutrons from the ^{7}Li(p,n)^{7}Be reaction with a 1.5-ns pulsed proton beam from the 3-MV Pelletron accelerator of the

---

* Present Address: *Panasonic Communications Co., Ltd.*
4-1-6 Minoshima, Hakata-ku, Fukuoka 812-8531, Japan

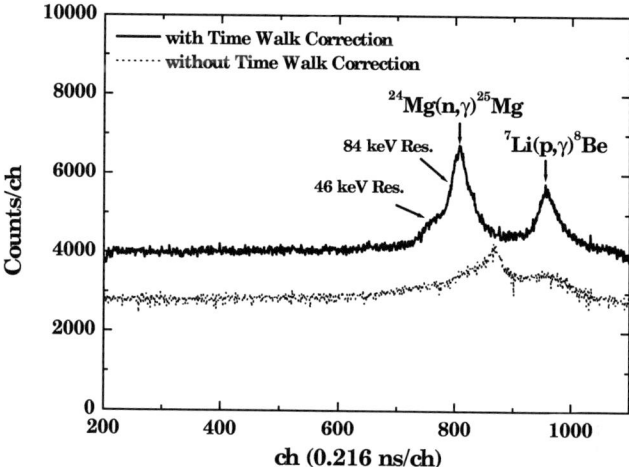

FIGURE 1. TOF spectra for the ^{24}Mg(n,γ)^{25}Mg reaction measurement with and without the time walk correction. In the corrected spectrum, the peaks due to the neutron-resonance capture by ^{24}Mg were well separated from the peak due to the ^{7}Li(p,γ)^{8}Be reaction.

Research Laboratory for Nuclear Reactors at the Tokyo Institute of Technology. The incident neutron spectrum on a capture sample was measured by means of a time-of-flight(TOF) method with a ^{6}Li-glass scintillation detector. The capture sample was a natural magnesium disk with a diameter of 90 mm and a thickness of 3 mm. A gold sample with a diameter of 90 mm and a thickness of 1 mm was used as a standard[3]. Capture γ rays were detected by a large anti-Compton HPGe spectrometer employing a TOF method. The HPGe spectrometer had a relative efficiency of 100 %, and was shielded against background γ rays and neutrons with potassium-free lead, borated polyethylene, cadmium and ^{6}LiH. Two-dimensional data on neutron TOF and γ-ray pulse height from the spectrometer were stored into a personal computer in a list mode.

The time resolution of the HPGe spectrometer was improved by using a relation between pulse height and time walk[1]. Figure 1 shows TOF spectra for the ^{24}Mg measurement with and without the time walk correction. In the corrected spectrum, the peaks due to the neutron-resonance capture by ^{24}Mg were well separated from the peak due to the ^{7}Li(p,γ)^{8}Be reaction in the neutron-production Li target. However, the 46-keV resonance peak could not be separated clearly from the 84-keV resonance peak.

RESULTS AND DISCUSSION

For the 46-keV d-wave resonance, twenty-one primary transitions were observed, and the corresponding partial capture kernels were derived, as shown in Table 1. We

TABLE 1. Partial capture kernels and relative intensities of the 46-keV resonance($3/2^+$) of ^{24}Mg

Final state (keV)	J^π	Partial capture kernel (eV) Present	Relative intensity (%) Present	Relative intensity (%) Bergqvist[5]
0	$5/2^+$	0.81±0.08	43.1±4.5	46.3±5.8
585.0	$1/2^+$	0.17±0.02	8.6±1.0	14.0±2.5
974.7	$3/2^+$	0.13±0.02	6.8±0.9	
1611.8	$7/2^+$	0.09±0.01†	5.0±0.7†	-
1964.6	$5/2^+$	0.02±0.01	0.9±0.3	-
2563.8	$1/2^+$	0.03±0.01	1.5±0.3	4.1±2.5
2737.7	$7/2^+$	0.07±0.01†	3.6±0.5†	2.5±1.7
3414.0	$3/2^-$	0.06±0.01	3.0±0.5	-
3907.7	$5/2^+$	0.05±0.01	2.6±0.5	-
3970.7	$7/2^-$	0.013±0.007†	0.7±0.3†	-
4276.8	$1/2^-$	0.04±0.01	2.1±0.4	-
4722.0	$5/2^+(3/2^+)$	0.07±0.01	3.7±0.6	8.3±3.3
5116.3	$1/2^-$	0.02±0.01	1.0±0.4	-
5520.9	$5/2$	0.03±0.01	1.3±0.5	-
5744	$(7/2,9/2^+)$	0.04±0.01†	2.2±0.7†	-
5860	$5/2^+$	0.04±0.01	2.0±0.6	-
5978.0	$7/2^+$	0.04±0.01†	2.2±0.6†	-
6082	$5/2^+$	0.03±0.01	1.5±0.6	-
6169	$3/2$	0.05±0.02	2.8±0.8	-
6570	$(1/2^+),3/2$	0.03±0.01	1.4±0.7	-
6678	$5/2$	0.07±0.02	3.9±1.1	-

† The same angular distribution was assumed as the dipole radiation.

TABLE 2. Total capture kernel of the 46-keV resonance of ^{24}Mg

Authors	Total capture kernel (eV)	Remarks
Present	1.888±0.092	Anti-Compton HPGe
Block[6]	0.8±0.2	Liquid scintillation tank
Weigmann[7]	0.848±0.007†	C_6F_6 scintillator
Mughabghab[8]	0.840±0.007	Evaluation
Sukhoruchkin[4]	0.902±0.007	Evaluation

† Corrected by Macklin and Winters[9]

tried to derive partial radiative widths by using the 46-keV resonance parameters in the evaluation by Sukhoruchkin[4], but reasonable widths were not obtained. It suggests that the re-evaluation of the total or neutron width should be performed. In Table 1, the relative intensities measured by Bergqvist et al.[5] with an NaI(Tl) detector are also shown.

The total capture kernel of the 46-keV resonance was obtained by summing up the partial capture kernels, and is compared in Table 2 with previous experiments[6, 7] and evaluations[4, 8]. The present result is about two times as large as the previous experiments.

TABLE 3. Partial capture kernels and radiative widths of the 84-keV resonance(3/2⁻) of ^{24}Mg

Final state		Partial capture kernel (eV)	Partial radiative width (eV)		
(keV)	J^π	Present	Present	Bergqvist[5]	Uchiyama[10]
0	5/2⁺	2.18±0.20	1.09±0.10	0.6	0.93±0.14
585.0	1/2⁺	8.28±0.74	4.14±0.37	1.88	3.98±0.57
974.7	3/2⁺	-	-	-	< 0.10
1611.8	7/2⁺	0.05±0.01†	0.02±0.01†	-	-
1964.6	5/2⁺	0.36±0.04	0.18±0.02	0.081	0.20±0.03
2563.8	1/2⁺	1.63±0.15	0.82±0.08	0.35	0.68±0.11
2737.7	7/2⁺	0.06±0.01†	0.03±0.01†	-	-
2801.5	3/2⁺	0.12±0.02	0.06±0.01	-	-
3414.0	3/2⁻	0.14±0.02	0.07±0.01	-	-
3907.7	5/2⁺	0.07±0.02	0.03±0.01	-	-
3970.7	7/2⁻	0.08±0.02†	0.04±0.01†	-	-
4276.8	1/2⁻	0.02±0.01	0.010±0.005	-	< 0.03
4722.0	5/2⁺	0.05±0.01	0.02±0.01	-	-
5012.3	7/2⁺	0.03±0.02†	0.02±0.01†	-	-
5116.3	1/2⁻	0.03±0.02	0.02±0.01	-	-
5474.8	1/2+	0.06±0.01	0.03±0.01	-	-
5520.9	5/2	0.06±0.02	0.03±0.01	-	-
5747.0	(3/2,5/2⁺)	0.04±0.02	0.02±0.01	-	-
5860	5/2⁺	0.04±0.02	0.02±0.01	-	-
6041.2	(7/2,11/2)⁺	0.04±0.02†	0.02±0.01†	-	-
6082	5/2⁺	0.06±0.03	0.03±0.01	-	-
6169	3/2	0.04±0.02	0.02±0.01	-	-
6362	3/2	0.09±0.03	0.05±0.01	-	-
6570	(1/2⁺,3/2)	0.11±0.04	0.06±0.02	-	-
6678	5/2	0.09±0.04	0.05±0.02	-	-
6777	1/2+	0.13±0.04	0.06±0.02	-	-

† The same angular distribution was assumed as the dipole radiation.

In the case of the 84-keV p-wave resonance, twenty-six primary transitions were observed, and the corresponding partial capture kernels and partial radiative widths were derived, as shown in Table 3. In the data analysis, the 84-keV resonance parameters evaluated by Sukhoruchkin[4] were also used. Previous experiments[5, 10] with an NaI(Tl) scintillator are also shown in Table 3.

The obtained total capture kernel and total radiative width are compared with previous experiments[6, 7, 10-12] and evaluations[4, 8] in Table 4.

CONCLUSION

The neutron capture γ rays from the 46- and 84-keV resonances of ^{24}Mg were measured with the HPGe spectrometer for the first time. Twenty-one and twenty-six primary transitions were observed for the 46- and 84-keV resonances, respectively. The corresponding capture kernels were derived for both the resonances, and the partial radiative widths were also obtained for the 84-keV resonance. The obtained total capture kernel of the 46-keV narrow resonance was about two times as large as previous experiments

TABLE 4. Total radiative width and the total capture kernel of the 84-keV resonance of ^{24}Mg

Authors	Total radiative width (eV)	Total capture kernel (eV)	Remarks
Present	6.93±0.39	13.8±0.8	Anti-Compton HPGe
Block[6]	6.3±0.9	-	Liquid scintillation tank
Nyström[11]	4.0±0.9	-	NaI(Tl)
Weigmann[7]	4.7±0.9†	9.0±1.9†	C_6F_6 scintillator
Allen[12]	4.4±0.6	-	Moxon-Rae
Uchiyama[10]	5.9±0.6	-	Anti-Compton NaI(Tl)
Mughabghab[8]	4.7±1.0	9.3±0.2	Evaluation
Sukhoruchkin[4]	5±1	10.0±2.0	Evaluation

† Corrected by Macklin and Winters[9]

and evaluations. As for the total radiative width of the 84-keV broad resonance, the present result was larger than evaluated values by about 40 %, but in agreement with the measurements by Block *et al.*[6] and Uchiyama *et al.*[10] within the experimental errors.

REFERENCES

1. S. Y. Lee, J. Hori, and M. Igashira, *J. Nucl. Sci. Technol.*, **36**, 719–737 (1999).
2. J. Hori, S. T. Park, S. Y. Lee, M. Igashira, and T. Ohsaki, *J. Nucl. Sci. Technol.*, **38**, 91–101 (2001).
3. ENDF/B-VI data file for ^{197}Au(MAT = 7925), evaluated by P. G. Young (1991).
4. S. I. Sukhoruchkin, Z. N. Soroko, and V. V. Deriglazov, *Low Energy Neutron Physics, subvolume B Tables of Neutron Resonance Parameters*, Landolt-Börnstein, New Series I/16B, Springer, Berlin, (1998).
5. I. Bergqvist, J. A. Biggerstaff, J. H. Gibbons and W. M. Good, *Phys. Rev.*, **158**, 1049–1062 (1967).
6. R. C. Block *et al.*, *Renseselaer Polytechnic Institute Report*, RPI-328-142 (1968).
7. H. Weigmann, R. L. Macklin, and J. A. Harvey, *Phys. Rev. C*, **14**, 1328–1335 (1976).
8. S. F. Mughabghab, M. Divadeenam, and N. E. Holden, *Neutron Cross Section* **1**, Part.A, BNL, Academic Press Inc, New York (1981).
9. R. L. Macklin and R. R. Winters, *Nucl. Sci. Eng.*, **78**, 110–111 (1981).
10. T. Uchiyama, M. Igashira and H. Kitazawa, *Phys. Rev. C*, **41**, 862–875 (1990).
11. G. Nyström, B. Lundberg, and I. Bergqvist, *Physica Scripta*, **4**, 95-99 (1971).
12. B. J. Allen, D. D. Cohen and F. Z. Company, *J. Phys. G*, **6**, 1173-1188 (1980).

Bound excited states in ^{27}F

Z. Elekes*, Zs. Dombrádi*, A. Saito[†], N. Aoi[**], H. Baba[†], K. Demichi[‡], Zs. Fülöp*, J. Gibelin[§], T. Gomi[‡], H. Hasegawa[‡], N. Imai[**], M. Ishihara[**], H. Iwasaki[†], S. Kanno[‡], S. Kawai[‡], T. Kishida[**], T. Kubo[**], K. Kurita[‡], Y. Matsuyama[‡], S. Michimasa[**], T. Minemura[**], T. Motobayashi[**], M. Notani[†], T.K Ohnishi[†], H.J. Ong[†], S. Ota[¶], A. Ozawa[∥], H.K. Sakai[‡], H. Sakurai[†], S. Shimoura[†], E. Takeshita[‡], S. Takeuchi[**], M. Tamaki[†], Y. Togano[‡], K. Yamada[‡], Y. Yanagisawa[**] and K. Yoneda[**]

*Institute of Nuclear Research of the Hungarian Academy of Sciences, P.O. Box 51, Debrecen H-4001, Hungary
[†]University of Tokyo, Tokyo 1130033, Japan
[**]The Institute of Physical and Chemical Research, 2-1 Hirosawa, Wako, Saitama 351-0198, Japan
[‡]Rikkyo University, 3 Nishi-Ikebukuro, Toshima, Tokyo 171, Japan
[§]Institut de Physique Nucléare, 15 rue Georges Clemenceau, 91406 Orsay, France
[¶]Kyoto University, Kyoto 606-8501, Japan
[∥]Tsukuba University, Tennoudai 1-1-1, Tsukuba-shi, Ibaraki 305-8571, Japan

Abstract. The ^{1}H(^{27}F,^{25,26,27}F) reactions have been studied at 40 MeV/nucleon average energy using a liquid hydrogen target. For ^{25}F, ^{26}F and ^{27}F nuclei, we have observed two γ-ray peaks each originating from the decay of two bound excited states. This is the first sign of the existence of bound excited states in ^{26,27}F. The presence of a single bound excited state in ^{27}F is a clear indication of a substantial change in the structure of the fluorine isotopes approaching the neutron dripline. The proposed second excited states in ^{25,26,27}F nuclei have no counterparts in either the psd or the $sdpf$ shell model calculations suggesting the appearance of nucler structure effects lying out of these model spaces.

Keywords: γ-spectroscopy, ^{25,26,27}F excited states, proton scattering, radioactive beam
PACS: 21.60.Cs; 23.20.Lv; 25.40.Ep; 27.30.+t; 29.30.Kv

INTRODUCTION

An intriguing problem is that the dripline of fluorine isotopes is located at least 6 neutrons farther than that of the oxygen isotopes. Whatever mechanism makes ^{31}F particle-bound; its traces should be visible in other fluorine nuclei, too. For instance, in ^{27}F the psd shell model [1], which can account for the properties of light fluorine nuclei, predicts the first excited state with spin 1/2$^+$ at 2.0 MeV energy, much higher than the neutron separation energy of 1.3(4) MeV. On the other hand, some shell breaking or dripline effects (continuum coupling, enhanced pairing) acting in ^{31}F can lower the energy of this state below the separation energy. According to recent Monte-Carlo shell model calculations it is enough to allow for the possibility of neutron cross shell excitations to have a bound excited state in ^{27}F [2]. To explore the traces of the mechanism which is expected to be responsible for the existence of ^{31}F in lighter fluorine nuclei, and to gain more information on its properties we have searched for bound excited states in ^{27}F.

EXPERIMENTAL

We have applied the (p,p') reaction in combination with γ-ray spectroscopy, which allows us to employ thick targets and low intensity radioactive beams. The experiment was carried out at the RIKEN Accelerator Research Facility. A 94 MeV/nucleon energy primary beam of ^{40}Ar with 60 pnA intensity hit an ^{181}Ta production target of 0.5 cm thickness. The reaction products were momentum and mass analyzed by the RIPS fragment separator. The secondary beam included neutron-rich O, F, Ne and Na nuclei with $A/Z \approx 3$. The total intensity was about 100 pps, while the fraction of individual isotopes varied in the range of 1-10% having a ^{27}F intensity of 4 pps on average. The identification of incident beam species was performed event by event by means of energy loss, time-of-flight (TOF) and magnetic rigidity ($B\rho$). The ^{27}F particles could be fully separated from other nuclei. The secondary beam was transmitted to a liquid hydrogen target of 30 mm diameter at F3. The average areal density of the hydrogen cooled down to 22 K was 210 mg/cm^2. The mean energy of ^{27}F isotopes was calculated at 39.6 MeV/nucleon. The position of the incident particles was determined by two PPACs placed at F3 upstream of the target. The scattered particles were detected and identified by a PPAC and a silicon telescope located about 80 cm downstream of the target. The telescope consisted of three layers of Si with thicknesses of 0.5, 0.5 and 1 mm. Each layer was made of a 2x2 matrix of detectors. The Z identification was performed by TOF-energy loss method where the TOF was taken between the secondary target and the PPAC. Isotope separation was carried out among the different fluorine isotopes by use of the ΔE-E method. To detect the de-exciting γ rays emitted by the inelastically scattered nuclei the DALI2 setup including 146 NaI(Tl) scintillator detectors surrounded the target.

FIGURE 1. Doppler-corrected spectra of γ rays emerging from ^{1}H(^{27}F,^{27}F) (a), ^{1}H(^{27}F,^{26}F) (b) and ^{1}H(^{27}F,^{25}F) (c) reactions. The solid line is the final fit including the spectrum curves from GEANT4 simulation and additional smooth polynomial backgrounds plotted as separate dotted lines for each nucleus. The insets in gray boxes show the *psd* shell model predictions.

ANALYSIS AND RESULTS

Fig. 1 plots the Doppler-corrected γ ray spectra for ^{27}F (a), ^{26}F (b) and ^{25}F (c) nuclei. First, the positions of the peak candidates (500, 750, 1200 keV for ^{27}F, 470, 660, 1300 for ^{26}F, 730, 1000, 1350, 1750 keV for ^{25}F) and their uncertainties were determined by fitting the spectra with Gaussian functions and smooth exponential backgrounds. After the peak positions have been determined they were fed into the detector simulation software GEANT and the resultant response curves plus smooth polynomial backgrounds were used to analyze the experimental spectra in terms of the significance of the peaks by taking the 2σ level as a criterion. According to this, there are two significant peaks at 727(22) and 1753(53) keV in the ^{25}F spectrum (Fig. 1 (c)). In the ^{26}F spectrum (Fig. 1 (b)), two peaks were found at 468(17) and 665(12) keV. The ^{27}F spectrum (Fig. 1 (a)) also shows two peaks at 504(15) and 777(19) keV.

DISCUSSION AND CONCLUSIONS

The experimental data can be compared with the predictions of the shell model calculations. The sd shell model [3] predicts the ground state of ^{25}F to be $5/2^+$, followed by a $1/2^+$ state at 911 keV and a $3/2^+$ one at 3373 keV. In ^{26}F, the members of the $\pi d_{5/2} \nu d_{3/2}$ multiplet give the lowest energy states starting with the 1^+ ground state and followed by the 2^+ at 681, the 3^+ at 1604 and the 4^+ state at 353 keV. In ^{27}F, a $5/2^+$ ground state is expected with the $1/2^+$ state as the first excited state at 1997 keV as mentioned earlier. Comparing the experimental data with these predictions, it is clearly seen that the energy of the 727 keV γ ray of ^{25}F and that of the 665 keV one in ^{26}F is fairly close to the predicted energies of the $1/2^+$ state in ^{25}F, and the 2^+ state in ^{26}F, respectively, and can be assigned to the decay of these states. On the other hand, both levels of ^{27}F and the second excited states of ^{25,26}F appear at too low energies independently whether they are constructed by placing the γ rays parallel or in cascade.

Extending the model space to the $sdpf$ shells [4], a lowering to 1.1 MeV of the $1/2^+$ excited state is calculated in ^{27}F. Although an excited state with a similar energy can be constructed by placing the two γ rays of ^{27}F on top of each other, on the basis of the expected decay properties, a state directly feeding the ground state is a more probable candidate for the spin 1/2 state. In spite of the $\sim$300 keV energy difference, the 777 keV transition may be a reasonable candidate for the decay of the $1/2^+$ state of the $sdpf$ shell model prediction. Thus, by allowing for breakdown of the N=20 neutron shell closure, half of the experimental results, namely the existence of the γ ray peaks with 700 keV energy in all the ^{25,26,27}F nuclei, may be explained [5].

The large energy deviation between at least one of the predicted and observed excited states suggests that these states intrude from a configuration outside of the model space, or the predicted energies strongly deviate from the reality due to some additional correlations not included in the models.

ACKNOWLEDGMENTS

We would like to thank the RIKEN Ring Cyclotron staff for their assist during the experiment. One of authors (Z. E.) is grateful for the JSPS Fellowship Program in RIKEN. The European authors thank the kind hospitality and support from RIKEN. The present work was partly supported by the Grant-in-Aid for Scientific Research (No. 1520417) by the Ministry of Education, Culture, Sports, Science and Technology and by OTKA T38404, T42733 and T46901. One of the authors (Zs. F.) thanks the support from Bolyai grant.

REFERENCES

1. E. K. Warburton, and B.A. Brown, *Phys. Rev.*, **C46**, 923 (1992).
2. Y. Utsuno, et al., *Phys. Rev.*, **C64**, 011301(R) (2001).
3. B. A. Brown, *http://www.nscl.msu.edu/ brown/sde.htm*
4. Y. Utsuno, et al., *Phys. Rev.*, **C60**, 054315 (1999).
5. Z. Elekes, et al., *Phys. Lett.*, **B599**, 17 (2004).

APPLICATION AND DEVELOPMENT
OF NUCLEAR TECHNIQUES

Thermal Neutron Capture Cross Sections Of The Palladium Isotopes

R.B. Firestone[1], M. Krtiáka[2], D.P. McNabb[3], B. Sleaford[3], U. Agvaanluvsan[3], T. Belgya[4] and Zs. Révay[4]

[1] Lawrence Berkeley National Laboratory Berkeley CA 94720
[2] Faculty of Mathematics and Physics, Charles University V Holešovickách 2, CZ-180 00 Prague 8, Czech Republic
[3] Lawrence Livermore National Laboratory, Livermore, California 94551
[4] Institute of Isotope and Surface Chemistry H-1525, Budapest, Hungary

Abstract. We have measured precise thermal neutron capture γ-ray cross sections cry for all stable Palladium isotopes with the guided thermal neutron beam from the Budapest Reactor. The data were compared with other data from the literature and have been evaluated into the Evaluated Gamma-ray Activation File (EGAF) [1]. Total radiative neutron capture cross-sections σ_γ can be deduced from the sum of transition cross sections feeding the ground state of each isotope if the decay scheme is complete. The Palladium isotope decay schemes are incomplete, although transitions deexciting low-lying levels are known for each isotope. We have performed Monte Carlo simulations of the Palladium thermal neutron capture deexcitation schemes using the computer code DICEBOX [2]. This program generates level schemes where levels below a critical energy E_{crit} are taken from experiment, and those above E_{crit} are calculated by a random discretization of an a priori known level density formula $\rho(E,J^\pi)$. Level de-excitation branching intensities are taken from experiment for levels below E_{crit} the capture state, or calculated for levels above E_{crit} assuming an a priori photon strength function and applying allowed selection rules and a Porter-Thomas distribution of widths. The advantage of this method is that calculational uncertainties can be investigated systematically. Calculated feeding to levels below E_{crit} can be normalized to the measured cross section deexciting those levels to determine the total radiative neutron cross-section σ^γ. In this paper we report the cross section measurements $\sigma_\gamma[^{102}Pd(n,\gamma)]=0.9\pm0.3$ b, $\sigma_\gamma[^{104}Pd(n,\gamma)]=0.6l\pm0.11$ b, $\sigma_\gamma[^{105}Pd(n,\gamma)]=2.1.1\pm1.5$ b, $\sigma_\gamma[^{106}Pd(n,\gamma)]=0.36\pm0.05$ b, $\sigma_\gamma[^{108}Pd(n,\gamma)(0)]=7.6\pm0.6$ b, $\sigma_\gamma[^{108}Pd(n,\gamma)(189)]=0.185\pm0.011$ b, and $\sigma_\gamma[^{110}Pd(n,\gamma)]=0.10\pm0.03$ b. We have also determined from our statistical calculations that the neutron capture states in ^{107}Pd are best described as $2^+[59(4)\%]+3^+[41(4)\%]$. Agreement with literature values was excellent in most cases. We found significant discrepancies between our results for ^{102}Pd and ^{110}Pd and earlier values that could be resolved by re-evaluation of the earlier results.

Keywords: Neutron capture on Palladium isotopes
PACS: 25.40.-h ,25.40.Lw

INTRODUCTION

Total radiative thermal neutron capture cross sections are usually determined by measuring either the neutron transmission rate through a target or the activation rate of radioactive product. Both methods require either knowledge of the neutron flux or use of a comparator material of well-known cross section. Transmission rates can be subject to significant corrections for neutron scattering and uncertainties in the target geometry. Activation measurements are usually more accurate, but they also require normalization to a comparator, typically gold, and they rely on the accuracy of the decay scheme normalization.

In this paper we report a new method for the determination of total radiative thermal neutron cross sections using prompt neutron capture γ-rays measured with guided neutron beams at the Budapest Reactor. Prompt γ-rays are measured with a high degree of sensitivity so that the cross section can be determined by both the total primary transition cross section deexciting the capture state and the total secondary transition cross section feeding the ground state. Complete decay schemes are available for very light isotopes, and complete measurements of secondary γ-ray cross sections feeding the ground state have been done for most isotopes of elements up to Fe. For heavier or low abundance isotopes, the measurements are generally incomplete.

For the more complex decay schemes of heavier elements, the contribution from unobserved continuum γ-rays feeding the ground state must be determined to obtain the total radiative cross section. In this paper we have used the Monte Carlo computer code DICEBOX [2] to calculate "complete" neutron capture decay schemes for the Palladium isotopes. These decay schemes are then normalized using measured transition cross sections deexciting low-lying levels in the Palladium isotopes to determine the total radiative cross sections.

EXPERIMENTAL

Neutron capture γ-ray cross sections for elemental targets with $Z = 1-84, 90, 92$ have been measured at the Budapest Reactor with the 2×10^6 ns^{-1} guided thermal neutron beam [3]. The target station is located far from the reactor where both primary and secondary γ-rays can be measured in low background conditions. Cross sections were measured using either stoichiometric compounds or accurately prepared mixtures containing the standard elements H, N, or Cl whose γ-ray cross sections are precisely known. The γ-ray cross sections were then accurately determined from their intensity relative to the standard γ-ray transitions of the comparators. An elemental target consisting of 1.6 g of PdCl$_2$ with a thickness of 0.4 gcm^{-2} was irradiated in the 2×10^6 ns^{-1}cm^{-2} guided thermal neutron beam at Budapest Reactor for 8753 seconds. Prompt gamma-rays from the target were detected with a 25% efficient, Compton-suppressed, HPGe detector. Counting efficiency was determined over the range of 50 keV to 10 MeV with radioactive sources and (n,γ) reaction gamma rays to a precision of better that 1% from 500 keV to 6 MeV and better than 3% at all energies [4]. The γ-ray spectra were analyzed using the Hypermet PC program [5]. A total of 202 γ-rays were assigned to the six Palladium isotopes ^{103}Pd, ^{105}Pd, ^{106}Pd, ^{107}Pd. ^{109}Pd, and ^{110}Pd on the basis of energy and intensity by comparison with data from the RNSDE [6] file. The γ-ray cross section data were sufficient to determine level deexcitation cross sections, $\sigma_{(level)}$, for 101 Palladium levels including at least one level from each isotope. These results are summarized in Table 1.

TABLE 1. Experimental neutron capture cross sections feeding excited states in Palladium isotopes. Ground state feedings for ^{106,109}Pd ere determined from experimental and theoretical side feedings to all levels below E_{crit} as described in the text.

Isotope	E(Level) (keV)	J^π	σ(level) (barns)	E_{crit} (keV)	Feeding (%)	σ_γ(This work) (barns)	σ_γ (Literature)* (barns)
^{103}Pd	118.736(17)	3/2+	0.51(14)	500	58(8)	0.9(3)	1.82(20)†
^{105}Pd	280.51(22)	3/2+	0.145(13)	680	30(8)	0.48(14)	
	306.25(3)	7/2+	0.040(8)		3.9(14)	1.02(42)	
	344.512(18)	1/2+	0.099(18)		1w)	0.66(31)	
	560.75(3)	3/2+	0.050(10)		6.5(15)	0.77(24)	
	644.53(4)	7/2-	0.063(6)		3.8(20)	1.7(9)	
	Average					0.61(11)	0.6(3)
^{106}Pd	0	0+	20.0(3)	2505	94.8(15)	21.1(15)	21.0(15)
^{107}Pd	115.74(12)	1/2+	0.095(9)	480	28(7)	0.34(9)	
	302.78(15)	5/2+	0.046(4)		8.5(30)	0.5420)	
	312.20(10)	7/2+	0.0244)		3.2(22)	0.8(5)	
	381.80(13)	3/2+	0.043(6)		11(2)	0.39(9)	
	471.21(24)	(3/2)+	0.024(5)		8(2)	0.30(10)	
	Average					0.36(5)	0.29(3)
^{109}Pd	0	5/2+	5.93(8)	350	78(6)	7.6(6)	7.6(4)
	188.990(10)	11/2-	0.185(11)		100	0.185(11)	0.18(3)
^{111}Pd	191.3(3)	[3/2]+	0.016(4)	200	18(5)	0.09(3)	
	195.1(2)	[3/2]+	0.019(8)		18(4)	0.11(5)	
	Average					0.10(3)	0.19(3)

* from reference [13] except where noted
† from reference [14]

DICEBOX CALCULATIONS

Theoretical statistical feedings to low-lying levels have been calculated using the computer code DICEBOX [2]. This code determines the theoretical uncertainty in the level feedings due to statistical fluctuations using an algorithm based on the extreme statistical model of nuclei. Below a critical energy E_{crit} the level scheme, i.e. energies, spins and parities of all levels and all depopulating transitions, is assumed to be known from experiment. Above this energy, an unknown set of levels is determined by a random discretization of a level density formula $\rho(E,J^\pi)$. The partial radiative width $\Gamma_{a\gamma b}$ γ-ray decay with an energy E_γ from a level a to a level b is assumed to vary with a Porter-Thomas distribution whose mean value is determined by the level density and a photon strength function $S(E\gamma)$ given by

$$\Gamma_{a\gamma b} = \frac{S(E_\gamma) \times E_\gamma^3}{\rho(E_a, J_a^\pi)}, \qquad (1)$$

where $\rho(E_a, J_a^\pi)$ is the level density at initial level a. Two choices of level density models are available with DICEBOX. The constant temperature model (CTF) [7] is given by

$$\rho(E_a, J_a^\pi) = f_J \frac{1}{T} e^{(E-E_0)/T} \qquad (2)$$

where T is the nuclear temperature and E_0 is the backshift. The back-shifted Fermi gas model (BSFG) [7-9] is given by

$$\rho(E_a, J_a^\pi) = \frac{f_J e^{2\sqrt{a(E-E_1)}}}{12\sqrt{2}\sigma a^{\frac{1}{4}}(E-E_1)^{\frac{5}{4}}} \qquad (3)$$

where a is the conventional shell-model level-density parameter and $E1$ is another back shift. The factor f_J represents the probability that a randomly chosen level has spin J [7] and is given by

$$f_J = \frac{2J+1}{2\sigma^2} e^{-(J+\frac{1}{2})^2/2\sigma^2} \qquad (4)$$

where $\sigma = 0.98 A^{0.29}$ for the CTF model and $\sigma = 0.298 A^{1/3} a^{1/4} (E-E_1)^{1/4}$ for the BSFG model. The parameters for both the CTF and BSFG models are determined from the known levels below E_{crit} and resonance data for levels above the neutron capture state. The CTF and BSFG models gave very similar results in these calculations.

The photon strength function $S^{XL}(E_\gamma)$ for multipolarities XL is assumed to be independent of spin and parity and is defined as

$$S^{XL}(E_\gamma) = \frac{1}{E_\gamma^{2L+1}} \frac{\langle \Gamma_{ab}^{(J)} \rangle_a}{\langle D_J \rangle} \qquad (5)$$

Here $\langle \Gamma_{ab}^{(J)} \rangle_{ab}$ presents the value of the partial radiation width $\Gamma_{ab}^{(J)}$ for a transition a → b, averaged over initial levels a with a fixed parity π and spin J. D_J is the average spacing between initial levels. Only the lowest-order multipolarities E1, M1, and E2 are included in these calculations. Numerous options for calculating $S^{XL}(E_\gamma)$ including single-particle, Axel and Brink [10], Kopecky and Chrien [11], Kadmenskij et al [12], and various other semi-empirical models are available in DICEBOX. The choice of model did not significantly affect our results.

DICEBOX constructs a complete, artificial decay scheme and then randomly simulates γ-ray cascades. Selection rules for different types of transitions are taken into account. One unique choice of a level scheme and all partial radiative widths is termed a nuclear realization. As there are an infinite number of possible nuclear realizations allowed by these assumptions, the simulated population of low-lying levels is subject to fluctuations. If the intensities of primary transitions to low-lying levels are known, as in the case of ^{106}Pd and ^{109}Pd, they are taken from experiment. Otherwise they are simulated and their values will differ from realization to realization.

The fraction of the total cross section feeding the observed levels in the Palladium isotopes assuming the indicated values of E_{crit} given in Table 1. The uncertainties in these

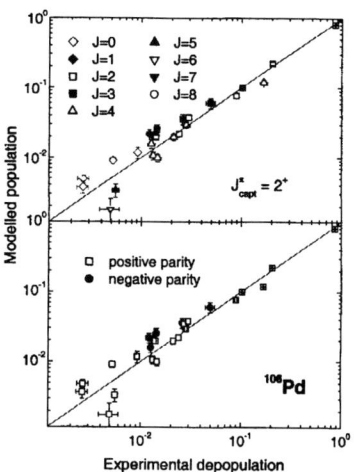

FIGURE 1. Experimental and statistical side feedings for ^{105}Pd(n,γ) assuming the capture state $J^\pi=2^+$.

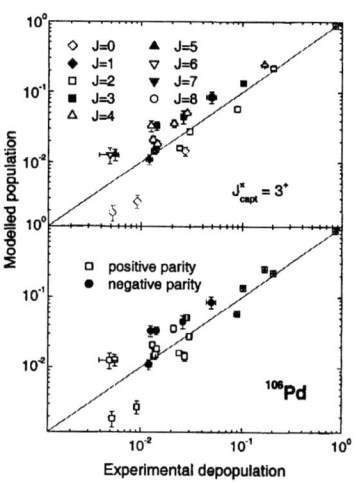

FIGURE 2. Experimental and statistical side feedings for ^{105}Pd(n,γ) assuming the capture state $J^\pi=3^+$.

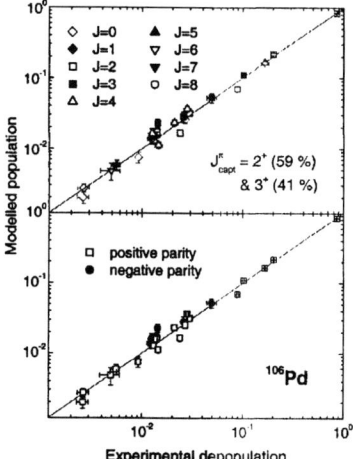

FIGURE 3. Experimental and statistical side feedings for ^{105}Pd(n, γ) assuming the capture state is 2+[59(4)%]+3+[41(4)%]

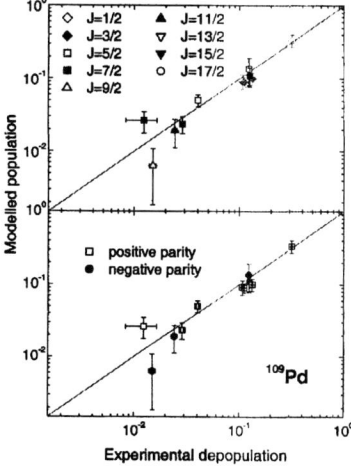

FIGURE 4. Experimental and statistical side feedings for ^{108}Pd(n,γ).

values were determined from the variations in the multiple realizations of each decay scheme. Complete γ-ray data were available for all levels below E_{crit} in ^{106}Pd and ^{109}Pd. For these cases we normalized the calculation by performing a least-squares fit of the calculated side feeding to each level from levels above E_{crit} to the experimental side feeding to each level determined by the intensity balance, $\sum \sigma_\gamma(in) - \sigma_\gamma(out)$, from levels below E_{crit}. The total feedings to the ground states of ^{106}Pd and ^{109}Pd, calculated from this normalization, are given in Table 1. The total radiative neutron capture cross sections σ_γ in Table 1 are determined simply by the ratio 100 x σ(level)/% Feeding.

RESULTS

The total radiative neutron capture cross sections for the Palladium isotopes derived from the experimental cross sections. $I_{\gamma+e}$, feeding the levels and the relative branching intensities calculated with DICEBOX feeding these levels is shown in Table 1. For ^{102}Pd(n,γ) we measured σ_γ=0.9±0.3 b based on the de-excitation cross section of the first excited state in ^{103}Pd. This value is considerably lower than σ_γ =1.82±0.20 b from Duncan and Krane [14]. That value was measured assuming $P_{357.4\gamma}$=0.0221±0.0007% for ^{103}Pd decay from reference [15]. A previous measurement by Zoller et al [16] gave $P_{357.4\gamma}$=0.0324±0.0016% leading to σ_γ=1.24±0.15 b which is in better agreement with our value. For ^{104}Pd(n,γ) we determined that σ_γ =0.61±0.11 b based on the average population of five levels in ^{105}Pd. This value is in good agreement with σ_γ=0.6±0.3 b from Mughabghab et al [13]. Similarly, For ^{106}Pd(n,γ) we determined that σ_γ=0.36±0.05 b from the average population of five levels in ^{107}Pd in agreement with σ_γ=0.29± 0.03 b from Ref. [13]. The experimental value σ_γ=0.1±0.03 b for ^{110}Pd(n,γ), based on the observation of two levels in ^{111}Pd, is lower than the compiled value of σ_γ=0.19±0.03 b which is from Sehgal et al [17]. Comparison of other cross sections measured in Ref. [17] indicates that they were typically about twice as large as the values compiled in Ref. [13]. Presumably this discrepancy resulted from an imprecise determination of the neutron flux in Ref. [17]. The Sehgal et al value can be renormalized by a factor of 0.54, necessary to correct their measurement of 14±2 b for the ^{108}Pd(n,γ) cross section, yielding σ_γ=0.10±0.02 b, which agrees with our result.

The values σ_γ=21.1±1.5 b for ^{105}Pd(n,γ) and σ_γ=7.6±0.6 b for ^{108}Pd(n,γ) were determined using data for many levels as described above. They are in excellent agreement with the compiled values from Ref. [13]. In addition, a new value has been determined for the cross section feeding the 4.7 min isomer in ^{109}Pd(n,γ) $\sigma_\gamma[^{108}$Pd(n,γ)(189)]=0.185±0.011 b, determined by measuring the 188.99-keV γ-ray cross section in equilibrium with the prompt transitions. Calculation of ^{106}Pd(n,γ) statistical feeding was complicated by the unknown relative contributions of capture states with spins 2^+ and 3^+. Figures 1 and 2 show comparisons of the experimental and statistical side feedings for each J^π assignment. Agreement between experiment and theory is poor in both cases. We then solved for the mixing fraction of capture state spins that minimizes the χ^2/f comparison between experiment and theory and determined that $2^+[59(4)\%]+3^+[41(4)\%]$ gave an excellent fit to the experimental and theoretical side feedings, as is shown in

Figure 3. The good agreement between experimental and statistical side feedings for ^{109}Pd, where the capture state is uniquely $1/2^+$, is shown in Figure 4.

CONCLUSIONS

New measurements of the total thermal neutron radiative cross sections for ^{102,104,105,106,108,110}Pd(n,γ) have been performed using prompt γ-ray cross sections to renormalize statistical model calculations using DICEBOX. These results had comparable precision to previous values. Our measurements provide independent confirmation of previous results and can be used to identify and resolve problems with earlier experiments. A large library of prompt neutron capture γ-ray data has been measured for all other stable elements and will be analyzed in the future to determine additional total radiative cross sections.

ACKNOWLEDGEMENTS

This work was supported by grants from the Director, Office of Science, Office of Basic Energy Sciences, of the U.S. Department of Energy under Contract No. DEAC03-76SF00098; the U.S. Department of Energy by University of California, Lawrence Livermore National Laboratory under Contract W-7405-Eng-48; NNSA Academic Alliance grant No. DE-FG03-03NA00076, and the Grant Agency of the Czech Republic under Contract 2OU03P 136

REFERENCES

1. R.B. Firestone, H.D. Choi, R.M. Lindstrom, G.L. Molnár, S.F. Mughabghab, R. Paviotti-Corcuera, Zs. Révay, V. Zerkin and C M Zhou, *Database of Prompt Gamma Rays from Slow Neutron Capture for Elemental Analysis*, International Atomic Energy Agency TECDOC, in press (2003)
2. F. Bečvář, Nucl. Instr. Meth. **A417**,434 (1998). 105, 113 (2000).
3. T. Belgya, Zs. Révay, B. Fazekas, I. Héjja, L. Dabolczi, G.L. Molnár, Z. Kis, J. Östör and Gy. Kaszk, Proc. 9th Intern. Symp. on Capture Gamma-Ray Spectroscopy and Related Topics, Budapest, Hungary, Oct. 8-12, G.L. Molnár, T. Belgya, Zs. Révay (Eds). Springer Verlag, Budapest, 826 (1997).
4. C.L. Molnár, Zs. Révay, and T. Belgya, Nucl. Instrum. Meth. Phys. Res. **A489**,140 (2002).
5. B. Fazekas, J. Östör, Z. Kis, G.L. Molnár, and A. Simanits, Proc. 9th Intern. Symp. on Capture Gamma-Ray Spectroscopy and Related Topics, Budapest, Hungary, Oct. 8- 12. G.L. MOM, T. Belgya, Zs. Révay (Eds). Springer Verlag, Budapest, 774 (1997).
6. Evaluated Nuclear Structure Data File, a computer file of evaluated experimental nuclear structure data maintained by the National Nuclear Data Center, Brookhaven National Laboratory.
7. A. Gilbert and A.G.W. Cameron, Can. J. Phys. **43**, 1466 (1965).
8. H.A. Bethe, Rev. Mod. Phys. **9**, 69 (1937).
9. W. Dilg, W. Schantl, H. Vonach, and M. Uhl. Nucl. Phys. **A217**,269 (1973).
10. P. Axel, Phys. Rev. **126**, 671 (1962).
11. J. Kopecky and R.E. Chrien, Nucl. Phys. **468**,285 (1987).
12. S.C. Kadmenskij, V.P. Markushev and V.I. Funnan. Sov. J. Nucl. Phys. **37**, 165 (1983).
13. S. F. Mughabghab, Thermal Neutron Capture Cross Sections, Resonance Integrals, and g-factors, INDC(NDS)-**440** (2003).
14. C.L. Duncan and K.S. Krane, Phys. Rev. **C71**.054322 (2005).
15. E.S. Macias, M.E. Phelps, D.G. Sarantites, and R.A. Meyer, Phys. Rev. **C14**,639 (1976).
16. W.H. Zoller, E.S. Macias, M.B. Perkal, and W.B. Walters, Nucl. Phys. **A130**,293 (1969).
17. M.L. Sehgal, H.S. Hans, and P.S. Gill, Nucl. Phys. **12**, 261 (1959).

Atomic-Nuclear Coupling Experiments

J. A. Becker

Lawrence Livermore National Laboratory
Livermore, CA 94550

Abstract. Atomic-nuclear coupling experiments are described, with an emphasis on recent experiments aimed at demonstrating the NEET mechanism in atomic nuclei. Upper limits for x-ray induced decay of the Hf-178 31-y isomer reported by Ahmad and his colleagues are presented, and these upper limits are contrasted with the positive reports of Collins and coworkers.

Keywords: atomic nuclear coupling, NEET, ^{235}U ^{189}Os, isomers.
PACS: 23.20.-g, 23.90.+W, 25.20.Dc, 27.70.+q

INTRODUCTION

Manipulating a nucleus with atomic probes and potentially gaining nuclear energy release at the cost of atomic energy is an attractive idea. Basic science interests include identification of the mechanisms and respective cross sections, and the prospect of "energy on demand" has a strong applied science interest. Familiar isotopes discussed in this context are illustrated in Figure 1. Interest in atomic-nuclear coupling has been renewed with recent experiments and theoretical work focused on induced decay of the 31-y isomer in ^{178}Hf at E_x = 2.4 MeV, initiated by x-ray irradiation, and by experiments done with the goal of observing "Nuclear Excitation by Electronic Transition" (NEET). The NEET process has been observed successfully in x-ray irradiation of ^{197}Au, while only cross-section limits have been obtained for ^{187}Os. Attempts have been made to excite the first excited state of ^{235}U and 77 eV ($t_{1/2}$ ~25 m) in a laser-induced plasma. These experiments together with theoretical calculations are discussed in the next sections. A "Workshop on Nuclear Isomers" organized by Prof. Yang Sun preceded CGS12. This workshop topics featured aspects of the many roles isomer studies have played in nuclear physics. Olivier Roig presented a poster at CGS12 describing superelastic scattering work, reporting an indirect measure of superelastic scattering for isomeric ^{177}Lu — 200 b.

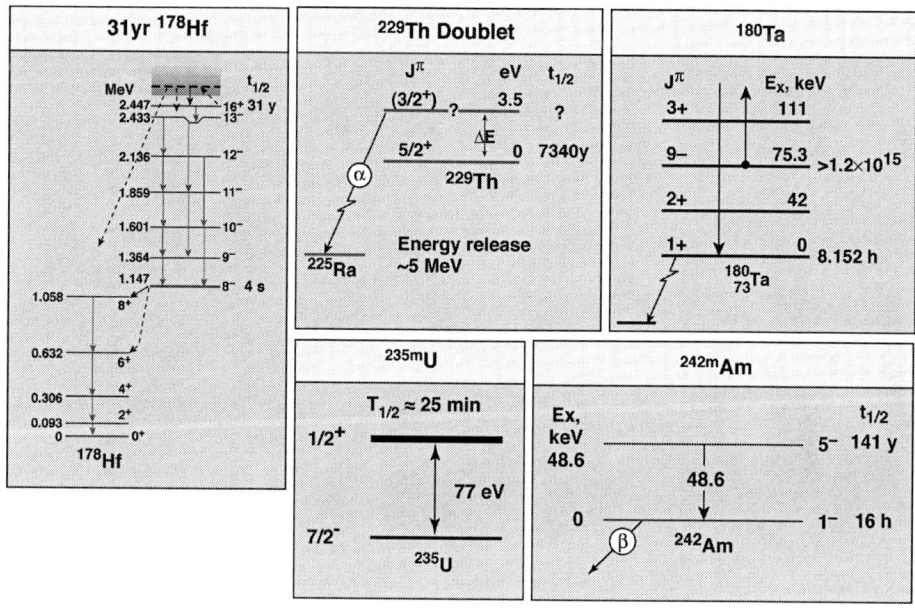

FIGURE 1. Nuclear isomers often mentioned in connection with atomic-nuclear coupling.

X-Ray Induced Decay of Nuclear Isomers

The 31-y isomer in ^{178}Hf (^{178m2}Hf) has attracted considerable interest, because of the potential high energy release and the isomer's long half life. ^{178m2}Hf is available in research quantities mixed with ^{178}Hf, and often other Hf isotopes, depending on the method of production. The positive reports of induced decay by Collins, et. al.,[1] (Citation is to the most recent work, where citations to earlier work may be found.) was quickly criticized as improbable because of the overwhelming E1 strength required by the reported cross section, coupled with the requirement of a nearby intermediate state with special properties. Verification experiments by Ahmad, el. al.,[2] and Roberts, Carroll, et. al.,[3] set upper limits for enhanced decay of the ^{178m2}Hf many orders magnitude below the continued positive reports of Collins and his coworkers. Results are compared in Figure 2. Belief in x-ray induced enhanced decay of the 2.4-MeV isomer with any practical cross section is contrary to scientific evidence. The discussion continues in popular literature.[4] Suggestions abound for practical applications.[5] Numerous popular press articles also are in evidence, and the discussion has spread abroad.

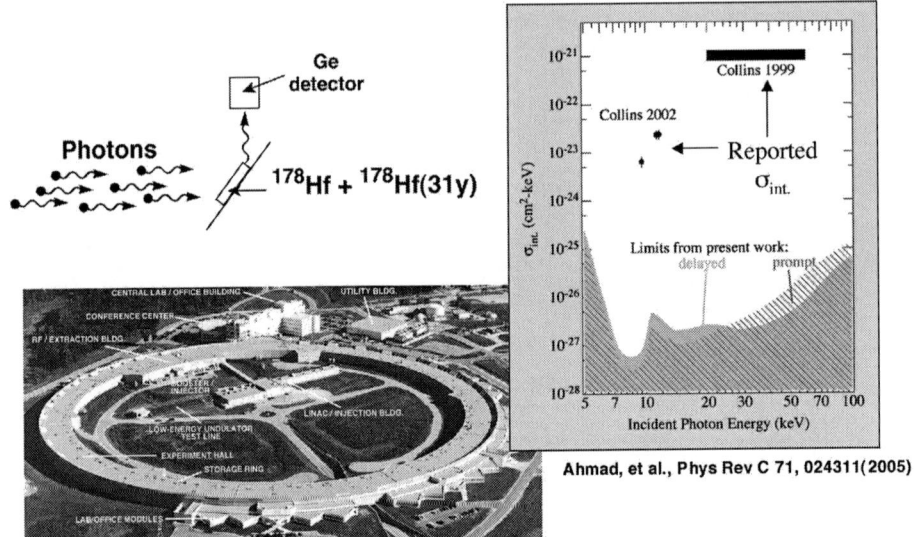

FIGURE 2. A cartoon of the experimental arrangement employed by Ahmad, et. al.,[2] upper limits to enhanced decay of isomeric [178]Hf obtained in their work, and a visual of the Advanced Photon Source. The upper limits come in 2 pieces: for enhanced decay through the 8⁻ isomeric state in [178]Hf, and for decay which bypasses that state. Ahmad, et al., used a "white" beam of incident x-rays. Roberts, Carroll et. al.,[3] have reported a cross sections limit of 10^{-25} keV-cm^2, covering (stepwise) 75% of a narrow region of incident x-ray energies 9.45 – 13.5 keV (not illustrated).

NEET

Morita[6] pointed to the possibility of NEET, suggesting NEET as a possibility of ^{235}U and ^{235m}U separation. Kishimoto[7] presented evidence for NEET in ^{197}Au, obtained in a daunting experiment at SP-ring8, The probability of NEET, reporting for the particular atomic excitation is P(NEET) = 5.0 ± 0.6 x 10^{-8}/k-vacancy, is good agreement with calculation ~ 4 x 10^{-8}/k-vacancy.[8,9] Configurations are given in Figure 3. Upper limits to the NEET cross section for ^{189}Os for populating the 69.537 keV 5/2⁻ excited state in ^{189}Os are reported by Ahmad, et. al.,[10] and by Akoi, et. al.,[11] as P(NEET) < 10^{-10}/k-vacancy, again in agreement with calculations of Tkalya and Harston. See Figure 4.

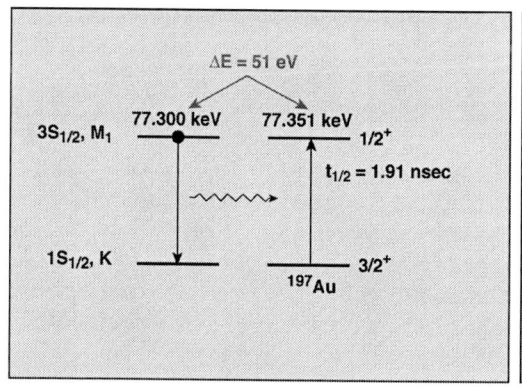 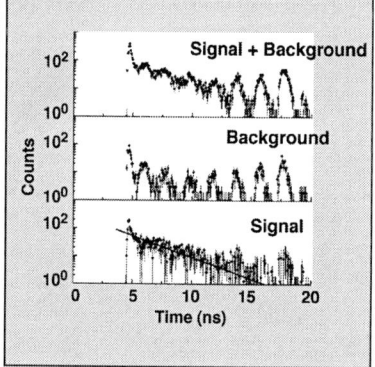

FIGURE 3. The relevant configuration in ^{197}Au for the NEET process, and the NEET signal observed by Kishimoto, et. al.[7]

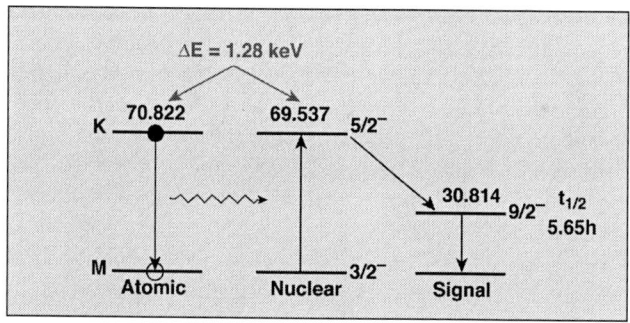

FIGURE 4. The relevant configurations in ^{189}Os for the NEET process, and the upper limits obtained in the experiments of (a) Ahmad, et. al.,[10] and (b) Akoi, et. al.,[11] and the calculations of (c) Tkalya[8] and (d) Harston[9] for P_{NEET}.

NEET in μ-Mesic Atoms

NEET is observed in μ-mesic atom studies. Here, because of the very high muonic atomic energies, the many continuum states enable the NEET condition of energy degeneracy and multiple matching. See, e.g., Bernow, et. al.[12] and Engfer[13]. The emphasis was on measurements of nuclear quadrupole moments, using NEET as a population mechanism.

NEET in a Plasma

Meot and his collaborators have attempted to populate ^{235m}U in a laser-induced plasma.[14] Elaborate calculations are required in order to set appropriate laser conditions so that the desired plasma is obtained. Morel, et. al.,[15] have shown that the laser conditions required for an observation is more than present day lasers can deliver. Meot and his collaborators are turning away from ^{235m}U, and they are planning an experiment in ^{201}Hg,[16] where calculations show the chances of success are higher. The laser pulse required is 1 ns long, and 100 Joules.

NEEC

Meot and his collaborators[16] are also mounting an experiment to search for Nuclear Excitation by Electron Conversion (NEEC) in ^{57}Fe, taking advantage of inverse kinetics and the energetic beams available at GANIL.

Dicke Radiation

Robert Dicke first used the term superradiance in 1954[17] to describe the emission of photons from a quantum state consisting of an ensemble of excited atoms coupled only by the electromagnetic field of the photon. This state maximizes when the number of excited states equals the number of ground states in the ensemble. The superradiant state emits photons at a rate proportional to the number of atoms in the ensemble squared. High isomer enrichment is required to enter the superradiant state An effort at Los Alamos is underway to develop a coherent photon source based on a Mössbauer crystal, specifically a Mössbauer crystal highly enriched (>10%) in ^{93m}Nb.[18] Fortunately, ^{93m}Nb is an isomer that can be obtained in high purity. It is produced in 89 percent abundance from the decay of ^{93}Mo and is produced in 95 percent abundance from ^{93}Zr decay. A single Mössbauer crystal is needed to achieve nuclear superradiance. The Mössbauer effect is necessary to permit the nuclei to be resonant with the emitted gamma ray. A crystal lattice is needed so that the nuclei will "feel" the electromagnetic field of the photon, in phase. Thus the superradiant state exists only along Bragg angles. Isomer collection is an issue, and crystal growing is difficult. An efficient crystal growing process was developed, and KNbF$_6$ crystals 2 mm in length have been grown. The attempt to grow K^{93m}NbF$_6$ crystals has stalled. This is a very exciting, long range research project.

ACKNOWLEDGMENTS

This work was funded in part by the U.S Department of Energy, and performed under the auspices of the U.S. Department of Energy by the University of California, Lawrence Livermore National Laboratory, Contract No. W-7405-Eng-48 (LLNL).

REFERENCES*

*Reference cited usually contains citations to the earlier work, which may be more appropriate in context.
1. C.B. Collins, et. al., Laser Physics 14, 154 (2004).
2. I. Ahmad, et. al., Phys. Rev. C 71, 024311 (2005).
3. J. J. Carroll, et. al., Laser Phys. Lett. 1,275 (2004).
4. B. Schwarzschild, Physics Today 57, 21 (May 2004); P. M. Walker and J. J. Carroll, Physics Today 58, 39 (June 2005).
5. D. Hambling, in NewScientist.com, 13 Aug. 2003.
6. M. Morita, Prog. Theor. Phys. 49, 1574 (1973).
7. S. Kishomoto, et. al., Phys. Rev. Lett. - 85, 1831 (2001).
8. E.V. Tkalya, in AIP Conference Proceedings 506, 486 (2000); E.V. Tkalya, Phys. Rev. C 71, 024606 (2005); E.V. Tkalya, Physics Uspekhi 48, 525 (2005).
9. M. Harston and J. J. Carroll, Laser Physics 15, 487 (2005).
10. I. Ahmad, et. al., Phys. Rev. C 61, R05304 (2000). See also the discussion by D. Gemmell, in AIP Conference Proceedings 652, 239 (2003).
11. K. Aoki, et. al., Hyperfine Int. 143, 111 (2002).
12. S. Bernow, et. al., Phys. Rev. Lett. 21, 457 (1968).
13. R. Engfer, in Proceedings of the International Conference on Nuclear Physics vol. 2, p. 437, Munich 1973, edited by J. de Boer and H.J. Mang, North Holland/Elsevier.
14. G. Claverie, et. al., Phys. Rev. C 70, 044303 (2004).
15. P. Morel, et. al., Phys. Rev. A 69, 0634214 (2004).
16. V. Meot, private communications (2005).
17. R. H. Dicke, Phys. Rev. 93(1) 99 (1954).
18. J. Wilhelmy, private communication (2005).

Distortion of pulse-height spectra of neutron capture gamma rays

A. Laptev[1], H. Harada[1], S. Nakamura[1], J. Hori[2],
M. Igashira[3], T. Ohsaki[3] and K. Ohgama[3],*

[1]*Japan Nuclear Cycle Development Institute, Tokai-mura, Naka-gun, Ibaraki 319-1194, Japan*
[2]*Research Reactor Institute, Kyoto University, Kumatori-cho, Sennan-gun, Osaka 590-0494, Japan*
[3]*Tokyo Institute of Technology, N1-26,2-12-1 O-okayama, Meguro-ku, Tokyo 152-8550, Japan*

Abstract. A distortion of pulse-height spectra of neutron capture γ-rays caused by γ-flash at neutron time-of-flight (TOF) measurement using a pulse neutron source has been investigated. Pulses from C_6D_6 detectors accumulated by flash-ADC were processed with both traditional analog-to-digital converter (ADC) and flash-ADC operational modes. A correction factor of γ-ray yields, due to baseline shift, was quantitatively obtained by comparing the pulse-height spectra of the two data-collecting modes. The magnitude of the correction factor depends on the time, which passed after γ-flash, and has complicated time dependence with a changing sign.

Keywords: Neutron capture; γ-rays; γ-flash; TOF method; Detector baseline; Flash-ADC.
PACS: 25.40.Lw; 29.30.Hs; 29.40.Mc; 29.85.+c

INTRODUCTION

The precise neutron capture cross-section measurement in wide neutron energy range is required for a wide area of applications ranging from nuclear astrophysics [1] to nuclear transmutation studies [2]. Neutron capture has been investigated at many neutron facilities in the world with a large amount of data having been already measured and published. Under measurement at pulse neutron source in TOF mode, any used γ-ray detector is significantly affected by strong γ-flash pulse originated when an accelerator beam strikes a neutron-producing target. The strong detector overload causes baseline distortion, which leads to distortion of the measured pulse-height. The later results in a systematic error of obtained capture cross-section value, which is based on γ-ray yield, by using traditional ADCs with fixed pulse-height bias. Recently, the flash-ADC has found wide applications in neutron cross-section measurements [3-5]. It gives an opportunity to study [6], quantitatively, the baseline shift effect on the measured capture cross-section and to investigate the magnitude of possible systematic error of many previous experiments. The result of this investigation is presented.

---

*Present address: Japan Nuclear Cycle Development Institute, O-arai-machi, Higashi-Ibaraki-gun, Ibaraki, 311-1393, Japan

EXPERIMENTAL SET-UP

The neutron capture cross-section study of Zr isotopes is carried out at the TOF facility of 46-MeV electron linac at the Research Reactor Institute of Kyoto University (KURRI). The used experimental set-up [6] is shown in Fig. 1. Bursts of fast neutrons were produced from the photo-neutron Ta target. The water tank surrounding the Ta target was used as a neutron moderator. The average current of electron beam was 40-44 μA, accelerator repetition rate was 200 Hz, and electron pulse width was 47 ns. A Pb-shadow bar of 10 cm length was placed in front of the Ta target to reduce the γ-flash generated by the target. The collimation system of B_4C, Li_2CO_3, Pb and paraffin shaped the neutron beam of 2 cm diameter in place of an investigated sample. The flight path of about 12 m was used. The powder sample of zirconium-92 oxide with 91.4% isotopic enrichment and 1 g element weight encapsulated in aluminum foil with size of 18×18 mm has been used for present measurement. Sample was placed between a pair of C_6D_6 liquid scintillate detectors with a total volume about 0.5 liter.

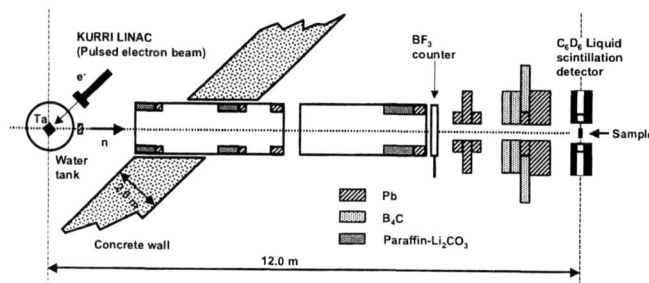

FIGURE 1. Experimental set-up.

BASELINE RECONSTRUCTION

The data acquisition system of this experiment is based on a 40 MHz flash-ADC [3] that corresponds to a 25 ns TOF-channel width (sampling interval). The flash-ADC is started by a pulse from the accelerator and digitizes detector signals over a TOF interval of about 3.3 ms, which covers over a neutron energy as low as 0.07 eV. To extract waveforms of separated pulses, the data accumulation code [3] is applied the digital filtering [7] to signal level yield of flash-ADC. On-line extracted pulse waveforms are recorded on the hard disk of a PC. Typical pulse width (FWHM) is 350-400 ns. Waveform examples of some incoming pulses measured in different regions of TOF spectrum are shown in Fig. 2. It is clearly seen that the baseline varies in the different TOF areas.

To estimate the magnitude of the baseline distortion, the baseline reconstruction has been done. For the reconstruction, the pre-histories of the pulse waveforms accumulated during an experiment have been used. The waveform pre-histories of all pulses were averaged for the accumulated data files and then the calculated baseline was smoothed by a method of adjacent averaging to reduce the statistical fluctuations. In Fig. 3, the reconstructed baseline $B(i_{TOF})$ dependence of the TOF channel number

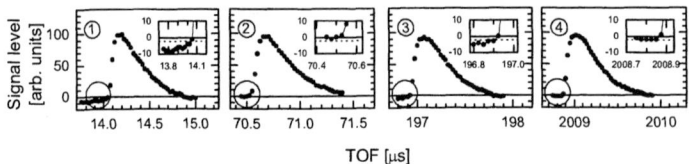

FIGURE 2. Waveforms of the C_6D_6 detector signal in different regions of TOF spectrum. Inserts show the pulse prehistory in more details; the dot line is the average level b_0 from the Fig. 3.

i_{TOF} for the neutron capture experimental data is shown. Obtained baseline has a significant shift at the beginning channels and then has fading fluctuations relative to some level, b_0. Distortion of the signal baseline is very strong and continues during a time up to 600 μs following the γ-flash. The value of the average level, b_0, is calculated by averaging the obtained baseline over TOF range of 800–3200 μs where all transient process, caused by γ-flash, is over:

$$b_0 = \overline{B(i_{TOF})}. \qquad (1)$$

The obtained b_0 corresponds to a zero value of the baseline at the measurement, and is shown by the dashed line in Fig. 3. Its small negative value reflects the current settings of the flash-ADC electronic module in the present experiment.

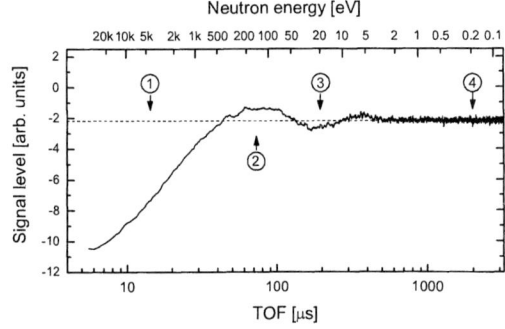

FIGURE 3. Baseline at the neutron capture experiment using C_6D_6 detector for ^{92}Zr sample. The dashed line is at -2.17 and is obtained as an average value of the baseline over TOF range of 800–3200 μs. The figures show a position of example waveforms from the Fig. 2.

CALCULATION OF PULSE-HEIGHT CORRECTION

The baseline distortion significantly affect measured pulse-height spectra of neutron capture γ-rays and results of the neutron capture cross-section measurements using data acquisition systems based on a traditional ADC with a fixed bias. To evaluate the magnitude of the correction, the reconstruction of the pulse-height spectra has been done. For spectra reconstruction, the method based on the operational principle of traditional ADC has been used. According to this principle, a digitizer yields the numerical value that corresponds to the pulse-height maximum. A two-dimensional "TOF × pulse-height" matrix was obtained.

For full simulation of the pulse processing by a data acquisition system based on an ADC with a fixed bias, the definite bias was applied to obtain γ-ray yields at every TOF channel. For calculations, this bias was chosen as a *bias* = 10 pulse-height channels relative to the level of the zero baseline, b_0. This value corresponds to a γ-ray energy of about 100 keV, which is typical for neutron capture cross-section measurements. Using this bias the convolution of "TOF × pulse-height" matrix has been done and the dependence of the integrated pulse-height spectra on the TOF channel number was obtained:

$$N_\Sigma(i_{TOF}) = \int_{bias+b_0}^{\infty} N(i_{TOF}, i_{PH}) di_{PH} . \qquad (2)$$

Here $N(i_{TOF}, i_{PH})$ is the constructed two-dimensional "TOF × pulse-height" matrix, $N_\Sigma(i_{TOF})$ is the integrated pulse-height spectra summarized from the chosen bias at every TOF channel, and i_{PH} is pulse-height channel. The value of $N_\Sigma(i_{TOF})$ does not take into account a baseline altering and therefore has systematic error at distortion of the baseline. However, by using the obtained information about the baseline it is possible to calculate an analogous value $N'_\Sigma(i_{TOF})$:

$$N'_\Sigma(i_{TOF}) = \int_{bias+B(i_{TOF})}^{\infty} N(i_{TOF}, i_{PH}) di_{PH} , \qquad (3)$$

which is free from such distortion, and to calculate a value of the relative correction that is needed to be introduced due to baseline distortion:

$$c(i_{TOF}) = \frac{N_\Sigma(i_{TOF}) - N'_\Sigma(i_{TOF})}{N'_\Sigma(i_{TOF})} . \qquad (4)$$

The value of correction, $c(i_{TOF})$, and its dependence on the TOF channel number, i_{TOF}, is presented in Fig. 4. The value of $N_\Sigma(i_{TOF})$ is entered directly into the expression for the measured cross-section of neutron capture. Its correction, $c(i_{TOF})$, is very important. The obtained correction factor is significantly large. For example, at TOF of 75 μs, E_n = 120 eV, $c \approx 0.05$, and at TOF of 15 μs, E_n = 4 keV, c = - 0.2. After a large negative value at initial 40 μs of TOF range, which corresponds neutron energy above 500 eV, the correction factor, $c(i_{TOF})$, has several maxima in the absolute value of magnitude with the different sign. These maxima are shown by dotted lines with the specified level in Fig. 4. The correction approximately vanishes around TOF of 600 μs, $E_n \approx 2$ eV, and its value averaging over the large TOF, more than 800 μs, is $\bar{c} \approx 0$.

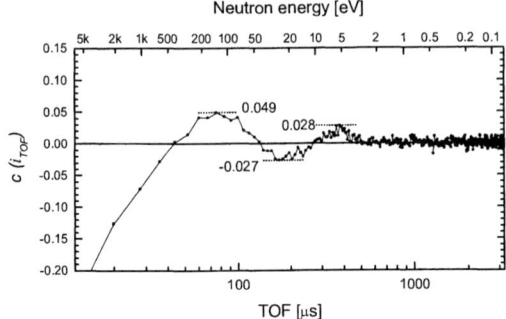

FIGURE 4. The TOF dependence of the correction $c(i_{TOF})$ for the summarized pulse-height spectrum caused by the baseline distortion for ^{92}Zr sample. The dotted lines with the specified level are magnitude maxima reached by the correction $c(i_{TOF})$.

CONCLUSION

The baseline shift effect on the neutron capture cross-section, measured by the TOF method using a pulsed neutron source, has been studied. It has been shown that a correction, due to baseline shift, has complicated energy dependence with sign changes and reaches magnitude of 20% at the beginning of time dependence of detector baseline distortion and about 5% at the first positive maximum at time about 75 µs after neutron burst. The baseline distortion continues up to the time of about 600 µs following the γ-flash. The demonstrated approach for calculation of a distortion of pulse-height spectra of neutron capture γ-rays can be used at any neutron capture experiment with a pulsed neutron source. Using this method, a correction factor for earlier measured cross-sections can be estimated after reproduction of the experimental conditions of past measurements. The obtained results give us insight into the possible systematic errors due to γ-flash of cross-sections measured in the past in the world.

REFERENCES

1. A. Aprahamian, K. Langanke and M. Wiescher, *Prog. Part. Nucl. Phys.* **54**, 535-613 (2005).
2. M. Igashira, M. Mizumoto, M. Oshima, A. Hasegawa, H. Harada, N. Yamano, H. Yamana, M. Baba, K. Kato, M. Sugawara, Y. Nagai, and K. Kawade, "A Nuclear Data Project in Japan" in *Int. Conf. on Nucl. Data for Sci. and Tech., Santa Fe, New Mexico, USA, 2004*, edited by R.C. Haight, M.B. Chadwick, T. Kawano, and P. Talou, AIP Conference Proceedings 769, American Institute of Physics, Melville, NY, 2005, pp. 601-604.
3. O. Shcherbakov, A. Donets, A. Evdokimov, A. Fomichev, T. Fukahori, A. Hasegawa, A. Laptev, V. Maslov, G. Petrov, S. Soloviev, Yu. Tuboltsev and A. Vorobyev, *J. Nucl. Sci. Tech.*, Suppl. **2**, 230-233 (2002).
4. O. Shcherbakov, K. Furutaka, S. Nakamura, H. Harada and K. Kobayashi, *Nucl. Instr. Meth.* A **517**, 269-284 (2004).
5. A.B. Laptev, A.Yu. Donets, A.V. Fomichev, A.A. Fomichev, R.C. Haight, O.A. Shcherbakov, S.M. Soloviev, Yu.V. Tuboltsev and A.S. Vorobyev, *Nucl. Phys.* A **734S**, E45-E48 (2004).
6. A.B. Laptev, H. Harada, S. Nakamura, J. Hori, M. Igashira, T. Ohsaki and K. Ohgama, *Nucl. Instr. Meth.* A **543**, 502-508 (2005).
7. D.D. Burgess and R.J. Tervo, *Nucl. Instr. Meth.* **214**, 431-434 (1983).

NUCLEAR ASTROPHYSICS
(R-PROCESS NUCLEOSYNTHESIS)

Nuclear Physics Issues of r-Process Nucleosynthesis

K.-L. Kratz

Institut für Kernchemie, Universität Mainz, Germany
HGF Virtual Institute for Nuclear Structure and Astrophysics, Mainz, Germany
Institute for Structure and nuclear Astrophysics, University of Notre Dame, USA

Abstract.
Nucleosynthesis theory predicts that about half of the chemical elements above iron are formed in explosive stellar scenarios by the r-process, i.e. a combination of rapid neutron captures, inverse photodisintegrations, and slower β-decays, β-delayed processes, as well as fission and possibly interactions with neutrinos. A correct modelling of this process, therefore, requires the knowledge of nuclear properties very far from stability and a detailed description of the astrophysical environments. With respect to nuclear data, after an initial period of measuring classical "waiting-point" nuclei with magic neutron numbers, recent investigations have paid special attention to shape transitions and the erosion of classical shell gaps with possible occurrence of new magic numbers. The status of experimental and theoretical nuclear data on masses and β-decay properties will be briefly reviewed, and consequences on the overall r-process matter flow up to the cosmochronometers ^{232}Th and ^{238}U will be discussed.

Keywords: Nucleosynthesis,Nuclear Data,r-Process,Cosmochronometers
PACS: 26.30.+k,26.50.+x,07.10.CV

INTRODUCTION

Historically, the field of nuclear astrophysics has always been concerned with the interpretation of observables (like specific signatures in luminosity, or elemental and isotopic abundance distributions; see e.g. [1, 2, 3, 4], and the description of the originating nucleosynthesis processes (see, e.g. [5, 6, 7]). Steadily improved observational instrumentation and spectroscopic detection methods, today allow detailed analyses of such observable features in the ejeta of various stellar events. And, moreover, progress in experimental and theoretical knowledge of nuclear-physics properties now offers increasingly meaningful comparisons between predicted and observed abundances, which yield valuable information about the required astrophysical conditions (for the r-process, see e.g. [8, 9, 10, 11]. In the present paper, I will focus on the main nuclear properties required for the classical rapid neutron-capture process. For sufficiently high neutron densities (n_n) and temperatures ($T_9=10^9$) during the main phase of an r-process, for which an (n,γ) $\rightleftharpoons$ (γ,n) equilibrium can be assumed, nuclear masses (in particular neutron separation energies, S_n) and gross β-decay properties (half-lives, $T_{1/2}$, and delayed neutron-emission probabilites, P_n) are of major interest. In this classical "waiting-point approximation" [5], the knowledge of only these three nuclear parameters would be sufficient to determine the whole set of r-process abundances for a given stellar condition in terms of n_n and T_9. Already in 1993, results from the Kratz – Thielemann collabora-

CP819 *Capture Gamma-Ray Spectroscopy and Related Topics: 12th International Symposium*
edited by A. Woehr and A. Aprahamian
© 2006 American Institute of Physics 0-7354-0313-9/06/$23.00

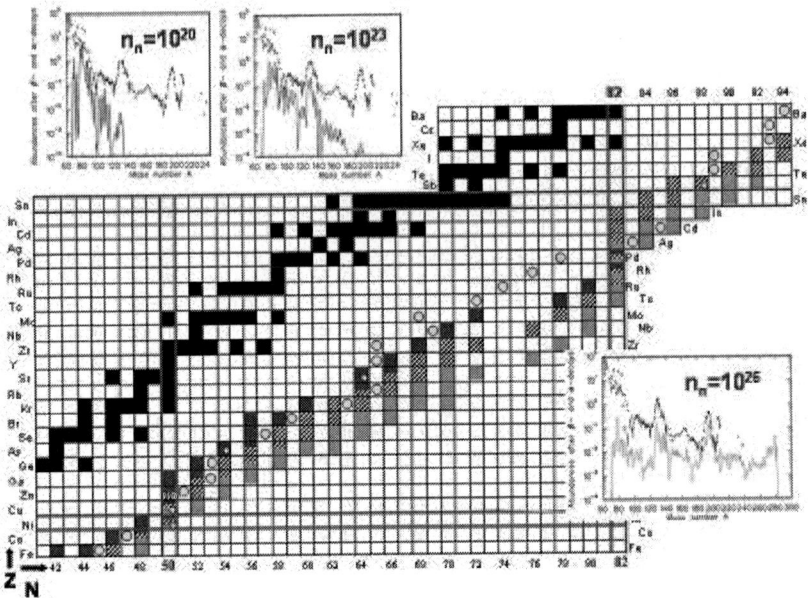

FIGURE 1. The r-process "boulevard" at freeze-out ($T_9=1.35$) between the Fe-seed and the $A \simeq 130\,N_{r,\odot}$ peak region for three selected neutron densities: dark squares $n_n=10^{20}$ cm^{-3}; hatched squares $n_n=10^{23}$ cm^{-3}, and grey squares $n_n=10^{26}$ cm^{-3}. The light-grey circles indicate the most neutron-rich isotope of each element, for which at least one integral β-decay property (a $T_{1/2}$ and / or a P_n-value) has been measured. The three inserts show the respective calculated r-abundances after decay back to stability, in comparison with the total $N_{r,\odot}$ distribution [14]. In these calculations, the mass formula ETFSI-Q [15] and the β-decay properties from the QRPA (GT+ff) model [17] have been used, together with all experimental data up to 2005 [19, 12, 16]. For further discussion, see text.

tion [8] (based on the first experimental information on N=50 and N=82 "waiting-point" nuclei), have indicated that this is, indeed, the case for conditions which reproduce the low-mass wings of the $A \simeq 80$ and $A \simeq 130$ peak regions of the solar-system isotopic r-process abundance ($N_{r,\odot}$) distribution. Although since that time, impressive experimental progress has been made (see, e.g. [12, 13]), still most of the relevant data have either not been determined or are not accessible to experiments at all. Hence, we are probing our knowledge of nuclear-structure properties when extrapolating from the known regions to the r-process "boulevard" near the neutron drip-line.

WHAT IS AN R-PROCESS ISOTOPE?

Since during the last few years, quite a number of experimental as well as theoretical publications have appeared that claim to deal with supposed r-process isotopes, it may be worth to define what neutron-rich nuclei should actually be called *"r-process isotopes"* in terms of the classical "waiting-point" concept of an r-process.

By now, among astrophysicists it is widely accepted that the observed isotopic $N_{r,\odot}$ pattern contains the fingerprints of the nuclear-physics properties of those isotopes lying in the late path of the r-process, just at the transition from the $(n,\gamma) \rightleftharpoons (\gamma,n)$ equilibrium to the "freeze-out" phase. As can easily be deduced from the nuclear Saha equation, for different n_n-components this transition occurs at different – constant – neutron separation energies (S_n) involving nuclei at different distance from β-stability. Already more than a decade ago, it has, for example, been shown by Kratz et al. [8] that for a chosen stellar temperature of $T_9=1.35$, these freeze-out conditions are reached at $n_n \simeq 10^{20}$ cm^{-3} for the $A\simeq80$ $N_{r,\odot}$ peak region, at $n_n \simeq 10^{23}$ cm^{-3} for the second $N_{r,\odot}$ peak at $A\simeq130$, and $n_n \simeq 10^{26}$ for the $A\simeq195$ peak region. Consequently, under the above conditions, different mass regions of the total $N_{r,\odot}$ distribution are produced, for which different individual isotopes for a given element lie in the respective r-process "path". The situation for the first part of the classical r-process, from the Fe-seed region up to just beyond the N=82 shell closure, is shown in Fig. 1. In order to obtain the observed $N_{r,\odot}$ distribution, the abundances of the individual n_n-components have to be summed up, finally resulting in an r-process *"boulevard"* with a width of several mass units for each element, rather than the historical, simplistic picture of a narrow *"path"*. For better illustration, let us chose strontium (Z=38), for which N=64 ^{102}Sr up to N=72 ^{110}Sr would be the (main) r-process isotopes for the above three n_n-conditions.

Furthermore, when talking about r-process isotopes, one should also consider their "importance" in terms of the initial and final (isotopic) abundances relative to the $N_{r,\odot}$ values. As can be deduced from Fig. 1 for the case of Sr chosen above, for the "low"-n_n freeze-out conditions of 10^{20} n/cm^3, the respective N=64 and N=66 isotopes contribute to about 10 % to the $N_{r,\odot}$ values, whereas the heavier ones (according to the Saha equation) are negligible. Similarly, for $n_n=10^{23}$ cm^{-3}, the abundances of the N=66 and N=68 Sr nuclides are important on the 10 % level, whereas the lower- and higher-N isotopes are negligible. For $n_n=10^{26}$, however, although the mass region between the first and second $N_{r,\odot}$ peak is still produced, the contribution of the N=70 and N=72 Sr isotopes is only on the per-mille level.

Finally, it should be pointed out that there are two types of r-process nuclei with different significance for the r-process. The even-neutron isotopes are the classical "waiting-point" nuclei. In each isotopic chain, the r-process matter flow is halted at such a nuclide (A,Z), where it must "wait" for its β-decay to the respective isobar of the next heavier element (A,Z+1). The odd-neutron r-process nuclides, on the other hand, determine the isotopic neutron-capture flow through their S_n values. They determine the position of the r-process path at nuclide (A,Z), when neutron capture (n,γ) into the next istotope (A+1,Z) is prevented by the dominating inverse photodisintegration (γ,n).

EXPERIMENTAL INFORMATION ON R-PROCESS NUCLEI

As has been discussed in detail elsewhere (see, e.g. [18, 10]), there are in principle two types of nuclear-structure information that may help to solve the r-process problems. First, as outlined before, there is the quite obvious need for nuclear masses and

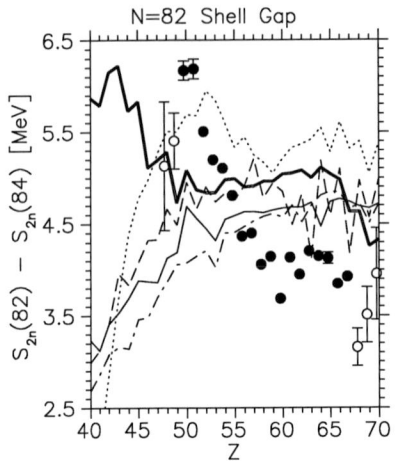 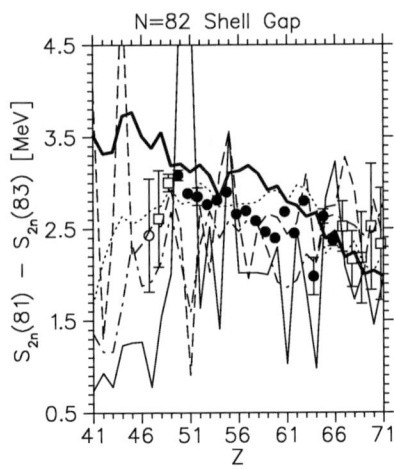

FIGURE 2. The N=82 shell gap as a function of Z. Theoretical mass predictions (FRDM, thick line [27]; HFB-2, dashed-dotted line [23]; HFB-8, thin line [24]; HFB-9 long-dashed line [25] and ETFSI-Q short-dashed line [15]) are compared to experimental values from NUBASE [19]. The data for Z=47,48 and 68-70 (open cirles) were deduced from combinations of measured and short-range extrapolated values.

the gross β-decay properties $T_{1/2}$ and P_n of those isotopes lying in the r-process boulevard. Second, also the evolution of shell structure from β-stability towards the neutron dripline must be understood. This requires detailed spectroscopic studies, in particular for neutron-magic "waiting-point" nuclei, as have been performed for N=50 ^{80}Zn [20] and N=82 ^{130}Cd [21]. For example, it originated from such detailed spectroscopic studies in the N=60 shape-transitional region, that already a decade ago Kratz and collaborators predicted that *"the calculated r-abundance hole in the A$\simeq$120 region"*, observed in most of the r-abundance calculations at that time, *"reflects ... the weakening of the shell strength ... below $^{132}Sn_{82}$."* [8]. This kind of "quenching" of the classical spherical shells with increasing isospin has later also been predicted by various theoretical models (among the first, see, e.g., the Hartree-Fock-Bogulyubov (HFB) ansatz of Dobaczewski et al. [22], and is meanwhile well established for the N=20 and N=28 shell closures and at least indicative for N=50 [8, 20] and N=82 [18, 21].

Nuclear Masses

A substantial number of theoretical approaches to calculate nuclear masses exist nowadays for various applications. The sophistication of these models is, however, quite different and should therefore be understood before they are used as nuclear input. The key question in the present case, i.e. r-process calculations, is: Are the masses calculated for unknown nuclei reliable, or do the model predictions diverge outside the experimental data set, to which their parameters were adjusted to? This question is best addressed, if published theoretical masses are compared to experimental data obtained

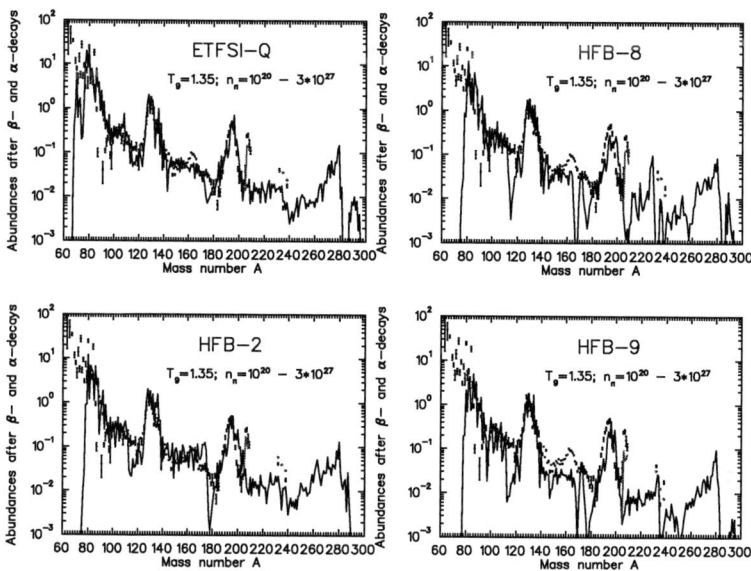

FIGURE 3. Comparison of calculated r-abundance distributions ($N_{r,calc}$) with the $N_{r,\odot}$ distribution using four different recent mass models, ETFSI-Q [15], HFB-2 [23], HFB-8 [24] and HFB-9 [25]. Apart from the S_n-values, all other nuclear-physics and astrophysics parameter were kept constant. For further discussion, see text.

after the publication of the calculations.

It has been claimed that the recent more sophisticated HFB mean-field mass models (see, e.g. [23, 24, 25]), that go beyond gross-theory (e.g. [26] and macroscopic-microscopic approaches (e.g. the Finite Range Droplet Model (FRDM) of Möller et al. [27]), would by virtue of their added microscopic complexity be physically more reliable and have better predictive power for unknown nuclei. However, recent tests ([28, 29, 30, 31]) have shown that this expectation has so far not been fulfilled. Moreover, there seem to be fundamental problems with the HFB models. As is discussed in detail in the topical review of Rikovska Stone [31], several parameters are clearly not treated in a *"selfconsistent"* and *"fully microscopic"* way, as claimed by the authors. Furthermore, there is no significant improvement of the quality of fits of the HF mass models to known data bases or new masses, with increasing sophistication and number of fitting parameters. This is most evident in the mass regions relevant to r-process calculations, i.e. the far-unstable regions of the neutron shell closures and shape transitions. Rikovska Stone even concludes that, despite the admirable effort of the Montreal–Brussels group, it is unlikely that the present HFB models will ever yield masses with a higher precision than about 0.7 MeV. Furthermore, the FRDM macroscopic-microscopic model (with its parameters fitted to the masses known in 1992), still is doing a better job with its best rms value of 633 keV for the 2003 mass evaluation [19] than any of the Extended Thomas-Fermi plus Strutinsky Integral (ETFSI), HFBCS and HFB mass models.

As an example for the present situation of mass predictions at magic neutron shells, which are of particular interest to the calculation of the $N_{r,\odot}$ peaks, Figure 2 gives comparisons of the experimental N=82 shell gap as a function of atomic number Z with several mass model predictions. In the left part of Fig. 2, the shell gap is defined as the difference of the two-neutron separation energies S_{2n} for the paired-neutron systems at N=82 and N=84. In the right part of Fig. 2, the S_{2n} differences of the odd-neutron systems N=81 and N=83 are shown. The message from both figures is unsatisfactory: **none** of the mass models is able to reproduce the overall experimental trend, and in particular the reduction of the shell gap on both sides of the doubly-magic ^{132}Sn nucleus. Furthermore, the recent HFB models show large odd-even-Z fluctuations of the gap in the odd-neutron S_{2n} presentation, in total disagreement with experiment. With respect to calculations of the $A \simeq 130$ $N_{r,\odot}$ peak, the region below Z=50 is of major interest. In this region, the FRDM shows an increase of the N=82 shell gap down to Z$\simeq$42, rather than the expected "quenching". On the other hand, all recent HFB models do not reproduce the pronounced extremum of the gap at Z$\simeq$50, but predict a rather structureless decrease of the shell gap down to Z$\simeq$40. The "best" trend is given by the ETFSI-Q [15] and the Duflo–Zuker [33] mass formulae, which both are, however, no real "models" but rather heuristically fine-tuned shell-structure approaches.

In Figure 3, comparisons of calculated r-abundance distributions ($N_{r,calc}$) with the solar-system $N_{r,\odot}$ pattern, using the mass models ETFSI-Q [15], HFB-2 [23], HFB-8 [24] and HFB-9 [25], are shown. In all four cases, the experimentally known masses from the 2003 mass evaluation of Audi et al. [19] have been superimposed. It is quite evident, that the use of the supposed "microscopic" HFB models does not improve the overall fits. On the contrary, compared to the ETFSI-Q fit the deviations become more pronounced, in particular for the two most recent HFB versions, which are recommended for astrophysics application by the authors [28]. In any case, the general consequences for the r-abundance calculations are clear: without the trend of the N=82 shell gap guided by the experimental masses around ^{132}Sn (note in particular the recent Q_β measurement of ^{130}Cd [21]), any realistic astrophysical calculation of the A$\simeq$130 peak would result in even larger deviations from the $N_{r,\odot}$ pattern, together with an incorrect description of the r-process matter flow through this most important bottle-neck region. It has been shown (see, e.g.[9, 10]), that this may have considerable consequences for the build-up of the heavier r-elements up to the 3rd peak region, and in turn will even influence the nucleosynthesis prediction of the actinide cosmochronometers leading to unreliable determinations of the age of our Galaxy [32].

Gross β-Decay Properties

Theoretically, the two gross β-decay parameters, the half-life ($T_{1/2}$) and the β-delayed neutron emission probability (P_n), are interrelated via their usual definition in terms of the so-called β-strength function (see, e.g. [27]). Although, taken separate each integral quantity has only a limited physical significance, combined they may well provide some first information on the underlying nuclear structure of far-unstable nuclei.

Over the years, several theoretical approaches have been applied to model β-decay (see, e.g. [34]). Like in the case of mass models, theoretical approaches predicting $T_{1/2}$

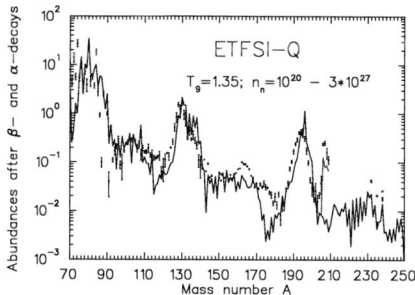
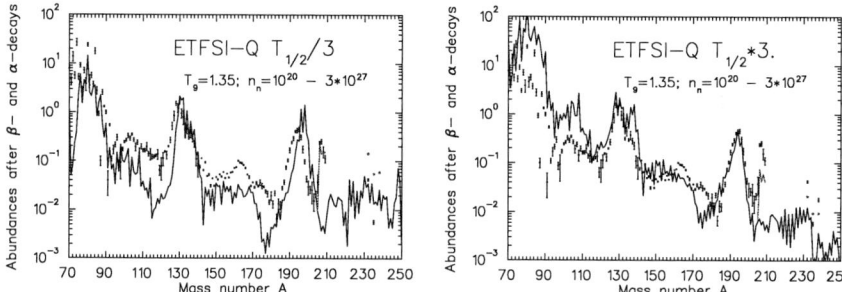

FIGURE 4. Comparison of calculated isotopic r-process abundances ($N_{r,calc}$) with the $N_{r,\odot}$ distribution for three different assumptions on the β-decay half-lives ($T_{1/2}$). Apart from the $T_{1/2}$, all other nuclear-physics and astrophysics parameters were kept constant. In the upper figure, the "standard" $N_{r,calc}$ result is shown when using the $T_{1/2}$ from our recent FRDM + QRPA(GT+ff) model calculations [17]. In order to test the (maximum) effect of the average $T_{1/2}$ uncertainties on the $N_{r,calc}$, in two subsequent calculations all $T_{1/2}$ have been divided by or multiplied with a factor three, respectively (see the two lower figures). The deviations in the resulting abundance distributions (most pronounced in the $A \simeq 80$, the rare-earth and the actinide regions) relative to the "standard" $N_{r,calc}$ pattern, clearly indicate the differences in the speed of the overall r-process matter flow introduced by the different $T_{1/2}$ sets. For further discussion, see text.

and P_n-values range from phenomenological, algebraic expressions up to large-scale microscopic models. Some of these approaches emphasize global applicability, whereas others seek self-consistency or the comprehensive inclusion of nuclear correlations. Unfortunately, to date none of these models incorporates all important ingredients. Most models still have strong limitations, e.g. they are restricted to GT-transitions, to spherical shapes, to even-even nuclei, or have too small shell-model spaces (as typical examples, see e.g. [27, 35, 36, 37, 38, 39]). Therefore, these models usually describe more or less successfully only small mass regions, if their locally fine-tuned parameters are kept fixed. Moreover, unfortunately the information made available in order to judge the physical reliability of these approaches is often very limited, and it is worrying to see no improvement in **global** calulations of gross β-decay properties for astrophysical applications. Therefore, for the time being these models are unsuitable as a basis for calculations of the full r-process.

There is so far only one consistent macroscopic-microscopic approach with global applicability of predicting a variety of nuclear properties which are needed for detailed dynamical r-process calculations (see, e.g. [40]): namely the FRDM combined with the quasi-particle random phase approximation (QRPA) of Möller et al. [27, 17]. Originally restricted to GT-transitions [27], the latest version of this model also includes first-forbidden (ff) corrections (FRDM + QRPA(GT+ff)), as well as a spreading of the single-particle GT-transitions according to experimental findings [17]. Analoguously, when adopting other mass models in our r-process calculations, such as the "quenched" version of ETFSI [15], we normally use theoretical β-decay quantities deduced from QRPA calculations with masses and shape parameters given by that particular model (e.g. ETFSI-Q + QRPA(GT+ff)).

The FRDM + QRPA(GT+ff) model shows an average error in predicting $T_{1/2} \leq 1$ s of about 3; accordingly, the average error for P_n-values is 3.5 (see Figs. 4 and 5 in [17]). Initially we thought that these uncertainty factors were satisfactory for astrophysical applications. Nevertheless, in order to check the (maximum) effect of the $T_{1/2}$ uncertainties on the calculated r-process abundances ($N_{r,calc}$) we have changed **all** $T_{1/2}$ values by a factor 3 in both directions, by keeping all other nuclear-physics quantities (S_n- and P_n-values) and astrophyscis parameters (T_9 and n_n-components, their time scales and weightings) constant (see, e.g. [10]). The results are shown in Fig. 4. The deviations in the respective r-abundance distributions (the two lower figures) relative to the "standard" $N_{r,calc}$ pattern (the upper figure) are most pronounced in the A$\simeq$80, the rare-earth and the actinide regions. They clearly indicate the differences in the speed of the overall r-process matter flow introduced by the different $T_{1/2}$ sets.

Progress in the above approaches certainly will ask for more than just the reproduction of known gross β-decay properties, but rather for a detailed prediction of the full "β-strength function" (S_β), at least for a number of selected "key" r-process isotopes. Experimentally, the determination of such S_β distributions requires full spectroscopic investigation, including quantites such as the Q_β value, the main GT- and low-lying ff-transitions with their excitation energies and log(ft) values. A recent, important example in this context has been the identification of several a priori unexpected features in the decay of N=82 ^{130}Cd [21, 12], which is presumably the most important "waiting-point" isotope for the **"main"** r-process. For the **"weak"** r-process, apart from N=50 ^{80}Zn [8, 20] the major bottle-neck for the r-matter flow through the A$\simeq$80 $N_{r,\odot}$ peak is doubly-magic ^{78}Ni, whose half-life has recently been determined at NSCL/MSU [41]. In Fig. 5, we show the predictions of the GT-decay scheme from QRPA calculations with two different nuclear potentials, i.e. a Folded-Yukawa and a standard Woods-Saxon potential. On the one hand, the apparent differences in the level schemes indicate the present model uncertainties. On the other hand, however, both decay schemes agree in the general outcome that (i) the $T_{1/2}$ is mainly determined by the GT-transition to a "low-lying" $J^\pi=1^+$ level in ^{78}Cu which has the $\nu p_{1/2} \otimes \pi p_{3/2}$ 2QP configuration, and (ii) the P_n value is dominated by the "high-lying" $J^\pi=1^+$ levels with $\nu g_{9/2} \otimes \pi g_{9/2}$ 2QP and $\nu f_{5/2} \otimes \pi f_{7/2}$ 4QP configurations. Experimentally, a full understanding of the decay scheme thus requires classical γ-spectroscopic studies as well as (high energy-resolution) delayed-neutron spectroscopy. Theoretically, it is easy to understand that

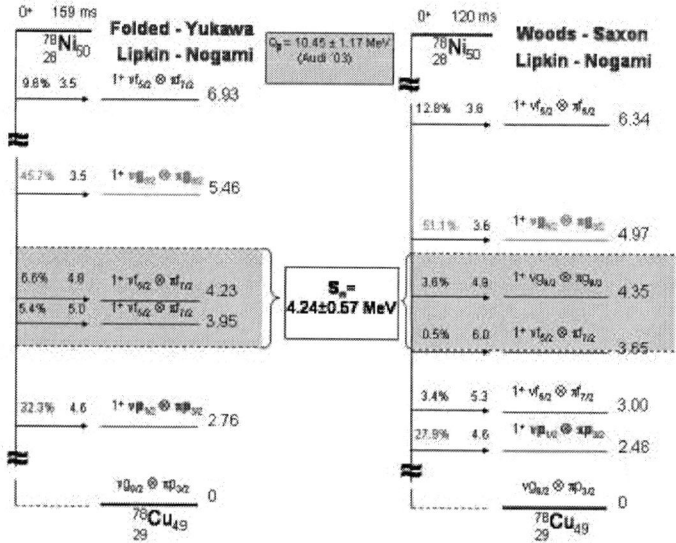

FIGURE 5. Predicted decay schemes of doubly-magic ^{78}Ni using the QRPA shell-model approach with a Folded-Yukawa (left part) and a Woods-Saxon potential (right part), and a Lipkin-Nogami pairing model. Both calculatoins are consistent in that the $T_{1/2}$ is determined by the energy and β-feeding of the "low-lying" 2QP $\nu p_{1/2} \otimes \pi p_{3/2}$ level. Analoguously, the P_n value is mainly determined by the $\nu g_{9/2} \otimes \pi g_{9/2}$ 2QP and $\nu f_{5/2} \otimes \pi f_{7/2}$ 4QP configurations. For further discussion, see text.

only shell-model approaches with a large model space, including the f- and g-shells, can give reliable answers. However, it is also evident from the above discussion that the correct $T_{1/2}$ *"number"* may fortuitously be obtained with a reduced model space when neglecting the GT-contribution from the ν g-shell and / or blocking the ν and π f-shell as inert core.

SUMMARY AND OUTLOOK

In summary, I have briefly reviewed the present status of the main nuclear-data needs of r-process nucleosynthesis. Both, progress in theoretical predictions as well as in experimental data is needed in order to further advance our knowledge. Apart from nuclear masses and gross β-decay properties, the importance of other nuclear quantities for full dynamic network calculations of the r-process(es) has to be checked. For example, neutron-capture rates are needed to correctly model the final "freeze-out" from the initial abundances during the decay back to stability (see, e.g. [40]). Moreover, fission barriers and fission fragment distributions have to be known in order to describe possible *"fission recycling"* in high neutron-density scenarios. Futhermore, neutrino reactions, which play an important role for determining the initial neutron excess in r-process models, may lead to a moderate processing of timescales and final r-abundances (see, e.g.

[10, 42]). Taken together, there is by all means hope to soon solve the problem of the *"origin of the heavy elements between Fe and Th, U"*, which has recently been considered number three of *The 11 Greatest Unanswered Questions in Physics* [43]).

REFERENCES

1. H.E. Suess and H.C. Urey, *Rev. Mod. Phys.*, **28**, 53 (1956).
2. W.H.G. Lewin et al., *Space Sci. Rev.*, **62**, 233 (1993).
3. K. Lodders, *ApJ*, **591**, 1220 (2003).
4. C. Sneden and J.J. Cowan, *Science*, **299**, 70 (2003).
5. E.M. Burbidge et al., *Rev. Mod. Phys.*, **29**, 547 (1957).
6. J.J. Cowan et al., *Phys. Rep.*, **208**, 267(1991).
7. G. Wallerstein et al., *Rev. Mod. Phys.*, **69**, 995 (1997).
8. K.-L. Kratz et al., *ApJ*, **403**, 216 (1993).
9. K.-L. Kratz et al., in *Origin of Elements in the Solar System: Implications of Post-1957 Observations* ed. by O. Manuel, Kluwer Academics / Plenum Publishers, Boston/Dordrecht, 2001, p. 119.
10. B. Pfeiffer et al., *Nucl. Phys.*, **A693**, 282 (2001).
11. K.-L. Kratz et al., *New Astron. Rev.*, **48**, 105 (2004).
12. K.-L. Kratz et al., *EPJ*, **A25 Supplement 1**, 633 (2005).
13. K.-L. Kratz et al., *AIP Conf. Proc.*, **769**, 1356 (2005).
14. F. Käppeler et al., *Rep. Prog. Phys.*, **52**, 131 (1989).
15. J.M. Pearson et al., *Phys. Lett.*, **B387**, 455 (1996).
16. B. Pfeiffer et al., *Prog. Nucl. Energy*, **41**, 39 (2002).
17. P. Möller et al., *Phys. Rev*, **C67**, 055802 (2003).
18. K.-L. Kratz et al., *Hyperfine Interactions*, **129**, 141 (2000).
19. G. Audi et al., *Nucl. Phys.*, **A729**, 3 (2003).
20. Kratz et al., *Phys. Rev.*, **C38**, 105 (1988).
21. I. Dillmann et al., *Phys. Rev. Lett.*, **91** 162503 (2003).
22. J. Dobaczewski et al., *Phys. Rev. Lett.*, **72**, 981 (1996).
23. S. Goriely et al., *ADNDT*, **77**, 311 (2001).
24. M. Samyn et al., *Phys. Rev.*, **C70**, 044309 (2004).
25. S. Goriely et al., *Nucl. Phys.*, **A750**, 425 (2005).
26. H. von Groote et al., *ADNDT*, **17**, 418 (1976).
27. P. Möller et al., *ADNDT*, **66**, 131 (1997).
28. J. M. Pearson in *"The r-Process: The Astrophysical Origin of the Heavy Elements"*, eds. Y.-Z. Qian et al., Proc. Inst. of Nucl. Theory, Vol. 13, World Scientific, 2004, p. 43.
29. D. Lunney et al., *Rev. Mod. Phys.*, **75**, 1921 (2003).
30. P. Möller, private communication 2005.
31. J. Rikovska Stone, *J. Phys. G: (Nucl. Part. Phys.)*, **31**, R211. (2005).
32. J.J. Cowan et al. *ApJ*, **521**, 194 (1999).
33. J. Duflo and A.P. Zuker, *Phys. Rev.*, **C52**, R23 (1995).
34. J. Engel in [28], p. 53
35. J. Engel et al., *Phys. Rev.*, **C60**, 014302 (1999).
36. G. Martinez-Pinedo, *Nucl. Phys.*, **A668**, 357c (2000), and refs. therein.
37. B.A. Brown et al., *Nucl. Phys.*, **A719**, 177c (2003), and refs. therein.
38. I.N. Borzov et al., *Phys. Rev.*, **C71**, 065801 (2005), and refs. therein.
39. T. Niksic et al., *Phys. Rev.*, **C71** 014308 (2005).
40. K. Farouqi et al., contrib. to this conference
41. P.T. Hosmer et al., *Phys. Rev. Lett.*, **94** 112501 (2005).
42. M. Terasawa et al., *ApJ.*, **608**, 470 (2004).
43. E. Haseltine, *Discover Magazine*, **23**, No.2 (2002).

Neutron captures and the r-process

K. Farouqi, K.-L. Kratz, B. Pfeiffer*,† and T. Rauscher, F.-K. Thielemann**

Institut für Kernchemie, Universität Mainz, Germany
†*HGF Virtual Institute for Nuclear Structure and Astrophysics, Mainz, Germany*
**Departement für Physik und Astronomie, Universität Basel, Switzerland*

Abstract. In order to study possible neutron-capture effects during an r-process, it is necessary to perform fully dynamical simulations. We have performed such calculations within the model of an adiabatically expanding high-entropy bubble of a SN II, using temperature-dependent reaction rates including the NON-SMOKER neutron-capture rates of Rauscher et al. [1].

Keywords: Nuclear astrophysics, r-process, neutrino-driven wind of type II SNe
PACS: 26.30.+k

THE ASTROPHYSICAL MODEL: NEUTRINO-DRIVEN WIND OF TYPE II SNE

During the final stages of the evolution of a massive $(8-25 M_\odot)$ star, an "iron" core forms in its central region and subsequently undergoes gravitational collapse. When the central density reaches nuclear matter density, the collapse stops abruptly to cause a "core bounce". A shock wave is created and starts to propagate outward. According to hydrodynamical calculations [3], this shock wave loses its entire kinetic energy within a few milliseconds to stall well inside the outer edge of the initial iron core, and no immediate disruption (a "prompt" explosion) of the star occurs. On a timescale from several tens of milliseconds to about half a second, the neutrinos streaming out from the new-born neutron star can deposit energy behind the standing accretion shock at a rate high enough to revive its outward motion and initiate the final explosion of the star. This is the neutrino-driven "delayed" explosion mechanism originally suggested by Wilson [4]. We have started our network calculations after the total photodisintegration of the matter above the nascent neutron star at $T_9 = 9$ with protons (p) and neutrons (n). The n-to-p-ratio is $Y_e = X_p = 1 - X_n$ with X_p and X_n being the mass fractions of protons and neutrons. Using the charged particle network of F.-K. Thielemann and the r-process code of C. Freiburghaus, combined with recent β-decay and neutron-capture rates, we were able to perform a detailed study of the α-process and of the subsequent r-process. Using the three parameters V_{exp} (expansion speed of the shock wave), $S \propto T^3/\rho$ (S entropy, T the temperature and ρ the matter density of the bubble) and Y_e, we could show that the above parameters have to fulfill specific conditions in order to make a subsequent r-process (see table 1).

TABLE 1. An example of parameters wich allow a subsequent r-process

V_{exp} (km/s)	τ_{exp} (ms)	Y_e	$S(K_B/\text{Baryon})$	Y_n/Y_{seed}
7500	35	0.49	$130 \leq S \leq 310$	$1 \leq \ldots \leq 150$
		0.47	$110 \leq S \leq 300$	$1 \leq \ldots \leq 150$
		0.45	$90 \leq S \leq 290$	$1 \leq \ldots \leq 150$
		0.43	$70 \leq S \leq 280$	$1 \leq \ldots \leq 150$
		0.41	$50 \leq S \leq 260$	$1 \leq \ldots \leq 150$

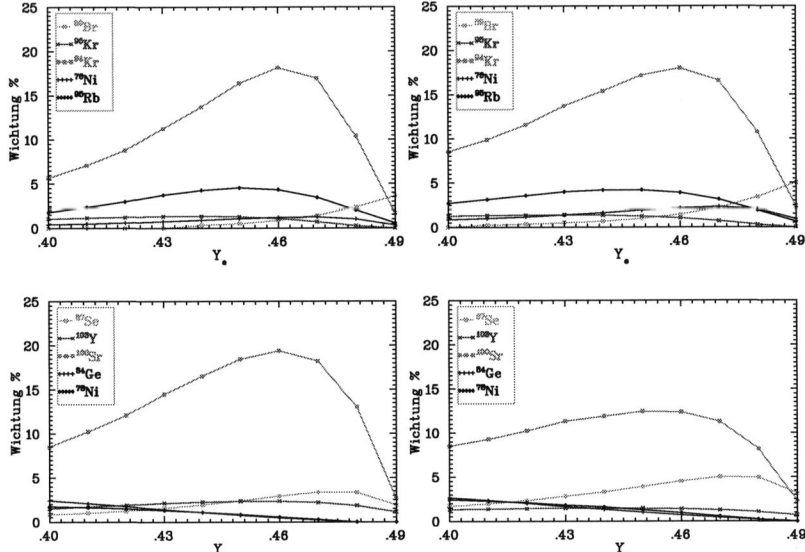

FIGURE 1. Seed nuclei distribution as a function of Y_e

SEED NUCLEI DISTRIBUTION AFTER AN α-RICH FREEZE-OUT

The maximum of the seed nuclei distribution after an α-rich freeze-out lies in the mass region between A=80 and 110 (see figure 1). This will avoid the bottle neck region at N=50, and provide a very fast r-process.

THE R-PROCESS

A first result for a simple superposition fitting five different mass regions in the high-entropy bubble with the corresponding entropies is shown in figure 2. The fit function $g(S_i)$ which we used is depending on the entropy S and on two other parameters X_1 and

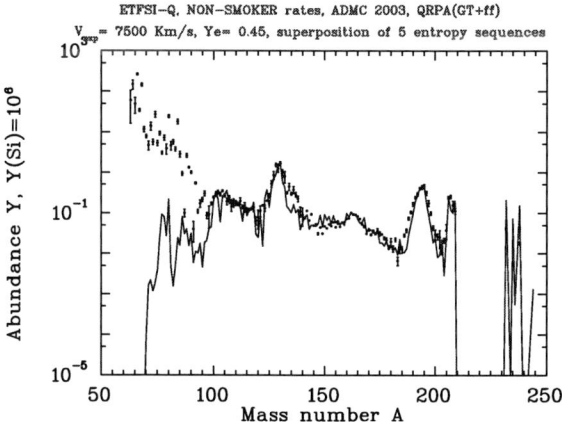

FIGURE 2. Superposition of 5 entropy sequences to reproduce the isotopic solar r-process pattern.

X_2, and the index i is the corresponding mass region:

$$g(S_i) = X_1 \cdot e^{-X_2 \cdot S_i}. \tag{1}$$

EFFECT OF THE NEUTRON CAPTURE IN THE R-PROCESS NUCLEOSYNTHESIS

In order to study the effect of the neutron capture rates we scaled the whole capture rates up and down by a factor of 100. The normal neutron capture rates are calculated at an energy of 30 keV. The result for the second and third peak at A=130 and 195 is shown by figure 3. As we expected, for the second peak there is no significant effect because of the very short freeze-out time, whereas for the third peak the effect is because of the longer freeze-out time "fatal". Therefore, a full network is needed if the region beyond A=140 wants to be studied.

THE ROLE OF β-DELAYED NEUTRONS

An other open question was whether the *beta*-delayed neutrons during and after the freeze-out will change the progenitor isotopic abundances or not. Figure 4 shows the answer for the second and third peaks. At A=130 mass region we see no significant difference between the two curves. Beyond A=140 it seems that the β-delayed neutrons are needed to shovel up the matter flow to the higher mass regions. At A=195 mass region we see also this effect. Without β-delayed neutrons the maximum of the red curve lies at A=194 and one unit is then needed to put the peak at the right place. This result was also expected because of the higher pn values of the progenitor nuclei in that mass region.

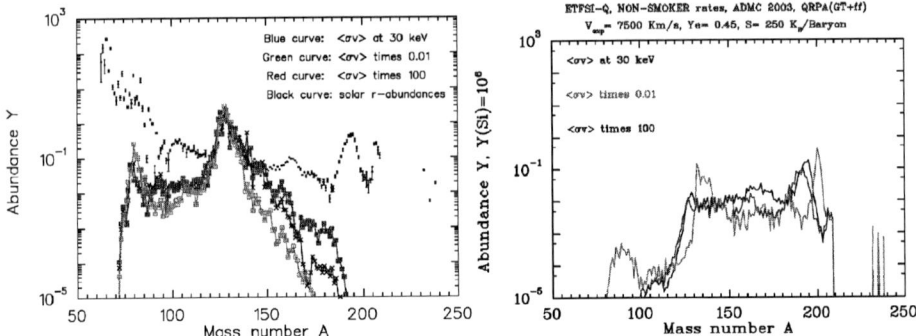

FIGURE 3. The neutron capture rates scaled up and down by a factor of 100.

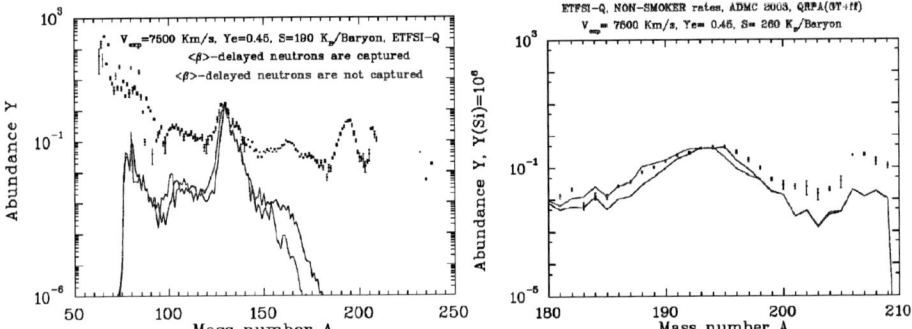

FIGURE 4. β-delayed neutrons and their roles in the r-process nucleosynthesis.

REFERENCES

1. T. Rauscher, T. and F.-K. Thielemann, Atomic data Nucl. Data Tables 75, 1 (2000)
2. C. Freiburghaus et al., *Ap. J.*, 516, 381 (1999)
3. Myra & Bludman, *ApJ*, 340, 384 (1989)
4. J. R. Wilson, *in Numerical Astrophysics* (1985)

Radioactive beam experiments relevant for the astrophysical r-process

H. Schatz

National Superconducting Cyclotron Laboratory, Michigan State University, East Lansing, MI 48824, USA
Dept. of Physics and Astronomy, Michigan State University, East Lansing, MI 48824, USA, Joint Institute for Nuclear Astrophysics, http://www.jinaweb.org

Abstract. Radioactive beam experiments play an important role in the quest for an understanding of the origin of the heavy elements in nature produced by the astrophysical rapid neutron capture process (r-process). I discuss recent progress in identifying the site and nature of the r-process and the reasons why nuclear physics experiments at existing and future radioactive beam facilities are needed to solve the r-process problem. Recent results and future plans for r-process related experiments at the National Superconducting Cyclotron Laboratory are also presented.

Keywords: r-process, radioactive beam experiments
PACS: i26.50.+x 23.40.-s

INTRODUCTION - THE CASE FOR THE NEED OF NUCLEAR PHYSICS IN THE R-PROCESS

The rapid neutron capture process (r-process) is one of the major nucleosynthesis processes in nature [1, 2]. It produces roughly half of the elements above the iron peak, and is largely responsible for the origin of elements such as xenon, gold, platinum, and uranium in nature. The fact that the site of the r-process and therefore its conditions and reaction sequences are still not known with certainty is one of the major open questions in science. This had been recognized for example, by the NRC study "Connecting Quarks with the Cosmos", which identified the question for the origin of the heavy elements as one of the 11 science questions for the 21st century at the intersection of astrophysics and particle physics [3].

A number of theoretical models have been proposed in the past as sites for the r-process. These include the neutrino driven wind off a forming neutron star in a core collapse supernova [4, 5], jets in core collapse supernovae [6, 7], neutron star mergers [8, 9], and outflows from rapidly accreting disks in collapsars powering gamma-ray bursts [10]. While all these models have been shown to exhibit some r-process within the respective parameter space, they all fall short of selfconsistently producing sufficient amounts of r-process elements in the proportions that would explain the solar r-process abundance pattern. Clearly more theoretical work is needed to overcome these issues. However, this alone does not solve the problem as even a self-consistent viable r-process model does not need to be the correct model realized in nature. In the end, experimental constraints are needed to guide theories and to identify the r-process site or sites.

So far, no clear smoking gun has been found that would pinpoint the type of astrophysical object responsible for the r-process. One possibility would be the detection of γ-rays from the decay of radioactive r-process nuclei in the ejecta of a galactic core collapse supernova (see for example [11]), but predicted gamma-ray fluxes are below current and near future sensitivity limits of gamma-ray observatories. Recently some evidence for the presence of non anthropogenic ^{244}Pu, an r-process isotope, has been found in deep sea manganese encrustations [12]. Only a single atom was found and the age range of the layer is with 1-14 Myr B.P. (before 1950) rather large. Nevertheless this finding is consistent with the recent detection of ^{60}Fe in a deep sea manganese crust that based on yield arguments has been interpreted as evidence of a nearby (10 Pc) core collapse supernova 2.8 Myr ago [13]. Taken together these results would favor core collapse supernovae as a site for the r-process, but more data are needed to draw final conclusions about a possible correlation of ^{244}Pu and ^{60}Fe in deep sea crusts.

Less direct constraints on the r-process site can be derived from observations related to the history of galactic chemical evolution of r-process elements or isotopes. One example are the abundance signatures of radioactive r-process isotopes that can be found in meteorites and that are sufficiently long-lived to carry some information about the history of their production such as the frequency of the nucleosynthesis event. Examples are ^{129}I and ^{182}Hf for both of which there is evidence of their presence in the early solar system from isotopic anomalies found in meteorites. Because of the different half-lives of ^{129}I (15.7 Myr) and ^{182}Hf (8.9 Myr) the final abundance levels depend differently on the r-process event frequency and ejection efficiency. It has therefore been argued that ^{129}I and ^{182}Hf must have been produced in different types of supernovae, ^{129}I in a low frequency type event, and ^{182}Hf together with the actinides in a high frequency event to explain the low ^{129}I and high ^{182}Hf abundances inferred for the presolar nebula [14]. However, observations of r-process elements in very metal poor stars seem to indicate that Hf is produced in the same r-process than Ba [16]. It might be difficult to find an r-process condition that produces all of ^{129}I but very little ^{138}Ba as the build up of the $A = 130$ peak will always result in some flow through the bottle-neck region (see for example Fig. 14 in [15]). At the very least this will require fine tuning of the r-process conditions.

The discovery of metal poor halo stars that are enriched in r-process elements and precision spectroscopy that allowed to measure in these stars accurate r-process abundance patterns for up to 28 elements [16] has also provided important observational clues (see [17] for a review). The material that these stars illuminate in their photosphere has been polluted by a single, or at most very few, r-process events. The fact that the A>130 elemental abundances observed in these stars, with the exception of the actinides [18, 19, 20, 21], agree with the r-process abundances inferred for the solar system by subtracting an s-process contribution has therefore been interpreted as a hint that the r-process sets in early, at low galactic metallicity levels, and produces a consistently similar pattern in each event for nuclei in the element number range from 56 to 79. Discrepancies between the stellar and solar r-process patterns for elements such as Mo or Ag have been interpreted as a hint for the existence of a second r-process producing extra amounts of lighter r-process elements at later times [22, 23]. Neutron capture element abundances for larger samples of stars over a wider range of metallicities that are available for some elements provide information about the history of r-process enrichment that can constrain event frequency and ejected masses. Recently Argast et al. [24] demonstrated that the observed enrichment and scatter in Eu and Ba abundances as a function of metallicity excludes neutron star mergers as the principle source of r-process elements but is consistent with a site related to core collapse supernovae. This conclusion depends on a comparison of observations with a model for galactic chemical evolution. In the future, ongoing surveys such as SEGUE (Sloan Extension for Galactic Understanding and Exploration) [25], will lead to the discovery of many more of such r-process enhanced metal poor stars greatly enhancing the amount of information available on the history of the r-process in our Galaxy.

However, observational constraints can at best point to the general type of astrophysical object responsible for the synthesis of certain r-process elements and do not necessarily reveal the actual r-process location, mechanism and conditions within such an object. One of the potentially most important pieces of information about the origin of the r-process elements are therefore the details of the r-process abundance pattern. An isotopic abundance pattern can be obtained from solar system abundances, once the contribution from the p- and the s-processes are subtracted [26]. Elemental abundance patterns of up to 28 elements have been obtained from r-process enriched metal poor stars [16]. These abundance patterns are sensitive to the conditions during the r-process. Neuton density and temperature directly determine the r-process path and therefore the final pattern. Extreme neutron to seed ratios lead to fission once nuclei around $A \approx 260-280$ are reached (see [27] for a recent review), which in turn leads to potentially distinctive signatures in the r-process pattern due to the production of fission fragments or, in the extreme case, due to fission cycling [28]. The timescale of the r-process freezeout [29] and the presence of neutrinos during the r-process [30, 31, 32] can further modify the final abundances. In principle, an analysis of the r-process abundance pattern can therefore yield not only further constraints for the astrophysical event responsible for r-process nucleosynthesis, but, in addition provides the only experimental constraint on the actual r-process environment within an astrophysical object such as a supernova. Therefore it seems unlikely that the r-process problem can be solved in a satisfactory way without dramatic improvement in our understanding of the underlying nuclear physics through experimental data.

Extracting any information about the r-process from observed abundance patterns requires knowledge of the underlying nuclear physics [15]. For example, without nuclear physics experiments demonstrating the existence neutron shell closures, the characteristic peaks in the solar abundance pattern [33] could not have been used to infer the rather extreme conditions needed for the r-process. Nuclear physics such as masses, beta-decay rates, branchings for beta-delayed neutron-emission, partition functions, fission rates and fragment distributions, neutrino interactions, neutron capture rates and through these quantities nuclear shapes, shell gap sizes, pairing interactions, etc all imprint themselves onto the final r-process abundances. Experimental constraints together with reliable theoretical predictions are needed in order to disentangle nuclear physics effects from astrophysics effects and to extract astrophysical information from the r-process abundance pattern. As an example it has been pointed out that the possibility of a significant reduction of the $N = 82$ and $N = 126$ shell gaps far from stability would lead to a better fit of the observed r-process abundances by r-process models in the mass regions right below the A=130 and A=195 abundance peaks [44]. However, experimental confirmation is essential to tell, whether this nuclear physics effect indeed solves the problem or whether an important piece of physics is missing in current r-process models. Recently first experimental evidence of such a quenching of the $N = 82$ shell gap far from stability has indeed been reported from radioactive beam experiments [43]. In the end, with accurate knowledge of the nuclear physics of the r-process and once the site of the r-process is identified, it might even be possible to use observed r-process abundance patterns to constrain the physics of the astrophysical site.

To illustrate the importance of nuclear data for an understanding of neutron capture processes it is useful to compare the r-process problem to the history of the s-process. The nuclei involved in the s-process are near stability and consequently a wide range of quite accurate experimental data has been accumulated over time. The site of the s-process was clearly identified early on through the discovery of unstable Tc in the envelopes of red giant stars [34]. Nevertheless the nature of the s-process in red giant stars remained unclear for a long time and only the details of the solar system abundance pattern together with decades of accurate laboratory measurements of neutron capture rates have finally confirmed the now widely accepted scenario of a sequence of He shell flashes [35]. These flashes produce a distinct sequence of neutron bursts that leads to the correct s-process flows at so called branching points where neutron capture rates and beta decay rates compete and therefore need to be accurately known. However, the exact stellar mixing process that leads to the formation of the ^{13}C pocket that is required to drive the main neutron source, the ^{13}C(α,n) reaction, is still not well understood. With ever increasing precision in neutron capture rate measurements it has now become possible to turn things around and use abundance observations in stars and presolar grains to constrain stellar mixing processes (for example [36]).

With the continuation of experimental r-process programs at existing radioactive beam facilities, and, most importantly, with the future availability of the Rare Isotope Accelerator (RIA), there is now a real opportunity to achieve similar progress in our understanding of the r-process. A large fraction of nuclei participating in the r-process will be reachable for experiments at RIA. The nuclear data obtained with RIA will therefore allow to extract critical experimental information about the r-process not only from the solar system abundances but also from the stream of observational data expected for r-process enhanced metal poor stars.

EXPERIMENTAL APPROACHES

Since the first measurement of the properties of a nucleus in the r-process (^{80}Zn [38], and ^{130}Cd [39] in 1986) around 30 r-process nuclei have been investigated experimentally mainly by measuring the ground state β-decay half-life (see Fig. 1). As Fig. 1 shows these are nuclei mostly around the $N = 50$ and $N = 82$ shell closures where the r-process path runs closer to stability. Currently a number of efforts are on the way at radioactive beam facilities worldwide to push the limits of experimental knowledge further into the path of the r-process. This includes the continuing program of decay studies of r-process nuclei at ISOLDE [40], most recently with measurements of the r-process nuclei ^{134,136}Sn [41], ^{138}Sn [42] and the decay spectroscopy of the r-process nucleus ^{130}Cd [43]. The latter result is of particular importance as the measured trend in isobaric mass differences indicates a reduction of the $N = 82$ shell gap far from stability, with important consequences for r-process modeling [44]. Decay spectroscopy of lighter r-process nuclei below Fe has been performed at GANIL [45] and is relevant for some isotopic anomalies found in meteorites that might originate from an r-process. In-flight fission of a ^{238}U beam has been demonstrated to be a powerful new production mechanism for r-process nuclei [46] and has been used at GSI to investigate decay properties of r-process nuclei in the Sb-Xe range [47].

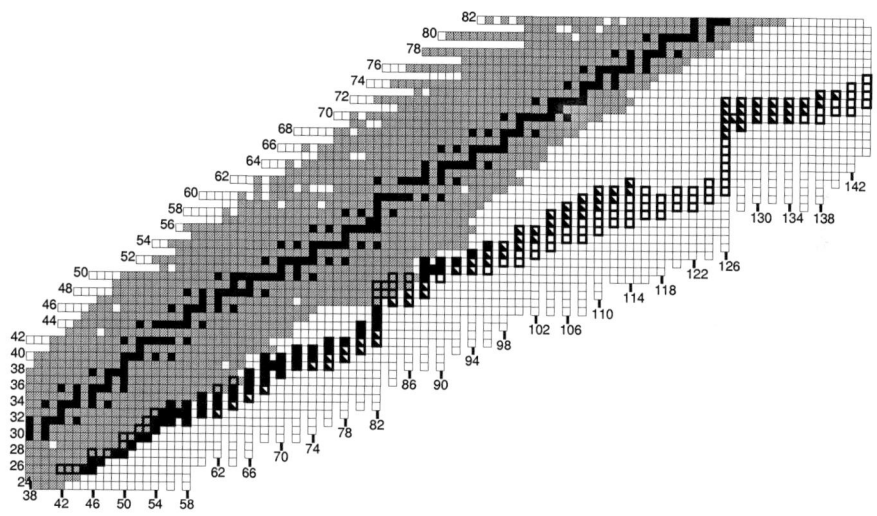

FIGURE 1. The path of the r-process on the chart of nuclides [20]. Framed are the nuclei that are major waiting points in the r-process flow and contribute significantly to the final abundances. Grey are nuclei for which at least a half-life is experimentally known, solid black are stable nuclei and r-process nuclei predicted to be within reach for at least a half-life measurement at the new NSCL Coupled Cyclotron Facility at MSU. Half-filled are r-process nuclei that are within experimental reach at the planned Rare Isotope Accelerator RIA.

A major step in the quest for experimental data on r-process nuclei has been the commissioning of the Coupled Cyclotron Facility at MSU's National Superconducting Cyclotron Laboratory (NSCL). Fig. 1 shows the reach of this facility for at least a half-life measurement. Clearly, a significant part of the r-process up to $A \approx 130$ is now within reach. First results from decay studies in the r-process at the NSCL are summarized below.

Mass measurements generally require higher beam intensities than decay studies. The masses of only 6 r-process nuclei are known experimentally, most of them near ^{132}Sn and all of them determined with the β endpoint technique, for example at Studsvik/OSIRIS [48] or ISOLDE [43]. In the future, high precision mass measurements with ion traps will be necessary to reduce systematic uncertainties. A first step in that direction has been made with the mass measurements of neutron rich fission products from a ^{252}Cf source with the Canadian Penning Trap at Argonne National Laboratory [49]. The results covering neutron rich nuclei in the range of Ba-Pr, still ≈ 8 mass units away from the r-process, show that some previous measurements indeed require significant revision.

In a complementary approach, fission fragment beams at GSI have been injected into the ESR storage ring to measure masses of extremely neutron rich nuclei using the isochroneous mass spectroscopy technique [50]. In a first experiment more than 40 new masses in the range of Ga-La have been determined and the r-process path was reached at Se [51].

Constraining neutron capture rates on nuclei far from stability poses still a greater challenge. One approach is the use of (d,p) neutron transfer reactions that can measure excitation energies and neutron spectroscopic factors that allow one to calculate more reliable neutron capture rates. First attempts to develop such a technique have been carried out at GANIL [52] and Oak Ridge [53]. The recent Oak Ridge experiments reach the main r-process path at ^{85}Se for the first time. Another possibility is Coulomb breakup of a radioactive beam that measures the rate of the (γ,n) reaction in the Coulomb field of a high Z nucleus. The detailed balance principle can then be used to constrain the neutron capture rate. This technique has been used successfully in several cases, for example the neutron capture on ^{14}C [54].

RECENT NSCL EXPERIMENTS

Recently a number of r-process related experiments have been carried out at the NSCL, taking advantage of the enhanced capabilities of the new Coupled Cyclotron Facility. The initial program focuses on β-decay properties. These experiments can be carried out with beam intensities as low as one nucleus per day, and allow one to obtain data on the most exotic nuclei within experimental reach today.

Radioactive beams of r-process nuclei are produced at the NSCL by fragmentation of stable, neutron rich beams at typical energies of 120 - 140 MeV/u. Important factors for the high exotic beam production capabilities of the facility are high primary beam currents, high beam energy that allows for the use of thicker production targets, and the large momentum acceptance of 5.5% of the A1900 superconducting fragment separator. For experiments with ^{136}Xe it becomes however necessary to restrict the momentum acceptance somewhat in order to block primary beam charge states. The A1900 produces a mixed radioactive beam that is transported to various experimental stations and typically contains on the order of a dozen species. Individual nuclei can then be identified event-by-event by measuring their momentum, charge, and velocity. Momenta are measured by tracking particles at the dispersive intermediate image of the A1900, charge numbers are obtained from energy loss measurements in a Si PIN diode, and velocities are determined from the time-of-flight between two plastic scintillators. For the β-decay experiments the identified nuclei are transported to the NSCL β counting system [55] and continuously implanted into a segmented (40x40 strips) double-sided silicon strip detector measuring time and location of the implantation. The same detector also registers the electrons emitted in the subsequent β-decay of the short lived nuclei allowing one to determine the decay time. With this setup decay properties of all species in the radioactive beam can be determined simultaneously.

The detector system can be surrounded with either the neutron detector NERO, or the γ-ray detector array SeGA. NERO is a neutron long counter using ^{3}He and BF$_3$ gas counters embedded in a polyethylene matrix that serves as a neutron moderator. The neutron detection efficiency is 25-37% for neutron energies up to 3 MeV. When used in combination with the β-counting system branchings for β-delayed neutron emission can be determined. With SeGA, typical γ-efficiencies of 5.3% at 1 MeV are reached allowing one to perform β decay-spectroscopy experiments with low secondary beam intensities.

In a first experiment we succeeded in measuring the half-life of the doubly-magic nucleus ^{78}Ni [56] - one of the last (besides ^{48}Ni) and most exotic of the 10 classical doubly-magic nuclei in nature with unknown properties. Before, just 3 ^{78}Ni nuclei had been identified in a pioneering experiment at GSI [57]. At the NSCL a total of 11 ^{78}Ni events were detected in 104 hours of beam time, using the fragmentation of a 15 pnA 140 MeV/u ^{86}Kr beam and taking advantage of the full momentum acceptance of the A1900. A half-life of 110^{+100}_{-60} ms was found, somewhat on the lower side compared to theoretical predictions typically ranging from around 100 ms [63] to 470 ms [60]. Together with the new half-life measurement for ^{77}Ni and the much improved values for ^{75}Ni and ^{76}Ni our measurements confirm that the trend of older more global models [60, 61] to overpredict half-lives in this region that is already seen for more stable nuclei continues into the path of the r-process. This is in line with more recent improvements of these models that all lead to a reduction of the predictions for Ni half-lives [62, 65]. Shell model calculations typically agree well with our measurements [63, 64]. However, the half-life of a nucleus is a quantity that depends on a number of nuclear structure features such as Q-value and the location of the beta-strength for various types of transitions in the daughter nucleus that can compensate each other. We are currently analyzing branchings for beta-delayed neutron emission, which will provide an additional constraint for the beta-strength above the neutron separation energy. This should allow to further test various model predictions and reveal further whether some of the agreements among models, and between models and experiment, are accidental.

The lower half-life for ^{78}Ni confirmed in our experiment leads to a substantial acceleration of the r-process due to the bottle-neck nature of ^{78}Ni. This will require significant readjustment of model parameters in r-process models that are characterized by a neutron capture flow through the $N = 50$ shell closure.

In two other NSCL experiments, fragmentation of a ^{136}Xe beam at 120 MeV/u was used to produce a beam of neutron rich Rh, Ag, and Pd nuclei close to the r-process path. One experiment used the β-counting station with SEGA to search for isomeric states and to perform β-decay spectroscopy. A first result was the measurement of the energy of the first 2^+ and 4^+ states in ^{120}Pd populated in the β-decay of ^{120}Rh [66]. In the other experiment the NERO detector was used to measure new β-decay half-lives and branchings for β-delayed neutron emission for neutron rich Tc, Rh, Ag, and Pd nuclei towards the r-process path [67].

SUMMARY

Experimental nuclear data on extremely neutron rich nuclei are needed together with advances in astronomy and theoretical astrophysics to answer the fundamental question of the origin of the heavy elements and the nature of the r-process. Tremendous progress is being made at existing radioactive beam facilities and experiments in the path of the r-process, albeit difficult, are now possible at least up to the Xe region. Together with the prospect of experiments with the planned Rare Isotope Accelerator (RIA) there is now for the first time the opportunity to address the nuclear physics side of the r-process problem in the coming decades.

ACKNOWLEDGMENTS

The author acknowledges support from NSF grants PHY 0110253 and PHY 0216783 (Joint Institute for Nuclear Astrophysics).

REFERENCES

1. J.J. Cowan, F.-K. Thielemann, and J.W. Truran, *Phys. Rep.*, , **208**, 267 (1991).
2. Y. Z. Qian, *Prog. Part. Nucl. Phys.*, **50**, 153 (2003).
3. Connecting Quarks with the Cosmos: Eleven Science Questions for the New Century (2003), Committee on the Physics of the Universe, National Research Council http://www.nap.edu/catalog/10079.html
4. S.E. Woosley and R.D. Hoffman *Ap. J.*, **395**, 202 (1992).
5. K. Takahashi, J. Witti and H.-Th. Janka, *A&A*, **286**, 857 (1994).
6. J.M. LeBlanc and J.R. Wilson *Ap. J.*, **161**, 541 (1970).
7. A.G.W. Cameron, *Ap. J.*, **562**, 456 (2001).
8. J.M. Lattimer et al. *Ap. J.*, **213**, 225 (1977).
9. S. Rosswog et al. *Astron. Astr.*, **341**,499 (1999).
10. R. Surman and G. C. McLaughlin, *Ap. J.*, **618**, 397 (2005).
11. Y.-Z. Qian, P. Vogel, and G. J. Wasserburg, *Ap. J.*, **506**, 868 (1998).
12. C. Wallner et al. *New Astr. Rev.*, **48**, 145 (2004).
13. K. Knie et al. *Phys. Rev. Lett.*, **93**, 171103 (2004).
14. Y.-Z. Qian, P. Vogel, and G. J. Wasserburg, *Ap. J.*, **494**, 285 (1998).
15. K.-L. Kratz et al. *Ap. J.*, **403**, 216 (1993).
16. C. Sneden et al. *Ap. J.*, **591**, 936 (2003).
17. J.W. Truran et al. *Publ. Ast. Soc. Pac.*, **114**, 1293 (2002).
18. R. Cayrel et al. *Nature*, **409**, 691 (2001).
19. J.J. Cowan et al. *Ap. J.*, **572**, 861 (2002).
20. H. Schatz et al. *Ap. J.*, **579**, 626 (2002).
21. S. Goriely and M. Arnould *Astron. Astr*, **379**, 1113 (2001).
22. B. Pfeiffer, U. Ott, and K.-L. Kratz *Nucl. Phys. A*, **688**, 575 (2001).
23. C. Travaglio et al. *Ap. J.*, **601**, 864 (2004).
24. D. Argast et al. *Astron. Astrophys.*, **416**, 997 (2004).
25. http://www.jinaweb.org/sdss2/
26. C. Arlandini et al. *Ap. J.*, **525**, 886 (1999).
27. I. V. Panov et al. astro-ph/0412654
28. P.A. Seeger, W.A. Fowler, and D.D. Clayton *Ap. J. Suppl.*, **11**, 121 (1965).
29. C. Freiburghaus et al. *Ap. J.*, **516**, 381 (1999).
30. Y.-Z. Qian et al. *Phys. Rev. C*, **55**, 1532 (1997).
31. Y.-Z. Qian *Ap. J.*, **569**, L103 (2002).
32. E. Kolbe, K. Langanke, and G.M. Fuller *Phys. Rev. Lett.*, **92**, 111101 (2004).
33. E. M. Burbidge, G. R. Burbidge, W. A. Fowler, and F. Hoyle, *Rev. Mod. Phys.*, **29**, 547 (1957).
34. P. W. Merrill, *Science*, **111** (1952) 484.
35. R. Gallino et al. *Ap. J.*, **497**, 388 (1998).
36. R. Reifarth et al. *Ap. J.*, **614**, 363 (2004).
37. M. Lugaro et al. *Ap. J.*, **593**, 486 (2003).
38. B. Ekstrom et al. *Phys. Scr.*, **34**, 614 (1986).
39. K.-L. Kratz et al. *Z. Phys. A*, **325**, 489 (1986).

40. K.-L. Kratz et al. *Hyperf. Int.* , **129**, 185 (2000).
41. J. Shergur et al. *Phys. Rev. C* , **65**, 4313 (2002).
42. K.-L. Kratz, private communication 2004.
43. I. Dillmann et al. *Phys. Rev. Lett.* , **91**, 2503 (2003).
44. B. Chen et al. *Phys. Lett. B* , **355**, 37 (1995).
45. O. Sorlin et al. *Nucl. Phys. A* , **719**, C193 (2003).
46. C. Donzaud et al. *EPJ A* , **1**, 407 (1998).
47. P. Santi et al. APS 2002 Fall Meeting of the DNP, abstract DE009.
48. B. Fogelberg et al. *Phys. Rev. Lett.* , **82**, 1823 (1999).
49. J. A. Clark et al. *Nucl. Phys. A* **746**, 342c (2004).
50. J. Stadlmann et al. *Phys. Lett. B* , **586**, 27 (2004).
51. M. Matos and C. Scheidenberger, private communication (2004).
52. O. Sorlin, private communication (2004).
53. J. S. Thomas et al. *Europ. Phys. Journ. A*, **25**, 371 (2005).
54. A. Horvath et al. *Ap. J.* , **570**, 926 (2002).
55. J.I. Prisciandaro et al. *Nucl. Intrum. Methods A* , **505**, 140 (2003).
56. P. Hosmer et al. *Phys. Rev. Lett.*, **94**, 112501 (2005).
57. M. Bernas et al. *Phys. Lett. B* , **415**, 111 (1997).
58. G. Audi *et al.* Nucl. Phys. A, **729**, 3 (2003).
59. J. Engel et. al. *Phys. Rev. C* , **60**, 014302 (1999).
60. P. Möller, J. R. Nix, and K.-L. Kratz, *At. Data Nucl. Data Tab.* , **66**, 131 (1997).
61. I. N. Borzov, S. Goriely, and J.M. Pearson, *Nucl. Phys. A* , **621**, 307c (1997).
62. P. Möller, B. Pfeiffer and K.-L. Kratz, *Phys. Rev. C* , **67**, 055902 (2003).
63. K. Langanke and G. Martinez-Pinedo *Rev. Mod. Phys.* , **75**, 819 (2003).
64. A. F. Lisetskiy, B. A. Brown, and M. Horoi, *Europ. Phys. Journ.*, **25**, 95 (2005), and A. F. Lisetskiy private communication.
65. I. N. Borzov, *Phys. Rev. C*, **71**, 5801 (2005).
66. W. B. Walters et al. *Phys. Rev. C*, **70**, 4314 (2004).
67. F. Montes et al. submitted.

Magnetic Dipole and Gamow-Teller Modes in Neutrino-Nucleus Reactions: Impact on Supernova Dynamics and Nucleosynthesis

P. von Neumann-Cosel*, T. Adachi†, A. Byelikov*, H. Fujita**, Y. Fujita†,
A. Heger‡, E. Kolbe§, K. Langanke¶, G. Martínez-Pinedo‖, A. Richter*,
A. Shevchenko* and Y. Shimbara†

*Institut für Kernphysik, Technische Universität Darmstadt, D-64289 Darmstadt, Germany
†Department of Physics, Osaka University, Toyonaka, Osaka 560-0043, Japan
**School of Physics, University of the Witwatersrand, Johannesburg 2050, South Africa
‡Theoretical Astrophysics Group, Los Alamos National Laboratory, Los Alamos, NM 87545, U.S.A
§Institut für Physik, Universität Basel, CH-4056 Basel, Switzerland
¶Gesellschaft für Schwerionenforschung, D-64291 Darmstadt, Germany
‖ICREA and Institut d'Estudis Espaicals de Catalunya, E-08034 Barcelona, Spain

Abstract. Some aspects of the importance of neutrino-induced reactions on nuclei within supernova physics are discussed. It is argued that important constraints on the experimentally unknown cross sections can be obtained from experimental studies of the nuclear response in selected cases. Examples are neutral-current induced reactions on fp-shell nuclei extracted from high-resolution inelastic electron scattering data providing the $M1$ strength distributions and the production of the exotic heavy, odd-odd nuclei ^{138}La and ^{180}Ta through charged-current reactions dominated by Gamow-Teller transitions. The Gamow-Teller strength can deduced from the (^{3}He,t) charge-exchange reaction at zero degree.

Keywords: Neutrino-induced reactions on nuclei, $M1$ and GT strengths, supernova dynamics, neutrino nucleosynthesis

INTRODUCTION

Inelastic neutral-current neutrino-nucleus scattering plays an important role in many astrophysical applications, including r-process nucleosynthesis [1, 2], the synthesis of certain elements like ^{10,11}B and ^{19}F during a supernova explosion by the v-process [3, 4] or for the detection of supernova neutrinos (see e.g. [5]). Particular relevance is expected in aspects of supernova physics [6, 7, 8]: *i)* for the neutrino opacities and thermalization during the collapse phase, *ii)* for the revival of the stalled shock in the delayed explosion mechanism, and *iii)* for explosive nucleosynthesis. In case *i)* neutrinos are produced by electron capture on nuclei and have progressively larger energies (up to 20 – 30 MeV) as the collapse proceeds. For cases *ii)* and *iii)* the neutrino spectra are approximately thermal with average energies $E_v = 15$ MeV (for v_e neutrinos) and 25 MeV (for v_μ and v_τ neutrinos).

Although inelastic neutrino-nucleus scattering is not yet considered in supernova simulations, their importance has been pointed out, in particular for nuclei in the iron ($A \approx 56$) mass range [7]. Except for the ground state transition to the $J = 1^+$, $T = 1$ state in ^{12}C at an excitation energy $E_x = 15.11$ MeV [9], currently no data for inelastic neutrino-nucleus scattering is available. However, for some iron-region nuclei precise data on the magnetic dipole ($M1$) strength distributions exist. It is shown that these data supply to a large extent the required information about the nuclear Gamow-Teller (GT) distribution and hence determine the inelastic neutrino-nucleus cross sections for supernova neutrino energies. It is also demonstrated that large-scale shell-model calculations agree quite well with the precise $M1$ data, thus validating the use of such models to determine the required cross sections for nuclei where no data exist, or at the finite-temperature conditions in a supernova.

Neutrino-induced reaction can make important contributions to the nucleosynthesis of specific isotopes [3]. Prime examples comprise ^{11}B and ^{19}F. These nuclides are produced by neutral-current reactions on ^{12}C and ^{20}Ne in the carbon and neon/oxygen shells of massive stars during supernova explosion, likely dominating over other sources for these isotopes. Since the weak interaction cross sections are so small, ν-nucleosynthesis can be effective only for isotopes with a large abundance difference of several orders of magnitude between mother and daughter nuclei. In general, neutral-current neutrino reactions of the type $(\nu, \nu'x)$, where x can be a neutron or a light charged particle, have been conjectured to be most important, while charged-current reactions contribute little to neutrino nucleosynthesis. However, this conclusion was based on very simplifying assumptions [3] for the neutrino-induced cross section for the heavier nuclides.

Recently, a refined analysis was performed which considerably changed the picture of the production of the rare, odd-odd isotopes ^{138}La and ^{180}Ta [4]. This study followed the neutrino nucleosynthesis of the light (e.g. ^{11}B, ^{19}F) and heavy (e.g. ^{138}La, ^{180}Ta) candidate nuclei in a self-consistent way in complete stellar evolution models that included the evolution of all isotopes up to bismuth from the time the star ignited central hydrogen burning through the supernova explosion [12]. The study of [4] also used improved cross sections for the neutrino-induced reactions on the key nuclides. Here, the neutrino-induced excitation functions for neutral- and charged-current reactions on ^{181}Ta, ^{180}Hf, ^{139}La, and ^{138}Ba were derived on the basis of RPA calculations [13].

These calculations predict that ^{138}La is made almost exclusively by the charged-current reaction ^{138}Ba$(\nu_e,e^-)^{138}$La (see also [14]). For ^{180}Ta the situation is slightly more complicated. Recent work has demonstrated that despite its isomeric nature s-process production cannot be excluded [15] and it may essentially account for close to 100 % of the solar abundance [16]. Also the γ-process has been identified as a possible source [17, 18]. The present results predict about half of the solar ^{180}Ta abundance to be made by the γ-process and half by the charged current reaction ^{180}Hf$(\nu_e,e^-)^{180}$Ta. Because the relevant neutrino energies are rather low the nuclear response is dominated again by the GT strength. The GT response can extracted from charge-exchange reactions at zero degree and we have set out to determine it in high-resolution (^{3}He,t) experiments on ^{138}Ba and ^{180}Hf populating ^{138}La and ^{180}Ta.

NEUTRAL-CURRENT INELASTIC NEUTRINO-NUCLEUS SCATTERING AND $M1$ RESPONSE IN NUCLEI

Because of the limited space in this volume, the interested reader is referred to Ref. [10], where the results of this study are presented (see also [11]).

NEUTRINO NUCLEOSYNTHESIS AND GT STRENGTH IN ^{138}LA AND ^{180}TA

As pointed out in the introduction, the latest calculations (see also [14]) predict the charged-current reaction ^{138}Ba(ν_e, e^-) to be the origin of ^{138}La and, at least partly, also of ^{180}Ta. While these results represent a considerable improvement, a large uncertainty in the quantitative estimates remains from the use of rather schematic RPA calculations to describe the response of the most important excitations, viz. GT and spin-dipole (SD) modes. Besides the need to introduce a quenching factor [13], a realistic description of the strength distributions requires the inclusion of complex configurations beyond the $1p-1h$ excitations considered in RPA. These correlations will fragment the GT strength more than described within RPA results and can in fact shift some strength over the relevant particle thresholds so that it will no longer contribute to the nucleosynthesis process. For example, excitation of GT transitions above the neutron threshold in ^{138}La at 7.5 MeV excitation energy will produce ^{137}La rather than ^{138}La.

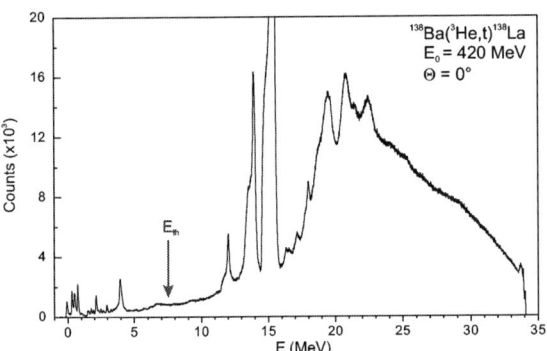

FIGURE 1. Spectrum of the ^{138}Ba$(^3$He,$t)$ reaction at $\Theta = 0° - 0.5°$. Above $E_x = 10$ MeV indicated by the vertical dashed line, the spectrum is suppressed by a factor of 10. There, the excitations are dominantly caused by background reactions on C and O due to the use of a (^{138}Ba+PVA) target. The neutron emission threshold in ^{138}La is indicated by an arrow.

To study this problem, a high-resolution measurement of the GT strengths in ^{138}La and ^{180}Ta using the ^{138}Ba$(^3$He,$t)$ and ^{180}Hf$(^3$He,$t)$ reactions at zero degrees has been performed recently at RCNP, Osaka. The properties of the $(^3$He,$t)$ reaction at $E = 140$ MeV/nucleon as a tool to measure detailed GT distributions have been well established in recent years, in which the proportionality of the zero degree cross sections has been established for the transitions with $B(GT) \geq 0.04$ from the comparison with the GT strengths from the isospin-analogous β decays [20, 21]. Utilizing recent advances in dispersion matching techniques [22], an energy resolution up to $\Delta E \approx 50$ keV (FWHM) in heavy nuclei can be realized. Figure 1 presents a spectrum for the the ^{138}Ba$(^3$He,$t)$ reaction at $\Theta = 0° - 0.5°$. Because barium is extremely oxydizing, the enriched ^{138}Ba material was dissolved in polyvenylalcohol (PVA) following the procedure described in [19]. This causes considerable background from C- and O-induced background reactions which completely dominate the spectrum for excitation energies above 10 MeV. However, we take advantage here of the large Q-value difference between the ^{138}Ba(^{3}He,t) and ^{12}C,^{16}O(^{3}He,t) reactions, respectively. Astrophysically, only the region below the neutron emission threshold (marked by an arrow in Fig. 1) is relevant because at higher excitation energies ^{137}La is produced, which decays to ^{137}Ba. In this energy range only two well-known transitions from the ^{13}C and ^{15}N target components contribute to the spectrum. Below $E_x \simeq 3.5$ MeV, resolved transitions are visible while the GT strength distribution in ^{138}La is smooth at higher energies.

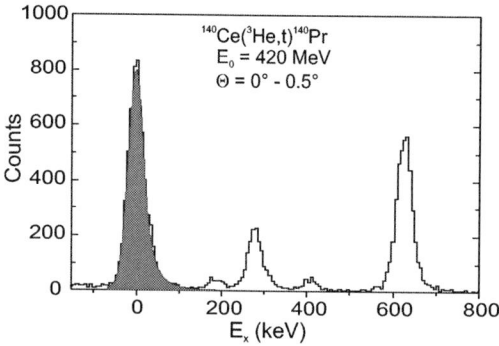

FIGURE 2. Spectrum of the ^{140}Ce(^{3}He,t) reaction at $\Theta = 0° - 0.5°$ and low excitation energies. A fit to determine the line content of the g.s. transition is indicated.

For an extraction of absolute GT strengths a normalization to known weak GT decays is necessary. The accumulated data indicate that this normalization is a smooth function of mass number A [23]. The normalization is achieved for the case of the ^{138}La by a measurement of the g.s. transition in the (^{3}He,t) reaction on the neighboring isotone ^{140}Ce. Since the final nucleus ^{140}Pr has a g.s. J^π value of 1^+, the $B(GT)$ value of the well measured EC decay [24] from this nucleus can be used. The result is presented in Fig. 2, where the low energy region in ^{140}Pr is displayed. Although the level density in such an odd-odd nucleus is large even at very low excitation energies (about 25 levels are known below 1 MeV [24]), the selectivity of the reaction leads to a limited number of observed transitions. The g.s. is well separated from excited states.

A preliminary $B(GT)$ strength distribution in ^{138}La is compared in Fig. 3 with that used in the stellar evolution calculations [4]. Up to $E_x = 3.2$ MeV, individual transitions are resolved; above the strength is shown in 100 keV bins. Clearly, the data exhibit a much more fragmented strength than predicted by the RPA results. This is particularly pronounced at excitation energies above 4 MeV. The astrophysical consequences are presently analyzed.

ACKNOWLEDGMENTS

The (^{3}He,t) experiments have been performed in a Cape Town / Darmstadt / Gent / Johannesburg / Münster / Osaka collaboration. I am particulary grateful for the contributions of T. Adachi, N. Botha, P. von Brentano, A. Byelikov, D. De Frenne, D. Frek-

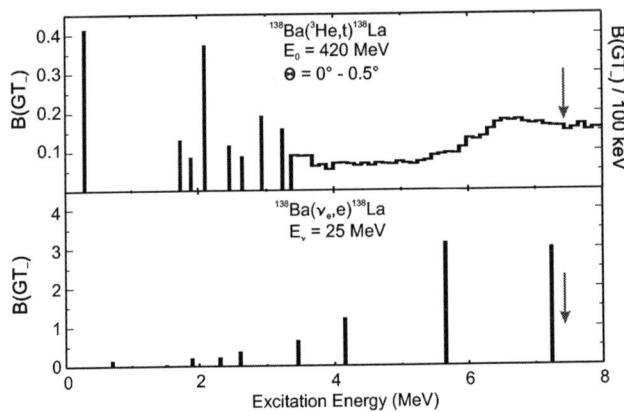

FIGURE 3. Comparison of experimental *GT* strength distribution in ^{138}La up to $E_x = 10$ MeV deduced from the present experiment with that used in the stellar evolution calculation calculations of [4].

ers, H. Fujita, Y. Fujita, E. Jacobs, A. Negret, N. Popescu, S. Rakers, A. Richter, A. Shevchenko, Y. Shimbara, and F.D. Smit. The astrophyiscal aspects are analyzed in collaboration with A. Heger, E. Kolbe, K. Langanke, and G. Martínez-Pinedo. This work has been supported by the DFG under contracts SFB 634 and 446 JAP 113/267/0-1.

REFERENCES

1. G. M.Fuller, and B. S.Meyer, *Astroph. J.*, **452**, 792 (1995).
2. W. C. Haxton, K. Langanke, Y. Z. Qian, and P. Vogel, *Phys. Rev. Lett.* **78**, 2694 (1997).
3. S.E. Woosley, D.H. Hartmann, R.D. Hoffman, and W.C. Haxton, *Astroph. J.*, **356**, 272 (1990).
4. A. Heger, et al., *Phys. Lett. B*, **606**, 258 (2005).
5. E. Kolbe, K. Langanke, G. Martínez-Pinedo, and P. Vogel, *J. Phys. G*, **29**, 2569 (2003).
6. S.W. Bruenn and W.C. Haxton, *Astroph. J.*, **376**, 678 (1991).
7. W.R. Hix, A. Mezzacappa, O.E. Bronson Messer, and S.W. Bruenn, *J. Phys. G*, **29**, 2523 (2003).
8. K. Langanke and G. Martínez-Pinedo, *Rev. Mod. Phys.*, **55**, 819 (2003).
9. L.B. Auerbach et al., *Phys. Rev. C*, **64**, 066501 (2001).
10. K. Langanke, G. Martínez-Pinedo, P. von Neumann-Cosel, and A. Richter, *Phys. Rev. Lett.*, **93**, 202501 (2004).
11. A. Juodagalvis, et al., *Nucl.Phys. A*, **747**, 87 (2005).
12. T. Rauscher, A. Heger, R.D. Hoffman, and S.E. Woosley, *Astroph. J.*, **576**, 323 (2002).
13. E. Kolbe, K. Langanke, and P. Vogel, *Phys. Rev. C*, **50**, 2576 (1994; *Nucl. Phys. A*, **652**, 91 (1999).
14. S. Goriely, M. Arnould, I. Barzov, and M. Rayet, *Astron. Astrophys.*, **375**, L35 (2001).
15. D. Belic et al., *Phys. Rev. Lett.*, **83**, 5242 (1999).
16. K. Wisshak et al., *Phys. Rev. Lett.*, **87**, 251102 (2001).
17. S.E. Woosley and W.M. Howard, *Astroph. J. Suppl.*, **36**, 285 (1978).
18. M. Rayet et al., *Astron. Astrophys.*, **298**, 517 (1995).
19. Y. Shimbara et al., *Nucl. Instrum. Meth. A*, **522**, 205 (2004).
20. Y. Fujita et al., *Phys. Rev. C*, **59**, 90 (1999).
21. Y. Fujita et al., *Phys. Rev. C*, **67**, 064312 (2003).
22. H. Fujita et al., *Nucl. Instrum. Meth. A*, **484**, 17 (2002).
23. Y. Fujita, private communication.
24. L.K. Peker, *Nucl. Data Sheets*, **73**, 261 (1994).

NUCLEAR STRUCTURE
(HIGH SPIN EXPERIMENTS)

Hyperdeformed rotational bands observed in the actinide region

A. Krasznahorkay*, M. Csatlós*, Y. Eisermann†, T. Faestermann**,
G. Graw†, J. Gulyás*, D. Habs†, M.N. Harakeh‡, R. Hertenberger†,
M. Hunyadi§, H.J. Maier†, Z. Máté*, O. Schaile†, P.G. Thirolf† and
H.J. Wirth†

*Institute of Nuclear Research of the Hungarian Academy of Sciences, P.O. Box 51, H-4001 Debrecen, Hungary
†Department f. Physik, Universität München, D-85748 Garching, Germany
**Technische Universität München, D-85748 Garching, Germany
‡Kernfysisch Versneller Instituut, 9747 AA Groningen, The Netherlands
§Institute of Nucl. Phys. **A671** (2000) 119. Nuclear Research of the Hungarian Academy of Sciences, P.O. Box 51, H-4001 Debrecen, Hungary

Abstract.
The fission probability of ^{234,236}U as a function of the excitation energy has been measured with high energy resolution using the $^{233,235}U(d,pf)$ reactions in order to study hyperdeformed (HD) rotational bands. Rotational band structures with a moment of inertia of $\Theta = 238 \pm 42$ $\hbar^2$/MeV and $\Theta = 217 \pm 38$ $\hbar^2$/MeV have been observed for ^{234,236}U, respectively, corresponding to hyperdeformed configurations. From the level density of the rotational bands the excitation energy of the ground state in the third minimum was determined to be $E_{III} = 3.1 \pm 0.4$ MeV and $E_{III} = 2.7 \pm 0.4$ MeV, for ^{234}U and for ^{236}U, respectively. The excitation energy of the lowest hyperdeformed transmission resonance and the energy dependence of the fission isomer population probability enabled the determination of the height of the inner fission barrier of ^{236}U $E_A = 5.15 \pm 0.20$ MeV and its curvature parameter $\hbar\omega_A = 1.2$ MeV. Using this new method the long-standing uncertainties in determining the height of the inner potential barrier in uranium isotopes could be resolved.

PACS: Pw21.10.Re, 24.30.Gd, 25.85.Ge, 27.90.+b

INTRODUCTION

The investigation of the superdeformed (SD, ratio of long to short axis 2:1) and hyperdeformed (HD, axis ratio 3:1) states started with the discovery of fission isomers and the study of their properties in the actinides. Later, it was continued with the search for SD and HD states in high-spin γ-spectroscopy [1, 2].

Recently, very effective, high resolution 4π γ-ray spectrometers like EUROBALL and GAMMASPHERE have been developed for studying these highly deformed states, which is one of the most vital fields in modern nuclear structure physics. Using these arrays many SD bands have already been studied, but no firm evidence could be obtained for any HD bands [3].

The theoretical description of fission isomerism is based on the introduction of shell corrections [4] to the smooth liquid drop potential energy surface, resulting in a double humped fission barrier with a second minimum containing SD states. Moreover, in the actinide region a third minimum in the potential energy (which contains HD states) was predicted already more than twenty years ago by Möller et al. [5]. According to recent calculations, in these nuclei the so-called third minimum of the potential barrier appears with deformation parameters $\beta_2 \approx 0.90$ and $\beta_3 \approx 0.35$ [4] and the depth is predicted to be much larger ($\Delta E \approx 3$ MeV [6]) than believed earlier [7].

The aim of our experiments was to resolve the structure of the vibrational resonances in the third minimum, to determine the J and K values as well as the fission decay properties of the HD excited states.

EXPERIMENTAL SETUP

The γ-spectroscopic studies of the SD states turned out to be very difficult in the actinide region because of the low cross sections and the high background produced by the fission fragments. However, at higher excitation energy sub-barrier transmission resonances appear in the fission probability at the position of quasi-bound states in the second and third minima of the potential barrier. By measuring the fission probability as a function of the excitation energy one can map these SD and HD states. Early attempts of studying these transmission resonances critically suffered from either a limited energy resolution [8] or statistical significance [9].

The first experimental evidence for HD states was found in a high resolution study of the transmission resonances in ^{236}U [10].

In this work we used (d,pf) reaction to determine the fission probability as a function of the excitation energy. The energy of the outgoing protons was analyzed with high energy resolution in coincidence with the fission fragments.

The experiments were carried out at the Munich Tandem Accelerator Laboratory using enriched (99 %) ≈ 30 μg/cm^2 thick ^{233}U (99.89 %) and 88 μg/cm^2 ^{235}U$_2$O$_3$ targets and deuteron beam energies of $E_d = 12$ MeV and 9.73 MeV, respectively.

The energy of the outgoing protons was analyzed by a Q3D magnetic spectrograph [11], which was set at $\Theta_L = 130^o$ and $\Theta_L = 125^o$ relative to the incoming beam axis and the solid angle was 10 msr [12]. The position of the analyzed particles in the focal plane was measured with a light-ion focal-plane detector of 1.8 m active length [13] and with a position sensitive light-ion focal-plane detector with individual cathode strip readout of 890 mm active length [14]. A line-width of ≤ 3 keV has been observed for elastic scattering of 20 MeV deuterons.

The fission fragments were detected by two position sensitive avalanche detectors (PSAD) [15, 16] consisting of two wire planes (with delay-line read-out), which determined the horizontal and vertical positions of the incident particles. The solid angle of the detectors was 30 % of 4π in the case of ^{234}U(d,pf) [12] and 4 % of 4π in the ^{236}U(d,pf) [17] experiment.

HYPERDEFORMED STATES IN ^{234}U

In order to investigate the HD bands the excitation energy was chosen between the energy of the inner and outer barriers of the second minimum. In this energy range the widths of the SD resonances in the second minimum should be much broader than those of the HD states due to the strong coupling to the normal deformed states. The widths of the HD states due to the higher outer barriers of the third minimum remain below the actual experimental resolution of ~5 keV.

Part of the proton spectrum measured in coincidence with the fission fragments is shown in Fig 1a).

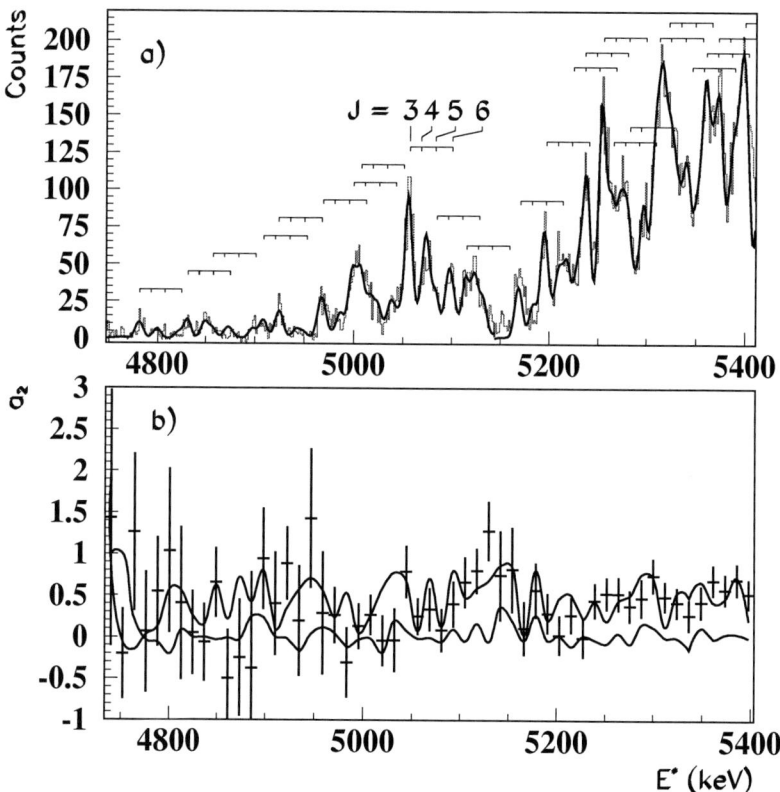

FIGURE 1. a) Part of the measured proton energy spectrum as a function of the excitation-energy for ^{234}U. The superimposed solid line shows a fit to the data using 24 rotational bands with common rotational parameter. b) Experimental fission-fragment angular-correlation coefficients (a_2) compared to the calculated ones using K=1 (upper curve) and K=3 (lower curve) for all of the bands.

Assuming overlapping rotational bands with the same moment of inertia, inversion parameter and intensity ratio for the members in a band, we fit our spectrum using simple Gaussians for describing the different band members in the same way as we did it previously [10]. The result of the fit is also shown in Fig. 1a).

Fission-fragment angular distributions were generated as a function of the excitation energy, normalized to the known (d,f) angular distribution and fitted with even Legendre polynomials (LP) up to fourth order. The a_2 angular distribution coefficient is shown in Fig. 1b) as a function of the excitation energy. In order to get information on the J and K values of the observed rotational bands, or to check our assumptions made for fitting the energy spectrum, the angular distribution coefficients of the fission fragments have been calculated and compared to the experimental ones. For more details see ref. [12].

HYPERDEFORMED STATES IN ^{236}U

The resonances at 5.27, 5.34 and 5.43 MeV had been previously identified as hyperdeformed resonances[10], however without resolving their rotational structure. This was achieved recently[17]. The measured high-resolution fission probability in terms of the excitation energy of the compound nucleus ^{236}U is shown in Fig. 2a,b). It was obtained by dividing the proton energy spectrum measured in coincidence with fission fragments by the smoothly varying proton spectrum from the (d,p) reaction.

The excitation energy region containing HD resonances was analyzed in two steps, beginning with the resonance structure above 5.2 MeV as shown in Fig. 2 b). We fit our experimental results with overlapping rotational bands with the same moment of inertia and intensity ratio for the members in a band [10, 12].

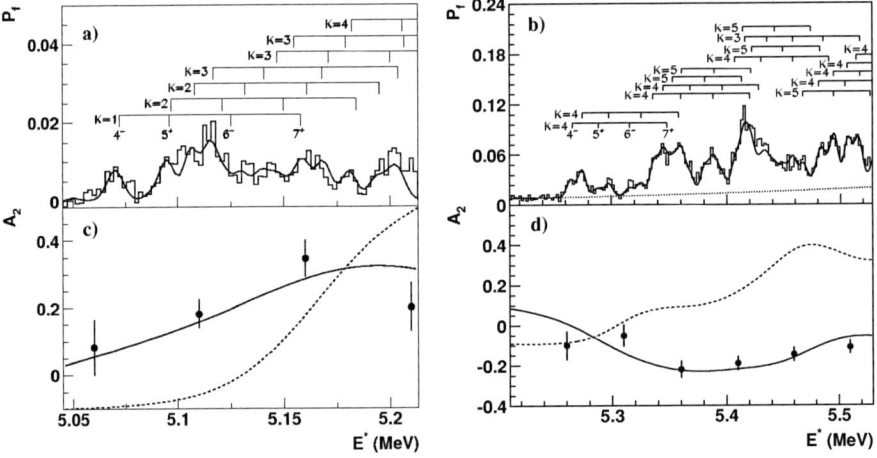

FIGURE 2. a,b)Fission probability (P_f) as a function of the excitation-energy for ^{236}U. The superimposed solid line shows a fit to the data using rotational bands. The picket fence structure s indicate the position of the band members used in the fit with K values as indicated for each band by left-sided numbers. The best fit has been obtained by using hyperdeformed rotational bands. c,d) Consistent description of the angular-correlation coefficient A_2 from Ref.[18] (data points) with the fit function from panel a,b) based on the distribution of K values as indicated in a,b).

Assuming alternating parity bands with J=K=3,4 and 5 band heads the fission probability between 5.2 and 5.5 MeV was fitted, resulting the fit curve superimposed on the data in Fig. 2b). The picket fence structures indicate the position of the rotational band members, whilst the numbers to the left denote the corresponding K values. With the series of 15 rotational bands shown in Fig. 2b) the angular correlation coefficients A_2 measured by Just et al. [18], data points in Fig 2d) also could be reproduced. The assumption of SD rotational bands with J=$2^+,4^+,6^+$ failed to reproduce the data as indicated by dashed line in Fig. 2d).

We deduced a rotational parameter of $\hbar^2/2\theta = 2.3 ^{+0.3}_{-0.5}$ keV and a moment of inertia of $\theta = 217 \pm 38\ \hbar^2/MeV$ from the data. This agrees nicely with the rotational parameter we obtained previously for ^{236}U ($\hbar^2/2\theta = 1.6 ^{+1.0}_{-0.4}$ keV)[10] and also with the values calculated by Shneidman et al.[19] who assumed dinuclear systems, suggesting the possibility of an exotic heavy clustering as predicted also by Ćwiok et al.[6] before.

In the second step of the analysis, the proton energy spectrum below 5.2 MeV was investigated, where no conclusive high-resolution data were so far available. The result of the analysis in the excitation energy region between 5.05 MeV and 5.2 MeV is shown in Fig.2 a), where the prompt fission probability is displayed together with the results of a fit by rotational bands similar to the procedure described above for the upper resonance region.

Assuming overlapping rotational bands in the second minimum with a spin sequence of $J^\pi = 2^+\text{-}4^+\text{-}6^+\text{-}8^+$ and a typical rotational parameter for superdeformed bands $\hbar^2/2\theta \approx 3.3$ keV, the experimental data could not be reproduced. The best fit to the data was obtained in the case of $\hbar^2/2\theta \approx 2.2$ keV, which suggested a hyperdeformed nuclear shape also with this assumption.

The rotational parameter derived from the best fit could be determined as $\hbar^2/2\theta = 2.4 ^{+0.6}_{-0.3}$ keV, corresponding to a hyperdeformed configuration. This result is in contrast to the old assumption that the decaying vibrational excitations originate from the superdeformed second minimum. The occurrence of a transmission resonance in the third minimum at 5.1 MeV requires a rather complete damping and an inner-barrier height E_A lower than 5.2 MeV.

ON THE DEPTH OF THE THIRD MINIMUM

The possibility of a complete spectroscopy in the hyperdeformed third minimum may open up the determination of the excitation energy of the ground state in the third minimum, i.e. the depth of the third minimum, exploiting the level density information. The density of $J = 3$ states in ^{234}U has been determined from our experimental data and calculated as a function of the excitation energy using the back-shifted Fermi-gas description with parameters determined by Rauscher et al. [20].

In order to estimate the depth of the third minimum we compared the experimentally obtained and calculated values. We assumed that the same parameterization of the level density formula is valid in the third minimum, as it was determined by Rauscher et al. [20] by fitting the level densities in the first minimum and which we have already checked also in the second minimum in the case of ^{240}Pu (for more details see ref. [21]). The result of the comparison is shown in Fig. 3a). From the comparison we found a value of 3.1 MeV for the energy of the ground state in the third minimum.

In order to get some estimate for the precision of the level distance analysis described

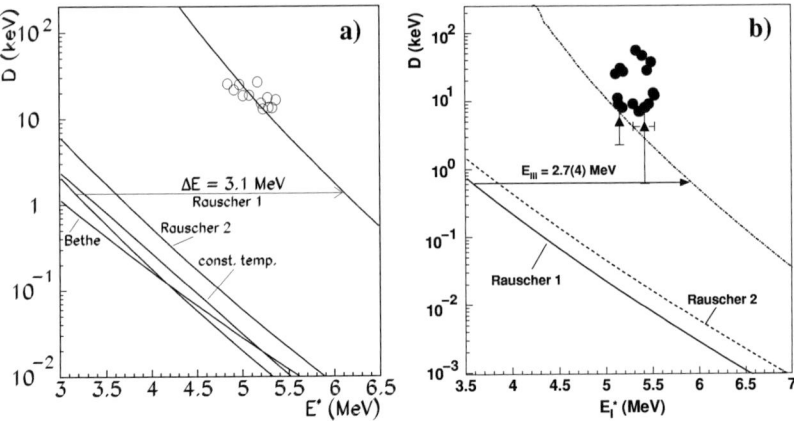

FIGURE 3. a) Average distances of the $J = 3$ levels in the third minimum of ^{234}U as a function of the excitation energy. The solid curves show calculated values by different formulas (see text for details), the circles correspond to experimental values. b) Distances of the J=5 spin states in the third minimum of ^{236}U as a function of the excitation energy.derived from the two resonance region shown in Fig. 2a),b). The solid curve ('Rausher 1') shows values calculated by parametrization of Rauscher et al.. The full points correspond to our experimental values. The dashed curve represents a calculation using the experimental rotational parameter instead of the rigid-rotor value.('Rausher 2'). The triangles indicate the average level density in the two resonance regions, corrected for unobserved J=5 states due to the K-dependent filtering bye the fission barrier. The dashed-dotted curve originated from shifting the curve labeled 'Rausher 1' through the corrected experimental level density data.

above we repeated the calculation of level distances by using the rotational parameter deduced in the present work. We used two other formulas to estimate the level distances, which were parameterized by von Egidy et al. [22]. From the uncertainties of the calculated and measured level distances the error of the energy determination was estimated to be 0.4 MeV.

Similarly to the case of ^{234}U, we determined the depth of the third minimum of ^{236}U by comparing the experimentally obtained level spacings of the J=5 members of the rotational bands of Fig. 2a) and 2b) with the calculated ones (Fig. 3b). The uncertainty of E_{III} composed of the uncertainty obtained from our analysis and the uncertainty introduced by the level density parametrization as quoted by the authors in ref. [20] was estimated to be ± 0.4 MeV.

Ćwiok et al. [6] predicted two different HD minima for both ^{234}U and for ^{236}U with very different β_λ ($\lambda = 3 - 7$) values. One of them had an octupole deformation parameter of $\beta_3 \approx 0.4$ and a minimum of $E_{III} = 3.5$ MeV and 3.8 MeV, respectively. The other was more reflection-asymmetric and had an octupole deformation parameter of $\beta_3 \approx 0.6$ and a minimum of $E_{III} = 2.7$ MeV and 2.4 MeV, respectively. The experimental values of $E_{III} = 3.1 \pm 0.4$ MeV obtained for ^{234}U and $E_{III} = 2.7 \pm 0.4$ MeV obtained in the present work were between the two predicted values with an error bar, which overlaps with both theoretical values.

CONCLUSION

In summary, we have measured the fission probability of ^{234}U and ^{236}U as a function of excitation energy with high energy resolution using the (d,pf) reaction. The rotational parameter obtained from fitting the energy spectrum was found to be $\hbar^2/2\theta = 2.1 \pm 0.2$ keV and 2.3 ± 0.4 keV for ^{234}U and ^{236}U, respectively, which is characteristic for the hyperdeformed nuclear shape. The level densities of the most strongly excited $J = 3$ states in ^{234}U and J=4 states in ^{236}U were compared to the prediction of the back-shifted Fermi-gas formula and the energy of the ground state in the third minimum was estimated to be $E_{III} = 3.1 \pm 0.4$ MeV and $E_{III} = 2.7 \pm 0.4$ MeV for ^{234}U and ^{236}U, respectively. These values agree well with the predicted ones [6, 19].

The hyperdeformed rotational-band structure observed in the rather low excitation energy region around 5.1 MeV independently supports our experimental finding of a rather deep third minimum, which is in agreement with theoretical predictions. We furthermore used the lowest transmission resonance in the third minimum to determine the height of the inner barrier E_A. In this way, we could answer the question of the so-called 'Thorium anomaly' in the region of Uranium isotopes.

Fig. 4 schematically shows the potential-energy surface of ^{236}U as a function of deformation (axis ratio) with the triple-humped fission barrier as deduced in this publication. Prominent new features are the lower inner barrier E_A and the large depth of the third minimum $(E_{B1}+E_{B2})/2 - E_{III}$.

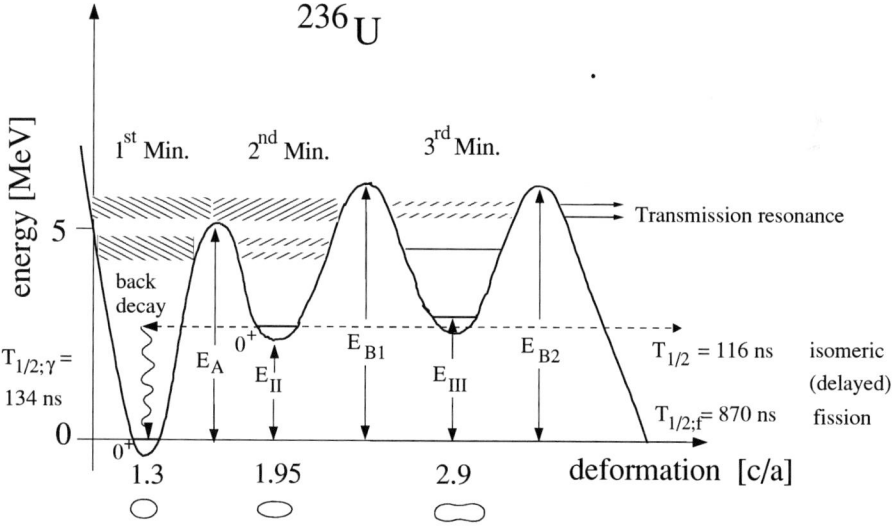

FIGURE 4. The triple-humped potential-energy surface of ^{236}U. Also damped class-I, class-II and class-III compound states are shown in the three minima. For strongly mixed class-I and class-II states, transmission resonances of class-III (HD) states occurred in fission.

Extrapolating the height of the inner barrier for even-even nuclei from the plutonium nuclei (~ 5.5 MeV) over the uranium nuclei (~ 5.0 MeV) one predicts this to be about 4.5 MeV for thorium nuclei. This represents a new trend in fission barriers for light actinides which is different from former assignments [23], but in good agreement with more recent theoretical predictions [7]. In the double-humped barrier picture the inner barrier for Th and Ra nuclei always had to be adjusted above the highest observed resonance, while now the existence of a deep third minimum only requires that the inner barrier lies above the highest resonance in the second minimum. For ^{232}Th this trend of the inner barrier is supported by the onset of the isomeric shelf at 4.6 MeV [24, 7]. The much lower inner barriers obtained allow a better understanding of the γ back-decay to the first minimum and lead to new predictions of lifetimes and yields of isomers in the second minimum of uranium isotopes.

Acknowledgment

The work has been supported by DFG under HA 1101/6-3 and 436 UNG 113/129/0, the Hungarian Academy of Sciences under HA 1101/6-1, and the Hungarian OTKA Foundation No. T038404.

REFERENCES

1. D. Habs, Nucl. Phys. **A502** (1989) 105c.
2. P.G. Thirolf and D. Habs, Prog. Part. Nucl. Phys. **49** (2002) 245.
3. H. Hübel., Acta Phys. Pol. **36** (2005) 1015.
4. V.M. Strutinsky, Nucl. Phys. **A95** (1967) 420.
5. P. Möller, S.G. Nilsson and R.K. Sheline, Phys. Lett. **B40** (1972) 329.
6. S. Ćwiok et al., Phys. Lett. **B322** (1994) 304.
7. W.M. Howard and P. Möller, At. Data Nucl. Data Tables **25** (1980) 219, and references therein.
8. J. Blons et al., J. Blons et al., Nucl. Phys. **A477** (1988) 231.
9. P. Glässel, H. Röser and H.J. Specht, Nucl. Phys. **A256** (1976) 220.
10. A. Krasznahorkay et al., Phys. Rev. Lett. **80** (1998) 2073.
11. H.A. Enge and S.B. Kowalsky, (in Proc. 3rd Int. Conf. on magnet technology, Hamburg (1970)).
12. A. Krasznahorkay et al., Phys. Lett. **B461** (1999) 15.
13. E. Zanotti et al., Nucl. Instr. Meth. **A310** (1991) 706.
14. H.F. Wirth, (PhD Thesis, TU Munich, (2001) Unpublished).
15. P.C.N. Crouzen, (PhD Thesis, Rijksuniversiteit Groningen, (1988) Unpublished).
16. M. Hunyadi, (PhD Thesis, Lajos Kossuth University Debrecen, (1999) Unpublished).
17. M. Csatlós et al., Phys. Lett **B615** (2005) 175.
18. M. Just, PhD thesis, Univ. Heidelberg (1978), unpublished.
19. T.M. Shneidman et al., Nucl. Phys. **A671** (2000) 119.
20. T. Rauscher, F.K. Thielemann, K.L. Kratz, Phys. Rev. **C 56** (1997) 1613.
21. M. Hunyadi et al., Phys. Lett. **B505** (2001) 27.
22. T. von Egidy, H.H. Smidt, A.N. Behkami, Nucl. Phys. **A481** (1988) 189.
23. S.B. Bjørnholm and J.E. Lynn, Rev. Mod. Phys. **52** (1980) 725.
24. V.E. Zhuchko et al., JETP Lett. **22** (1975) 118.

High Spin Bands in the A ~ 130 Nuclei: A "Non-chiral" Explanation

Rościsław Kaczarowski

The Andrzej Sołtan Institute for Nuclear Studies, 05-400 Świerk, Poland

Abstract. Pairs of close-lying, high-spin rotational bands with similar properties are frequently observed in the odd-odd nuclei in the A ~ 130 nuclei region, especially in the vicinity Z ~ 57 and N ~ 75. New phenomenological approach based on independent additivity of both signature sequences in the involved rotational bands built on single-quasiparticle proton and neutron states in the odd-A nuclei is proposed in the aim to understood structure of both observed bands built on the two-quasiparticle states. Softness of the core is approximately taken into account. The predictions based on this approach are compared with experimental data in the case of ^{132}La. The proposed signature selection rule explains some properties of the γ-transitions between these bands.

Keywords: Enter Keywords here.
PACS: 21.60.-n, 23.20.Lv, 27.60.+j

INTRODUCTION

Recent experimental investigations of odd-odd nuclei in the A ~ 130 nuclei region, especially in the nuclei with proton and neutron numbers close to Z ~ 57 and N ~ 75, have resulted in observation of systematic pairs of close-lying, high-spin rotational bands with similar properties built on nuclear state consisting the $h_{11/2}$ odd-proton and $h_{11/2}$ odd-neutron [1-8]. The nature of these bands is not yet fully understood. They were interpreted as a "chiral bands" (predicted and described in Refs. [9,10], in the framework of a core-particle-hole coupling model (CPHC) [4] and in the framework of the pair-truncated shell model (PTSM) (Ref. [12] and references mentioned therein). On the other hand new mean field calculations exclude a possibility of existence of "chiral" bands below critical frequency value estimated as $\hbar \omega_{crit} \approx$ 0.5 – 0.6 MeV [11], which is in contrast to previous mean filed calculations [10].

The aim of this paper is to examine a relatively simple and strictly phenomenological approach based on assumption that the signature sequences of two rotational bands in odd-Z and odd-N bands add independently. The signatures of the two sequences of a rotational band in odd-A nucleus are defined here as $\alpha = (-1)^{I-1/2} \cdot \frac{1}{2}$. In the case of two strongly coupled proton and neutron bands with no signature splitting in the odd-A nuclei, only one two-quasiparticle band with $\Delta I=1$ spin sequence in odd-odd nucleus is formed as a result of coupling. When one of the odd-proton or odd-neutron bands has non-zero signature splitting, two $\Delta I=1$ bands in odd-odd nucleus are expected, while in a special case when signature splitting exists in both proton and neutron rotational bands, one may expect even four similar rotational bands with $\Delta I=2$ spin sequence in odd-odd nucleus built on two-quasiparticle configurations arising from all four combinations of signature sequences in the proton and neutron rotational bands ($\pi_+\nu_+$, $\pi_+\nu_-$, $\pi_-\nu_+$ and $\pi_-\nu_-$).

The occurrences of such bands, so called "semidecoupled" and "doublydecoupled", respectively, have been already observed and interpreted in odd-odd heavy transitional nuclei (Re, Ir) bands by A.J. Kreiner et al (see e.g. Refs. [13-15]).

To check these predictions in the A ~ 130 nuclei region, a new phenomenological approach using the modified additivity rule (soft-core additivity) has been used in the case of ^{132}La nucleus for comparison of the observed and predicted values for several experimental observables, such as moments of inertia, routhians, alignments and energies of rotational levels. The obtained results will be presented. The experimental data used in these calculations have been taken from Refs. [4,1] (^{132}La), Ref. [16] (^{131}La) and Ref. [17] (^{131}Ba).

HIGH SPIN BANDS IN THE A ~ 130 NUCLEI

In the discussed A~130 nuclei region and for moderate quadrupole deformation $\beta_2 \approx$ 0.2 the Fermi level for Z~57 is expected to be closed to the ½⁻[550], $h_{11/2}$ and 3/2⁻[541], $h_{11/2}$ Nilsson configurations while neutron Fermi level for N~75 lies in close proximity to the high-K, 9/2⁻[514], $h_{11/2}$ configuration. Rotational bands built on these configurations are observed in neighboring odd-Z and odd-N, respectively, nuclei (see e.g. Refs. [16, 17]).

The signature splitting observed in the rotational band built on the 9/2⁻[514], $h_{11/2}$ configuration in ^{131}Ba is not predicted by the Coriolis calculations using model described in Ref. [18] where couplings of all configurations originating from the $h_{11/2}$ neutron orbital have been taken into account. Therefore the observed splitting may be due to the triaxial shape of ^{131}Ba nucleus in this configuration. This is not unexpected, because many of the highest-K configurations in a given orbital show similar staggering. The rotational bands built on the K=11/2⁻ proton state of the $h_{11/2}$ orbital parentage in heavy transitional nuclei are (Re, Ir) are a typical examples, but also in the A~130 nuclei region such staggering is observed in the rotational bands built on the highest-K configurations, even on configurations of the low-j, $d_{5/2}$ and $g_{7/2}$ orbitals as well as the high-j, $h_{11/2}$ orbital (see e.g. ^{149}Pm nucleus [19]). Staggering of the rotational band levels and oscillations of intraband B(M1)/B(E2) values are considered as a typical signature of triaxial deformation of nucleus. However, both these effects occur also frequently in axial nuclei.

The rotational levels of bands built on low-K, $h_{11/2}$ proton configurations are strongly mixed due to the Coriolis interaction, what leads to formation of decoupled rotational band structure where favored part has negative signature $\alpha_{fav}=(-1)^{j-1/2}\cdot\frac{1}{2} = -1/2$ (Ref. [20]). Coriolis calculations using the model described in Ref. [18], which takes into account coupling of all $h_{11/2}$ proton configurations, show that due to large negative decoupling parameter of the ½⁻[550] configuration (the value of a=-5.8 is expected in the framework of the Nilsson model with deformation parameters ε_2=0.20 and ε_4=-0.03), the lowest energy state in the favored branch has a spin value much greater than value of K, usually I = 7/2⁻ or 11/2⁻ in this nuclei region (depending on exact a value for a given nucleus), while in the unfavored branch the lowest energy state has I = 5/2⁻ or 9/2⁻. In the odd-odd nuclei coupling of the 9/2⁻[514] odd neutron to both signatures of the ½⁻[550] proton rotational band results in two band heads with the highest possible value of K=5 (according to the Nordheim weak rule), and with band-head spins I=6 for favored band

and I=5 for the unfavored one. However, again due to large decoupling parameter value for the K=1/2 proton configuration, the lowest rotational levels may have spin values I = 8^+ or 10^+ for the lower, favored band and I = 7^+ or 9^+ for the higher, unfavored band.

In such way the systematic appearance of two high-spin, ΔI=1 rotational bands with similar properties ("sister bands") in this nuclei region acquires an apparent explanation.

Taking into account the signature splitting observed in the rotational band built on the ν9/2⁻[514] configuration in odd-N, ^{131}Ba nucleus, one may expect even four high-spin rotational bands with ΔI=2 spin sequences in ^{132}La. However, if the even-even core of this nucleus has stable prolate deformation due to the well known strong prolate-deformation driving properties of the odd-proton occupying the high-j, K=1/2⁻, [505] configuration then the oblate and/or the γ-deformation driving properties of the 9/2⁻[514] odd-neutron, placed in the prolate potential well, may be cancelled in the odd-odd ^{132}La nucleus. Under these assumed circumstances, it is possible that the signature splitting due to the odd-neutron may disappear in, presumably prolate, ^{132}La and, consequently, only two bands built on 2-quasiparticle states resulting from the coupling of the 9/2⁻[514] neutron with both signatures of the 1/2⁻[505] proton configuration would be observed in this nucleus, in agreement with existing experimental data.

Soft-core Additivity Rule

The additivity rule, commonly used to interpret properties of the odd-odd nuclei, states that a value of a given additive observable S, such as alignment and routhian values, in the rotational bands of odd-odd, Z+1, N+1, A+2 nucleus can be approximately estimated knowing the values of this observable in the respective rotational bands in odd-Z, A+1 nucleus, as well as in the odd-N, A+1 nucleus

$$S_{Z+1,N+1}^{A+2}(\omega) = S_{Z+1}^{A+1}(\omega) + S_{N+1}^{A+1}(\omega) \qquad \text{i)}$$

where ω denotes a given rotational frequency.

However, one should bear in mind that alignment as well as routhian values in a rotational band build on the given configuration are calculated in relation to collective rotation of the core with kinematical moment of inertia defined by Harris parameters J_0 and J_1 used in the formula $J(\omega) = J_0 + J_1 \omega^2$.

In the case of "partially" additive observables (where differences between the value of the observable in the odd-A nucleus and the respective value for the even-even core are assumed to be additive), such as moment of inertia and rotational level energies, the additivity rule can be expressed as

$$S_{Z+1,N+1}^{A+2}(\omega) = S_{Z,N}^{A}(\omega) + dS_{Z+1}^{A+1}(\omega) + dS_{N+1}^{A+1}(\omega) \qquad \text{ii)}$$

where

$$dS_{Z+1}^{A+1}(\omega) = S_{Z+1}^{A+1}(\omega) - S_{Z,N}^{A}(\omega) \quad \text{and} \quad dS_{N+1}^{A+1}(\omega) = S_{N+1}^{A+1}(\omega) - S_{Z,N}^{A}(\omega) \qquad \text{iii)}$$

and where the quantity $S^A_{Z,N}$ is the value of observable in the even-even core.

In this simple approach an influence of the p-n interaction on properties of the final odd-odd nucleus is neglected, as well as changes of the even-even nucleus core properties (deformation parameters, pairing, etc.) due to presence odd-proton, odd-neutron and both of them. Therefore-even ^{130}Ba nucleus can not be considered as a good choice of the core for ^{131}Ba and ^{131,132}La nuclei.

However, in the transitional nuclei region where nuclei are soft in relation to changes of deformation parameters (including non-axial deformation parameter γ) and where, as well, pairing strength can change considerably when odd-nucleons are added, the approximation discussed above becomes too crude. It is obvious that a proper treatment of the core is essential for a description of "soft" nuclei. Therefore, in order to take into account these changes of the soft core parameters, one should try to find out a good approximation of the core parameters J_0 and J_1 not only separately for odd-Z, odd-N and odd-odd nuclei of interest, but also separately for each involved proton and neutron configuration. Obviously this presents a considerable challenge and some common criteria are necessary to choose a set of reasonable core parameters for considered odd-A, odd-odd and even-even nuclei of interest.

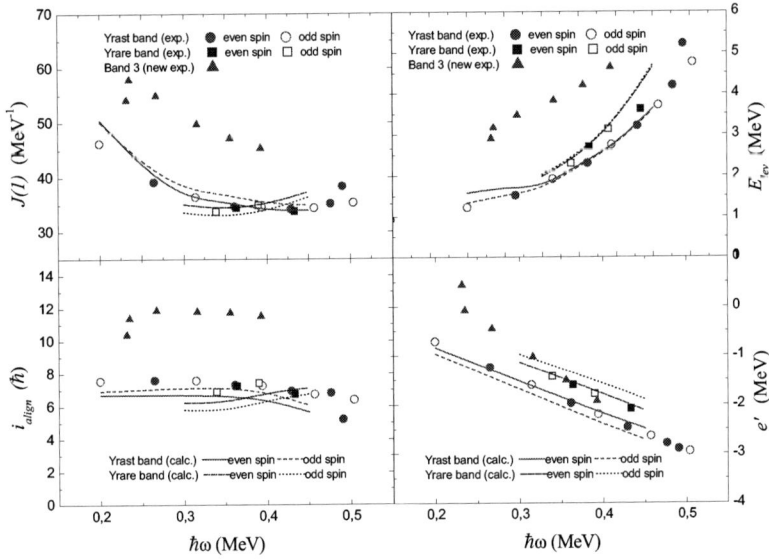

FIGURE 1. (Color online) Comparison of the experimental and predicted by soft-core additivity rule values of kinematical moment of inertia $J(1)$, alignment i, routhians e' and level energies in rotational bands in ^{132}La. The experimental values are shown with filled and open symbols, for (initial) levels with even- and odd-spin values, respectively. The calculated values are given by curves. The data for the recently observed third rotational band in ^{132}La [1] are also included (triangles, olive online).

In this paper one of the several possible criteria of choice of individual cores has been tested, namely, the request of constant value of the alignment for the first three members of a energy-favored rotational band sequences in odd-A nuclei, i_{al}=const. This condition has been used in order to calculate J_0 and J_1 core parameters for rotational band build on the 1/2⁻[505] proton configuration in ^{131}La and for rotational band build on the 9/2⁻[514] neutron configuration in ^{131}Ba. For the odd-odd ^{132}La the mean core was used, i.e. the used J_0 and J_1 parameters were mean values of these parameters in ^{130,132}Ba and ^{132,134}Ce neighboring nuclei, calculated from condition i_{al}=0 for the first three rotational levels in the respective ground state bands. The chosen core parameters are shown in Table 1. The experimental data used in the calculations have been taken from Refs. [4,1] (^{132}La), Ref. [16] (^{131}La) and Ref. [17] (^{131}Ba) and Ref. [21] (even-even nuclei).

TABLE 1. The values J_0 and J_1 parameters used in the calculations

Nucleus and configuration	J_0	J_1
^{131}La, 1/2$^-$[505]	10.76	66.42
^{131}Ba, 9/2$^-$[514]	12.10	34.65
^{132}La, High-spin bands	5.75	66.64

The comparison of experimental values of moments of inertia, alignments, routhians and rotational level energies as a function of rotational frequency in high spin rotational bands in ^{132}La with the respective values calculated using the soft-core approach is shown in Fig. 1. The experimental values for the third, recently discovered [1], rotational band in this nucleus evidently show the very different character of this band. The high values of the kinematical moment of inertia, alignment and level energies indicate, that this band is probably built on the four-quasiparticle state. No calculations for this band were done.

A new signature selection rule

Additional arguments supporting an interpretation of the two lowest high-spin bands in ^{132}La, denoted as Band 1 and Band 2 in Fig.2, as rotational bands resulting from the coupling, respectively, of negative- and positive signature sequence members of the π1/2$^-$ [505] rotational band with members of the ν9/2$^-$[514] rotational band follow from the behavior of inter-band quadrupole, E2, electromagnetic transition between these two bands.

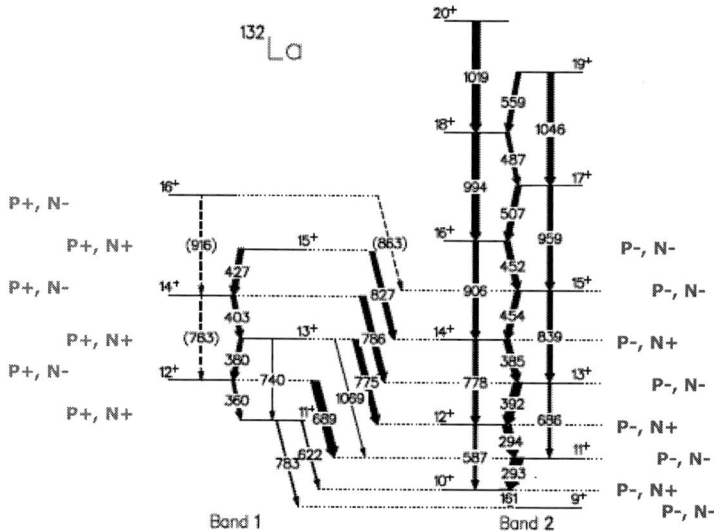

FIGURE 2. (Color online) Partial level scheme of ^{132}La. Level and γ-energies were taken from Ref. [1]. Symbols $P+$, $P-$, $N+$ and $N-$ denote the respective signatures of the proton and neutron components of the two-quasiparticle wave function of given rotational state.

In the partial level scheme of these bands in ^{132}La established earlier [1,4] and shown here in Fig. 2 one may easily note either absence or very low intensity of several energy-favored E2 transitions between bands, e.g. between states with spin I in rotational Band 1 to the states with spin $I-2$ in rotational Band 2. Similar absences or weakness of the analogous inter-band E2 transitions have been observed also in pairs of the positive-parity rotational bands built on the $\pi\frac{1}{2}^-[505]\circ\nu 9/2^-[514]$ configuration in other neighboring odd-odd nuclei, e.g. ^{130}Cs, ^{134,136}Pm [5] and in ^{138}Eu [2], what strongly suggests a quite probable existence of common γ-transition selection rule responsible for hindrance of the discussed γ-transition.

If proton-neutron interaction is sufficiently weak, the proton and neutron components in the wave functions of rotational states are relatively independent and the interference between these components is rather small. Therefore one may assume that these components retain their "internal" signature values. Under this assumption pairs of these "internal" proton and neutron signatures can be attributed to each member of the rotational band in odd-odd nucleus, as shown in Fig.2, where P+, P-, N+ and N- denote, respectively, positive or negative the internal proton (neutron) signature. One may also note, that the observed γ-transitions take place only between pairs of rotational states when no more than one odd particle changes its "internal" signature. The γ-transitions between states requiring the change of both "internal" signatures of the involved proton and neutron are either non-existent or weak. These observations may be summarized as a proposed new *signature selection rule* for electromagnetic transition in the odd-odd nuclei:

i. the γ-transitions requiring the change of only one "internal" signature are allowed:
(P+, N+) → (P-, N+) and (P+, N-) → (P-, N-)

ii. the γ-transitions requiring the change of both "internal" signatures are forbidden, or, at least, hindered:
(P+, N+) → (P-, N-) and (P+, N-) → (P-, N+).

The above rule provides an evident explanation of the discussed above anomalies in intensities of inter-band E2 transitions between the $\pi h_{11/2} \circ \nu h_{11/2}$ bands observed in the A ~ 130 nuclei region.

SUMMARY AND CONCLUSIONS

It has been shown that a relatively simple and strictly phenomenological approach based on assumption that the signature sequences of two rotational bands in odd-Z and odd-N bands add independently allows to explain in a simple and apparent way, without any additional assumptions, an existence of two high-spin "sister" rotational bands built on the $\pi h_{11/2} \circ \nu h_{11/2}$ configurations in the A~130 transitional nuclei region. The experimental values and behavior of the moment of inertia, alignment, routhians and rotational level energies as a function of rotational frequency in both lowest rotational bands in the odd-odd ^{132}La nucleus agree reasonably well with the ones deduced from experimental data using the described soft-core additivity approach. However, the energy splitting between the bands is smaller than predicted by using the additivity rules. The proposed signature selection rule explains in an evident way the absence or weak intensity of certain inter-band E2 transition, expected to be strongly favored by their high energy.

ACKNOWLEDGMENTS

The author wishes to thank Professor Stefan Frauendorf for fruitful discussions and for the careful reading and commenting the manuscript and Professor Umesh Garg for helpful comments and remarks.

REFERENCES

1. J. Srebrny,et al, *Acta Phys. Pol. B* **36**, 1063, (2005)
2. A.A. Hecht *et al*, *Phys. Rev C.* **68**, 054310 (2003).
3. T. Koike *et al,,* *Phys. Rev. Lett.* **86**, 971 (2001).
6. A.A. Hecht *et al*, *Phys. Rev. C* **63**, 051302(R) (2001).
7. D.J. Hartley *et al*, *Phys. Rev. C* **64**, 031304(R) (2001).
8. R.A. Bark, *et al*, *Nucl. Phys.* **A691**, 577 (2001).
9. S. Frauendorf and J. Meng, *Nucl. Phys.* **A617**, 131 (1997).
10. V.I. Dimitrov, S. Frauendorf, and F. Dönau, *Phys. Rev. Lett.* **84**, 5732 (2000).
11. P. Olbratowski, J. Dobaczewski, J. Dudek and W. Płóciennik, *Phys Lett.* **93**, 052501 (2004).
12. K. Higashiyama, N. Yoshinaga, and K. Tanabe, *Phys. Rev. C* **72**, 024315 (2005)
13. A.J. Kreiner, D.E. Di Gregorio, A.J. Fendrik, J. Davidson, M. Davidson, *Phys. Rev. C* **29**, 1572 (1984).
14. A.J. Kreiner J. Davidson, M. Davidson, D. Abriola, C. Pomar, P. Thieberger, *Phys. Rev. C* **36** 2309 (1987).
15. A.J. Kreiner, *Phys. Rev. C* **38** 2486 (1988).
16. L. Hildingsson, C.W. Beausang, D.B. Fossan, R. Ma, E.S. Paul, W.F. Piel, Jr., and N. Xu, *Phys. Rev. C* **39**, 471 (1989).
17. R. Ma, Y. Liang, E.S. Paul, N. Xu, D.B. Fossan, L. Hildingsson, R.A. Wyss, *Phys. Rev. C* **41**, 717 (1990).
18. R. Kaczarowski, *Comp. Phys. Comm* **13**, 63 (1977).
19. M.A. Jones *et al,,* *Nucl. Phys.* **A669**, 119 (2000).
20. F.R. Xu, W. Satuła, R. Wyss, *Nucl. Phys.* **A669**, 119 (2000).
21. R.B. Firestone, F. Chu, M. Baglin, "*Table of Isotopes*", Lawrence Berkeley National Laboratory, University of California, 8th edition, 1999.

Nuclear Structure and Octupole Collectivity in Nucleus ^{122}Cs

I.M.Govil

Department of Physics, Panjab University, Chandigarh-India

Abstract : The nuclear structure of ^{122}Cs has been studied using ^{107}Ag (^{19}F,p3n)^{122}Cs fusion reaction at beam energy of 93 MeV. Fifteen new transitions belonging to ^{122}Cs have been observed extending the level scheme up to $E_x \sim 7$ MeV and J $\sim 28\hbar$. We have performed the linear polarization measurements using the Clover detectors to assign the unknown spins and parities of a few bands. We have also observed a few linking transitions between negative and positive parity bands indicating the octupole collectivity in this nucleus. The band structure of this nucleus is discussed in the frame work of the cranked Hartree-Fock-Bogoliubov model and the microscopic Projected Hartree- Fock model.

Keywords: Nuclear structure, γ-ray, octupole collectivity, polarization.
PACS: 21.60.Ev;23.20.Lv;27.60+j

INTRODUCTION

The odd-odd nuclei in the mass region (50<Z<58) exhibit some of the recently known phenomena of signature inversion and chiral symmetry [1-10]. The physics underlying signature inversion has been attributed to the triaxiality, the p-n interaction [11] and the quadrupole pairing [12]. In the case of the Cs isotopes, as the neutron number decreases towards the mid-shell (N=66), the neutron Fermi level also lowers into the low Ω $h_{11/2}$ orbitals and therefore, the possibility of the residual interactions increases between the valence protons and the neutrons . In this mass region octupole collectivity has also been observed in ^{114}Xe [13], that is partly because of the proton and the neutron orbitals, $h_{11/2}$ and $d_{5/2}$ having $\Delta l=3$ are near the Fermi surface. The knowledge about the structure of the odd-odd nuclei in this region is still fragmentary because these nuclei have higher level densities leading to a much more complex energy spectra than those of the even-even or odd-A nuclei. A major problem of the odd-odd nuclei is the lack of the spin and parity assignments, which is usually compounded by the observation of many unlinked bands. The purpose of the present paper is to explore the band structure of the ^{122}Cs nucleus. In the earlier investigations [4,14] the level structure of ^{122}Cs has been studied using only six HPGe detectors, and in these measurements, spin and parity for some of the observed rotational bands were not assigned or confirmed. In the present work, we have made an attempt to establish the unknown spin-parity of these bands using the linear polarization measurements. Due to the new spin and parity assignment, we could also explore some linking transitions between negative and positive parity bands indicating the possibility of octupole collectivity in this nucleus. The results are compared with the cranked Hartree-Fock-Bogoliubov model and the microscopic projected Hartree- Fock model.

EXPERIMENTAL DETAILS

High spin states in the ^{122}Cs nucleus were populated using the ^{107}Ag(^{19}F,p3n)^{122}Cs fusion evaporation reaction at beam energy of 93 McV. The beam was provided by the 14UD Pelletron facility at TIFR, Mumbai India. An isotopically enriched 1mg/cm^2 thick ^{107}Ag target on a 10mg/cm^2 thick Au backing was used. The de-exciting gamma rays were detected using the Indian National Gamma Array [15,16] consisting of eight clover detectors in conjunction with a 14-element NaI(Tl) multiplicity filter. The photo peak efficiency for the array was 1.6 %. The detectors were coplanar and placed at 60°, 90°, 120°, 150°, 210°, 250°, 285°, and 325° with respect to the beam direction. A total of about 200 million triple or higher-fold coincidence events were recorded in the experiment. The efficiency and energy calibration were performed with the standard γ-ray ^{152}Eu and ^{133}Ba radioactive sources. After gain-matching, the coincidence events were sorted into symmetric and asymmetric matrices for detailed off line analysis. The data is analyzed using both RADWARE [17] and IUCSORT [18] computer programs. The multipolarity of the de-exciting γ-ray were deduced from the observed γ-γ angular correlation and R$_{DCO}$ measurements.

Multipolarity and spin assignments were further corroborated by the γ-ray linear polarization measurements. The use of the Clover detectors facilitated polarization measurements [19-22]. The individual crystals are considered as a scatterer and the two adjacent crystals as the observers, within a single Clover detector. Two asymmetric polarization matrices were constructed by placing events along one axis which contained the energy recorded in any one of the detector, while the other axis corresponded to the energy scattered in a perpendicular or parallel segments of the Clover with respect to the beam axis. From the projected spectra, the number of γ-rays with scattering axis perpendicular (N$_{per}$) and parallel (N$_{par}$) to the emission plane were obtained for a given γ-ray. From these spectra, the asymmetry parameter Δ$_{asym}$ was obtained using the relation

$$\Delta_{asym} = \frac{aN_{per} - N_{par}}{aN_{per} + N_{par}} \qquad (1)$$

The correction parameter **a** due to the asymmetry of the present experimental configuration has been deduced using radioactive sources. An electric transition results in a positive value for the Δ$_{asym}$ whereas a negative value corresponds to a magnetic transition. A near-zero value is indicative of a possible admixture. As we use the data from detectors at all angles in the reaction plane, we may not perform the quantitative polarization measurements. However, we may get the useful information about the nature of γ-ray of interest.

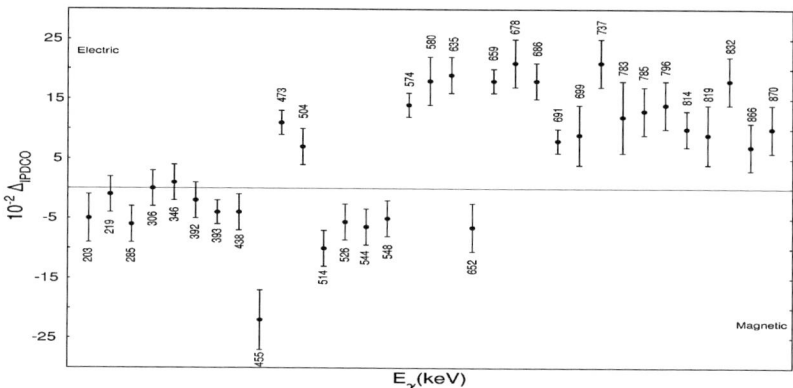

FIGURE 1. Experimental γ-ray asymmetry parameter Δasym, from polarization measurements plotted for γ-ray transitions of ^{122}Cs. A positive value corresponds to an electric transition and a negative value indicates the magnetic transition. The quoted errors are due to background subtraction and fitting.

The polarization sensitivity is measured by measuring the asymmetry parameter (Δ_{asym}) and verifying the Multipolarity for the well known E2 and M1 transitions. Fig.1 illustrates the results of these measurements. The parities of the bands C, D and E have been assigned for the first time in the present work. A positive sign of the Δ_{asym} characterize the E1/E3 nature of the 678 keV transition linking the negative parity band B with the positive parity band A.

RESULTS & DISCUSSION

The proposed level scheme of ^{122}Cs from the present work is shown in Figure 2, where the ordering of transitions are based on relative γ-ray intensities and γ-γ coincidence using proper gates. The level scheme has been established up to Ex ~ 7 MeV and angular momentum up to 28$\hbar$. All the new transitions reported by us have been confirmed by reverse gates and multiple gates belonging to a particular band.

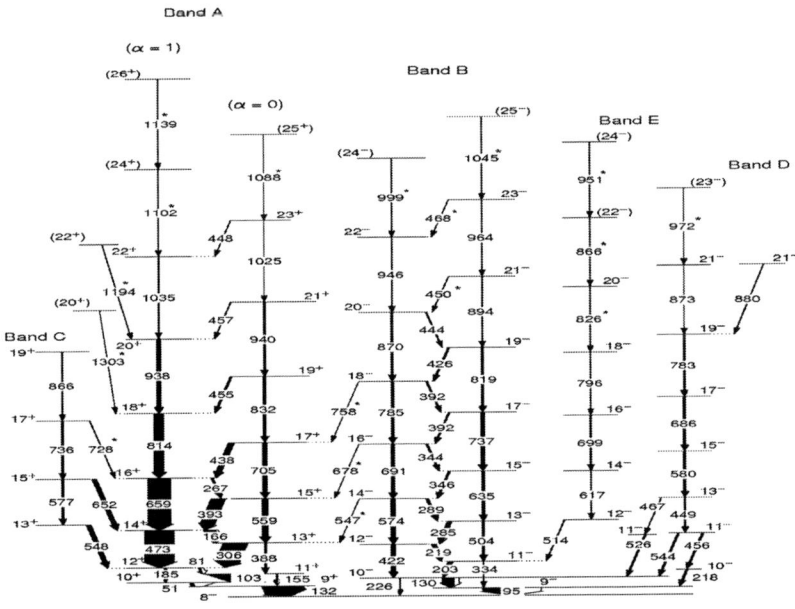

FIGURE 2. Level scheme for ^{122}Cs populated in ^{107}Ag(^{19}F, p3n)^{122}Cs reaction. Newly observed transitions are marked with an asterisk. The energies are marked within ±1 keV. The spin and parity assignments, given in parentheses, are tentative.

In addition to the transitions reported earlier by Moon et al. [4], about 15 new transitions have been identified in the present work. These transitions are indicated by asterisk(*) in the level scheme. The gated spectra with 132 keV γ-ray in Fig.3 indicates six new transitions belonging to band A, of these three transitions viz 1102 keV (24+→ 22+), 1139 keV (27+→ 25+), 1194 keV (22+→ 20+) and 1303 keV (20+→ 18+) belong to α = 0 (positive signature) of band A while one transitions viz 1088 keV (25+ → 23+), belong to α = 1 (negative signature of band A) and one linking transitions viz 728 keV (17+→ 16+) between band C and band A.

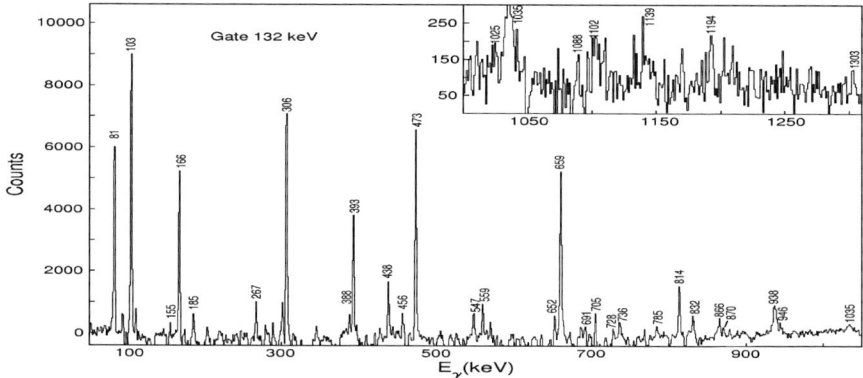

FIGURE 3. γ-γ coincidence spectrum with gate on 132 keV transition.

Further the gated spectra with 95 keV transition indicates the presence of three new transitions 826 keV (20⁻→ 18⁻), 866 keV (22⁻ → 20⁻) and 951 keV (24⁻→ 22⁻) belonging to band E and one transition of 1045 keV belonging to band B. In the spectrum gated with 946 keV γ-ray transition of band B shown in Fig. 4, one may clearly see two new linking transitions of 678 keV and 758 keV between negative parity band B and positive parity band A.

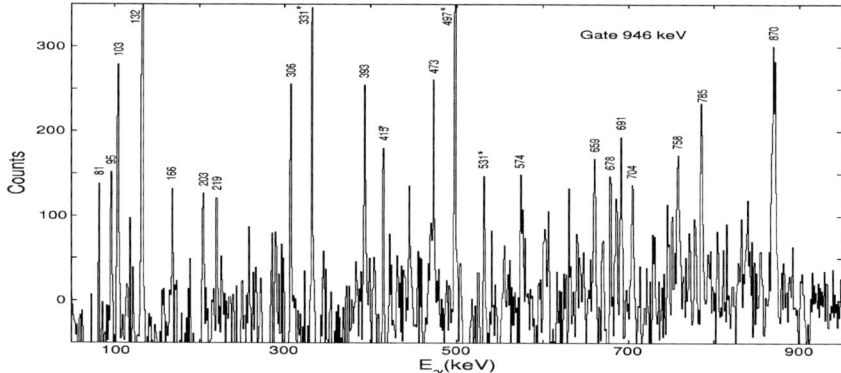

FIGURE 4. γ-γ coincidence spectrum with gate on 946 keV transition. The transitions marked with asterisk belong to the neighbouring nuclei and the background.

As mentioned above we have observed three linking transitions viz 547, 678, 758 keV between negative parity band B and positive parity band A, indicating these to be either E1 or E3 nature. Recently there have been reports on observation of higher multipole (E3) electric transitions by Angelis et al. [13] in this region. Due to the strong mixing of $\nu d_{5/2}$ orbital (l=2) in negative parity band B and the $\nu h_{11/2}$ configuration (l=5) in band A, there is a possibility of Δl=3 octupole transition between these two bands. However, it would be of interest to confirm the Multipolarity of these transitions by lifetime measurements.

The band structure and electromagnetic properties for this nucleus have been compared with the microscopic Hartree-Fock model with angular momentum Projection [23]. The experimental data has also been compared with the Woods-Saxon Cranked Shell Model (CSM) calculations [24]. The results are found to be in good agreement with theory.

SUMMARY

The level structure of the Z=55 odd-odd ^{122}Cs nucleus has been studied with eight Clover detectors array using ^{19}F + ^{107}Ag fusion evaporation reaction. Fifteen new transitions belonging to this nucleus have been identified extending the level scheme up to excitation energy of 7 MeV and spins up to $8\hbar$. Linear polarization measurements with the Clover detectors as polarimeter were performed to assign unknown spin and parity of some of the transitions. We have also reported some new linking E1 or E3 transitions between the negative and positive parity bands, indicating the octupole collectivity in this nucleus. The observed level scheme is compared with the microscopic Projected Hartree-Fock calculation and the results are found to be in a reasonably good agreement, which provides useful information about the dominant microscopic configurations of these bands. It would be interesting to explore the level structure of this nucleus further, to search for the possible existence of chirality and octupole collectivity.

ACKNOWLEDGMENTS

The authors would like to thank the accelerator crew for providing an excellent beam from 14UD Pelletron at TIFR, Mumbai. The financial support from UGC, and CSIR is gratefully acknowledged.

REFERENCES

1. B. Cederwall et al. Nucl. Phys. A542, (1992)454.
2. T. Komatsubara et al. Nucl. Phys. A557 (1993)419c.
3. J. F. Smith et al. Phys. Rev. C58 (1998)3237.
4. C.-B. Moon , T. Komatsubara and K. Furuno, Nucl. Phys. A674(2000)343.
5. S. Frauendrof and Jie Meng, Nucl. Phys. A617, (1997)131.
6. T. Koike, Phys. Rev. C 67,044319(2003).
7. K. Starosta et al. Phys. Rev. Lett. 86 (2001)971.
8. F. Liden et al. Nucl. Phys. A550 (1992)365.
9. R. Zheng, S. Zhu, N. Cheng, and J. Wen , Phys. Rev. C64, 014313(2001).
10. Y. Liu, Y. Ma, H. Yang, and S. Zhou, Phys. Rev. C52 (1995)2514.
11. T. Tajima, Nucl. Phys. A572 (1994)365.
12. F. R. Xu, W. Satula, R. Wyss, Nucl. Phys. A669(2000)119.
13. G. de Angelis et al. Phys. Lett. 535B, 93 (2002).
14. Yong-Name U et al. J. Phys. G 31 (2005)B1-B6.
15. S. Lakshmi et al. Phys. Rev. C66, 0141303(R)(2002).
16. P. Datta et al. Phys. Rev. C67, 014325 (2003).
17. D. C. Radford , Nucl. Instrum. Methods Phys. Res.A361 (1995)297.
18. N. S. Pattabiraman ,S. N.Chintalapudi and S. S. Ghugre , Nucl. Instrum. Methods Phys. Res. A526, 432 (2004).
19. G. Duchtene F. A. beck, P. J. twin, G. de France et al., Nucl. Instrum. Methods Phys. Res.A 432, 90(1999).
20. K. Starosta, Nucl. Instrum. Methods Phys. Res.A 423, 16(1999).
21. Ch. Droste, Nucl. Instrum. Methods Phys. Res. A 337, 430(1999).
22. S.Ray et al. Phys. Rev. C69, 0543114 (2004)
23. C. R. Praharaj, J. Phys. G14(1988) 843.
24. W.Nazarewicz et.al. Nucl. Phys. A 435 (1985) 397

Rotational Damping, Ridges, and the Quasi–continuum of γ Rays in ^{152}Dy

T. Lauritsen*, R.V.F. Janssens*, T.L. Khoo*, I. Ahmad*, M.P. Carpenter*,
F.G. Kondev*, C.J. Lister*, E.F. Moore*, D. Seweryniak*, S. Zhu*,
T. Døssing[†], B. Herskind[†], A.M. Heinz[**], D.G. Jenkins[‡], R.M. Clark[§],
P. Fallon[§], A.O. Macchiavelli[§], D. Ward[§], G. Lane[¶], P. Chowdhury[‖],
A. Korichi[††], A. Lopez–Martens[††] and A.J. Larabee[‡‡]

*Argonne National Laboratory, Argonne, Illinois 60439, USA.
[†]Niels Bohr Institute, DK–2100, Copenhagen, Denmark.
[**]A.W. Wright Nuclear Struct Lab, Yale University, New Haven, CT 06520, USA.
[‡]Department of Physics, University of York, Heslington, York YO10 5DD, UK.
[§]Lawrence Berkeley National Laboratory, Berkeley, California 94720, USA.
[¶]Department of Nuclear Physics, Australian National University, Canberra ACT 0200, Australia.
[‖]University of Massachusetts, Lowell, MA 01854, USA.
[††]C.S.N.S.M, IN2P3-CNRS, bat 104-108, F-91405 Orsay Campus, France.
[‡‡]Greenville College, Greenville, IL 62246, USA.

Abstract.
The quasi–continuum of γ rays from the feeding and decay of superdeformed and normal bands in the nucleus ^{152}Dy have been extracted in 1 – and 2 – dimensional spectra. The E_γ–E_γ correlations in the latter reveals strong ridges associated with superdeformed and normal states in this nucleus. The entry distributions for normal and superdeformed bands were also extracted. A Monte Carlo model was developed that *simultaneously* describes all the quasi–continuum and ridge spectra as well as the feeding intensity of the superdeformed bands. Through the calculation of the continuum of γ rays at finite temperature, the rotational damping widths in the normal and superdeformed wells were extracted.

Keywords: Superdeformation, Heavy Ion Fusion, High Spin Spectroscopy, Continuum Spectroscopy, Rotational Damping, Entry Distribution.
PACS: 23.20.Lv, 23.20.En, 27.70.+q, 27.80.+w

INTRODUCTION

In selected regions of the chart of nuclei, shell effects lead to an excited minimum associated with a large, prolate deformation with a major to minor axis ratio of ~2:1 [1, 2, 3]. The properties of the excitations occurring in this superdeformed (SD) minimum continue to be a subject of much interest. The goals of this work were to understand [i] the feeding (and decay) of normal deformed (ND) and SD bands in the A~150 mass region, [ii] the properties of excited SD and ND states above the respective yrast lines as revealed by the associated quasicontinuum (QC) γ–ray spectra and [iii] the γ–γ correlations.

The rotational damping is described by three parameters: Γ_μ, Γ_{rot} and I_{nar} [4, 5, 6, 7, 8]. The narrow rotational damping width, Γ_μ, describes the interval over which a basis state mixes with other basis states, due to the residual interaction, to form compound states. In the decay of an initial state at spin I, Γ_{rot} characterizes the spread over the many final states at spin I-2, through collective E2 matrix elements. Finally, I_{nar} describes the

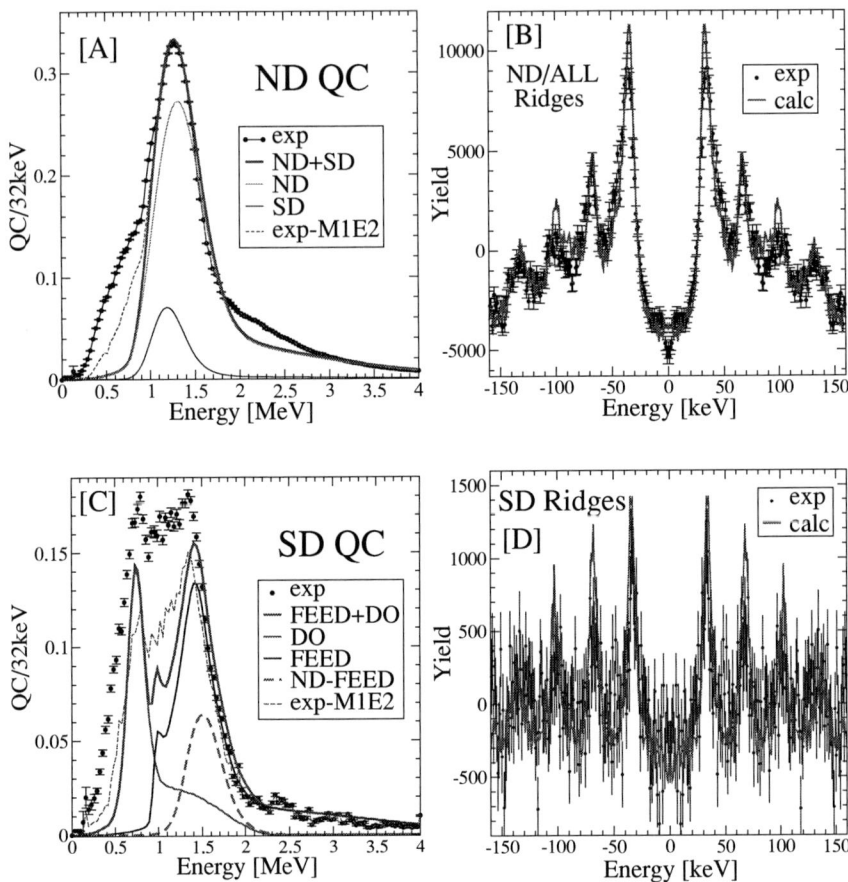

FIGURE 1. [A]([C]) The QC spectra of γ rays and [B]([D]) γ–γ ridges when pairwise coincidence gates are set on combinations of clean ND (SD band 1) lines in ^{152}Dy [10]. The lines show the results of MC calculations, which reproduce the spectra. See text for details.

relative strength of the narrow fraction. The three rotational damping width parameters were extracted by reproducing the measured QC and ridge spectra in Fig. 1 with a Monte Carlo (MC) calculation described in detail in [9].

THE MEASURED QC AND RIDGE SPECTRA

With clean, pairwise coincidence gates placed on ND and SD transitions, both 1–dimensional (1D) γ spectra and 2–dimensional (2D) γ-γ matrices were extracted from the data. Background spectra and matrices were also collected, so that a background subtraction [11] could be performed. The 1D spectra were then unfolded [12], corrected for efficiency, normalized to multiplicity, and the discrete peaks removed. The result is the QC of γ rays from the cooling of the nucleus, which precedes the population of discrete states near the yrast line. The QC spectra are shown in Figs. 1A and 1C. The latter includes the γ rays from both the feeding (FEED) and decay–out (DO) of SD band 1. The 2D matrices were unfolded in 2D before being 'COR subtracted' [13], so that there were no net counts left. After discrete peaks (and stripes) have been removed in these matrices, a projection was made on the cross–diagonal. These projections, which show the ridges, are presented in Figs. 1B and 1D for the ND and SD gated matrices, respectively.

ND and SD bands were populated in ^{152}Dy using the reaction ^{108}Pd(^{48}Ca,4n)^{152}Dy at 191 MeV (mid target). The ^{48}Ca beam was delivered by the 88–inch cyclotron facility at the Lawrence Berkeley National Laboratory and the target consisted of a stack of two $\sim$0.4 mg/cm^2 self-supporting ^{108}Pd foils. The γ rays were measured with the Gammasphere (GS) array, comprised of 100 Compton suppressed germanium detectors. As described in Ref. [10], events associated with the detecting the decay of the 86 ns, 17^+ yrast isomer [14] on a stopper foil.

THE ENTRY DISTRIBUTION

The starting points for the γ cascades followed in a MC code come from a *measured* 2D entry distribution, i.e. the distribution in the spin and energy plane that a nucleus occupies after the last particle has been evaporated. This distribution was measured by recording the sum–energy energy and number of detectors that fire in GS when gates are placed on clean combinations of ND lines (at moderate spins). The response of GS was measured using a ^{88}Y source and a MC procedure was used to unfold and determine the true total energy and multiplicity distribution of the γ cascades. The properties of the QC (i.e., multipolarity decomposition) allows for the multiplicity to be translated into spin. The entry distribution, when gates are placed on ND lines, is shown in Fig. 2A, along with the ND and SD yrast lines. The mean entry spin and energy were measured to be 49.6(1.0)$\hbar$ and 28.4(6) MeV, which are in excellent agreement with the mean values obtained from the 1D QC spectrum analysis: 49.0(1.0)$\hbar$ and 29.3(7) MeV.

The SD entry distribution can be measured in the same way by placing clean pairs of gates on transitions in the SD band 1 of ^{152}Dy. This distribution is shown in Fig. 2B. The mean entry spin and energy are 56.6(1.5)$\hbar$ and 29.2(8) MeV. Compared to the ND (I,E) entry distribution in Fig. 2A, the SD entry distribution is clearly moved towards higher spins and lower excitation energy with respect to the SD yrast line. Obviously, the part of the entry distribution below the lowest of the two yrast lines is unphysical and is attributed to uncertainties in the measurement of the entry distribution.

MONTE CARLO CALCULATIONS

A MC calculation is used to generate γ-ray spectra, which *simultaneously* reproduce both the 1D and 2D spectra; i.e., the QC spectrum and the ridges shown in Fig. 1. The MC simulations include both the γ rays feeding and deexciting the SD well. In the decay-out of the SD well, there is a significant E2 component (around 0.7 MeV in Fig. 1C). In contrast, the decay-out spectrum in the A$\sim$190 region is dominated by statistical high-energy E1 transitions [15]. As was mentioned in the introduction, the MC calculation contains the rotational damping parameters Γ_μ, Γ_{rot}, and I_{nar} in the SD and

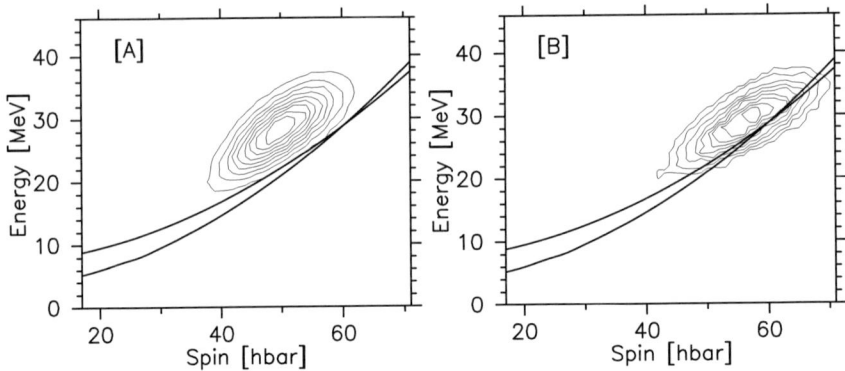

FIGURE 2. The measured entry distributions when gates are set on pairwise ND transitions [A] and SD transitions [B]. Also shown are the ND and SD yrast lines in ^{152}Dy.

ND wells [4, 5, 6, 7, 8]. Reproducing the measured QC spectra and ridges through the MC calculations allows for a measurement of the rotational damping in the two wells.

In this presentation, the (renormalized) rotational damping functions of Refs. [4, 5, 6] are used in the MC code. Without varying the strength of the theoretical functions, it is not possible to reproduce the QC and ridge spectra. Thus, the ansatz in these calculations is that the strengths need to be modified for simultaneous fits to the QC and ridge spectra in Fig. 1. For each γ–decay step, a total of eight decay widths are calculated: namely those of quadrupole E2 transitions ($\Delta I = 2$) and statistical E1 transitions ($\Delta I = -1, 0, 1$) in both the ND and SD wells. If the nucleus is SD at the time of decay, the four decay widths in the ND well are attenuated by the penetration probability through the barrier that separates the ND and SD states. The barrier penetration is small if the SD state is deep in the SD well, and near unity at the top of the barrier. On the other hand, if the nucleus is in a state of the ND well, the four calculated decay widths in the SD well are attenuated correspondingly. The subsequent decay path is selected in a MC fashion from values of the eight decay widths. If a ND decay width is selected, the decay occurs in the ND well and vice versa.

For E2 transitions, the wide or narrow components are first selected based on I_{nar}. An additional Gaussian deviation around the mean γ transition energy is then added, the width of which is based on the input rotational damping values Γ_μ or Γ_{rot}. This is actually done before the final E2 decay widths are calculated, since the E^5 term in the decay width formula strongly favors E2 decays where the (mostly) wide component adds to the decay γ–ray energy, giving rise to an important *extra cooling* of the γ cascades.

In the simulations, the single γ rays of the cascades are binned, forming the calculated QC spectra displayed in Fig. 1A. For the ND (which include SD precursor states) QC spectrum, shown in Fig. 1A, all MC γ cascades are used no matter which well the γ cascades were trapped in. For the SD QC spectrum, Fig. 1C, only cascades trapped in the SD well, which happens about separating the SD well from the ND well is adjusted to get the observed population of the SD band in the MC calculations. In the SD feeding region, ~ 42–$66\hbar$, the height of the barrier is ~ 1.3 MeV.

Quantity	ND	SD
Γ_μ	16.8 keV	5 keV
Γ_{rot}	176 keV	220 keV

Tab. 1. The mean values of the distribution of rotational damping parameters that are able to reproduce the observed QC and ridges spectra in the MC calculation.

The calculated γ–γ coincidences in the cascades are binned in 2D γ–γ matrices after correcting for the experimental detector efficiency. The simulated 2D γ–γ coincidence matrices have peaks and 'stripes' removed (even though discrete peaks are not included in the model), and are COR subtracted, *exactly* as done for the experimental data. Subsequently, the calculated ridge spectra are extracted by projecting on the cross-diagonal from the *same* γ–energy regions as for the experimental matrices. Figures 1B and 1D present the calculated ridges obtained in this way with the same selection of cascades as for the measured QC spectrum.

CONCLUSIONS AND FUTURE

It is remarkable how well the ND and SD QC – and ridge – spectra can be reproduced using the MC calculations. The model seems to contain all the physics necessary to describe the cooling of the ^{152}Dy nucleus. The theoretical Γ_μ, Γ_{rot} and I_{nar} rotational damping functions must in general be renormalized by large factors to reproduce the data. Work is in progress to improve the calculations by using a more realistic level density formula at high spins.

The analysis presented here is in progress and all results should be considered preliminary. This work was supported in part by the U.S. Dept. of Energy, Office of Nuclear Physics, under Contract No. W–31–109–ENG–38 and DE–AC03–76SF00098, and the Danish Natural Science Foundation.

REFERENCES

1. P. J. Twin *et al.*, *Phys. Rev. Lett.* **57**, 811 (1986).
2. P. Nolan *et al.*, *J. Phys G* **11**, L17 (1985).
3. R. V. F. Janssens, and T. L. Khoo, *Annu. Rev. Nucl. Part. Sci.* **41**, 321 (1991).
4. B. Lauritzen, T. Døssing, R.A. Broglia, *Nucl. Phy.* **A457**, 61 (1986).
5. T. Døssing, and B. Herskind, *Acta Phy.* **32**, 2565 (2001).
6. M. Matsuo *et al.*, *Phys. Lett.* **B465**, 1 (1999).
7. T. Døssing *et al.*, *Phys. Rep.* **268**, 1 (1996).
8. B. Mottelson, *Proceedings of the international seminar on the fontiers of nuclear spectroscopy*, World Scientific, 1993.
9. T. Lauritsen *et al.*, *Phys. Rev. C*, to be published (2005).
10. T. Lauritsen *et al.*, *Phys. Rev. Lett.* **88**, 042501 (2002).
11. B. Crowell *et al.*, *Nucl. Instrum. Methods.* **A355**, 575 (1995).
12. D. C. Radford *et al.*, *Nucl. Instrum. Methods* **A258**, 111 (1987).
13. O. Andersen *et al.*, *Phys. Rev. Lett.* **43**, 687 (1979).
14. T. L. Khoo *et al.*, *Phys. Rev. Lett.* **41**, 1027 (1978).
15. T. Lauritsen *et al.*, *Phys. Rev.* **C62**, 044316 (2000).

Multinucleon Transfer Reactions to Study Single-Particle Evolution in Se Isotopes

[1]P.H.Regan, [1]G.A.Jones, [1]Zs.Podolyák, [2]N.Yoshinaga,
[3]K.Higashiyama, [4]G.deAngelis, [4,5]Y.H.Zhang, [4]A.Gadea, [6]C.A.Ur
[4]M.Axiotis, [6]D.Bazzacco, [7]R.Broda, [8]D.Bucurescu, [6]E.Farnea,
[1]W.Gelletly, [8]M.Ionescu-Bujor, [8]A.Iordachescu, [4]Th.Kröll,
[1]S.D.Langdown, [6]S.Lenzi, [6]S.Lunardi, [4]N.Marginean, [4]T.Martinez,
[9]N.Medina, [6]R.Menegazzo, [4]D.R.Napoli, [10]B.Quintana, [11]B. Rubio,
[4]C. Rusu, [12]R.Schwenger, [4]D.Tonev, [1,4]J.J.Valiente Dobon,
[13]W. von Oertzen,

[1]*Dept. of Physics, School of Electronics and Physical Sciences, Guildford, Surrey, UK*
[2]*Department of Physics, Saitama University, Saitama City, 338-8570, Japan*
[3]*Department of Physics, University of Tokyo, Hongo, Tokyo, 113-0033, Japan*
[4]*INFN Laboratori Nazionali di Legnaro,Legnaro, Italy*
[5]*Institute of Modern Physics, CAS, Lanzhou, People's Republic of China*
[6]*INFN, Sezione di Padova, Padova, Italy*
[7]*Institute of Nuclear Physics, Krakow, Poland*
[8]*National Institute for Physics and Nuclear Engineering, Bucharest, Romania*
[9]*Instituto de Fisica, Universidade de Sao Paulo, Sao Paulo, Brazil*
[10]*Grupo de Fisica Nuclear, Universidade de Salamanca, Spain*
[11]*Instituto de Fisica Corpuscular, Valencia, Spain*
[12]*Institut für Kern und Hadronenphysik, Forschungszentrum Rossendorf, Dresden, Germany*
[13]*Freie Universität Berlin, Fachbereich Physik, Berlin, Germany*

Abstract. This contribution reports on the results of an experiment to study the near-yrast states in selenium isotopes around N=50, following their population in thick-target, multinucleon transfer reactions between an ^{82}Se beam and a ^{192}Os target. Discrete decays from states with spins of 20ℏ and higher are clearly identified in the Os-like fragments. Information on the size of the energy gap between the $f_{5/2}$ and $p_{3/2}$ proton orbitals can be inferred from the spectroscopy of the N=48 isotone, ^{82}Se. The experiment results for the level schemes for ^{80,82}Se derived in the current work are compared with restricted-basis shell model calculations, which give a good description of the yrast sequences of these nuclei using a basis space limited to excitations in the ($p_{3/2}$, $p_{1/2}$, $g_{9/2}$) neutron and ($f_{5/2}$, $p_{3/2}$, $p_{1/2}$) proton orbitals.

Keywords: gamma-ray spectroscopy, energy levels, shell model calculations, deep-inelastic reactions.
PACS: 21.10.-k , 21.60.Cs, 21.70.Hi , 27.50.+e

INTRODUCTION

The study of neutron-rich nuclei is of significant current interest. Of particular focus is the question of the robustness of magic numbers at N,Z=28,40 and 50 and the appearance of sub-shell closures at nucleon numbers 32 and 34 with neutron/proton number evolution. Single-particle 'monopole' energy shifts between specific orbitals can result in a rearrangement of the traditional ordering of sub-shells associated with near-spherical nuclei close to stability [1,2], which in turn results in the appearance/removal of new/traditional magic numbers. Recent reports following initial studies with projectile fragmentation [3,4] and deep-inelastic reactions [5,6,7] have suggested a possible sub-shell closure for neutron number N=32 (and perhaps N=34), associated with a large single particle gap between the neutron $p_{3/2}$ and $p_{1/2}$ orbitals. In this case, it has been proposed that a weakening of the proton-neutron $\pi 1f_{7/2} \rightarrow \nu 1f_{5/2}$ interaction and a significant $\nu 2p_{3/2} \rightarrow \nu 2p_{1/2}$ spin-orbit splitting causes the N=32 shell closure by causing the p-shell neutron orbitals to be lowered in energy with respect to the $f_{5/2}$ states, until they are inverted compared to the near-stable ordering (in which it is expected that the $f_{5/2}$ lies below the $p_{3/2}$ and $p_{1/2}$ respectively).

Monopole energy shifts can be inferred from both direct measurement of single-particle energies in odd-A nuclei or alternatively, by identifying configuration specific decays in the decay sequences at higher spins [8]. The current work is aimed at the investigation of the proton energy levels above the Z=28 shell closure for N~50. This work also allows information on the evolution of the N=50 shell closure for neutron-rich nuclei. The near-stable selenium isotopes, with Z=34 and N~50 are usually associated with a full $f_{5/2}$ proton sub-shell in their lowest lying, spherical configuration. Higher-spin states which can not be simply made in the available neutron valence space require excitations from the proton $f_{5/2}$ orbital into the $p_{3/2}$ level. The size and consistency of this Z=34 sub-shell closure (i.e., the $\nu f_{5/2} \rightarrow \nu p_{3/2}$ gap) is one of the focuses of the current work.

EXPERIMENTAL DETAILS, DATA ANALYSIS & RESULTS

The nuclei of interest were populated following heavy-ion multi-nucleon transfer reactions using 'thick target' measurement [9,10]. The beam of ^{82}Se ions of energy 460 MeV was provided by the tandem XTU and superconducting LINAC ALPI at the Laboratori Nazionali di Legnaro, Italy. The ions were incident on an isotopically enriched (97.8%) ^{192}Os target of thickness 50 mg/cm^2. The target was then backed with a 0.2mm layer of Ta which was thick enough to stop all the subsequent reaction products. Reaction gamma rays were measured using the GASP spectrometer [11] which consists of 40 Compton-suppressed coaxial hyperpure germanium detectors and a 72 element inner BGO ball for gamma-ray multiplicity and calorimetry measurements. Events consisting of at least three Compton suppressed germaniums and 2 BGO elements which fired in prompt (~30 ns) coincidence within an event time of approximately 1µs were written to tape for subsequent off-line analysis. The typical on-target beam current during the experiment was approximately 2 particle nA, which resulted in a master-trigger rate of approximately 4kHz over the course of the six day run. Further details and results of this experiment have been presented in references [9,10,12,13].

The data were sorted in offline analysis into standard Radware [14] symmetric γ-ray coincidence matrices and cubes. Examples of sums of double-gated gamma-ray triples spectra from the cube analysis are shown in figures 1 and 2 for Se and Os nuclei respectively. The gamma-ray information on the Os isotopes has been reported elsewhere [9], but we note again the high-spin population on the near-target nuclei in the current work. Discrete states with angular momenta of 20ℏ and higher are observed in $^{186\rightarrow190}$Os, highlighting the effectiveness of multinucleon transfer reactions for the population of near-yrast states for near-stable nuclei.

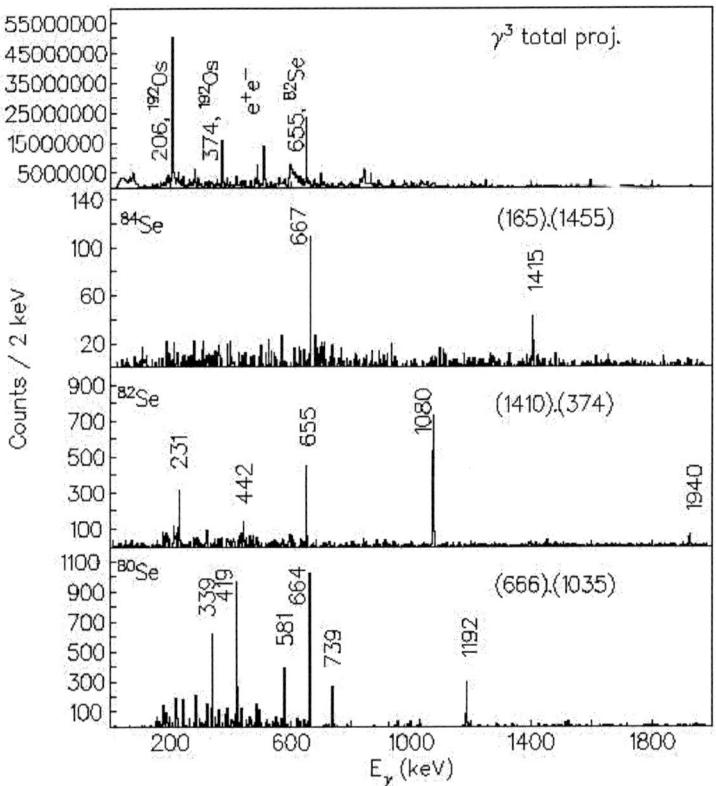

FIGURE 1. Total projection of triples γ-ray coincidence cube (upper panel) and double-gated γ-ray gates on transitions in ^{80,82,84}Se from the current work.

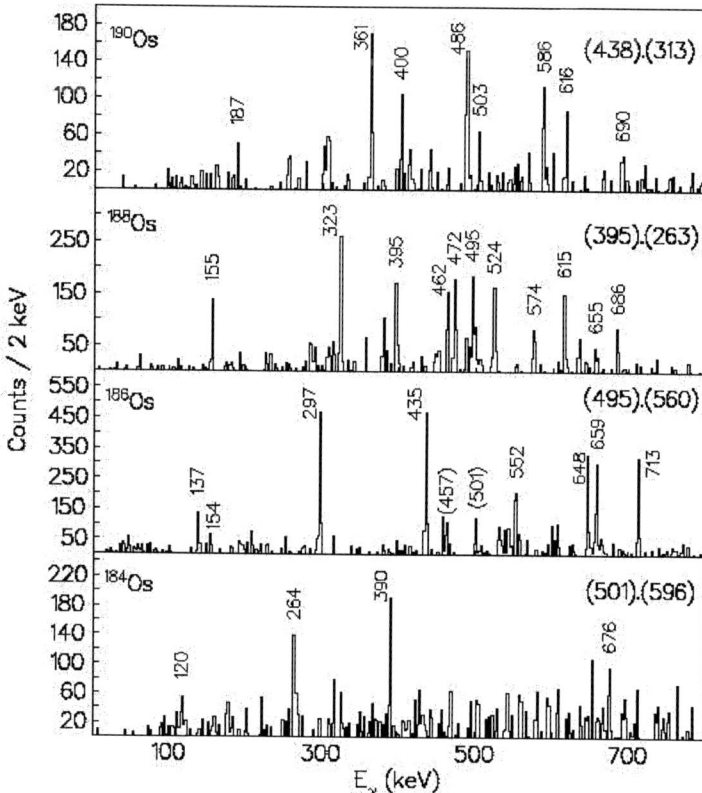

FIGURE 2. Double-gated triples gamma-ray spectra on transition in Os isotopes observed in the current work. Note that decays from discrete states with spins of 20ℏ and higher are observed in these cascades.

The decay schemes associated with ^{80,82,84}Se deduced from the current work are shown in figure 3. These data are consistent with previously reported studies of ^{80,82}Se performed using the EUROBALL array following population in fusion-fission reactions [15,16,17]. States up to spin 8ℏ have also been identified previously in the N=48 isotone ^{82}Se following deep-inelastic reactions [18,19], with a nanosecond isomer associated with the $(\nu g_{9/2})^{-2}$, $I^\pi = 8^+$ seniority isomer reported at E_x=3519 keV. (Such 'seniority' isomers are a standard feature throughout the N=48 isotonic chain [18]). Three further gamma rays are observed in the current work in cascade above the 3519 keV state. The most intense is the 1940 keV transition which feeds directly into the 8^+ isomeric level. While definite spin/parity assignments were not possible for these states in the current work, comparison with both shell model calculations (see below), and other N=48 isotones [20,21] strongly suggests that the 1940 keV transition represents the yrast $10^+ \rightarrow 8^+$ decay in ^{82}Se. There is also some evidence for an isomeric decay with a half-life in the 10ns range from the 2365 keV level in ^{80}Se. This has also been identified in fusion-fission work between a ^{178}Hf beam and ^{27}Al target with GAMMASPHERE [22]. The data on the N=52 isotone, ^{84}Se, presented here has been reported in ref [10] and is consistent with other reports from both deep-inelastic [23,24] and fusion-fission data [25].

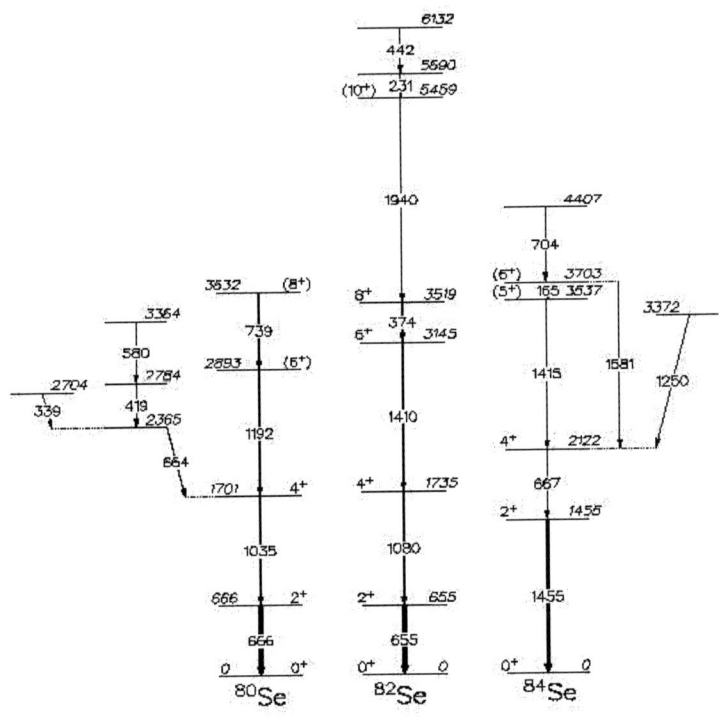

FIGURE 3. Partial level schemes for ^{80,82,84}Se as deduced in the current work

SHELL MODEL CALCULATIONS

Full shell model calculations have been performed for ^{80,82}Se assuming a simple valence space made up from neutrons occupying $p_{3/2}$, $p_{1/2}$ and $g_{9/2}$ orbitals, while the proton valence space was restricted to the $p_{3/2}$, $p_{1/2}$ and $f_{5/2}$ levels. The predicted energy levels from these calculations are compared with the experimental data determined for these nuclei in the current work, in figures 4 and 5. In general, the shell model calculations reproduce the experimental levels rather well. The large energy gap predicted between the yrast 10^+ and 8^+ levels in ^{82}Se can be understood simply by the increase in valence space required to allow states of this high angular momentum. Up to spins of $8\hbar$, near yrast, low-seniority states can be constructed primarily via excitations in the two neutron $g_{9/2}$ holes in the N=50 shell. Higher spins require extra proton excitations, and since in the lowest excitation one would expect a full, close $f_{5/2}$ proton shell for the spherical Z=34 Se isotopes, this corresponds to excitations between the $f_{5/2}$ and $p_{3/2}$ proton configurations.

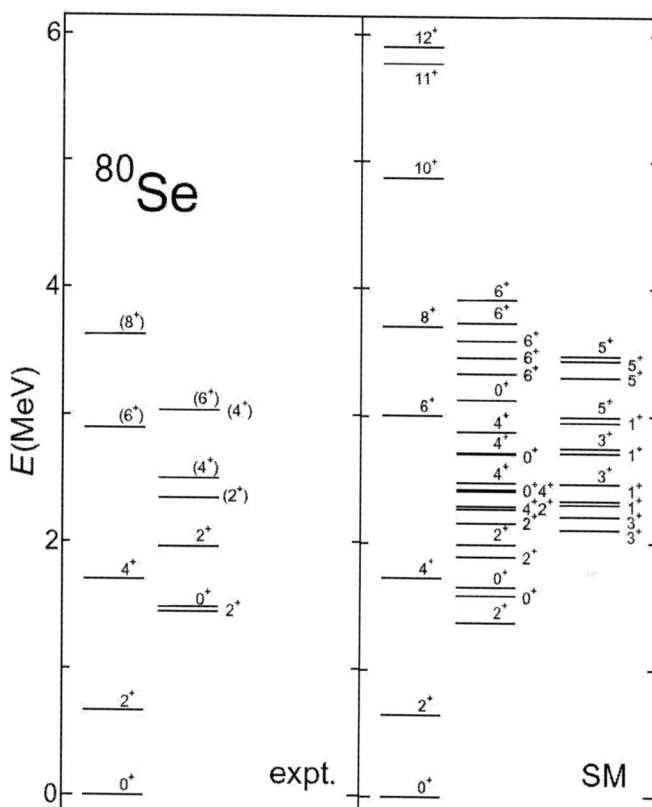

FIGURE 4. Comparison between experimental and shell model levels in ^{80}Se.

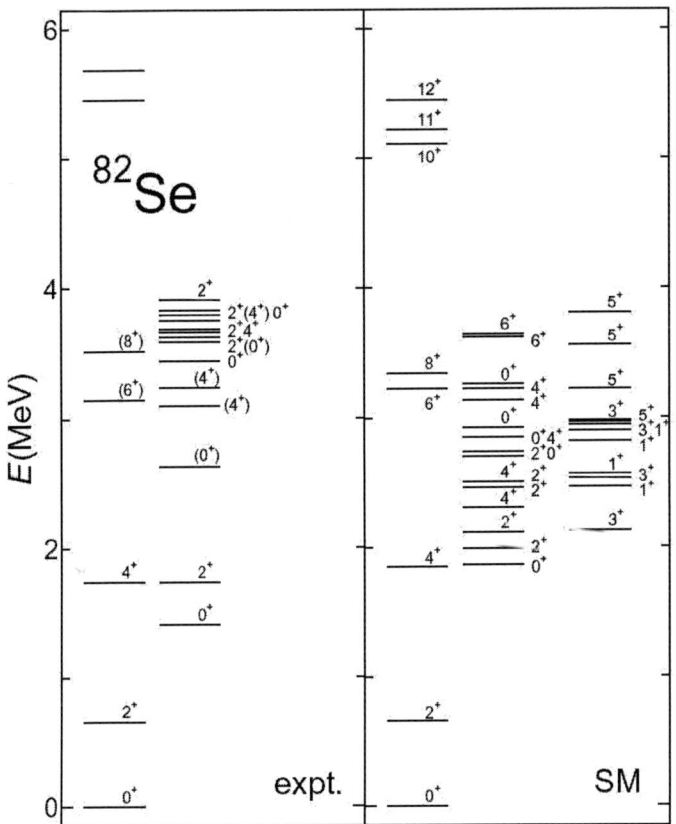

FIGURE 5. Comparison between experimental and shell model levels in ^{82}Se.

SUMMARY AND FUTURE WORK

In summary, the near-yrast states of ^{80,82,84}Se have been studied following population in heavy-ion collisions between an ^{82}Se beam and a thick ^{192}Os target. The results are compared with restricted basis shell model calculations which reproduce the energy observed energy levels rather well and provide a mechanism for measuring the $f_{5/2} \rightarrow p_{3/2}$ neutron single-particle splitting in this region.

ACKNOWLEDGMENTS

This work is supported by the Engineering and Physical Sciences Research Council (UK), EU Contract Nos. ERBFMGECT 980110 and HPRI-CT-1999-00083 and a Marie Curie grant of the European Community program IHP under Contract No. HPMF-CT-2002-02018.

REFERENCES

1. B.A.Brown, *Prog. Part. Nucl. Phys*. **47** (2001) 517.
2. T. Otsuka et al., Phys. Rev. Lett., **87** (2001) 082502
3. D.-C. Dinca et al., *Phys. Rev.* **C71**, 041302(R) (2005)
4. S.N.Liddick et al., *Phys. Rev.* **C70**, 064303 (2004)
5. S.N. Liddick et al., *Phys. Rev. Lett.* **92**, 072502 (2004)
6. B. Fornal et al., *Phys. Rev.* **C70**, 064304 (2004)
7. R.V.F. Janssens et al., *Phys. Lett.* **546B**, 55 (2002)
8. A. F. Lisetky et al., *Phys. Rev.* **C70,**, 044314 (2004)
9. Zs. Podolyak et al., *Int. J. Mod. Phys*. **E13**, 123 (2004)
10. Y.Zhang et al., *Phys. Rev.* **C70**, 024301 (2004).
11. D. Bazzacco, *Proceedings of the International Conference on Nuclear Structure at High Angular Momentum*, Ottawa, 1992, Vol. II. p. 376, Report No. AECL 10613
12. G.A. Jones et al., *Acta Phys. Pol.* **B36**, 1323 (2005)
13. G.A. Jones et al., *J. Phy.s.* **G31**, S1891 (2005)
14. D.C. Radford, *Nucl. Inst. Meth. Phys. Res.* **A361**, 297 (1995)
15. M.G. Porquet et al., Proc.2nd Intern.Workshop Nuclear Fission and Fission-Product Spectroscopy, Seyssins, France, p.212 (1998); AIP Conf. Proc. **447** (1998)
16. J.K.Tuli, *Nucl. Data. Sheets* **98**, 209 (2003)
17. B. Singh, *Nucl. Data Sheets* **105**. 223 (2005)
18. T. Ishii et al., *Eur. Phys. J.* **A13**, 15 (2002)
19. A. Makishima et al., *Phys. Rev.* **C59**, R2331 (1999)
20. E.K. Warburton et al., *Phys. Rev.* **C31**, 1184 (1985)
21. E.K. Warburton et al., *J.Phys.* **G12**, 1017 (1986)
22. G.A. Jones, *private communication*
23. S. Lundardi, *Acta Phys. Pol.* **B36**, 1301 (2005)
24. G. deAngelis, *Nucl. Phys.* **A751**, 553c (2005)
25. A. Prevost et al., *Eur. Phys. J* **A22**, 391 (2004)

NUCLEAR STRUCTURE
(THEORY)

Nuclear Mean Field from the Shell-Model Point of View

N. A. Smirnova, A. De Maesschalck, K. Heyde

Department of Subatomic and Radiation Physics, University of Ghent, Proeftuinstraat 86, 9000 Ghent, Belgium

Abstract.
  Some of the intriguing questions related to the properties of exotic nuclei are how their shell structure changes with respect to that of nuclei close to the line of β-stability and what mechanisms are responsible for the observed changes. We discuss the variations of the nuclear mean field in a series of isotopes from proton-rich to very neutron-rich species studied within the nuclear shell model. We make use of the approximate expression for the shell-model centroids of the single-particle states and apply it to explore the evolution of the mean field in the series of F, Cu and Sb isotopes. The approximation allows us to treat heavy systems which at present cannot be studied by exact diagonalization. The changes of the single-particle energies are shown to be mainly due to the monopole part of the effective two-body interaction. The comparison between different schematic and realistic forces is performed in an attempt to understand the role of specific terms acting within the effective interaction.

Keywords: Single-particle level structure, monopole shift, shell model, effective interactions
PACS: 21.10.Pc,21.10.Jx,21.60.Cs

INTRODUCTION

The nuclear mean field is a property that follows from the nucleon-nucleon (n-n) force which gives rise to saturation properties inside the atomic nucleus and the Pauli principle. Assuming existence of a well behaved effective n-n potential, one can derive the nuclear mean field within the Hartree-Fock (HF) theory. In particular, we will note that if one simplifies the interaction to a schematic form $V = V_0 \delta(\vec{r} - \vec{r}')$, the local Hartree potential,

$$U(\vec{r}) = \int V(\vec{r},\vec{r}')\rho(\vec{r}')d\vec{r}', \qquad (1)$$

will behave like the nuclear mass density $\rho(\vec{r})$.

  As was recognized long ago [1, 2], experimental information on odd-A nuclei in the region of β-stability indicates that the mean field can be well approximated by a harmonic oscillator potential with a strong spin-orbit component and an attractive $\hat{l}^2$ term, i.e.

$$h = h_{\text{h.o.}} + v_{ll}\hat{l}^2 + v_{ls}\hat{l}\cdot\hat{s}. \qquad (2)$$

With appropriate strengths, such an independent particle model explains the known shell gaps that give rise to increased stability at N,Z=2,8,20,28,(40),50,82,126 and the spin and parity assignments of the ground states of most stable odd-A nuclei. However, this shell structure of the single-particle energy spectra will change for nuclei situated far from the region of β-stability (neutron-rich and proton-rich nuclei).

The phenomenological potentials are certainly not applicable for carrying out extrapolations to nuclear systems at extreme values of isospin, because its parameters have been determined for the limited patch of stable and almost stable nuclei. Using the HF or Hartree-Fock-Bogoliubov (HFB) method, and starting from well tested Skyrme functionals, it has been shown [3, 4] that the local part of the mean-field exhibits a more diffuse surface region in neutron-rich nuclei as compared to that in nuclei near stability (more like the shape of a Woods-Saxon potential). It has also been noted [3] that one can actually simulate this transition rather well by gradually turning off the $v_{ll}l^2$ term in the one-body potential (2).

Using effective two-body forces that allow to determine the nuclear average potential through HF(B) theory, one runs into problems when moving far away from the region of stable nuclei. Within self-consistent calculations, one needs a two-body force, which in turn, generates the mean field and the energy spectra. Since there is no *ab-initio* knowledge of how the n-n force changes with N and/or Z (isospin dependence), the applications to the regions very far from stability are still limited, and various studies of the nuclear matter at extreme ratios of neutrons to protons (e.g. neutron matter) can be very helpful.

In this contribution we attack the problem through studies of the effect that specific components of the force, such as spin-exchange $(\vec{\sigma}_1 \cdot \vec{\sigma}_2)$, tensor $S_{12} = (\vec{\sigma}_1 \times \vec{\sigma}_2)^{(2)} \cdot Y^{(2)}(\hat{r})$, two-body spin-orbit, etc, have on changes in the single-particle energy spectra in correcting the mean-field to the lowest order.

In the next parts, we shall try to understand a number of the observed variations in single-particle energies (in as much as they are well determined experimentally for regions where Z (N) number is just adjacent to a 'closed' shell) calculating the contribution of the residual proton-neutron interaction mainly to the monopole field. Comparison of theoretical predictions with the experimental data, as well as comparison of the results obtained with realistic and specific schematic forces helps to elucidate the driving mechanism for these changes.

SHELL-MODEL APPROACH

Evolution of the shell structure when moving to neutron-rich or proton-rich nuclei can be most clearly seen in the energy spectra of odd-A isotopes (isotones) having an extra proton (neutron) outside a closed proton (neutron) core. Then, within the shell-model framework, the proton (neutron) self-energy correction has a relatively simple expression, containing the monopole part of the proton-neutron interaction and the pairing correlations among valence neutrons (protons). For the case of a proton interacting with neutrons, spread over a given number of orbitals by a pairing force, the modified (or effective) single-proton energy can be obtained as [5, 6, 7]:

$$\tilde{\varepsilon}_{j_\pi} = \varepsilon_{j_\pi} + \sum_{j_v} \overline{E}_{j_\pi j_v}(2j_v+1)v_{j_v}^2 . \tag{3}$$

Here $v_{j_v}^2$ denote the occupation probabilities of the corresponding valence neutron orbitals (j_v), over which the summation is performed, and $\overline{E}_{j_\pi j_v}$ is the average proton-

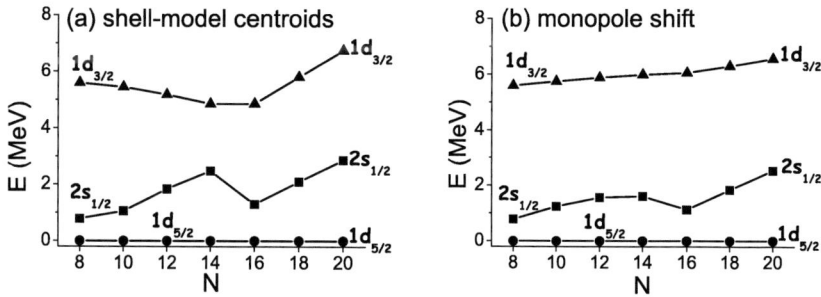

FIGURE 1. F-isotopes: (a) shell-model centroids of the single-particle states (5); (b) effective single-particle energies obtained from eq. (3).

neutron interaction energy for a given proton and a given neutron orbital,

$$\overline{E}_{j_\pi j_\nu} = \frac{\sum_J \langle j_\pi j_\nu; J|V|j_\pi j_\nu; J\rangle (2J+1)}{\sum_J (2J+1)}. \quad (4)$$

Note that only the monopole part of the effective two-body interaction V gives a non-zero contribution to (4). Thus, the evolution of the effective single-proton (single-neutron) energies as a function of the neutron (proton) number due to the described mechanism is usually referred to as the monopole shift.

One can easily see that under the assumption of the normal filling of neutron orbitals, expression (3) coincides with the standard definition of the effective single-particle energies as single-nucleon separation energies obtained from the diagonal part of the shell-model monopole Hamiltonian [8, 9, 10, 11]. Detailed studies of the pairing correlations in addition to the monopole interaction have been also performed in Ref. [12].

In spite of its simplicity, expression (3) represents a very good approximation to the full single-particle centroids as obtained from the complete diagonalization of the shell-model Hamiltonian:

$$E(nlj) = \frac{\sum_k S^k_{nlj} E^k_{nlj}}{\sum_k S^k_{nlj}}, \quad (5)$$

where S^k_{nlj} are spectroscopic factors of the state with the energy E^k_{nlj}. As an example, the single-proton states of (sd)-shell obtained from (5) and those obtained by the approximate formula (3) are compared in Fig. 1. The USD interaction [13] was used and the diagonalization was performed with the shell-model code ANTOINE [14] (this code was used throughout our work). To get $v^2_{j_\nu}$ in eq. (3), the BCS equations for the pure pairing force of $G = 19/A$ MeV were solved. The agreement is seen to be very good. Similar agreement holds for the Cu-isotopes, for which the importance of the BCS correlations in the neutron system is clearly present (see Fig. 2(c–f) in Ref. [16]).

This result provides a direct confirmation that the monopole part of the proton-neutron effective interaction together with the pairing correlations gives the major contribution

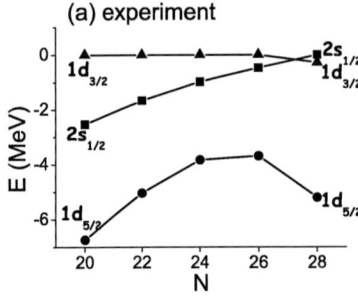 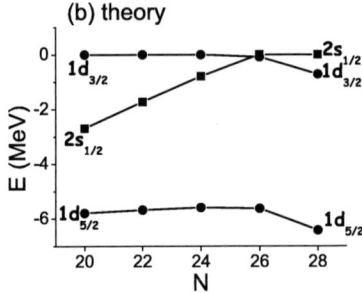

FIGURE 2. Variation of single-particle states in K-isotopes: (a) experimental centroids; (b) monopole shift (3).

to the shift of the shell-model single-particle centroids in nuclei adjacent to semi-magic ones. Thus, the approach can be used to study the evolution of the shell structure throughout the nuclear chart, including the regions where the model space is too big for performing exact diagonalization.

THEORY VERSUS EXPERIMENT

There exists a significant amount of experimental data accumulated on the shifts of the single-particle states in different mass regions [17, 18] and new data on more and more exotic systems are under way [19, 20, 21, 22, 23]. To compare theoretical predictions with experimental data one would like at best to possess information on the full single-particle centroids, which means the knowledge of both energies and spectroscopic factors of the states of a certain J^π. This condition prohibits a systematic detailed comparison between theoretical predictions and experimental data, since at present typically only few isotopes or isotones in a given chain are characterized by extensive enough data on the single-particle spectroscopic strength. In Ref. [16] we discuss the situation for Cu-nuclei.

In Fig. 2 we compare the experimental single-particle centroids for K-isotopes [24, 25] and the monopole shift results (3) using the effective interaction for $(sdpf)$-shell model space from Ref. [26] and the pure pairing force with $G = 19/A$ MeV. The gradual shift of the proton $2s_{1/2}$ state with neutrons filling $1f_{7/2}$ orbital is well reproduced.

In Fig. 3, the experimental energies of the lowest states of a given spin and parity in Sb-isotopes [24, 19, 27] are compared with the monopole shift predictions using the realistic effective interaction for $(1g2d3s1h_{11/2})$-shell model space adjusted in Ref. [28] on the basis of the G-matrix [29] and a pure pairing force with $G = 23/A$ MeV. The experimental lowest $5/2^+$, $7/2^+$ and $11/2^-$ states are assumed to contain a dominant contribution of the proton $2d_{5/2}$, $1g_{7/2}$ and $1h_{11/2}$ single-particle configurations. The theoretical calculations start from the known proton single-particle and neutron single-hole energies in ^{133}Sb [19] and ^{131}Sn [27], respectively. The encouraging result is that

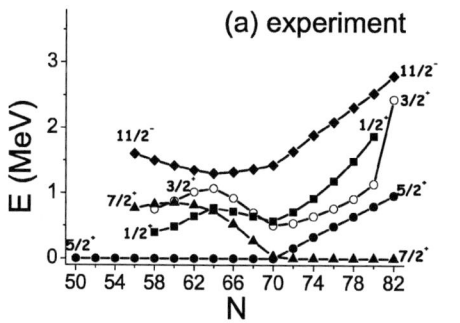 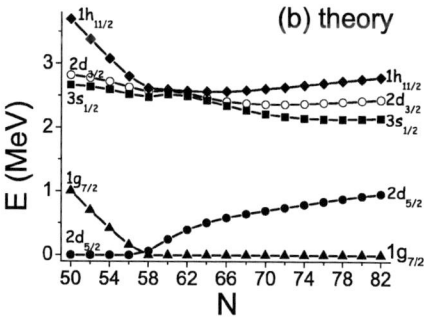

FIGURE 3. Variation of single-particle states in Sb-isotopes: (a) experimental lowest states; (b) monopole shift (3).

the realistic interaction reproduces well the overall dropping of $1g_{7/2}$ as neutrons fill the $1h_{11/2}$ orbital, although the crossing with the $2d_{5/2}$ state is calculated to be around ^{109}Sb. One of the ways to improve is to take into account a similar inversion in neutron single-particle energies used in the solution of the BCS equations which is known to take place (the neutron $2d_{5/2}$ state is the lowest in light Sn-nuclei, while $1g_{7/2}$ is the lowest in heavy Sn). The increase of the energy of the proton $1h_{11/2}$ compared to that of the $1g_{7/2}$ state in heavy Sb-nuclei can be noted as well.

The thorough analysis of experimental data reveals a number of interesting features characteristic for almost all series of isotopes and isotones. As has been concluded [17] from the single-particle spectra of odd-A Cu, Sb-isotopes and $N = 29, 51, 83$ isotones, the proton-neutron interaction matrix elements of the type $\langle j_> j'_<; J|V|j_> j'_<; J \rangle$ or $\langle j_< j'_>; J|V|j_< j'_>; J \rangle$ (here $j_> = l + 1/2$, $j_< = l - 1/2$) are regularly more attractive than those of the type $\langle j_> j'_>; J|V|j_> j'_>; J \rangle$ or $\langle j_< j'_<; J|V|j_< j'_<; J \rangle$[1]. This property of the nuclear force has been recognized earlier [30] in the studies of the mechanism responsible for the origin of nuclear deformation. A similar trend has been revealed from experimental and theoretical studies of light neutron-rich nuclei and a particular reason for that, namely, the dominant effect of $(\vec{\sigma} \cdot \vec{\sigma})(\vec{\tau} \cdot \vec{\tau})$-term in the central part of the effective interaction has been proposed [31].

Indeed, the monopole shift calculations [32] in different regions of isotopes and isotones on the basis of realistic interactions well confirm the enhanced attraction between $j_>$-$j_<$ ($j_<$-$j_>$) proton-neutron states. The question now is whether we may, on the quantitative basis, determine which particular component of the effective interaction is responsible for these particular shifts?

---

[1] Important to note that in this comparison, the reference and migrant orbitals should be characterized by the same principle quantum number n. Otherwise, the matrix elements will be simply suppressed by geometry, which is the case for the Bi-isotopes (relative shift of the proton $1h_{9/2}$ and $2f_{7/2}$ orbitals).

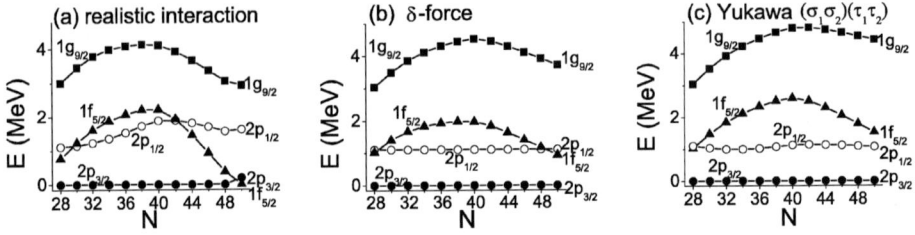

FIGURE 4. Monopole shift (3) in Cu-isotopes from (a) the realistic interaction [28], (b) δ-force (6) with $V_{eff} = 400$ MeV, and (c) the central interaction (7) with $A = C = D = 0$, $F = 1$, $V_0 = -10$ MeV, $\mu = 0.7$ fm^{-1}.

REALISTIC INTERACTION VERSUS SCHEMATIC FORCES

We have performed systematic comparison of the monopole shifts obtained with realistic and two schematic effective forces (see Ref. [32] for details):

1. Zero-range δ-force with spin-exchange

$$V_\delta(1,2) = -V_{eff}\delta(\vec{r}_1 - \vec{r}_2)(1 - \alpha + \alpha\vec{\sigma}_1 \cdot \vec{\sigma}_2); \quad (6)$$

2. The most general central potential with finite-range radial dependence

$$V(1,2) = V(r)\left(A + C(\vec{\tau}_1 \cdot \vec{\tau}_2) + D(\vec{\sigma}_1 \cdot \vec{\sigma}_2) + F(\vec{\sigma}_1 \cdot \vec{\sigma}_2)(\vec{\tau}_1 \cdot \vec{\tau}_2)\right). \quad (7)$$

The parameters in these expressions have been adjusted to the experimental data in odd-odd nuclei from the corresponding mass region. We used the Yukawa-type potential $V(r) = -V_0 \exp(-\mu r)/(\mu r)$ in (7), but other finite-range dependences (exponential, Gaussian, etc) yield very similar results.

Fig. 4 shows the results obtained for Cu-isotopes using the realistic interaction [28], δ-force and the $(\vec{\sigma} \cdot \vec{\sigma})(\vec{\tau} \cdot \vec{\tau})$-term of the central potential (the latter has the biggest matrix elements between the states of interest as was discussed in Ref. [31] and therefore produces the largest variations). Obviously, none of the schematic forces allows to reproduce the observed steep decrease of the $1f_{5/2}$ state relative to the $2p_{3/2}$ state in neutron-rich Cu-isotopes. Similarly, schematic forces (6) and (7) fail to reproduce the crossing of the proton $1g_{7/2}$ and $2d_{5/2}$ states in Sb-isotopes, well described by the realistic interaction [32].

Another interesting remark is that the δ-force and the spin-isospin independent part of the central potential (7) provide no relative shifts of spin-orbit partners, while the isospin-dependent terms of the central potential affect them in a very subtle way (note, that the pure spin-exchange part in (6) and (7) does not contribute to the average matrix element (4) at all).

These results have motivated us to study other possible ingredients present in the effective n-n force, such as the tensor force

$$V_T(1,2) = V(r)(\vec{\sigma}_1 \times \vec{\sigma}_2)^{(2)} \cdot Y^{(2)}(\hat{r})(v_0 + v_\tau(\vec{\tau}_1 \cdot \vec{\tau}_2)), \quad (8)$$

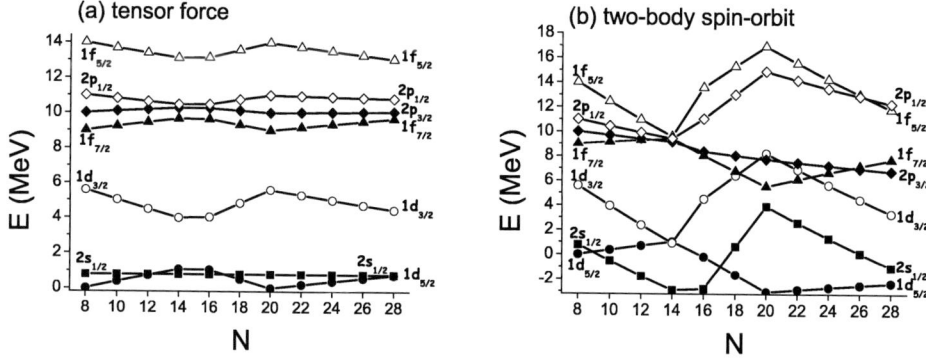

FIGURE 5. Variations of the proton single-particle states produced (a) by the tensor force (8) with $V_0 = 100$ MeV, $\mu = 0.7$ fm^{-1}, $v_0 = 0$, $v_\tau = 1$, and (b) by the two-body spin-orbit interaction (9) with $V_0 = -50$ MeV, $\mu = 0.7$ fm^{-1}, within $sdpf$-shell model space.

and the nonlocal two-body spin-orbit interaction

$$V_{lS}(1,2) = V(r)(\vec{l}\cdot\vec{S}). \tag{9}$$

Fig. 5 shows the results of the shifts of the proton single-particle states obtained with (8) and (9) in $(sdpf)$-model space under the assumption of a normal filling for neutrons (for arbitrarily chosen strengths). The particular role of these interactions has been discussed by T. Otsuka [33] as well. Indeed, it is easy to notice that using the appropriate sign of the strength of these forces, the observed variations (the enhanced attraction of proton-neutron $j_<$-$j'_>$ or $j_>$-$j'_<$ pairs) are well accounted for. However, we want to stress that it is the central part of the interaction which governs the major behavior of single-particle levels in the nuclear mean-field and in order to determine reliably the role of non-central forces one should have a confident estimate on their relative strength from nuclear spectroscopic properties. This work is in progress.

OUTLOOK

Systematic calculations of the monopole shifts in nuclei adjacent to the semi-magic ones are performed within the shell-model framework. The main mechanism for the variation of the single-particle states when moving away from the stable to highly unstable nuclei as originating from the monopole part of the proton-neutron effective interaction and the pairing between alike valence particle is well confirmed. The nuclei considered in detail here represent a particular case of many-nucleon systems; however, we believe that the essential results obtained on the evolution of the shell structure will remain valid in the more general case. The realistic interaction reproduces well the experimental data, which signal the specific behavior of the proton-neutron $j_<$-$j'_>$ or $j_>$-$j'_<$ pairs. To understand this phenomenon a thorough comparison of the results obtained using realistic and schematic interactions is performed. It is shown that, although the δ-force and the most general central potential (with the dominant part generated by $(\vec{\sigma}\cdot\vec{\sigma})(\vec{\tau}\cdot\vec{\tau})$-term) allows sometimes to obtain an overall trend, they fail to reproduce the experimentally observed effects in detail. The non-central interactions, such as tensor force and/or non-local two-body spin-orbit force may lead to the desired behavior, but their relative contribution still has to be determined.

ACKNOWLEDGMENTS

N.A.S. thanks E. Caurier and F. Nowacki from IReS (Strasbourg) for making available some of the interactions used here, and for useful discussions. We thank B. R. Barrett and R.F. Casten for their interest and fruitful discussions. This work was supported by the Inter-University Attraction Poles (IUAP) under project P5/07. K.H. is grateful to the FWO-Vlaanderen for financial support.

REFERENCES

1. M. G. Mayer, *Phys. Rev.*, **75**, 1969–1970 (1949).
2. O. Haxel, J. H. D. Jensen, H. E. Suess, *Phys. Rev.*, **75**, 1766 (1949).
3. J. Dobaczewski, I. Hamamoto, W. Nazarewicz, J. A. Sheikh, *Phys. Rev. Lett.*, **72**, 981–984 (1994).
4. J. Dobaczewski, W. Nazarewicz, et al, *Phys. Rev.*, **C53**, 2809–2839 (1996).
5. A. L. Goodman, *Nucl. Phys.*, **A287**, 1–12 (1977).
6. R. A. Sorensen, *Nucl. Phys.*, **A420**, 221–236 (1984).
7. K. Heyde, J. Jolie, J. Moreau et al, *Nucl. Phys.*, **A466**, 189–226 (1987).
8. R.K.Bansal, and J.B.French, *Phys. Lett.*, **11**, 145–148 (1964).
9. A. Poves, and A. P. Zuker, *Phys. Rep.*, **70**, 235–314 (1981).
10. T. Otsuka, M. Honma,T. Mizusaki et al, *Prog. Part. Nucl. Phys.*, **47**, 319–400 (2001).
11. E. Caurier, G. Martínez-Pinedo, F. Nowacki, A. Poves, A. P. Zuker, *Rev. Mod. Phys.*, **77**, 427–488 (2005).
12. A. P. Zuker, *Nucl. Phys.*, **A576**, 65–108 (1994).
13. B. A. Brown, and B. H. Wildenthal, *Ann. Rev. Nucl. Part. Sci.*, **38**, 29–66 (1988).
14. E. Caurier, shell model code ANTOINE, IReS, Strasbourg, 1989–2002
15. E. Caurier, and F. Nowacki, *Acta Phys. Polon.*, **30**, 705–714 (1999).
16. N. Smirnova, A. De Maesschalck, A. Van Dyck, K. Heyde, *Phys. Rev.*, **C69**, 044306 (2004).
17. S. Franchoo, Ph. D. thesis, University of Leuven (1999).
18. S. Franchoo et al, *Phys. Rev. C*, **64**, 054308 (2001).
19. M. Sanchez-Vega, B. Fogelberg, H. Mach et al, *Phys. Rev. Lett.*, **80**, 5504–5507 (1988).
20. J. Shergur et al, *Nucl. Phys.*, **A682**, 493c–497c (2001).
21. H. De Witte, Ph.D. thesis, University of Leuven (2004).
22. J.S. Thomas et al, *Phys. Rev.*, **C71**, 021302 (2005).
23. J.P Schiffer et al, *Phys. Rev. Lett*, **92**, 162501 (2005).
24. http://www.nndc.bnl.gov
25. P. Doll, G.J. Wagner, K.T. Knöpfle, G. Mairle, *Nucl. Phys.*, **A263**, 210–236 (1976).
26. J. Retamosa, E. Caurier, F. Nowacki, and A. Poves, *Phys. Rev.* **C55**, 1266–1274 (1997).
27. B. Fogelberg, H. Gausemel, K.A. Mezilev, et al, *Phys. Rev.*, **C70**, 034312 (2004).
28. F. Nowacki, Ph.D. thesis, IReS, Strasbourg, 1996; F. Nowacki, private communication.
29. M. Hjorth-Jensen, T.T.S. Kuo, and E. Osnes, *Phys. Rep.*, **261**, 125–270 (1995).
30. P. Federman, and S. Pittel, Phys. Rev. C **20**, 820 (1979).
31. T. Otsuka, R. Fujimoto, Y. Utsuno et al, *Phys. Rev. Lett.*, **87**, 082502 (2001).
32. N. Smirnova, A. De Maesschalck, K. Heyde, in preparation.
33. T. Otsuka, talk at the Conf. 'Limits 2004', ANL (2004).

Softness of doubly-magic ^{78}Ni and related topics

A. F. Lisetskiy[1]*†, B. A. Brown* and H. Schatz*†

National Superconducting Cyclotron Laboratory, Michigan State University, East Lansing, Michigan 48824-1321
†*Joint Institute for Nuclear Astrophysics,Michigan State University, East Lansing, Michigan 48824-1321*

Abstract. Results of shell model calculations for ^{76}Ni and ^{78}Ni with newly derived effective interaction for the $f_{5/2}p_{3/2}p_{1/2}g_{9/2}$ and the $f_{7/2}f_{5/2}p_{3/2}p_{1/2}g_{9/2}$ model spaces are presented. The calculated spectra of ^{76}Ni in different spaces are compared and the role of the ^{56}Ni core breaking is discussed. Calculated half-life and branching ratios for beta-decay of ^{78}Ni and the effect associated with the $f_{7/2}$ orbital are analyzed.

Very neutron-rich nuclei play an important role in the astrophysical rapid neutron capture process (r-process) which is thought to be responsible for the origin of about half of the heavy element beyond iron. However the exact path of this reaction network, which is determined by the details of the nuclear structure, is not well known due to the sparse experimental information for neutron-rich regions. In this context the nuclear chart region in a vicinity of ^{78}Ni is of very special interest, since the ^{78}Ni is the only doubly-magic nucleus that represents an important waiting point in the r-process, where the reaction sequence halts to wait for the decay of the nucleus. Furthermore the structure of doubly magic ^{78}Ni may help to answer the question about persistence of classical magic numbers or appearance of new ones at exotic values of isospin. Experimental investigations of neutron-rich nuclei have greatly advanced the last decade providing access to many new regions of the nuclear chart. Very recent studies have established beta decay properties of neutron-rich nickel isotopes up to ^{78}Ni [1], have provided new information on spectra of nickel isotopes up to ^{76}Ni [2] and copper isotopes up to ^{77}Cu [3, 4] and have yielded the B(E2) values for the radioactive nuclei ^{78,80,82}Ge finding evidence for the persistence of the $N = 50$ closed shell [5].

Experimental progress is accompanied by theoretical developments as well. Recently we have derived new effective interaction for the $f_{5/2}p_{3/2}p_{1/2}g_{9/2}$ model space from a fit to updated experimental data for Ni isotopes from A= 57 to A=78 and N= 50 isotones from ^{79}Cu to ^{100}Sn [6]. The new interaction represents $T = 1$ part of the interaction for the full proton-neutron $f_{5/2}p_{3/2}p_{1/2}g_{9/2}$ model space covering more than 200 nuclei of the nuclear chart triangle formed by doubly-magic ^{56}Ni, ^{78}Ni and ^{100}Sn nuclei. The $T = 1$ part of the interaction allowed us to explore many interesting phenomena like disappearance of seniority isomers and emergence of vibration-like collectivity in

---

[1] Gesellschaft für Schwerionenforschung mbH, Darmstadt, Germany

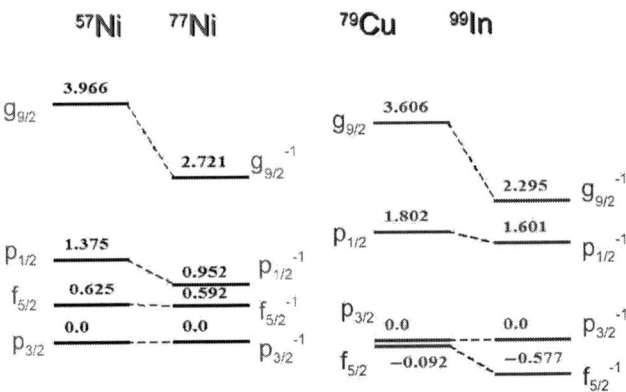

FIGURE 1. Calculated one-particle levels for ^{57}Ni and ^{79}Cu and one-hole i inverted levels for ^{77}Ni and ^{99}In.

$A = 70-76$ nickel isotopes caused by unusually large two body matrix element (TBME) for the $g^2_{9/2}$ configuration with $J^\pi = 2^+$ [7]. Further progress towards a construction of the $T = 0$ part of the effective interaction has been achieved very recently [8]. We have extended the pool of fitted experimental data by including Cu isotopes and those of Zn, Ga, Ge, As, Se, Br and Kr isotopes which are characterized by relatively small Hamiltonian matrixes.

In present contribution we discuss the possible ways to understand the enhancement of the residual interaction in the $J^\pi = 2^+$ state for the $g_{9/2}$ orbital and to illustrate the effect of the ^{56}Ni core breaking on the calculated half-life of ^{78}Ni.

The effective interaction for the $f_{5/2}p_{3/2}p_{1/2}g_{9/2}$ configurational space is specified uniquely in terms of interaction parameters consisting of four single-particle energies and 169 TBMEs. The starting point for the fitting procedure was the derived $T = 1$ interaction for nickel isotopes [6] and the $T = 0$ part of realistic G-matrix interaction based on the Bonn-C NN potential. Using Linear combination Method for fitting procedure we found that convergence of the χ^2 in the first iteration is achieved already at 14 linear combinations and we have chosen this as a reasonable number for all following iterations. We performed several iterations until the eigen-energies converged. The values of the interaction parameters are adjusted to fit 450 energy levels for Ni, Cu and some of Zn, Ga, Ge, As, Se, Br, Kr isotopes and all of N=50 isotones [8]. The new studies resulted in significant modification of single-particle energies as compared to our previous results [6]. New calculated spectra of one-particle and one-hole states for corner-stone nuclei in the considered configurational space are shown in Figure 1. We succeeded to describe nickel isotopes between ^{68}Ni and ^{78}Ni nuclei very well with new effective interaction, however the excitation energies of 2^+ states in corresponding Valence Mirror Symmetry (VMS) partner $A = 90-98$ $N = 50$ isotones are predicted to be too low. This problem has been discussed by us in [6]. The 2^+ states in Ni isotopes lay lower than in $N = 50$ isotones, while an opposite is expected for the same structures with decrease of mass number. The origin of this difference is mainly in $(g^2_{9/2})_J$ TBMEs: effective inter-

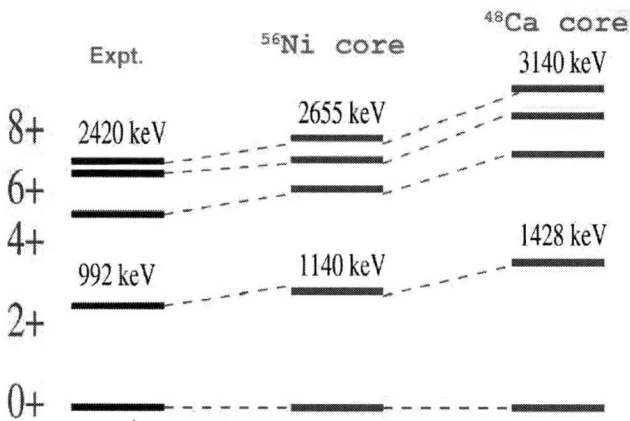

FIGURE 2. Calculated and experimental spectra of ^{76}Ni. For the calculations with the ^{48}Ca core only one proton excitations from the $f_{7/2}$ orbital are taken into account.

action for nickel isotopes in $J = 2$ and $J = 4$ states has to be by 300 keV more attractive than for corresponding $N = 50$ isotones. The best illustrative example of the observed difference is the two-hole $g_{9/2}^{-2}$ spectra of ^{76}Ni and ^{98}Cd [2].

One may expect that the Z=28 shell gap near ^{78}Ni is relatively weak as compared to the N= 50 shell gap near ^{100}Sn. To test the effect of cross-shell (Z=28) excitations we performed calculations in expanded configurational space allowing one proton excitations from $f_{7/2}$ orbital to the $f_{5/2}p_{3/2}p_{1/2}g_{9/2}$ space. This means that the ^{48}Ca nucleus is an inert core instead of ^{56}Ni. We combined our new interaction with the part of the GXPF1 interaction [9] that involves the $f_{7/2}$ orbital. The lacking $f_{7/2}g_{9/2}$ TBME's were modeled by the SDI. The $f_{7/2}$ orbital was put at 4 MeV below the $p_{3/2}$ orbital. Results of the calculations are shown in Figure 2. Our calculations indicate that ^{56}Ni core breaking does not lower the energy of the 2+ state but pushes it and other $g_{9/2}^{2}$ states up. Proton $f_{7/2}$ excitations contribute considerably to pairing but do not change much interaction in quadrupole or higher multipolarity channels. Since the part of the interaction that involves $f_{7/2}$ orbital has been taken from the GXPF1 interaction which works perfectly in a vicinity of ^{56}Ni, i.e. near the $N = Z$ line, one may conclude that the interaction has to be modified for the neutron-rich region. Namely, the quadrupole part of the interaction has to be stronger or/and pairing has to be weaker.

The expansion of the space discussed above helps to explore the role of the $f_{7/2}$ degree of freedom in beta-decay of neutron-rich nickel isotopes. The calculations performed with our new interaction in the $f_{5/2}p_{3/2}p_{1/2}g_{9/2}$ space yield good description of Q-values, neutron separation energies, beta-delayed neutron emission rates and half-lives. However, the half-lives of $^{74-78}$Ni isotopes are reproduced using very strong quenching ($\alpha_q = 0.35$) for the GT matrix elements. The distribution of the GT strength for ^{78}Ni in this case is shown in Figure 3 (left). The calculations in enlarged space with the ^{48}Ca core yield good description of the beta-decay properties too, however the half-lives are

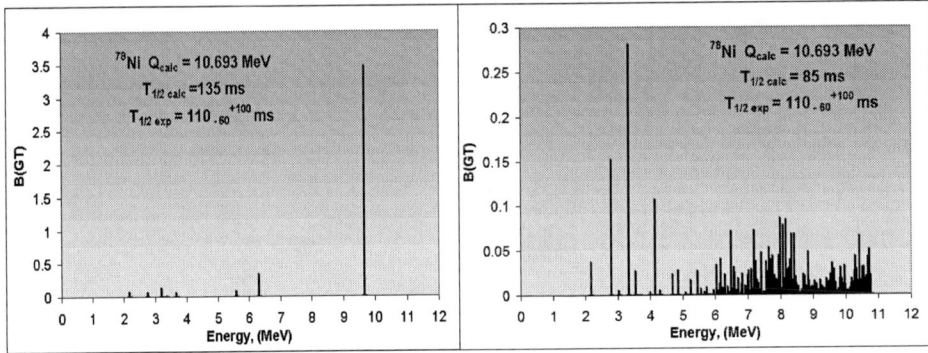

FIGURE 3. Distribution of the GT strength for beta-decay of ^{78}Ni calculated within the $f_{5/2}p_{3/2}p_{1/2}g_{9/2}$ space with the ^{56}Ni inert core (left) and within the $f_{7/2}+f_{5/2}p_{3/2}p_{1/2}g_{9/2}$ space with the ^{48}Ca core (right). The value of quenching factor $\alpha_q = 0.35$ ($\alpha_q = 0.7$) for GT matrix elements was used for the case shown on the left (right) panel.

described now with a standard value of quenching factor $\alpha_q = 0.7$. As an example the distribution of the GT strength for ^{78}Ni is shown in Figure 3 (right). One may conclude that the $f_{7/2}$ orbital plays an important role. On the other hand, present calculations show that reliable estimations of half-lives and other beta-decay observables may be obtained in smaller $f_{5/2}p_{3/2}p_{1/2}g_{9/2}$ space for nuclei which can not be yet calculated in space which includes the $f_{7/2}$ shell due to restricted computational power. Large-scale shell model calculations with many-nucleon excitations from the $f_{7/2}$ orbital and are required in order to get better insight into the problems discussed in present contribution. We acknowledge support from NSF Grant No. PHY-0244453.

REFERENCES

1. Hosmer, P.T., et al., *Phys. Rev. Lett*, **94**, 112501 (2005).
2. Mazzocchi, C., et al., *Phys. Lett. B*, **622**, 45 (2005).
3. Van Roosbroeck, J., et al., *Phys. Rev. C* **69**, 034313, (2004).
4. Ressler, J.J., et al, in preparation.
5. Padilla-Rodal, E., et al., *Phys. Rev. Lett*, **94**, 122501 (2005).
6. Lisetskiy, A., Brown, B.A., Horoi, M., and Grawe, H., *Phys. Rev. C*, **70**, 044314 (2004).
7. Horoi, M., Lisetskiy, A. F., Brown, B. A., Proceedings of the Conference on "Nuclei at the Limits", Argonne National Laboratory July 26-30, 2004, *AIP Conf. Proc.* **764** 170 (2005).
8. Lisetskiy, A., Brown, B., Horoi, M., in preparation.
9. Honma. M, Otsuka. T, Brown. B. A., and Mizusaki. T., *Phys. Rev. C* **65**, 061301(R), (2002).

Regular and chaotic nuclear vibrations

P. Cejnar, M. Macek, P. Stránský, M. Kurian

Institute of Particle and Nuclear Physics, Charles University, 18000 Prague, Czech Republic

Abstract. We investigate classical limit of the geometric collective model and interacting boson model at zero angular momentum. Both models exhibit complex interplay between regular and chaotic modes of motions. Types of classical orbits are shown to affect properties of quantal spectra.

INTRODUCTION

While it is assumed that atomic nuclei at low energies exhibit a mixture of regular and chaotic dynamics, the principal mechanism that controls the transition between both types of motions still remains mostly unknown. Several possible sources of irregularity—diffuseness of the nuclear surface [1], spin-orbit coupling [2], deformations of higher multipolarities [3]—can be identified already on the mean-field level, but the most substantial effects probably result from residual interactions, which, however, are too difficult for an explicit analysis.

A viable alternative to the attempts to solve the nuclear many-body problem in its full complexity is the use of simplified models that capture only some essential dynamical features of nuclei in various regimes. Since residual interactions are responsible for highly correlated collective modes of motions (complementary to single-particle modes), one may try to search signatures of regularity and chaos in the known elementary collective models, such as the Geometric Collective Model (GCM) [4, 5] or the Interacting Boson Model (IBM) [6]. This approach was pioneered by Alhassid *et al.* [7] and by Paar *et al.* [8] using the IBM and later followed by Cejnar and Stránský [9] using the GCM. In spite of the the apparent simplicity of both IBM and GCM approximations, surprising variability and complexity of possible types of collective modes was discovered. It turns out that nuclear collective motions belong to the most interesting cases of dynamics on the border between order and chaos.

COLLECTIVE HAMILTONIANS

In this contribution we present some new results obtained for the *classical* limit of both the IBM and GCM. We use the GCM Hamiltonian in a truncated form,

$$H_{\text{GCM}} = \frac{\sqrt{5}}{2K}[\pi \times \pi]^{(0)} + \sqrt{5}A[\alpha \times \alpha]^{(0)} - \sqrt{\frac{35}{2}}B[[\alpha \times \alpha]^{(2)} \times \alpha]^{(0)} + 5C\left([\alpha \times \alpha]^{(0)}\right)^2, \quad (1)$$

where $\alpha^{(2)}$ stands for the spherical tensor of quadrupole shape variables (now playing the role of generalized coordinates) and $\pi^{(2)}$ for the associated tensor of momenta.

Higher-order kinetic and potential terms are neglected. Although the Hamiltonian has four adjustable parameters (K, A, B, and C), the possibility to independently scale energy, coordinates and time implies that in the classical case there is only one essential control parameter (while the others can be set to unity).

The tensor of quadrupole shape variables depends only on five independent real coordinates, which can be associated with two shape parameters

$$\alpha_0^{(2)} \equiv x = \beta \cos\gamma \quad , \quad \sqrt{2}\,\text{Re}\,\alpha_{\pm 2}^{(2)} \equiv y = \beta \sin\gamma. \tag{2}$$

and three Euler angles $(\theta_1, \theta_2, \theta_3)$ characterizing the orientation of the principal frame of deformation relative to the laboratory frame. If the angular momentum $L = -i\sqrt{10}[\alpha \times \pi^*]^{(1)}$ is set to zero, the principal frame is at rest and the Euler angles become irrelevant. In this case, the system becomes effectively 2-dimensional, the configuration space given by (x,y) or the corresponding polar coordinates (β, γ) that coincide with the Bohr deformation parameters. The $L=0$ Hamiltonian reduces to

$$H_{\text{GCM\_0}} = \underbrace{\frac{1}{2K}(\pi_x^2 + \pi_y^2)}_{\pi^2 = \pi_\beta^2 + \left(\frac{\pi_\gamma}{\beta}\right)^2} + \underbrace{A(x^2+y^2) + B(x^3 - 3y^2 x) + C(x^2+y^2)^2}_{A\beta^2 + B\beta^3 \cos 3\gamma + C\beta^4}, \tag{3}$$

which represents just a 2D oscillator with quadratic, cubic, and quartic potential terms.

The IBM Hamiltonian (in the quantum form) employed here reads as

$$\hat{H}_{\text{IBM}} = \underbrace{\hbar\omega}_{1} \left[\frac{\eta}{N}\hat{n}_d - \frac{1-\eta}{N^2}(\hat{Q}_\chi \cdot \hat{Q}_\chi)\right], \tag{4}$$

where N is the total number of bosons, $\hat{n}_d$ the d-boson number operator, and $\hat{Q}_\chi^{(2)} = s^\dagger \tilde{d} + d^\dagger \tilde{s} + \chi[d^\dagger \times \tilde{d}]^{(2)}$ the quadrupole operator. Two control parameters $\eta \in [0,1]$ and $\chi \in [-\frac{\sqrt{7}}{2}, +\frac{\sqrt{7}}{2}]$ define an extended Casten triangle [10]. The classical limit of Hamiltonian (4) can be obtained via the procedure elaborated in Ref.[11]. It makes use of Glauber coherent states $|\alpha\rangle \propto \exp(\alpha_s s^\dagger + \sum_\mu \alpha_\mu d_\mu^\dagger)|0\rangle$, where complex coefficients $\{\alpha\}$ define a set of 6+6 classical-like coordinates and momenta. Reduction of the phase space to 10 dimensions (as in the GCM) is reached by fixing the average number of bosons $\langle N \rangle \equiv \langle \alpha|\hat{N}|\alpha\rangle = |\alpha_s|^2 + \sum_\mu |\alpha_\mu|^2$. The classical Hamiltonian $H_{\text{IBM}} = \langle \alpha|\hat{H}_{\text{IBM}}|\alpha\rangle$ is then obtained in the limit $\langle N \rangle \to \infty$ after some appropriate transformations of variables $\{\alpha\}$. The result is similar to Eq.(1), but contains additional position-dependent terms in the kinetic energy. With the angular momentum $\hat{L} = \sqrt{10}[d^\dagger \times \tilde{d}]^{(1)}$ set to zero, the procedure again leads to just a 2D configuration space. The $L=0$ classical counterpart of Hamiltonian (4) is given explicitly in Ref.[12].

REGULAR AND CHAOTIC REGIMES

Although numerous Hamiltonians similar to that in Eq.(3) were studied in the past, the GCM and IBM at $L=0$ number among the most striking examples of 2D systems

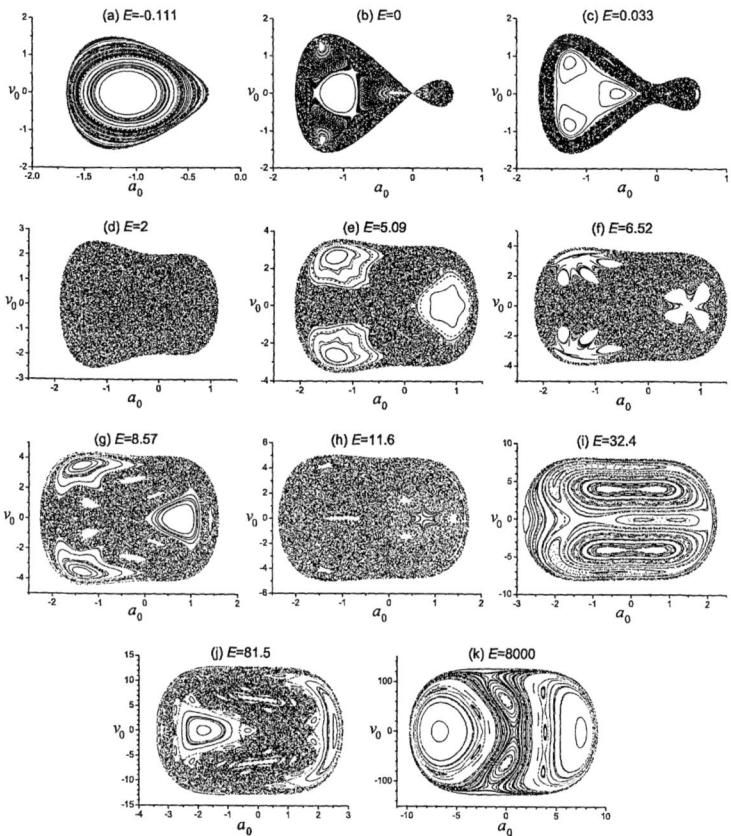

FIGURE 1. Poincaré phase-space sections corresponding to Hamiltonian (3) with $K = -A = C = 1$ and $B = 1.09$ for various values of energy (panels a–k). In each section we show 30000 passages of 52 randomly chosen trajectories through the plane $y = 0$. Horizontal and vertical axes, respectively, represent x and π_x at the moment of crossing (π_y for each point is determined by the remaining values).

with mixed regular and chaotic dynamics. A surprising feature is an oscillatory, fine-structured dependence of the degree of chaos on energy and control parameters [9], which is in contrast to mostly monotonous behaviors observed in the other related systems (see, e.g., the Hénon-Heiles system [13] or the one in Ref.[14]).

To illustrate the variability of nuclear collective vibrations, we show in Fig.1 the evolution of Poincaré phase-space sections of a fixed GCM Hamiltonian with increasing energy. Let us stress that Hamiltonian (3) produces regular dynamics at very low as well as at very high energies—see panels (a) and (k), respectively. While the low-E regime is approximated by a deformed harmonic oscillator, motions at high E are dominated by the quartic potential which is also integrable. It is in between these limits where the cubic term of the potential leads to the complex interplay of various regular and chaotic modes. Note that regular phase-space domains correspond to organized areas in Poincaré sections (with individual points falling on smooth curves) whereas the chaotic domains exhibit themselves as homogeneously (ergodically) filled areas (see panel d).

A particularly amazing feature is the emergence of an island of complete regularity at $E = 32.4$ (panel i) from the sea of chaos at lower and higher energies. While the reason for this peculiar behavior remains unknown, we basically understand the mechanism which breaks the regularity in the *low-E* (harmonic-oscillator like) limit (in Fig.1 this happens somewhere between the first two panels). It can be shown [15] that chaos sets in at approximately that value of energy for which the $x \times y$ area accessible to classical motions changes the shape from fully convex to partly concave. This way of generating chaos is similar to billiard systems, but we have to stress that in our case we deal with a smooth potential with no hard-wall reflections.

As demonstrated in Ref.[9], an oscillatory dependence of the degree of chaos is observed also with varying control parameters of the GCM Hamiltonian. In particular, when departing from the integrable $B = 0$ regime (corresponding to γ-soft nuclei), one passes an interval where chaos decreases considerably below the surrounding level (with a fine structure depending on energy). This region seems to be related to the "arc of regularity" reported by Alhassid and Whelan inside the IBM Casten triangle [7]. Work is in progress [15] on the task to extend these GCM observations to nonzero angular momenta, when the full 5D configuration space must be taken into account.

THE INTEGRABLE CASE

In the rest of this contribution we will deal with the IBM. Instead of continuing the discussion of classical chaos, however, we will focus on the completely *integrable* regime of Hamiltonian (4) with $\chi = 0$ [16] (similar to the GCM with $B = 0$). This corresponds to the transition between O(6) ($\eta = 0$) and U(5) ($\eta = 1$) dynamical symmetries with unbroken O(5) dynamical symmetry, implying that the seniority v is a good quantum number all the way. The classical $L = 0$ IBM Hamiltonian in this case reads as

$$H_{\text{IBM\_0}} = \frac{\eta}{2}\pi^2 + (1-\eta)\beta^2\pi^2 + \frac{5\eta - 4}{2}\beta^2 + (1-\eta)\beta^4, \tag{5}$$

where we immediately mark the existence of a second motional integral, π_γ, which is uniquely related to the O(5) Casimir invariant (in the classical limit $C_2[O(5)] \propto \pi_\gamma^2$ for $L = 0$). Hamiltonian (5) represents motions (with just an unusual expression for kinetic energy) in an axially-symmetric Mexican-hat potential. (Note that due to an N-dependent scaling of the phase space the physical domain is restricted to $\beta \leq \sqrt{2}$.) On the first sight it might seem that there is nothing new to be learned by modelling the dynamics in such a trivial case, but the opposite is true.

Classical orbits generated by Hamiltonian (5) can be classified by the ratio

$$R = \frac{T_\gamma}{T_\beta} = \frac{\langle \omega_\beta \rangle}{\langle \omega_\gamma \rangle} \tag{6}$$

of periods associated with γ and β oscillations (or the inverse ratio of the respective average angular frequencies). If R is rational the orbit is periodic, while trajectories with

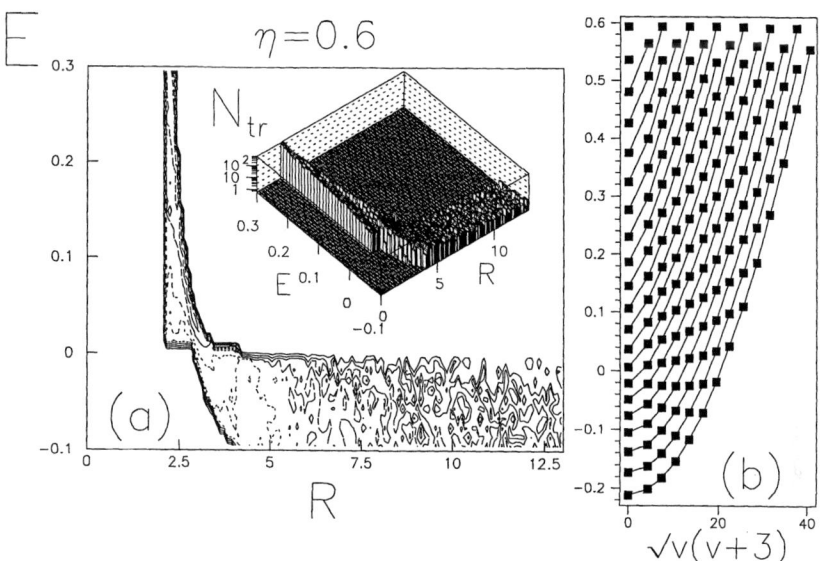

FIGURE 2. (a) Distribution of trajectories over the ratio (6) as a function of energy for Hamiltonian (5) with $\eta = 0.6$. The histogram was calculated using a sample of about 50000 generated orbits. (b) The corresponding $N = 40$ lattice of $L = 0$ quantum states characterized by energy E and seniority v.

irrational R never close. The distribution of the ratio R for orbits with various energies is shown in Fig.2(a) for $\eta = 0.6$. We see that at $E = 0$ the motions undergo an abrupt change: the orbits with $R \in [2,3]$ arise on the $E > 0$ side but those with $R > 3$ die out as E increases (the latter phenomenon is due to a certain type of bifurcations, when two types of orbit with a given R "annihilate"[16]). While the orbits at low energies can be characterized by $R \geq R_{min} > 3$, at very high energies only values close to $R \approx 2$ survive.

This has rather interesting consequences. Motions with large ratios R form flower-like orbits zigzagging inside the circular valley of the Mexican-hat potential, but those with $R \approx 2$ form bouncing-ball orbits that traverse across the $\beta \approx 0$ region as in the spherical oscillator case. The former orbits are typical for the O(6) limit of Hamiltonian (5), whereas the latter ones characterize the U(5) limit. We see that the transition between both types of motions happens (from the classical viewpoint) around $E \approx 0$.

Does this observation have any practical implications on the quantum level? Fig.2(b) shows a lattice of the IBM $L = 0$ states for $N = 40$ in the plane energy vs. seniority. For $E > 0$ the lattice resembles the U(5) limit—it is essentially the lattice associated with the spherical isotropic 2D harmonic oscillator [cf. Eq.(5) with $\eta = 1$], although certain differences follow from the fact that the usual O(2) angular momentum is replaced here by the $L = 0$ projection of the O(5) invariant. On the other hand, the $E < 0$ lattice is essentially that of the O(6) limit. The crossover between both types of spectra happens in the region $E \approx 0$, where also the classical orbits change their forms (panel a).

It is evident (panel b) that at $E \approx 0$ states with different seniorities become nearly degenerate, which results in a huge bunching pattern of $L = 0$ levels observed in $\eta \in$

[0.3, 0.8] spectra across the O(6)-U(5) transition [16]. This phenomenon is explained via semiclassical trace formulas [17], describing oscillatory components of the level density in terms of the associated classical periodic orbits. The clue is the fact that the $\beta = 0$ local maximum of the potential in Eq.(5) generates orbits with periods $T \to \infty$ (for $E \to 0$) that yield a diverging contribution to the Berry-Tabor trace formula. On a deeper level, this anomalous bundle of orbits and the corresponding "defect" in the grid of quantum states can be accommodated in a recently developed framework of so-called *monodromy* (see, e.g., Ref.[18] and refs. therein). It is likely that future elaboration of this concept within the IBM will shed more light on the mechanism leading to redistributions of levels between the O(6) and U(5) multiplets.

We hope you are convinced now that sometimes it is worth pretending that nuclei are classical objects...

ACKNOWLEDGMENTS

P.C. thanks J. Jolie, R. Casten, and J. Dobeš for many stimulating discussions. This work was supported by the Czech Ministry of Education under Project MSM 0021620834.

REFERENCES

1. R. Arvieu, F. Brut, J. Carbonell, and J. Touchard, Phys. Rev. A **35**, 2389 (1987).
2. P. Rozmej and R. Arvieu, Nucl. Phys. **A545**, C497 (1992).
3. W.D. Heiss, R.G. Nazmitdinov, and S. Radu, Phys. Rev. Lett. **72**, 2351 (1994); Phys. Rev. C **52**, 3032 (1995); W.D. Heiss and R.G. Nazmitdinov, Phys. Rev. Lett. **73**, 1235 (1994).
4. G. Gneuss, U. Mosel, and W. Greiner, Phys. Lett. **30B**, 397 (1969); **31B**, 269 (1970); G. Gneuss and W. Greiner, Nucl. Phys. **A171**, 449 (1971).
5. J.M. Eisenberg and W. Greiner, *Nuclear Theory*, Vol. 1, *Nuclear Models* (North-Holland, Amsterdam, 1987).
6. F. Iachello and A. Arima, *The Interacting Boson Model* (Cambridge University Press, Cambridge, England, 1987).
7. Y. Alhassid, A. Novoselsky, and N. Whelan, Phys. Rev. Lett. **65**, 2971 (1990); Y. Alhassid and N. Whelan, Phys. Rev. C **43**, 2637 (1991); Phys. Rev. Lett. **67**, 816 (1991); Y. Alhassid and A. Novoselsky, Phys. Rev. C **45**, 1677 (1992); Y. Alhassid and D. Vretanar, Phys. Rev. C **46**, 1334 (1992); N. Whelan and Y. Alhassid, Nucl. Phys. **A556**, 42 (1993).
8. V. Paar and D. Vorkapić, Phys. Lett. B **205**, 7 (1988); Phys. Rev. C **41**, 2397 (1990); V. Paar, D. Vorkapić, and A.E.L. Dieperink, Phys. Rev. Lett. **69**, 2184 (1992).
9. P. Cejnar and P. Stránský, Phys. Rev. Lett. **93**, 102502 (2004).
10. E. López-Moreno and O. Castaños, Phys. Rev. C **54**, 2374 (1996); J. Jolie, R.F. Casten, P. von Brentano, and V. Werner, Phys. Rev. Lett. **87**, 162501 (2001).
11. R.L. Hatch and S. Levit, Phys. Rev. C **25**, 614 (1982).
12. Y. Alhassid, in *Perspectives for the Interacting Boson Model*, ed. R.F. Casten *et al.* (World Scientific, Singapore, 1994), p. 591.
13. M. Hénon and C. Heiles, Astron. J. **69**, 73 (1964).
14. T.H. Seligman, J.J.M. Verbaarschot, and M.R. Zirnbauer, Phys. Rev. Lett. **53**, 215 (1984).
15. P. Stránský, M. Kurian, and P. Cejnar, to be published.
16. S. Heinze, P. Cejnar, J. Jolie, and M. Macek, nucl-th/0504016; M. Macek, P. Cejnar, J. Jolie, and S. Heinze, nucl-th/0504017 (submitted to Phys. Rev. C).
17. M.C. Gutzwiller, J. Math. Phys. **12**, 343 (1971); M.V. Berry and M. Tabor, Proc. R. Soc. Lond. A**349**, 101 (1976).
18. K. Efstathiou, M. Joyeux, and D.A. Sadowskií, Phys. Rev. A **69**, 032504 (2004).

CONTINUUM SHELL MODEL, REACTIONS AND GIANT RESONANCES

Vladimir Zelevinsky* and Alexander Volya[†]

*NSCL and Department of Physics and Astronomy, Michigan State University, East Lansing, MI 48824-1321, USA
[†]Department of Physics, Florida State University, Tallahassee, FL 32306-4350, USA

Abstract. We give a short review of our recent work developing the continuum shell model based on the effective non-Hermitian Hamiltonian. The unified description of discrete spectra and nuclear reactions is the main advantage of the approach. We comment on the first applications to loosely bound nuclei and pygmy resonances.

Keywords: continuum shell model, decay, resonance, exotic nuclei, collective excitations
PACS: 24.10Cn, 42.50.Fx, 89.75.Fb

1. The broad title of this presentation does not mean that we will try to fully cover all those subjects. Our goal is to briefly outline a promising approach [1, 2] aimed at unifying the description of nuclear phenomena in discrete spectrum, continuum, and, what might be currently the most important, on the borderline between them. The boundary layer contains the wealth of physics of marginally stable many-body systems. Nuclei far from the valley of stability give the best example of such systems, and the experiments of the last decade demonstrate new exciting features. However, other areas, from exotic baryons to quantum dots and future quantum computers, reveal many examples of similar phenomena [3].

2. Let us start with stable systems. Quantum many-body theory usually recommends to find a suitable, advisably self-consistent, mean field and determine the quasiparticle excitations. The residual interaction between the quasiparticles leads to collective effects and, on the other hand, to quantum chaos. In nuclei we can restrict the valence space in a reasonable way, find the effective interaction corresponding to this truncation and solve either the full shell model or simpler microscopic models, − with great success in describing experimental data. The starting point is the effective Hermitian Hamiltonian containing the mean field and interaction terms, $H = H^\circ + V$.

Of course, the realistic mean field determines the single-particle spectrum that at some energy comes to unbound orbitals. Correspondingly, above the threshold excitation energy many-body states are unbound and have finite lifetime. The conventional shell model (SM) determines the excited states as if they were stable neglecting coupling to the continuum. Here another part of nuclear science, the theory of nuclear reactions, enters the game. The unbound states appear in reaction theory as intermediate in the processes of excitation and deexcitation. For relatively large lifetime, they are seen as narrow resonances in reactions. Regrettably, the structure and reactions are separated by a historical barrier of language and tools that is not easy to overcome.

Marginally stable, or open, systems require this barrier to be destroyed. Loosely bound nuclei live in the proximity of the continuum, they may have only few (or none) discrete excited states. Almost any real or virtual excitation brings such a system to the continuum. And this is the physics we like to study, — halos and skins, unusual shell structure, metastable shapes and cluster states, new collective modes etc. We need a theoretical apparatus to unify the description of bound and continuum states, structure and reactions, especially on their borderline.

3. Among many attempts moving in this direction, an attractive possible solution is provided by the continuum shell model (CSM) [4, 5, 6] based on the effective non-Hermitian Hamiltonian. One can (in principle, exactly but in practice approximately) eliminate the part of Hilbert space related to reaction channels and find the effective Hamiltonian $\mathcal{H}$ acting only in the intrinsic space of the shell model. The price one pays is in new features of the effective Hamiltonian — it is non-Hermitian (unbound states decay) and explicitly contains running energy E as a parameter. This operator in the intrinsic space of many-body states $|1\rangle$ is given by

$$\mathcal{H}(E) = H + \Delta(E) - \frac{i}{2} W(E), \qquad (1)$$

where H is the standard SM Hamiltonian, the two new terms, which appeared as a result of elimination of channel states, are given by matrix elements,

$$\Delta_{12}(E) = \text{P.v.} \int dE' \sum_c \frac{A_1^c(E') A_2^{c*}(E')}{E - E'}, \qquad (2)$$

and

$$W_{12}(E) = 2\pi \sum_{c(\text{open})} A_1^c(E) A_2^{c*}(E), \qquad (3)$$

The new parameters here are the amplitudes A_1^c which are proportional to the matrix elements of the full original Hamiltonian between the SM states $|1\rangle$ and continuum states $|c;E\rangle$ in the channel c at energy E.

The Hermitian dispersive part $\Delta(E)$ reflects virtual off-shell excursions to and from continuum. Here all, closed and open, channels contribute on equal footing. The principal value integral (2) only weakly depends on energy and at this stage we assume that it is simply included into the SM part H. This is convenient since it guarantees that for bound states our solution will exactly coincide with that found in the traditional SM. In the future many-body effects coming from Δ deserve to be studied. The anti-Hermitian absorptive part W describes real on-shell decays and contains only contributions of channels c open at given energy. At energy E below threshold energy E_c^{th} in the channel c, the amplitudes A_1^c vanish. This energy dependence is crucial for the borderline phenomena.

4. The gap between the structure and reactions is bridged because the same matrix elements A_1^c determine the scattering processes. The amplitude T^{ba} of the reaction $a \to b$ is given by

$$T^{ba}(E) = \sum_{12} A_1^{b*}(E) \left(\frac{1}{E - \mathcal{H}} \right)_{12} A_2^a. \qquad (4)$$

The scattering matrix that determines the observable cross sections can be found as

$$S^{ba} = s_b^{1/2} \left(\delta^{ba} - iT^{ba} \right) s_a^{1/2}, \qquad (5)$$

where the *s*-factors include potential phase shifts and contributions of remote channels not included explicitly in calculations. The unitarity of the scattering matrix is ensured by construction since all couplings with the continuum are included in the propagator of eq. (4) with the same amplitudes A which give the entrance and exit of the process.

The poles of the scattering matrix are the eigenvalues of $\mathscr{H}$,

$$\mathscr{E}_\alpha(E) = E_\alpha(E) - \frac{i}{2}\Gamma_\alpha(E), \tag{6}$$

which are located in the lower part of the complex energy plane. They represent resonances depending on real energy of the process and can be obtained also by the diagonalization of $\mathscr{H}$. The result displays a complicated pattern of cross sections with interfering resonances, energy-dependent centroids and widths and non-exponential decay.

5. The first realistic applications of CSM [1, 2] allowed us to obtain a good description of loosely bound nuclei. A special attention should be paid to the parameterization of the decay amplitudes A_1^c which are not present in the usual SM. At this stage we took into account one-body and two-body channels. The near-threshold behavior is mainly determined by the available phase space volume. The one-body decays are calculated in terms of the mean field, while the two-body channels include the emission of correlated pairs that is in fact responsible for the finite life time of certain even-even isotopes. Due to the crucial influence of threshold locations one first has to find them from solving the analogous problem for the daughter nuclei. This leads to the necessity to consider the isotope chains in their entirety starting, for example, from a very stable double-magic nucleus. In Fig. 1 we show the results of the CSM calculations for the ^{18}O nucleus; more results for the entire chain of oxygen isotopes are reported in [2]. Emphasizing the natural connection between the bound states and reaction calculations in CSM, the figure includes both the level scheme and the scattering cross section computed within the same formalism and using the same interaction. The predictions of many resonance positions and widths are waiting for their experimental testing.

6. New collective dynamics [5, 8] follows from the factorized character of the anti-Hermitian part (3) dictated by unitarity. When typical intrinsic widths, $\gamma_1^c = |A_1^c|^2$, are small compared with the level spacing D, we see narrow isolated resonances in this channel. However, when $\gamma \geq D$, the resonances overlap, and a sharp redistribution of widths occurs with segregation of broad states, accumulating a large fraction of the summed widths of all overlapping states, from long-lived trapped states deprived of their widths. This transition looked at in the time domain separates direct processes from those going via compound states. Here the coupling to the continuum defines a natural doorway basis [8], and the number of short-lived states is equal to the number of open channels with overlapped resonances. This phenomenon is analogous to the super-radiance in quantum optics [3], where the individual atoms coupled through their common radiation field create the coherent broad state. In many-body systems, the intrinsic levels are coupled through the common decay channels.

The analogs of super-radiance are found in atomic and molecular physics, condensed matter and particle physics. Neutron resonances in heavy nuclei, seemingly studied already in all details for numerous applications, provide a good example of hidden new physics. Here an envelope seen as a broad single-particle resonance corresponds to the super-radiant state while narrow neutron resonances are trapped long-lived states.

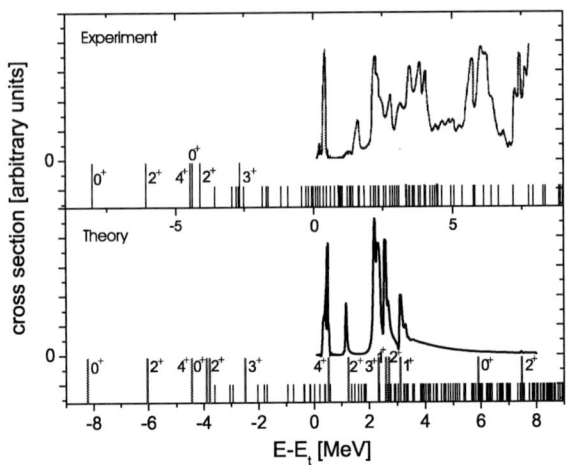

FIGURE 1. Comparison of experimental (upper plot) and theoretical (lower plot) neutron scattering cross section in ^{18}O. The set of shell model states and states observed in experiment [7] are shown below the cross section curve. States marked with longer lines correspond to positive parity and can be identified with those in the sd-shell model valence space.

Statistics of neutron widths and level spacings was one of the cornerstones of application of random matrix theory as a limit of extreme quantum chaos. However, based on the effective Hamiltonian (1) we can assert [5] that the level repulsion at small distances should be absent because of the finite width of resonances. This is hard to test at low energies. However, the distortion of spectral statistics becomes more noticeable [9, 10] along with the growth of energy and resonance widths. The statistics of complex poles was never reliably extracted from the data although it should be possible for neutron resonances and for microwave cavities. The Ericson fluctuations also should be modified since well known theory [11] did not take into account super-radiant coherence of resonances and related effects.

7. The collective dynamics through coupling with continuum coexist with traditional collectivity caused by coherent multipole-multipole interactions accessible by standard approaches and leading to collective vibrations and giant resonances. While giant resonances collectivize the multipole strength shifted for isovector modes up from particle-hole energy, the interaction via continuum collectivizes the width of the particle-hole excitations. Thus, "normal" giant resonances should coexist with super-radiant resonances around unshifted particle-hole energy [12, 13]. This is the mechanism of emergence of pygmy-branches of giant resonances. They are expected to become more pronounced in weakly bound nuclei where the role of the continuum increases. In Fig. 2 the schematic picture showing the formation of giant resonance is presented. The upper part of the figure is a result of the scattering cross section calculation where fragmentation of reso-

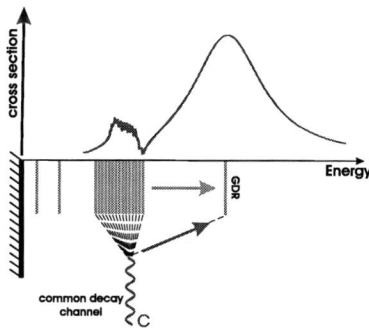

FIGURE 2. Schematic figure showing formation of giant and pygmy resonances, the upper part shows the neutron scattering cross section.

nance strength is transparent.

Currently the work is in progress and we expect many interesting advances in parallel to new experimental tests.

We acknowledge the support from the NSF grant PHY-0244453 and from DOE, grant DE-FG02-92ER40750.

REFERENCES

1. A. Volya and V. Zelevinsky, Phys. Rev. C **67**, 54322 (2003).
2. A. Volya and V. Zelevinsky, Phys. Rev. Lett. **94** (2005) 052501.
3. A. Volya and V. Zelevinsky, in *Nuclei and Mesoscopic Physics*, ed. V. Zelevinsky, AIP Conference Proceedings **777**, 2005, p. 229.
4. C. Mahaux and H.A. Weidenmüller, *Shell Model Approach to Nuclear Reactions* (North Holland, Amsterdam, 1969).
5. V.V. Sokolov and V.G. Zelevinsky, Nucl. Phys. **A504**, 562 (1989).
6. I. Rotter, Rep. Prog. Phys. **54**, 635 (1991).
7. V. McLane, C. Dunford, and P. Rose, *Neutron Cross Section Curves*, vol. 2 (Academic Press, Boston, 1988).
8. V.V. Sokolov and V.G. Zelevinsky, Ann. Phys. (N.Y.) **216**, 323 (1992).
9. S. Mizutori and V.G. Zelevinsky, Z. Phys. **A346**, 1 (1993).
10. F.M. Izrailev, D. Saher, and V.V. Sokolov, Phys. Rev. E **49**, 130 (1994).
11. T. Ericson and T. Mayer-Kuckuk, Ann. Rev. Nucl. Sci. **16**, 183 (1966).
12. V.V. Sokolov and V.G. Zelevinsky, Fizika (Zagreb) **22**, 303 (1990).
13. V.V. Sokolov, I. Rotter, D.V. Savin, and M. Müller, Phys. Rev. C **56**, 1031, 1044 (1997).

Correlation of Nuclear Level Densities and Masses

T. von Egidy* and D. Bucurescu[†]

*Technische Universität München, D-85748 Garching, Germany
[†]Horia Hulubei National Institute of Physics and Nuclear Engineering, R-76900 Bucharest, Romania

Abstract. Level density parameters for 310 nuclei between ^{19}F and ^{251}Cf have been determined for the Back Shifted Fermi Gas and the Constant Temperature models, by fitting both low energy discrete levels and the s-wave neutron resonace spacings. The determined parameters have been found well correlated with quantites available from the mass tables, and simple formulas are proposed on this basis, which can be used to extrapolate to nuclei far from stability.

Keywords: Nuclear level densities
PACS: 21.10.-k, 21/10/Ma

INTRODUCTION

The nuclear level densities (LD) are very important in many applications, since they represent one of the main ingredients in statistical model calculations of nuclear reaction cross sections. A special item is how to calculate level densities for nuclei far from stability, where they cannot be determined experimentally. Theoretical (microscopic) approaches are difficult to use in applications, and are not very reliable. On the other hand, experimental level densities can be determined only for a limited number of nuclei, and it is a problem how to extrapolate such data to unstable, far away nuclei [1].

Most of the experimental determinations of the nuclear level densities are based on the direct counting of the observed nuclear states at low energy, and/or of the neutron resonances at energies close to the neutron binding energy. For most of the applications, such data are evaluated with simple two-parameter models, such as the back-shifted Fermi gas (BSFG) model [2, 3] or the constant temperature(CT) model [2]. Such models do not take into account many details of the nuclear interactions (like shell effects, pairing, etc.) and therefore the empirical (fitted to data) parameters exhibit rather complicated behaviours. There are many such determinations in the literature [1], but the problem of the extrapolation of the empirically determined parameters to far away nuclei persists also in this case, since there is no consistent systematics of these parameters.

In the present work, we first determine a new set of phenomenological LD parameters for such simple models as the BSFG and the CT model, by fitting the latest data on low-excitation energy levels *and* neutron resonance spacings at the neutron binding energies for 310 nuclei. We study then how these model parameters correlate with different observables which are both experimentally known for a large number of nuclei (larger than that in our set) and more reliably calculated theoretically or extrapolated from the existing data. We find excellent correlations with quantities which can be derived from the mass tables, and on this basis we propose simple formulas which describe the main features of our empirical LD data. These formulas are proposed as reliable means for the extrapolation to unknown nuclei. A detailed report of these results is under way [4].

EMPIRICAL LEVEL DENSITY PARAMETERS

To evaluate the experimental LD data, we have used three models. The first is the usual back-shifted Fermi gas model (BSFG) [2], which has as parameters a (the level density parameter) and E_1 (an excitation energy shift, or backshift). In this version, a is independent of the excitation energy. However, microscopic calculations show that the shell effects must be damped at higher excitation energies. On this basis, a phenomenological expression has been proposed [5, 6], $a(U,Z,N) = \tilde{a}[1 + \frac{S(Z,N)-\Delta}{U-E_2} f(U-E_2)]$, which will be refered to in the following as the BSFG with energy dependence of the a parameter (BSFG-ED). Here, $f(U-E_2) = 1 - e^{-\gamma(U-E_2)}$ and $S(Z,N) = M_{exp} - M_{LD}$ is the "shell correction", with M_{exp} the experimental mass, and M_{LD} a mass calculated with a macroscopic, liquid drop formula. For the later, we use the formula of Pearson, which does not include shell, pairing, or deformation effects [8]. The parameters of the BSFG-ED model are $\tilde{a}$, and the backshift E_2. Finallly, we have used the constant temperature (CT) model, having as parameters the temperature T and a backshift E_0 [2].

The set of nuclei for which LD parameters were determined, comprises 310 nuclei between ^{19}F and ^{251}Cf. For each nucleus, and each of these models, the two parameters were determined with the procedure of Ref. [7], by a fit to the newest discrete low-lying levels in a certain spin and energy window (data considered to be complete), *and* the average neutron resonance LD data. We emphasize that in our fit procedure the two parameteres were taken free in the fit. In many other determinations, the backshifts E_i were considered to be due only to the pairing, and consequently they were effectively 'eliminated' by relating them to the neutron and proton pairing energies, either from experimental tables or from an average formula.

The empirically determined parameters have, generally, a complicated behaviour. For the BSFG and CT models they can be seen in Figs. 1 and 2. Some empirical observations are the following. The parameters a, $\tilde{a}$, and T do not show notable dependence on the type of nucleus (even-even, odd-A or odd-odd). a shows a strong dependence on shell effects, having large variations at the shell closures, as observed in many other previous, similar parameter sets. Similarly, T shows strong variations at the shell closures. The parameter $\tilde{a}$ has no more shell effects.

The back-shift energies E_1, E_2, and E_0 have an even more complicated behaviour, and in addition present a dependence on the type of the nucleus. However, the behaviour of these parameters is found very similar for all three models.

CORRELATIONS OF THE LEVEL DENSITY PARAMETERS WITH SHELL AND PAIRING EFFECTS

The empirical parameters were compared to different other quantities, obtained from atomic masses and other tabulated experimental nuclear properties, in search of useful

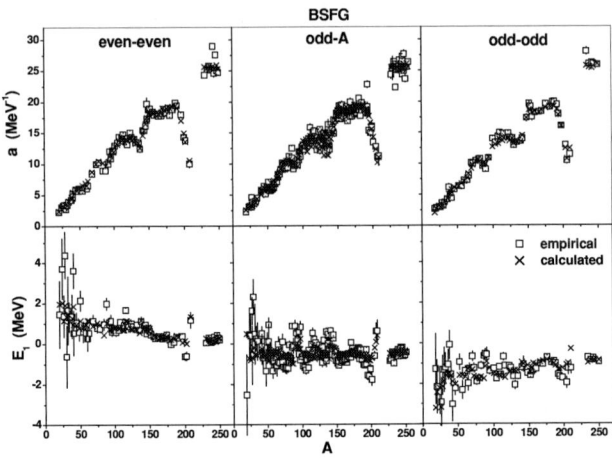

FIGURE 1. The empirical parameters a and E_1 of the BSFG model, and their fits with the formulas from Table 1.

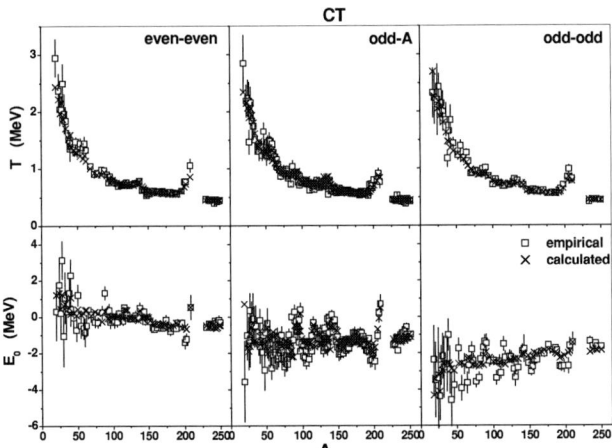

FIGURE 2. The empirical parameters T and E_0 of the CT model, and their fits with the formulas from Table 1.

correlations which explain different characteristics of these data. Then, it was tried to reproduce the level density parameters by simple formulas which use these quantities. Even-even, odd and odd-odd nuclei were frequently treated separately. Linear, quadratic, logarithmic, etc. dependencies were tried, but simple functions with few variables were preferred. The best results of these careful evaluations are presented below. Generally, the obtained normalized χ^2 values lie between 2 and 4. Besides the mass, which determines the shell correction S, we have found that very useful quantities, which correlate well with the backshifts E_i are the deuteron pairing energy P_d (as defined in the mass tables [9, 10]), and the derivative of the shell correction $S(Z,N)$ with respect to the mass

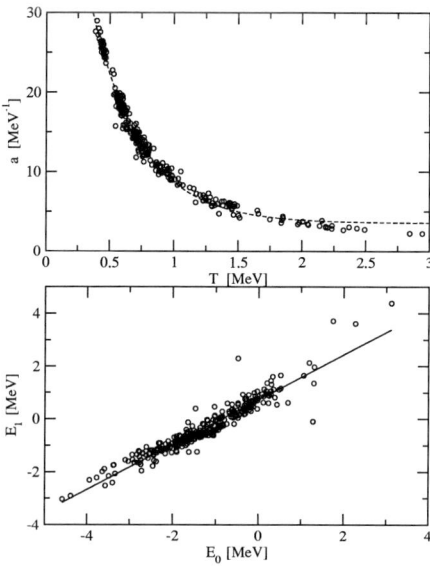

FIGURE 3. Correlations between the parameters of the BSFG and CT models. An exponential fit (for a versus T relation) and a linear one (for E_1 versus E_0) are shown.

A, which we calculate numerically from the mass tables.

As observed long time ago [2], and in many more recent publications, the level density parameter a (Fig. 1) shows an almost linear dependence on A and a clear correlation with the shell correction S. The parameter T (Fig. 2), which has an average dependence on mass as $A^{-2/3}$ [7], can be also well correlated with the shell correction. The parameter $\tilde{a}$ has an almost linear behaviour with mass, as proposed in Ref. [5].

The S values in our case show slightly different curves for even-even, odd and odd-odd nuclei. Consequently a pairing correction has to be applied: $S' = S - \Delta$. Various Δ were introduced in the fit procedures of a and T. The best results were obtained with $\Delta = +0.5P_d$; 0; $-0.5P_d$ for the even-even, odd-A, and odd-odd nuclei, respectively, where P_d is the deuteron pairing energy.

As it can be seen also in Figs. 1 and 2, the back-shift curves are quite different for even-even, odd and odd-odd nuclei. The curves for the even-even nuclei are relatively smooth, while the odd-odd nuclei exhibit more fluctuations and odd nuclei have strong variations. Therefore a pairing correction has to be introduced. Again, comparison of various pairing corrections showed that the use of P_d is the best choice. In addition, some structures in the back-shifts E_i which remain still unaccounted for, were found well correlated with the derivative of the shell correction $S(Z,N)$ with respect to mass.

The formulas adopted in this work are presented in Table 1, and some results of the fit can be seen in Figs. 1 and 2. The corresponding parameters for the BSFG and BSFG-ED are practically the same.

TABLE 1. Resume of the formulas proposed for the description of the phenomenological model parameters. The quantities P_d, $S(Z,N)$, and $dS(Z,N)/dA$ are given in MeV, therefore the dimensions of the different p_i constants in the formulas are such that a, $\tilde{a}$ are in MeV^{-1}, and T, E_i are in MeV, respectively. Note that for the compactness of the table we use the same notation, p_i, for the fitted parameters in the different formulas, but, for example in the case of the BSFG, the p_i constants in the formula of a and of E_1, respectively, have different meanings (they are factors for different quantities and have completely different values).

Model	Formula		Parameters in the formula			Type of nuclei
		p_1	p_2	p_3	p_4	
BSFG	$a/A = p_1 + p_2 S'(Z,N) + p_3 A$	0.127(1)	4.98(13)×10^{-3}	−8.95(53)×10^{-5}		all
	$E_1 = p_1 − 0.5P_d + p_4 \frac{dS(Z,N)}{dA}$ $= p_2 − 0.5P_d + p_4 \frac{dS(Z,N)}{dA}$ $= p_3 + 0.5P_d + p_4 \frac{dS(Z,N)}{dA}$	−0.468(30)	−0.565(23)	−0.231(39)	0.438(83)	e-e o-e,e-o o-o
BSFG-ED	$\tilde{a}/A = p_1 + p_3 A$	0.127(1)		−9.05(53)×10^{-5}		all
	$E_2 = p_1 − 0.5P_d + p_4 \frac{dS(Z,N)}{dA}$ $= p_2 − 0.5P_d + p_4 \frac{dS(Z,N)}{dA}$ $= p_3 + 0.5P_d + p_4 \frac{dS(Z,N)}{dA}$	−0.477(31)	−0.577(23)	−0.231(40)	0.442(87)	e-e o-e,e-o o-o
CT	$T \cdot A^{2/3} = p_1 + p_2 S'(Z,N) + p_3 S'(Z,N)^2$	17.45(6)	−0.51(4)	0.051(10)		all
	$E_0 = p_1 − 0.5P_d + p_2 \frac{dS(Z,N)}{dA}$ $= p_3 − P_d + p_4 \frac{dS(Z,N)}{dA}$ $= p_1 + 0.5P_d + p_2 \frac{dS(Z,N)}{dA}$	−1.23(3)	0.32(14)	−1.42(3)	0.84(14)	e-e o-e,e-o o-o

$S(Z,N) = M_{exp} − M_{LD}$, $S'(Z,N) = S(Z,N) − \Delta$; $\Delta = (+0.5, 0, −0.5) P_d$ for e-e, odd, o-o, respectively.

$P_d = \frac{1}{4}(−1)^{Z+1} [S_d(A+2, Z+1) − 2S_d(A,Z) + S_d(A−2, Z−1)]$; ($S_d$ - deuteron separation energy) [9, 10]

RELATION BETWEEN THE BSFG AND CT MODEL PARAMETERS

[!] The phenomenological parameters a and T, and the backshifts E_1 and E_0, of the BSFG and CT models, respectively, show very simple relationships, which have not been noticed before. This is shown in Fig. 3, where it can be seen that the relation a versus T is almost exponential, while that between E_1 and E_0 is practically linear. Thus, for not too high excitation energies, the two models are found to be equivalent.

CONCLUSIONS

A first result of our investigation is a new set of nuclear level density parameters for 310 nuclei between ^{19}F and ^{251}Cf, for three two-parameter models (BSFG, BSFG-ED, and CT), by fitting the newest complete experimental data for both the low-energy excited states and the level density at the neutron binding energy. The empirical LD model parameters for all nuclei have been described by simple formulas with few parameters (Table 1). It is remarkable that these formulas describe also the main features of the back-shift energies E_i which were determined from the data as free fit parameters.

The proposed formulas involve only quantities which can be taken or derived from the mass tables [9], so in essence our formulas calculate the level densities from nuclear masses. The formulas do not depend on any model; in extrapolating them to nuclei far from stability one relies only on the methods of extrapolation developed for the nuclear masses [11]. This means also that the coarse structure of the nuclei at higher excitations is mainly determined by the mass of the ground state. The choice of these formulas has been completely empirical. Although they might not be the "most appropriate" functional forms (from the theoretical point of view), they highlight interesting connections between the level densities and the mass table.

The present method of calculating level density parameters for nuclei for which there are no experimental data available, not from some local systematics and/or by some arbitrary extrapolation, but by using the formulas proposed above, seems to be a reasonable way of extrapolation to nuclei far from stability. The BSFG and CT models, respectively, may be safely used up to the neutron binding energy or slightly higher energies. The BSFG-ED model may be recommended up to higher energies (∼15-20 MeV), since it contains the recommended damping of the shell effects with increasing excitation energy. Finally, smooth correlations have been found between the BSFG and CT model parameters.

REFERENCES

1. A.V. Ignatyuk, IAEA-TECDOC-1034, IAEA, Vienna, 1998, p.65
2. A. Gilbert and A.G.W. Cameron, Can. J. Phys. **43**, 1446 (1965); P.J. Brancazio and A.G.W. Cameron, Can. J. Phys. **47**, 1029 (1969).
3. W. Dilg, W. Schantl, H. Vonach, and M. Uhl, Nucl. Phys. **217**, 269 (1973)
4. D. Bucurescu T. von Egidy and D. Bucurescu , Phys. Rev. **C**, in press.
5. A.V. Ignatyuk, G.N. Smirenkin, and A.S. Tishin, Sov. J. Nucl. Phys. **21**, 255 (1975)
6. A.V. Ignatyuk, K.K. Istekov, and G.N. Smirenkin, Sov. J. Nucl. Phys. **29**, 450 (1979)
7. T. von Egidy, H.H. Schmidt, and A.N. Behkami, Nucl. Phys. **A481**, 189 (1988)
8. J.M. Pearson, Hyp. Interactions **132**, 59 (2001)
9. G. Audi, A.H. Wapstra, and C. Thibault, Nucl. Phys. **A729**, 337 (2003)
10. The AME2003 files, retrieved from http://csnwww.in2p3.fr/amdc
11. A.H. Wapstra, G. Audi, and C. Thibault, Nucl. Phys. **A729**, 129 (2003)

Effective Interactions and Operators in Nuclei within the No-Core Shell Model

Bruce R. Barrett*, Petr Navratil[†], Ionel Stetcu* and James P. Vary[†,**,‡]

*Department of Physics, University of Arizona, P.O. Box 210081, Tucson, Arizona 85721
[†]Lawrence Livermore National Laboratory, Livermore, P.O. Box 808, California 94551
**Department of Physics and Astronomy, Iowa State University, Ames, Iowa 50011
[‡]SLAC, MS 81, 2575 Sand Hill Rd, Menlo Park, California 94025

Abstract. We review the application of effective operator formalism to the *ab initio* no core shell model (NCSM). For short-range operators, such as the nucleon-nucleon potential, the unitary-transformation method works extremely well at the two-body cluster approximation and good results are obtained for the binding energies and excitation spectra of light nuclei ($A \leq 16$). However, for long-range operators, such as the radius or the quadrupole moment, performing this unitary transformation at the two-body cluster level does not include the higher-order correlations needed to renormalize these long-range operators adequately. Usually, such correlations can be obtained either by increasing the order of the cluster approximation or by increasing the model space. We will discuss the difficulties of these approaches as well as alternate possible solutions for including higher-order correlations in small model spaces.

Keywords: shell model, nuclear spectra, electromagnetic transitions
PACS: 21.60.Cs, 23.20.-g, 23.20.Js

INTRODUCTION

In the last few years, significant advances in theoretical methods and computer power have allowed the description of the low-lying states in light nuclei from first principles, *i.e.*, starting from realistic nucleon-nucleon (NN) interactions that fit the experimental phase-shifts [1], and theoretical three-body forces [2]. The no-core shell model (NCSM) is a particularly flexible method that allows all the *A* nucleons to interact and preserves all the symmetries of the original Hamiltonian [3, 4]. The main tool is a unitary transformation [5, 6] of the Hamiltonian from the infinite space to a model space, which allows a diagonalization in a many-body basis. In the lowest cluster approximation, the unitary transformation approach works extremely well for binding energies and excitation spectra of light ($A \leq 16$) and selected medium-mass nuclei [7].

Besides spectra, other properties of the nuclear states are of interest, as they impose a strong test on the theoretical wave functions. In particular, we concentrate on electromagnetic transitions. The same unitary transformation used to compute the effective interaction can be used in order to obtain effective operators in the model spaces used to diagonalize the effective Hamiltonian. We found that, in the lowest approximation, the unitary transformation has little effect on long-range observables [8], such as quadrupole moments and radii. In retrospect, these results can be understood as long-range observables "feel" effects due essentially to all *A* nucleons; the two-body cluster approximation, however, accounts for short-range correlations and cannot accommodate long-range correlations.

In this paper, we shortly review the theory of effective operators in the framework of the unitary transformation approach and discuss applications to light nuclei, as well as possible avenues for including more higher-order correlations in small model spaces.

THEORETICAL APPROACH

For over fifty years, the state of the art method for a microscopic description of nuclear states has been the phenomenological shell model, where on top of the single-particle states one adds a residual interaction and diagonalizes in a restricted model space, assuming an inert core. Work by Arima and Horie [9] and later extended by Bertsch [10] introduced the core polarization concept into the nuclear shell model. Core polarization is a method to induce long-range correlations in a nucleus, and its effect was an improved agreement of the nuclear spectrum with the experiment. Without entering into details concerning issues like convergence, we have to point out that, while reasonable results have been obtained for the nuclear spectra, applications of the same many-body techniques to other observables, such as the magnetic dipole and electric quadrupole, never achieved the same level of success. In particular, the empirical effective charges of $1.5e$ for the proton and $0.5e$ for the neutron have remained a puzzle for decades.

In the NCSM, we use a unitary transformation in order to include correlations left out by the truncation of the model space. The transformed Hamiltonian

$$\mathcal{H} = e^{-S} H e^{S}, \qquad (1)$$

preserves the spectrum of the initial Hamiltonian. Moreover, if the antihermitian operator S is determined by the condition $P\mathcal{H}Q = 0$, that is, the transformed Hamiltonian does not couple the model, or P, space with the excluded, or Q, space, then the effective interaction in the model space is energy independent. Furthermore, if one determines the operator S so that the additional decoupling condition $Q\mathcal{H}P = 0$ is fulfilled, it can be shown that the effective operators determined by the transformation

$$\mathcal{O} = e^{-S} O e^{S} \qquad (2)$$

are also energy independent [5, 11]. Formally, the operator S can be written by means of another operator ω as $S = \text{arctanh}(\omega - \omega^{\dagger})$, where the new operator fulfills $Q\omega P = \omega$. Hence, one obtains the energy-independent effective Hamiltonian in the model space P

$$H_{eff} = P\mathcal{H}P = \frac{P+P\omega^{\dagger}Q}{\sqrt{P+\omega^{\dagger}\omega}} H \frac{P+Q\omega P}{\sqrt{P+\omega^{\dagger}\omega}}, \qquad (3)$$

and, analogously, any observable can be transformed to the P space as [5, 11]

$$O_{eff} = P\mathcal{O}P = \frac{P+P\omega^{\dagger}Q}{\sqrt{P+\omega^{\dagger}\omega}} O \frac{P+Q\omega P}{\sqrt{P+\omega^{\dagger}\omega}}. \qquad (4)$$

The transformed operator preserves the same symmetries of the initial operator, e.g., the tensor character, as the transformation is a spin-isospin scalar. However, we have to

TABLE 1. $B(E2)$, in $e^2 fm^4$, and relative kinetic energy expectation value, in MeV, for selected nuclei and model spaces, using the bare and effective operators computed in the two-body cluster approximation.

Nucleus	Observable	Model Space	Bare operator	Effective operator
^{6}Li	$B(E2, 1^+0 \to 3^+0)$	$2\hbar\Omega$	2.647	2.784
^{6}Li	$B(E2, 1^+0 \to 3^+0)$	$10\hbar\Omega$	10.221	-
^{6}Li	$B(E2, 2^+0 \to 1^+0)$	$2\hbar\Omega$	2.183	2.269
^{6}Li	$B(E2, 2^+0 \to 1^+0)$	$10\hbar\Omega$	4.502	-
^{10}C	$B(E2, 2_1^+0 \to 0^+0)$	$4\hbar\Omega$	3.05	3.08
^{12}C	$B(E2, 2_1^+0 \to 0^+0)$	$4\hbar\Omega$	4.03	4.05
^{4}He	$\langle g.s.|T_{rel}|g.s.\rangle$	$8\hbar\Omega$	71.48	154.51

point out that the transformation is much more involved for a non-scalar observable than for a scalar one. Indeed, because general tensor operators can couple different angular momentum states, one has to use different transformation operators ω in Eq. (4).

Once one obtains ω, the problem is solved. However, this is a task as difficult as solving the full A-body problem, as the exact solution is an A-body operator, regardless of the rank of the interaction with which one starts. Therefore, one makes the cluster approximation, in which one finds ω for the problem involving a nucleons ($a < A$), and then uses this solution to compute the effective Hamiltonian and operators in the model space via Eqs. (3) and (4). This interaction, which reproduces exactly the spectrum of the a-body problem, is then used to compute the solution to the A-body problem. There are two important convergence properties. Thus, keeping the model space fixed and increasing the cluster size, one converges to the exact solution. The same convergence to the exact solution can be also obtained by keeping the cluster size fixed and increasing the model spaces. In principle, because of these properties, one expects to observe larger renormalization in smaller model spaces and for small cluster sizes, or, in other words, the renormalization effects are more visible in small model spaces for a given cluster approximation.

RESULTS AND DISCUSSION

In a recent publication [8], we have investigated the effect of the renormalization of electromagnetic operators in realistic calculations for p-shell nuclei. Because our first goal was a qualitative understanding of the influence of effective operators and not a highly accurate description of the experimental data, we have used only two-body, high precision NN interactions, leaving out the three-body forces for the time being. This also motivates the use of rather small model spaces and of the two-body cluster approximation, given that, as expected from the convergence properties of effective operators mentioned at the end of the last section, in such conditions, larger renormalization effects are expected.

In calculations of several nuclei, we saw very little effect of the effective operators in the results for the $B(E2)$, as illustrated in Table 1. One may find this somehow surprising, as previous investigations in the framework of the NCSM [12] successfully obtained the correct effective proton and neutron phenomenological charges, which haunted the nuclear structure community for so long. However, the main difference is that the calculation in Ref. [12] included up to six-body correlations in the case of ^{6}Li, i.e., the nucleus chosen for this investigation. Comparison of the two results already suggests that higher-order clusters can play an important role in the renormalization of the $E2$ operator.

In contrast with the $E2$ operator, the kinetic energy operator is well-renormalized at the two-body cluster level, as seen in Table 1. The difference can be understood in terms of the character of the two operators. The kinetic energy is short range, while the quadrupole is long range; at the two-body cluster level, the unitary transformation renormalizes mainly the short-range core of the interaction, leaving unchanged the long range part. Hence, in order to account for long-range correlations in the two-body cluster approximation, one needs to enlarge the model space; an illustration of this approach can be seen in Table 1, where the $B(E2)$ value obtained in $10\hbar\Omega$ is significantly larger than the one calculated in $2\hbar\Omega$. Although for very light nuclei increasing the model spaces is not a problem, the use of a large model space quickly becomes numerically intractable for heavier systems. Because the utilization of very large model spaces is impractical, if not impossible, for all but the lightest nuclei ($A \leq 10$), it is worthwhile to develop new methods to introduce more correlations in a smaller model space. Such methods should also provide a better understanding of the underlying physics. Possible such methods would include, among others,

1. performing large model-space calculations for light nuclei, and then explicitly truncating these results into a very small model space, as done in Ref. [12] for ^{6}Li, thereby building into the final results all the A-body correlations, and
2. calculating the effective Hamiltonian in a large model space and using its matrix elements as input into a many-body perturbation theory calculation of a new, further renormalized effective Hamiltonian in a much smaller model space.

Work is in progress to implement such approaches.

ACKNOWLEDGMENTS

B.R.B and I.S. acknowledge partial support by NFS grant PHY0244389. The work was performed in part under the auspices of the U. S. Department of Energy by the University of California, Lawrence Livermore National Laboratory under contract No. W-7405-Eng-48. P.N. received support from LDRD contract 04-ERD-058. J.P.V. acknowledges partial support by USDOE grant No DE-FG-02-87ER-40371.

REFERENCES

1. R. B. Wiringa, V. G. J. Stoks and R. Schiavilla, Phys. Rev. C **51**, 38 (1995);R. Machleidt, F. Sammarruca and Y. Song, Phys. Rev. C **53**, R1483 (1996).

2. S. A. Coon, M. D. Scadron, P. C. McNamee, B. R. Barrett, D. W. E. Blatt, and B. H. J. McKeller, Nucl. Phys. **A317**, 242 (1979); S.A. Coon and H. K. Han, Few Body Systems 30, 131 (2001).
3. P. Navrátil, J. P. Vary, W. E. Ormand, and B. R. Barrett, Phys. Rev. Lett. **87**, 172502 (2001).
4. P. Navrátil, J. P. Vary, and B. R. Barrett, Phys. Rev. Lett. **84**, 5728 (2000); Phys. Rev. C 62, 054311 (2000).
5. S. Okubo, Prog. Theor. Phys. **12**, 603 (1954).
6. J. Da Providencia and C. M. Shakin, Ann. of Phys. **30**, 95 (1964); K. Suzuki and S.Y. Lee, Prog. Theor. Phys. **64**, 2091 (1980); K. Suzuki, Prog. Theor. Phys. **68**, 246 (1982); K. Suzuki and R. Okamoto, Prog. Theor. Phys. **70**, 439 (1983); K. Suzuki, Prog. Theor. Phys. **68**, 1999 (1982); K. Suzuki and R. Okamoto, Prog. Theor. Phys. **92**, 1045 (1994).
7. S. Popescu, S. Stoica, J.P. Vary and P. Navratil, to be published.
8. I. Stetcu, B. R. Barrett, P. Navrátil, and J. P. Vary, Phys. Rev. C **71**, 044325 (2005).
9. A. Arima and H. Horie, Prog. Theor. Phys. (Kyoto) **12**, 623 (1954); Phys. Rev. 99,778 (1955).
10. G. F. Bertsch, Nucl. Phys. **74**, 234 (1965).
11. P. Navrátil, H. Geyer, and T. T. S. Kuo, Phys. Lett. **B 315**, 1 (1993).
12. P. Navrátil, M. Thoresen, and B. R. Barrett, Phys. Rev. C **55**, R573 (1997).

NUCLEAR REACTIONS

Precision Gamma-Ray Spectroscopy at the Institut Laue Langevin

H.G. Börner

Institut Laue-Langevin, F-38042 Grenoble, France

Abstract. Currently the Institut Laue-Langevin follows two main experimental issues in the ILL nuclear physics studies. In the first one thermal neutron capture is used to excite nuclei up to the binding energy. Subsequently emitted gamma rays are characterized with ultra-high precision using the crystal spectrometers *GAMS*. In the second one, neutron-induced fission allows the study of the structure of heavy neutron-rich isotopes using the *LOHENGRIN* recoil mass separator. Most of today's measurements at ILL are done using the in-pile target arrangements of these two instruments, but some work is also carried out with neutrons at external positions fed by neutron guides. We will present an overview on the current status of the various experimental approaches and discuss the results and applications of a selected set of measurements.

INTRODUCTION

The Institut Laue-Langevin (ILL) is Europe's leading research facility for fundamental research using neutrons. The ILL operates the brightest neutron source in the world, reliably delivering intense neutron beams to 40 unique scientific instruments. The Institute welcomes 1500 visiting scientists per year to carry out world class research in solid state physics, biology, chemistry and nuclear and particle physics. Funded by its three founder members, France, Germany and the United Kingdom, the ILL has also signed scientific collaboration agreements with seven other European countries.

ULTRA-HIGH RESOLUTION GAMMA-RAY SPECTROSCOPY

At the high-flux reactor one can obtain very high specific activities when exploiting the flux of 5×10^{14} neutrons/cm^{2}s available at the in-pile target facilities. This allowed us to aim for the installation of gamma-ray spectrometers with the highest possible energy resolution. The crystal spectrometers GAMS [1,2] have been extensively used to address a large variety of different fields with an emphasis to nuclear-structure studies. Originally they were mainly used to construct level schemes with the aim to do that as completely as possible. A well known example is ^{168}Er [3] where - within the spin window which can be reached in neutron capture - a complete level scheme was established up to an excitation energy of about 2.5 MeV. Another example is ^{196}Pt [4], the level scheme of which was the first one to be associated with the O(6) limit of the IBM. To date we use the ppm resolving power mainly to observe the tiny Doppler shifts obtained when a nucleus recoils after the emission of gamma rays. This allows us to determine lifetimes of excited states - in the pico to femto second region - populated after thermal neutron capture. The method, called Gamma Ray Induced

Doppler broadening (GRID), is in detail discussed in [5].

Examples for lifetime studies concern for instance the observation of mixed-symmetry states in ^{54}Cr [6], the investigation of multiphonon excitations in medium-heavy nuclei such as ^{114}Cd [7], the determination of quadrupole-octupole coupled states in ^{144}Nd [8], the test of selection rules predicted for ^{196}Pt by O(6) [9], the search for multiphonon excitations in heavy deformed nuclei such as ^{168}Er [10,11] or ^{178}Hf [12] and the very recent measurements [13] of lifetimes in the critical point nucleus ^{152}Sm (see for instance [14]).

The resolution which is now obtained with flat crystals has been pushed to its limits and corresponds to that predicted by dynamical diffraction theory – at the expense of detection efficiency. A geometry, using bent crystals, would allow one to obtain much higher detection efficiency but the resolution is not yet comparable. One of the aims for the future will therefore remain the development of bent crystals with improved resolving power. In the past this development was mainly limited due to manpower problems. In parallel improvements of precision continue to be a great challenge at these instruments. Efforts will continue to approach the 10ppb range.

SPECTROSCOPY OF NEUTRON-RICH NUCLEI

The LOHENGRIN recoil mass separator [15] for unslowed fission products has contributed now for 30 years to research in fission and spectroscopy of very neutron rich nuclei. Basically, fission fragments are analysed on their flight path through a magnetic and an electric field. This technique ensures a high mass resolution with a clear separation between fragments with different mass numbers, or more precisely with different ratios of mass number (A) to ionic charge (q). The source for the spectrometer is a thin target of fissile material which is placed close to the core of the reactor in a high thermal neutron flux of almost 10^{15} n/(cm²·s). Fragments with the same ratio A/q are focussed onto parabolas. The high fission count rates allow searchs for very rare fission events. Fragment yields have been analysed down to the level of 10^{-8}/fission and have e.g. led to the discovery of new neutron-rich isotopes in Fe and Co. The useful range of atomic masses available at the separator for gamma-ray spectroscopy studies, is approximately 70 to 170. These limits are set only by yield distributions from fission. Various fissile targets are available, from ^{229}Th to ^{251}Cf, and targets can be chosen to optimise the production rate for a particular nucleus of interest. The kinetic energy of the fragments passing through the separator varies between about 50 and 110 MeV. With a flight path of 23m the separation time is around 1 or 2 μs.

Fission fragments arriving at the focal point of the spectrometer will have the same A/q ratio but will have different kinetic energies. A selection of ionisation chambers (to measure the kinetic energies) are available at Lohengrin, each optimised to a different application. The different applications include - besides fission studies - high-efficiency gamma-ray and conversion-electron spectroscopy and gamma-ray angular distribution studies (an example for an experimental set up is shown in figure 1).

FIGURE 1. *A detector set-up at the Lohengrin focal point with two 60% Ge-detectors and one Si-detector in position.*

Detection of conversion electrons is performed by two adjacent silicon detectors placed a few millimetres behind a 12 micrometer thick mylar foil, which stops the fission fragments. The kinetic energy of the fission fragment can be tuned, with the separator, so that they come to rest within the last few micrometers of the mylar foil, allowing high-resolution conversion-electron spectroscopy. The overall conversion-electron detection efficiency is around thirty percent. The lower limit of detection is around 15 keV, which is very useful when searching for isomeric transitions, which are often of low energy.

For gamma-ray detection currently two clover detectors are now available and can be placed face-to-face, perpendicular to the beam, close to the focal point of the beam to give the maximum amount of solid-angle coverage possible. Figure 2 demonstrates the quality of data obtained to date. The gamma-ray multiplicity in all experiments is generally low, allowing few detectors to be used. In collaboration with the Studsvik group an array of barium-fluoride detectors allows fast-timing measurements to be performed on excited nuclear-states with lifetimes in the range ten picoseconds to nanoseconds. For isomers with microsecond lifetimes the correlation between the arrival time of the fission fragment, in the ionisation chamber, and the detection of a gamma ray, or conversion electron, is sufficient to measure the lifetime of the excited state.

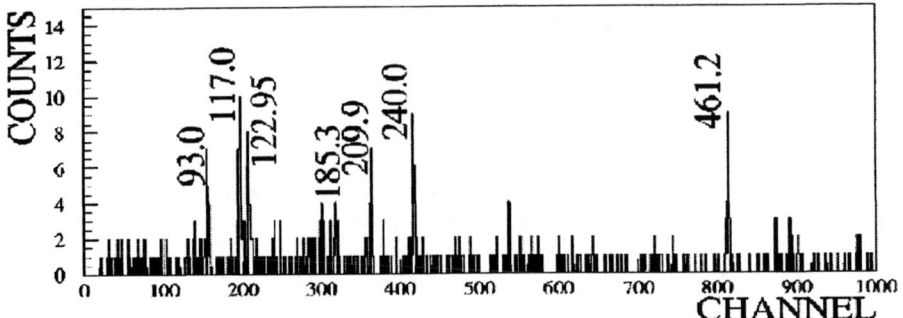

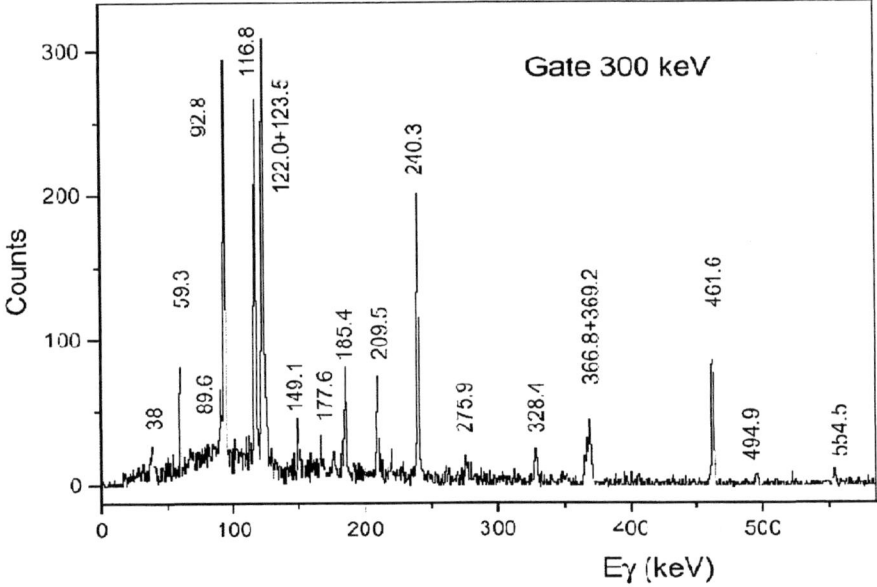

FIGURE 2. *Top: Gamma rays from ^{96}Rb in 1999. Bottom: Gamma rays from ^{96}Rb in 2005.*

The best technique to obtain information on excited nuclear states of neutron-rich nuclei far from stability, at Lohengrin, is to observe the decay of states below microsecond isomers. The transit time of approximately one to two microseconds through the separator currently allows isomeric states with lifetimes as short as about half a microsecond to be observed. The unique character of isomeric states give valuable nuclear structure information. Isomeric states are present across the whole landscape of the nuclear chart resulting from both single particle effects (such as yrast spin traps) and collective motion (*K*-isomers and shape isomers). Putting a time window of several tens of microseconds between the detection of the arrival of a fission fragment and the detection of a gamma ray, plus the ability to select individual masses (or in favourable cases individual isotopes) gives very clean spectra. The cleanliness of these spectra allows transitions to be identified and assigned to nuclei with just tens of gamma-ray photo peak counts. The same is true for conversion electrons.

Isomers are useful not only because they allow measurements of the properties of weakly produced nuclei but because these states have unique a character among many other excited states. Isomers are generally at higher angular momentum than ground states, giving isomer spectroscopy of odd-odd nuclei a distinct advantage over beta decay spectroscopy, where only low-spin states are populated, as intermediate-spin states can be accessed. New isomers were discovered around the mass 100 region including ^{94}Y, ^{96}Rb and ^{106}Nb [16]. In the case of ^{96}Rb this is the first measurement of any excited states in this very neutron-rich nucleus. Measurements of many isomers in the region of the neutron-rich doubly-magic nucleus ^{132}Sn have also been performed recently. New isomeric spin $19/2^+$ states were found in the isotopes $^{125, 127, 129}$Sn [17]. An isomeric state of spin $23/2^+$ was discovered in ^{129}Sn also [18]. Previous to this work the highest spin states known in these heavy, odd-mass Sn nuclei were of spin $11/2^-$. Production of these nuclei and states would be very difficult at most other existing nuclear physics facilities. Comparison of these states with theoretical interpretations of low-lying negative-parity states shows some unexpected discrepancies for Sn isotopes (figure 3 shows a spectrum obtained for mass 129) close to the doubly-magic nucleus ^{132}Sn. Concerning the antimony isotopes around ^{132}Sn, a new isomer was observed in ^{130}Sb with spin 13^+ [19], and isomers in $^{131, 133}$Sb [19][20] were also investigated. New 10^+ isomers were observed in ^{132}Te, ^{134}Xe [22] and together with measurements of the isomer in ^{130}Te [21], show a strong increase in the B(E2) strength from Sn to Te isotopes. Small configuration mixings involving proton 2^+ configurations in the 8^+ and 10^+ levels, due to the *p-n* interaction, can have strong effects on the B(E2) values, but negligibly perturb the energies of these states. More recently microsecond isomers in the In and Cd isotopes, in the mass range A=123 to 130, were investigated at LOHENGRIN, through thermal-neutron induced fission reactions of Pu targets [23]. It was found that most likely no μs isomers exist in ^{126,128}Cd with half-lives longer than 0.5 μs. The level schemes of odd-mass $^{123-129}$In and new measurements of the μs half-lives of the odd-odd $^{126-130}$In were established. It was also found that the M2 transitions are much faster in the In isotopes than in the proton-magic Sn isotopes.

SPECTROCOPY AT A NEUTRON-GUIDE

At present, the fission reaction, induced by thermal neutrons on a fissile target placed close to the core of a reactor, is certainly one of the most effective methods for producing extremely rare medium-heavy to heavy neutron rich isotopes isotopes. The cross sections for this type of reaction are generally large: For example $\sigma_f \sim$ 1000 barn for the reaction n_{th} +^{241}Pu. As a result, thanks to ILL's extremely high neutron flux, $\Phi_n \sim 10^{15}$ n/cm^2/s, it is possible to produce up to 10^{12} fissions/s with only 1 mg of fissile material if placed close to the reactor core. As discussed above, in the past these exceptional properties have been successfully used with the spectrometer Lohengrin.

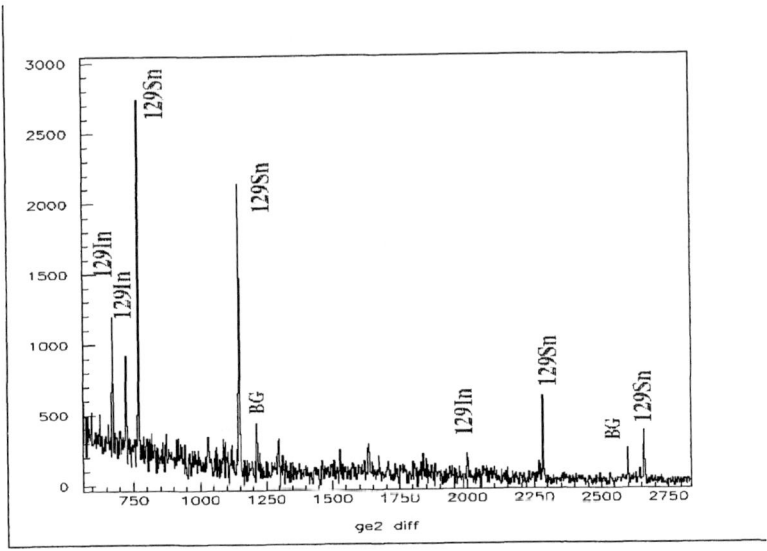

FIGURE3. *Example of a spectrum obtained for mass 129*

As it takes about a microsecond for the fission products to travel from the sample to the Lohengrin detectors, Lohengrin is ideally situated for the investigation of micro- to millisecond isomers and their decay, using its good mass resolution. A complementary (and up to now not yet efficiently exploited) approach is to use an external beam of thermal neutrons which allows prompt fission products to be studied. The ILL's neutron guides produce high-flux beams of Φ_n around 10^9 - 10^{10} n/cm^2/s, of very high purity (no γ noise and no fast neutrons from the reactor core). Neutrons are capable of inducing fission reactions on fissile targets placed outside the reactor, in a sufficiently clean environment to be able to study the prompt and delayed gamma rays emitted by the fission products. Using a very high-efficiency gamma detector array with a granularity sufficient to detect intermediate multiplicity events (~10), it is possible to obtain information on the structure of nuclei inaccessible by other currently available methods, such as spontaneous fission. Experiments of this kind are proposed for several neutron guide positions. In this way, Lohengrin and these new experimets should provide us with complementary information and make the ILL an important centre for the study of exotic nuclei. Using a neutron guide, different fissile targets, ranging from Th to Cf, can be irradiated, making it possible to study regions of nuclei inaccessible by spontaneous fission. As an example, the reaction n_{th} + ^{241}Pu makes it possible to reach even nuclei in the vicinity of ^{78}Ni (which are inaccessible with spontaneous fission sources). The use of different fissile targets should also help to identify the fission products observed (different yields). Finally, on the guide it is possible to perform irradiations over long periods, corresponding to several reactor cycles, something which cannot be done on an accelerator. A target mass as low as ~ 1 to 10 μg of ^{241}Pu, for instance, will be sufficient. First such measurements were

FIGURE 4. *Top view on a set-up used at the neutron guide H113*

already carried out successfully at ILL during 2004/2005 (one of the experiments carried out recently, is shown in figure 4).

REFERENCES

1. M.S. Dewey et al., *Nucl. Instr. Meth.* **A284** (1989) 151
2. E.G. Kessler et al., *Nucl. Instr. Meth.* **A457** (2001) 187
3. H.G. Börner et al., *Phys. Rev. Lett.* **66** (1991) 691
4. J. A. Cizewski et al, *Nucl. Phys.* **A323** (1979) 349
5. H.G. Börner and R. Krücken, *Nucl. Phys. News Int.* Vol. 10, No. 3 (2000) 11
6. K.P. Lieb et al., *Phys. Lett.* **B215** (1988) 50
7. R.F. Casten et al., *Phys. Lett.* **B297** (1992) 19
8. S.J. Robinson et al., *Phys. Rev. Lett.* **73** (1994) 412
9. H.G Börner et al., *Phys. Rev.* **C42** (1990) R2271
10. L. Genilloud et al., *Phys Rev.* **C62**, 034313 (2000)
11. H. Lehmann et al., *Phys. Rev.* **C 57** (1998) 569
12. A. Aprahamian et al., *Phys. Rev.* **C 65**(2002)031301(R)
13. N.V. Zamfir et al., *Phys. Rev.* **C 65**(2002)067305
14. R.F. Casten and N.V. Zamfir, *Phys. Rev. Lett.* **87**, 052503 (2001)
15. E. Moll et al., *Kerntechnik* 19 (1977) 374
16. J. Genevey et al., *Phys. Rev.* **C 59** (1999) 82-89.
17. J.A. Pinston et al., *Phys. Rev.* **C 61** (2000) 024312.
18. J. Genevey et al., *Phys. Rev.* **C 65** (2002) 034322.
19. J. Genevey et al., Eur. Phys. J. **A 9** (2000) 191-195.
20. J. Genevey et al., *Eur. Phys. J* **A 7** (2000) 463-465.
21. J. Genevey et al., *Phys. Rev.* **C 63** (2001) 05315.
22. J. Genevey et al., *Phys. Rev.* **C 67** (2003) 054312.
23. A. Scherillo et al., *Phys. Rev.* **C70** (2004) 054318
24. G. Smith et al. Private communication

Radiative Capture versus Coulomb Dissociation

Henning Esbensen

Physics Division, Argonne National Laboratory, Argonne, Illinois 60439

Abstract. Measurements of the Coulomb dissociation of ^{8}B have been used to infer the rate of the inverse radiative proton capture on ^{7}Be. The analysis is usually based on the assumptions that the two processes are related by detailed balance and described by E1 transitions. However, there are corrections to this relation. The Coulomb form factors for the two processes, for example, are not identical. There are also E2 transitions and higher-order effects in the Coulomb dissociation, and the nuclear induced breakup cannot always be ignored. While adding first-order E2 transitions enhances the decay energy spectrum, the other mechanisms cause a suppression at low relative energies. The net result may accidentally be close to the conventional first-order E1 calculation, but there are differences which cannot be ignored if accuracies of 10% or better are needed.

Keywords: Coulomb dissociation, radiative capture
PACS: 25.40.LW, 25.70.De,26.65.+t

INTRODUCTION

The radiative proton capture on unstable nuclei, $a+p \to b+\gamma$, can be extremely difficult to measure at low energy so it is important to develop alternative methods that can provide the necessary information. One method is to produce a secondary beam of the product nucleus b and then measure the Coulomb dissociation reaction, $b+\gamma \to a+p$, in the Coulomb field from a high-Z target. The basic idea of this method is that the two reactions are related by detailed balance. Thus by measuring one of the two reactions one should be able to infer information about the other [1]. However, there are corrections to the Coulomb dissociation that have to be considered before the inverse radiative capture cross section can be reliably determined.

An excellent example of a radiative proton capture reaction that has been measured both directly and indirectly using the Coulomb dissociation method is the ^{7}Be$+p \to ^8$B$+\gamma$ reaction, which is of great interest to solar neutrino physics. By comparing the results it was realized [2] that there are significant differences. Thus the S factor, $S(E_{rel})$, extracted from Coulomb dissociation experiments is slightly steeper when plotted as a function of the relative energy, E_{rel}, of the proton and ^{7}Be, and the S factor extrapolated to zero relative energy, $S_{17}(0)$, is about 10% smaller than obtained in recent direct capture measurements. In an effort to understand what causes these differences, we have examined the assumptions that are commonly made in the analysis of Coulomb dissociation experiments [3]. The basic assumptions are

- the Coulomb dissociation is determined by first-order dipole (E1) transitions,
- the dipole field can be calculated in the so-called far-field approximation.

We find that the far-field approximation is inaccurate for ^{8}B, and that E2 transitions, nuclear processes, and higher-order effects cannot always be ignored as discussed below.

SEMICLASSICAL DESCRIPTION OF DISSOCIATION

We have used a semiclassical description of the dissociation of ^{8}B in the Coulomb and nuclear fields from a target nucleus. The relative motion of projectile and target is simulated by classical Coulomb trajectories, whereas the relative motion of the proton and the ^{7}Be core is treated quantum mechanically. The ground state wave function is modeled as the eigenfunction of a simple two-body Hamiltonian, which includes the proton-core Coulomb potential, and a Wood-Saxon type nuclear potential [4, 5] adjusted to produce the proton separation energy of 137 keV. The two-body wave function is evolved in time in the Coulomb and nuclear fields of the target nucleus by solving the time-dependent Schrödinger equation numerically. All calculations are based on the same two-body model so that one can clearly see the effect of various approximations. All calculations were performed for a ^{8}B beam energy of 52 MeV/n on a Pb target, which are the experimental conditions in Ref. [6].

Far-field approximation. The first application of the model is to test the validity of the far-field (FF) approximation that is commonly made in the multipole expansion of the Coulomb field from the target nucleus. In the FF approximation it is assumed that the distances r_p and r_c of the proton and the ^{7}Be core, respectively, from the center-of-mass of ^{8}B are smaller than the distance R between ^{8}B and the target nucleus. The two intrinsic distances are related by $r_p = 7r_c$, so while the condition for the FF approximation is usually fulfilled for the core distance r_c, it is not always fulfilled for the proton distance r_p, because of the weak binding and the long radial tail of the two-body ground state wave function. The correct or unrestricted multipole expansion of the Coulomb field is

$$V_{\text{Coul}}(\mathbf{r}_x, \mathbf{R}) = \sum_{\lambda\mu} \frac{4\pi Z_x Z e^2}{2\lambda+1} \frac{r_<^\lambda}{r_>^{\lambda+1}} Y_{\lambda\mu}^*(\hat{r}_x) Y_{\lambda\mu}(\hat{R}), \qquad (1)$$

where $r_< = \min(r_x, R)$, $r_> = \max(r_x, R)$, $Z_x e$ is the charge and $\mathbf{r}_x$ is the position of the proton (or the ^{7}Be core) with respect to the target, and Ze is the charge of the target nucleus. This expression should be corrected for the finite size of the target. In the FF approximation one assumes that $r_> = R$.

The failure of the FF approximation is illustrated in the left panel of Fig. 1 for a fixed impact parameter of 20 fm with a Pb target at 52 MeV/n. The first-order E1 and E2 decay energy spectra obtained in the FF approximation (dashed curves) are compared to the unrestricted first-order E0, E1, and E2 spectra (solid curves). The latter calculations were based on the unrestricted multipole expansion, Eq. (1), and were corrected for the finite size of the target. It is seen that the FF approximation is poor at low E_{rel}, in particular for E2 transitions.

Dynamic effects. The decay energy spectra obtained in various approximations are compared in the right panel of Fig. 1. The top dashed curve is the first-order result obtained in the FF approximation; it is just the sum of the two dashed curves shown in Fig. 1. The next (solid) curve is the unrestricted first-order result obtained as the sum of the unrestricted E0, E1, and E2 multipole spectra shown in Fig. 1. This curve is suppressed at low relative energies compared to the first-order FF spectrum. The suppression is relatively insensitive to the beam energy but it depends on impact parameter and affects the dissociation probability out to very large impact parameters as shown in Ref. [5].

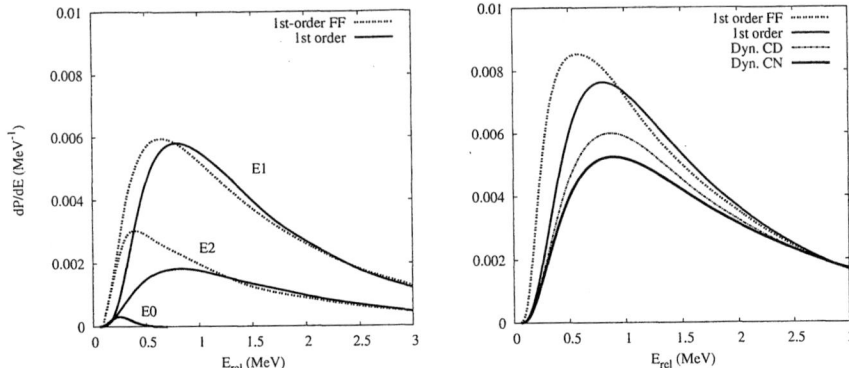

FIGURE 1. Decay energy spectra for ^{8}B at 52 MeV/n and an impact parameter of 20 fm with a Pb target. The left panel shows the first-order E0, E1 and E2 spectra obtained with the unrestricted multipole expansion of the Coulomb field (solid curves), and in the far-field (FF) approximation (dashed curves). The right panel compares the summed first-order spectra with the dynamic calculations of the Coulomb dissociation (Dyn. CD) and the Coulomb-nuclear induced dissociation (Dyn. CN).

In decreasing order, the third curve in the right panel of Fig. 1 shows the dynamic Coulomb dissociation spectrum (Dyn. CD). It is suppressed compared to the second curve (1st order) due to higher-order Coulomb-induced processes. In Ref. [5] it was demonstrated that the dominant part of the relative suppression of the dissociation probability, the so-called Barkas factor, has the approximate form

$$B \approx -\frac{Ze^2}{E}\frac{\text{Constant}}{\sqrt{b^2+a^2}}, \qquad (2)$$

in terms of the target charge Ze, the beam energy E, and the impact parameter b. The parameter a was found to be about 37 fm. The expression (2) shows that the absolute suppression is of order Z^3. Such a term is caused by an interference of first- and second-Born amplitudes, and it is the result of an interesting interplay between E1 and E2 transitions [7]. In fact, the Z^3 suppression would vanish if there were no E2 transitions. This Z^3 effect is also known as the Barkas effect in atomic physics, where it accounts for the difference in the stopping powers of protons and antiprotons [8].

The lowest solid curve in the right panel of Fig. 1 shows the further reduction that occurs in the decay energy spectrum when the nuclear proton-target interaction is also included in the dynamic calculation (Dyn. CN). It is clear that the first-order FF approximation is a very poor approximation in the example shown here, compared to the complete calculation. The first-order E1 FF approximation (shown in the left panel) is accidentally in much better agreement with the full calculation, but there are differences. Thus the full dynamic calculation is suppressed by about 10% at $E_{rel} = 0.25$ MeV and enhanced by 10% at $E_{rel} = 1$ MeV (see Fig. 4 of Ref. [3]). This trend explains why Coulomb dissociation experiments, when analyzed in the first-order E1 FF approximation, produce a steeper slope of $S_{17}(E_{rel})$ than obtained in direct capture measurements.

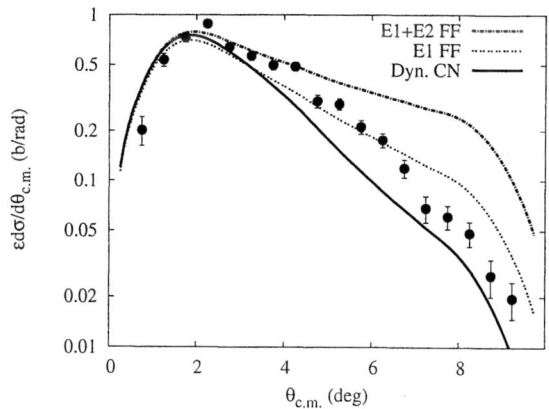

FIGURE 2. Comparison to the *RIKEN* experiment [6] for the relative energy cut: $0.5 \leq E_{rel} \leq 0.75$ MeV. The calculations were corrected for the experimental efficiency.

The only way to make the first-order E1 FF approximation a reliable tool is to increase the beam energy and increase the impact parameter. Another possibility is to argue that the E2 strength of ^{8}B is very small. If that is the case, then the dynamic Z^3 effect would vanish, as explained above, and the FF approximation would only affect the E1 spectrum. It is therefore of great interest to know the actual E2 strength of ^{8}B.

E2 strength from momentum distributions. One major difference between the radiative proton capture on ^{7}Be and the Coulomb dissociation of ^{8}B is that the radiative capture is almost entirely due to E1 transitions at low energies, below the M1 resonance, whereas E2 transitions must play some role in the Coulomb dissociation. However, the actual E2 strength of ^{8}B has been controversial [9, 10]. The effect of E2 transitions has been seen in measurements of the momentum distribution of ^{7}Be fragments [11], where the interference of E1 and E2 amplitudes produces an asymmetry, but the E2 strength that was needed to fit the data was only half of the model prediction.

One reason the extracted E2 strength is so small is that the analysis was based on the far-field approximation. Thus it is possible to reproduce the data without any adjustment of the E2 strength when the correct or unrestricted Coulomb form factors are used. This was recently demonstrated by Mortimer *et al.* [12] in DWBA calculations that were based on essentially the same structure model as used here and in Ref. [11]. Higher-order processes may reduce the asymmetry, but it is not clear by how much. In any case, it appears that the E2 strength that is needed to reproduce the measured asymmetry is at least as strong as - and maybe even stronger than - predicted by the structure model.

E2 strength from decay energy spectra. The analysis of measured decay energy spectra, on the other hand, suggests that the E2 strength is very small and essentially consistent with zero [6, 10]. The analysis performed in Ref. [6] was based on DWBA calculations, so it is of interest to see what is the effect of higher-order processes. This is illustrated in Fig. 2, where the measured angular distribution [6] for the relative energy cut: $0.5 \leq E_{rel} \leq 0.75$ MeV, is compared to our semiclassical calculations [3].

All calculations were corrected for the experimental efficiency. It is seen that the first-order E1+E2 FF calculation has a very poor shape in comparison to the data, whereas the first-order E1 FF calculation has a better shape. This is consistent with Refs. [6, 10].

The full dynamic calculation is shown by solid curve in Fig. 2. By comparing it to the first-order E1+E2 FF approximation it is seen that nuclear and higher-order processes, combined with unrestricted Coulomb form factors, have a very large effect at large scattering angles. The full calculation may have an even better shape in comparison with the data than predicted by the first-order E1 FF approximation. The most important observation is, however, that an analysis based on the first-order E1 or E1+E2 far-field approximation is unreliable, in particular at large scattering angles. The conclusion of Refs. [6, 10] that the E2 strength must be very small is therefore not justified.

CONCLUSIONS

The calculations presented here demonstrate that the extraction of radiative proton capture rates from Coulomb dissociation experiments is very difficult when high precision is needed. The conventional analysis, which is based on first-order E1 transitions in the far-field approximation (detailed balance) is inaccurate by up to 10%, and sometimes even more, depending on the kinematics of the measurement. To improve the analysis it is necessary to use unrestricted Coulomb form factors and include realistic E2 transitions as well as nuclear and higher-order processes. Relativistic effects, which have not been considered here, should also be implemented in the Coulomb excitation.

It is interesting to note that the extraction of radiative neutron capture rates from Coulomb dissociation experiments is simpler. The FF approximation is justified, the E2 strength is small, and the Z^3 effect will therefore also be small.

ACKNOWLEDGMENTS

This work was supported by the U.S. Department of Energy, Office of Nuclear Physics, under Contract No. W-31-109-ENG-38.

REFERENCES

1. G. Baur, C. A. Bertulani, and H. Rebel, Nucl. Phys. A **458**, 188 (1986).
2. A. R. Junghans et al., Phys. Rev. C **68**, 065803 (2003).
3. H. Esbensen, G. F. Bertsch, and K. A. Snover, Phys. Rev. Lett. **94**, 042502 (2005).
4. H. Esbensen, and G. F. Bertsch, Nucl. Phys. A **600**, 37 (1996).
5. H. Esbensen and G. F. Bertsch, Phys. Rev. C **66**, 044609 (2002).
6. T. Kikuchi et al., Eur. Phys. J. A **3**, 213 (1998).
7. Alexander Volya and Henning Esbensen, Phys. Rev. C **66**, 044604 (2002).
8. L. H. Andersen, P. Hvelplund, H. Knudsen, S. P. Möller, J. O. Pedersen, E. Uggerhöj, K. Elsener, and E. Morenzoni, Phys. Rev. Lett. **62**, 1731 (1989).
9. K. Langanke and T. D. Shoppa, Phys. Rev. C **49**, R1771 (1994); C **51**, 2844 (1995).
10. Moshe Gai and Carlos A. Bertulani, Phys. Rev. C **52**, 1706 (1995).
11. B. Davids et al., Phys. Rev. Lett. **81**, 2209 (1998).
12. J. Mortimer, I. J. Thompson, and J. A. Tostevin, Phys. Rev. C **65**, 064619 (2002).

Spectroscopy studies using two-nucleon knockout

J. A. Tostevin

Department of Physics, School of Electronics and Physical Sciences, University of Surrey, Guildford GU2 7XH, United Kingdom

Abstract. Two-proton removal reaction cross sections, from ^{208}Pb at 1 GeV/nucleon, are estimated as an example of the direct population of (high-spin) seniority-2 isomeric states, here in ^{206}Hg. Nucleon removal by both the stripping and diffractive mechanisms is considered. The cross sections in this specific (test) case are significant and can provide direct two-nucleon removal predictions of isomeric ratios.

Keywords: Direct reactions, two-proton knockout, seniority isomers, neutron-rich spectroscopy
PACS: 21.10.Jx, 24.10.-i, 25.70.-z, 27.80.+w

INTRODUCTION

One-nucleon, and selected two-nucleon knockout reactions from intermediate energy beams proceed as sudden, direct reactions [1, 2, 3]. When these are combined with coincident gamma-ray, or other final-state-selective detection of the reaction residues, the partial cross section measurements provide a demanding test of modern nuclear structure model predictions of one- and two-nucleon configurations in nuclei. To date, direct two-nucleon knockout reactions have been considered using uncorrelated, partially correlated (cluster) and fully-correlated (shell model) structure model descriptions, together with eikonal reaction theory [2, 3]. Recent applications, to light and medium-mass nuclei, include two-proton removal from neutron-rich ^{28}Mg [2, 3], ^{44}S [4] and ^{54}Ti [5] and two-neutron removal from the neutron-deficient ^{34}Ar, ^{30}S and ^{26}Si [6]. The eikonal reaction theory is able to include both the elastic (diffraction) and inelastic (stripping) nucleon-removal mechanisms.

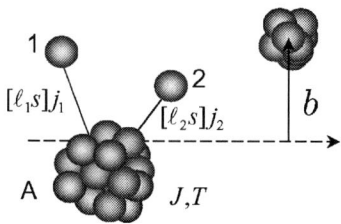

FIGURE 1. Schematic of the uncorrelated nucleons description of the two-nucleon knockout reaction.

Here, as a test case for heavier nuclei, we investigate the direct, two-proton removal cross sections from ^{208}Pb(0^+). Final state J^π selectivity is assumed and we consider the population of specific isomeric states in the ^{206}Hg residue. We assume, see Figure 1, the removal of the pair of nucleons from Hartree-Fock single-particle states ϕ_{j_1} and ϕ_{j_2}. Thus, except for the spatial correlations due to the nucleons being bound to a common core, and those arising from antisymmetry and angular momentum coupling, the two protons are assumed to be uncorrelated. As was discussed in [2, 3], the stripping (inelastic breakup) cross section is then, with $\hat{J}^2 = (2J+1)$,

$$\sigma_{j_1 j_2}(J) = \frac{1}{\hat{J}^2} \sum_M \int d\vec{b} \, |\mathscr{S}_c|^2 \, \langle \overline{[\phi_{j_1} \otimes \phi_{j_2}]_{JM}} | (1 - |\mathscr{S}_1|^2)(1 - |\mathscr{S}_2|^2) | \overline{[\phi_{j_1} \otimes \phi_{j_2}]_{JM}} \rangle. \quad (1)$$

The integral is carried out over all projectile center-of-mass (cm) impact parameters $\vec{b}$, see Figure 1. The $\mathscr{S}_i$ ($i = 1, 2, c$) are the eikonal S-matrices [1] describing the interactions of the two nucleons (1,2) and the A-body residue, or core c, with the target. Each $\mathscr{S}_i$ is a function of its own impact parameter b_i and is assumed to be spin-independent. This expression reflects the stripping mechanism where the residue interacts at most elastically with the target, survives the collision, and escapes to infinity; reflected by $|\mathscr{S}_c|^2$. The two removed nucleons interact inelastically with the target and are absorbed from the elastic channel; seen by the product of their absorption probabilities $\prod_i (1 - |\mathscr{S}_i|^2)$. Here we have made the (sudden) adiabatic and the eikonal (forward scattering) approximations, both expected to be excellent at the energies of interest. We have also made the spectator-core approximation, that dynamical excitation of the core during the collision can be neglected, as are the effects of recoil of the heavy mass A residue, by setting $b_c = b$.

Eq. (1) has been applied to ^{28}Mg and other systems [2, 3]. Additional (diffractive) cross section enters the terms in the eikonal theory with the form

$$\bar{\sigma}_{j_1 j_2}(J) = \frac{1}{\hat{J}^2} \sum_M \int d\vec{b} \, |\mathscr{S}_c(b)|^2 \, \langle \overline{[\phi_{j_1} \otimes \phi_{j_2}]_{JM}} | |\mathscr{S}_1|^2 (1 - |\mathscr{S}_2|^2) | \overline{[\phi_{j_1} \otimes \phi_{j_2}]_{JM}} \rangle, \quad (2)$$

where only one nucleon (here 2) is absorbed. These describe all processes where both nucleon 1 and the core emerge from the collision. These diffraction contributions are included here and are discussed in detail elsewhere [7]. The cross section for diffractive removal of both nucleons is negligible at the energy of interest. Reference to figure 2 suggests that indirect population of the ^{206}Hg final states of interest, by one-proton removal to states in the ^{207}Tl continuum and proton evaporation, is also expected to be negligible.

DIRECT TWO-PROTON KNOCKOUT FROM ^{208}PB

We calculate two-proton knockout cross sections from a 1.0 GeV per nucleon ^{208}Pb beam incident on a ^{9}Be target. The required core- and proton-target S-matrices were calculated from the core and target point nucleon densities using the optical limit of Glauber's multiple scattering theory [8, 9]. A zero-range nucleon-nucleon (NN) effective interaction was assumed with strength determined, in the usual way [10], by the free pp and np cross sections. The real-to-imaginary ratios of the forward scattering NN amplitudes were taken to be zero. The neutron and proton densities in ^{206}Hg were given by Hartree-Fock calculations [11], using the Skyrme SkP interaction [12], with rms radii

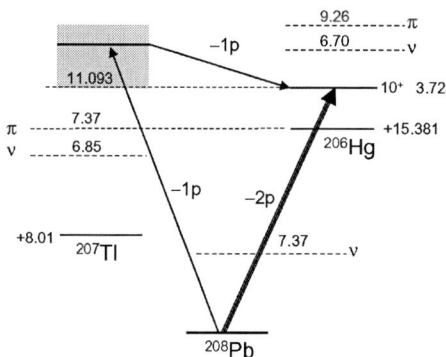

FIGURE 2. Schematic of the direct two-proton knockout reaction to the ^{206}Hg (10^+) state, and the associated neutron and proton thresholds. Non-direct population of ^{206}Hg bound states, by one-proton removal to excited ^{207}Tl followed by proton evaporation, involves intermediate states above the (lower) neutron evaporation threshold and is expected to be small.

of 5.606 fm and 5.445 fm, respectively. The density of the ^{9}Be target was assumed to be of Gaussian form with rms matter radius of 2.36 fm. However, analyses of one-nucleon knockout reveal that calculations show little sensitivity to the details of these radial forms, e.g. [13].

The ^{206}Hg final states of interest, Figure 3, are the (10^+) (3.723 MeV), (8^+) (3.623 MeV), (7^-) (2.466 MeV) and 5^- (2.102 MeV) [seniority-2] two-proton hole states. These have dominant components with $\pi[0h_{11/2}^{-2}]$, $\pi[0h_{11/2}^{-2}]$, $\pi[0h_{11/2}^{-1}1d_{3/2}^{-1}]$ and $\pi[0h_{11/2}^{-1}2s_{1/2}^{-1}]$, respectively. The ^{208}Pb ground state to ^{206}Hg ground state two proton separation energy is 15.381 MeV, Figure 2. Hence the separation energies to these $J^\pi(E^*)$ excited states are $[15.381 + E^*]$ MeV. These yield $S_{2p}(0h_{11/2}^2) \approx 19.0$ MeV, $S_{2p}(0h_{11/2}1d_{3/2}) = 17.85$ MeV and $S_{2p}(0h_{11/2}2s_{1/2}) = 17.50$ MeV. We take, $S_p(0h_{11/2}) = 9.5$ MeV, $S_p(1d_{3/2}) = 8.35$ MeV, $S_p(2s_{1/2}) = 8.0$ MeV, in agreement with the one proton separation energy and the ^{207}Tl spectrum.

The bound state proton-core single-particle wave functions were calculated in Woods-Saxon potential wells with a conventional diffuseness parameter, $a = 0.70$ fm. The potential radius parameters, r_0, were adjusted (for each single particle orbital) to reproduce the rms radius of the Hartree-Fock calculation [11]. These give $r_0 = 1.336$ fm, 1.282 fm and 1.323 fm for the $1d_{3/2}$, $0h_{11/2}$ and $2s_{1/2}$ states. The strength of the binding potentials were adjusted to reproduce the physical separation energies, as above. A 6.0 MeV spin-orbit potential, with the same geometry parameters as for the central potential, was included. Thus, both the Coulomb and centrifugal barriers experienced by the removed protons are included fully. Our results for the knockout of uncorrelated proton pairs are collected in Table 1.

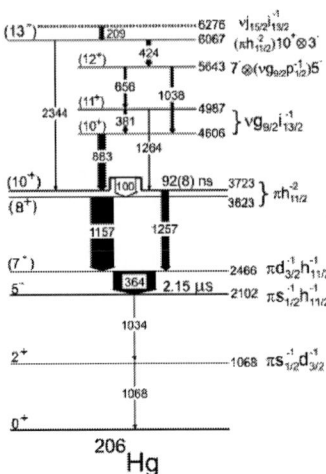

FIGURE 3. States in ^{206}Hg [14], showing the (10^+), (8^+), (7^-) and 5^- two-proton-hole states, their excitation energies, and their dominant configurations.

As the two removed nucleons are assumed to be uncorrelated, and assuming that the valence proton structure in ^{208}Pb is $[j]^N$, several results follow. The first is that the calculated (unit) cross sections for removal of a proton pair are as given in Table 1. This sets the scale for the anticipated cross sections. Based on an assumed $[j]^N$ configuration this predicts an inclusive cross section of $N(N-1)\sigma_{jj}(J)/2$. However, it also follows that this cross section yield will be shared between different core final states with associated strengths $S(J^\pi)$. For removal of a pair from a $|[j]^N, 0^+\rangle$ occupied sub-shell, N even, the inclusive cross section will be distributed between final states J^π according to the square of the coefficients of fractional parentage and, explicitly [15],

$$S(J^\pi \neq 0^+) = \frac{N(N-1)}{2}\left[\frac{2(N-2)}{(N-1)}\frac{(2J+1)}{(2j-1)(2j+1)}\right], \quad (3)$$

where $\sum_J S(J^\pi) = N(N-1)/2$. This yields $S(10^+) = 21$, $S(8^+) = 17$, being $S(J^\pi) =$

TABLE 1. Calculated two-proton removal cross sections from ^{208}Pb at 1.0 GeV/nucleon. The theoretical stripping cross sections, $\sigma_{j_1 j_2}(J)$ [Eq. (2)] and the total diffractive contributions, $\tilde{\sigma}_{j_1 j_2}(J)$, are shown. Also shown are the scaling factors $S(J^\pi)$ for each J^π final state and the resulting theoretical partial cross sections σ_J.

J^π	2p configuration	$\sigma_{j_1 j_2}(J)$ (mb)	$\tilde{\sigma}_{j_1 j_2}(J)$ (mb)	$S(J^\pi)$	σ_J (mb)
(10^+)	$[0h_{11/2}]^2$	0.0082	0.0016	21	0.21
(8^+)	$[0h_{11/2}]^2$	0.0073	0.0015	17	0.15
(7^-)	$[0h_{11/2}, 1d_{3/2}]$	0.0153	0.0038	15	0.29
5^-	$[0h_{11/2}, 2s_{1/2}]$	0.0133	0.0034	11	0.18

$2J+1$, due to the filled $[0h_{11/2}]^{12}$ sub-shell, and with $\sum_J S(J^\pi) = 66$. Thus, in excess of half the inclusive cross section from proton pair removal from the $[0h_{11/2}]^{12}$ sub-shell is expected to populate the (seniority-2) (10^+) and (8^+) final states.

The corresponding inclusive cross section for knockout of a pair from different sub-shells, e.g. a $[j_1]^{N_1}[j_2]^{N_2}$ configuration, is $N_1 N_2 \sigma_{j_1 j_2}(J)$. The analogous distribution of this strength among J^π final states is also discussed in [15]. For the case of protons removed from a pair of *filled* sub-shells, as arises for both the (7^-) ($[0h_{11/2}]^{12}[1d_{3/2}]^4$) and 5^- ($[0h_{11/2}]^{12}[2s_{1/2}]^2$) states, these are simply $S(J^\pi) = (2J+1)$. These weights were used to compute the (10^+) (3.723 MeV), (8^+) (3.623 MeV), (7^-) (2.466 MeV) and 5^- (2.102 MeV) direct two-proton removal cross sections, shown in Table 1.

SUMMARY AND CONCLUSIONS

We have calculated two-proton removal cross sections, from ^{208}Pb at 1.0 GeV/nucleon on a ^{9}Be target, populating the assumed (10^+) (3.723 MeV), (8^+) (3.623 MeV), (7^-) (2.466 MeV) and 5^- (2.102 MeV) seniority-2, two-proton hole states in ^{206}Hg. This important (test) case predicts significant cross sections from the direct two-proton knockout mechanism and a ratio of the isotopic yields for the $(10^+) : 5^-$ states of $0.21 : 0.83$. This ratio is in line with very preliminary experimental indications [16]. Other systems of interest, and with final states expected to be populated cleanly via the direct two-proton knockout mechanism, include $\pi[0g_{7/2}]^2$ and $\pi[0h_{11/2}]^2$ proton-pair removal from ^{136}Xe and a ^{206}Hg secondary beam, respectively.

The financial support of the United Kingdom Engineering and Physical Sciences Research Council (EPSRC) through Grant No. EP/D003628, and extended discussions with Drs P.H. Regan, Zs. Podolyák, and P.D. Stevenson are gratefully acknowledged.

REFERENCES

1. P.G. Hansen and J.A. Tostevin, *Annu. Rev. Nucl. Part. Sci.* **53**, 219 (2003).
2. D. Bazin *et al.*, *Phys. Rev. Lett.* **91**, 012501 (2003).
3. J.A. Tostevin, G. Podolyák, B.A. Brown, and P.G. Hansen, *Phys. Rev. C* **70**, 064602 (2004).
4. J. Fridmann *et al.*, *Nature* **435**, 922 (2005).
5. A. Gade, R.V.F. Janssens *et al.*, in preparation.
6. K. Yoneda *et al.*, in preparation.
7. J.A. Tostevin, *Diffractive contributions to direct two-nucleon knockout reactions*, in preparation.
8. R.J. Glauber, in *Lectures in Theoretical Physics*, ed. by W.E. Brittin (Interscience, New York, 1959), Vol.1, p315.
9. J.A. Tostevin, *Nucl. Phys. A* **682**, 320c (2001).
10. L. Ray, *Phys. Rev. C* **20**, 1857 (1979).
11. P.D. Stevenson, private communication, 2005.
12. J. Dobaczewski, H. Flocard and J. Treiner, *Nucl. Phys. A* **422**, 103 (1984).
13. A. Gade *et al.*, *Phys. Rev. Lett.* **93**, 042501 (2004).
14. B. Fornal *et al.*, *Phys. Rev. Lett.* **87**, 212501 (2001).
15. Norman K. Glendenning, *Phys. Rev.* **137**, B102 (1965).
16. Zs. Podolyák *et al.*, private communication, 2005.

Influence of the projectile description on breakup calculations

Pierre Capel* and Filomena Nunes[†]

*TRIUMF, 4004 Wesbrook Mall, Vancouver, B.C., Canada V6T 2A3
[†]National Superconducting Cyclotron Laboratory and Department of Physics and Astronomy, Michigan State University, East Lansing, Michigan 48824

Abstract. Coulomb dissociation of ^{11}Be and ^{8}B are studied theoretically with the aim of analyzing the influence of the projectile description on the calculations. Both projectiles are modeled as two-body systems, and various potentials with different geometry are considered for both of them. As expected, their breakup reactions are found to be peripheral. The calculations depend on the initial bound state (through its asymptotic normalization coefficient) and also on the distorted waves that describe the continuum (mainly through the phase shifts). The relative importance of both aspects varies from one projectile to the other and depends on the physical constraints imposed.

Keywords: halo nuclei, potential description, dissociation, ^{11}Be, ^{8}B
PACS: 24.10.-i, 25.60.Gc, 25.60.-t, 27.20.+n

1. INTRODUCTION

Coulomb breakup is probably one of the most used tools to study the structure of halo nuclei [1]. In that reaction, the halo nucleons dissociate through the interaction with a heavy target. Spectroscopic information can be inferred by comparing experimental data with the results of accurate theoretical models. Several models have been developed, or extended to this end [2]: adiabatic approximation [3], continuum discretized coupled channel method (CDCC) [4, 5], time-dependent technique [6–8],...

These reaction models usually rely on a simple two-body description of the projectile: a pointlike halo loosely bound to a structureless core. The interaction between the two bodies is simulated by a central potential, usually of Woods-Saxon form factor, plus a spin-orbit coupling term. The depth of the potential is adjusted to reproduce the energy of the bound states and some resonances. However, the geometry of the potential is by no means unique, and variations of its radius and/or its diffuseness may affect significantly the calculated cross sections.

Due to the very low binding energy of the projectile, the reaction is expected to be mostly peripheral. Therefore, one usually assumes the breakup cross section to be proportional to the square of the asymptotic normalization coefficient (ANC). However, several questions remain open: is the reaction sufficiently peripheral for a dependence on the ANC? To what extend is it true when couplings in the continuum are not negligible? How does the final-state interaction affect the results? In this talk, we present the results of calculations of the Coulomb breakup of ^{11}Be and ^{8}B. Using different potentials to describe each nucleus, we focus on the sensitivity of the calculations on the projectile description.

In Sec. 2, we present a brief description of the reaction models to be used. The various potentials chosen to describe the projectiles are listed in Sec. 3. The results of the calculation of the ^{11}Be breakup on lead are summarized in Sec. 4, and those of the dissociation of ^{8}B on ^{58}Ni are presented in Sec. 5. The conclusions are drawn in the final section.

2. THEORETICAL FRAMEWORK

We are interested in the breakup of a projectile P on a target T. The projectile is seen as a two-body system: a pointlike and structureless fragment f of mass m_f and charge $Z_f e$ loosely bound to a structureless core c of mass m_c and charge $Z_c e$. The target is simulated by a structureless particle of mass m_T and charge $Z_T e$.

The internal structure of the projectile is simulated by the Hamiltonian H_0

$$H_0 = -\frac{\hbar^2}{2\mu}\Delta + V_{cf}(\vec{r}), \qquad (1)$$

TABLE 1. Parameters of the ^{10}Be-n potentials, R_0 is parametrized as $r_0 A_c^{1/3}$.

Potential	V_{leven} (MeV)	V_{lodd} (MeV)	V_{LS} (MeV fm^2)	a (fm)	r_0 (fm)
V1	62.52	39.74	21.0	0.6	1.2
V2	66.325	38.37	12.44	0.5	1.2
V3	58.905	40.025	27.68	0.7	1.2
V4	71.28	49.015	29.95	0.6	1.1
V5	55.25	32.515	12.86	0.6	1.3
V6	59.05	59.05	0	0.62	1.236

where $\mu = m_c m_f/(m_c + m_f)$ is the c-f reduced mass, and $\vec{r}$ is the relative coordinate of the fragment to the core.

The potential which simulates the core-fragment interaction is composed of a point-sphere Coulomb potential, a nuclear central part, and a spin-orbit coupling term:

$$V_{cf}(\vec{r}) = V_C(r) - V_l f(r, R_0, a) + \vec{L} \cdot \vec{I} V_{LS} \frac{1}{r} \frac{d}{dr} f(r, R_0, a), \qquad (2)$$

where $f(r, R_0, a) = \left[1 + \exp\left(\frac{r - R_0}{a}\right)\right]^{-1}$ is the usual Woods-Saxon form factor. The orbital momentum of the c-f relative motion is denoted by $\vec{L}$, and $\vec{I}$ is the fragment spin. The core spin is set to zero. The depths of V_{cf} are adjusted to reproduce the bound states of the system and some of its resonances. The interactions between the projectile constituents and the target are simulated by optical potentials chosen in the literature (see Refs. [5, 8] for details).

In this framework, the breakup reaction of the two-body projectile onto the target is approximated by a three-body scattering problem. We solve this problem using either the time-dependent technique described in Ref. [8] or the CDCC model of Ref. [5], depending on applicability.

The time-dependent description of the reaction relies on the semiclassical approximation [9] in which the projectile-target relative motion is approximated by a classical trajectory. Along that trajectory, the projectile experiences a time-dependent potential that simulates the Coulomb and nuclear fields of the target. The wave function describing the internal structure of the projectile is therefore the solution of a time-dependent Schrödinger equation. Starting with the projectile in its ground state, the equation is solved numerically using the evolution algorithm described in Ref. [8].

The CDCC method consists in a fully quantal technique in which the three-body wave function is expanded over the basis of the eigenstates of H_0 (1). The continuum is included in the calculation through a discretization into energy bins. Each bin wave function is obtained by summing all the positive-energy eigenstates of H_0 included in the bin. The three-body Schrödinger equation can then be reduced to a one dimensional coupled channel equation. This equation is solved in the same manner as in Ref. [5].

3. CORE-FRAGMENT POTENTIALS

As done in previous works [6–8, 10], we describe ^{11}Be as a neutron loosely bound to a ^{10}Be core in its 0^+ ground state. The depths of the potential are adjusted to reproduce the energies of the low-lying states of ^{11}Be. The shell inversion observed between the bound states is reproduced using a parity-dependent depth V_l. The $\frac{1}{2}^+$ ground state is modeled by a $1s1/2$ state, the $\frac{1}{2}^-$ excited state by a $0p1/2$ state, and the first $\frac{5}{2}^+$ resonance is reproduced in the $d5/2$ wave.

In order to study the sensitivity of our calculations to the potential choice, we develop five sets of parameters that reproduce the physical states mentioned above (see Table 1). The first potential (V1) has been devised in Ref. [10]. The next four (V2 to V5) have been obtained by varying either the diffuseness or the radius of V1. We also use a sixth potential (V6) developed by Fukuda et al. to analyze their experiment [11]. It reproduces only the ground state energy and does not contain a spin-orbit coupling term.

The ^{8}B nucleus is described by the usual two-body system: a proton loosely bound to a ^{7}Be core [5, 12]. The internal structure of the core is neglected, and its spin is set to zero in the calculations. The 2^+ ground state of ^{8}B is assumed to be a pure $0p3/2$ proton single-particle state.

In order to study the sensitivity of the calculations to the description of the projectile, five ^{7}Be-p potentials are considered (see Table 2). They are obtained by varying the diffuseness of a simplified version of the potential developed

TABLE 2. Parameters of the ^{7}Be-p potentials, R_0 is parameterized as $r_0 A_c^{1/3}$. The Coulomb radius is $R_C = 1.3 A_c^{1/3}$ fm.

Potential	V_l (MeV)	V_{LS} (MeV fm^2)	a (fm)	r_0 (fm)
T1	45.23	19.59	0.40	1.25
T2	44.98	19.59	0.52	1.25
T3	44.47	19.59	0.60	1.25
T4	43.50	19.59	0.70	1.25
T5	42.28	19.59	0.80	1.25

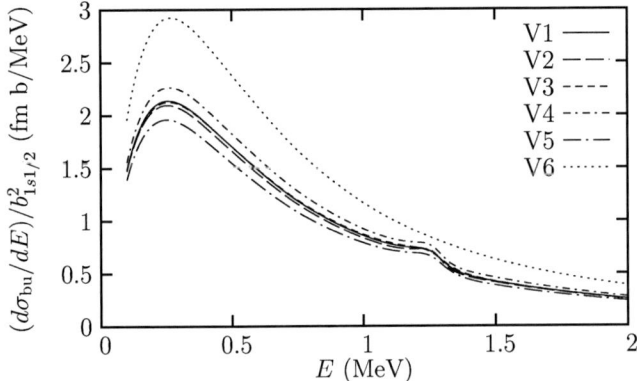

FIGURE 1. Sensitivity to the potential geometry: breakup cross section of ^{11}Be on ^{208}Pb at 69 MeV/nucleon as a function of energy. The cross sections are divided by the square of the single particle ANC $b_{1s1/2}$. Calculations are performed with the different ^{10}Be-n potentials listed in Table 1.

by Esbensen and Bertsch [12] (potential T2). While the spin-orbit term is kept unchanged for all potentials, the depth of the central part V_l is adjusted to reproduce the 137 keV binding energy of the bound state.

4. BREAKUP OF ^{11}BE ON ^{208}PB

The breakup of ^{11}Be on ^{208}Pb at 69 MeV/nucleon has recently been measured at RIKEN [11]. From the analysis of their experiment, Fukuda et al. extracted a spectroscopic factor of 0.7 for the $|^{10}\text{Be}(0^+) \otimes s1/2\rangle$ configuration in the ground state. To evaluate the sensitivity of this figure to the ^{10}Be-n potential geometry, we perform the calculations using the six potentials of Table 1. We use the time-dependent technique of Ref. [8]. The breakup cross section is displayed in Fig. 1 as a function of the relative energy between the ^{10}Be and the neutron after breakup. To remove the major dependence on the ANC of the bound state $b_{1s1/2}$, the cross sections are divided by the square of the ANC.

The total cross sections obtained with potentials V1–V6 all exhibit the same shape—if one excepts the small bump at 1.3 MeV due to the $\frac{5}{2}^+$ resonance, which is not reproduced by V6. However, they strongly differ in amplitude. For example, V6 leads to a cross section larger by 40 % than V1–V5. Moreover, even though they have been adjusted on the same levels, potentials V1–V5 leads to variation in the cross section as large as 20 %. It appears that the variations differ strongly from one partial wave to the other. While the $p3/2$ contributions bear most of the difference between V1–V5, the $p1/2$ contributions for those five potentials are very similar to one another. On the other hand the major difference between V6 and the other five lies in the $p1/2$ contribution. This indicates that breakup calculations depend

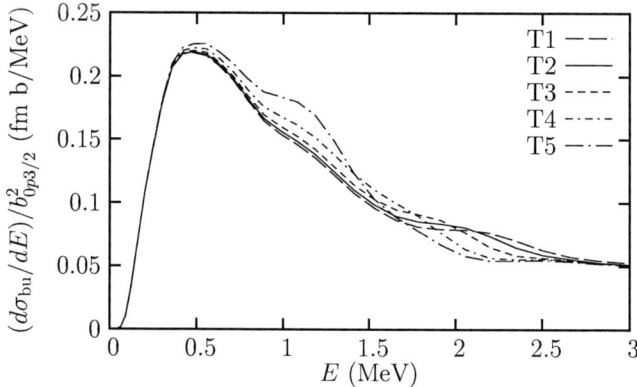

FIGURE 2. Sensitivity to the potential geometry: breakup cross section of ^{8}B on ^{58}Ni at 25.75 MeV as a function of energy. The cross sections are divided by the square of the single particle ANC $b_{0p3/2}$. Calculations are performed with the different ^{7}Be-p potentials listed in Table 2.

not only on the ground-state wave function, but also on the description of other partial waves.

Besides these calculations, we perform a test using plane waves for the positive-energy states of the projectile, instead of the distorted waves obtained with the potentials of Table 1. The ground state is still described using potentials V1–V6. In that case, the breakup cross section is found to be exactly proportional to $b^2_{1s1/2}$. This result suggests that, as expected, the breakup reaction is very peripheral. It also indicates that the description of the continuum has a significant influence on the calculation.

The results of Fig. 1 are reproduced qualitatively using a simple first-order approximation [9] using a pure E1 Coulomb potential to model the projectile-target interaction. This shows that the major difference in the breakup cross section can be explained by the direct transition from the initial $1s1/2$ bound states to the continuum states in the p waves. The couplings in the continuum play thus a minor role. This also indicates that the sensitivity of the breakup calculation to the potential choice can be qualitatively understood by comparing the distorted waves obtained with the potentials of Table 1. This analysis shows us that potentials V1–V6 lead to significant differences in the $p3/2$ phase shifts, which explain the large variations observed in that contribution. In the $p1/2$ partial wave, however, potentials V1–V5 exhibit very similar phase shifts, because they have all been adjusted to reproduce the $\frac{1}{2}^-$ excited state of ^{11}Be in that partial wave. This explains the similarity in their $p1/2$ contribution to the cross section. Potential V6, which does not reproduce the ^{11}Be excited state, gives a very different $p1/2$ phase shift. This difference is responsible for the larger $p1/2$ contribution of potential V6 compared to the others.

5. BREAKUP OF ^{8}B ON ^{58}NI

In order to extend our analysis to proton-halo nuclei, we perform a similar analysis for the breakup of ^{8}B on ^{58}Ni at 26 MeV. This reaction was measured at Notre Dame [13], and analyzed within the CDCC framework [5]. We conduct our calculations in the same manner as in that reference, except that we consider the five ^{7}Be-p potentials of Table 2. The corresponding breakup cross sections are plotted in Fig. 2 as a function of the energy. The contributions of the $p3/2$ and $p1/2$ partial waves are also depicted. As in the ^{11}Be case, the cross sections are divided by the square of the ANC of the initial bound state $b_{0p3/2}$.

Contrarily to the previous case, the breakup cross section is here approximately proportional to $b^2_{0p3/2}$. This is explained by the relative proportionality of the dominant $p3/2$ contributions—as well as the s and d ones—with the square of the ANC. Above 1 MeV, all cross sections exhibit a bump, whose loctaion varies with the potential choice. This bump is due to the presence of an unfitted resonance in the $p1/2$ partial wave, which distorts that contribution.

Therefore, the potential dependence of the calculation varies from one partial wave to the other as in the ^{11}Be case.

As for ^{11}Be, when the calculations are performed with pure Coulomb waves to describe the continuum, the breakup cross sections are found to be exactly proportional to $b_{0p3/2}^2$. This, again, indicates that the reaction is peripheral and depends on the description of the continuum. An analysis of the influence of the ^{7}Be-p potentials on the distorted waves enables us to understand qualitatively this dependence. First, we observe that all the potentials lead to very similar $p3/2$ phase shifts. This is due to the fact that T1–T5 are all adjusted to reproduce the binding energy of the ^{8}B ground state in that partial wave. This similarity explains that the dominant $p3/2$ is approximatively proportional to the square of the ANC. Since the energy of the $p1/2$ resonance changes with the potential choice, the continuum wave functions in that partial wave strongly differ from one potential to the other. This explains the strong dependence of that contribution to the potential choice.

6. CONCLUSION

In this talk, we have presented an analysis of the sensitivity of the breakup calculations to the projectile description for both ^{11}Be and ^{8}B. This analysis indicates that breakup calculations are sensitive to the potential choice. This sensitivity is larger in the ^{11}Be case than in the ^{8}B case. We have shown that it is not only due to variations in the ground state wave function, but also to the description of the continuum. Therefore, the breakup reaction probes not only the initial bound state of the projectile, but also its continuum.

Since the reaction is very peripheral, the sensitivity to the potential is due to the subsequent variation in the asymptotic part of the wave functions, i.e. the ANC and the phase shifts. We have observed that when potentials are adjusted to reproduce loosely-bound states or low-lying resonances, they lead to similar phase-shifts. However, when the potentials are not constrained, variations in their geometry may lead to significant differences in the breakup cross sections.

When extracting spectroscopic information form breakup experiments, one should be particularly cautious to the influence the description of the continuum has on the calculations. The potential should be constrained, as much as possible, by experimental data. If the data are scarce, the sensitivity of the extracted values to the potential geometry should be evaluated.

ACKNOWLEDGMENTS

P.C. acknowledges the support from the Natural Sciences and Engineering Research Council of Canada (NSERC). F.M. acknowledges the support from the NSCL at Michigan State University and from the National Science Foundation grant PHY-0456656.

REFERENCES

1. I. Tanihata, *J. Phys. G*, **22**, 157 (1996).
2. J. Al-Khalili, and F. M. Nunes, *J. Phys. G*, **29**, R89 (2003).
3. J. A. Tostevin, S. Rugmai, and R. C. Johnson, *Phys. Rev. C*, **57**, 3225 (1998).
4. M. Kamimura, M. Yahiro, Y. Iseri, H. Kameyama, Y. Sakuragi, and M. Kawai, *Prog. Theor. Phys. Suppl.*, **89**, 1 (1986).
5. J. A. Tostevin, F. M. Nunes, and I. J. Thompson, *Phys. Rev. C*, **63**, 024617 (2001).
6. T. Kido, K. Yabana, and Y. Suzuki, *Phys. Rev. C*, **50**, R1276 (1994).
7. H. Esbensen, G. F. Bertsch, and C. A. Bertulani, *Nucl. Phys. A*, **581**, 107 (1995).
8. P. Capel, D. Baye, and V. S. Melezhik, *Phys. Rev. C*, **68**, 014612 (2003).
9. K. Alder, and A. Winther, *Electromagnetic Excitation*, North-Holland, Amsterdam, 1975.
10. P. Capel, G. Goldstein, and D. Baye, *Phys. Rev. C*, **70**, 064605 (2004).
11. N. Fukuda, T. Nakamura, N. Aoi, N. Imai, M. Ishihara, T. Kobayashi, H. Iwasaki, T. Kubo, A. Mengoni, M. Notani, H. Otsu, H. Sakurai, S. Shimoura, T. Teranishi, Y. X. Watanabe, and K. Yoneda, *Phys. Rev. C*, **70**, 054606 (2004).
12. H. Esbensen, and G. F. Bertsch, *Nucl. Phys. A*, **600**, 37 (1996).
13. V. Guimarães, J. J. Kolata, D. Peterson, P. Santi, R. H. White-Stevens, S. M. Vincent, F. D. Becchetti, M. Y. Lee, T. W. O'Donnell, D. A. Roberts, and J. A. Zimmerman, *Phys. Rev. Lett.*, **84**, 1862 (2000).

NUCLEAR STRUCTURE

Distribution of the GT strength starting from the ground state of ^{14}N

A. Negret*, T. Adachi†, C. Bäumer**, A.M. van den Berg‡, G.P.A. Berg‡, P. von Brentano§, D. Frekers**, D. De Frenne*, K. Fujita¶, Y. Fujita†, E. W. Grewe**, P. Haefner**, K. Hatanaka¶, M. Hunyadi‡, M.A. de Huu‡, H. Johansson‖, E. Jacobs*, Y. Kalmykov††, K. Kawabata‡‡, A. Korff**, K. Nakanishi¶, P. von Neumann-Cosel††, T. Ogama†, L. Popescu*, S. Rakers**, A. Richter††, N. Ryezayeva††, Y. Sakemi¶, A. Shevchenko††, Y. Shimbara†, Y. Shimizu¶, A. Tamii‡‡, M. Uchida‡‡, H.J. Wörtche‡ and M. Yosoi‡‡

*Vakgroep Subatomaire en Stralingsfysica, Universiteit Gent, B-9000 Gent, Belgium
†Department of Physics, Osaka University, Toyonaka, Osaka 560-0043, Japan
**Institut für Kernphysik, Westfälische Wilhelms-Universität Münster, D-48149 Münster, Germany
‡Kernfysisch Versneller Instituut, Rijksuniversiteit Groningen, NL-9747 AA Groningen, The Netherlands
§Institut für Kernphysik,Universität zu Köln, 50937 Köln, Germany
¶Research Center for Nuclear Physics, Osaka University, Ibaraki, Osaka 567-0047, Japan
‖Gesellschaft für Schwerionenforschung mbH, Darmstadt, Germany
††Institut für Kernphysik, Technische Universität Darmstadt, D-64289 Darmstadt, Germany
‡‡Department of Physics, Kyoto University, Sakyo, Kyoto 606-8224, Japan

Abstract.
Two charge-exchange reactions were performed at intermediate energies in order to investigate the Gamow-Teller strength distribution starting from ^{14}N$_{g.s.}$ in β^+ and β^- directions: ^{14}N$(d,^2$He$)^{14}$C and ^{14}N$(^3$He$,t)^{14}$O. We describe here a few delicate details of the experimental techniques used in these investigations.

INTRODUCTION

The charge-exchange reactions performed at intermediate energies offer the possibility to study the collective excitations in nuclei [1]. The investigation of the most simple excitation of the spin and isospin, called Gamow-Teller (GT) (by analogy to the similar β-decay), represents an opportunity to study a few basic features of the nuclear force due to the very simple excitation mechanism. Unfortunately, the limited resolution of (p,n) or (n,p) - type charge-exchange reactions made difficult in the past the distinction of the GT strength distribution over individual final levels. With the improved resolution obtained in the (n,p)-type $(d,^2$He$)$ reaction and in the (p,n)-type $(^3$He$,t)$ reaction this goal is now achievable [2, 3].

We report here on a particular case where the very good resolution obtained in these two reactions makes a step forward possible in the understanding of a long-standing problem related to the β-decay of ^{14}C and ^{14}O to the ground state of ^{14}N. In both cases, the initial J^π is 0^+ and the initial isospin T is 1 while for ^{14}N, $J_f^\pi=1^+$ and $T_f=0$. Although fast GT transitions would be expected, in both cases they are suppressed by a few orders of magnitude: $\log(ft)_{^{14}C\to^{14}N}=9.0$ and $\log(ft)_{^{14}O\to^{14}N}=7.3$ [4].

Two experiments were performed in order to investigate the GT strength distribution B(GT) starting from the ground state of ^{14}N among final states in ^{14}C and ^{14}O. For the $^{14}N(d,^2He)^{14}C$ experiment performed at KVI, Groningen (Nl) with $E_d=171$ MeV, the resolution was ≈ 170 keV and for the $^{14}N(^3He,t)^{14}O$ experiment performed at RCNP, Osaka (Jp) at $E_{3He}=420$ MeV, the obtained resolution was ≈ 35 keV.

EXPERIMENTAL TECHNIQUES

The two experiments presented here are rather different. Although both use spectrometer detecting techniques, the focus in the two cases is on different issues.

We should note that in both cases the spin and isospin selection rules allow only the excitation via GT transitions of $T=1$ final states with $J^\pi=0^+$, 1^+ or 2^+. It was therefore expected that the spectra obtained for the two final mirror nuclei ^{14}C and ^{14}O are very similar if the isospin symmetry holds for this system.

The $(d,^2He)$ reaction

The $^{14}N(d,^2He)^{14}C$ reaction was investigated at KVI, making use of the AGOR Cyclotron, the Big Bite Spectrometer (BBS) [5] and the EuroSuperNova (ESN) [6] detector. In principle, the $(d,^2He)$ reaction is difficult because ^2He is an unbound system. It represents two protons in a $^1S_0$ state. For the success of the experiment the two correlated protons have to be detected simultaneously in the focal plane of the spectrometer over a huge proton background generated by deuterium breakup processes. The readout system of the ESN detector makes this possible [7]. The limited momentum acceptance of BBS acts as a filter for the $^1S_0$ wave of the two protons.

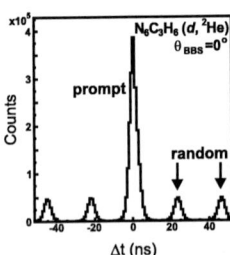
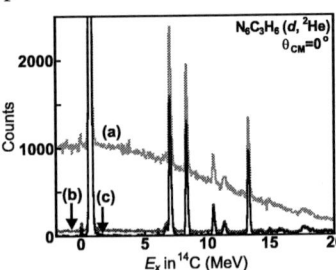

FIGURE 1. **Left panel**: the time difference of the two protons detected in the focal plane. **Right panel**: the background subtraction technique based on the spectrum shown in the left panel: (a) represents the initial energy spectrum, (b) is the spectrum gated on the prompt protons and (c) represents the final spectrum obtained after the subtraction of the random events from spectrum (b).

The background subtraction is mainly based on the relative timing of the two protons in in the focal plane (Fig. 1, left panel). Gating on the prompt peak and then subtracting the random contribution the energy spectrum can be cleared as shown in the right panel of Fig. 1.

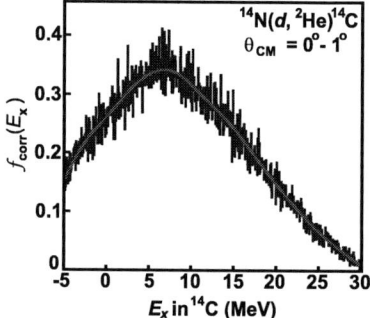

FIGURE 2. Correction function for the ^{14}N$(d,^2$He$)^{14}$C reaction. The dark spectrum is the result of the Monte Carlo simulation and the light line is a smoothed version of it, used as correction function.

Another delicate step in the $(d,^2$He$)$ data analysis is the correction for the efficiency of detecting two particles in the same focal plane. This correction was performed by a Monte Carlo simulation of the experiment [8] where the gates imposed by the spectrometer acceptance and detector were taken into account. Fig. 2 shows the correction function applied to the energy spectrum in order to obtain the cross sections. However, this method implies an uncertainty of the absolute values of the cross sections of of the order of $\approx 20\%$, estimated from repeated runs of the simulation with various experimental gates. We should note that this uncertainty does not propagate in the B(GT) values since for that purpose an additional calibration is needed, as will be shown.

The ^{14}N$(^3$He$,t)^{14}$O reaction

The ^{14}N$(^3$He$,t)^{14}$O reaction was investigated at RCNP. There, the entire experimental setup and effort of the experimental team is dedicated to the improvement of the resolution of the final spectra. We used the Ring Cyclotron, the WS-course beam line [9] and the Grand Raiden Spectrometer [10] together with a focal plane detector.

The technique used to improve the resolution is based on the matching of dispersion of the beam with the dispersion of the spectrometer. The WS-course beam line is dedicated to this purpose. For the improvement of the energy resolution the beam spot is dispersed in the horizontal plane (the bending plane of the spectrometer; this represents the *lateral dispersion matching*. For the reconstruction of the horizontal scattering angle also the *angular dispersion matching* was achieved [11]. The *over-focus mode* was needed for the good reconstruction of the vertical scattering angle [12].

A special attention was paid during the analysis to the energy resolution of the spectra. The dependences of the vertical position in the focal plane of the particles on their horizontal and vertical scattering angle were corrected with polynomial functions.

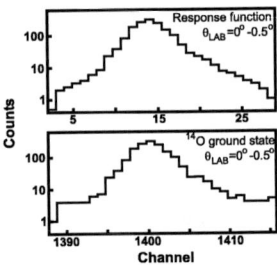

FIGURE 3. The response function of the spectrometer in the ^{14}N(^{3}He,t)^{14}O reaction at scattering angles $\leq 0.5°$ (upper panel) deduced from the shape of the ground state peak (lower panel).

The final spectra were fitted, with functions representing the convolution of the natural response of the detection system (deduced from the shape of the ground state peak as shown in Fig. 3) with Breit-Wigner functions of variable widths. This procedure allowed the determination of the natural width of most of the excited levels [13].

RESULTS AND CONCLUSIONS

The analysis procedure allows the construction of the angular distributions and the deduction of the B(GT) distributions in ^{14}C and ^{14}O.

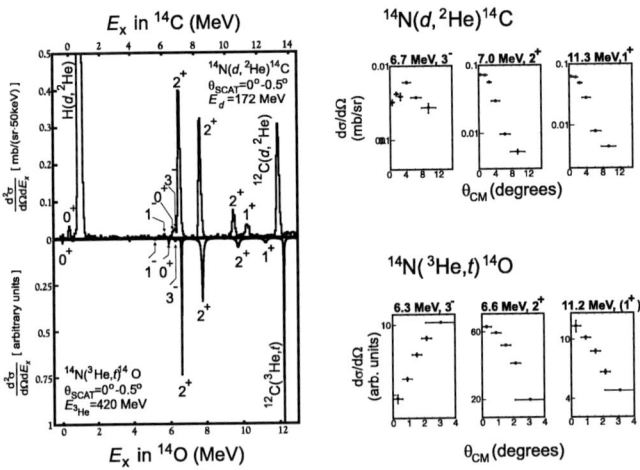

FIGURE 4. Spectra at very forward scattering angles (left panel) and a few angular distributions (right panel) for the two described reactions.

The two final spectra for very forward scattering angles are compared in the left panel of Fig. 4. They were used for the determination of the GT strength to each individual final level. In order to deduce the B(GT)s a calibration value is needed in both cases. Because the transitions to the final ground states are very suppressed and have unusual angular distributions, they cannot be used for calibration. Therefore, the proportionality factor between the cross sections and B(GT)s was deduced, in both cases, from the transitions starting from ^{12}C which was also present in the target (to ^{12}B$_{g.s.}$ and ^{12}N$_{g.s.}$, respectively), for which the log(ft)s are well-known [14]. However, the extrapolation of these proportionality factors from ^{12}C to ^{14}N involves an uncertainty of the absolute B(GT) values of the order of 30%, based on Ref. [15] but also on our previous experience [16]. However, the relative values (i.e. the B(GT) distributions) are not affected by this uncertainty.

The data were divided in several angular binnings in order to build the angular distributions necessary for identification the GT (ΔL=0) transitions. Due to calibration uncertainties, the cross sections deduced from the (^{3}He,t) experiment are expressed in arbitrary units. However this is not an obstacle since the shape of the angular distribution is sufficient to determine the angular momentum transferred and because for the B(GT) values a calibration is required. The right panel of Fig. 4 shows the angular distributions for a few corresponding states in the final nuclei. Apart from the ground states, all positive-parity final states have a clear ΔL=0 behaviour and we considered them as GT states.

The angular distributions and the comparison of the two final spectra allows also the assignment of several J^π values [13].

The consequences of our results regarding the B(GT) distributions and the implications on the structure of the ^{14}N, ^{14}C and ^{14}O nuclei deserve special attention. They are still under analysis. A forthcoming paper is forseen to address these aspects based on the results of the analysis presented here and comparing these results with various theoretical approaches.

REFERENCES

1. Rapaport, J., and Sugarbaker, E., *Annu. Rev. Nucl. Part. Sci.*, **44**, 109–151 (1994).
2. Hagemann, M. et al.., *Phys. Rev. C*, **71**, 014606 (2005).
3. Fujita, Y. et al., *Phys. Rev. Lett.*, **92**, 062502 (2004).
4. Ajzenberg-Selove, F., Kelley, J., and Nesarja, C., *Nucl. Phys. A*, **523**, 1–196 (1991).
5. van den Berg, A., *Nucl. Instrum. Methods Phys. Res. A*, **99**, 637–640 (1995).
6. Hagemann, M. et al., *Nucl. Instrum. Methods Phys. Res. A*, **437**, 459–470 (1999).
7. Rakers, S. et al.., *Nucl. Instrum. Methods Phys. Res. A*, **481**, 253–261 (2002).
8. Hagemann, M., Ph.D. thesis, Ghent University (2001), unpublished.
9. Wakasa, T. et al., *Nucl. Instrum. Methods Phys. Res. A*, **482**, 79–93 (2002).
10. Fujiwara, M. et al., *Nucl. Instrum. Methods Phys. Res. A*, **422**, 484–488 (1999).
11. Fujita, H. et al., *Nucl. Instrum. Methods Phys. Res. A*, **482**, 17–26 (2002).
12. Fujita, H. et al., *Nucl. Instrum. Methods Phys. Res. A*, **469**, 55–62 (2001).
13. Negret, A. et al., *Phys. Rev. C*, **71**, 047303 (2005).
14. Ajzenberg-Selove, F., and Kelley, J., *Nucl. Phys. A*, **506**, 1–158 (1990).
15. Taddeucci, T., Goulding, C., Carey, T., Byrd, R., Goodman, C., Gaarde, C., Larsen, J., Horen, D., Rapaport, J., and Sugarbaker, E., *Nucl. Phys. A*, **469**, 125–172 (1987).
16. Rakers, S. et al., *Phys. Rev. C*, **65**, 044323 (2002).

Microscopic Calculations for Waiting-Point Nuclei*

K. P. Drumev, C. Bahri and J. P. Draayer

Department of Physics and Astronomy, Louisiana State University, Baton Rouge, LA, 70803

Abstract. Shell-model calculations for upper fp-shell nuclei using realistic interactions are reported. Valence nucleons beyond the N=28=Z core are considered to fill levels of the normal parity upper fp-shell and the unique parity configurations that consists either of the $g_{9/2}$ level or the whole gds-shell. These two cases are handled within a standard M-scheme approach and an SU(3) picture, respectively. Results for low-lying energy spectra, single-particle occupancies and symmetry properties of the eigenstates are reported. Various truncations are considered that key on the number of nucleon pairs allowed to occupy the unique-parity space. The calculations demonstrate the importance of the unique-parity space to the structure of upper fp-shell nuclei.

Keywords: SU(3) shell model, upper fp-shell, collective levels
PACS: 21.60.Fw, 27.50.+e, 21.10.Re

INTRODUCTION

The nuclear shell model has been applied successfully in a description of various aspects of nuclear structure, in large part because it is based on a minimum number of assumptions. Although direct diagonalization of the Hamiltonian matrix in the full Hilbert space would be desirable, the dimensionality of such a space is often too large to allow calculations of this type to be done. In the upper fp-shell, for example, we are unable to do this for some waiting-point nuclei like ^{68}Se and ^{72}Kr, which are important for the rp-process in nucleosynthesis [1]. Recently, in order to relax this restriction dramatically, various stochastic approaches, for instance, the Shell-Model Monte Carlo method [2], have been investigated. Alternatively, algebraic models using the symmetry properties of the systems under investigation have been developed [3].

The role of the intruder levels that penetrate down into lower-lying shells in atomic nuclei has been the focus of many studies and debates. These levels are found in heavy nuclei where the strong spin-orbit interaction destroys the underlying harmonic oscillator symmetry of the nuclear mean-field potential. They are important and have to be included in the model space if experimentally observed states of higher spin or opposite parity are to be described. In this contribution we report on calculations that consider the occupancy of these levels, their contribution to the nuclear deformation, and the role they play in the overall dynamics of the system.

---

* Support provided by the U.S. National Science Foundation under grant No: PHY-0140300 and PHY-0500291

RESULTS FROM A REALISTIC INTERACTION

First, we carried out M-scheme shell-model calculations for the ^{64}Ge and ^{68}Se nuclei in the pf$_{5/2}$g$_{9/2}$ model space, assuming that the occupancy of the f$_{7/2}$ orbital was frozen [4]. This was done with different cuts of the full model space. The Hamiltonian we used was a G-matrix with a phenomenologically adjusted monopole part [5]. Also, a renormalized version of this interaction in the pf$_{5/2}$ space was introduced for describing beta decay [6]. Energy spectra for these two nuclei with both interactions are shown in Figure 1 where a comparison is made between those using the G-matrix and its renormalized version. Both describe the experimental data well.

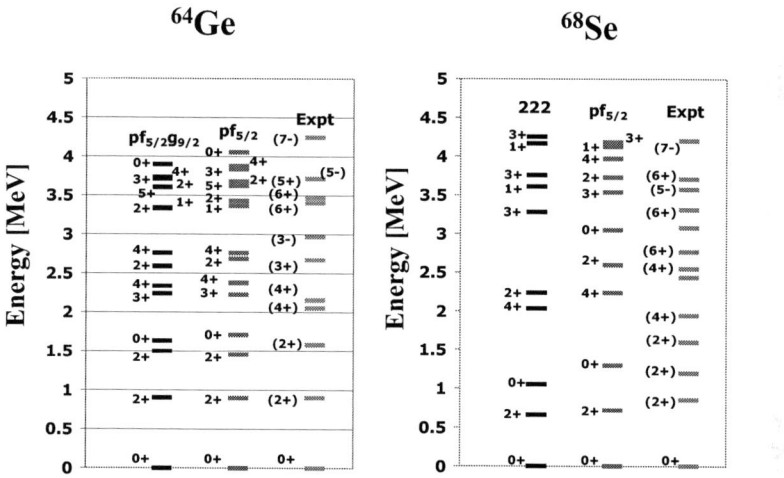

FIGURE1. Energy spectra for ^{64}Ge and ^{68}Se. The basis that is used in the G-matrix calculation for ^{68}Se is labeled by a sequence of three numbers, TPN, where T denotes the maximum total number of particles allowed in the intruder level and P and N denote the same for the protons and neutrons, respectively.

In Figure 2, results for the single-particle occupancies of states from the ground-state (g.s.) band of ^{64}Ge are shown. The lower (solid) bars represent the occupation numbers in the pf$_{5/2}$g$_{9/2}$-space calculation from the many-particle basis states which have no particles in the intruder level while the upper (gray) bars refer to those that count the occupancy when the intruder level is active. It is obvious that including configurations with at most two particles (protons or neutrons) in the intruder is enough to describe them with the same success as when we use the full space. Similar behavior is observed in the beta and gamma bands of the nucleus.

Next, results for the pseudo-SU(3) symmetry in the states of different bands in ^{64}Ge are shown in Figure 3 using the renormalized version of the interaction in the pf$_{5/2}$ space. The distribution of the second order Casimir operator C_2 of SU(3) in the g.s. and gamma bands indicates contribution of 50-60% for the leading SU(3) representation, which is an indication that the SU(3) symmetry is quite good. These observations suggest that the use of a symmetry-adapted, truncated set of basis states will give us better results for nuclei like ^{68}Se but in such cases the intruder level will play an even more significant role.

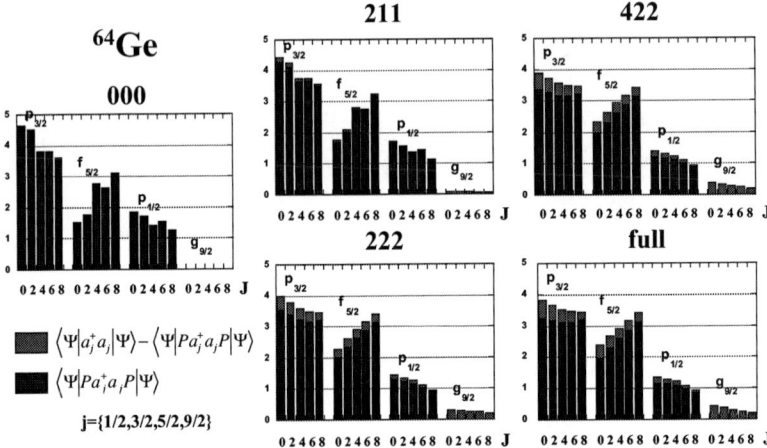

FIGURE 2. Different cuts and comparisons of the single-particle occupancies for the g.s. band of ^{64}Ge. Again, the basis that is used in each case is labeled by a sequence of three numbers, TPN, where T denotes the maximum total number of particles allowed in the intruder level and P and N denote the same for the protons and neutrons, respectively.

RESULTS IN THE SU(3) MODEL WITH ACTIVE INTRUDER LEVELS

Until recently, SU(3) shell-model calculations – real SU(3) for light nuclei and pseudo-SU(3) for heavy – have been performed in either one (protons and neutrons filling the same shell, e.g. the ds-shell) or two (protons and neutrons filling different shells, e.g. for rare earth and actinide nuclei) spaces only. Results for low-energy spectra of even-even and odd-mass heavy deformed nuclei have been published over the years [7]. Their B(E2) transition strengths, including both scissors and twist modes and their fragmentation have been successfully but only qualitatively described [8].

This simplified picture has now been extended by explicitly taking intruder levels into account [9]. The model space consists of two parts for each type of particles – protons and neutrons – a normal parity pseudo-shell and a unique parity shell, composed of the higher orbitals with opposite parity. The many-particle basis states

$$|\{\{a_{\pi N},a_{\pi U}\}\rho_\pi(\lambda_\pi\mu_\pi),S_\pi;\{a_{\nu N},a_{\nu U}\}\rho_\nu(\lambda_\nu\mu_\nu),S_\nu\}\rho(\lambda\mu)kL,S;JM\rangle \quad (1)$$

are built as SU(3)-coupled states with a well-defined particle number and total angular momentum where $a_{\sigma\tau} = \{N_{\sigma\tau}[f_{\sigma\tau}]\alpha_{\sigma\tau}(\lambda_{\sigma\tau}\mu_{\sigma\tau}),S_{\sigma\tau}\}$ are the basis-state labels for the four spaces in the model (σ stands for N or U and τ - for π or ν).

The Hamiltonian

$$H = \sum_{\sigma,\tau}(H_{sp}^{\sigma\tau} - GS^{\sigma\tau\dagger}S^{\sigma\tau}) - \frac{\chi}{2}Q.Q - \sum_{\substack{\sigma\neq\sigma'\\\tau}}GS^{\sigma\tau\dagger}S^{\sigma'\tau} \quad (2)$$

contains spherical Nilsson single-particle energies

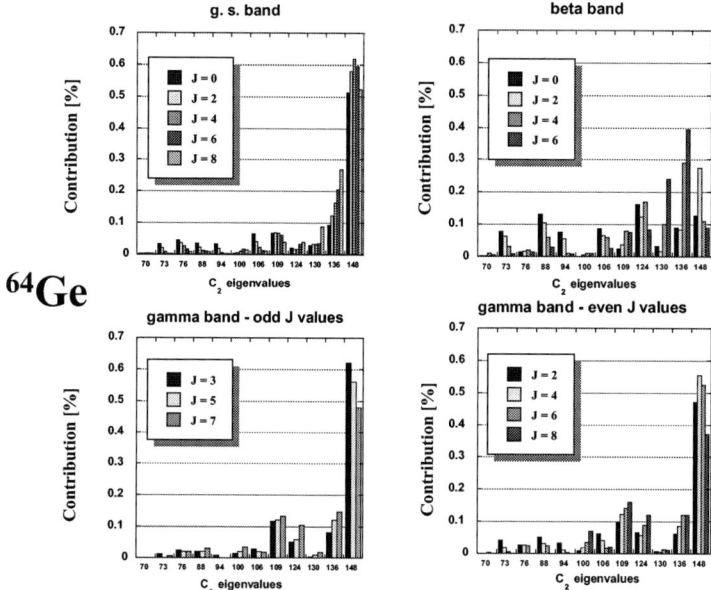

FIGURE 3. Pseudo-SU(3) content of the low-lying states in different bands of ^{64}Ge using the renormalized version of the G-matrix interaction.

$$H_{sp}^{\sigma\tau} = \sum_i \hbar\omega\{\eta_i + 3/2\} - k\hbar\omega\{2\vec{l}_i\cdot\vec{s}_i + \mu\vec{l}_i^{\,2}\} \tag{3}$$

for protons and neutrons as well as the quadrupole-quadrupole and pairing interactions. The single-particle terms together with the proton and neutron pairing interaction mix the SU(3) basis states allowing for a realistic description of the energy spectra. Most of the parameters we use for ^{64}Ge(^{68}Se) in the Hamiltonian were fixed from the systematics [10]:

$\hbar\omega = 41/A^{1/3} = 10.25(10.04)$

$k_{\pi U} = k_{\nu U} = 0.0367(0.0367)$ $\qquad \mu_{\pi U} = \mu_{\nu U} = 0.0568(0.0568)$

$\chi = 22/A^{5/3} = 0.0214(0.0194)$ $\qquad G = 15/A = 0.234(0.221)$

Those used for the single-particle terms in the pseudo spaces were taken to be consistent with single-particle energies in the G-matrix interaction used above:

$k_{\pi N} = k_{\nu N} = 0.011(0.011)$ $\qquad \mu_{\pi N} = \mu_{\nu N} = 1.056(1.056)$

The calculations were carried out using a set of basis states with (pseudo-)spin zero and one proton and neutron configurations. Since the most important configurations are those with highest spatial symmetry [11], we ignored all configurations that involve an odd number of particles in any of the four spaces and only consider the interplay between those having zero or two protons and/or two neutrons in the unique space. Then, from all the possible couplings we chose those irreducible representations with the highest value for the second order Casimir operator of (pseudo-)SU(3).

For our choice of coefficients in the Hamiltonian, the configurations with no particles in the unique space lie lowest and determine the structure of the lowest energy eigenstates. For ^{64}Ge, the other two groups of basis states (with two and four particles in the unique space) start to play a role above excitation energies of about 5 MeV. Also, since the spin-orbit interaction does not mix states from these groups it does not have any effect on the low-lying states. The mixing between configurations with different number of particles, although small, tracks back to pair scattering between the normal and unique parity spaces.

In Figure 4a one can see good agreement of the calculated results with experiment for the first states from the g.s and gamma bands of ^{64}Ge. For all the states calculated energies lie lower than the theoretical predictions with the G-matrix interaction. For ^{68}Se (Figure 4b) the SU(3) results suggest a shifting in order between the second J=2 and first J=4 states. This agrees with the order found in experiment.

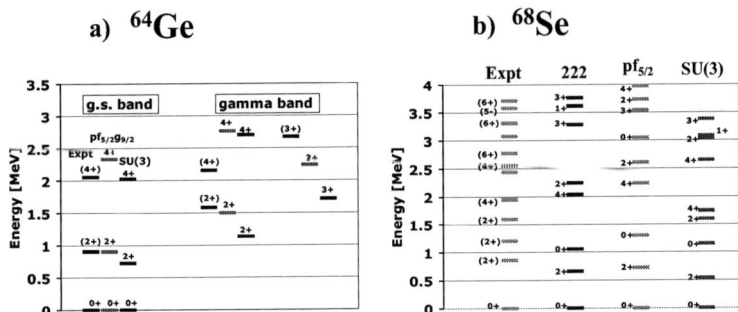

FIGURE 4. Results for the excitation spectra of ^{64}Ge and ^{68}Se from the SU(3) shell model compared with the experiment and the G-matrix results.

In conclusion, results from calculations in an extended SU(3), symmetry-adapted basis show that it is possible to obtain good results for nuclei of the upper fp-shell by considering only highest symmetry configurations.

REFERENCES

1. H. Schatz et al., Phys. Repts. **294**, 167 (1998).
2. S. E. Koonin, D. J. Dean, and K. Langanke, Phys. Repts. **577**, 1 (1996).
3. J. P. Elliott, Proc. Roy. Soc. London, Ser. A **245**, 128 (1958); **245**, 562 (1958); R. D. Ratna Raju, J. P. Draayer, and K. T. Hecht, Nucl. Phys. **A202**, 433 (1973).
4. M. Honma, T. Mizusaki, and T. Otsuka, Phys. Rev. Lett. **77**, 3315 (1996).
5. E. Caurier, F. Novacki, A. Poves, and J. Retamosa, Phys. Rev. Lett. **77**, 1954 (1996).
6. P. Van Isacker, O. Juillet and F. Novacki, Phys. Rev. Lett.. **82**, 2060 (1999).
7. C. Vargas, J. G. Hirsch and J. P. Draayer, Nucl. Phys. **A690**, 409 (2001); **A697**, 655 (2002); G. Popa, J. G. Hirsh and J. P. Draayer, Phys. Rev. C **62**, 064313 (2000); C. Vargas, J. G. Hirsch, T. Beuschel and J. P. Draayer, Phys. Rev. C **61**, 031301 (2000).
8. C. Vargas, J. G. Hirsch and J. P. Draayer, Phys. Lett. B **551**, 98 (2003).
9. C. Bahri et al., to be published.
10. P. Ring and P.Schuck, The Nuclear Many-Body Problem, Springer, Berlin, 1979; M. Dufour, A. P. Zuker, Phys. Rev. C **54**, 1641 (1996).
11. J P. Draayer, K. J. Weeks and K. T. Hecht, Nucl. Phys. **A381**, 1 (1982); J P. Draayer and K. J. Weeks, Ann. Phys. (N.Y.) **156**, 41 (1984); O. Castanos, J. P. Draayer and Y. Leschber, *ibid.* **180**, 290 (1987).

Level densities of iron isotopes and low-energy enhancement of γ-strength function

A.V. Voinov*, S.M. Grimes*, U. Agvaanluvsan†, E. Algin**, T. Belgya‡,
C.R. Brune*, M. Guttormsen§, M.J. Hornish¶, T.N. Massey¶, G.E.
Mitchell‖, J. Rekstad§, A. Schiller†† and S. Siem§

*Department of Physics and Astronomy, Ohio University, Athens, OH 45701
†Lawrence Livermore National Laboratory, L-414, 7000 East Avenue, Livermore, CA 94551
**Department of Physics, Osmangazi University, Meselik, Eskisehir, 26480 Turkey
‡Institute of Isotope and Surface Chemistry, Chemical Research Centre HAS, P.O.Box 77, H-1525 Budapest, Hungary
§Department of Physics, University of Oslo, N-0316 Oslo, Norway
¶Department of Physics and Astronomy, Ohio University, Athens, OH 45701
‖North Carolina State University, Raleigh, NC 27695
††NSCL, Michigan State University, East Lansing, MI 48824

Abstract. The neutron spectrum from the ^{55}Mn$(d,n)^{56}$Fe reaction has been measured at $E_d = 7$ MeV. The level density of ^{56}Fe obtained from neutron evaporation spectrum has been compared to the level density obtained from Oslo-type ^{57}Fe$(^3He, \alpha\gamma)^{56}$Fe experiment [1]. The good agreement supports the recent results [1, 2] including the low-energy enhancement in the γ-strength function for iron isotopes. The new level density function allowed us to investigate an excitation energy dependence of this enhancement, which is shown to increase with increasing excitation energy.

Keywords: level density, γ-strength function, evaporation spectra
PACS: 21.10.Ma, 24.30.Gz, 25.40.Lw, 25.55.Hp, 27.40.+z

INTRODUCTION

The unusual low-energy enhancement of the γ-strength function below the particle separation threshold has been found recently in Oslo-type experiments, first for ^{56}Fe and ^{57}Fe nuclei [1] then for the set of molybdenum isotopes [3]. This result contradicts an existing understanding based on different extrapolations of the tail of the giant dipole resonance function towards the low-energy region.

Although it has been proven that the Oslo method works well, the question remains about the applicability of the Axel-Brink hypothesis upon which the method is based. The hypothesis assumes that the γ-strength function depends on the energy of the γ-transition only and not on the excitation energy of initial and final states. It contradicts the modern understanding of damping properties of the giant resonance width, which should depend on the temperature of the final states [4, 5]. The Axel-Brink hypothesis allows one to factorize the first generation γ-matrix $P(E_i, E_\gamma)$ measured in Oslo experiments [6, 7] as

$$P(E_i, E_\gamma) \propto \rho(E_i - E_\gamma) \mathcal{T}(E_\gamma). \tag{1}$$

Here, ρ is the nuclear level density (NLD), $\mathcal{T}(E_\gamma)$ is the radiative transmission coefficient connected to the γ-strength function as $f(E_\gamma) \propto \mathcal{T}/E_\gamma^3$ and E_i is the energy of an initial state. The ρ and $\mathcal{T}$ functions are determined by the iterative procedure [8] through the adjustment of each data point of these two functions until a global χ^2 minimum with the experimental $P(E_i, E_\gamma)$ matrix is reached. Furthermore, the Oslo method assumes that the γ-decay pattern from any excitation energy is independent if it is directly populated by a nuclear reaction or by a nuclear reaction followed by one or more γ-rays. In order to address all of the above concerns, the level density of ^{56}Fe has been measured independently from the particle evaporation spectrum of the ^{55}Mn$(d,n)^{56}$Fe reaction. Such spectra are described by the simple model based on the Hauser-Feshbach formalism implying that the shape of the particle spectra depends on the level density of final nuclei and transmission coefficients of outgoing particles. Because transmission coefficients are directly connected to the capture cross section of inverse reactions and can be obtained experimentally, the level density can be deduced from spectra.

In this work the level density obtained from the neutron evaporation spectrum is compared to one obtained from the Oslo type ^{57}Fe$(^3$He$,\alpha\gamma)^{56}$Fe experiment. Results of the latter experiment can be found in recent publications [1, 2]. In this work, we will concentrate solely on the ^{55}Mn$(d,n)^{56}$Fe experiment.

EXPERIMENTS AND METHODS

The experiment has been performed with a 7 MeV deuteron beam provided by John Edwards Laboratory Tandem Accelerator at Ohio University. A self-supporting 1μm manganese foil has been used as a target. The energy of outgoing neutrons is determined by a time-of-flight technique using a 7 m flight path, and the neutrons are detected in liquid scintillator NE213 detectors. A pulse width of 3 ns provided the energy resolution of $\sim$100 and 800 keV for 1- and 14-MeV neutrons, respectively. The detector efficiencies were measured by using the neutron "standard" spectrum obtained at an angle of 120° from the ^{27}Al(d,n) reaction on a thick, stopping Al target at $E_d = 7.44$ MeV [9]. The swinger beam facility has been used for the measurement of the angular distribution of neutrons from 20 to 150 degrees.

The procedure to extract the level density from evaporation spectra was proposed in Ref.[10]. It is based on the Hauser-Feshbach theory of compound nuclear reactions according to which the level density is determined by both the transmission coefficients of outgoing particles and the level density of the residual nucleus. In order to calculate the transmission coefficients, ten neutron optical model potentials proposed in the RIPL-2 [11] data base have been tested. All of them have been found to give the same result (the same shape of neutron evaporation spectra) within $\sim$15% over the range of 1-14 MeV of neutron energy. Finally, the potential of D.Wilmore and Hodgson has been adopted. In order to extract the level density of ^{56}Fe, the following procedure has been used: 1) The Fermi-gas level density model is chosen to calculate the neutron evaporation spectrum. The parameters of the model have been adjusted to reproduce the experimental spectrum as closely as possible. 2) The model level density has been improved by a binwise renormalization according to the expression:

$$\rho(E) = \rho(E)_{\text{input}} \frac{(d\sigma/d\varepsilon)_{\text{meas}}}{(d\sigma/d\varepsilon)_{\text{calc}}} \quad (2)$$

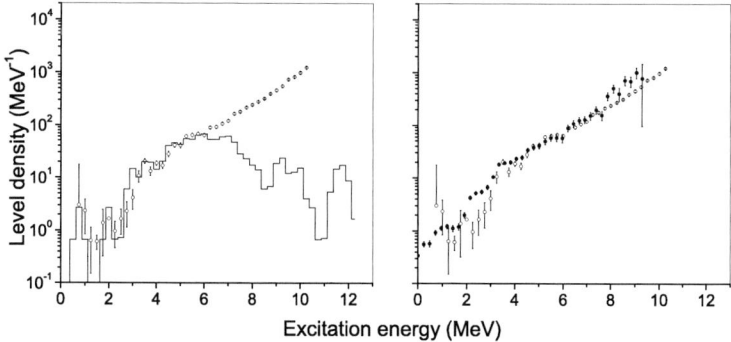

FIGURE 1. The comparison of the level density extracted from neutron evaporation spectra (open circles) with discrete level density (left panel) and with level density obtained from Oslo-type experiment (right panel)

The extracted level density $\rho(E_x)$ for the ^{56}Fe nucleus is shown in Figure 1. The absolute normalization of this function has been obtained by fitting to the density of discrete levels below 6 MeV of an excitation energy. One can see the good structural agreement between these two level densities up to about 6 MeV of excitation energy beyond which the discreet level density is incomplete. Figure 2 also shows the comparison between level densities obtained from both the current and Oslo experiments [2]. One can see the good general agreement which provides increased confidence to both methods. Moreover, the important result is that the step at 3 MeV obtained from the Oslo experiment discussed in the Ref.[2] is well reproduced by the current data. Also, the step at 6 MeV of excitation energy is now seen to be real as well because it is reproduced by both experiments. These steps have been interpreted as a result of the breaking of cooper pairs. The disagreement between Oslo and current data at excitation energies below 2.5 MeV can be understood as there are too few levels available at this excitation energy. It results in a larger fluctuation of primary γ-transitions due to poor spin and parity averaging, thereby making population of some of these levels more preferable in the Oslo experiment.

Thus, in spite of the fact that these two methods rely on different underlying assumptions, different nuclear reactions and different mathematical techniques to extract NLD, they have very much in common, namely the dominance of the statistical mechanism of nuclear reactions in both cases. It also serves as a double check of these two methods.

γ-STRENGTH FUNCTION OF ^{56}FE

The γ-strength functions for ^{56}Fe and ^{57}Fe have been obtained from the Oslo-type experiment in Ref. [1]. The striking feature of these functions is the increase of γ-strength in the region below 3 MeV. The main drawback of the Oslo method is the assumption of

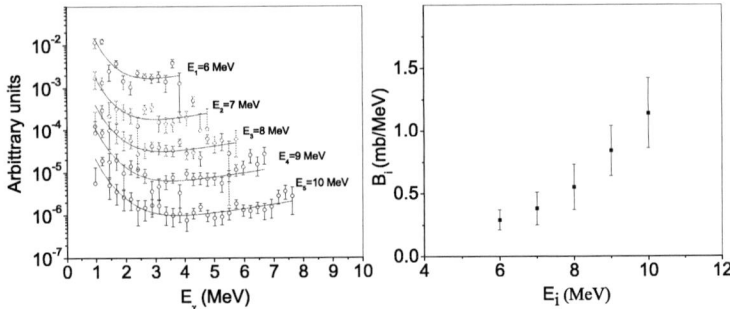

FIGURE 2. Left: the γ-strength function extracted from the Oslo experiment using the level density from neutron evaporation spectra of the ^{55}Mn$(d,n)^{56}$Fe reaction; right: B_i coefficients obtained from the fit

Axel-Brink hypothesis, which prevents us from the study of the temperature dependence of γ-strength function. Such temperature dependence follows from the temperature dependence of giant resonance widths caused by different damping mechanisms debated in literature.

In this work, we can use the first generation matrix $P(E_i, E_\gamma)$ obtained from Oslo experiments and the level density from the ^{55}Mn$(d,n)^{56}$Fe reaction to get the γ-strength functions corresponding to different excitation energies of initial states:

$$f(E_\gamma, E_i) = \frac{1}{2\pi} \frac{N(E_i) P(E_i, E_\gamma)}{\rho(E_i - E_\gamma) E_\gamma^3}, \quad (3)$$

where $E_i = E_\gamma + E_f$ is the energy of the initial state of the γ-transition. Because of unknown normalization $N(E_i)$, it is possible to investigate only the shape of the function (3) at different E_i but not its absolute magnitude. We assume that it can be described by the combination of the low energy tail of the giant electric dipole resonance (GEDR) which is described by simple Lorentz formula, and the soft pole, for which the power function has been chosen, as it was used in [1]. In this case, the γ-strength function is reproduced by the following expression:

$$f(E_\gamma, E_i) = \frac{A_i}{3\pi^2 \hbar^2 c^2} \left(\frac{\sigma_0 E_\gamma \Gamma^2}{(E_\gamma^2 - E^2)^2 + E_\gamma^2 \Gamma^2} + \frac{B_i}{E_\gamma^C} \right) \quad (4)$$

where A_i, B_i and C are fitting parameters, and E=16.8 MeV and $\Gamma = 4$ MeV are the position and width of the GEDR. The parameters B_i determine the relative contribution of the soft pole to the γ-strength at different E_i. The result of the fit is shown in the Figure 2. We have observed that the parameter B_i increases gradually as E_i increases, thereby indicating the increase of the soft pole contribution.

SUMMARY AND CONCLUSIONS

The level density of the ^{56}Fe nucleus extracted from two different experiments are in good agreement with each other. It indicates the consistency of these two methods and the possibility to apply both methods to an investigation of a broad range of nuclei. It helps to eliminate most of the systematic errors associated with these methods, including the unknown contribution of direct processes in particle evaporation spectra and the assumption of the Axel-Brink hypothesis in the Oslo method.

The γ-strength function for the ^{56}Fe isotope obtained in Ref.[1] has now been extracted by taking into account the level density from the neutron evaporation spectrum. It allows one to investigate the γ-strength function at a different excitation energy of initial states. We confirm the presence of the low-energy enhancement at all excitation energies from 5 to 10 MeV. The contribution of the soft pole increases with increasing excitation energy. This fact might indicate the temperature dependence of γ-strength caused by the damping properties of giant resonances.

ACKNOWLEDGMENTS

The authors acknowledge support from the following organizations: a) National Nuclear Security Administration under the Stewardship Science Academic Alliances program through DOE Research Grants No. DE-FG03-03-NA00074 and No. DE-FG03-03-NA00076; b) Norwegian Research Council; c) DOE research grants No. W-7405-ENG-48 and DE-FG02-97-ER41042; d) EU5 Framework Programme under Contract No. HPRI-CN-1999-00099.

REFERENCES

1. A. Voinov, E. Algin, U. Agvaanluvsan, T. Belgya, R. Chankova, M. Guttormsen, G.E. Mitchell, J. Rekstad, A. Schiller, and S. Siem, *Phys.Rev. Lett.* **93**, 142504-1 – 142504-4 (2004).
2. A. Schiller et al., *Phys. Rev. C* **68** 054326-1 – 054326-6(2003).
3. M.Guttormsen, R.Chankova, U.Agvaanluvsan, E.Algin, L.A.Bernstein, F.Ingebretsen, T.Lonroth, S.Messelt, G.E.Mitchell, J.Rekstad, A.Schiller, S.Siem,A.C.Sunde, A.Voinov, and S. Ødegård, *Phys. Rev. C* **71**, 044307-1 – 044307-7 (2005).
4. S.G. Kadmenskiĭ, V.P. Markushev, and V.I. Furman, *Yad. Fiz.* **37**, pp. 277 –283 (1983), [*Sov. J. Nucl. Phys.* **3**, 165 – 171 (1983)].
5. S.F. Mughabghab and A.A.Sonzogni, *Phys. Rev. C* **65**, 044620-1 – 044620-5 (2002).
6. P. Axel, *Phys. Rev.* **126**, 671 –683 (1962).
7. D.M. Brink, *Ph.D. thesis*, Oxford University, 1955.
8. A. Schiller, L. Bergholt, M. Guttormsen, E. Melby, J. Rekstad, and S. Siem, *Nucl. Instrum. Methods Phys. Res.* **A447**, 498 – 511 (2000).
9. T.N. Massey, S. Al-Quraishi, C.E. Brient, J.F. Guillemette, S.M. Grimes, D. Jacobs, J.E. O'Donnell, J. Oldendick and R. Wheeler, *Nuclear Science and Engineering* **129**, 175 – 179 (1998).
10. H.Vonach, *Proceedings of the IAEA Advisory Group Meeting on Basic and Applied Problems of Nuclear Level Densities, Upton, NY, 1983*, BNL Report No.BNL-NCS-51694, 1983, pp.247 – 290.
11. Reference Input Parameter Library (RIPL-2), URL http://www-nds.iaea.org/RIPL-2
12. D.Wilmore and P.E.Hodgson, *Nucl. Phys.* **55**, 673 – 694 (1964)

Spin-isospin excitations from the ground-state of ^{64}Ni

L. Popescu[1*], T. Adachi[†], C. Bäumer[**], G.P.A. Berg[2 ‡],
A.M. van den Berg[‡], P. von Brentano[§], D. Frekers[**], D. de Frenne[*],
K. Fujita[¶], Y. Fujita[†], E.W. Grewe[**], P. Haefner[**], K. Hatanaka[¶],
M. Hunyadi[‡], M. de Huu[‡], E. Jacobs[*], H. Johansson[∥], A. Korff[**],
A. Negret[1*], K. Nakanishi[¶], P. von Neumann-Cosel[††], S. Rakers[**],
N. Ryezayeva[††], Y. Sakemi[¶], A. Shevchenko[††], Y. Shimbara[¶], Y. Shimizu[¶],
H. Simon[∥], Y. Tameshige[†], A. Tamii[¶], M. Uchida[‡‡], H.J. Wörtche[‡] and
M. Yosoi[‡‡]

Vakgroep Subatomaire en Stralingsfysica, Universiteit Gent, B-9000 Gent, Belgium
†*Department of Physics, Osaka University, Toyonaka, Osaka 560-0043, Japan*
**Institut für Kernphysik, Westfälische Wilhelms-Universität Münster, D-48149 Münster, Germany*
‡*Kernfysisch Versneller Instituut, Rijksuniversiteit Groningen, NL-9747 AA Groningen, The Netherlands*
§*IKP Köln University, Köln, Germany*
¶*RCNP, Osaka University, Ibaraki, Osaka 567-0047, Japan*
∥*Gesellschaft für Schwerionenforschung mbH, D-64291 Darmstadt, Germany*
††*Institut für Kernphysik, Technische Universität Darmstadt, Germany*
‡‡*Dept. Phys., Kyoto University, Sakyo, Kyoto 606-8224, Japan*

Abstract. Spin-isospin (Gamow-Teller) excitations in ^{64}Cu and ^{64}Co have been studied using (^{3}He,t) and ($d,^2$He) charge-exchange reactions on ^{64}Ni. As the isospin of the ^{64}Ni ground-state is $T_0=4$, states with $T=3$, 4 and 5 in ^{64}Cu are excited via the (^{3}He,t) reaction and states with $T=5$ in ^{64}Co via ($d,^2$He). If we assume that the nuclear interaction is charge symmetric, the $T=5$ states in ^{64}Cu should appear at corresponding excitation energies (if corrected for the Coulomb displacement) and with similar strengths as the $T=5$ states in ^{64}Co. As in the ^{64}Cu spectrum the $T=5$ states are very weakly excited, only by combining the results of the two complementary experiments one can estimate the Gamow-Teller strength starting from ^{64}Ni in a consistent way.

INTRODUCTION

Similar to β-decay, the charge-exchange (CE) reactions replace a proton in a nucleus by a neutron (the inverse reaction of β^- decay) or a neutron by a proton (the inverse of β^+ decay) (see fig. 1). Although the charge exchange processes involve strong interactions, the matrix elements that enter into the reaction rates are the same as in the case of

---

[1] Permanent address: NIPNE, Atomistilor 407, P.O.Box MG-6, Bucharest-Magurele, Romania
[2] Present address: Dep of Physics, University of Notre Dame, 225 Neuwland, Science Hall, Notre Dame, IN 46446-5670

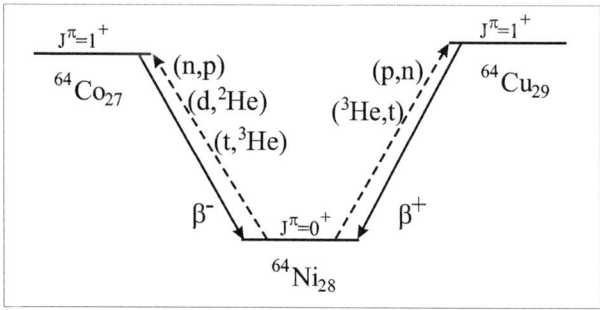

FIGURE 1. Schematic drawing of GT transitions between ^{64}Co, ^{64}Ni and ^{64}Cu, ground states either via β-decay, or via different CE reactions.

the β-decay mediated by the weak interaction. The main advantage of CE reactions over β-decay is that the obtainable information is not limited by the Q-value to lower energy final states: a complete picture of the spin-isospin transitions up to high excitation energies can be given.

CHARGE-EXCHANGE REACTIONS ON ^{64}NI

Spin-isospin (Gamow-Teller) excitations in ^{64}Co and ^{64}Cu have been studied using $(d,^2\text{He})$ and $(^3\text{He},t)$ CE reactions on ^{64}Ni.

The proportionality between the intermediate incident energy $(d,^2\text{He})$ reaction cross section extrapolated to zero momentum transfer $(q=0)$ and the Gamow-Teller (GT) transition strength (B_{GT}^+) was shown experimentally [1, 2]. This CE reaction, by its spin-isospin-flip selection character [3], is an excellent tool in the study of GT transitions. We have studied the $^{64}\text{Ni}(d,^2\text{He})^{64}\text{Co}$ reaction at the KVI, Groningen, using a 170 MeV deuteron beam of the AGOR accelerator. The ejectiles were momentum analyzed with the Big-Bite Spectrometer (BBS) and the EuroSuperNova detection system [3]. An energy resolution of about 110 keV was achieved, enabling the separation of several individual states (see fig. 2 a)). The $\Delta L=0$ assignment is based on an angular distribution analysis.

Also the proportionality between the intermediate incident energy $(^3\text{He},t)$ reaction cross section extrapolated to $q=0$ and the GT transition strength (B_{GT}^-) was shown experimentally [4]. The $^{64}\text{Ni}(^3\text{He},t)^{64}\text{Cu}$ experiment was performed at the Grand Raiden Spectrometer, RCNP, Osaka, with a 420 MeV ^{3}He beam. An unprecedent good energy resolution of 35 keV was achieved. The low and intermediate excitation energy region could be decomposed into individual peaks (see fig. 2 b)) and the $\Delta L=0$ transitions identified via an angular distribution analysis.

Both transitions from ^{64}Co ground-state (g.s.) and from ^{64}Cu g.s. to ^{64}Ni g.s. are allowed GT β transitions, with $(\log ft)_{64Co->64Ni} = 4.3$ and $(\log ft)_{64Cu->64Ni} = 5$ as determined in β-decay studies [5]. Therefore an absolute calibration is available in both cases for connecting the reaction cross sections with the transition strengths.

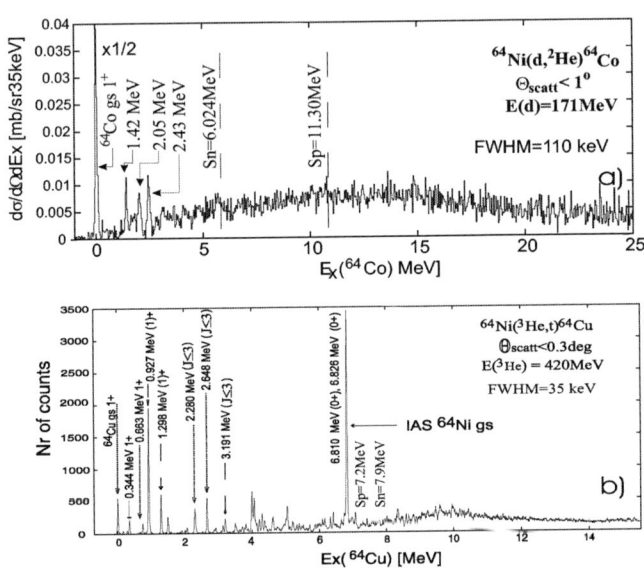

FIGURE 2. a) ^{64}Co and b) ^{64}Cu spectra obtained in the $(d,^2\text{He})$ and $^{64}\text{Ni}(^3\text{He},t)$ experiments. As the difference between the intensities of the ^{64}Co g.s. peak and the rest of the spectrum was considerable, this peak was scaled by a factor of 1/2. The large peak in the ^{64}Cu spectrum b) corresponds to the IAS of ^{64}Ni g.s. Neutron and proton separation energies are also indicated.

THE B_{GT} DISTRIBUTION OVER DIFFERENT ISOSPIN COMPONENTS

In a nucleus, the z-component of the isospin is $T_z = \frac{1}{2}(N-Z)$ and the isospin of the ground and excited states must be $T \geq |T_z|$. The isospin nature of the states excited via CE and inelastic scattering (IE) reactions is discussed in what follows, where we denote by T_0 the z-component of the target isospin.

In terms of one-particle one-hole excitations, the isospin of the quantum of excitation characterizing the isospin-flip transitions will be $\tau = 1$ with an eigenvalue $\mu_\tau = 0, \pm 1$. Therefore the isospin excitations give rise to a triplet of states with $T = T_0 - 1, T_0, T_0 + 1$.

Fig. 3 shows the possible isospin components excited by different CE and IE reactions starting from a nucleus with ground state isospin $T = T_z = T_0$:

- via the $(^3\text{He},t)$ reaction the z-component of the isospin of the final nucleus is $T_z = T_0 - 1$, therefore all three isospin components mentioned above are excited.
- via the $(d,^2\text{He})$ reaction the final nucleus has $T_z = T_0 + 1$ and, as for the final states $T \geq T_z$, the $T_0 + 1$ are the only states that can be excited.
- an IE reaction, as (p,p'), will excite T_0 and $T_0 + 1$ states in the target nucleus.

In fig. 3 the dashed lines connect the states with the same isospin in different nuclei (with the same A but different T_z): the so called Isobaric Analogue States (IAS).

All the low-lying energy levels in the $(^3\text{He},t)$ spectrum have the minimum isospin $(T_0 - 1)$, while the first $T_0 + 1$ state occurs at excitation energies well above the threshold

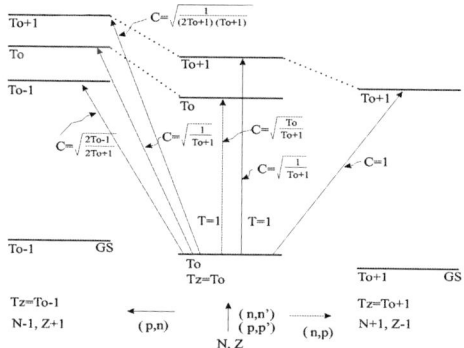

FIGURE 3. Schematic representation of the isospin transitions starting from a nucleus with ground-state isospin $T_0=T_z=(N-Z)/2$. The corresponding Clebsch-Gordon coefficients (C) for different transitions are indicated. Picture taken from [6]

for proton emission [7]. Due to the charge independence of the nucleonic interactions, the binding energy for the lowest state with any given isospin $T > T_z$ can be obtained. The difference in binding energy for states in the same nucleus (so the same A and T_z) and having different values of T is given by the symmetry energy.

In the absence of the coupling between the isospin of the vibration τ and T_0, the intensity ratios of the different isospin components are given by the square of the corresponding Clebsch-Gordan (CG) coefficients for those components:

$$C_{CG}^2 = |\langle T_i \, T_{z_i} \, \tau \, \mu_\tau \, | \, T_f \, T_{z_f} \rangle|^2 \qquad (1)$$

However, for nuclei with neutron excess, like ^{64}Ni, an isovector potential is created in the nucleus which acts on the isospin of the vibration, τ, so that the states with low total isospin are favored. In the presence of this coupling, the fundamental shift produced by the neutron excess is [7]:

$$E_{coupling} = \frac{V_1}{A}(\tau T_0) = \frac{V_1}{2A}[T(T+1) - T_0(T_0+1) - \tau(\tau+1)] \qquad (2)$$

whereby V_1 is of the order of 100 MeV. It is clear that the interaction energies given by eq. 2 become quite large, even for moderate values of T_0.

Table 1 gives the calculated values of the square of the CG coefficients and the displacement energies (calculated using eq. 2) for the three isospin components excited in ^{64}Cu via the (^{3}He,t) reaction. The difference in energies gives the relative positions of the centroids and the intensity ratios represents the ratios of the summed strengths for different isospin components. The amount of the GT strength exhausted by each isospin component is also indicated. As can be observed, most of the strength is exhausted by the $T_0 - 1$ component, that forms the giant GT resonance. The $T_0 \longrightarrow T_0 + 1$ transitions are very much suppressed, only 2.2% of the GT strength going to $T_0 + 1$ states. But the transitions to $T_0 + 1$ component can be studied in detailed in the (d,^{2}He) reaction, where it is the only component excited. Combining the results of the two complementary ex-

TABLE 1. Relative intensities (as C_{CG}^2) and positions of different isospin components excited in the ^{64}Ni(^{3}He,t)^{64}Cu reaction. The C_{CG}^2 as indicated in [6] were used (see fig 3). The relative positions were calculated by using eq. 2. The % of the total $B(GT)$ exhausted by each component is also given.

T	$T_0-1=3$	$T_0=4$	$T_0+1=5$
C_{CG}^2	$\frac{7}{9}$	$\frac{1}{5}$	$\frac{1}{49}$
% of $\Sigma B(GT)$	77.8%	20%	2.2%
$E_{coupling}$ [MeV]	-7.21	-1.56	6.25

periments, the GT strength starting from the g.s. of ^{64}Ni can be estimated in a consistent way.

To complete this study, spin-isospin excitations of $T=T_0=4$ and $T=T_0+1=5$ states in ^{64}Ni will be investigated using the ^{64}Ni(p,p') reaction.

These experiments on ^{64}Ni are part of a systematic study on the IAS in A = 58, 60, 62, 64 Cu, Ni and Co nuclei.

The final analysis of the experimental data is still ongoing. The results will be published as an extended paper.

These experiments were performed at RCNP, Osaka University and at KVI, Groningen. The authors are grateful to the accelerator group of RCNP and KVI for providing a high-quality ^{3}He and d beam. The Ghent group acknowledges support from the FWO-Flanders. L.P., T.A. and A.N. acknowledge support from the 21st Century COE program "Toward a new basic science" of Graduate School of Science, Osaka University. We also acknowledge support of the European Community "Access to Research Infrastructure action of the Improving Human Potential Programme"

REFERENCES

1. S. Rakers, e. a., *Physical Review C*, **65**, 044323 (2002).
2. E.W. Grewe, e. a., *Physical Review C*, **69**, 064325 (2004).
3. S. Rakers, e. a., *Nuclear Instrumentations and Methods A*, **481**, 252 (2002).
4. M. Fujiwara, e. a., *Physical Review C*, **59**, 90–100 (1999).
5. Firestone, R., and Shirley, V., editors, *Table of Isotopes*, Willey, New York, 1996, eight edn.
6. Satchler, G., *Direct Nuclear Reactions*, Oxford University Press, 1983.
7. Bohr, A., and Mottelson, B., *Nuclear Structure, Volume I and II*, World Scientific, 1998, second edn.

POSTER SESSION

α-stripping Reactions with Exotic Nuclei: $^{12}C(^{7}Be,^{3}He)^{16}O$

H. Amro[1], F.D. Becchetti[1], Yu Chen[1], H. Jiang[1], M.Ojaruega[1], H. C. Griffin[1], J.J. Kolata[2], B. B. Skorodumov[2], J.D. Hinnefeld[3], G. Peaslee[4]

[1]Department of Physics, University of Michigan, Ann Arbor, Michigan 48109-1120, USA
[2]Department of Physics, University of Notre Dame, Notre Dame, Indiana 46556-5670, USA
[3]Department of Physics, Indiana University-SB, South Bend, Indiana 46634-7111 USA
[4]Department of Physics and Engineering, Hope College, Holland, Michigan 49422-9000USA

Abstract. The $^{12}C(^{7}Be,^{3}He)^{16}O$ reaction has been studied at $E(^{7}Be)=34$ MeV. At this energy this Reaction exhibits a high selectivity for populating known α-cluster states in ^{16}O ($J^{\pi} = 0^{+},3^{-},2^{+},1^{-},4^{+}$, and 6^{+}). The angular distributions for the $0^{+},3^{-},2^{+},1^{-}$, and 4^{+} levels are obtained at forward angles. Likewise, large reaction rates were measured for the ^{3}He transfer channel, i.e. $^{12}C(^{7}Be,\alpha)^{15}O$.

INTRODUCTION

The study of alpha-particle transfer reactions on *p*-shell and *sd*-shell nuclei is important for our understanding of the structure of these nuclei and for the analysis of the helium-burning process in nuclear astrophysics. α-particle reduced widths of bound and nearly-bound states in principle can be extracted in α-stripping reactions such as (^{7}Li,t) and (^{6}Li,d)^{1,2}. The study of such reactions provides complimentary information to that btained from (α,α) scattering and (α,γ) capture reactions. The pronounced selectivity of such reactions and forward peaking of the angular distributions of their rates are Indicative of direct α-transfer reaction mechanism. A new α-transfer reaction, ($^{7}Be,^{3}He$) which is analogous to (^{7}Li,t) reaction may prove to be better-suited since it has a better reaction kinematics and perhaps reduced projectile breakup in the excitation Regions of interest[2].

EXPERIMENTAL RESULTS

The ^{7}Be beam was produced via the $^{6}Li(^{3}He,P)^{7}Be$ production reaction. The 37 MeV ^{6}Li primary beam was provided by the FN tandem accelerator at the University of Notre Dame. The ^{7}Be beam, 10^{5}/s, at 34 MeV was collected and focused using the UM/UND *TwinSol* RNB facility to the scattering chamber. Five ΔE-E telescopes were mounted in this chamber, covering an angular range 15-75°, to collect reaction products. A thick, ~ 1 mg/cm^{2}, self supporting natural carbon target was used together with a natural gold target. An example of a $^{12}C(^{7}Be,^{3}He)^{16}O$ reaction spectrum at 15° is shown in Figure 1.

The ^{3}He- and the α-groups, corresponding to α-transfer and ^{3}He-transfer to ^{12}C, respectively, are clearly visible in this Figure. Due to angular and energy resolution limitations in this experiment, the $2^+,1^-$ and $0^+,3^-$ resonances could not be resolved as can be seen in the inset of Figure 1. Reliable reaction rates could not be extracted for the 16.3 MeV resonance in ^{16}O observed in the measurement due to a strong beam related ^{3}He group that contaminated our data. Angular distributions of the reaction rates for the sum of the $2^+,1^-$ and $0^+,3^-$ resonances are extracted and presented in Figure 2 along with the reaction rates for ^{12}C(^{7}Li,t)^{16}O of reference [2]. Reliable comparison between the two reactions cannot be done at this time. A future high resolution ^{12}C(^{7}Be,^{3}He)^{16}O measurement with TOF information is planned to allow for more precise results.

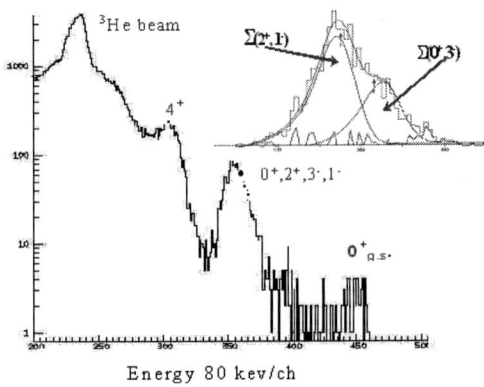

FIGURE 1. ^{3}He spectrum obtained at 15°. Known α-cluster states in ^{16}O are indicated by their J^π.

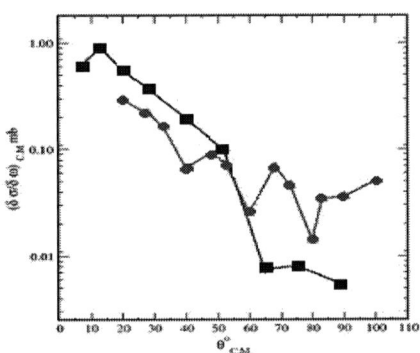

FIGURE 2. ^{12}C(^{7}Be,^{3}He)^{16}O angular distributions for the $\Sigma(2^+,1^-)$ resonances in ^{16}O. The data are from the current work (square symbols) and for ^{12}C(^{7}Li,t)^{16}O from ref.[2] (circles).

REFERECCES

1. M. E. Cobern *et. al.* Phys. Rev. **C14**, 491 (1976).
2. F.D. Becchetti *et. al.* Nucl. Phys. **A305**, 293 (1978).

Structural Ambiguities In ^{114}Cd

D. Bandyopadhyay[1,2], P. E. Garrett[1], S. R. Lesher[2], C. Fransen[2], N. Boukharouba[2], M. T. McEllistrem[2], and S.W. Yates[2,3]

[1] *Department of Physics, University of Guelph, Guelph, Ontario, Canada, N1G2W1.*
[2] *Department of Physics and Astronomy, University of Kentucky, Lexington, KY 40506-0055.*
[3] *Department of Chemistry, University of Kentucky, Lexington, KY 40506-0055.*

Abstract. Properties of low-spin states in ^{114}Cd have been studied through the (n,n'γ) reaction. Gamma-ray angular distributions and excitation functions have been used to characterize the decays of the excited levels, and level lifetimes have been obtained with the Doppler-shift attenuation method. Data have been compared with the theoretical calculations following the Interacting Boson Model. Multiphonon states have been explored on the basis of their decay scheme and collective nature of the transitions.

Keywords: ^{114}Cd, Energy levels, Multiphonon states.
PACS: 21.10.Hw, 21.10.Re, 23.20.Lv, 27.60.+j

In the cadmium isotopes, the presence of additional 0^+ and 2^+ states near the 2-phonon triplet made the phonon structure ambiguous. It has been found that even a firm assignment of two phonon structures is a challenge [1]. Here, we present some interesting features observed while higher phonon states in ^{114}Cd were sought.

Three-phonon structures have been observed in $^{108-112,116-120}$Cd isotopes [2-6]. In Ref. [1], Casten et al. suggested the 3-phonon structure in ^{114}Cd nucleus studied through GRID technique [1]. In contrast, Mheemeed et al. suggested the 0_4^+ and 2_4^+ of this proposed quintuplet as the higher-lying members of 2p4h excitations [7].

We have studied low-lying states of ^{114}Cd at the University of Kentucky 7 MV accelerator facility through the inelastic scattering of fast neutrons. The γ-ray excitation functions of the levels in ^{114}Cd were measured over the range of neutron energies from 1.9 to 3.8 MeV in 0.1 MeV steps, and γ-ray angular distributions were measured for 3 neutron energies of 2.5, 3.0, and 3.5 MeV. Lifetimes were measured following the Doppler Shift Attenuation Method and transition strengths were calculated from the measured lifetimes, branching ratios and multipole mixing ratios. The detailed results may be found somewhere else [8].

Systematics of 3-phonon levels over the Cd isotopes support the 0^+_4 state at 1859 keV as the member of the 3-phonon multiplet. Surprisingly, no strong *E2* transitions have been observed from this state to the 2-phonon 2^+ state at 1210 keV questioning the validity of its 3-phonon origin. However, a transition has been observed from this level to the intruder 2^+ level at 1364 keV. One reason could be that the expected 650-keV transition is difficult to observe in the presence of a strong 651-keV transition from 1210-keV level to 558-keV level.

The 3^+_1 level at 1865 keV has been assigned as the member of quartet of 3-phonon states, while the 3^+_2 state at 2205 keV could not be explained in the multiphonon scenario. Considerable *E2* branching has been observed from this state to the 3-

phonon states, but the 4-phonon octet should not exhibit a 3^+ member [9]. Theoretical calculations suggest this state as an intruder excitation. A strong E0 transition from this level to the 3^+_1 state also indicates a very different configuration from the 3^+_1 3-phonon state [7]. Therefore, we have assigned this state as a member of a quasi-γ-band. The question arises what will be the band head of this γ-band? The 3_2^+ level decays to both the 1842-keV and 2048-keV 2^+ states. IBM calculations suggest normal and intruder 2^+ excitations at 1.9 and 1.8 MeV respectively. The expected "V" shaped behaviour in the systematics also supports the 1842-keV level as the intruder state and 2048-keV level as a multiphonon excitation in conflict with the assignment of Casten et al [1].

Fragmented hexadecapole excitations have been suggested in ^{112}Cd and ^{110}Cd [10]. The spin of the 2152-keV level has been assigned as 3^+ or 4^+ [11]. We confirm the spin of this state as 4^+. Considering the systematics, the possibility of the origin of this state as a hexadecapole excitation cannot be ruled out.

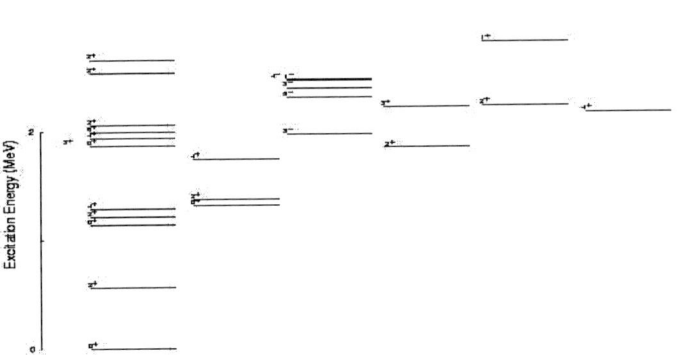

FIGURE 1. Possible multiphonon interpretation of levels in ^{114}Cd.

References

1. R. F. Casten et al., Phys. Lett. **297B**, 19 (1992).
2. A. Aprahamian et al., Phys. Rev. Lett. **59**, 535 (1987).
3. Y. Wang et al., Phys. Rev. C **67**, 064303 (2003).
4. P. E. Garrett et al., Phys. Rev. C **64**, 024316 (2001).
5. F. Corminboeuf et al., Phys. Rev. Lett. **84**, 4060 (2000).
6. A. Gade et al., Phys. Rev. C **66**, 034311 (2002).
7. A. Mheemeed et al., Nucl. Phys. **412A**, 113 (1984).
8. D. Bandyopadhyay et al., to be published.
9. A. Bohr and B. R. Mottleson, Nuclear structure, Vol II.
10. M. Pignanelli et al., Nucl. Phys. **540A**, 27 (1992).
11. J. Blachot, Nuclear Data Sheets **97**, 593 (2002).

^{207}Pb(n,2nγ)^{206}Pb Cross-Section Measurements by In-Beam Gamma-Ray Spectroscopy

P. Baumann[1], C. Borcea[2,3], E. Jericha[4], S. Jokić[5], M. Kerveno[1],
S. Lukić[1,5], L.C. Mihailescu[2], A. Pavlik[6], A.J.M. Plompen[2], G. Rudolf[1]
and the n_TOF collaboration

[1] *IReS, IN2P3, Strasbourg, France*
[2] *EC, JRC, IRMM, Geel, Belgium*
[3] *NIPNE, Bucharest, Romania*
[4] *Technische Universität Wien, Wien, Austria*
[5] *INN Vinča, Belgrade, Serbia and Montenegro*
[6] *Universität Wien, Wien, Austria*

Abstract. ^{207}Pb(n,2nγ)^{206}Pb cross section were measured for incident neutron energies between 6 and 20 MeV with the white neutron beam produced at GELINA. The γ-ray production cross section for the main transition (803 keV, $2^+ \to 0^+$) in ^{206}Pb is compared to results obtained at Los Alamos and to the TALYS and EMPIRE-II code predictions.

Keywords: Neutron cross-sections; Gamma-ray production; Time-of-flight; Enriched lead
PACS: 25.40.-h, 24.60.Dr, 29.30.Kv

INTRODUCTION

Innovative systems for energy production and/or waste transmutation require new nuclear data in a wide neutron energy range. For (n,xn) reactions with x>1 at E_n > 6-8 MeV, experimental data are very rare and evaluated data show large discrepancies. Thus, within the European Commission n_TOF-ND-ADS project, a program to measure (n,xn) cross-sections was initiated. In particularly, it is of great importance to know all properties of stable lead isotopes as they are involved in materials (target in Pb/Bi) for Accelerator Driven Systems (ADS). This paper presents ^{207}Pb(n,2nγ)^{206}Pb cross-section measurements for incident neutron energies between 6 and 20 MeV produced by the white neutron beam at GELINA (IRMM, Geel).

EXPERIMENTAL PROCEDURE AND RESULTS

The ^{207}Pb(n,2nγ)^{206}Pb measurements were performed, in parallel with (n,n'γ) measurements, by the detection of prompt γ rays in HPGe detectors. The main difficulty was the presence of a strong γ flash which arrives only a few μsec before the fastest neutrons (2.5 μsec for 20 MeV neutrons). For the reactions corresponding to high-energy neutrons, pile up of pulses from the γ flash together with events from neutron-induced γ rays may occur. These pile ups increase significantly the dead time, which can only be avoided at the expense of a strong deterioration of the energy resolution. To solve these problems we have developed a data acquisition system based on digital pulse processing techniques[1]. Signals from the preamplifier of the HPGe crystals were sampled at a rate of 65 MS/s using 14-bit Flash ADC modules. The samples were analysed on line by a built-in-Field-Programmable Gate Array (FPGA). For the energy determination, the trapezoïdal shaping technique was used. The algorithm for time determination consisted in a digital timing filter followed by a digital constant fraction discriminator. The lead sample (90g) enriched to 92% in ^{207}Pb was placed at 200 m from the neutron source. For the ^{207}Pb(n,n'γ) measurements, conventional electronics[2] was employed, while for the (n,2nγ) measurements two HPGe detectors (70% and 100% rel. eff.) among the three used, were also read out with digital electronics. The neutron flux was measured with a ^{235}U fission-chamber.

Figure 1 shows the cross-section data between 6 and 20 MeV neutron energies for the main γ transition (803 keV, $2^+ \rightarrow 0^+$) in ^{206}Pb. We compare our preliminary data to the results obtained by Vonach et al.[3] and to the TALYS[4] and EMPIRE-II[5] code predictions. They present a slight difference in shape and a difference in values of about 20 % with the code predictions and the previous data. Using model calculations[4,5], we have deduced the total (n,2n) cross-section values. These values are in very good agreement with the data of Frehaut et al.[6].

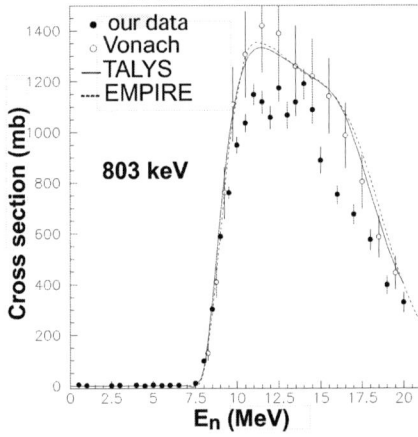

FIGURE 1. Cross section for the $2^+ \rightarrow 0^+$ (803 keV) transition in ^{206}Pb and comparison to the data of Vonach and to code predictions. Contributions of ^{206}Pb(n,n'γ) and ^{208}Pb(n,3nγ) are subtracted.

REFERENCES

1. S. Lukić, PhD Thesis, Strasbourg University (2004), and P. Medina et al., Electronic Tech. Group, IReS.
2. L.C. Mihailescu et al., Nucl. Inst. & Meth. in Phys. Res. A 531, 375 (2004).
3. H. Vonach et al., Phys. Rev. C50, 1952 (1994).
4. A. J. Koning et al., "TALYS : Comprehensive nuclear reaction modelling", Proc. of the Int. Conf. on Nuclear Data for Science and Technology, Santa Fe, USA, Sept. 26-Oct. 1, 2004 (AIP Conf. Proc., Vol. 769) p. 1154.
5. M. Herman, "EMPIRE-II", http://www-nds.iaea.org/empire/ and I. Raškinytė and H. Leeb, n_TOF report.
6. J. Frehaut et al., Proc. on « Symp. on Neutron Cross Section from 10-50 MeV » May 12, 1980, Upton, LI, USA.

Anomalous neutron radiative capture in ^{197}Au revisited

M. Krtička *, F. Bečvář*, M. Heil†, F. Käppeler†, R. Reifarth†,**,
I. Tomandl‡, F. Voss† and K. Wisshak†

Charles University, Faculty of Mathematics and Physics, 180 00 Prague, Czech Republic
†*Forschungszentrum Karlsruhe, Postfach 3640, D-76021 Karlsruhe, Germany*
**Los Alamos National Laboratory, Los Alamos, New Mexico 87545, USA*
‡*Nuclear Physics Institute of the Czech Academy of Sciences, 250 68 Řež, Czech Republic*

Abstract. Behavior of the photon strength function of ^{198}Au is treated using a novel approach based on studying γ cascades following the thermal and keV neutron capture in ^{197}Au.

Keywords: Neutron capture, γ cascades, photon strength function, pygmy resonance
PACS: 21.10.Pc, 21.10.Ma, 23.20.Lv 24.30.Gd

Although the observation of unusual behavior of neutron capture in heavy nuclei, including ^{197}Au, dates from 1953 [1], the existing experimental data on this phenomenon seem still problematic. In particular, regarding the ^{197}Au(n,γ)^{198}Au reaction, it is not quite clear whether the photon strength function $S_\gamma(E_\gamma)$, governing the emission of capture γ rays, displays near energy of ≈ 5.5 MeV the so-called pygmy resonance, or a mere irregularity having nothing to do with any nuclear vibrational mode. The pygmy resonance in ^{198}Au has been observed, e.g., in Ref. [2], while more recent data from the average resonance neutron capture in ^{197}Au [3] do not show any its traces. In this context we note that during last years the attention has been turned to exotic $E1$ soft vibrational modes of heavy nuclei, see, e.g., Ref. [4]. One of such modes may be represented by out-of-phase oscillation of the neutron excess and the $N = Z$ core [5].

In order to get new data on S_γ we studied γ cascades following the neutron capture in ^{197}Au. We measured *singles* capture γ-ray spectra for individual γ multiplicities m. These measurements were undertaken at neutron energies of 90-100 keV using the 4π BaF$_2$ γ calorimeter [7] installed at the Karlsruhe Van de Graaff accelerator. In addition, with the aid of the sum-coincidence setup at the Řež reactor [8], we undertook measurements of spectra of γ rays belonging to *all* Two-Step Cascades (TSCs) initiating at the thermal neutron capturing state of the ^{197}Au+n system and ending at preselected levels in ^{198}Au. These TSC spectra were retrieved for 15 well-resolved levels in ^{198}Au below 810 keV. Examples of the spectra from both measurements are shown in Fig. 1.

All spectra were compared with Monte Carlo predictions based on the validity of the statistical model. The γ cascades were generated using the DICEBOX algorithm [10]. The response of the γ calorimeter was simulated with the aid of the GEANT code [9]. For generating γ cascades three options for S_γ were considered: (i) the temperature-dependent expression for S_γ proposed by Kadmenkij, Markushev and Furman (KMF), Ref. [11]; (ii) the sum of the KMF strength function and a pygmy resonance term characterized by $\Gamma_{PY} = 1.7$ MeV, $E_{PY} = 5.9$ MeV and an optimum choice of the peak

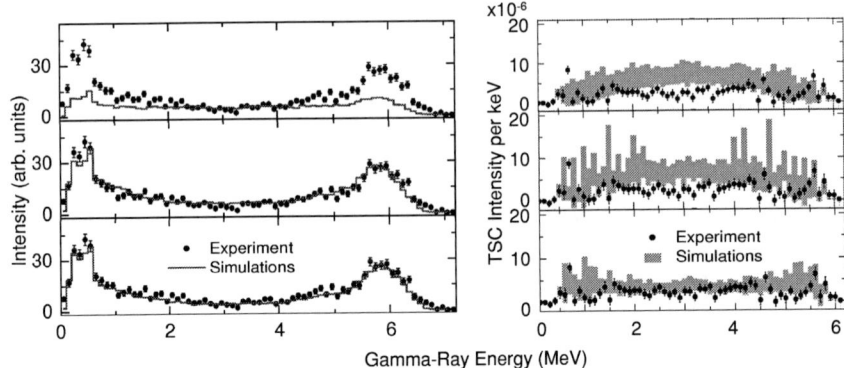

FIGURE 1. Left part: a comparison of the measured singles γ-ray spectrum for $m = 2$ with its DICE-BOX/GEANT simulations (capture of neutrons with $E_n = 90$-100 keV). Right part: an analogous comparison for the spectrum of TSCs populating the 381 keV, $J^\pi = 3^+$ level in ^{198}Au (DICEBOX simulations, the thermal neutron capture). The upper, middle and lower panels refer, respectively, to options (i-iii) for S_γ. Shaded areas represent simulated TSC intensities together with their rms Porter-Thomas uncertainties.

photoabsorption cross section $\sigma_{PY} = 8$ mb; (iii) the KMF strength function multiplied by a depression factor Q which takes a constant value of 0.2 for energies $E_\gamma < 4$ MeV and then it increases linearly, reaching at $E_\gamma = 5.8$ MeV a limiting value $Q = 1$. In the KMF expression for S_γ the Migdal factor $\sqrt{(1+2f_1'/3)/(1+2f_0')}$ was each time adjusted to reproduce the known value of $\Gamma_\gamma = 139$ meV for the 4.9 eV neutron resonance of ^{197}Au.

As is seen from Fig. 1, it is unlikely that S_γ of ^{198}Au exhibits a pygmy resonance. This conclusion is corroborated by our detailed analysis which shows that option (iii) leads to almost ideal predictions of all the TSC spectra measured and that this is not the case for options (i-ii). If a pygmy resonance exists, its σ_{PY} will be evidently <8 mb, i.e. much smaller than the value of 30 mb known from Ref. [2]. The neutron capture in ^{197}Au seems, indeed, anomalous, but it is still not clear whether the $E1$ photon strength is abruptly depressed or only locally redistributed at energies near 5.5 MeV.

This work has been partly supported by the Grant Agency of the Czech Republic under contract No. 202/03/P136.

REFERENCES

1. G. A. Bartholomew, B. B. Kinsey, *Canad. J. Phys.* **79**, 1025 (1953).
2. M. Igashira *et al.*, *Nucl. Phys.* **A457**, 301 (1986).
3. S. F. Mughabghab, C. L. Dunford, *Phys. Lett.* **B 487**, 155 (2000).
4. J. Enders *et al.*, *Acta Physica Polonica B* **36**, 1077 (2005).
5. R. Mohan, M. Danos, L. C. Biedenharn, *Phys. Rev. C* **3**, 1790 (1971).
6. S. Joly, D. M. Drake, L. Nilsson, *Phys. Rev. C* **20**, 2072 (1979).
7. K. Wisshak *et al.*, *Nucl. Instr. Methods A* **292**, 595 (1990).
8. J. Honzátko *et al.*, *Nucl. Instr. Methods A* **376**, 434 (1996).
9. M. Heil *et al.*, *Nucl. Instr. Methods A* 459, **229** (2001).
10. F. Bečvář, *Nucl. Instr. Methods A* **417**, 434 (1998).
11. S. G. Kadmenskij, V. P. Markushev, V. I. Furman, *Sov. J. Nucl. Phys.* **37**, 165 (1983).

New method for the determination of accurate gamma-ray intensities for the $^{14}N(n,\gamma)^{15}N$ high energy standard

T. Belgya

Institute of Isotopes, Chemical Research Centre, HAS H-1525 Budapest, Hungary)

Abstract. The $^{14}N(n,\gamma)^{15}N$ reaction is a primary source of high energy gamma-rays for use in calibration of detectors for other neutron-capture reactions. The gamma ray-intensities of ^{15}N produced by thermal neutron capture and the gamma-ray detection efficiency function have been simultaneously determined from gamma peak areas alone using the basic principle of an intensity balance. A least square fit was made to a new type of intensity balance calculation, combined with traditional efficiency fitting of radioactive sources. This latter ensures the compatibility with low energy efficiencies, while providing an unbiased efficiency function for higher (up to 10 MeV) gamma-ray energies. The calculation is based on the assumption that the ^{15}N decay scheme is complete. From the internal consistency of the resulting intensities, it is believed that they are more accurate than previously published values. The same is true for the derived efficiency function.

INTRODUCTION

Accurate determination of the full-energy-peak efficiency (FEPE) calibration of modern germanium gamma-ray detectors in the energy range of 0.1-10 MeV is a long-standing problem[1-3]. To determine the FEPE one needs to have radioactive sources with precisely known intensities. Unfortunately the commonly available, well calibrated sources only go up to a gamma energy of 2.7 MeV (^{24}Na source). To determine intensities for higher energy sources, one would need a good efficiency calibration and vice versa, so Raman et al. hit the nail on the head when they characterized this problem as the chicken-egg dilemma[1].

To solve this problem, We have decided to work out a new methodology for FEPE determination, which is independent from any previous results, using the thermal neutron capture reaction. First we applied this methodology to determine the relative gamma ray intensities from the $^{14}N(n,\gamma)^{15}N$ reaction.

MEASUREMENTS & DATA EVALUATION

The neutron-capture experiments using a mass of 1.194 g of urea-d (ND_2COND_2) were performed at the PGAA-NIPS facilities of the Budapest Research Reactor[4]. The $^{14}N(n,\gamma)^{15}N$ spectrum was accumulated for about 264,000 s with a count rate of about 43 cps and used for the new method. To obtain reliable results for the low energy efficiency, we included efficiency data determined for each recommended peak from measurements performed on radioactive sources. We used commercial ^{60}Co, ^{133}Ba, ^{152}Eu and ^{207}Bi standards and home made ^{110m}Ag and ^{182}Ta sources.

All of the measured spectra used in this work were evaluated with the Hypermet PC program package[5] developed in our department. This program provided peak areas and positions, as well as their statistical uncertainties.

THE NEW METHOD AND RESULTS

Let us consider a decay scheme of a nucleus with a number of levels n, up to and including the capture state. In the capture reaction the gamma source is the capture state while the drain is the ground state. We will call a decay cascade any decay path which connects the capture state to the ground state with a consecutive series of two or more non-parallel transitions. First we prove that the sum of gamma-ray intensities crossing any single line –the so called crossing line – between any two neighboring states is a constant C. The crossing intensity means the intensity of a gamma ray, which crosses a crossing line, and the crossing intensity sum (CIS) is the sum of all such crossing intensities for a given crossing line. The CIS represent the continuity equations for this discrete problem.

It is easy to see that any one decay cascade going from the capture state to the ground state will increase each of the n-1 crossing intensity sums by 1 (the intensity of one photon is 1). Thus if we repeat the capture excitation process C times then the value of each of the n-1 crossing intensity sums will be C. The experimental intensity $I_{i,j}$ of a certain gamma ray de-exciting level i and feeding level j can be expressed with the inverse FEPE $\varepsilon^{-1}(E_{i,j})$ and the peak area $A_{i,j}$ as

$$I_{i,j} = A_{i,j}\varepsilon^{-1}(E_{i,j})$$

The peak areas are corrected for the self absorption in the sample and for the electron conversion. To find out the unknown intensities and the detector efficiency function simultaneously, we can utilize the CIS method. The crossing intensity sum for crossing line with index of f (f=1, 2, ..., n-1) is

$$S_f = \sum_{\substack{i>f \\ j \leq f}} I_{i,j} = \sum_{\substack{i>f \\ j \leq f}} A_{i,j}\varepsilon^{-1}(E_{i,j}) = C$$

Instead of solving the direct CIS equations, which can be over determined we minimize the following least square function, which is set up using the inverse efficiency for both the gamma lines of the radioactive sources and for the crossing intensity sums

$$X^2 = \sum_f (S_f - C)^2/\sigma_f^2 + \sum_m \frac{\left(\varepsilon_m^{-1*} - \varepsilon^{-1}(E_{i,j})\right)^2}{\sigma_m^2}$$

The least square function consists of two terms, the first represents the CIS principle and the other uses radioactive sources to fix the low energy efficiency, where there are no strong transitions in the ^{14}N(n,γ) reactions. In this way we can obtain a functional representation of the detector efficiency and determine the intensities at the same time. In our concrete case the intensities $I_{i,j}$ belong to the transitions de-exciting ^{15}N, the $\varepsilon^{-1}(E_{i,j})$ is the unknown inverse FEPE function, the ε_m^{-1*} is the mth inverse detector efficiency value determined from one of the corresponding radioactive source gamma-lines and σ_m^2 is its uncertainty. The σ_f^2 values have to be calculated from the uncertainties of the corrected peak areas $A_{i,j}$.

The least square equation was then solved by a computer program we have developed for this problem. We found that above 2.7 MeV our resulting efficiency curve values were smaller than the efficiency curve obtained from ^{15}N intensities published in Jurney et al.[2] and the difference reached 6% at about 8 MeV. This difference can be demonstrated by the ratios of the intensities of the strongest ^{15}N transitions in Fig. 1.

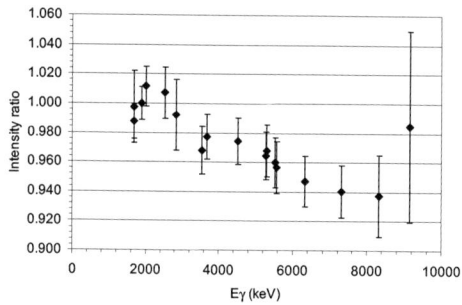

FIGURE 1. Ratios of intensities for the strongest ^{14}N capture gamma rays. Intensities of Jurney et al.[2] were divided by intensities from this work. The two data sets were normalized to the 1884 keV intensity.

The influence of these new nitrogen intensities is substantial for many nuclear data, including capture cross sections, gamma-ray production cross sections and derived quantities. Validation of these new findings and development of the least square function by including correlations between intensities are in progress.

The author is grateful for the support of the Bolyai scholarship of the Hungarian Academy of Sciences.

REFERENCES

1. S. Raman, C. Yonezawa, H. Matsue, H. Iimura, and N. Shinohara, Nucl. Instr. & Methods **A 454** (2-3), 389-402 (2000).
2. E.T. Jurney, J.W. Starner, J.E. Lynn, and S. Raman, Phys. Rev. **C 56**, 118-134 (1997).
3. T. J. Kennett, W. V. Prestwich, and J. S. Tsai, Nucl. Instr. and Methods **A 249**, 366-378 (1986); G. L. Molnár, Z. Révay, and T. Belgya, Nucl. Instr. & Methods **A 489** (1-3), 140-159 (2002).
4. Z. Révay, T. Belgya, Z. Kasztovszky, J. L. Weil, and G. L. Molnár, Nucl. Instr. & Methods **B 213**, 385-388 (2004).
5. B. Fazekas, T. Belgya, L. Dabolczi, G. Molnar, and A. Simonits, Journal of Trace and Microprobe Techniques **14** (1), 167-172 (1996).

Simultaneous measurement of (n,γ) and (n,fission) cross sections with the DANCE 4π BaF$_2$ array

T.A. Bredeweg[*], M.M. Fowler[*], J.A. Becker[†], E.M. Bond[*],
M.B. Chadwick[*], R.R.C. Clement[†], E.-I. Esch[*], T. Ethvignot[**],
T. Granier[**], L.F. Hunt[*], R.A. Macri[†], J.M. O'Donnell[*], R.S. Rundberg[*],
J.M. Schwantes[*], J.L. Ullmann[*], D.J. Vieira[*], J.B. Wilhelmy[*],
J.M. Wouters[*], C.-Y. Wu[†] and J.E. Yurkon[‡]

[*]*Los Alamos National Laboratory, Los Alamos, NM, 87545, USA*
[†]*Lawrence Livermore National Laboratory, Livermore, CA, 94550, USA*
[**]*CEA-DAM, BP 12, Bruyères-le-Châtel, 91680, France*
[‡]*National Superconducting Cyclotron Laboratory, East Lansing, MI, 48824, USA*

Abstract. Neutron capture cross section measurements on many of the actinides are complicated by low-energy neutron-induced fission, which competes with neutron capture to varying degrees depending on the nuclide of interest. Measurements of neutron capture on ^{235}U using the Detector for Advanced Neutron Capture Experiments (DANCE) have shown that we can partially resolve capture from fission events based on total photon calorimetry (i.e. total γ-ray energy and γ-ray multiplicity per event). The addition of a fission-tagging detector to the DANCE array will greatly improve our ability to separate these two competing processes so that improved neutron capture and (n,γ)/(n,fission) cross section ratio measurements can be obtained. The addition of a fission-tagging detector to the DANCE array will also provide a means to study several important issues associated with neutron-induced fission, including (n,fission) cross sections as a function of incident neutron energy, and total energy and multiplicity of prompt fission photons. We have focused on two detector designs with complementary capabilities, a parallel-plate avalanche counter and an array of solar cells.

Keywords: fission-tagging, neutron capture, neutron induced fission
PACS: 29.40.-n, 25.85.-w

INTRODUCTION

The competition between (n,γ) and (n,fission), and its dependence on incident neutron energy is of great interest to Stockpile Stewardship and nuclear attribution as well as for advanced reactor design and the advanced fuel cycle initiative. For the actinides, there is essentially nothing known on the properties of prompt fission γ-rays as a function of incident neutron energy. Also, many of the existing cross section measurements are quite old and may be subject to systematic errors.

Two classes of experiments, each with specific goals can be envisaged. Both require a compact detector design capable of fitting inside the DANCE array. The first class would be high resolution measurements of both the capture and fission cross sections over the range of neutron energies available at the Lujan Center. This would require a detector with very high detection efficiency for fission fragments to cleanly separate

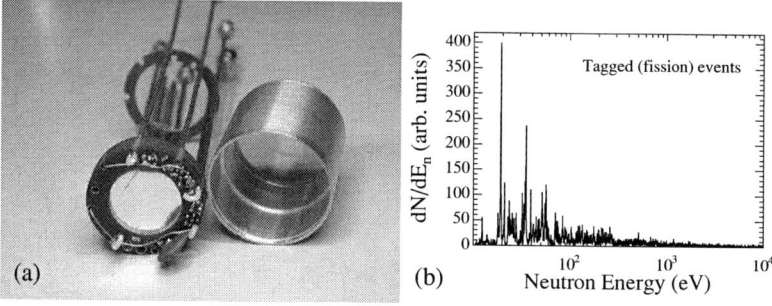

FIGURE 1. Panel (a): PPAC fission-tagging detector along side a standard DANCE target holder. Panel (b): ^{235}U(n,fission) neutron energy spectrum for PPAC tagged (fission) events.

capture from fission events. The second class would be a study of fission specific phenomena for neutron-induced fission as a function of neutron energy. It may also be possible to study these quantities as a function of the fission fragment mass asymmetry and orientation of the scission axis. These measurements would require a detector that provides some degree of angular information and has reasonable energy resolution for fission fragments.

INITIAL TEST RESULTS

Both of the designs mentioned above have recently been tested inside the DANCE array. In both cases the results for tagged (fission) events have been consistent with expectations for fission fragment detection efficiency, energy resolution and timing. Panel (a) of Fig. 1 shows the PPAC detector next to a standard DANCE target holder. Panel (b) is the neutron energy spectrum for "tagged" ^{235}U(n,fission) events obtained using the PPAC detector. The target was a thin electro-deposit of 93 μg/cm^2 ^{235}U onto a 220 μg/cm^2 aluminized mylar foil. Initial results this PPAC test have shown a 90+% efficiency for detecting at least one fission fragment, however, more work is needed to fully characterize the detectors performance. These tests have also shown that both fission-tagging detector designs provide clean discrimination against α-particles from the target material, which is an important consideration for many of the actinides. A suite of production measurements of neutron capture and neutron induced fission on $^{241,243\&242m}$Am are planned for the Fall 2005 LANSCE run cycle.

ACKNOWLEDGMENTS

Work performed under the auspices of the U.S. Department of Energy by the University of California, Los Alamos National Laboratory (W-7405-ENG-36) and Lawrence Livermore National Laboratory (W-7405-ENG-48). Work benefited from use of Los Alamos Neutron Science Center (W-7405-ENG-36). Work benefited from use of the National Superconducting Cyclotron Laboratory supported by NSF grant No. PHY-0110253.

Classical chaos in the interacting boson model

Michal Macek, Pavel Cejnar

Institute of Particle and Nuclear Physics, Charles University, Prague

Abstract. We explore classical dynamics of quadrupole deformations within the space of control parameters of the interacting boson model at $L = 0$. Poincaré sections are used to visualize regions of regular and chaotic motion in the phase space. Changes of stability of certain trajectories are observed while crossing the Alhassid-Whelan arc of regularity in the parameter space. Periodic orbits are related to the observed fluctuations of level density.

We study the classical limit of the IBM Hamiltonian $\hat{H} = \frac{\eta}{N}\hat{n}_d - \frac{1-\eta}{N^2}(\hat{Q}_\chi \cdot \hat{Q}_\chi)$ at zero angular momentum (see also Ref.[1]) with the aim (i) to analyze details of dynamics in regular and chaotic regions of the parameter space and (ii) to search for effects related to strong fluctuations of the level density. Note that the model shows fully regular dynamics in its dynamical symmetries and along the O(6)-U(5) transition, but semiregular behavior was also detected along the Alhassid-Whelan "arc" $\chi = \chi_{\text{reg}}(\eta)$ inside the Casten triangle [2] (the explanation of this phenomenon still being unknown). Moreover, anomalous bunchings of levels observed in various parameter regions—namely along the O(6)-U(5) transition [3] and along the semiregular arc [4]—indicate exceptional influence of some stable periodic orbits (relationship given by Gutzwiller and Berry-Tabor trace formulas or their generalizations [5]).

To visualize regular and chaotic motions in the phase space, we use the method based on Poincaré sections—see an example in Fig. 1. We have performed an extensive scan of classical dynamics for different control parameters (η, χ) and energies, and created a large collection of phase-space pictures [6]. In general, the disclosed interplay of chaotic and regular features is rather complicated. One can, nevertheless, recognize three basic types of low-energy behaviors corresponding to parameter regions below, at, and above the semiregular arc: (i) Between the SU(3)-U(5) edge and the arc, stable γ-vibrations represent the dominant low-energy periodic motion, while β-vibrations are unstable. (ii) At about $\chi = \chi_{\text{reg}}(\eta)$, the stability properties of both kinds of vibrations are exchanged. (iii) In the direction toward the O(6)-U(5) edge from the arc, both modes become unstable and chaotic behavior prevails.

At $\chi = \chi_{\text{reg}}(\eta)$, we observe a regular island in the Poincaré sections at $E \approx 0$ which disappears at lower and higher energies as well as in the neighboring regions of the Casten triangle. This is shown in Fig. 2. The periodic trajectory inside the island represents vibrations predominantly in the γ-direction that seem to be responsible (through semiclassical trace formulas) for the $E \approx 0$ level-density fluctuation observed along the semiregular arc (see Fig. 9 in Ref.[4]). Note that a similar level bunching in the O(6)-U(5) region was subject to a detailed study in Ref.[3].

The reader is encouraged to visit our web page [6] where a more complete (and updated) collection of results is shown.

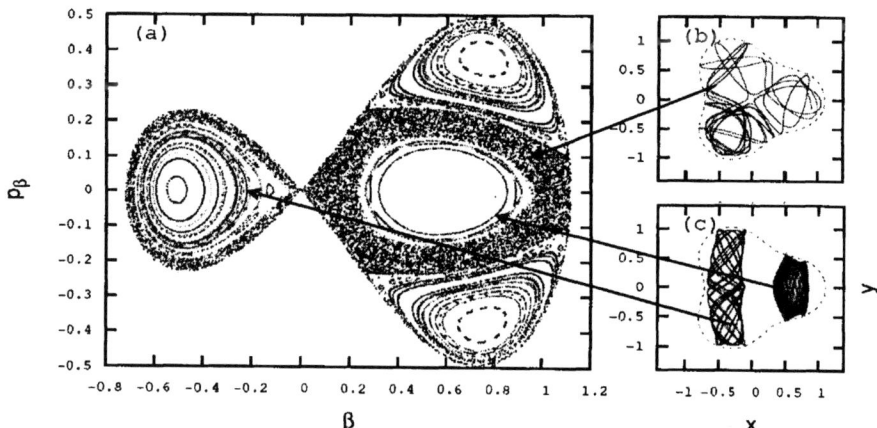

FIGURE 1. Poincaré section of the IBM classical phase space at $\eta = 0.7, \chi = -0.75$ (semiregular arc) and $E = 0$. Panel (a) shows a plane cutting the 4D phase space at $\gamma = 0$, points mark passages of 120 trajectories through the plane. Examples of individual trajectories are given in the rhs panels. Points left behind by regular trajectories lie on topological circles, while chaotic trajectories create an ergodic "sea". The picture demonstrates the coexistence of regular and chaotic domains in the phase space.

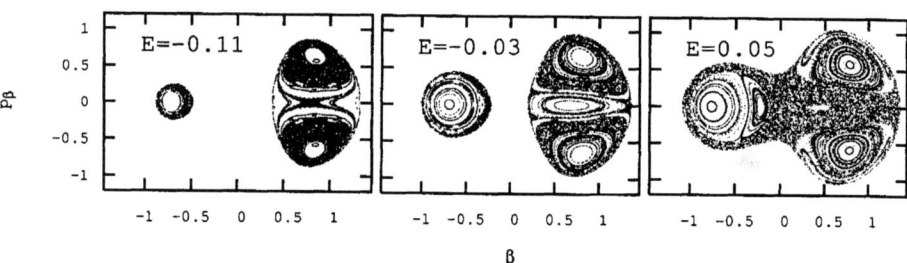

FIGURE 2. Poincaré sections in the semiregular arc at $\eta = 0.5$ for three different energies close to $E = 0$. The central panel shows a regular island around $\beta \approx 0.7, p_\beta = 0$. The existence of a stable central periodic orbit inside this island is presumably responsible for the $E \approx 0$ level bunching pattern observed in the semiregular arc. At lower or higher energies, the trajectory is unstable—see the lhs and rhs panels.

REFERENCES

1. P. Cejnar et al., contribution in these Proceedings.
2. Y. Alhassid and N. Whelan, Phys. Rev. Lett. **67**, 816 (1991).
3. S. Heinze, P. Cejnar, J. Jolie, and M. Macek, nucl-th/0504016; M. Macek, P. Cejnar, J. Jolie, and S. Heinze, nucl-th/0504017 (submitted to Phys. Rev. C).
4. P. Cejnar and J. Jolie, Phys. Rev. E **61**, 6237 (2000).
5. H.-J. Stöckmann, *Quantum Chaos. An Introduction* (Cambridge University Press, UK, 1999).
6. http://www-ucjf.troja.mff.cuni.cz/~geometric.

Measurement of Disintegration Rates and Absolute γ-ray Intensities

Daniel J. DeVries and Henry C. Griffin

Department of Chemistry, University of Michigan, Ann Arbor, MI 48109

Abstract. The majority of practical radioactive materials decay by modes that include γ-ray emission. For questions of "how much" or "how pure," one must know the absolute intensities of the major radiations. We are using liquid scintillation counting (LSC) to measurements of disintegration rates, coupled with γ-ray spectroscopy to measure absolute γ-ray emission probabilities. Described is a study of the ^{227}Th chain yielding absolute γ-ray intensities with ~0.5% accuracy and information on LSC efficiencies.

Keywords: Th-227; Gamma-ray intensity; Liquid scintillation counting

Gamma-ray spectroscopy is perhaps the most common technique for characterizing radioactive materials. Liquid scintillation analysis is a widely used method for determining disintegration rates of α and β- emitting samples. Coupling these techniques can result in a powerful method for determining absolute γ-ray emission probabilities for an entire decay chain. The uncertainties associated with measurements obtained with this method depend greatly on the accuracy and precision of the detection efficiencies used. Using liquid scintillation counting to determine the disintegration rate of α samples is attractive because it is a 4π counting method. Alpha decays are essentially mono-energetic, with decay energies of several MeV, resulting in essentially 100% detection efficiency [1]. LSC for β- decay count rate analysis is more problematic, due to the continuous distribution of β- particle energies, resulting in detection efficiencies of less than 100%.

A 36% Princeton Gamma-Tech HPGe detector was used, employing a standard counting geometry with a sample holder giving reproducible (<0.2% variation) detection efficiencies. Gamma detection efficiencies are based on ^{60}Co as a primary standard, and uncertainties of ~1% at 100 keV and <0.5% at 1 MeV are inferred [2,3]. A Packard Tri-Carb 1900TR liquid scintillation analyzer was used for LSC work. Commercial sources of ^{3}H and ^{14}C gave LSC detection efficiencies of 61.0% and 96.7% respectively.

Solvent extraction based on TTA (thenyl-trifluoroacetone) was used to isolate and purify ^{227}Th from ^{227}Ac. The time of the last effective separation (last removal of daughter nuclei) was recorded, since after this point the purified sample undergoes simple growth and decay. Several independent characterizations of the γ sample, prepared from the stock solution, were made by γ-ray spectroscopy. Three LSC samples were prepared from the stock solution through a dilution step, used reduce the uncertainty in small mass measurements, and each was counted 50 times over a period of ~10 days to determine their absolute disintegration rates. Growth and decay calculations were used to find the count rates of ^{227}Th, ^{223}Ra, and their descendants for each sample. This total count rate was then fit to the observed count rate by minimizing the sum of the squares of the residuals to find the optimal values for the initial activities of ^{227}Th, ^{223}Ra, and also to determine the β- detection efficiency of the LSC system. Each optimized activity was divided by its mass to give three independent determinations of the activity per gram of the stock solution at the time of last effective separation. Table 1 shows the results of the fitting process.

TABLE 1. Initial stock solution disintegration rates of ^{227}Th and ^{223}Ra, and the β^- detection efficiency, obtained from the fitting method; including statistical uncertainties. The reduced χ^2 for each fit is also shown.

	Th CPM/g at T=0	Ra CPM/g at T=0	Beta Efficiency	Reduced χ^2
Sample-1	675218 ± 870	329 ± 293	0.9890 ± 0.0036	2.82
Sample-2	672732 ± 812	587 ± 235	0.9901 ± 0.0029	1.86
Sample-3	672904 ± 867	452 ± 264	0.9911 ± 0.0032	2.08
	673579 ± 490	**518 ± 160**	**0.9901 ± 0.0018**	

When using an indiscriminant counting system like LSC, with growth and decay calculations to determine disintegration rates, purity of the sample is paramount. Extra ^{223}Ra activity, not removed during purification, would cause the fitting method to return a high initial ^{227}Th disintegration rate. Extra activity due to ^{211}Pb or ^{211}Bi contamination would have a similar effect, though only on the first few counts due to their shorter half-lives. Purity of the stock solution was verified two ways. First, the initial ^{223}Ra activity, computed from the fitting of LSC data and shown in Table 1, indicated an activity of 518±160 CPM/g. Second, γ analysis of the 269 keV radium photo peak from the first γ count (134 min live time, started 90 min after the last effective separation) yielded a disintegration rate of 3459±273 DPM/g. Growth and decay calculations gave a rate of 2546±2 DPM/g, based on decay of the initial thorium activity. The excess 913±297 DPM/g due to radium is in agreement with the LSC determined amount, and when combined, give a weighted average of 722±155 DPM/g due to radium (<0.15% of the thorium rate). Confident in its purity, the initial ^{227}Th disintegration rate was then applied to the γ counting samples, and yielded the absolute intensities shown in Table 2.

TABLE 2. The last three γ counts showing absolute intensities for select transitions in the chain. Each count is an independent measurement of the absolute γ intensity for each transition. Intensities are based on the absolute disintegration rate of the stock solution, as determined by LSC, the photo peak statistics, and the γ detection efficiency. Uncertainties are purely statistical and do not account for γ detection efficiencies.

Nuclide	Energy(keV)	Count # 6	Count # 5	Count # 4	Weighted Average	NDS Intensity
Th-227	236.23	12.510 ± 0.015	12.481 ± 0.014	12.480 ± 0.014	12.481 ± 0.007	11.2 ± 0.6 [4]
Th-227	300.22	2.143 ± 0.005	2.147 ± 0.004	2.138 ± 0.004	2.145 ± 0.002	2.0 ± 0.2 [4]
Ra-223	269.70	13.373 ± 0.024	13.431 ± 0.025	13.372 ± 0.030	13.397 ± 0.014	13.90 ± 0.28 [6]
Ra-223	324.11	3.658 ± 0.012	3.659 ± 0.013	3.689 ± 0.016	3.665 ± 0.007	3.99 ± 0.07 [6]
Rn-219	271.46	10.800 ± 0.021	10.833 ± 0.022	10.839 ± 0.027	10.825 ± 0.012	10.80 ± 0.22 [6]
Rn-219	402.01	6.610 ± 0.017	6.620 ± 0.019	6.663 ± 0.023	6.644 ± 0.010	6.59 ± 0.22 [6]
Po-215	439.04	0.053 ± 0.003	0.058 ± 0.003	0.056 ± 0.004	0.056 ± 0.002	0.04 ± [6]
Pb-211	405.04	3.998 ± 0.014	4.017 ± 0.014	4.013 ± 0.018	4.009 ± 0.008	3.78 ± 0.05 [7]
Pb-211	831.85	3.436 ± 0.016	3.423 ± 0.017	3.436 ± 0.021	3.434 ± 0.010	3.52 ± 0.05 [7]
Bi-211	351.23	13.672 ± 0.025	13.854 ± 0.027	14.100 ± 0.033	13.976 ± 0.015	12.91 ± 0.11 [7]
Corr. Bi-211	351.23	13.107 ± 0.096	13.066 ± 0.133	12.970 ± 0.189	13.068 ± 0.070	12.91 ± 0.11
Po-211/Tl-207	569.21	0.009 ± 0.002	0.010 ± 0.002	0.015 ± 0.003	0.011 ± 0.001	0.007 ± 0.001 [5]
Po-211/Tl-207	897.66	0.272 ± 0.005	0.267 ± 0.005	0.273 ± 0.007	0.271 ± 0.003	0.266 ± 0.020 [5]

The internal agreement within each set of counts is good; ~75% are within 1σ of the weighted average, and typical uncertainties are ~0.2%. Deviations from the literature values show no systematic effects, with ratios of these values to NDS differing by ~4%. The uncorrected values for ^{211}Bi show a clear trend; ascribed to a weak 0.85% intense (relative to 100% for the 236 keV γ-ray) 350.5 keV γ-ray from ^{227}Th [6]. This weak γ-ray was used to subtract the thorium contribution from the 211Bi photo peak for each count. The results, shown in Table 2 as

Corrected Bi-211, agree internally and with the literature, showing no statistically significant trend. The 12.481±0.007 reported here for the intensity of the 236 keV γ-ray in ^{227}Th does not agree with the most recent NDS evaluation of 17.5±1.4 [6], but does agree with the previously established intensity shown in Table 2. Agreement of this work with published intensities suggests that LSC is capable of reliably following disintegration rates of samples with several daughters over a period of days.

REFERENCES

1. W.J. McDowell, "Alpha Counting and Spectrometry Using Liquid Scintillation Methods," NAS-NS-3116, USDOE (1986).
2. E. Schönfeld et al., Int. J. Appl. Radiat. Iso. 56 (2002) 215.
3. H.C. Griffin and C.S. Sumithrarachchi, Trans. Amer. Nucl. Soc. 85 (2001) 234.
4. C. Maples, Nucl. Data Sheets 22 (1977) 275.
5. M.J. Martin, Nucl. Data Sheets 70 (1993) 315.
6. E. Browne, Nucl. Data Sheets 65 (1992) 669.
7. A. Artna-Cohen, Nucl. Data Sheets 63 (1991) 79.

New Band Mechanism of Doubly-Odd Nuclei Around Mass 130

Koji Higashiyama[1], Naotaka Yoshinaga[2]

[1] *Department of Physics, University of Tokyo, Hongo, Tokyo 113-0033, Japan*
[2] *Department of Physics, Saitama University, Saitama City 338-8570, Japan*

Keywords: Chopsticks configurations; Pair-truncated shell model, 134La calculated levels
PACS: 21.60.Cs, 21.60.Ev, 27.60.+j

The band structure with the $\nu h_{11/2} \otimes \pi h_{11/2}$ configuration in the doubly-odd nucleus ^{134}La is investigated within the framework of a pair-truncated shell model (PTSM). In the model, the shell-model basis states of doubly-odd nuclei are constructed by adding an unpaired neutron and an unpaired proton to the even-even core, which consists of angular momentum zero (*S*) and two (*D*) collective pairs. We use the pairing plus quadrupole type interaction, whose force strengths are determined so as to reproduce the energy levels of the even-even nuclei in the mass $A \approx 130$ region [1].

In Fig. 1, the measured spectrum based on the $\nu h_{11/2} \otimes \pi h_{11/2}$ configuration for ^{134}La is compared with the theoretical result. The energy levels for the yrast states are almost perfectly reproduced, except that in our calculation the 8_1^+ state is located between the 9_1^+ and 10_1^+ states. Concerning the yrare states, our theoretical result also provides a successful description of the energy levels, though only four levels are observed experimentally. We also calculate the ratios $B(M1;I \rightarrow I-1)/B(E2;I \rightarrow I-2)$ for the yrast states, and obtain a good agreement with experiment.

The partial level scheme of ^{134}La constructed from the theoretical results of the *M*1 and *E*2 transitions is shown in Fig. 1 The solid arrows indicate *E*2 transitions, and the dotted arrows indicate *M*1 transitions. The numerals on the right side of the *E*2 transitions denote the $B(E2;I \rightarrow I-2)$ values (in $e^2 b^2$), and those beneath the *M*1 transitions denote the $B(M1;I \rightarrow I-1)$ values (in μ_N^2). The states within four $\Delta I=2$ *E*2 bands with the bandhead states of the 8_1^+, 9_1^+, 10_1^+ and 11_1^+ are connected by the strong *E*2 transitions to the same members of the $\Delta I=2$ *E*2 bands, and by the strong *M*1 transitions to the states in the neighboring $\Delta I=2$ *E*2 bands.

In search of the microscopic origin of the electromagnetic properties, we analyze the reduced matrix elements of *M*1 operators, and compare the $B(M1)$ values for two-nucleon system of one neutron and one proton both in the same $0h_{11/2}$ orbital with those of actual calculations. We also calculate the effective angles between the angular momenta of the neutrons and protons in the $0h^{11/2}$ orbitals. It turns out that the closing of two angular momenta enhances the $B(M1)$ values, and the odd-spin yrast states (I) ($I \geq 11$) have a fully aligned configuration of two angular momenta with angular momentum 11. Similarly, the even-spin yrast states (I-1), the odd spin states (I-2) and the even spin states (I-3) are formed mainly with the configurations of angular momentum 10, 9 and 8, respectively. Schematic illustrations of these configurations (chopsticks configurations) are presented below each $\Delta I=2$ $E2$ band in Fig.~2. Since quadrupole collectivity plays an important role in describing the $\Delta I=2$ $E2$ bands, we conclude that the level scheme of ^{134}La arises from the chopsticks configurations of the unpaired nucleons in the $0h_{11/2}$ orbitals, weakly coupled with the quadrupole collective excitations of the even-even part of the nucleus. Similar situations are also obtained in ^{132}Cs, ^{130}Cs and ^{132}La. The detailed results and their analyses are presented in Refs. [3,4]

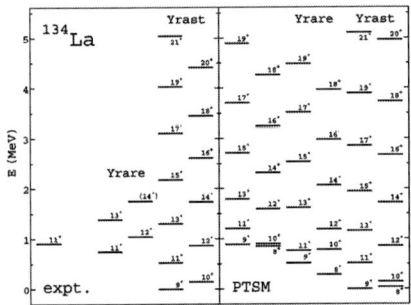

FIGURE 1. Comparison of the experimental energy spectrum (expt.) with the PTSM result (PTSM). The experimental data are taken from Ref.[2].

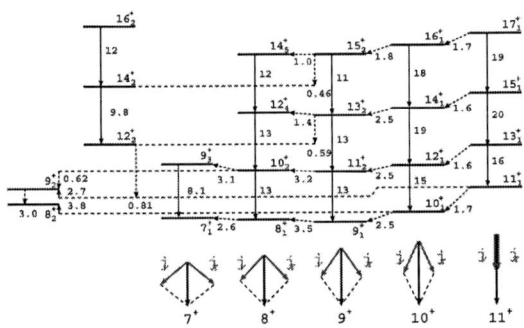

FIGURE 2. Partial level scheme of ^{134}La suggested by the PTSM calculation.

REFERENCES

1 N. Yoshinaga and K. Higashiyama, *Phys. Rev.* **C69**}, 054309 (2004).
2 R. A. Bark et al., *Nucl. Phys.* **A691**, 577 (2001).
3 K. Higashiyama and N. Yoshinaga, *Prog. Theor. Phys.* 113, 1139 (2005);
4 K. Higashiyama, N. Yoshinaga, and K. Tanabe, *Phys. Rev.* **C72**, 024315 (2005).

Measurement of Neutron Capture Cross Section for Natural Palladium

J. Hori[1], H. Yashima[1], T. Oishi[2], W. Takahashi[2], M. Baba[2], K. Nakajima[1]

[1] *Research Reactor Institute, Kyoto University, Kumatori-cho, Sennan-gun, Osaka 590-0494, Japan*
[2] *Cyclotron and Radioisotope Center, Tohoku University, Aoba, Aramaki, Aoba-ku, Sendai 980-8578, Japan*

Abstract. The neutron capture cross section of natural palladium has been measured relative to $^{10}B(n,\alpha)$ standard cross section by the neutron time-of-flght (TOF) method in the energy range of 0.01 eV to 10 keV using a detection assembly of $Bi_4Ge_3O_{12}$ (BGO) scintillators and a 46 MeV electron linear accelerator (linac) at the Research Reactor Institute, Kyoto University (KURRI). The relative cross section has been normalized at 0.0253 eV to the reference value estimated by using the cross sections of stable palladium isotopes in the JENDL-3.3 library. The present results were compared with the evaluated values in the JENDL-3.3.

Keywords: Neutron capture cross section, TOF method, Natural palladium
PACS: 25.40.Lw; 29.30.Hs; 29.40.Mc

INTRODUCTION

The precise neutron capture cross sections for long-lived fission products (LLFPs) are required for the research and development of the nuclear transmutation technology. Palladium-107 is one of the most important LLFP nuclei. It is, however, not easy to prepare the highly enriched ^{107}Pd sample. In order to make a correction for the isotope impurities in the sample precisely, it is necessary to verify the capture cross section data of the stable palladium isotopes, which would exist in the sample as impurities. In this work, the measurement of the capture cross section of natural palladium was carried out.

EXPERIMENTAL PROCEDURE

The measurement has been performed by the neutron TOF method using the 46 MeV linac at the KURRI. A natural palladium metal disk of $1.8 \times 1.8 \times 0.1$ cm^3 was used as a sample. Prompt gamma-rays from the capture sample were detected with a total absorption spectrometer consisting of twelve BGO scintillators [1], which located at 12.7 ± 0.02 m from the neutron source. Pulsed neutrons were produced from a water-cooled Ta target in a water tank as a neutron moderator. The linac was operated with two different modes: one was for the measurement at low energy region below 20 eV with a repetition rate of 50 Hz, a pluse width of 3 µs, averaged current of 78 µA, and the other is for the measurement at high energy regions above about 2 eV with a repetition rate of 200 Hz, a pluse width of 47 ns, averaged current of 45 μA. The signals from the BGO spectrometer were fed into the time digitizer though the amplifier and the discriminator. The start time of the time digitizer was triggered by the linac electron burst and the dwell time was set to 2.0 or 0.125 μs with a spectrum length of 8192 channel. The data taking with a digital signal processing (DSP) technique was also applied and the timing for each signal was derived though the analysis of the signal waveform. The background measurement has been made without the sample and the background level has been confirmed by using the notch-filters of In, Co and Mn and by the block off method with a thick ^{10}B plug (4.54 g/cm^2). In addition, the background due to scattered neutron has been experimentally estimated by using a graphite sample and the correction function for neutron scattering and/or self-shielding was calculated by using the Monte Carlo code MCNP-4C.

RESULTS AND DISSCUSSION

The TOF spectra with the time digitizer and the DSP technique were compared in high energy region. Though the baseline undershoot was usually observed in about 20 μs from the gamma burst, the current comparison confirmed that the data taking with the time digitizer was correctly performed in the energy region below 10 keV as shown in Fig. 1. The relative measurement of the ^{nat}Pd(n,γ) reaction has been normalized at 0.0253 eV to the evaluated value in JENDL-3.3. Figure 2 shows the current result comparing with the evaluated cross sections in JENDL-3.3. For the integrated cross section of a few resonances, the discrepancies were observed by a factor of 0.8~3.5.

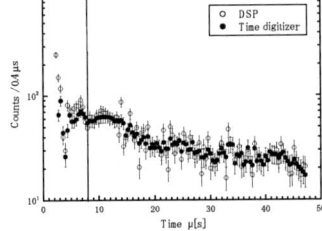

FIGURE 1. Comparison of TOF spectra with the time digitizer and the DSP technique.

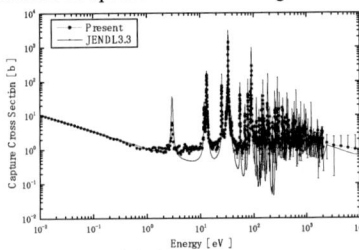

FIGURE 2. Comparison of the JENDL-3.3 with the current measurement for the ^{nat}Pd(n,γ) reaction.

REFERENCES

1. S. Yamamoto et al., *Nucl. Instr. Meth.*, **A249**, 484 (1986).

Systematic Measurement of keV-neutron Capture Cross Sections and Capture Gamma-ray Spectra of Sn Isotopes

J. Nishiyama*, M. Igashira*, T. Ohsaki*, G. N. Kim[†], W. C. Chung** and T. I. Ro**

Research Laboratory for Nuclear Reactors, Tokyo Institute of Technology, 2-12-1-N1-26 O-okayama, Meguro-ku, Tokyo 152-8550, Japan
[†]*School of Physics and Energy Science, Kyungpook National University, 1370 Sankyok-dong, Buk-gu, Daegu 702-701, Korea*
**Department of Physics, Dong-A University, 840 Hadan 2-dong, Saha-gu, Busan 604-714, Korea*

Abstract. The capture cross sections and capture γ-ray spectra of ^{117,119}Sn were measured in an incident neutron energy region from 10 to 100 keV and at 570 keV, using a 1.5-ns pulsed neutron source by the ^{7}Li(p,n)^{7}Be reaction and a large anti-Compton NaI(Tl) γ-ray spectrometer. A pulse-height weighting technique was applied to observed capture γ-ray pulse-height spectra to derive capture yields. The capture cross sections of ^{117,119}Sn were obtained with the error of about 5% by using the standard capture cross sections of ^{197}Au. The present cross sections were compared with previous experimental data and the evaluated values in JENDL-3.3 and ENDF/B-VI. The capture γ-ray spectra of ^{117,119}Sn were derived by unfolding the observed capture γ-ray pulse-height spectra.

Keywords: neutron capture, cross section, gamma-ray spectra, nuclear transmutation, tin 126, tin 117, tin 119, anti-Compton NaI(Tl) spectrometer, time-of-flight method, pulse-height weighting technique, experimental data
PACS: 25.40.Lw, 27.60.+j, 29.30.Kv

INTRODUCTION

The neutron capture cross sections of long-lived fission products (LLFPs) are important physical quantities for the study on the transmutation of radioactive nuclear wastes. The nuclide ^{126}Sn is one of the LLFPs. However, there is no experimental data for ^{126}Sn, because the preparation of high-purity sample is difficult and, moreover, γ-ray radiation from a sample causes a serious background.

On the other hand, keV-neutron capture cross sections and capture γ-ray spectra of stable Sn isotopes contain important information useful for the theoretical evaluation of the capture cross sections of ^{126}Sn. Thus, we have started a systematic measurement of keV-neutron capture cross sections and capture γ-ray spectra of stable Sn isotopes. In the present contribution, the results for ^{117,119}Sn are shown.

EXPERIMENTAL PROCEDURE

The capture cross sections and capture γ-ray spectra of ^{117,119}Sn were measured in the incident neutron energy region from 10 to 100 keV and at 570 keV, using the

3-MV Pelletron accelerator of the Research Laboratory for Nuclear Reactors at the Tokyo Institute of Technology. Pulsed keV neutrons were produced from the ^{7}Li(p,n)^{7}Be reaction with a 1.5-ns bunched proton beam from the accelerator. The pulse-repetition rate was 4 MHz. The ^{117,119}Sn samples were highly enriched metal plates, and the net weight of each sample was about 1 g. Capture γ rays were detected with a large anti-Compton NaI(Tl) spectrometer [1] by means of a time-of-flight method.

DATA PROCESSING AND RESULTS

A pulse-height weighting technique [2] was applied to the observed capture γ-ray pulse-height spectra to obtain capture yields. Using the standard capture cross sections of ^{197}Au [3], the capture cross sections of ^{117,119}Sn were derived with the error of about 5%. The present results of ^{119}Sn are compared in Fig. 1 with previous experimental data [4, 5] and the evaluated values in JENDL-3.3 [6] and ENDF/B-VI [7]. The capture γ-ray spectra were derived by unfolding the observed capture γ-ray pulse-height spectra. The capture γ-ray spectrum of ^{119}Sn in the incident neutron energy region from 15 to 100 keV is shown in Fig. 2. The strong primary transitions from the capture states to the ground and first excited states are observed around 8 and 9 MeV, respectively.

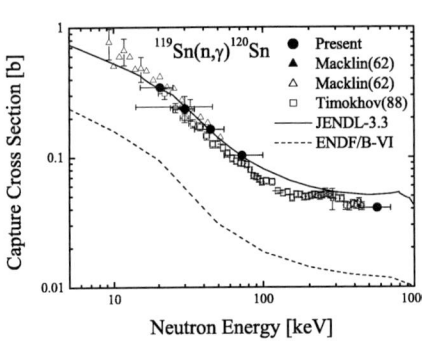
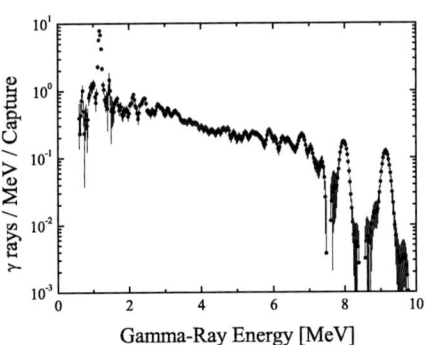

FIGURE 1. Neutron capture cross sections of ^{119}Sn in the keV region

FIGURE 2. Capture γ-ray spectrum of ^{119}Sn in the incident neutron energy region from 15 to 100 keV

REFERENCES

1. S. Mizuno, M. Igashira and K. Masuda, *J. Nucl. Sci. Technol.*, **36**, 493-507 (1999).
2. R.L. Macklin and J.H. Gibbons, *Phys. Rev.*, **159**, 1007-1012 (1967).
3. ENDF/B-VI data file for ^{197}Au(MAT=7925), evaluated by P. G. Young (1991).
4. R.L. Macklin, T. Inada and J.H. Gibbons, *Nature.*, **194**, 1272 (1962); EXFOR11958.006 (1976), EXFOR11981.005 (1976).
5. V.M. Timokhov et al., *Sov. J. Nucl. Phys.*, **50**, 375-380 (1989); EXFOR40960.008 (1988).
6. JENDL-3.3 data file for ^{119}Sn(MAT=5046), evaluated by JNDC FP Nuclear Data W.G. (1990).
7. ENDF/B-VI data file for ^{119}Sn(MAT=5046), evaluated by R. E. Schenter and F. Schmittroth (1974).

$^{18}F(\alpha,p)^{21}Ne$ Reaction: Neutron Source For r-Process In Supernovae

H.-Y. Lee[1], M. Beard[1], H.-W. Becker[2], M. Couder[1], A. Couture[1], J. Görres[1], L. Lamm[1], P. LeBlanc[1], S. O'Brien[1], A. Palumbo[1], E. Stech[1], E. Strandberg[1], W. Tan[1], C. Ugalde[1,*], M. Wiescher[1]

[1]*Department of Physics, University of Notre Dame, Notre Dame, IN, USA*
[2]*Dynamitron Tandem Laboratory, Ruhr-Universität, Bochum, Germany*

Abstract. The reaction rate of $^{18}F(\alpha,p)^{21}Ne$ has been studied using the inverse reaction $^{21}Ne(p,\alpha)^{18}F$. This has been measured by the activation method in the energy range of the relevant Gamow window. Experimental results will be discussed and compared with the results of Hauser-Feshbach calculations and previous measurements.

Keywords: r-process nucleosynthesis, supernovae, activation method, ^{18}F, ^{21}Ne
PACS: 25.40.Lw, 26.30.+k, 27.30.+t, 29.25.-t

INTRODUCTION

Recent observations of the abundance distribution of heavy elements in metal poor old stars suggest an indication of the existence of more than one r-process site [1,2] proposing a different site for the synthesis of light r-process nuclei. This led to a reexamination of previously proposed models for r-process sites which originally did not provide sufficient neutron flux for the production of the very heavy elements. One of these models is the r-process nucleosynthesis in the supernova shock traveling through the He-rich shell of the pre-supernova star [3]. In this scenario, the ^{14}N from preceding CNO burning is converted to ^{18}F, via the reaction $^{14}N(\alpha,\gamma)^{18}F$ and subsequent β^+-decay forms ^{18}O, which produce neutrons by $^{18}O(\alpha,n)$ or $^{18}O(\alpha,\gamma)^{22}Ne(\alpha,n)$. The statistical model predicted that the reaction of $^{18}F(\alpha,p)^{21}Ne$ can compete with the β^+-decay at high helium density and temperatures in the shock front. Since $^{21}Ne(\alpha,n)$ has a positive Q-value, it has been identified as a stronger neutron source than $^{22}Ne(\alpha,n)$. Because no published data on the $^{18}F(\alpha,p)^{21}Ne$ reaction are available, we have measured the total cross section of the inverse reaction $^{21}Ne(p,\alpha,)^{18}F$ using an activation method.

EXPERIMENTAL SETUP

The experiment was performed at the KN and FN accelerators at the Nuclear Structure Laboratory at the University of Notre Dame in an energy range of 2.5 MeV to 3.5 MeV with p-beam currents of 2-30 µA. The targets were made by implantation of ^{21}Ne (0.27%

---
*Current address: TUNL,University of North Carolina, Chapel Hill, NC

natural abundance). The backings are thick gold layers on copper disks to reduce background reactions and to have good thermal conductivity for efficient cooling. Implantation was done with 2 different energies, 150 keV and 400 keV to achieve a homogenous ^{21}Ne distribution .This was monitored using the 768 keV resonance in the ^{21}Ne(p,γ) reaction [5]. During the activation experiment, the ^{18}F activity was measured by counting 511 keV annihilation γ rays in coincidence. The detection system consisted of two Ge clover detectors mounted face to face [6] with an absolute coincidence counting efficiency of 2.63%, which was measured with a weak ^{22}Na source.

RESULTS AND CONCLUSIONS

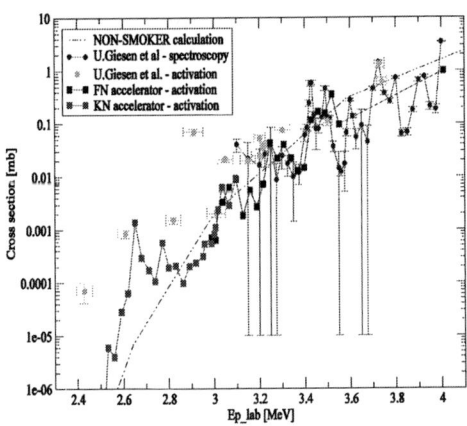

Figure 1. The total cross section (preliminary) of 21Ne(p,α)18F before 18O(p,n)18F yield subtraction

The preliminary total cross section [see Fig.1] showed that the measurements appears to be in good agreement with the Hauser-Feshbach calculation, and the current data (solid square) overlap with previous spectroscopy data, corrected for the target thickness of ~23 keV. At energies of 2.43 MeV, 2.82 MeV, and 2.91 MeV[4], the current data show no existence of resonances in ^{21}Ne(p,α)^{18}F. The observed yield at 2.65 MeV reflects a resonance of ^{18}O(p,n)^{18}F [7], which can also contribute to the ^{18}F activity above the threshold of 2.57 MeV. Since the existence of a thin oxygen layer on the target surface is unavoidable, it is necessary to subtract the yield of ^{18}O(p,n)^{18}F from ^{21}Ne(p,α) ^{18}F, especially below 3 MeV, when the both yields are comparable. For the final analysis, we plan to measure oxygen content on all targets and correct for the contamination.

REFERENCES

1. C.Seden, Nature **409**, 673 (2001)
2. J.W.Truran, J.J.Cowan, in the proceedings of the 10th workshop on 'Nuclear Astrophysics' Ringberg Castle, eds.W.Hillebrandt, E.Müller, MPA/P12, P.64 (2000)
3. J.W.Truran, J.J.Cowan, A.G.W.Cameron, Astrophys. J. **222**, L63 (1978)
4. U. Giesen, Diploma Thesis, Universitat Münster (1987)
5. W.H.Schulte, et al., Nucl. Instr. and Meth. B**64**, 383 (1992)
6. S.Dababneh, et al., Nucl. Instr. and Meth. A**517**, 230 (2004)
7. J.K.Bair, Phys. Rev. C**8**, 120 (1973)

Search for Multiphonon and Mixed-Symmetry States in ^{127}I

S. Mukhopadhyay*, J.N. Orce*, S.N. Choudry*, V. Varadarajan*, A. Kumar*, M.T. McEllistrem* and S.W. Yates*,†

*Department of Physics and Astronomy, University of Kentucky, Lexington, Kentucky 40506-0055
†Department of Chemistry, University of Kentucky, Lexington, Kentucky 40506-0055

Abstract.
The low-spin structure of ^{127}I has been studied using the (n,n'γ) reaction at neutron energies ranging from 1.2 to 2.3 MeV and the (n,n'γγ) reaction at 3 MeV. New excited levels and γ-rays have been identified by using excitation functions and coincidence data. Lifetimes were also determined from an angular distribution experiment. Future work involves determining multipolarities and spin assignments in order to obtain $B(M1)$ and $B(E2)$ values.

Keywords: Transition strengths, multiphonon states, mixed-symmetry states, odd-A nuclides
PACS: 21.10.Re; 21.10.Tg; 21.60.Fw; 23.20.-g; 23.20.En; 23.20.Gq

INTRODUCTION

In the triaxial $O(6)$ region, mixed-symmetry (*MS*) states have only been identified in $^{128}_{54}$Xe$_{74}$ [1]. Little is known about how vibrational anharmonicities and MS states are modified by the presence of an odd particle. In the present work, we search for multiphonon and *MS* states built on the 5/2$^+$ ground state and the 7/2$^+$[404] and 3/2$^+$[422] Nilsson orbitals of the neighboring odd-mass nucleus, $^{127}_{53}$I$_{74}$. The low-spin structure of this nucleus has been studied through the (n,n'γ) inelastic scattering reaction using the 7 MeV accelerator at the University of Kentucky. Excitation functions (1.2 - 2.3 MeV), angular distribution (2 MeV) and γ-γ coincidence (3 MeV) measurements were performed. Pulsed-beam techniques were used to reduce background, with beam pulses separated by 533 ns and bunched to about 1.5 ns. This approach allowed the use of the time-of-flight techniques for background suppression by gating on the appropriate prompt time windows [2]. The collected γ-rays were detected using a BGO Compton-suppressed 55% HPGe spectrometer with 2.0 keV resolution. These measurements have led to an expanded decay scheme and the determination of lifetimes in the fs range for ^{127}I.

The combination of excitation functions and γ-γ coincidence data is a very powerful tool for constructing a level scheme. From excitation functions, energy thresholds help to resolve doublets or triplets of γ-rays and allow the determination of direct decays of high-energy γ-rays to either the ground state or low-lying single-particle excitations. The left panel of Fig. 1 shows the relative cross-section as a function of neutron energy of the proposed 828 keV γ-ray depopulating the 1479 keV level. Additional coincidence analysis (shown on the right panel of Fig. 1) confirms the γ-ray placement in the ^{127}I level scheme. Furthermore, lifetimes were measured through the Doppler-shift

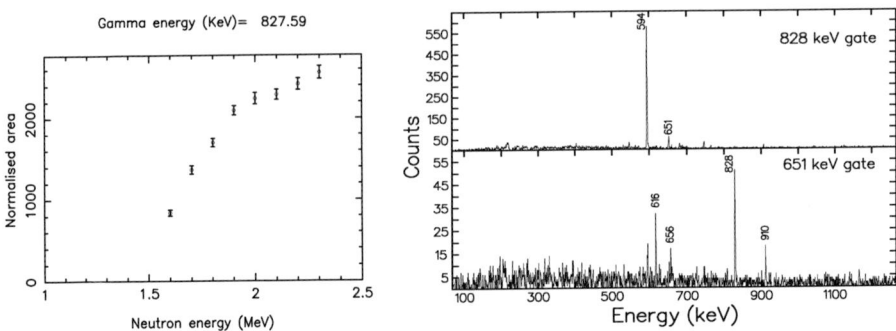

FIGURE 1. The left panel shows the cross-section versus neutron energy for the proposed 828 keV γ-ray depopulating the 1479 keV level. The right panel shows background-substracted coincidence spectra gated on the 828 and 651 keV γ-rays, confirming such a placement.

attenuation method following the (n,n′γ) reaction [3].

In principle, the states of an odd-A nucleus can be described in terms of a valence nucleon weakly or intermediately coupled to the excited states of the neighbouring even-even vibrational core [4]. In the coupling limit, the coupling Hamiltonian can be treated as a perturbation, whereas intermediate coupling also allows for mixing between several single-particle states coupled to core excitations. The angular momentum of the core, J_{core} and single-particle states, J_{sp}, couple to form a multiplet of states with a total angular momentum [5, 6] given by, $|J_{sp} - J_{core}| \leq J_{core} \otimes J_{sp} \leq |J_{sp} + J_{core}|$.

An even more complex scenario might be encountered in odd-A nuclei with respect to even-A nuclides. In the odd-A case the interplay of MS, multiphonon, intruder and single-particle degrees of freedom may strongly perturb the nuclear wavefunctions. However, multiphonon and MS states can be identified from their different decay signatures. Whereas the former preferentially decay through strongly collective $E2$ transitions following the $\Delta n_\lambda = \pm 1$ selection rule for vibrational states, the latter exhibit strong $M1$ dipole transitions to the phonon excitations and weak $E2$ transitions to the ground state. Future work involves determining multipolarities and the determination of $B(M1)$ and $B(E2)$ values. This material is based upon work supported by the U.S. National Science Foundation under Grant No. PHY-0354656.

REFERENCES

1. I. Wiedenhöver et al., Phys. Rev. C **56**, R2354 (1997).
2. M. T. McEllistrem, *Nuclear Research with Low Energy Accelerators*, Academic Press, New York (1967), p. 167; P. E. Garrett, N. Warr and S. W. Yates, J. Res. Natl. Stand. Technol. **105**, 141 (2000).
3. T. Belgya, G. Molnár and S.W. Yates, Nucl. Phys. **A607**, 43 (1996).
4. K. Heyde and P. J. Brussaard, Nucl.Phys. A**104**, 81 (1967).
5. A. de-Shalit, Phys. Rev. **122** 1530 (1961).
6. R. E. Anderson et al., Phys. Rev. C **15**, 123 (1977).

^{106}Cd and ^{112}Sn: Alpha-Induced Cross Section Measurements For The Astrophysical P-Process

A.Palumbo[1], J.Görres[1], H.-Y. Lee[1], M.Wiescher[1], W.Rapp[1], N.Özkan[2], R.T.Güray[2], G.Efe[2], Gy.Gyürky[3], Zs. Fülop[3], E.Somorjai[3]

1 University of Notre Dame, Notre Dame, Indiana, 46556, USA
2 Kocaeli University, Department of Physics, 41380 Umuttepe, Kocaeli, Turkey
3 Institute of Nuclear Research (ATOMKI), P.O. Box 51 H-4001 Debrecen, Hungary

Abstract. The simulation of p-process nucleosynthesis requires thousands of reaction rates. These rates are typically calculated with the statistical Hauser Feshbach Model (HF-Model) which depends critically on the reliability of the optical nucleus potential. In particular, the predicted (α,γ) cross sections at low energies (below 20 MeV) are sensitive to the choice of the alpha potential. Dscrepancies have been observed between the experimental cross sections and the theoretical values (NON-SMOKER). Currently, only a few α-induced reaction rates have been measured. To explore the trend ^{106}Cd(α,γ)^{110}Sn and ^{112}Sn(α,γ)^{116}Te cross sections were measured at the Notre Dame FN Tandem accelerator. First results are presented and discussed.

Keywords: Astrophysical p-process, capture reactions, optical potential;
PACS: 26.30+k, 26.50+x, 27.60+j;

INTRODUCTION

The p-process is responsible for the synthesis of the neutron deficient stable nuclei above Fe. These isotopes are called "p-nuclei" and their abundance accounts typically for less than 1% of the total isotopic abundances. The p-process is an astrophysical scenario that is not well understood [1]; according to present interpretation, p-nuclei are formed through photodisintegration reactions of mainly the (γ,n) (γ,α) and (γ,p) types on seed r- or s- nuclei. These photodisintegration processes require high temperatures of $T_9 = 2-3$. Sufficient abundance of seed material must be present and the reaction must freeze-out before too heavy of a photoerosion takes place (leaving Fe nuclei as the end product) [2]. Possible production sites where the above criteria are met include Type II and Type I supernovae.

Experiment and First Results

Only four α-capture reactions on p-nuclei had previously been measured: ^{70}Ge [3], ^{96}Ru [4], ^{112}Sn [5], ^{144}Sm [6]. To probe the HF predictions for α capture near the Z=50 closed shell, we have performed two independent measurements of ^{106}Cd(α,γ)^{110}Sn and a first study of ^{112}Sn(α,γ)^{116}Te. For these measurements we used the activation method. In this technique, a stable sample is bombarded with a charged particle beam producing a radioactive species whose characteristic γ-rays are then measured.

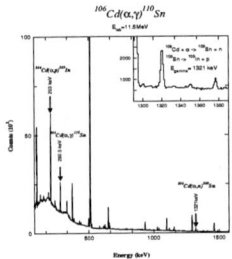 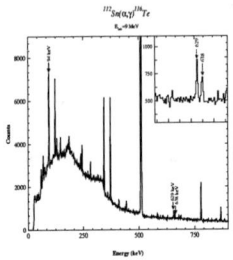

FIGURE 1. γ-ray spectra of ^{110}Sn and ^{116}Te

Highly enriched ^{106}Cd and ^{112}Sn targets were activated in the energy range of 8-12 MeV in 0.5 MeV increments at the ATOMKI cyclotron and the Notre Dame FN tandem accelerator.

After irradiation, the γ-rays of the radioactive targets were measured using two Clover detectors in close geometry (9 mm between detector and target) surrounded by 5 cm lead bricks to improve detector efficiency and low energy background reduction.

The experimental S-factor data are displayed in figure 2 as function of alpha energy. The results from both facilities show good agreement. The experimental S-factor values for ^{112}Sn(α,γ)^{116}Te do also agree with previously published results [5]. Comparison with Hauser Feshbach predictions of the S-factor curve indicates deviations towards lower energy where the theoretical values become larger. This is particularly noteworthy in the case of ^{106}Cd(α,γ)^{110}Sn with deviations of up to a factor of three in the lower energy range.

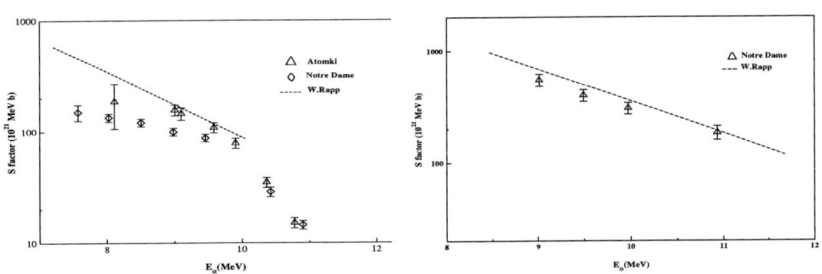

FIGURE 2. The S-factor of ^{106}Cd(α,γ)^{110}Sn (measured at Notre Dame and Atomki) is shown in the left panel. The right panel displays the S-factor for ^{112}Sn(α,γ)^{116}Te. The dashed lines show Hauser Feshbach (HF) predictions for the two reactions.

REFERENCES

1. M. Rayet et al., Astro. Astrophys. 298, 517-527 (1995).
2. M. Arnould and S. Goriely, Phys. Rep. 384 Issue 1-2 (2003) 1.
3. Zs. Fülop, A.Z. Kiss, E. Somorjai et al., Z. Phys. A 355 (1996) 203.
4. W. Rapp, H.J. Brede, M. Heil, Nucl. Phys. A688 (2001) 427.
5. N. Özkan, A. StJ. Murphy et al., Nucl. Phys. A 710 (2002) 469.
6. E. Somorjai, Zs. Fülop, A.Z. Kiss et al., Astron. Astrophys. 33 (1998) 1112

$K = 0^+$ bands in Gadolinium nuclei

G. Popa[*], A. Aprahamian[†], A. M. Bruce[**], J. G. Hirsch[‡] and J. P. Draayer[§]

[*]*The Ohio State University, Mansfield, OH 44906*
[†]*University of Notre Dame, Notre Dame, IN 46556*
[**]*University of Brighton, Brighton, BN2 4GJ, United Kingdom*
[‡]*Instituto de Ciencias Nucleares, UNAM, 70-543 México 04510 DF, México*
[§]*Louisiana State University, Baton Rouge, LA 70803*

Abstract. The recent $^{160}Gd(p,t)^{158}Gd$ experiment[1] has raised again the question about the nature of the low-lying excited 0^+ states in deformed nuclei. New (p,t) experiments have resolved typically about ten low-lying excited 0^+ states[2]. Calculations using the IBA model [3], the projected shell model [4] and the quasiparticle-phonon model [5] have answered partially this question and showed the complicated nature of these states.
We study the energy levels, their wave function content, and the corresponding $B(E2)$ transitions in $^{156,158,160}Gd$ nuclei, in the framework of the pseudo-SU(3) model. The theory uses a realistic Hamiltonian with single-particle energies and monopole pairing interactions in addition to the usual quadrupole-quadrupole term, with the pairing and quadrupole-quadrupole strengths taken from systematics. The calculations are carried out with a set of basis states with pseudo-spin zero proton and neutron configurations. The calculated $B(E2)$ values suggest that the first excited 0^+ state might have a mixed rotational and collective vibrational character, where the latter – quasiparticle type excitations – are not considered in this calculation.[1]

Keywords: $K = 0^+$, collective levels, Gadolinium
PACS: 21.60.Fw, 21.10.Re, 27.70.+q

MODEL SPACE AND HAMILTONIAN

We calculated the low-energy spectra and the corresponding $B(E2)$ transitions in $^{156,158,160}Gd$ nuclei within the framework of the pseudo-SU(3) model. The building blocks of the model are the pseudo-SU(3) proton and neutron states with pseudo spin zero. The many-particle states are built as pseudo-SU(3) coupled states with a well-defined particle number and total angular momentum [6, 7].

The Hamiltonian includes spherical Nilsson single-particle terms for the protons and neutrons, $H_p^{sp} + H_n^{sp}$, the quadrupole-quadrupole ($Q \cdot Q$) and pairing interactions, H_p^P and H_n^P, with interaction strengths taken from systematics, as well as four 'rotor-like' terms which preserve the SU(3) symmetry. The parameters (a, b, a_{sym}, a_3) of the 'rotor-like' part of the Hamiltonian are used for fitting the band-head energy of the first excited $K^\pi = 0^+$, $K^\pi = 2^+$, and $K^\pi = 1^+$ states, and the moment of inertia of the ground band. Parameter values are close to those used in the description of neighboring even-even and odd-A nuclei [8, 9]. The single-particle terms together with the proton and neutron pairing interaction mix the SU(3) basis states allowing for a realistic description of the

---
[1] A second footnote in the abstract

TABLE 1. Experimental energies for the excited $K^\pi = 0^+$ and $K^\pi = 2^+$ states in four Gadolinium nuclei. Calculated and experimental values for two $B(E2)$ transitions in three Gadolinium nuclei are also included.

nucleus	^{154}Gd	^{156}Gd	^{158}Gd	^{160}Gd
Energy [MeV] of $K^\pi = 0_2^+$	0.68	1.05	1.19	1.33
Energy [MeV] of $K^\pi = 2_2^+$	0.99	1.15	1.19	0.99
Exp. $B(E2; 0_2^+ \to 2_{g.s.})$ [W.u]		4.18 → 10.97 *	1.2†	
Th. $B(E2; 0_2^+ \to 2_{g.s.})$ [W.u]		0.5	0.08	0.07
Exp. $B(E2; 2_{K=2}^+ \to 0_{g.s.})$ [W.u]		2.73 → 4.25	3.4(3)	3.80(22)
Th. $B(E2; 2_{K=2}^+ \to 0_{g.s.})$ [W.u]		6.4	6.8	8.1

* Experimental data for ^{156}Gd extracted from [11].
† Experimental data for $^{158,160}Gd$ extracted from [10].

inter-band $B(E2)$ transitions.

RESULTS AND ANALYSIS

The energies of the excited $K^\pi = 0^+$, and $K^\pi = 2^+$ states in even-even nuclei display interesting behavior. The first one increases from ^{154}Gd to ^{160}Gd, while the latter increases from 0.99 MeV in ^{154}Gd to 1.19 MeV in ^{158}Gd, and then decreases back to 0.99 MeV in ^{160}Gd (see Table 1). The experimental and calculated $B(E2; 0_2^+ \to 2_{g.s})$ and $B(E2; 2_{K=2}^+ \to 0_{g.s.})$ transitions are also given in Table 1. The calculated $B(E2)$ values for the first transition are smaller than the experimental ones by about a factor of ten for all three nuclei. The experimental $B(E2)$ value decreases by about a factor of five from ^{156}Gd to ^{158}Gd. This trend is reproduced by the calculations. However, the calculated $B(E2)$ values for the latter transitions are over estimated by about a factor of two for all three nuclei. Having reproduced the correct trend in the $B(E2)$ values for these transitions in all three nuclei, we conclude that the missing $B(E2)$ strength from the excited 0^+ state may come from other excitations and configurations, that are not considered in this model.

REFERENCES

1. S. R. Lesher et.al., Phys. Rev. C **66**, 051305(R) (2002).
2. H.-F. Wirth et.al., Phys. Rev. C **69**, 044310 (2004).
3. N. V. Zamfir et.al., Phys. Rev. C **66**, 057303 (2002).
4. Y. Sun et.al., Phys. Rev. C **68**, 061301(R) (2003).
5. N. Lo Iudice et.al., Phys. Rev. C **70**, 064316 (2004).
6. T. Beuschel, J. G. Hirsch, and J. P. Draayer, Phys. Rev. C **61** 54307 (2000).
7.] G. Popa, J. G. Hirsch and J. P. Draayer, Phys. Rev. C **62** 064313 (2000).
8. J. P. Draayer, G. Popa and J. G. Hirsch, Acta Phys. Pol. 32, 2697 (2001).
9. C. Vargas, J. G. Hirsch, T. Beuschel, J. P. Draayer, Phys. Rev. C **61** 31301 (2000).
10. G. Audi and A. H. Wapstra, Nucl. Phys. A **595** (4)(1995).
11. R. C. De Haan, private communication and U. Notre Dame, Ph. D. Thesis (2001).

Nuclear Reaction and Structure Databases of the National Nuclear Data Center

B. Pritychenko*, R. Arcilla*, M.W. Herman*, P. Obložinský*,
D. Rochman*, A.A. Sonzogni*, J.K. Tuli* and D.F. Winchell*

*National Nuclear Data Center, Brookhaven National Laboratory, Upton, NY 11973-5000, U.S.A.

Abstract. The National Nuclear Data Center (NNDC) collects, evaluates, and disseminates nuclear physics data for basic research and applied nuclear technologies. In 2004, the NNDC migrated all databases into modern relational database software, installed new generation of Linux servers and developed new Java-based Web service. This nuclear database development means much faster, more flexible and more convenient service to all users in the United States. These nuclear reaction and structure database developments as well as related Web services are briefly described.

Keywords: Nuclear Data
PACS: 29.85.+c

The National Nuclear Data Center coordinates the US Nuclear Data Program (US-NDP) and Cross Section Evaluation Working Group (CSEWG), maintains and contributes to the nuclear structure (ENSDF, NSR, XUNDL) and nuclear reaction (ENDF, CSISRS, CINDA) databases along with several derived databases (NuDat, MIRD), and publishes Nuclear Data Sheets journal and Nuclear Wallet Cards booklet [1, 2].

The NNDC has been providing remote electronic access to its databases and other information since 1986. In order to improve the quality of nuclear data services and take advantage of latest software and hardware developments, NNDC started nuclear database migration project in 1999. New Web interfaces were integrated with Sybase relational databases and Nuclear Data Portal http://www.nndc.bnl.gov was launched on April 19, 2004. A brief description of the major nuclear data services is presented below.

Evaluated Nuclear (reaction) Data File: The ENDF reaction database contains recommended data from the United States ENDF/B-VI.8 library. ENDF Web interface also provides access to the other evaluated nuclear reaction libraries: JEFF-3.1, JENDL-3.3, BROND, and CENDL.

Evaluated Nuclear Structure Data File: The ENSDF evaluated nuclear structure and decay database, as of August 22, 2005, contains recommended data for 2929 nuclides, organized in over 15451 individual datasets.

Nuclear Science References: NSR is an indexed bibliography of nuclear physics papers and reports containing over 1.8×10^5 nuclear science articles, indexed according to content.

Nuclear structure and decay Data: NuDat 2.1 contains evaluated (recommended) nuclear structure and decay data extracted from ENSDF and Nuclear Wallet Cards for all-known nuclei, with about 1.40×10^5 levels, 2.04×10^5 γ-ray energies, *etc*.

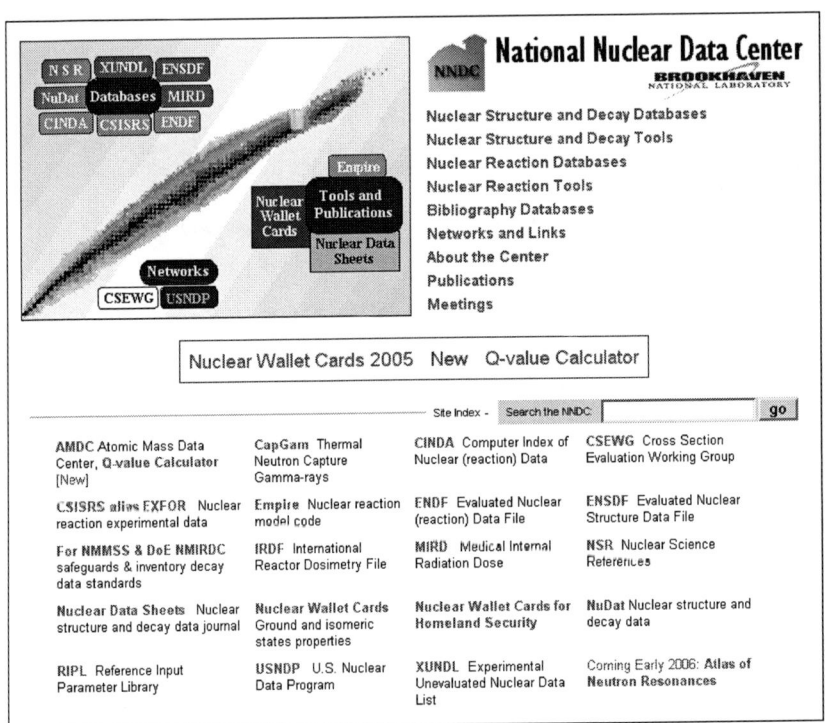

FIGURE 1. Front page of the NNDC Web Services: Nuclear Data Portal (www.nndc.bnl.gov). All elements of the page, including graphic images, are hyperlinked.

The new Web-based nuclear data retrieval system (shown in Figure 1) generated overwhelming response from nuclear data users. This system produced a two-fold increase in the NNDC Web data retrievals compared with the previous service. If this trend continues, the forecast for 2005 will be 720 K retrievals. It is expected that NNDC nuclear data retrieval system will meet growing demands in the future.

ACKNOWLEDGMENTS

This work is supported by the Office of Nuclear Physics, Office of Science of the U.S. Department of Energy, under contract no. DE-AC02-98CH10886 with Brookhaven Science Associates, LLC.

REFERENCES

1. National Nuclear Data Center, *http://www.nndc.bnl.gov* (accessed August 22, 2005).
2. B. Pritychenko *et al*, Proc. Int. Conf. on Nuclear Data for Science and Technology, Santa Fe, September 26 - October 1, 2004, AIP **769** (2005) 132.

Nuclear Structure Properties of Neutron Rich Ge-Br Isotopes in the Astrophysical r-Process

M. Quinn[1], A. Aprahamian[1], A. Woehr[1], P. Mantica[2], J. Pereira Conca[2], H. Schatz[2], S. Hennrich[3], and K.-L. Kratz[3]

[1] *Department of Physics, University of Notre Dame, Notre Dame, IN, USA*
[2] *National Superconducting Cyclotron Laboratory, Michigan State University, East Lansing, Michigan, USA*
[3] *Institut fur Kernchemie, Universitaet Mainz, Mainz, Germany*

Keywords: Nuclear Astrophysics, Beta Decay
PACS: 23.40.-s 25.70.Mn 28.60.+s

The astrophysical r-process is responsible for synthesis of roughly half of the elements heavier than iron. In spite of this significance, there are many uncertainties regarding the site of the r-process and the neutron-rich nuclei involved. Studying these nuclei presents a challenge, as they lie far from the valley of stability. Nuclear properties such as β decay half-lives and β-delayed neutron emission probabilities are critical inputs for r-process models. The neutron rich Ge-Br isotopes are in the region just after the N=50 bottle neck in the "classical" r-process, or may serve as seed material for the high entropy neutrino-wind r-process.

Neutron rich nuclei play an important role in both nuclear astrophysics and nuclear structure.[1] Central to the study of nuclear structure are the shape of a nucleus and how that shape changes from spherical to deformed. In some regions there is a smooth change in deformation as nucleons are either added or subtracted, and yet in other regions this transition is very rapid. The Ge-Br isotopes to be measured lie between the N=56 sub-shell closure and the onset of deformation at N=60, just below the Sr-Zr region, for which the most pronounced transition from spherical to deformed ground-state shapes have been observed.

Neutron rich Ge-Br isotopes will be studied at the NSCL at Michigan State University. Production will be by fragmentation of a 120 MeV/u ^{136}Xe beam on a Be target. The A1900 fragment separator will block unwanted species produced in this reaction. The transmitted nuclei will be implanted in a dual-sided silicon detector, which is part of the Beta Counting System detector. Implanted nuclei will be identified by the ΔE-time of flight method. Beta decays will then be detected by a single-sided silicon detector, which is also part of the Beta Counting System. Beta-delayed neutrons will be detected by the NERO neutron counter, which will surround the Beta Counting System. This will provide measurements of the β decay half-lives and β-delayed neutron emission probabilities. Then, SEGA germanium detector array will be used to detect gamma rays from decays of nuclei implanted in the Beta Counting System. The first 2^+ and 4^+ levels will be measured, determining the deformation of these nuclei.

TABLE 1. QRPA calculations of half-lives and β delayed neutron emission probabilities for spherical and deformed nuclear shapes. The data are taqken from the proposal to the NSCI/MSU experiment 03034 and references therein. [2]

Isotope	$T_{1/2}$ [ms] spherical	P_n [%]	$T_{1/2}$ [ms] deformed	P_n [%]	$T_{1/2}$ [ms] experiment	P_n [%]
Ge-85	832	8.3	186	5.4	540(50)	14(3)
Ge-86	627	31.5	177	5.4		
Ge-87	364	33.9	56.7	6.3		
Ge-88	171	69.4	45.8	6.7		
As-86	834	19.4	286	11.7	945(8)	26(7)
As-87	739	46.7	238	82.9	560(110)	17.5(25)
As-88	445	32.1	70.7	41.9		
As-89	218	77.0	63.0	93.3		
As-90	21.1	8.9	22.8	42.5		
As-91	61.1	92.2	33.2	95.7		
Se-89	1646	0.5	137	0.6	410(40)	7.8(25)
Se-90	724	0.6	141	1.1		
Se-91	39.3	0.2	37.6	1.3	270(50)	21(10)
Se-92	137	2.3	62.3	2.7		
Se-93	24.0	14.5	51.6	7.1		
Se-94	39.0	3.7	48.2	23.9		
Br-94	33.4	14.2	113	56.6	70(20)	68(16)
Br-95	53.2	93.8	70.2	79.1		
Br-96	19.2	31.9	36.7	56.2		
Br-97	20.2	97.2	42.4	92.0		

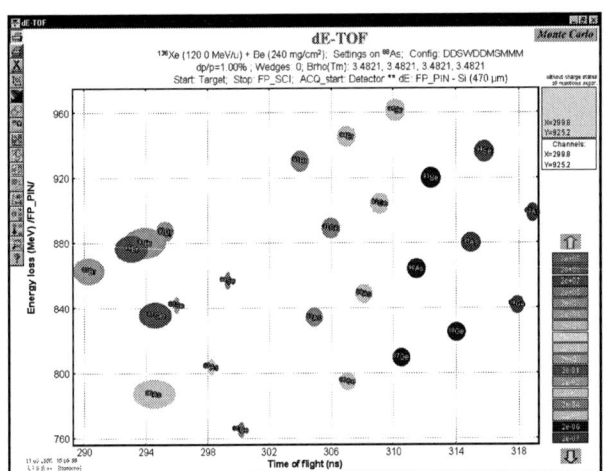

FIGURE 1. LISE simulation of the ΔE – time of flight spectrum for particle identifcation.

REFERENCES

1. A. Aprahamian, K.Langanke, and M. Wiescher. Prog. Part. Nucl. Phys., **54**:535 (2005).
2. A. Woehr *et al.*, Proposal to the experiment 03034 at NSCL/MSU (2003) and references there in

Thermal Neutron Capture and Thermal Neutron Burn-up of K isomeric state of ^{177m}Lu: a way to the Neutron Super-Elastic Scattering cross section

O. Roig[a] G. Bélier[a], V. Méot[a], J. Aupiais[c], J.-M. Daugas[a], Ch. Jutier[c], G. Le Petit[c], A. Letourneau[b], F. Marie[b], P. Romain[a] and Ch. Veyssiere[d]

[a] *CEA/DIF/DPTA Service de Physique Nucléaire, BP 12, 91680 Bruyères-le-Châtel, France*
[b] *CEA/DSM/DAPNIA Service de Physique Nucléaire, CE Saclay, 91191 Gif-sur-Yvette, France*
[c] *CEA/DIF/DASE Service de Radio-analyse, Chimie et Environnement, BP 12, 91680 Bruyères-le-Châtel, France*
[d] *CEA/DSM/DAPNIA Service d'Ingénierie des Systèmes, CE Saclay, 91191 Gif sur Yvette, France*

Abstract. Thermal neutron radiative capture and burn-up measurements of the K isomeric state in ^{177}Lu form part of an original method to indirectly obtain the neutron super-elastic scattering cross section at thermal energy. Neutron super-elastic scattering, also called neutron inelastic acceleration, occurs during the neutron collisions with an excited nuclear level. In this reaction, the nucleus could partly transfer its excitation energy to the scattered neutron.

INTRODUCTION

As nuclear isomers are nuclear excited states of long half-life, they have a potentiality as energy traps [1]. Many experiments have attempted to de-excite isomers [2,3]. However, the induced deexcitation of isomers comes up against an antagonism: the higher the isomer half life, the more difficult the deexcitation is. The induced deexcitation of K isomers [1] may be different because their half-life is not only due to the spin difference but also to the K difference. Quantum number K is the projection of the total nuclear spin on the symmetry axis of a deformed nuclei. At low energy, the K-hindrance factors are well reproduced by the empirical rule $100^{\Delta K - \lambda}$ (where λ is the transition multipolarity [4]) although large deviations were observed in a few cases [5]. At higher energy, the situation is quite different. In fact, the more the excitation energy increases, the less K is a good quantum number. At neutron separation energy, the K-hindrance is expected to play no part [6]. The observation of so-called neutron super-elastic scattering on ^{152m}Eu [7] and ^{180m}Hf [8] confirms this point of view. Although super-elastic cross sections were measured, the emitted neutron spectra were never observed. Pertinent nuclear structure information on the compound nucleus wave function could be extracted from these neutron lines.

From this point of view, the nucleus ^{177m}Lu is a promising candidate. It is a K isomer with a half-life of 160.44 days [9,10], a 970 keV excitation energy and the various expected super-elastic scattering neutron transitions (116keV, 125keV, 334keV, ...). These properties are encouraging to carry out an experiment in order to detect the scattered neutron and to measure their energies (Figure 1).

METHOD

Before performing such a difficult experiment, properties of ^{177}Lu and its neighboring nuclei allow to get the value of the super elastic cross section by an indirect method subtracting the radiative capture cross section from the burn-up cross section at thermal neutron energy.

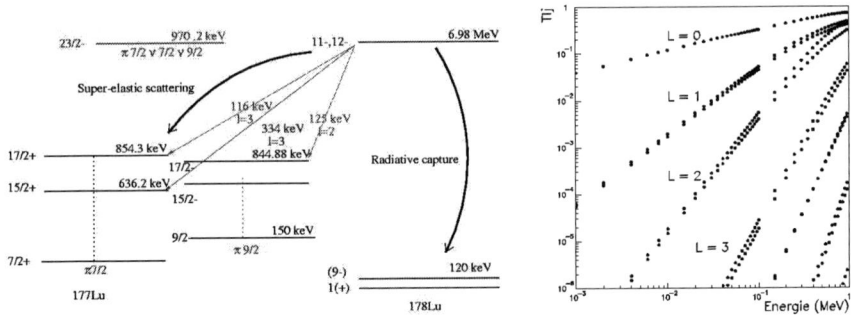

FIGURE 1. (a) Scheme of super-elastic scattering and radiative capture on ^{177m}Lu. (b) Transmission coefficients calculated by optic model (P. Romain) for ^{177m}Lu.

In the thermal neutron flux of a reactor, the burn-up cross section can be written:

$$\sigma_{\text{burn-up}} = \sigma_{n,\gamma}^{\text{therm.}} + \int_{E_{Cd}} \varphi \cdot \sigma_{n,\gamma}^{\text{resonance}} \cdot dE + \sigma_{SE}^{\text{therm.}} + \int_{E_{Cd}} \varphi \cdot \sigma_{SE}^{\text{resonance}} \cdot dE \quad (1)$$

The ^{177m}Lu isomer was produced at Institute Laüe Langevin in Grenoble by thermal neutron irradiation [11,12] of a ^{176}Lu powder which was highly enriched (99.993%) in Bruyères-le-Châtel. Nanograms of Lutetium were extracted by chemical separation before producing the isomeric targets by a direct deposit method.

The radiative capture cross section was determined by measuring the ^{178}Lu activation by performing γ-ray spectroscopy of irradiated ^{177m}Lu targets. In this case, irradiation time and cooling time are short. ^{178m}Lu can be identified by the measurement of the half-life of its γ-rays. These measurements were achieved in various neutron flux at different temperature. Then, according to the Westcott convention, the thermal radiative capture cross section has been extracted. For a neutron velocity of 2200 m.s^{-1}, this value is 418±26 b13. For the burn-up cross section, the measurement procedure is a γ-spectroscopy of ^{177m}Lu both before and after a long irradiation time. Activities during both periods are determined as a function of time. The difference between the decay of the ^{177m}Lu initial sample without irradiation and the measured activity after irradiation gives the rate of the burn-up and the cross section. A first value of 590 b [12] was obtained. These measurements need several values in various neutron flux in order to obtain a thermal burn-up cross section on the basis of Westcott convention. The difference between these two thermal cross sections could be explained by the existence of ^{177m}Lu de-excitation via (n,n') super-elastic scattering. If the latter cross section is sufficiently high, it allows us to contemplate a direct observation of accelerated neutrons.

REFERENCES

1. P.M. Walker and G. D. Dracoulis, *Nature* 399, 35-40 (1999).
2. Ch. Briançon et al., 8th Intern. Symp. on Capture γ - Ray Spectroscopy, Fribourg, Ed. World Scientific (1994) p. 192.
3. L. Lakosi and T.C. Nguyen, *Nucl. Phys. A* **697**, 44 (2002).
4. K.E.G. Löbner, *Phys. Lett. B***26**, 369 (1968).
5. P.M. Walker et al., *Phys. Rev. Lett.* **67**, 433 (1991).
6. A. Bohr and B.M. Mottelson, Ed. Benjamin, New York (1975).
7. I.A. Kondurov, E.M. Korotkikh and Yu. V. Petrov, *JETP Lett.* **31**, 20 (1980).
8. I.A. Kondurov, E.M. Korotkikh, Yu. V. Petrov and G.I. Shuljak, *Phys. Lett. B* **106**, 383 (1981).
9. M. Jorgensen, O.B. Nielsen and G. Sidenius, *Phys. Lett.* **1**, 321 (1962).
10. G.D. Dracoulis et al., *Phys. Lett. B* **584**, 22 (2004).
11. L. Maunoury et al., *Nucl. Phys. A* **701**, 286c (2002).
12. O. Roig et al., *Nucl. Instrum. Methods. A* **521**, 5-11 (2004).
13 G. Bélier et al., *Phys. Rev. C*, to be published (2005).

γ-ray detection possibilities in the European Recoil Separator for Nuclear Astrophysics

D.Schürmann*, A.Di Leva*, L.Gialanella[†], D.Rogalla**, F.Strieder*, N.De Cesare[‡], A.D'Onofrio**, G.Imbriani[†], R.Kunz*, A.Ordine[†], V.Roca[†], C.Rolfs*, M.Romano[†], F.Schümann*, F.Terrasi** and H.-P.Trautvetter*

*Institut für Experimentalphysik III, Ruhr-Universität Bochum, Bochum, Germany
[†]Dipartimento di Sc. Fisiche, Università Federico II, Napoli and INFN, Napoli, Italy
**Dipartimento di Sc. Ambientali, Seconda Università di Napoli, Caserta and INFN, Napoli, Italy
[‡]Dipartimento di Sc. della Vita, Seconda Università di Napoli, Caserta and INFN, Napoli, Italy

The remarkable experimental efforts to study the radiative capture reaction $^{12}C(\alpha,\gamma)^{16}O$ over the last decades [1–9] focused on the observation of the capture γ-rays, including one experiment [4] that used the coincident detection of the ^{16}O recoils. Due to the low cross section and various background sources depending on the exact nature of the experiments, γ-ray data with useful but still inadequate precision were limited to center-of-mass energies $E_{cm} = E = 1.0$ to 3.2 MeV.

A new experimental approach has been undertaken at the 4 MV Dynamitron tandem accelerator in Bochum, called ERNA = European Recoil separator for Nuclear Astrophysics [10–14]. In this approach, the reaction is initiated in inverted kinematics, $^4He(^{12}C,\gamma)^{16}O$, i.e. a ^{12}C ion beam is guided into a windowless ^4He gas target. The recoil separator suppresses the intense ^{12}C beam; the ^{16}O recoils in a selected charge state are then counted in a ΔE-E telescope placed at the end of the beam line. ERNA is designed to study the reaction over the energy range $E = 0.7$ to 5.0 MeV. The detection of the ^{16}O recoils allows, for the first time, a direct measurement of the total cross section of $^{12}C(\alpha,\gamma)^{16}O$, including possible non radiative transitions. A detection system is installed around the gas target in order to observe also γ-rays in coincidence with the ^{16}O recoils: different detectors are available (HPGe, NaI and BaF$_2$), in order to optimize the specifications of the detection system to the experimental needs. The coincident measurement allows the identification of the different amplitudes involved in the radiative capture, where the coincidence condition suppresses the background in the γ-spectra by several orders of magnitude. Details about the γ-ray detection setup will be given elsewhere [15]. Recently the total cross section of the $^{12}C(\alpha,\gamma)^{16}O$ reaction in the energy range $E = 1.9$ MeV to $E = 4.9$ MeV was measured with ERNA [16]. As part of these measurements a new resonance was found in $^{12}C(\alpha,\gamma)^{16}O$ at an energy $E = 4.888$ MeV, corresponding to a known 0^+ state ($\Gamma = 1.5 \pm 0.5$ keV) reported in [17]. Fig. 1 shows a γ-spectrum collected on top of the resonance using a HPGe detector placed at 90° with respect to the beam axis. A spectrum collected about 30 keV off-resonance, normalized to the integrated beam current, is also shown. The cascade

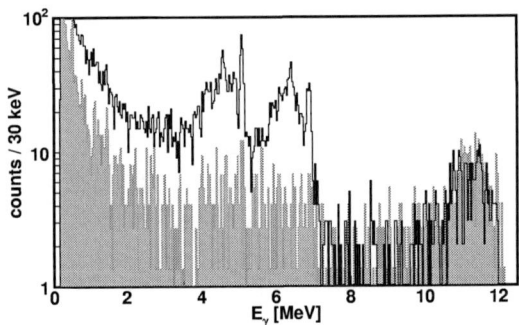

FIGURE 1. γ-spectrum at E = 4.892 MeV (solid line) with a HPGe detector at $\Theta_{lab} = 90°$. The shaded (light grey) curve shows the spectrum 30 keV off resonance and demonstrates, that the ground state transition at $E_\gamma = 12$ MeV is not originating from this 0^+ resonance.

transitions are significantly reduced while the observed yield of the ground state transition is unchanged. This demonstrates that the ground state transition at the energy of the 0^+ resonance is produced by the contribution of the tails of the neighbouring resonances underlying the 0^+ resonance. Therefore, the γ-decay of the resonance proceeds mostly through the $E_x = 6.92$ MeV state – 66 ± 15 % normalized to the observed number of resonant reactions – with a 15 ± 10 % branching to the $E_x = 7.12$ MeV state and a weak branching to the $E_x = 6.13$ MeV ($J^\pi = 3^-$) state. Although this resonance is not of direct interest for nuclear astrophysics, its discovery in the radiative channel is a good example for the sensitivity of the ERNA setup for γ-spectroscopy, that in the future will allow a clean identification of the different capture amplitudes in $^{12}C(\alpha,\gamma)^{16}O$.

REFERENCES

1. P.Dyer, and C.A.Barnes, *Nucl.Phys.A* **233**, 495 (1974).
2. K.-U.Kettner, et al., *Z.Phys.A* **308**, 73 (1982).
3. A.Redder, et al., *Nucl.Phys.A* **462**, 385 (1987).
4. R.M.Kremer, et al., *Phys.Rev.Lett.* **60**, 1475 (1988).
5. J.M.L.Quellet, et al., *Phys.Rev.C* **54**, 1982 (1996).
6. G.Roters, et al., *Eur.Phys.J.A* **6**, 451 (1999).
7. D.Rogalla, *Diploma Thesis*, Ruhr-Universität Bochum, 1997.
8. L.Gialanella, et al., *Eur.Phys.J.A* **11**, 357 (2001).
9. R.Kunz, et al., *Phys.Rev.Lett.* **86**, 3244 (2001).
10. D.Rogalla, et al., *Nucl.Instr.Meth.A* **437**, 266 (1999).
11. D.Rogalla, et al., *Eur.Phys.J. A* **6**, 471 (1999).
12. D.Rogalla, et al., *Nucl.Instr.Meth.A* **513**, 573 (2003).
13. L.Gialanella, et al., *Nucl.Instr.Meth.A* **522**, 432 (2004).
14. D.Schürmann, et al., *Nucl.Instr.Meth.A* **531**, 428 (2004).
15. D.Schürmann, et al. (2005), `to be published`.
16. D.Schürmann, et al. (2005), `to be published`.
17. D.R.Tilley, H.R.Weller, and C.H.Cheves, *Nucl.Phys.A* **564**, 1 (1993).

The Radiative Strength Function Using the Neutron-Capture Reaction on ^{94,95}Mo

S.A. Sheets*,†, U. Agvaanluvsan**, M. Krticka‡, G.E. Mitchell*,†, J.A. Becker**, F. Becvar‡ and the DANCE Collaboration§

*North Carolina State University, Raleigh, NC
†Triangle University Nuclear Laboratory, Durham, NC
**Lawrence Livermore National Laboratory, Livermore, CA
‡Charles University, Prague, Czech Republic
§Los Alamos National Laboratory, Los Alamos, NM

Abstract. We investigate the energy and multiplicity distributions of gamma-rays following neutron capture in ^{94,95}Mo. By comparing simulated gamma cascades following radiative capture with experimental spectra we try to learn something about the photon strength function at energies below the separation energy. The point of this comparison is to try and confirm recent experiments that found an anomalous low energy enhancement in the photon strength function for molybdenum isotopes.

Keywords: Photon Strength Function, Electromagnetic transitions
PACS: 25.40.Lw, 23.20.-g

INTRODUCTION

The γ-decay strength function (γSF) for a γ-ray emission of multipolarity XL is defined as the average reduced partial radiation width per unit energy interval of resonance spacing D:

$$f_{XL}(E_\gamma) = \frac{<\Gamma_{XL}(E_\gamma)>_{avg}}{E_\gamma^{2L+1} D}. \qquad (1)$$

In other words, the γSF gives an average probability of emission of a gamma ray with energy E_γ and multipolarity XL. The γSFs are important inputs for compound nucleus calculations of capture cross sections, γ-ray production spectra, and isomeric state populations.
Recently, an anomalous enhancement has been seen in the γSF in iron and molybdenum isotopes below gamma ray energies of 4 MeV [1][2]. At this time there exist no theoretical model which can explain this low energy enhancement of the γSF. A goal of this work is to further investigate this low energy enhancement using the (n,γ) reaction measured with the DANCE detector array.

METHOD

Measurements are performed using the DANCE detector. DANCE is a highly segmented BaF$_2$ array which makes it suitable for measuring γ-cascades. Different models of the

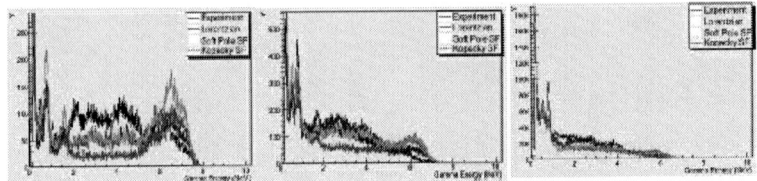

FIGURE 1. Gamma energy distribution for two, three, and four step cascades for ^{94}Mo

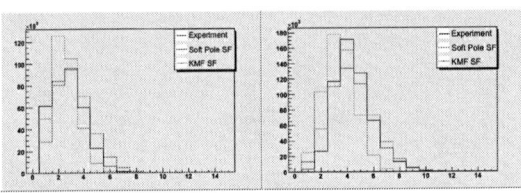

FIGURE 2. Multiplicity distribution for ^{94}Mo on the left and ^{95}Mo on the right

γSF will give different γ-ray multiplicity and γ-ray energy distributions. Our goal is to compare the experimental multiplicity and energy distributions to simulated distributions to infer the form of the γSF.

The monte carlo program DICEBOX is used to simulate gamma cascades for different models of the γSF and the level density. These cascades are given as input into a GEANT simulation of the DANCE detector response. Fig. 1. gives the result of a comparison between the experimental and simulated spectra for ^{94}Mo. By setting neutron energy gates around the resonances and sum energy gates around the Q-value we eliminate most of the background to get a clean capture gamma spectra. So far the multiplicity energy cascades are not well reproduced by simulated spectra.

In Fig. 2 we compare the multiplicity distribution to two models of the γSF. The first model is the commonly used KMF model [3]. The second is the "soft-pole" strength function given by [2] to explain the low energy enhancement of the γSF. In both ^{94,95}Mo the "soft-pole" strength function better represents the experimental data.

CONCLUSION

Although the analysis is still in its preliminary stage we have found evidence that the enhancement seen in [1][2] exists. We expect further work will clarify the nature of the low energy behavior of the γSF in ^{94,95}Mo.

REFERENCES

1. A. Voinov, and et al., *Phys. Rev. Lett* **93**, 142504 (2004).
2. M. Guttormsen, *Phys. Rev. C* **71**, 044307 (2005).
3. S. Kadmenskii, V. Markushev, and V. Furman, *Sov. J. Nucl. Phys.* **27**, 165 (1983).

Multichannel R-matrix Analysis of CNO Cycle Reactions

E. C. Simpson*, R. E. Azuma†, M. Wiescher*, A. Champagne**, P. Bertone**, H. -P. Trautvetter‡, J. Görres* and C. Ugalde**

Department of Physics, University of Notre Dame, Notre Dame, IN
†*University of Toronto, Toronto, Canada*
**University of North Carolina, Chapel Hill, NC*
‡*Institut für Experimentalphysik III, Ruhr-Universität Bochum, Bochum, Germany*

Abstract. The CNO cycle is the main process for hydrogen burning in stars somewhat larger than the Sun. The reaction cross sections at Gamow energies are typically in the femto to pico-barn range and are consequently very difficult to measure experimentally. The CNO reaction rates are based on extrapolations of experimental data from higher energies. We have developed a multi-channel R-matrix code to provide a new tool for fitting experimental data and making extrapolations to lower energies in all reaction and scattering channels. This approach does not only allow for a more reliable extrapolation, it also provides insight and guidance for the next generation of low energy experiments.

Keywords: R-matrix, CNO Cycle
PACS: 25.40.Lw Radiative capture - 26.20.+f Hydrostatic stellar nucleosynthesis

THE ^{14}N(p,γ)^{15}O REACTION

The ^{14}N(p,γ)^{15}O reaction determines the overall rate of the CNO cycle and hence energy production in stars larger than 1.5 solar masses. It is also plays a role in determining the age of globular clusters. Recently this reaction was measured at LUNA (Laboratory for Underground Nuclear Astrophysics, Gran Sasso, Italy,) [1] and LENA (Laboratory for Experimental Nuclear Astrophysics, North Carolina, USA) [2]. In this analyis the ^{14}N(p,p)^{14}N scattering channel has also been included [3]. Preliminary fits, which include the data of [4], are shown in Figure 1.

Our preliminary results are in reasonable agreement with results from the literature (see Table 1). Further analysis is underway to determine the effect of the elastic scat-

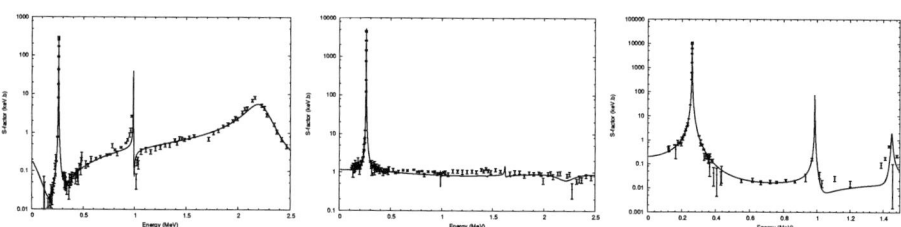

FIGURE 1. Preliminary R-matrix fits for γ transitions to the ground, 6.79 MeV and 6.18 MeV states.

TABLE 1. Summary of S(0) values (in units of keV b)

Transition to (MeV)	Ref. [4]	Ref. [5]	Ref. [1]	Ref. [2]	Present
0.00	1.55±0.34	$0.08^{+0.13}_{-0.06}$	0.25±0.06	0.49±0.08	0.14
6.18	0.14±0.05	$0.06^{+0.11}_{-0.02}$	$0.06^{+0.11}_{-0.02}$ *	0.04±0.01	0.20
6.79	1.41±0.02	1.63±0.17	1.35±0.05	1.15±0.05	1.17

* Taken from Ref. [5]

tering channel on the low energy extrapolation. Measurements of the subthreshold state lifetime [6] may offer some additional constraint on the S(0) in the ground state transition. S_{total} measurements may constrain the S(0) further still.

THE ^{15}N(p,γ)^{16}O AND ^{15}N(p,α)^{12}C REACTIONS

The branching between the first and second CNO cycles is defined by the ^{15}N(p,γ)^{16}O [7] and ^{15}N(p,α)^{12}C [8, 9, 10] reactions. We are reanalysing these reactions incorporating the ^{12}C(α,α)^{12}C [11] channel. Figure 2 shows preliminary fits.

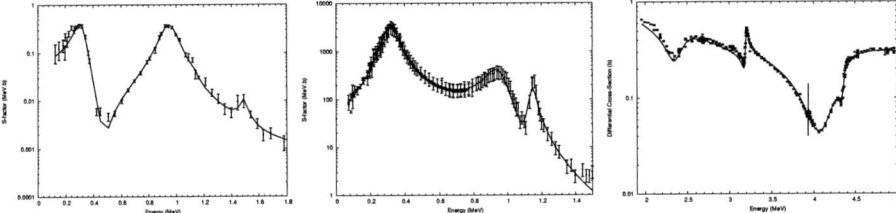

FIGURE 2. Preliminary R-matrix fits for ^{15}N(p,γ)^{16}O, ^{15}N(p,α)^{12}C and ^{12}C(α,α)^{12}C

Currently only data up to the proton threshold (5 MeV) has been considered in the ^{12}C(α,α)^{12}C channel, though higher energy data is available. As the α exit channels share R-matrix parameters, the reliability of the extrapolations in the ^{15}N(p,α)^{12}C should be improved. This in turn should improve the reliability in the γ exit channel.

REFERENCES

1. A. Formicola et. al., *Physics Letters B* **591**, 61 (2004).
2. R. C. Runkle et. al., *Physical Review Letters* **94**, 082503 (2005).
3. A. J. Ferguson et. al, *Elastic Scattering of Protons by Nitrogen*, Atomic Energy of Canada Report PD-261 (1956).
4. U. Schröder et. al., *Nuclear Physics A* **467**, 240 (1987).
5. C. Angulo, and P. Descouvement, *Nuclear Physics A* **690**, 755–768 (2001).
6. P. F. Bertone et. al., *Physical Review Letters* **87**, 152501 (2001).
7. C. Rolfs, and W. S. Rodney, *Nuclear Physics A* **235**, 450 (1974).
8. A. Redder et. al., *Zeitschrift für Physik A* **305**, 325 (1982).
9. J. L. Zyskind et. al., *Nuclear Physics A* **320**, 404 (1979).
10. A. Schardt et. al., *Physical Review* **86**, 527 (1952).
11. P. Tischhauser et. al., *Physical Review Letters* **88**, 072501 (2002).

Spectroscopic structure of exotic ^{19}Na. Astrophysics implication.

B.B. Skorodumov*, G.V. Rogachev[†,*], P. Boutachkov*, A. Aprahamian*, J.J. Kolata*, L.O. Lamm*, M. Quinn* and A. Woehr*

ISNAP, University of Notre Dame, Notre Dame, IN 46556
[†]*Physics Department, Florida State University, Tallahassee, FL 32306*

Abstract. The structure of proton-rich isotope ^{19}Na was studied in resonance elastic scattering of a radioactive ^{18}Ne beam on a proton target using inverse-kinematics method [1]. Only one state of ^{19}Na (the second excited state) was observed. A combined R-matrix and potential model analysis was performed. The spin-parity assignment of the second excited state was confirmed to be $1/2^+$. We show that the position of the $1/2^+$ state significantly affects the astrophysically important ^{18}Ne(2p,γ) reaction rate through that state but the totalreaction rate remains unchanged 2p-capture on ^{18}Ne proceeds mostly via the ground and first excited states in ^{19}Na at stellar temperatures.

PACS: 21.10.-k, 24.30.-v

INTRODUCTION

The structure of light exotic isotopes is currently one of the major topics in nuclear physics. Properties of light neutron-deficient isotopes are not as well known as the properties of their neutron-rich mirrors. One such example is ^{19}Na, a proton drip-line isotope which is unbound with respect to proton emission by only 321 keV. The main goal of this work was to extend our knowlegde of the structure of ^{19}Na to higher excitation energies (up to 2.7 MeV) using the resonance elastic scattering of a radioactive beam of ^{18}Ne on protons (the upper limit in excitation energy is implied by the maximum possible energy of ^{18}Ne beam that can be produced at ISNAP Ref. [2]). The structure of low lying resonances in ^{19}Na also has some astrophysical interest. It was shown in Ref. [3] that at certain stellar temperature and density conditions two-proton capture reaction on ^{18}Ne can play an important role in bridging the waiting point nucleus ^{18}Ne and provide continuous flow between the CNO and the FeNi mass region in the rp-process. The ^{18}Ne(2p,γ) reaction rate was recalculated taking into account new information on the excitation energy of the second excited state.

RESULTS

The experiment was carried out at the TwinSol radioactive nuclear beam facility Ref. [2] of ISNAP (Institute of Structure and Nuclear Astrophysics) at the University of Notre Dame. A beam of ^{18}Ne was produced via the ^{3}He(^{16}O,^{18}Ne)n reaction and used later for elastic scattering on polyethylene (CH$_2$) target. The Thick-Target Inverse-Kinematics (TTIK) technique Ref. [1] was used to obtain an excitation function for resonance elastic scattering. In the conversion of raw spectra to the absolute cross section we used the measured ratio of ^{18}Ne to the intensity of the primary beam ^{16}O. The kinematic relations and energy losses of protons in the target were taken into account in the transformation to the c.m. excitation functions. Monte Carlo simulations of the experimental energy resolution were performed for several excitation energies and angles. The typical experimental resolution at 165° was found to be about 30 keV at c.m. energies above 1 MeV. The precision of the excitation energies is about 20 keV, defined mainly by the uncertainties in the specific energy losses of ^{18}Ne ions in the CH_2 target. The absolute cross section values are obtained with uncertainties of 15%.

Only one state, at an excitation energy of 0.74 MeV, was observed. Its spin and parity were confirmed to be $1/2^+$, Ref. [4]. A potential model approach was used to obtain the single particle spectroscopic factor of this resonance. We show that the state has a reasonably pure single-particle structure $(2s1/2)^1$, in agreement with shell-model calculations Ref. [5]. The absence of another low-lying state (observed at an excitation energy of 2.37 MeV in the mirror nucleus ^{19}O) in the ^{18}Ne+p excitation function is indirect confirmation of the shell-model prediction that this is a $9/2^+$ state built on an excited states of ^{18}Ne. As such, it would be very narrow and only weakly populated in elastic scattering.

We did not observe a broad peak at 2.4 MeV, found in Ref. [6], which indicates that this peak is unrelated to elastic scattering and may be the result of sequential 2p decay from the higher lying excited states of ^{19}Na (as argued in Ref. [6]). The potential model with potential parameters obtained in this work was used to predict the excitation energy and width of a $3/2^+$ single particle state in ^{19}Na. This state should be well in the range of the experimental data of work Ref. [6], but there is no indication of this state. Additional measurements at higher excitation energies (and different angles) are needed to locate this resonance or understand the reason for the disappearence of this state.

The ^{18}Ne(2p,γ) astrophysical reaction rate was recalculated using the new experimental information on the excitation energy of the $1/2^+$ state. The (2p,γ) reaction rate due to this state is actually 100 times lower than it was suggested in previous work Ref. [3], however, due to the fact that the ^{18}Ne(2p,γ) reaction at stellar temperatures is dominated by the ground and the first excited state this finding did not make significant impact on the total reaction rate.

REFERENCES

1. K. P. Artemov et.al. *Sov. J. Nucl. Phys.*, 52, 406, 1990.
2. M. Y. Lee et. al. *NIM, A*, 422, 536, 1999.
3. J. Gorres et. al. *Phys. Rev. C.*, 51, 392, 1995.
4. C. Angulo et. al. *Phys. Rev. C.*, 67, 014308, 2003.
5. A. Volya and V. Zelevinsky. *Phys. Rev. Lett.*, 94, 052501, 2005.
6. F. Oliveira, et. al. *Eur. Phys. J. A*, 24, 237, 2005

Development Of A Prompt Gamma-ray Analysis Combined With Multiple Gamma-ray Detection

Y. Toh, M. Oshima, M. Koizumi, A. Osa and A. Kimura

Japan Atomic Energy Research Institute, Tokai, Ibaraki 319-1195,Japan

Abstract. By applying the multiple gamma ray detection method to PGA, the interference from strong gamma ray can be reduced, therefore quantification limits of trace elements are improved significantly. MPGA detector system is constructed at the guide-hall of JRR-3M in JAERI. Several standard samples were measured by MPGA detector system.

Keywords: MPGA, neutron capture, prompt gamma-ray
PACS: 01.50.Pa, 23.20.Lv, 29.30.Kv

INTRODUCTION

Prompt gamma ray analysis (PGA) is a rapid and non-destructive technique[1]. The chemical form and shape of the sample are relatively unimportant. However, since many prompt gamma-rays are emitted by the excited compound nuclei by neutron capture, these interfere with each other.

Most of nuclei emit two or more prompt gamma rays simultaneously in the neutron capture reaction. In multiple prompt gamma ray analysis (MPGA), the prompt gamma rays are simultaneously measured by two or more sets of gamma ray detectors, these gamma rays are reconstructed in the pair of two prompt gamma rays, and it is added to the two dimensional spectrum which sets two axes as the energy for every events. The gamma-ray distribution corresponding to the pair of the gamma ray which are emitted simultaneously will be obtained. By analyzing this two-dimensional gamma-ray peak, interference has the tendency to decrease. MPGA needs not only the knowledge of energies of capture gamma-rays, but also nuclear structure data. While the energy is usually accurate, many of the intensity and the excited level structure are of poor knowledge. Complete sets of these nuclear data are required.

DETECTOR SYSTEM

MPGA detector system is constructed at the guide-hall of JRR-3M in Japan Atomic Energy Research Institute(JAERI). This system consists of three Clover Ge detectors with BGO Compton suppressor. Detectors geometry is optimized by GEANT4[2]. The sample-detector distance is approximately 5cm. The absolute efficiency of the system is about 8% for 1.0 MeV gamma-ray. The neutron flux of a C-2-3 port is about $1.0 \times 10^{6-8}$ neutron/cm^{2}sec and the beam energy is about 3.1meV. The data acquisition hardware consist of the VME system crate that contains Energy, Timing and Coincidence modules. DSP(Digital Signal Processing) and FPGA(Field programmable Gate Array) are used in the VME modules. The DAQ system can be easily upgraded to accommodate more detectors (Maximum 256 channel)[3]. Automatic liquid nitrogen filling system and automatic sample changing system and helium gas delivery system are installed.

PERFORMANCE TEST

Food, alloy and rare and toxic metal etc. were measured by MPGA detector system. Counting rate of room background is about 50 count/sec. Background from beam duct, target holder and air were suppressed by neutron collimator(enriched LiF tile) and replacing air nitrogen with helium, therefore total background is low enough for MPGA. Figure1 shows the two dimensional spectrum of standard 50μg cadmium sample which is sealed in FEP (fluorinated ethylene propylene) film. Measuring time is 600 seconds and event rate is about 1.5k count/sec. Cadmium emits some prompt gamma rays, such as 558, 576, 651 and 725 keV, simultaneously. The coincident gamma ray pair of 558 and 651 keV has the strongest intensity. The quantification limits and accuracy of analytical results mainly depend on the matrix of sample and nuclear data, respectively. We plan to develop a database for typical samples and elements.

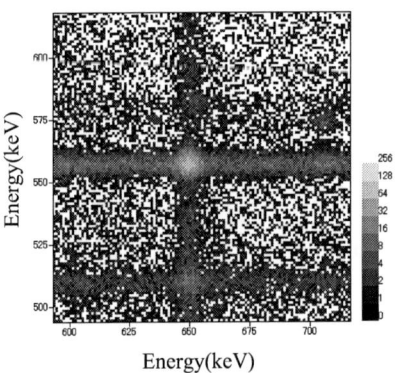

FIGURE 1. The spectrum of standard cadmium sample (50μg).

ACKNOWLEDGMENTS

We would like to express the gratitude to the crew of the JRR-3M for providing the neutron beam. This study is supported by Industrial Technology Research Grant Program in 03A52003c from New Energy and Industrial Technology Development Organization (NEDO) of Japan.

REFERENCES

1. R.E. Chrien, in: Neutron Radiative Capture, ed. B.J. Allen, I. Bergqvist, R.E. Chrien, D. Gardner, W.P. Poenitz, 1984. Pergamon Oxford, pp. 187.
2. S. Agostinelli, J. Allison, K. Amako, J. Apostolakis, H. Araujo, P. Arce, M. Asai, D. Axen, S. Banerjee, G. Barrand, et al., 2003. *Nucl. Instr. Meth. A***506**, 250-303(2003)
3. A.Kimura, Y. Toh, M. Koizumi, A. Osa, M. Oshima, J. Goto and M. Igashira, "Development of a Data Acquisition system for a multiple gamma-ray detection method" in *International Conference on Nuclear Data for Science & Technology - ND2004*, edited by R C. Haight et al., AIP Conference Proceedings 769, American Institute of Physics, Melville, NY, 2005, pp. 792-795.

Systematic Studies of Odd Isotopes in the Vicinity of the Closed Shell Z=50

J. Honzátko[*], I. Tomandl[*], V. Bondarenko[†], T. von Egidy[‡],
H.-F. Wirth[‡], R. Hertenberger[§], Y. Eisermann[§], G. Graw[§],
D. Bucurescu[¶] and V. Yu. Ponomarev[#]

[*]*Nuclear Physics Institute, 250 68 Řež, Czech Republic*
[†] *University of Latvia, Kengaraga 8, Riga, LV–1063, Latvia*
[‡] *Technische Universität München, D-85748 Garching, Germany*
[§] *Ludwig-Maxmilians Universität München, D-85748 Garching, Germany*
[¶]*Institute of Physics and Nuclear Engineering, 76900 Bucharest, Romania*
[#]*Institut für Kernphysik, Technische Universität Darmstadt, D-64289 Darmstadt, Germany*

Abstract. The isotopes in vicinity of the closed shell Z=50, the long chain of odd Te isotopes from ^{119}Te to ^{131}Te, ^{125}Sn and ^{123}Sn, have been investigated by means of (d,p), (d_{pol},p), (d_{pol},t), (^{3}He,α) and (n,γ) reactions. The experimental data were interpreted within the framework of Interacting Boson Fermion Model (IBFM) and Quasiparticle Phonon Model (QPM).

INTRODUCTION

The long chain of available Te isotopes provides a nice laboratory for the detailed investigation of evolution of various nuclear properties going from ^{119}Te, neutron midshell with N=67, to ^{131}Te, almost closed shell N=79, which approaches the doubly magic nucleus ^{132}Sn. In order to address this challenging task we studied all available odd Te isotopes, ^{119}Te [1], ^{121}Te [2], ^{123}Te [3], ^{125}Te [4], ^{127}Te [5], ^{129}Te [6] and ^{131}Te [7]. Recently, we have extended this systematics with a study of ^{125}Sn and ^{123}Sn.

EXPERIMENTAL PROCEDURES AND RESULTS

To obtain extensive and complete experimental information for all studied nuclei we combined results from light particle induced transfer and thermal neutron capture reactions, which provided mutually complementary data. The thermal neutron capture measurements, single γ spectra as well as $\gamma - \gamma$ coincidences, were performed at the light-water reactor LWR-15 at Řež near Prague [8]. The products

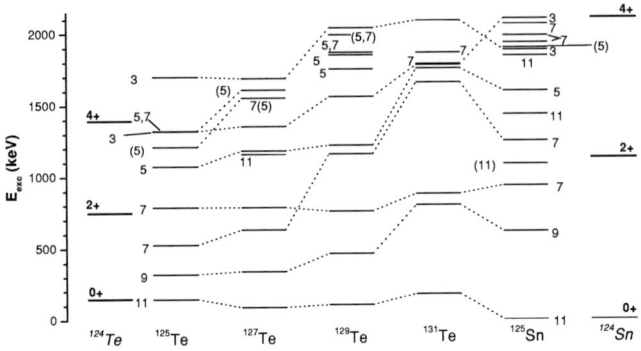

FIGURE 1. Part of the systematic of experimental negative-parity levels in the heavy odd-mass Te nuclei compared to those in ^{125}Sn. Double spin values are given for odd nuclei.

from light particle induced transfer reactions were analyzed with the Q3D magnetic spectrograph [9] at the Tandem Accelerator Laboratory in Garching.

Our recent results on ^{125}Sn have brought data for the investigation of changes of various nuclear properties passing from the Te chain to the Z=50 shell. One example of the extension of the Te systematic is given in Fig. 1. Similarly to heavy Te isotopes, we have obtained the evidence for the very dominant contribution of the direct capture mechanism in the reaction ^{124}Sn(n,γ)^{125}Sn and have observed the whole γ-decay intensity from the capture state. We have identified 99 % of the ground and isomer state population and 88 % of the capture state depopulation. This small incompleteness of the later value is caused by the presence of the very low-energy primary transitions. The extensive (d,p) data from the polarized beam experiment up to excitation energies of 5.3 MeV enable us to disclose the fragmentation of $2f_{7/2}$, $2f_{5/2}$, $3p_{3/2}$, $3p_{1/2}$ shells over three orders of magnitude. On the other hand, we can see that the population of the first $11/2^-$ state is weaker than in the Te isotopes. This can be explained by much smaller contribution of the $p_{3/2}$ component in the configuration of the low-spin members of the $h_{11/2}$ family of states in ^{125}Sn.

REFERENCES

1. D. Bucurescu et al., *Nucl. Phys.* **A674**, 11 (2000).
2. D. Bucurescu et al., *Nucl. Phys.* **A672**, 21 (2000).
3. V. Bondarenko et al., *Nucl. Phys.* **A673**, 85 (2000).
4. J. Honzátko et al., *Nucl. Phys.* **A645**, 331 (1999).
5. J. Honzátko et al., *Nucl. Phys.* **A756**, 249 (2000).
6. H.-F. Wirth et al.,*Nucl. Phys.* **A716**, 3 (2003).
7. I. Tomandl et al., *Nucl. Phys.* **A717**, 149 (2003).
8. J. Honzátko et al., *Nucl. Instr. Methods* A **376**, 434 (1996).
9. M. Löffler et al., *Nucl. Instr. Methods* **111**, 1 (1973).

Mass of the lowest $T = 2$ state of ^{32}S

S. Triambak*,†, A. García*, E. G. Adelberger*, G. J. P. Hodges*, H. E. Swanson*, S. A. Hoedl*, S. K. L. Sjue* and A. L. Sallaska*

*Physics Department, University of Washington, Seattle, Washington 98195
†Department of Physics, University of Notre Dame, Notre Dame, Indiana 46556

Abstract. We determined the mass of the lowest $T = 2$ state in ^{32}S with unprecedented accuracy and precision ($\Delta M/M \approx 10^{-5}$). The state was produced via the ^{31}P(p,γ) reaction and the energies of the de-excitation γ rays measured with high accuracy. This, together, with a recent measurement of the mass of ^{32}Ar provides a stringent test of the Isobaric Multiplet Mass Equation (IMME) for the $A = 32$ multiplet. A significant violation of the IMME is observed.

Keywords: Isobaric Multiplet Mass Equation, Doppler shift, isopsin mixing
PACS: 21.10.-k, 25.40.Ny

INTRODUCTION

If the nucleon-nucleon interaction is expressed in terms of isospin as a two-body charge dependent force, then the charge dependent Hamiltonian can be written as the sum of an isoscalar, an isovector and an isotensor of rank 2 to form an isospin multiplet, the masses of whose members are related by the Isobaric Multiplet Mass Equation [1] :

$$M(T_z) = a + bT_z + cT_z^2 \qquad (1)$$

where $T_z = \frac{N-Z}{2}$.

A recent measurement of the ^{32}Ar mass made the lowest $A = 32$, $T = 2$ multiplet the most precisely measured quintet. But the mass of the $T_z = 0$ member of the multiplet stands on a weak footing. On one hand, a ^{31}P(p,α) study yielded an excitation energy of 12049(2) keV [2], while an earlier ^{31}P(p,γ) measurement obtained a significantly different excitation energy of 12045.0(4) keV [3] with no published details about the difficult task of determining the excitation energy to such high precision. In this paper we present results of a remeasurement of the excitation energy of the lowest $T = 2$ state of ^{32}S using the ^{31}P(p,γ) reaction. This provides us with the most stringent test of the Isobaric Multiplet Mass Equation.

EXPERIMENTAL PROCEDURE AND DATA ANALYSIS

The ^{31}P(p,γ) measurement was performed using the University of Washington FN tandem accelerator. An $\approx 6\mu$A, 3.285 MeV proton beam was bombarded on an implanted ^{31}P water-cooled target to populate the state of interest. Two independent measurements were done at different times using 50% HPGe detectors. Firstly, data were taken using two detectors at ≈ 60 cm from the target at $\pm 90°$ to the beam and in a later measurement data were taken with a single detector at 15.8 cm from the target and 0° to the beam. The γ ray energies were calibrated using a ^{56}Co source present at all times during data acquisition and ^{27}Al(p,γ) and ^{35}Cl(n,γ) capture radiation. Gamma rays with well known energies up to ≈ 10 MeV from ^{27}Al(p,γ) were produced by bombarding a 15μA, 992 keV proton beam on a 20$\mu g/cm^2$ thick ^{27}Al target, while neutrons for the latter were produced by impinging a 600 nA, 1.912 MeV proton beam on a thick ($\approx 500\mu g/cm^2$) Li$_2$O target to initiate the ^{7}Li(p,n) reaction. These were further moderated by a 4 cm thick paraffin slab before capturing onto a large volume of NaCl to produce γ rays with well known energies up to ≈ 8.5 MeV.

The data showed that in spite of using a temperature controlled electronics rack small gain-shits were observed which had to be corrected for. In additon, careful Monte Carlo calculations were performed to simulate Doppler shifts on γ energies. Doppler effects on γ rays from the (p,γ) reactions were simulated using the half-life of the decaying state to randomly generate decay times from an exponential distribution and further taking into consideration the energy loss by the nucleus during that lifetime and the target-detector misalignment. For γ rays from the ^{35}Cl(n,γ) reaction the neutron angular distribution was fed into a program that simulated neutron scattering on paraffin and capture on ^{35}Cl to provide us with the eventual recoil velocities of the nuclei prior to γ emission. ADC non-linearities and incident γ-ray angle dependence on peak centroids were studied carefully. The points in the non-linear region were not used for gain shift correction or energy determination. The incident γ ray angle dependence on peak centroids was a relatively insignificant effect.

Peak centroids were obtained by fitting the γ peaks with a Gaussian folded with a delta function and two low energy exponential tails.

RESULTS AND CONCLUSIONS

There is excellent agreement between γ energies derived from the $\pm 90°$ detectors and the Doppler corrected γ energies from the 0° detector. We obtain an excitation energy of 12048.26(30) keV which agrees with Ref. [2] with higher precision but is $\approx 8\sigma$ higher than the value reported by Antony et. al. On using the best available results for the masses of other four members of the $A = 32, T = 2$ multiplet we observe a significant violation of the IMME with $Q(\chi^2, \nu) = 0.0003$. However, excellent agreement with the IMME is obtained if the excitation energy of the lowest $T = 2$ state in ^{32}S is ≈ 2.5 keV lower than our result. It is possible that isospin mixing with an existing narrow $T = 0$, $J^\pi = 0^+$ level ≈ 118 keV below the $T = 2$ state shifts the state upward by ≈ 2.5 keV.

REFERENCES

1. E. P. Wigner, *Proceedings of the Robert A. Welch Conferences on Chemical Research, Houston* (Robert Welch Foundation, Houston, 1957), Vol. 1.
2. J. F. Wilkerson *et. al.*, Nucl. Phys. A **549**, 223, (1992).
3. M. S. Antony *et al.* in *Proceedings of the International Conference on Nuclear Physics, Berkeley, 1980* (Lawrence Berkeley Laboratory, Berkeley, CA, 1980), Vol. 1.

The Complete (n,γ) Level Scheme of ^{124}Te

T. von Egidy*, H.-F. Wirth*, I. Tomandl[†] and J. Honzatko[†]

Physik-Department, Technische Universität München, D-85748 Garching, Germany
[†]*Nuclear Physics Institute, CZ-25068 Rez, Czech Republic*

Abstract. Using the (n,γγ) reaction a level scheme of ^{124}Te was established with more than 150 levels. The level density and population properties are investigated.

Keywords: ^{123}Te(n,γγ)^{124}Te, E_n = thermal, ^{124}Te levels deduced
PACS: 21.10.-k, 21.10.Ma

EXPERIMENTS AND RESULTS

Recent investigations of Te isotopes using the (n,γ) reaction by the Rez-Munich-Riga-Bucharest collaboration demonstrated that rather complete nuclear level schemes can be constructed. Also the level scheme of ^{124}Te is complete up to 3 MeV [1]. However, it was on the other hand still quite incomplete, because only 8% of the capture state depopulation and only 77% of the ground state population was established. This is surprising and therefore a challenge for experimental nuclear physicists, since the neutron capture cross section of ^{123}Te is very large (418b). Consequently we studied the ^{123}Te(n,γ)^{124}Te reaction with γ − γ coincidences at the research reactor of Rez near Prague. More than 150 levels were established. All strong γ transitions were placed in the level scheme, 475 transitions by coincidence. The level scheme is supposed to be complete in the spin range 0 - 4 up to 3.1 MeV (less than 5% missing or wrongly placed levels). The nuclear structure of ^{124}Te has been carefully investigated by several authors [2, 3].

LEVEL DENSITY

Above 3 MeV the level density increases dramatically and only a small fraction of the levels is observed in the (n,γ) reaction. This is demonstrated in Fig.1 (left) where the cumulative number of levels is plotted as function of the level energy. It was shown in several publications [4] (and references therein) that the exponential increase of the level density can be described with only two free parameters. We applied the fit procedure of Ref. [4] to the first 61 levels including the resonance density at the neutron binding energy of 9.4 MeV and obtained for the backshifted Fermi gas (BSFG) and the constant temperature (CT) models the parameters given in Fig.1. The BSFG model reproduces the level density very well in particular keeping in mind that also a point at 9 MeV determines the two parameters. The CT model does not work so well.

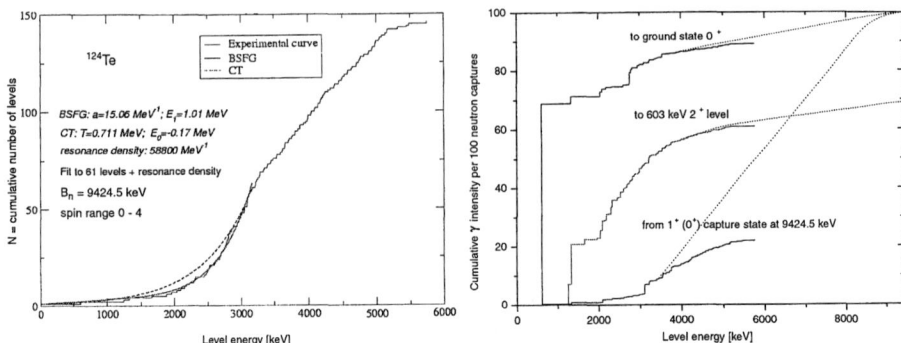

FIGURE 1. Left: Cumulative number of levels as a function of the excitation energy compared with fits of the BSFG (backshifted Fermigas) and CT (constant temperature) models. Right: Cumulative intensity of γ lines coming from the neutron capture state at 9424.5 keV and of γ lines populating the ground state or the first 2^+ level. The dotted lines show the very rough extrapolated behaviour.

POPULATION PROPERTIES

A very interesting feature of the present (n,γ) reaction is the flux of γ intensity from the neutron capture state to the ground state. This is illustrated in Fig.1 (right). The lower curve is the cumulative γ intensity depopulating the capture state. It is shown that only 20% of the depopulation is observed. This means that 80% goes via weak transitions to numerous levels above 3.5 MeV. The dotted line indicates the probable extrapolation to the final 100% at 9.4 MeV. A detailed theoretical calculation has to be performed including the fragmentation of E1 transition strength in order to explain this behaviour.

The second curve shows the cumulative γ intensity populating the 603 keV 2^+ level. The total observed populating intensity is 61%, while the depopulating intensity is 69%. About 8% of the population is missing and indicated by the dotted line.

The third curve displays the cumulative γ intensity to the ground state. Comparing the ground state with the first 2^+ level, it was assumed that the total observed ground state population is 90%. The largest population is given by the $2^+ - 0^+$ transition of 69%.

REFERENCES

1. R. Georgii et al., *Nucl. Phys.* **A592** (1995) 307.
2. C. Doll et al., *Nucl. Phys.* **A672** (2000) 3.
3. N. Warr et al., *Nucl. Phys.* **A636** (1998) 379.
4. T. von Egidy and D. Bucurescu, *Phys. Rev.* **C**, submitted.

Nature of One- and Two-Phonon Mixed Symmetry States in ^{92}Zr and ^{94}Mo from High-Resolution Electron and Proton Scattering

P. von Neumann-Cosel[*], N.T. Botha[†], O. Burda[*], J. Carter[**],
R.W. Fearick[†], S.V. Förtsch[‡], C. Fransen[§], H. Fujita[**,‡], M. Kuhar[*],
A. Lenhardt[*], R. Neveling[‡], N. Pietralla[¶], V.Yu. Ponomarev[*], A. Richter[*],
E. Sideras-Haddad[**], F.D. Smit[‡] and J. Wambach[*]

[*]*Institut für Kernphysik, Technische Universität Darmstadt, D-64289 Darmstadt, Germany*
[†]*Physics Department, University of Cape Town, Rondebosch 7700, South Africa*
[**]*School of Physics, University of the Witwatersrand, Johannesburg 2050, South Africa*
[‡]*iThemba LABS, PO Box 722, Somerset West 7129, South Africa*
[§]*Institut für Kernphysik, Universität zu Köln, D-50937 Köln, Germany*
[¶]*Department of Physics & Astronomy, SUNY, Stony Brook, NY 11794-3800, USA*

Abstract.
High-resolution inelastic electron (performed at the S-DALINAC) and proton (performed at iThemba LABS) scattering experiments on ^{92}Zr and ^{94}Mo with emphasis on $E2$ transitions are presented The measured form factors and angular distributions provide a measure for the F-spin purity, respectively the isovector nature, of the proposed one-phonon mixed symmetry states and furthermore provide a sensitive test of a possible two-phonon character of excited 2^+ states.

The concept of symmetric and mixed-symmetric quadrupole phonons as building blocks of the low-energy structure in spherical nuclei has been thoroughly investigated in recent years. A particularly interesting case are the $N = 52$ isotones. The nucleus ^{94}Mo has been studied in great detail and candidates for the mixed-symmetry one-phonon [1] and most of the two-phonon states [2, 3, 4] have been established based on characteristic decay pattern. This approach also provides a good description of ^{96}Ru [5] but seems to fail for ^{92}Zr indicating a substantial breaking of F-spin in the latter [6, 7].

We present high-resolution inelastic electron scattering (performed at the S-DALINAC) and proton scattering (performed at iThemba LABS) experiments on ^{92}Zr and ^{94}Mo with emphasis on $E2$ transitions populating 2^+ states. The energy resolution $\Delta E \approx 30$ keV (FWHM) was sufficient to resolve all experimentally known 2^+ states below $E_x = 4$ MeV. The measured form factors and angular distributions represent a sensitive test of a possible two-phonon character of excited states by comparison to QPM, shell model, and IBM-2 predictions. The results are complementary to the experimental approaches used so far [1–7] by measuring the one-phonon content in the wave functions of two-phonon candidates. The comparison of (e,e′) and (p,p′) results furthermore provides a measure for the F-spin purity, respectively the isovector nature, of the proposed one-phonon mixed symmetry states.

Figure 1 presents as examples the (e,e′) form factors and (p,p′) angular distributions of the transitions populating the symmetric and mixed-symmetry one-phonon states in

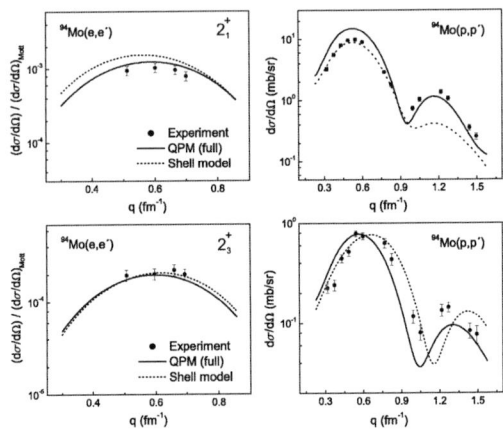

FIGURE 1. Excitation of the symmetric (2_1^+) and mixed-symmetry (2_3^+) one-phonon states in ^{94}Mo in high-resolution (e,e′) and (p,p′) experiments compared to QPM and shell-model calculations.

^{94}Mo. These are compared to QPM [8] (solid lines) and shell-model (dashed lines) calculations with the interaction described in [7]. The description within the QPM is satisfactory except for a slight over prediction of the (p,p′) cross sections for the transition to the 2_1^+ state. However, there is a systematic uncertainty of 20 – 30% depending on the choice of the effective target-nucleon interaction [9]. The results shown here were obtained with a G matrix parameterization derived from the Paris potential. The shell-model results show some deficiencies at higher momentum transfers tested in the proton scattering data, where correlations outside the valence space become important. Overall, the phonon character of the transitions to the 2_1^+ (symmetric) and 2_3^+ (mixed symmetric) states is confirmed by this analysis. The data furthermore provide detailed information about the purity of the proposed two-phonon states.

ACKNOWLEDGMENTS

This work has been supported by the DFG under contracts SFB 634 and 445 SUA 113/6/0-1 and by the South African NRF.

REFERENCES

1. C. Fransen et al., *Phys. Lett. B*, **508**, 219 (2001).
2. N. Pietralla et al., *Phys. Rev. Lett.*, **83**, 1303 (1999).
3. N. Pietralla et al., *Phys. Rev. Lett.*, **84**, 3775 (2000).
4. C. Fransen et al., *Phys. Rev. C*, **67**, 024307 (2003).
5. H. Klein et al., *Phys. Rev. C*, **65**, 044315 (2002).
6. V. Werner et al., *Phys. Lett. B*, **550**, 140 (2002).
7. C. Fransen et al., *Phys. Rev. C*, **71**, 054304 (2005).
8. N. Ryezayeva, et al., *Phys. Rev. Lett.*, **89**, 272502 (2002).
9. F. Hofmann, et al., *Phys. Lett. B*, **612**, 165 (2005).

CGS-12 Special Workshop on Isomers

Sunday, September 4th 2005

Chair: GEORGE DRACOULIS

10:00	(10) min	ANI APRAHAMIAN	Welcome and general remarks
10:10	(40+5) min	PHIL WALKER	Nuclear Isomers: Stepping Stones to the Unknown
10:55	(30+5) min	JAMES CARROLL	Status in the Search for the Triggered Depopulation of Nuclear Isomers
11:30	(30+5) min	ANTON TONCHEV	Nuclear Isomers as a Probe for Nuclear Structure and Application Studies
12:05	(30 + 5) min	ADAM HAYES	Violations of K-Conservation in ^{178}Hf

12:40 Lunch Break

Chair: UMESH GARG

13:50	(30+5) min	PARTHA CHOWDHURY	K-Isomers : Navigating an Approximate Symmetry through Spin, Isospin and Oscillator Shells
14:25	(30+5) min	YANG SUN	Challenge in the Theoretical Description of K-Isomers
15:00	(30+5) min	ROBERT GRZYWACZ	Isomer Studies near ^{78}Ni and ^{100}Sn

15:35 Coffee Break

Chair: YANG SUN

16:10	(30+5) min	PADDY REGAN	New Opportunities for Isomer Physics using the Stopped RISING Fragmentation Set-up at GSI
16:45	(30+5) min	ART CHAMPAGNE	Isomers and Nuclear Astrophysics

CGS-12 Conference Program

Monday, September 5th 2005

Nuclear Structure I (Isomers and Gamma Spectroscopy)

Chair: RICK CASTEN

8:30	(10+5) min	ANI APRAHAMIAN	Welcome and general remarks
8:45	(10+5) min	PROF. ANTHONY HYDER ASSOCIATE VICE PRESIDENT OF GRADUATE STUDIES AND OFFICE OF RESEARCH	Welcome in Behalf of Notre Dame
9:00	(25+5) min	GEORGE DRACOULIS	Isomers and Aspects of Nuclear Structure
9:30	(25+5) min	JONATHAN BILLOWES	Isomer and Ground State Properties of Yttrium and Scandium Isotopes Measured by Laser Spectroscopy
10:00	(15+5) min	NORBERT PIETRALLA	In-Beam γ-Spectroscopy of Low-Spin Mixed-Symmetry States of ^{138}Ce with Gammasphere in Singles-Mode
10:20	Coffee Break		

Nuclear Structure II : Dedicated to the Memory of D.D. Warner

Chair: ANI APRAHAMIAN

11:00	(25+5) min	RICK CASTEN	Emerging Collectivity in Nuclei, Shape Coexistence, and Proton-Neutron Interactions
11:30	(25+5) min	PIET VAN ISACKER	Symmetry and Wigner Energy in Nuclei
12:30	(25+5) min	TAKAHARU OTSUKA	Varying Shell Structure of sd-pf Nuclei
13:00	(25+5) min	ALISON BRUCE	Light Proton Drip Line Nuclei Studied Using Secondary-Fragmentation Reaction at GSI
13:30	Lunch break		

Nuclear Astrophysics I (CN0 Cycle)

Chair: MICHAEL WIESCHER

14:30	(25+5) min	ART CHAMPAGNE	Seeing the Stars and Capturing Gamma-Rays at LENA
15:00	(15+5) min	HANNS-PETER TRAUTVETTER	Low Energy Underground Study of $^{14}N(p,\gamma)^{15}O$ Differential and Total Cross Section at LUNA
15:20	(15+5) min	RALF PLAG	An Independent Measurement of the $^{12}C(\alpha,\gamma)^{16}O$ Cross Section with the Karlsruhe 4π BaF$_2$ Detector
15:40	(15+5) min	AARON COUTURE	$^{19}F(p,\gamma)$: Putting a Lid on the CNO Cycle

16:00   Coffee break

Fundamental Physics

Chair: ALAN WUOSMAA

16:30	(15+5) min	VICTOR FLAMBAUM	Parity and Time Reversal Violation in Atomic Nuclei and Test of the Standard Model
16:50	(15+5) min	PIL-NEYO SEO	Measurement of Parity-Violating Gamma-ray Asymmetry in Radiative Neutron Capture on Hydrogen at LANSCE
17:10	(15+5) min	RAYMOND MOREH	Scattering of keV Neutrons and the Problem of Quantum Entanglement
17:30	(15+5) min	GARY MITCHELL	Random Matrix Thermodynamics
17:50	(15+5) min	MARCUS MCELLISTREM	Isospin Symmetry of Isobaric Triplets along the $N \sim Z$ Line
18:10	(15+5) min	SADIA CHOUDRY	Study of Charge Symmetry in ^{18}O and ^{18}Ne

18:30   Dinner Break

Data for Nuclear- and Astrophysics Applications with Reception (Room 210 - 214)

Chair: RICK FIRESTONE

19:30	(15+5) min	FILIP KONDEV	Nuclear Structure and Decay Data Evaluation: Current Status and Future Perspectives
19:50	(15+5) min	ALEXANDER SUKHARUK	An Updated Library of Reaction Rates for the Astrophysical rp-Process: The Importance of Nuclear Structure Effects
20:10	(15+5) min	IRIS DILLMANN	(n,γ) Cross Sections of Light p Nuclei: Towards an Updated Experimental Database for the s and p-Process and KADoNiS- An Experimental Online Database for the s and p Processes
20:30	(15+5) min	BORIS PRITYCHENKO	Nuclear Data Resources for Capture Gamma-Ray Spectroscopy and Related Topics: Nuclear Reaction and Structure Databases of the National Nuclear Data Center
20:50	(15+5) min	KIM GUINYUN	Measurements of Neutron Capture Cross

Sections for Gd Isotopes in the Energy Region from 10 keV to 90 keV

Tuesday, September 6th 2005

Nuclear Masses and Shell Model

Chair: GRANT MATHEWS

8:30	(25+5) min	ALEJANDRO FRANK	*Nuclear Forecasting as Pattern Recognition: Can we Predict Nuclear Masses?*
9:00	(25+5) min	JUHA ÄYSTÖ	*Recent Developments in Mass Measurements of Neutron-Rich Nuclei*
9:30	(15+5) min	M.B. GOMEZ HORNILLOS	*Mass Measurements with the CSS2 & CIME Cyclotrons @ GANIL*
9:50	(15+5) min	MILAN MATOS	*Recent Achievements and Future of Mass Measurements with Fast Radioactive Beams*
10:10	(15+5) min	BRUCE BARRETT	*Effective Interactions and Operators in Nuclei within the No-Core Shell Model*

10:30 Coffee break

Nuclear Astrophysics II (Charged Particle Reactions and Reaction Rates)

Chair: ART CHAMPAGNE

11:00	(15+5) min	WANGPENG TAN	*Threshold States in ^{19}Ne and the CNO Breakout Reaction $^{15}O(\alpha,\gamma)^{19}$Ne*
11:20	(25+5) min	HENDRIK SCHATZ	*Nuclear Astrophysics with fast Radioactive Beams*
11:50	(15+5) min	YANG SUN	*Nuclear Isomers: Structures and Applications*
12:10	(15+5) min	GYORGY GYURKY	*Radiative Capture Reactions and α-Elastic Scattering on ^{106}Cd for the Astrophysical p-Process*
12:30	(25+5) min	ALBERTO MENGONI	*Reaction Rates for Stellar Nucleosynthesis*

13:00 Lunch break

Nuclear Reactions I (Reactions with Neutron-Rich Nuclei)

Chair: JERRY WILHELMY

14:30	(25+5) min	JOLIE CIZEWSKI	*Neutron-Transfer Reactions with Exotic Neutron-Rich Beams:*
15:00	(15 +5) min	FILOMENA NUNES	*Do Halos Exist on the Drip Line of Deformed Nuclei?*
15:20	(15 +5) min	PLAMEN BOUTACHKOV	*Doppler Shift as a Tool for Studies of Resonant*

| 15:40 | (15+5) min | Emil Betak | (p,n) Reactions with RIBs, Spectroscopy of ^{7}He Model Calculations of Radiative Capture of Nucleons in MeV Region |

| 16:00 | | Coffee break | |

Nuclear Structure III (0^+ States)

Chair: Gabriela Popa

16:30	(25+5) min	Shelly Lesher	The Nature of 0^+ States in ^{158}Gd
17:00	(15+5) min	Nicola Lo Iudice	Pairing Versus Quadrupole Collectivity of Low-Lying 0^+ States in Deformed Nuclei
17:20	(15+5) min	Naotaka Yoshinaga	Recent Progress on the 0^+ Dominance
17:40	(15+5) min	Paul Garrett	Gamma-Ray Spectroscopy at TRIUMF-ISAC
18:00	(15+5) min	Bronwyn Hyland	Precision Branching Ratio Measurement of ^{62}Ga

| 18:20 | Break | | |

| 19:30 | Poster Session | | During the poster session, there is a reception with food and drinks. The reception starts at 19:00 hours. |

There will be a prize of $500 awarded to the best poster presented by a postdoctoral fellow or a graduate student. A selection committee of 3 distinguished scientists will be judging the posters. The prize is funded by **Prof. Steven Yates** of Kentucky University for CGS-12 in honor of the memories of **Jean Kern, Raman Subramanian** and **Gabor Molnar** (Founder's Award) and it will be awarded at the banquet.

Wednesday, September 7th 2005

Nuclear Astrophysics III (s-Process Nucleosynthesis)

Chair: KARL-LUDWIG KRATZ

8:30	(25+5) min	MICHAEL HEIL	The s Process and its Relation to Explosive Nucleosynthesis
9:00	(15+5) min	ANA ALPIZAR-VINCENTE	Measurement of Neutron Capture Cross Section of ^{62}Ni in the keV Region
9:20	(15+5) min	ROBERT HATARIK	^{102}Pd(n,γ) cross section measurement using DANCE
9:40	(15+5) min	STEFANO MARRONE	Measurement of the ^{139}La(n,γ) Cross Section at n_TOF
10:00	(15+5) min	CESAR DOMINGO	Measurement of the Resonance Capture Cross Section of ^{204,206}Pb and Termination of the s-Process
10:20	Coffee break		

Neutron Capture Reactions I

Chair: HANS BÖRNER

11:00	(15+5) min	UNDRAA AGVAANLUVSAN	The Radiative Strength Function Using the Neutron-Capture Reaction on ^{151,153}Eu
11:20	(15+5) min	TAMAS BELGYA	Determination of Thermal Neutron Capture Cross Sections Using Cold Neutron Beams at the Budapest PGAA-NIPS Facilities
11:40	(15+5) min	STEPHAN WALTER	Method for (n,γ) Cross Section Measurements on Unstable Isotopes
12:00	(15+5) min	ROBERT RUNDBERG	Neutron Capture Cross Sections of ^{236}U and ^{237}U
12:20	(15+5) min	DANIEL CANO-OTT	Neutron Capture Cross Section Measurements at n_TOF of ^{237}Np, ^{240}Pu and ^{243}Am for the Transmutation of Nuclear Waste
12:40	(15+5) min	RON NELSON	Neutron-Induced Gamma-Ray Cross Section Standards and ^{56}Fe Inelastic Scattering
13:00	Lunch break		

15:00	Boarding of busses in front of McKenna Hall
15:15	Departure of busses

Conference Excursion to Chicago and Banquet
The conference banquet will take place on board a ship and will include an architectural tour of the city of Chicago from the Chicago River.

Please bring your conference nametag to the excursion!

Thursday, September 8th 2005

Nuclear Structure IV

Chair: TORBEN LAURITSEN

8:30	(25+5) min	JAN JOLIE	Level Dynamics of Yrare States
9:00	(15+5) min	VOLKER WERNER	New Findings for Mixed-Symmetry States
9:20	(15+5) min	FRANTISEK BECVAR	Independent Evidence for M1 Scissors Resonances Built on the Levels in the Quasicontinuum of ^{163}Dy
9:40	(15+5) min	STEVE YATES	Complex Two-Phonon Structures in ^{93}Nb
10:00	(15+5) min	ANTON TONCHEV	Low-Energy Dipole Modes of Excitation Below the Neutron Separation Energy
10:20	(15+5) min	ADAM HECHT	Search for Enhanced Alpha Preformation in the N=Z+1 nuclei ^{113}Ba, ^{109}Xe, ^{105}Te

10:40 Coffee break

Neutron Capture Reactions II and Nuclear Structure

Chair: DENIS DE FRENNE

11:00	(15+5) min	CHRIS ANGELL	Photo-Induced Population of the $h_{11/2}$ Isomeric States in (γ,n) Reactions
11:20	(15+5) min	VLADIMIR SOLOVYEV	General Approach to Materials Classification Using Neutron Analysis Techniques
11:40	(15+5) min	MASAYUKI IGASHIRA	Measurement of Gamma Rays from keV-Neutron Capture Reaction on Zr-90, 94
12:00	(15+5) min	TOSHIRO OHSAKI	Measurement of Capture Gamma Rays from the 46- and 84-keV Neutron Resonances of ^{24}Mg
12:20	(15+5) min	KIM LISTER	Doorway States in the $^{12}C(^{12}C,\gamma)^{24}$Mg Radiative Capture Process
12:40	(15+5) min	ZOLTAN ELEKES	Bound Excited States in ^{27}F

13:00 Lunch break

Application and Development of Nuclear Techniques

Chair: JAMES KOLATA

14:30	(15+5) min	RICK FIRESTONE	Thermal Neutron Capture Cross Sections of the Palladium Isotopes
14:50	(25+5) min	JOHN BECKER	Recent experiments on Atomic-Nuclear Coupling
15:20	(15+5) min	ROBERT RUNDBERG	The Detector for Advanced Neutron Capture Measurements at LANSCE
15:40	(15+5) min	ALEXANDER LAPTEV	Distortion of Pulse-Height Spectra of Neutron Capture Gamma Rays

16:00 Coffee break

Nuclear Astrophysics IV (r-Process Nucleosynthesis)

Chair: FRANZ KÄPPELER

16:30	(25+5) min	KARL-LUDWIG KRATZ	Nuclear-Physics Issues of r-Process Nucleosynthesis
17:00	(15+5) min	KHALIL FAROUQUI	Neutron Captures and the r-Process
17:20	(15+5) min	PETER VON NEUMAN-COSEL	Magnetic Dipole and Gamow-Teller Modes in Neutrino-Nucleus Reactions: Impact on Supernova Dynamics and Nucleosynthesis
17:50	(15+5) min	MOSHE GAI	How Accurately Do We Know the Cross Section of the $^7Be(p,\gamma)^8Be$ Reaction?

18:10 Dinner break

Physics and Film

In the evening, there is the possibility to see movies as a part of the Physics & Film Series (sponsored by JINA and the Physics Department at the performing arts center of the University of Notre Dame).

19:00 Metropolis (1926)

20:00 The Andromeda Strain (1970)

Friday, September 9th 2005

Nuclear Structure VI (High Spin Experiments)

Chair: NORBERT PIETRALLA

8:30	(25+5) min	ATTILA KRASZNAHORKAY	Hyperdeformed Rotational Bands Observed in the Actinide Region
9:00	(15+5) min	SLAWEK KACZAROWSKI	High Spin Bands in the A ~ 130 Nuclei: A "Non-Chiral" Explanation
9:20	(15+5) min	INDRAMANI GOVIL	Nuclear Structure and Octupole Collectvity in ^{122}Cs Nucleus
9:40	(15+5) min	TORBEN LAURITSEN	Rotational Damping, Ridges and the Quasicontinuum of γ rays in ^{152}Dy
10:00	(25+5) min	PADDY REGAN	Multinucleon Transfer Reactions to Study Nuclear Structural Evolution

10:30 Coffee break

Nuclear Structure VII (Theory)

Chair: STEFAN FRAUENDORF

11:00	(25+5) min	NADYA SMIRNOVA	Nuclear Mean Field from the Shell Model Point of View
11:30	(15+5) min	ALEXANDER LISETSKIY	Softness of Doubly-Magic ^{78}Ni and Related Topics
11:50	(25+5) min	PAVEL CEJNAR	Regular and Chaotic Nuclear Vibrations: Monodromy, Bifurcations, Regular Islands...
12:20	(15+5) min	VLADIMIR ZELEVINSKY	Shell Model, Nuclear Reactions and Giant Resonances
12:40	(15+5) min	TILL VON EGIDY	Correlation of Nuclear Level Densities and Masses

13:00 Lunch break

Nuclear Reactions II

Chair: FILOMENA NUNES

14:30	(25+5) min	HANS BÖRNER	Precision Gamma-Ray Spectroscopy at the Institut Laue Langevin
15:00	(15+5) min	HENNING ESBENSEN	Radiative Capture versus Coulomb Dissociation Experiments
15:20	(15+5) min	JEFFREY TOSTEVIN	Spectroscopy and Correlation Studies Using Two-Nucleon Knockout
15:40	(15+5) min	PIERRE CAPEL	Influence of the Projectile Description on Breakup Calculations

16:00   Coffee break

Nuclear Structure VIII

Chair: ALISON BRUCE

16:30	(15+5) min	ALEXENDRU NEGRET	GT strength starting from $^{14}N_{g.s.}$
16:50	(15+5) min	KALIN DRUMEV	Microscopic Calculations for Waiting-Point Nuclei
17:10	(15+5) min	ALEXANDER VOINOV	Level Density and Unusual Low Energy Enhancement of γ-Strength in Iron Isotopes
17:30	(15+5) min	LUCIA-ANA POPESCU	Spin-Isospin Excitations From The Ground-State of ^{64}Ni
17:50			Closing Remarks

20:00   "A Universe of Dreams" Ensemble Galilei with NPR's Neal Conan at the DeBartolo Perfoming Arts Center.
A performance of music, poetry and stories by the Ensemble Galilei with projected images from the Hubble Space Telescope.

End of Conference

Have a Safe Trip Home!

LIST OF PARTICIPANTS

Agvaanluvsan, Undraa
Lawrence Livermore National Laboratory
5456 Arlene Way
Livermore CA 94550
USA
agvaanluvsan1@llnl.gov

Almaraz-Calderon, Sergio
University of Notre Dame
Department of Physics
225 Nieuwland Science Hall
Notre Dame IN 46556
USA
salmaraz@nd.edu

Alpizar-Vicente, Ana
20B Verde Ridge
Los Alamos NM 87544
USA
aalpizar@mines.edu

Amro, Hanan
University of Michigan/University of Notre Dame
225 Nieuwland Sciene Hall
Notre Dame IN 46556
USA
ahamro@nd.edu

Angell, Chris
2822 Pickett Rd.
#123
Durham NC 27705
USA
cangell@tunl.duk.edu

Aprahamian, Ani
University of Notre Dame
Department of Physics
225 Nieuwland Science Hall
Notre Dame IN 46556
USA
aapraham@nd.edu

Aysto, Juha
University of Jycas Kyla
Department of Physics
P.O. Box 35 (YFL)
FIN-40014
FINLAND
juha.aysto@phys.jyu.fi

Baldwin, Harriet
University of Notre Dame
Center for Continuing Education
Notre Dame IN 46556
USA
baldwin.1@nd.edu

Barrett, Bruce
University of Arizona
P.O. Box 210081
1118 E. 4th St.
Tucson AZ 85721-0081
USA
bbarrett@physics.arizona.edu

Baumann, Paule
IReS-IN2P3-CNRS
Bat 24 23, rue du Loess BP 28
Strasbourg BAS-RHIN 67037
FRANCE
paule.baumann@ires.in2p3.fr

Beard, Mary
University of Notre Dame
Department of Physics
225 Nieuwland Science Hall
Notre Dame IN 46566
USA
mbeard@nd.edu

Becker, John
LLNL
7000 East Avenue
MS L-414
Livermore CA 94501
USA
jabecker@llnl.gov

Becvar, Frantisek
Charles University in Prague
V Holesovickach 2
Prague 8 CZ-180 00
CZECH REPUBLIC
becvar@mbox.troja.mff.cuni.cz

Belgya, Tamás
Inst. of Isotopes CRC HAS
Konkoly Thege u. 29-33
Budapest - 1525
HUNGARY
belgya@iki.kfki.hu

Bell, Elizabeth
LANL
1281 B Vallecita Dr
Santa Fe NM 87501-8879
USA
ebell@lanl.gov

Berg, Georg
University of Notre Dame
Department of Physics
46556
USA
gberg@nd.edu

Betak, Emil
Institute of Physics SAS
Dubravska cesta 9
Bratislava SLOVAKIA 84511
SLOVAKIA
betak@savba.sk

Billowes, Jonathan
University of Manchester
Schuster Building
Brunswick Street
Manchester M13 9PL
ENGLAND
j.billowes@manchester.ac.uk

Boerner, Hans
Institut Laue Langevin
6 rue Horrowiz
Grenoble ISERE 38042
FRANCE
borner@ill.fr

Boutachkov, Plamen
University of Notre Dame
Department of Physics
225 Nieuwland Science Hall
Notre Dame In 46556
USA
pboutach@nd.edu

Bredeweg, Todd
Los Alamos National Laboratory
PO Box 1663, MS J514
Los Alamos NM 87545
USA
toddb@lanl.gov

Bruce, Alison
University of Brighton
101 Compton Rd.
Brighton BN1 5AL
ENGLAND
alison.bruce@brighton.ac.uk

Cano Ott, Daniel
CIEMAT
Avda. Complutense 22
Madrid MADRID 28040
SPAIN
daniel.cano@ciemat.es

Capel, Pierre
TRIUMF
4004 Wesbrook Mall
Vancouver BC V6T2A3
CANADA
pierre.capel@centraliens.net

Casten, Richard
Yale University
PO Box 208124
WNSL
New Haven CT 06520-8124
USA
rick@riviera.physics.yale.edu

Cejnar, Pavel
IPNP, Charles University
V Holesovickach 2
Prague CZ-18000
Czech Republic
cejnar@ipnp.troja.mff.cuni.cz

Champagne, Art
Univ. of North Carolina
Dept. of Physics and Astronomy
Phillips Halll, CB3255
Chapel Hill NC 27599
USA
aec@tunl.duke.edu

Choudry, Sadia
University of Kentucky
Department of Physics and Astronomy
Chemistry-Physics Bldg. Rm 177
Lexington KY 40506-0055
USA
snchou0@pa.uky.edu

Cizewski, Jolie
Rutgers University
Department of Physics & Astronomy
136 Frelinghuysen Road
Piscataway NJ 08854-8019
USA
cizewski@physics.rutgers.edu

Coshow, Suzanne
University of Notre Dame
Department of Physics
225 Nieuwland Science Hall
Notre Dame IN 46556
USA
scoshaw@nd.edu

Couder, Manoel
University of Notre Dame
Department of Physics
225 Nieuwland Science Hall
Notre Dame IN 46566
USA
mcouder@nd.edu

Couture, Aaron
University of Notre Dame
Department of Physics
225 Nieuwland Science Hall
Notre Dame IN 46556
USA
acouture@nd.edu

De Frenne, Denis
University Ghent
Proeftuinstraat, 86
Gent O-VLAANDEREN 9000
BELGIUM
denis.defrenne@UGent.be

Dillmann, Iris
Forschungszentrum Karlsruhe, IK
Hermann-von-Helmholtz-Platz 1
Eggenstein-Leopoldshafen BW 76344
GERMANY
iris.dillmann@ik.fzk.de

Domingo-Pardo, César
CSIC-Univ. Valencia
Apdo Correos 22085
Valencia VALENCIA 46071
SPAIN
cesar.domingo.pardo@cern.ch

Dracoulis, George
Australian National University
Dept. Nuclear Physics
R.Phys.S.E.
Canberra A.C.T. 200
AUSTRALIA
george.dracoulis@anu.edu.au

Drumev, Kalin
Lousiana State University
202 Nicholson Hall
Baton Rouge LA 70803
USA
kalin@epscor.phys.lsu.edu

Elekes, Zoltan
ATOMKI
Bem ter 18/c.
Debrecen H-4026
HUNGARY
elekes@atomki.hu

Esbensen, Henning
Argonne Nat. Lab.
Physics Division, Bldg. 203
Argonne National Laboratory
Argonne IL 60439
USA
esbensen@phy.anl.gov

Falahat, Sascha
University of Notre Dame
Department of Physics
225 Nieuwland Science Hall
Notre Dame In 46556
USA
sfalahat@nd.edu

Fang, Lei
Yale University
6970 Orange St., Apt. 203
New Haven CT 6511
USA

Farouqi, Khalil
Institut für Kernchemie, Uni Mainz
Fritz-Strassmann-Weg 2
Mainz RP D-55128
GERMANY
farouqi@uni-mainz.de

Firestone, Richard
Lawrence Berkeley National Laboratory
1 Cyclotron Rd
MS 88R0192
Berkeley CA 94720
USA
rbf@lbl.gov

Flambaum, Victor
Argonne National Laboratory
Building 203
9700 S. Cass Ave.
Argonne IL 60439
USA
flambaum@phys.unsw.edu.au

Frank, Alejandro
UNAM
Circuito Exterior CU
Apdo Postal 70-543
Mexico City DF 4510
MEXICO
frank@nucleares.unam.mx

Frauendorf, Stephen
University of Notre Dame
Department of Physics
225 Nieuwland Science Hall
Notre Dame IN 46556
USA
sfrauend@nd.edu

Furutaka, Kazuyoshi
1700-1-a-104, Sirakata
Tokai-mura, Naka-gun Ibaraki-Ken 319-1106
JAPAN
furutaka@jb3.so-net.ne.jp

Gai, Moshe
UConn/Yale
LNS at Avery Point
1084 Shennecossett Rd.
Groton CT 6340
USA
moshe.gai@yale.edu

Garg, Umesh
University of Notre Dame
225 Nieuwland Science Hall
Notre Dame IN 46530
USA
garg@nd.edu

Garrett, Paul
University of Guelph
Dept. of Physics
Guelph ON N0B1S0
CANADA
pgarrett@physics.uoguelph.ca

Gasques, Leandro
University of Notre Dame
Department of Physics
225 Nieuwland Science Hall
Notre Dame IN 46556
USA
lgasques@nd.edu

Goerres, Joachim
University of Notre Dame
Department of Physics
225 Nieuwland Science Hall
Notre Dame IN 46556
USA
jgoerres@nd.edu

Gomez Hornillos, Maria Belen
University of Liverpool
Oliver Lodge Lab.
Oxford Street
Liverpool MERSEYSIDE L69 7ZE
ENGLAND
mbg@ns.ph.liv.ac.uk

Govil, Indra Mani
Physics Department
Panjab University
Chandigarh-160014 U.T 160014
INDIA
imgovil@pu.ac.in

Griffin, Henry
Univ. of Mich.
Chemistry 1055
Univ. of Mich.
Ann Arbor MI 48109-1055
USA
hcg@umich.edu

Gyurky, Gyorgy
Institute of Nuclear Research (ATOMKI)
Bem ter 18/c
Debrecen HAJDU-BIHAR 4024
HUNGARY
gyurky@atomki.hu

Haight, Robert
Los Alamos National Laboratory
MS H855
Los Alamos NM 87545
USA
haight@lanl.gov

Hammer, Wolfgang
University of Notre Dame
Department of Physics
225 Niewland Science Hall
Notre Dame IN 46556
USA
mail@jwhammer.de

Hatarik, Robert
LANL
20B Verde Ridge
Los Alamos NM 87544
USA
rhatarik@mines.edu

Hecht, Adam
Univ. MD, Argonne Nat Lab
415 W. Fullerton Parkway, 1205
Chicago IL 60614
USA
hecht@phy.anl.gov

Heil, Michael
FZK
Hermann-von-Helmholtz-Platz 1
Eggenstein-Leopoldshafen BW 76344
GERMANY
michael.heil@ik.fzk.de

Higashiyama, Koji
University of Tokyo
7-3-1 Hongo
Bunkyo-ku TOKYO 113-0033
JAPAN
higashi@nt.phys.s.u-tokyo.ac.jp

Hori, Jun-ichi
Kyoto University
Asashiro Nishi, Kumatori-cho
Sennan-gun OSAKA 590-0494
JAPAN
hori@rri.kyoto-u.ac.jp

Hyder, Anthony
University of Notre Dame
Graduate School
Main Building
Notre Dame IN 46566
USA

Hyland, Bronwyn
University of Guelph
Dept. of Physics, Gordon St
Guelph ON N1G 2W1
CANADA
bhyland@uoguelph.ca

Igashira, Masayuki
Tokyo Institute of Technology
2-12-1-N1-26 O-okayama
Meguro-ku Tokyo 152-8550
JAPAN
iga@nr.titech.ac.jp

Jolie, Jan
Institut fuer Kernphysik
University of Cologne
Zulpicherstr. 77
Cologne NW 50939
GERMANY
jolie@ikp.uni-koeln.de

Jungman, Susanne
University of Notre Dame
Department of Physics
225 Nieuwland Science Hall
Notre Dame IN 46556
USA
susanne.jungmann@web.de

Kaczarowski, Roscislaw
Soltan Institute for Nuclear Studies
Dept. of Nuslear Spectroscopy
05-400 Swierk
POLAND
rKaczarowski@bigfoot.com

Kaeppeler, Franz
FZK
Hermann-von-Helmholtz-Platz 1
Eggenstein-Leopoldshafen BW 76344
GERMANY
franz.kaeppeler@ik.fzk.de

Kahl, David
Beloit College
USA

Kim, Guinyun
Kyungpook National Univ.
1370 Sankyok-dong
Buk-gu
Daegu DAEGU 702-701
KOREA
gnkim@knu.ac.kr

Kondev, Filip
Argonne National Laboratory
9700 S. Cass Avenue
Argonne IL 60439
USA
kondev@anl.gov

Konecki, Graham
University of Notre Dame
Department of Physics
225 Nieuwland Science Hall
Notre Dame IN 46556
USA
gkonecki@alumni.nd.edu

Krasznahorkay, Atilla
ATOMKI
P.O. Box 51
H-4001 Debrecen
HUNGARY
kraszna@atomki.hu

Kratz, Karl-Ludwig
Institut für Kernchemie
Fritz-Strassmann-weg 2
Mainz RP 55128
GERMANY
klkratz@uni-mainz.de

Lamm, Larry
University of Notre Dame
Department of Physics
225 Nieuwland Science Hall
Notre Dame IN 46556
USA
llamm@nd.edu

Laptev, Alexander
Japan Nucl. Cycle Development Inst.
4-33, Muramatsu, Tokai-mura, Naka-gun
Tokai-mura Ibaraki 319-1194
JAPAN
laptev@jnc.go.jp

Lauritsen, Torben
Argonne National Lab
9700 S Cass Ave
Argonne IL 60517
USA
torben@anl.gov

LeBlanc, Paul
University of Notre Dame
Department of Physics
225 Niewland Science Hall
Notre Dame IN 46556
USA
pleblanc@nd.edu

Lee, HyeYoung
University of Notre Dame
Department of Physics
225 Nieuwland Science Hall
Notre Dame IN 46556
USA
hlee2@nd.edu

Lesher, Shelly
K.U. Leuven
IKS
Celestijnenlaan 200D
B-3001 Leuven
BELGIUM
shelly.lesher@fys.kuleuven.ac.be

Li, Tao
Univeristy of Notre Dame
Department of Physics
225 Nieuwland Science Hall
Notre Dame IN 46556
USA
tli1@nd.edu

Lisetskiy, Alexander
Michigan State University
National Superconductivity Cyclotron Laboratory
East Lansing MI 48824-1231
USA

Lister, C. J. (Kim)
Argonne National Laboratory
9700 S. Cass Ave
Argoone IL 60439
USA
Lister@anl.gov

Lo Iudice, Nicola
Università di Napoli Federico II
Dipartimento di Scienze Fisiche, Monte S
Napoli I-80126
ITALY
loiudice@na.infn.it

Marrone, Stefano
INFN
via Orabona 4
BARI I-70125
ITALY
stefano.marrone@ba.infn.it

Matos, Milan
Michigan State University
National Superconductivity Cyclotron Laboratory
East Lansing MI 48824-1231
USA

McEllistrem, Marcus
University of Kentucky
Chem-Phys. Rm. 177
Lexington KY 40506-0055
USA
marcus@uky.edu

Mengoni, Alberto
IAEA - Nuclear Data Section
P.O.Box 100
Wagramer Strasse 5
Vienna 1400
AUSTRIA
a.mengoni@iaea.org

Mitchell, Gary
North Carolina State University
Dept of Physics
Raleigh NC 27695
USA
mitchell@tunl.duke.edu

Moreh, Raymond
R.P.I & Ben-Gurion University
110 Eighth St, NES Build I-21, R.P.I
Troy NY 12180
USA
moreh@bgumail.bgu.ac.il

Mukhopadhyay, Sharmistha
University of kentucky
2077 Fontaine Road
Lexington KY 40502
USA
shamuk23@yahoo.co.in

Negret, Alexandru
Ghent University
Proeftuinstraat 86
Gent EAST FLANDERS 9000
BELGIUM
AlexandruLiviu.Negret@UGent.be

Nelson, Ronald
LANL
217 Sereno Dr
Santa Fe NM 87501
USA
ronelson99@mindspring.com

Neumann-Cosel, Peter von
Institut fur Kernphysik
TU Darmstadt
Darmstadt HE 64289
GERMANY
vnc@ikp.tu-darmstadt.de

Nishiyama, Jun
Tokyo Institute of Technology
2-12-1 N1-26 O-okayama
Meguro-ku TOKYO 1528550
JAPAN
jun-nishiyama@nr.titech.ac.jp

Nunes, Filomena
Michigan State University
Cyclotron Lab
East Lansing MI 48824
USA

O'Brien, Shawn
University of Notre Dame
225 Nieuwland Science Hall
University of Notre Dame
Notre Dame IN 46556
USA
sobrien4@nd.edu

Ohsaki, Toshiro
Tokyo Institute of Technology
N1-26, 2-12-1
O-okayama
Meguro-ku TOKYO 152-8550
JAPAN
tohsaki@nr.titech.ac.jp

Otsuka, Takaharu
Dept. of Physics, University of Tokyo
Hongo
Bunkyo-ku TOKYO 113-0033
JAPAN
otsuka@phys.s.u-tokyo.ac.jp

Palumbo, Annalia
University of Notre Dame
Department of Physics
225 Nieuwland Science Hall
Notre Dame IN 46556
USA
apalumbo@nd.edu

Pietralla, Norbert
SUNY at Stony Brook
Nuclear Structure Lab
Dept.of Physics & Astronomy
Stony Brook NY 11794
USA
Norbert.Pietralla@stonybrook.edu

Plag, Ralf
Forschungszentrum Karlsruhe, IK
Hermann-von-Helmholtz-Platz 1
Eggenstein-Leopoldshafen BW 76344
GERMANY
iris.dillmann@ik.fzk.de

Popa, Gabriela
University of Notre Dame
Department of Physics
225 Nieuwland Science Hall
Notre Dame IN 46556
USA
gpopa@nd.edu

Popescu, Lucia-Ana
Ghent University
Proeftuinstraat 86
Ghent 9000
BELGIUM
LuciaAna.Popescu@UGent.be

Pritychenko, Boris
901 Skyline Drive
Coram NY 11727
USA
pritychenko@bnl.gov

Quinn, Matthew
University of Notre Dame
Department of Physics
225 Nieuwland Science Hall
Notre Dame IN 46556
USA
mquinn1@nd.edu

Rao, Madhu
University of Notre Dame
Department of Physics
225 Nieuwland Science Hall
Notre Dame In 46556
USA

Regan, Patrick
University of Surrey
School of Electronics & Physics Science
University of Surrey
Guildford SURREY GU2 7XH
ENGLAND
p.regan@surrey.ac.uk

Roig, Olivier
CEA/DPTA/SPN
BP 12
Bruyeres-Le-Chatel 91680
FRANCE
olivier.roig@cea.fr

Rundberg, Robert
Los Alamos National Laboratory
Bikini Atoll Road, SM-30
MS - J514
Los Alamos NM 87545
USA
rundberg@lanl.gov

Sakharuk, Alexander
Michigan State University
National Superconductivity Cyclotron Laboratory
1 Cyclotron Laboratory
East Lansing MI 48824-1321
USA
sakharuk@nscl.msu.edu

Schuermann, Daniel
Ruhr-Universität Bochum
Universitätsstr 150
Bochum NW 44780
GERMANY
schuermann@ep3.rub.de

Seo, Pil-Neyo
North Carolina State University
Dept. of Physics, Campus Box 8202
Mail 110, North Carolina State Universit
Raleigh NC 27695
USA
pilneyo_seo@ncsu.edu

Sheets, Steven
TUNL
2327 37th St Apt B
Los Alamos NM 87544
USA
sasheets@tunl.duke.edu

Simpson, Edward
University of Notre Dame
Department of Physics
225 Niuewland Sciene Hall
Notre Dame IN 46556
USA
esimpso1@nd.edu

Skorodumov, Boris
University of Notre Dame
Department of Physics
225 Niewland Science Hall
Notre Dame IN 46556
USA
bskorodo@nd.edu

Smirnova, Nadya
University of Ghent
Proeftuinstraat 86
Ghent B-9000
BELGIUM
Nadya.Smirnova@UGent.be

Solovyev, Vladimir
Saint Gobain Crystals
12345 Kinsman Rd
Newbury OH 44065
USA
vsolovyev@gmail.com

Stech, Edward
University of Notre Dame
Department of Physics
225 Nieuwland Science Hall
Notre Dame IN 46556
USA
estech@nd.edu

Strandberg, Elizabeth
University of Notre Dame
Department of Physics
225 Niuewland Sciene Hall
Notre Dame IN 46566
USA
emcnassa@nd.edu

Sun, Yang
University of Notre Dame
Department of Physics
225 Nieuwland Science Hall
Notre Dame IN 46566
USA
ysun@nd.edu

Tan, Wanpeng
University of Notre Dame
Department of Physics
225 Nieuwland Science Hall
Notre Dame IN 46556
USA
wtan@nd.edu

Teymurazyan, Artur
University of Notre Dame
Department of Physics
225 Nieuwland Science Hall
Notre Dame IN 46556
USA
atymura@nd.edu

Toh, Yosuke
Japan Atomic Energy Research Institute
Tokai
Shirakata,Shirane2-4
Naka IBARAKI 319-1195
JAPAN
toh@jball4.tokai.jaeri.go.jp

Tomandl, Ivo
NPI Academy of Sciencies
NPI Rez
Rez near Prague 250 68
CZECH REPUBLIC
tomandl@ujf.cas.cz

Tonchev, Anton
TUNL/Duke Univ.
BOX 9008
Durham NC 27708-0308
USA
tonchev@tunl.duke.edu

Tostevin, Jeffrey
University of Surrey
Department of Physics
Guildford SURREY GU2 7XH
ENGLAND
j.tostevin@surrey.ac.uk

Tostevin, Lena
University of Surrey
Department of Physics
Guildford SURREY GU2 7XH
ENGLAND
j.tostevin@surrey.ac.uk

Trautvetter, Hanns Peter
Ruhr-Universität Bochum
Fakultaet f. Physik und Astronomie
Bochum 44780
GERMANY
trautvetter@ep3.ruhr-uni-bochum.de

Triambak, Smarajit
University of Notre Dame
4225 9th Ave NE
Apt 15
Seattle WA 98105
USA
smarajit@u.washington.edu

Ullmann, John
Los Alamos National Laboratory
MS H855
LANSCE-NS
Los Alamos NM 87545
USA
ullmann@lanl.gov

Van Isacker, Pieter
GANIL
BP 55027
Caen CALVADOS F-14076
FRANCE
isacker@ganil.fr

Vaz, Pedro
ITN/DPRSN
Estrada Nacional 10
P-2686-953Sacavem
PORTUGAL
pedrovaz@itn.mces.pt

Voinov, Alexander
Ohio University
Accelerator Lab. Department of Physics
and Astronomy, OU
Athens OH 45701
USA
voinov@ohio.edu

von Egidy, Till
Technical University Munich
James Franck Strasse
D-85748 Garching D-85748
GERMANY
egidy@ph.tum.de

Walker, Phil
University of Surrey
Department of Physics
Guildford SURREY GU2 7XH
ENGLAND
p.walker@surrey.ac.uk

Walter, Stephan
Forschungszentrum Karlsruhe, IK
Hermann-von-Helmholtz-Platz 1
Eggenstein-Leopoldshafen BW 76344
GERMANY
iris.dillmann@ik.fzk.de

Wang, Xiaofeng
University of Notre Dame
Department of Physics
225 Nieuwland Science Hall
Notre Dame IN 46556
USA
xwang13@nd.edu

Werner, Volker
Yale University
PO Box 208124
New Haven CT 06520-8124
USA
paula.fox@yale.edu

Wiescher, Michael
University of Notre Dame
Department of Physics
225 Nieuwland Science Hall
Notre Dame In 46556
USA
mwiesche@nd.edu

Wilhelmy, Jerry
LANL
1753 Laukahi Street
Honolulu HI 96821
USA
j_wilhelmy@lanl.gov

Woehr, Andreas
University of Notre Dame
Department of Physics
225 Nieuwland Science Hall
Notre Dame IN 46556
USA
awoehr@nd.edu

Yates, Steven
University of Kentucky
Department of Chemistry
305 Chemistry-Physics Building
Lexington KY 40506-0055
USA
yates@uky.edu

Yoshinaga, Naotaka
Saitama University
Department of Physics
Saitama 338-8570
JAPAN
yoshinaga@phy.saitama-u.ac.jp

Zelevinsky, Vladimir
National Cyclotron Laboratory
Michigan State University
210 Cyclotron Lab
East Lansing MI 48824-1321
USA
zelevinsky@nscl.msu.edu

Author Index

A

Abbondanno, U., 283, 288, 318
Adachi, T., 432, 535, 550
Adelberger, E. G., 607
Aerts, G., 283, 288, 318
Agvaanluvsan, U., 295, 389, 545, 597
Ahmad, I., 459
Ahmed, M. W., 340
Ahn, T., 11
Ai, H. C., 35
Ai, J., 24
Albers, D., 105
Algin, E., 545
Alpizar-Vicente, A. M., 273, 278, 295
Álvarez, H., 283, 288, 318
Alvarez-Velarde, F., 283, 288, 318
Ammar, Z., 233
Amon, L., 35
Amro, H., 24, 557
Andreoiu, C., 105, 249
Andreyev, A., 249
Andriamonje, S., 283, 288, 318
Andrzejewski, J., 283, 288, 318
Angell, C. T., 350, 363
Aoi, N., 383
Aprahamian, A., 43, 196, 221, 587, 591, 601
Arcilla, R., 589
Arima, A., 244
Ashley, S. F., 249
Assimakopoulos, P., 283, 288, 318
Audouin, L., 283, 288, 318
Aupiais, J., 593
Austin, R. A. E., 249
Avez, B., 216
Axiotis, M., 464
Azuma, R. E., 599

B

Baba, H., 383
Baba, M., 577
Badurek, G., 283, 288, 318
Bahri, C., 540
Bai, X., 176
Ball, G. C., 105, 249
Bandyopadhyay, D., 101, 249, 559
Bardayan, D. W., 209
Barea, J., 151
Barrett, B. R., 504
Baumann, P., 283, 288, 318, 561
Bäumer, C., 535, 550
Bazzacco, D., 464
Beard, M., 186, 581
Beausang, C., 24
Becchetti, F. D., 221, 557
Becker, H.-W., 581
Becker, J. A., 249, 295, 323, 396, 568, 597
Bečvář, F., 283, 288, 295, 318, 345, 563, 597
Belgya, T., 138, 300, 389, 545, 565
Bélier, G., 593
Berg, G. P. A., 535, 550
Berthoumieux, E., 283, 288, 318
Bertone, P., 599
Běták, E., 226
Blackmon, J. C., 209
Blank, B., 159
Block, R. C., 86
Bond, E. M., 312, 568
Bondarenko, V., 605
Borcea, C., 561
Börner, H. G., 511
Boswell, M., 350
Botha, N. T., 611
Boukharouba, N., 233, 559
Boutachkov, P., 221, 601
Bowman, J. D., 81
Bredeweg, T. A., 273, 278, 295, 312, 568
Bricault, P., 105
Broda, R., 464
Brown, B. A., 483
Brown, T. B., 233
Bruce, A. M., 71, 587
Brune, C. R., 545
Bucurescu, D., 464, 498, 605
Burda, O., 611
Bychowski, J. P., 221
Byelikov, A., 432

C

Caballero, L., 159
Cakirli, R. B., 35
Calviño, F., 283, 288, 318
Cano-Ott, D., 283, 288, 318
Capel, P., 528
Capote, R., 283, 288, 318
Carlini, R. D., 81
Carpenter, M. P., 11, 459
Carrillo de Albornoz, A., 283, 288, 381
Carter, J., 611
Casten, R. F., 24, 35, 49
Cejnar, P., 487, 570
Cennini, P., 283, 288, 318
Chadwick, M. B., 568
Chakrawarthy, R. S., 249
Champagne, A., 599
Chan, S., 249
Chartier, M., 159
Chautard, F., 159
Chen, N., 176
Chen, Y., 221, 557
Chepel, V., 283, 288, 318
Chiaveri, E., 283, 288, 318
Choudry, S. N., 96, 101, 583
Chowdhury, P., 459
Chubarian, G., 221
Chung, W. C., 133, 579
Chupp, T. E., 81
Churchman, R., 105, 249
Cizewski, J. A., 209
Clark, R. M., 459
Clement, R. R. C., 295, 568
Cline, D., 24
Colonna, N., 283, 288, 318
Coombes, H., 249
Cortes, G., 283, 288, 318
Couder, M., 186, 191, 581
Couture, A., 186, 191, 283, 288, 318, 581
Covrig, S. D., 81
Cox, J., 283, 288, 318
Cross, D., 105
Csatlós, M., 439

D

Dabaghyan, M., 81
Dahlfors, M., 283, 288, 318
Daly, J., 191
Danon, Y., 86
Dashdorj, D., 323
Daugas, J.-M., 593
David, S., 283, 288, 318
Davids, C. N., 355
deAngelis, G., 464
De Cesare,, 595
De Frenne, D., 535, 550
de Huu, M. A., 535, 550
De Maesschalck, A., 475
Demichi, K., 383
Demonchy, C. E., 159
Devlin, M., 323
DeVries, D. J., 572
DeYoung, P. A., 221
Di Leva, A., 595
Dillmann, I., 123, 283, 288, 318
Dolfini, R., 283, 288, 318
Dombrádi, Zs., 383
Dombsky, M., 105
Domingo-Pardo, C., 283, 288, 318
D' Onofrio, A., 595
Døssing, T., 459
Draayer, J. P., 540, 587
Dracoulis, G. D., 3
Dridi, W., 283, 288, 318
Drumev, K. P., 540
Dugnet, T., 216
Duran, I., 283, 288, 318

E

Efe, G., 201, 585
Eisermann, Y., 439, 605
Eleftheriadis, C., 283, 288, 318
Elekes, Z., 201, 383
Elliot, A., 118
Embid-Segura, M., 283, 288, 318
Esbensen, H., 518
Esch, E.-I., 273, 278, 295, 568
Ethvignot, T., 323, 568

F

Faestermann, T., 439
Fallon, P., 459
Farnea, E., 464
Farouqi, K., 419
Fearick, R. W., 611
Ferrant, L., 283, 288, 318
Ferrari, A., 283, 288, 318
Ferreira-Marques, R., 283, 288, 318
Finlay, P., 249
Firestone, R. B., 138, 389
Fisker, J. L., 118, 196
Fitzpatrick, C. R., 35
Fitzpatrick, L., 283, 288, 318
Folden, III, C. M., 295
Förtsch, S. V., 611
Fotiades, N., 323
Fowler, M. M., 568
Frais-Koelbl, H., 283, 288, 318
Frank, A., 151
Fransen, C., 233, 340, 559, 611
Freedman, S. J., 81
Freeman, S. J., 35
Frekers, D., 535, 550
Fujii, K., 283, 288, 318
Fujita, H., 432, 611
Fujita, K., 535, 550
Fujita, Y., 432, 535, 550
Fülöp, Zs., 201, 383, 585
Furman, W., 283, 288, 318

G

Gadea, A., 464
Gai, M., 171
Galaviz, D., 201
Gallino, R., 283, 288, 318
García, A., 607
Garnsworthy, A. B., 35
Garrett, P. E., 105, 249, 323, 559
Gelletly, W., 464
Gentile, T., 81
Georgiev, G., 159
Geppert, C., 105
Gericke, M. T., 81
Gerl, J., 24
Gialanella, L., 595
Gibelin, J., 383

Gillibert, A., 159
Gillis, R. C., 81
Goldberg, V. Z., 221
Gomez Hornillos, M. B., 159
Gomi, T., 383
Goncalves, I., 283, 288, 318
Gonzalez-Romero, E., 283, 288, 318
Görres, J., 186, 191, 201, 581, 585, 599
Goverdovski, A., 283, 288, 318
Govil, I. M., 454
Gramegna, F., 283, 288, 318
Granier, T., 323, 568
Graw, G., 439, 605
Greene, G. L., 81
Greife, U., 273, 278
Grewe, E. W., 535, 550
Griesmayer, E., 283, 288, 318
Griffin, H. C., 557, 572
Grimes, S. M., 545
Grinyer, G. F., 105, 249
Guerrero, C., 283, 288, 318
Gulyás, J., 439
Gunsing, F., 283, 288, 318
Guo, B., 176
Güray, R. T., 201, 585
Gurdal, G., 35
Guttormsen, M., 545
Gyürky, Gy., 201, 585

H

Haas, B., 283, 288, 318
Habs, D., 439
Hackman, G., 105, 249
Haefner, P., 535, 550
Haight, C., 278
Haight, R. C., 273, 283, 288, 295, 312, 318
Hanemaayer, V., 105
Hannant, C. D., 233
Harada, H., 402
Harakeh, M. N., 439
Hasegawa, H., 383
Hatanaka, K., 535, 550
Hatarik, R., 273, 278, 295
Hayes, A. B., 24
Hecht, A. A., 24, 355
Heger, A., 432

Heil, M., 123, 181, 265, 283, 288, 307, 318, 345, 563
Heinz, A. M., 24, 35, 355, 459
Heinze, S., 331
Hemingray, S., 118
Hennrich, S., 591
Herman, M. W., 128, 589
Herrera-Martinez, A., 283, 288, 318
Herskind, B., 459
Hersman, F. W., 81
Hertenberger, R., 439, 605
Heyde, K., 475
Higashiyama, K., 464, 575
Hinnefeld, J. D., 557
Hirsch, J. G., 151, 587
Hodges, G. J. P., 607
Hoedl, S. A., 607
Hoffman, D. C., 295
Honzátko, J., 605, 609
Hori, J., 402, 577
Hornish, M. J., 545
Hoteling, N., 355
Howell, C. R., 350
Hughes, R., 24
Hunt, L. F., 312, 568
Hunyadi, M., 439, 535, 550
Hyland, B., 105, 249

I

Igashira, M., 133, 283, 288, 318, 373, 378, 402, 579
Ikenaga, D., 378
Illes, E., 249
Imai, N., 383
Imbriani, G., 595
Ino, T., 81
Ionescu-Bujor, M., 464
Iordachescu, A., 464
Isaev, S., 283, 288, 318
Ishihara, M., 383
Iwasaki, H., 383

J

Jacobs, E., 535, 550
Jacquot, B., 159
Janssens, R. V. F., 11, 24, 459

Jenkins, D. G., 459
Jericha, E., 283, 288, 318, 561
Jiang, H., 557
Johansson, H., 535, 550
Johnson, M. S., 209
Jokić, S., 561
Jolie, J., 96, 331
Jones, G. A., 35, 249, 464
Jones, G. L., 81
Jones, K. L., 209
Jurado, B., 159
Jutier, Ch., 593

K

Kaczarowski, R., 447
Kadi, Y., 283, 288, 318
Kalmykov, Y., 535
Kandes, M., 81
Kanno, S., 383
Käppeler, F., 123, 181, 283, 288, 307, 318, 345, 563
Karadimos, D., 283, 288, 318
Karamanis, D., 283, 288, 318
Karwowski, H. J., 350, 363
Kawabata, K., 535
Kawai, S., 383
Kelley, J. H., 350, 363
Kerveno, M., 283, 288, 318, 561
Ketlerov, V., 283, 288, 318
Khoo, T. L., 459
Kim, G. N., 133, 579
Kimura, A., 603
Kishida, T., 383
Kiss, G. G., 201
Kneissl, U., 340
Koehler, P., 283, 288, 318
Kohstall, C., 340
Koizumi, M., 603
Kolata, J. J., 221, 557, 601
Kolbe, E., 432
Koltick, D. S., 368
Kondev, F. G., 113, 459
Konovalov, V., 283, 288, 318
Korff, A., 535, 550
Korichi, A., 459
Kossionides, E., 283, 288, 318
Kozub, R. L., 209
Krasznahorkay, A., 439

Kratz, K.-L., 409, 419, 591
Kretschmer, A., 201
Kröll, Th., 464
Kronenberg, A., 312
Krtiáka, M., 389
Krtička, M. M., 283, 288, 295, 318, 345, 563, 597
Kruizenga, A., 118
Kubo, T., 383
Kuhar, M., 611
Kulp, W. D., 249
Kumar, A., 583
Kunz, R., 595
Kurian, M., 487
Kurita, K., 383

L

Lamboudis, C., 283, 288, 318
Lamm, L. O., 186, 221, 581, 601
Lane, G., 459
Langanke, K., 432
Langdown, S. D., 464
Laptev, A., 402
Larabee, A. J., 459
Lassen, J., 105
Lauritsen, T., 459
Lauss, B., 81
Lavoie, J. P., 105
LeBlanc, P., 186, 581
Lecesne, N., 159
Lee, H.-Y., 186, 191, 201, 581, 585
Leeb, H., 283, 288, 318
Lenhardt, A., 611
Lenzi, S., 464
Le Petit,, 593
Lépine-Szily, A., 159
Lesher, S. R., 96, 101, 233, 559
Leslie, J. R., 105, 249
Letourneau, A., 593
Leuschner, M. B., 81
Li, J., 340
Li, Z., 176
Lian, G., 176
Lindote, A., 283, 288, 318
Linnemann, A., 96, 340
Lisetskiy, A. F., 483
Lister, C. J., 11, 24, 355, 459
Liu, W., 176

Lo Iudice, N., 239
Lopes, I., 283, 288, 318
Lopez-Martens, A., 459
López Vieyra, J. C., 151
Lozano, M., 283, 288, 318
Lozowski, W. R., 81
Lukić, S., 283, 288, 318, 561
Lunardi, S., 464

M

Macchiavelli, A. O., 459
Macek, M., 487, 570
Macri, R. A., 295, 568
Mahurin, R., 81
Maier, H. J., 439
Mantica, P., 591
Marganiec, J., 283, 288, 318
Marginean, N., 464
Marie, F., 593
Marques, L., 283, 288, 318
Marrone, S., 283, 288, 318
Martinez, T., 318, 464
Martínez-Pinedo, G., 432
Mason, M., 81
Massey, T. N., 545
Mastinu, P., 283, 288, 318
Masuda, Y., 81
Máté, Z., 201, 439
Matoš, M., 164
Matsuyama, Y., 383
Mattoon, C., 249
Mazzocchi, C., 355
McCutchan, E. A., 35
McEllistrem, M. T., 96, 101, 233, 559, 583
McKay, C. J., 96
McNabb, D. P., 138, 389
Medina, N., 464
Melconian, D., 105
Menegazzo, R., 464
Mengoni, A., 257, 283, 288, 318
Méot, V., 593
Merrick, M., 233
Meyer, D. A., 24
Michimasa, S., 383
Mihailescu, L. C., 561
Milazzo, P. M., 283, 288, 318
Minemura, T., 383

657

Mitchell, G. E., 91, 295, 545, 597
Mitchell, G. S., 81
Mittig, W., 159
Molnár, G. L., 138
Moore, E. F., 24, 459
Moreau, C., 283, 288, 318
Moreh, R., 86
Morton, A. C., 105, 249
Mosconi, M., 283, 288, 318
Motobayashi, T., 383
Mukhopadhyay, S., 101, 583
Müller, S., 340
Muto, S., 81
Mynk, M., 96

N

Nakajima, K., 577
Nakamura, S., 402
Nakanishi, K., 535, 550
Nann, H., 81
Napiorkowski, P., 24
Napoli, D. R., 464
Navratil, P., 504
Negret, A., 535, 550
Nelson, R. O., 323
Neuman, M., 86
Neveling, R., 611
Neves, F., 283, 288, 318
Nishiyama, J., 579
Nitsche, H., 295
Nogueira, T. N., 91
Notani, M., 383
Nunes, F. M., 216, 528

O

Oberhummer, H., 283, 288, 318
Obložinský, P., 128, 589
O'Brien, S., 186, 283, 288, 318, 581
O'Donnell, J. M., 273, 278, 295, 312, 568
Ogama, T., 535
Ohgama, K., 373, 402
Ohnishi, T. K., 383
Ohsaki, T., 133, 373, 378, 402, 579
Oishi, T., 577
Ojaruega, M., 557

Ong, H. J., 383
Orce, J. N., 96, 101, 583
Ordine, A., 595
Orr, N., 159
Osa, A., 603
Oshima, M., 283, 288, 318, 603
Ota, S., 383
Otsuka, T., 65
Ozawa, A., 383
Özkan, N., 201, 585

P

Page, S. A., 81
Pain, S. D., 209
Palombo, J., 355
Palumbo, A., 186, 201, 581, 585
Pancin, J., 283, 288, 318
Papachristodoulou, C., 283, 288, 318
Papadopoulos, C., 283, 288, 318
Paradela, C., 283, 288, 318
Pardo, R. C., 24
Parker, W., 295
Pato, M. P., 91
Patronis, N., 283, 288, 318
Pavlik, A., 283, 288, 318, 561
Pavlopoulos, P., 283, 288, 318
Pearson, C. J., 105, 249
Pearson, M., 105
Peaslee, G. F., 221, 557
Penttilä, S. I., 81
Pereira Conca, J., 591
Perrot, L., 283, 288, 318
Petkov, P., 96
Pfeiffer, B., 419
Phillips, A. A., 105, 249
Pietralla, N., 11, 340, 611
Pignatari, M., 265
Pinayev, I. V., 340
Pitz, H. H., 340
Plag, R., 123, 181, 283, 288, 307, 318
Plompen, A. J. M., 283, 288, 318, 561
Plukis, A., 283, 288, 318
Poch, A., 283, 288, 318
Podolyák, Zs., 464
Politi, G., 159
Ponomarev, V. Yu., 605, 611
Popa, G., 587
Popescu, L., 535, 550

Pretel, C., 283, 288, 318
Pritychenko, B., 128, 589

Q

Qian, J., 35
Quesada, J., 283, 288, 318
Quinn, M., 221, 591, 601
Quintana, B., 464

R

Rainovski, G., 11
Rakers, S., 535, 550
Ramsay, W. D., 81
Rapp, W., 585
Rauscher, T., 118, 123, 201, 283, 288, 318, 419
Regan, P. H., 35, 249, 464
Reifarth, R., 273, 278, 283, 288, 295, 307, 318, 345, 563
Rekstad, J., 545
Ressler, J. J., 249
Révay, Zs., 138, 300, 389
Richter, A., 432, 535, 611
Ro, T. I., 133, 579
Roca, V., 595
Rochman, D., 589
Rogachev, G. V., 221, 601
Rogalla, D., 595
Roig, O., 593
Rolfs, C., 595
Romain, P., 593
Romano, M., 595
Rosetti, M., 283, 288, 318
Rousseau, M., 159
Roussel-Chomaz, P., 159
Rubbia, C., 283, 288, 318
Rubio, B., 464
Rudolf, G., 283, 288, 318, 561
Rullhusen, P., 283, 288, 318
Rundberg, R. S., 273, 278, 295, 312, 568
Rusu, C., 464
Ryezayeva, N., 535, 550

S

Saito, A., 383
Sakai, H. K., 383
Sakemi, Y., 535, 550

Sakharuk, T., A., 118
Sakurai, H., 383
Salgado, J., 283, 288, 318
Sallaska, A. L., 607
Santra, S., 81
Sarazin, J. J., 249
Sarchiapone, L., 283, 288, 318
Sartorelli, J.C., 91
Savajols, H., 159
Savran, D., 340
Savvidis, I., 283, 288, 318
Schaile, O., 439
Schatz, H., 118, 423, 483, 591
Scheck, F., M., 340
Schiller, A., 545
Schlegel, Ch., 24
Schumaker, M. A., 249
Schumaker, M.A., 105
Schümann, F., 595
Schürmann, D., 595
Schwantes, J. M., 295, 312, 568
Schwarzenberg, J., 249
Schwenger, R., 464
Seo, P.-N., 81
Seweryniak, D., 24, 355, 459
Sharapov, E. I., 81
Sheets, S. A., 295, 597
Shergur, J., 355
Shevchenko, A., 432, 535, 550
Shimbara, Y., 432, 535, 550
Shimizu, Y., 535, 550
Shimoura, S., 383
Shirikova, N. Yu., 239
Shu, N., 176
Sideras-Haddad, E., 611
Siem, S., 545
Simon, H., 550
Simon, M. W., 24
Simpson, E. C., 599
Sjue, S. K. L., 607
Skorodumov, B. B., 221, 557, 601
Sleaford, B., 389
Sleaford, B. W., 138
Smirnova, N. A., 475
Smit, F. D., 611
Smith, K., 118
Smith, M. B., 249
Smith, M. S., 209
Smith, T. B., 81
Snow, W. M., 81

Solovyev, V. G., 368
Somorjai, E., 201, 585
Sonnabend, K., 201
Sonzogni, A. A., 128, 589
Srebrny, J., 24
Stech, E., 186, 191, 581
Stedile, M., 340
Stephan, C., 283, 288, 318
Stetcu, I., 504
Stoyer, M., 355
Strandberg, E., 186, 191, 581
Stránský, P., 487
Strieder, F., 595
Sun, Y., 30, 196
Sushkov, A. V., 239
Svensson, C. E., 105, 249
Swanson, H. E., 607
Szentmiklósi, L., 300

T

Tagliente, G., 283, 288, 318
Tain, J. L., 283, 288, 318
Takahashi, W., 577
Takeshita, E., 383
Takeuchi, S., 383
Tamaki, M., 383
Tameshige, Y., 550
Tamii, A., 535, 550
Tan, W. P., 186, 191, 581
Tassan-Got, L., 283, 288, 318
Tavora, L., 283, 288, 318
Teng, R., 24
Terlizzi, R., 283, 288, 318
Terrasi, F., 595
Thielemann, F.-K., 118, 123, 419
Thirolf, P. G., 439
Thomas, J. S., 209
Thompson, N. J., 35
Togano, Y., 383
Toh, Y., 603
Tomandl, I., 345, 563, 605, 609
Tonchev, A. P., 340, 350, 363
Tonev, D., 464
Tornow, W., 340, 350, 363
Tostevin, J. A., 523
Trautvetter, H.-P., 595, 599
Triambak, S., 607
Tsoneva, N., 350

Tufaile, A. P. B., 91
Tuli, J. K., 113, 589

U

Uberseder, E., 186
Uchida, M., 535, 550
Ugalde, C., 186, 191, 581, 599
Ullmann, J. L., 273, 278, 295, 312, 568
Ur, C. A., 464

V

Valiente-Dobón, J. J., 105, 249, 464
van den Berg, A. M., 535, 550
Van Isacker, P., 57, 151
Vannini, G., 283, 288, 318
Varadarajan, V., 101, 583
Vary, J. P., 504
Vaz, P., 283, 288, 318
Velázquez, V., 96, 151
Ventura, A., 283, 288, 318
Vetter, K., 24
Veyssiere, Ch., 593
Vieira, D. J., 273, 278, 295, 312, 568
Villamarin, D., 283, 288, 318
Villari, H., 159
Vincente, M., 283
Vincente, M. C., 288, 318
Vlachoudis, V., 283, 288, 318
Vlastou, R., 283, 288, 318
Voinov, A. V., 545
Volya, A., 493
von Brentano, P., 96, 340, 535, 550
von Egidy, T., 498, 605, 609
von Garrel, H., 340
von Neumann-Cosel, P., 432, 535, 550, 611
von Oertzen, W., 464
Voss, F., 283, 288, 318, 345, 563

W

Waddington, J. C., 249
Walker, P. M., 16, 249
Walter, S., 283, 288, 307, 318, 340
Walters, W. B., 355

Wambach, J., 611
Wang, B., 176
Ward, D., 459
Warr, N., 233
Watters, L. M., 249
Weller, H. R., 340
Wendler, H., 283, 288, 318
Werner, V., 35, 96, 340
Wiescher, M., 118, 186, 191, 196, 201, 283, 288, 318, 581, 585, 599
Wilburn, W. S., 81
Wilhelmy, J. B., 273, 295, 312, 568
Wilk, P., 295
Williams, S. J., 35, 249
Winchell, D. F., 589
Winkler, R., 35
Wirth, H.-F., 605, 609
Wirth, H. J., 439
Wisshak, K., 181, 283, 288, 318, 345, 563
Woehr, A., 221, 591, 601
Wollersheim, H. J., 24
Wong, J., 249
Wood, J. L., 249
Woods, P. J., 355
Wörtche, H. J., 535, 550
Wouters, J. M., 273, 278, 295, 312, 568
Wu, C. Y., 24, 295
Wu, C.-Y., 568
Wu, Y. K., 340

Y

Yamada, K., 383
Yan, S., 176
Yanagisawa, Y., 383
Yashima, H., 577
Yates, S. W., 96, 101, 233, 583
Yates, S.W., 559
Yoneda, K., 383
Yoshinaga, N., 244, 464, 575
Yosoi, M., 535, 550
Younes, W., 323
Yuan, V., 81
Yurkon, J. E., 568

Z

Zelevinsky, V., 493
Zeng, S., 176
Zerkin, V., 128
Zganjar, E. F., 249
Zhang, Y. H., 464
Zhao, Y. M., 244
Zhu, H., 81
Zhu, S., 11, 355, 459
Zilges, A., 201